T0131448

Linearized Plane Gravitational Wave

$$ds^2 = -dt^2 + dx^2 + dy^2 + dz^2 + h_{\alpha\beta}dx^\alpha dx^\beta$$

where (rows and columns in t, x, y, z order)

$$h_{\alpha\beta}(t,z) = \begin{pmatrix} 0 & 0 & 0 & 0 \\ 0 & f_+(t-z) & f_\times(t-z) & 0 \\ 0 & f_\times(t-z) & -f_+(t-z) & 0 \\ 0 & 0 & 0 & 0 \end{pmatrix}$$

for a wave propagating in the z-direction.

Friedman–Robertson–Walker Cosmological Models

$$ds^2 = -dt^2 + a^2(t)\left[d\chi^2 + \left\{\begin{matrix} \sin^2\chi \\ \chi^2 \\ \sinh^2\chi \end{matrix}\right\}(d\theta^2 + \sin^2\theta d\phi^2)\right], \quad \left\{\begin{matrix} \text{closed} \\ \text{flat} \\ \text{open} \end{matrix}\right\}.$$

$$ds^2 = -dt^2 + a^2(t)\left[\frac{dr^2}{1-kr^2} + r^2(d\theta^2 + \sin^2\theta d\phi^2)\right], \quad \begin{pmatrix} k=+1, \text{ closed} \\ k=0, \text{ flat} \\ k=-1, \text{ open} \end{pmatrix}.$$

THE GEODESIC EQUATION

- Lagrangian for the Geodesic Equation of a test particle

$$L\left(\frac{dx^\alpha}{d\sigma}, x^\alpha\right) = \left(-g_{\alpha\beta}(x)\frac{dx^\alpha}{d\sigma}\frac{dx^\beta}{d\sigma}\right)^{1/2}$$

where σ is an arbitrary parameter along the world line $x^\alpha = x^\alpha(\sigma)$ of the geodesic.

- Geodesic equation for a test particle (coordinate basis)

$$\frac{d^2x^\alpha}{d\tau^2} = -\Gamma^\alpha_{\beta\gamma}\frac{dx^\beta}{d\tau}\frac{dx^\gamma}{d\tau} \quad \text{or} \quad \frac{du^\alpha}{d\tau} = -\Gamma^\alpha_{\beta\gamma}u^\beta u^\gamma$$

where τ is the proper time along the geodesic and $u^\alpha = dx^\alpha/d\tau$ are the coordinate basis components of the four-velocity so that $\mathbf{u}\cdot\mathbf{u} = -1$. The Christoffel symbols $\Gamma^\alpha_{\beta\gamma}$ follow from Lagrange's equations or from the general formula (8.19). The geodesic equation for light rays takes the same form with τ replaced by an affine parameter and $\mathbf{u}\cdot\mathbf{u} = 0$.

- Conserved Quantities

$$\boldsymbol{\xi}\cdot\mathbf{u} = \text{constant}$$

where $\boldsymbol{\xi}$ is a Killing vector, e.g., $\xi^\alpha = (0, 1, 0, 0)$ in a coordinate basis where the metric $g_{\alpha\beta}(x)$ is independent of x^1.

Gravity

An Introduction to Einstein's General Relativity

Einstein's theory of general relativity is a cornerstone of modern physics. It also touches upon a wealth of topics that students find fascinating – black holes, warped spacetime, gravitational waves, and cosmology. This ground-breaking text, reissued by Cambridge University Press, helped to bring general relativity into the undergraduate curriculum, in order to make this fundamental theory accessible to virtually all physics majors. One of the pioneers of the "physics-first" approach to the subject, renowned relativist James B. Hartle recognized that there is typically not enough time in a short introductory course for the traditional, math-first, approach to the subject. In his text, he provides a fluent and accessible physics-first introduction to general relativity that uses a minimum of new mathematics and begins with the essential physical applications. This market-leading text is ideal for a one-semester course for undergraduates, requiring only introductory mechanics as a prerequisite.

James B. Hartle was educated at Princeton University and the California Institute of Technology where he completed a Ph.D. in 1964. He is currently Professor of Physics at the University of California, Santa Barbara. His scientific work is concerned with the application of Einstein's relativistic theory of gravitation (general relativity) to realistic astrophysical situations, especially cosmology.

Professor Hartle has made important contributions to the understanding of gravitational waves, relativistic stars, and black holes. He is currently interested in the earliest moments of the Big Bang, where the subjects of quantum mechanics, quantum gravity, and cosmology overlap. He has visited Cambridge often since 1971 and collaborated closely with Stephen Hawking over many years, most notably on their famous "no boundary proposal" for the origin of the universe.

Professor Hartle is a member of the US National Academy of Sciences, a fellow of the American Academy of Arts and Sciences, and a member of the American Philosophical Society. He is a founder and past director of the Institute for Theoretical Physics in Santa Barbara. He was awarded the 2009 Einstein Prize of the American Physical Society for his work on gravitational physics.

Gravity

An Introduction to Einstein's General Relativity

JAMES B. HARTLE
University of California, Santa Barbara

CAMBRIDGE
UNIVERSITY PRESS

Shaftesbury Road, Cambridge CB2 8EA, United Kingdom

One Liberty Plaza, 20th Floor, New York, NY 10006, USA

477 Williamstown Road, Port Melbourne, VIC 3207, Australia

314–321, 3rd Floor, Plot 3, Splendor Forum, Jasola District Centre, New Delhi – 110025, India

103 Penang Road, #05–06/07, Visioncrest Commercial, Singapore 238467

Cambridge University Press is part of Cambridge University Press & Assessment, a department of the University of Cambridge.

We share the University's mission to contribute to society through the pursuit of education, learning and research at the highest international levels of excellence.

www.cambridge.org
Information on this title: www.cambridge.org/9781316517543

DOI: 10.1017/9781009042604

This book was previously published by Pearson Education Inc., 2003
Reissued, with corrections, by Cambridge University Press 2021 (version 2, February 2024)

Printed in Great Britain by CPI Group (UK) Ltd, Croydon CR0 4YY

A catalogue record for this publication is available from the British Library

ISBN 978-1-316-51754-3 Hardback

Additional resources for this publication at www.cambridge.org/hartle

To Mary Jo

Contents

Boxes

Preface

Einstein's relativistic theory of gravitation—general relativity—will shortly be a century old. At its core is one of the most beautiful and revolutionary conceptions of modern science—the idea that gravity *is* the geometry of four-dimensional curved spacetime. Together with quantum theory, general relativity is one of the two most profound developments of twentieth-century physics.

General relativity has been accurately tested in the solar system. It underlies our understanding of the universe on the largest distance scales, and is central to the explanation of such frontier astrophysical phenomena as gravitational collapse, black holes, X-ray sources, neutron stars, active galactic nuclei, gravitational waves, and the big bang. General relativity is the intellectual origin of many ideas in contemporary elementary particle physics and is a necessary prerequisite to understanding theories of the unification of all forces such as string theory.

An introduction to this subject, so basic, so well established, so central to several branches of physics, and so interesting to the lay public is naturally a part of the education of every undergraduate physics major. Yet teaching general relativity at an undergraduate level confronts a basic problem. The logical order of teaching this subject (as for most others) is to assemble the necessary mathematical tools, motivate the basic defining equations, solve the equations, and apply the solutions to physically interesting circumstances. Developing the tools of differential geometry, introducing the Einstein equation, and solving it is an elegant and satisfying story. But it can also be a long one, too long in fact to cover both that and introduce the many contemporary applications in the time that is typically available for an introductory undergraduate course.

Gravity introduces general relativity in a different order. The principles on which it is based are discussed at greater length in Appendix D, but essentially the strategy is the following: The simplest physically relevant solutions of the Einstein equation are presented *first*, without derivation, as spacetimes whose observational consequences are to be explored by the study of the motion of test particles and light rays in them. This brings the student to the physical phenomena as quickly as possible. It is the part of the subject most directly connected to classical mechanics, and requires the minimum of new mathematical ideas. The Einstein equation is introduced later and solved to show how these geometries originate.

A course for junior or senior level physics students based on these principles and the first two parts of this book has been part of the undergraduate curriculum at the University of California, Santa Barbara for over twenty-five years. It works.

Acknowledgments

It will be disappointing if my colleagues in gravitational physics find anything new here. It would mean that they have not studied the classic texts of Landau and Lifshitz (1962), Misner, Thorne and Wheeler (1970), Taylor and Wheeler (1963), Wald (1984), and Weinberg (1972) on which this exposition relies so heavily. I have not acknowledged individual points of indebtedness to these works. I do so generally here.

I am especially grateful to Roger Blandford, Ted Jacobson, Channon Price, Kip Thorne, and Bob Wald, who read early versions of the entire book and provided helpful advice on its overall structure in addition to numerous corrections. The book has benefited from the comments and criticism of my colleagues that have taught from preliminary versions of it over the years. Vernon Barger, Omer Blaes, Doug Eardley, Jerome Gauntlett, Gary Horowitz, Clifford Johnson, Shawn Kolitch, Rob Myers, Thomas Moore, Stan Peale, Channon Price, and Kristin Schleich have my gratitude in this regard. Many colleagues commented constructively on individual chapters. Lars Bildsten, Omer Blaes, Peter D'Eath, Doug Eardley, Wendy Freedman, Daniel Holz, Gary Horowitz, Scott Hughes, Robert Kirshner, Lee Lindblom, Richard Price, Peter Saulson, Bernard Schutz, David Spergel, Joseph Taylor, Michael Turner, Bill Unruh, and Clifford Will have my particular thanks for their help. I am grateful to Eric Adelberger, Neil Ashby, Matt Colless, Francis Everitt, Andrea Ghez, John Hall, Jim Moran, Michael Perryman, Wolfgang Schleich, Tuck Stebbins, Dave Tytler, and Jim Williams for assistance with some of the boxes and figures. Instructive exchanges on particular points with Dave Arnett, Peter Bender, Dieter Brill, J. Richard Gott, Jeanne Dickey, Andrew Fabian, Jeremy Gray, Gary Gibbons, Wick Haxton, Gordon Kane, Angela Olinto, and Roger Penrose were useful. In lists this long it is inevitable that I have left somone out. To them I offer my appologies and my hope for a second printing.

The help in providing many of the figures is acknowledged in the individual figure captions.

Students too numerous to mention pointed out mistakes, typos, and arguments that lacked clarity, and I would like to thank Joe Alibrandi, Maria Cranor, Ian Eisenman, William Kaufmann, Bill Paxton, and Taro Sato for their particular contributions and perspectives.

Thanks are due to Esther Singer and Reta Benhardt, who typed the lecture notes on which this book was based. But special thanks are due to Thea Howard for the electronic typesetting of the manuscript in its various and many stages and for the electronic drafting of almost all of the line figures.

Special mention and thanks are due to Leonard Parker who wrote the *Mathematica* programs for computing curvature on the book website, to Lee Lindblom, who computed the stellar models in Chapter 24, and to Matt Hansen, who read the manuscript in all its later stages, each time correcting mistakes and providing valuable suggestions.

My acknowledgments would not be complete without thanking the staffs at Addison-Wesley headed by Adam Black, Integre Technical Publishing under

Leslie Galen, and Scientific Illustrators with George Morris for their cooperation, flexibility, advice, and patience.

The book is dedicated to my wife Mary Jo, for her unstinting support in so many ways, selfless flexibility in the face of deadlines, and boundless tolerance for too-optimistic estimates of when it would be finished.

James Hartle
June, 2002

Preface to the Reissue

It is a source of great satisfaction for me to see my text reissued by Cambridge University Press.

Gravity is the force that governs our universe on the largest scales of space and time according to the laws of Einstein's general theory of relativity.

General relativity may seem mathematically complex but its underlying physical ideas are remarkably simple. Mass curves four-dimensional spacetime. Free mass moves in curves in that spacetime on paths of extremal length. The applications of general relativity to understand phenomena such as pulsars, quasars, the expansion of the universe, the big bang, and gravitational waves are conceptually simple and a source of fascination to scientists and the public at large. They are phenomena that undergraduate physics majors see on TV and read about in the press. General relativity is central to modern physics and to science in general.

When I arrived in Santa Barbara in 1966 as a new assistant professor, I instituted a course for undergraduate physics majors where they could learn something of the foundations of this beautiful, profound, and essential subject, something of its real-world astrophysical applications, and something of its experimental tests. The notes I wrote for the course were the basis for *Gravity*, written years later.

The idea was simple: teach the basic physics first, stress the real-world applications, and leave the mathematical framework for later. This way, also, a course could be taught by many faculty, not just specialists in general relativity. It worked! The result years later was *Gravity*, which has been widely used around the world. (For more, see Appendix D on Pedagogical Strategy later in the book.)

It remains for me to extend my thanks once again to all those who contributed to *Gravity* originally (acknowledged above), and now to the staff at Cambridge University Press under Vince Higgs for reissuing it. I am also grateful to Donald DeLand, formerly of Integre Technical Publishing, who was instrumental in sourcing the archived print files.

James Hartle
January, 2021

Figure Credits

Credits for figures in Boxes with multiple figures are listed in the order they appear.

Figure 1.1, Adapted from a figure prepared by Clifford Will for Hartle et al. 1999.

Figure 1.3, Courtesy of Paul Scowen and Jeff Hester (Arizona State University), and the Mt. Palomar Observatories.

Figure 1.4, Simulation and image courtesy of Rob Hynes.

Figure 1.5, Courtesy of NASA Jet Propulsion Laboratory.

Figure 1.6, Courtesy of the Boomerang Collaboration.

Box 2.1, Courtesy of Randall Rickleffs, McDonald Observatory.

Box 2.1, Courtesy of James Williams and the Jet Propulsion Laboratory, California Institute of Technology, Pasadena, California.

Box 2.1, Courtesy of James Williams and the Jet Propulsion Laboratory, California Institute of Technology, Pasadena, California.

Box 2.2, Courtesy of the Boomerang Collaboration.

Box 4.1, Courtesy of John Hall, from Brillet and Hall 1979.

Box 4.2, The author learned of these diagrams from G.W. Gibbons. The example shown is taken from Marey 1885.

Box 4.3, Courtesy of John Biretta, Space Telescope Science Institute. Data from Biretta, Moore and Cohen 1996.

Figure 6.2, Courtesy of Eric Adelberger, University of Washington.

Figure 6.3, NASA.

Box 6.2, Courtesy of Robert Nelson.

Box 7.2, Drawing of Kip Thorne by Matthew Zimet, reproduced from Thorne 1994 with the permission of W.W. Norton, Inc.

Figure 10.1, Adapted from Vessot and Levine 1979.

Figure 10.3, From Campbell and Trumpler 1923.

Figure 10.5, Courtesy of E. Fomalont and the National Radio Astronomy Observatory.

Figure 10.6, Courtesy of E. Fomalont, National Radio Astronomy Observatory.

Figure 10.7, From Shapiro et al. 1977, courtesy of Irwin Shapiro.

Figure 11.4, Courtesy of Tony Tyson, Bell Laboratories, Lucent Technologies and NASA from W. Colley, J.A. Tyson, and E. Turner 1996.

Figure 11.6, Courtesy of Kim Griest and the MACHO collaboration.

Figure 11.8, Courtesy of Chris Reynolds from data in Y. Tanaka et al., 1995.

Figure 11.9, Courtesy of the NAIC-Arecibo Observatory, a facility of the NSF, and David Parker.

Figure 11.10, Courtesy of J.H. Taylor, Princeton University.

Figure 13.1, Simulation and image courtesy of Rob Hynes.

Figure 13.2, Courtesy of Jerome Orosz from Orosz et al. 1996.

Figure 13.3, Courtesy of J.M. Moran and L. J. Greenhill based on data in Miyoshi et al., 1995 and Herrnstein et al., 1999.

Figure 13.4, Courtesy of Andrea Ghez, UCLA from data in Ghez, et al., 2002.

Figure 13.5, Courtesy of Chris Carilli and Rick Perley, NRAO

Box 14.1, Courtesy of Francis Everitt, GP-B project.

Box 15.1, Adapted from Thorne et al. 1986.

Figure 16.5, Courtesy of Caltech/LIGO.

Figure 17.1, Courtesy of Peter Challis, Harvard-Smithsonian Center for Astrophysics and AURA.

Figure 17.2, Space Telescope Science Institute and NASA.

Figure 17.3, Courtesy of NASA Goddard Space Flight Center and the COBE Science Working Group from data in Fixsen et al. 1996 and Mather et al. 1999.

Figure 17.4, From M. Roberts 1988 courtesy of M. Roberts based on data from Cram, Roberts, and Whitehurst 1980.

Figure 17.5, Courtesy of Barbara Carter, Harvard-Smithsonian Center for Astrophysics.

Figure 17.8, Courtesy of Michael Perryman from the Hipparcos Astrometric Satellite project of the European Space Agency.

Figure 17.9, Courtesy of Wendy Freedman, Carnegie Observatories from data in Persson et al. 2003.

Figure 17.10, Courtesy of Peter Challis, Harvard-Smithsonian Center for Astrophysics, Space Telescope Science Institute, and NASA.

Figure 17.11, Courtesy of Wendy Freedman, Carnegie Observatories from data in Freedman, et al. 2001.

Figure 17.12, Courtesy of NASA Goddard Space Flight Center and the COBE Science Working Group from data in Bennett et al. 1996.

Figure 17.13, Courtesy of Steve Maddox, Will Sutherland, George Efstathiou, and John Loveday.

Figure 17.14, Courtesy of John Peacock and the 2dF Galaxy Redshift Survey Team.

Box 19.1, Courtesy of David Tytler and Nao Suzuki, University of California, San Diego.

Figure 19.2, Courtesy of Adam Riess and the High-Z Supernova Search Team.

Figure 19.4, Courtesy of M. Tegmark from an analysis described in Wang, Tegmark and Zaldarriga 2001.

Figure 23.3, Courtesy of J.H. Taylor from data in J.H. Taylor and J.M. Weisberg 1989 and further unpublished updates.

Figure 24.4, Computation and graph courtesy of Lee Lindblom, California Institute of Technology.

Figure 24.5, Computation and graph courtesy of by Lee Lindblom, California Institute of Technology.

Figure 24.6, Prepared by Lee Lindblom, California Institute of Technology, from data in B. Harrison, K. Thorne, M. Wakano, and J.A. Wheeler 1965 and P. Glendenning 1985.

Figure 24.7, Computation and graph courtesy of Lee Lindblom, California Institue of Technology.

Organizational Notes

The pedagogical principles that guided the writing of this book are explained in Appendix D. However the following notes may be immediately useful in navigating the text:

- **Boxes:** The boxes contain material that illustrates or expands on the basic material in the text. Sometimes this is a qualitative explanation of a related phenomenon or idea, sometimes a description of a relevant experiment. Sometimes these are expositions that require a knowledge of physics beyond the basic mechanics and special relativity that is assumed in the text. *It is not necessary to understand the boxes to understand the text.*
- **Problems:** The labels on the problems mean the following:

  A = More algebra needed than most problems.
  B = Refers to a discussion in a Box.
  C = More challenging than most problems.
  E = Asks for an order of magnitude estimate in contrast to a calculation.
  N = Requires some computer work.
  P = Requires some aspect of physics outside the prerequites assumed to this text, e.g., electromagnetism.
  S = Straightforward (in the author's opinion.)

  A problem with no labels is just an ordinary problem, referring to the text, of average difficulty, etc.
- ***Mathematica* Programs:** Several *Mathematica* programs are provided for computing curvature quantities for general metrics, orbits, and cosmological models. These can be downloaded from the website below.
- **Website:** A website containing current information about the book can be found at the time of writing at:

  `www.cambridge.org/hartle .`

  This includes current errata, notebook files for the *Mathematica* programs, supplementary discussion (Web supplements), and links to other sites that were useful at the time of writing.
- **A few symbols:**

  $\equiv$ defined to be
  $\approx$ approximately equal to
  $\sim$ of order of magnitude
  $\rightarrow$ asymptotically approaches
  $\odot$ the Sun
  $\oplus$ the Earth

PART I

Space and Time in Newtonian Physics and Special Relativity

The major phenomena of gravitational physics are briefly described and the idea that the geometry of space and time is a physical question is introduced. Essential elements of Newtonian physics and special relativity are reviewed. Tools for describing the geometry of spacetime are developed.

CHAPTER 1

Gravitational Physics

Gravity is one of the four fundamental interactions. The classical theory of gravity—Einstein's general relativity—is the subject of this book. General relativity is central to the understanding of frontier astrophysical phenomena such as black holes, pulsars, quasars, the final destiny of stars, the big bang, and the universe itself. General relativity is also concerned with the minute departures of the orbits of the planets from the laws of Newton and is a necessary ingredient in the operation of the Global Positioning System used every day. As one of the fundamental forces, gravity is central to the quest for a unified theory of all interactions; many of the ideas for these "final theories" originate in general relativity.

Gravitational physics is thus a two-frontier science. Its important applications lie at both the largest and smallest distances considered in contemporary physics. On the largest scales, gravitational physics is linked to astrophysics and cosmology. On the smallest scales it is tied to quantum and elementary particle physics. These two frontiers become one at the big bang, where the whole of the observable universe today is compressed into the smallest possible volume. This introductory text treats only the *classical* (nonquantum) theory of gravity whose direct applications are mostly on large distance scales, but the ideas and methods developed here reemerge in different guises at the frontier of the very small. This introduction gives a brief survey of some of the phenomena for which classical general relativity is important.

The origins of general relativity can be traced to the conceptual revolution that followed Einstein's introduction of special relativity in 1905. Newton's centuries-old gravitational force law is inconsistent with special relativity. According to Newton's law, two bodies of mass m_1 and m_2 attract one another with a gravitational force whose magnitude is

$$F_{\text{grav}} = \frac{Gm_1m_2}{r_{12}^2}, \tag{1.1}$$

where r_{12} is the distance between them, and G is Newton's gravitational constant 6.67×10^{-8} dyn $\cdot$ cm^2/g^2. The Newtonian gravitational force acts instantaneously. The force on one mass depends on the position of the second at the same time. However, instantaneous interaction is prohibited in special relativity where no signal can travel faster than the speed of light. Newtonian gravity can therefore only be an approximation to a yet more fundamental theory.

In 1915, Einstein's quest for a relativistic theory of gravity resulted not in a new force law or a new theory of a relativistic gravitational field, but in a profound conceptual revolution in our views of space and time. Einstein saw that the experimental fact that all bodies fall with the same acceleration in a gravitational field led naturally to an understanding of gravity in terms of the curvature of the four-dimensional union of space and time—*spacetime*. Mass curves spacetime in its vicinity, and the trajectories along which all masses fall are the straight paths in this curved spacetime. In Newtonian theory the Sun exerts a gravitational force on the Earth and the Earth moves around the Sun in response to that force. In general relativity the mass of the Sun curves the surrounding spacetime, and the Earth moves on a straight path in that curved spacetime. Gravity is geometry.

The remainder of this chapter briefly introduces some phenomena in the universe for whose understanding general relativity is important. A few properties of the gravitational interaction that help to explain when gravity is important can already be seen from the Newtonian gravitational force law (1.1):

- Gravity is a universal interaction in Newtonian theory between all mass, and, since $E = mc^2$, in relativistic gravity between all forms of energy.
- Gravity is unscreened. There are no negative gravitational charges to cancel positive ones, and therefore it is not possible to shield (screen) the gravitational interaction. Gravity is always attractive.
- Gravity is a long-range interaction. The Newtonian force law is a $1/r^2$ interaction. There is no length scale that sets a range for gravitational interactions as there are for strong and weak interactions.
- Gravity is the weakest of the four fundamental interactions acting between individual elementary particles at accessible energy scales. The ratio of the gravitational attraction to the electromagnetic repulsion between two protons separated by a distance r is

$$\frac{F_{\text{grav}}}{F_{\text{elec}}} = \frac{Gm_p^2/r^2}{e^2/(4\pi\epsilon_0 r^2)} = \frac{Gm_p^2}{(e^2/4\pi\epsilon_0)} \sim 10^{-36}, \tag{1.2}$$

  where m_p is the mass of the proton and e is its charge.

These four facts explain a great deal about the role gravity plays in physical phenomena. They explain, for example, why, although it is the weakest force, gravity governs the organization of the universe on the largest distance scales of astrophysics and cosmology. These distance scales are far beyond the subatomic ranges of the strong and weak interactions. Electromagnetic interactions *could* be long range were there any large-scale objects with net electric charge. But the universe is electrically neutral, and electromagnetic forces are so much stronger than gravitational forces that any large-scale net charge is quickly neutralized. Gravity is left to govern the structure of the universe on the largest scales.

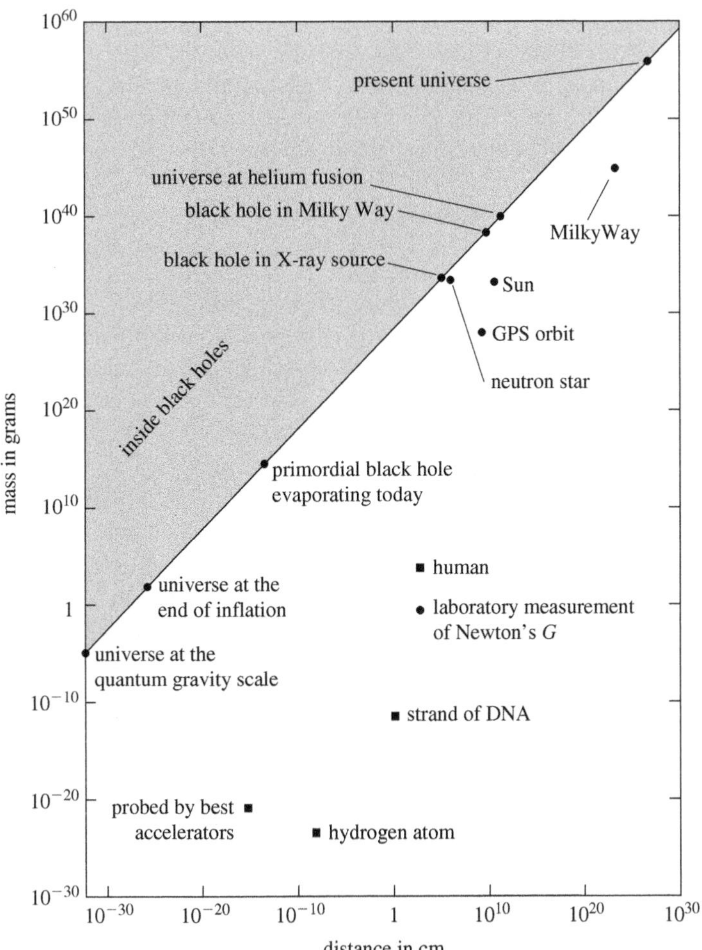

FIGURE 1.1 Gravitational physics deals with phenomena on scales of distance and mass ranging from the microscopic to the cosmic—the largest range of scales considered in contemporary physics. There are phenomena for which gravity is important over this whole range of scales that are shown on this plot of characteristic mass M vs. characteristic distance R. Representative ones are indicated by circles. Other illustrative phenomena where gravitation plays little role are shown by squares. Phenomena above the diagonal line are unobservable, because they take place inside black holes. Phenomena close to the diagonal line $2GM = c^2R$ are the ones for which relativistic gravity is important. The largest scales are the frontier of astrophysics; the smallest are those of elementary particle physics. The smallest distance shown ($\sim 10^{-33}$ cm) is the Planck length marking the boundary between classical and quantum gravity. Scales referring to the universe at various moments in its history denote the size of the volume that light could travel across since the big bang and the mass inside that volume if the universe always had the expansion rate it had at that moment.

This book is not concerned with all phenomena for which gravity is important but rather with phenomena for which *relativistic* gravity is important. Newtonian gravity, for instance, is adequate for understanding the internal structure of the Sun. It turns out that relativistic gravity becomes important for an object of mass M and size R only when the characteristic dimensionless ratio formed with the velocity of light c,

$$\frac{GM}{Rc^2}, \tag{1.3}$$

is a significant fraction of unity. Figure 1.1 shows a range of phenomena in the universe and their characteristic values of M and R. The ones closest to the line $2GM = c^2R$ are the ones for which relativistic gravity is most important. We now describe a few of these in more detail.

Precision Gravity in the Solar System

By the measure (1.3) the Earth is not a very relativistic system: $GM_{\oplus}/c^2R_{\oplus} \sim 10^{-9}$. (The astronomical symbol for the Earth is $\oplus$.) Yet such is the precision required in clocks at the heart of the Global Positioning System (GPS) (Figure 1.2) that it would fail in about half an hour were the effects of general relativity not taken into account in their operation (Chapter 6).

For the Sun ($\odot$), $GM_{\odot}/c^2R_{\odot} \sim 10^{-6}$. General relativistic effects on the orbits of the planets are therefore small, but they are detectable in precise observations. For example, the precise amount by which the position of the Mercury's closest approach to the Sun shifts in each orbit is a test of general relativity. General

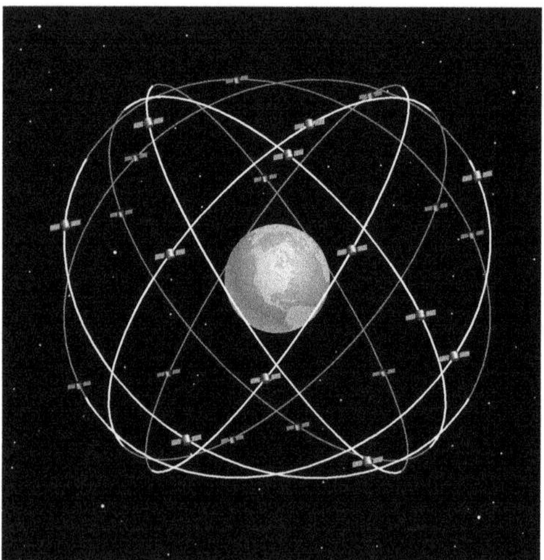

FIGURE 1.2 The configuration of satellites for the Global Positioning System, for which the tiny effects of general relativity are important.

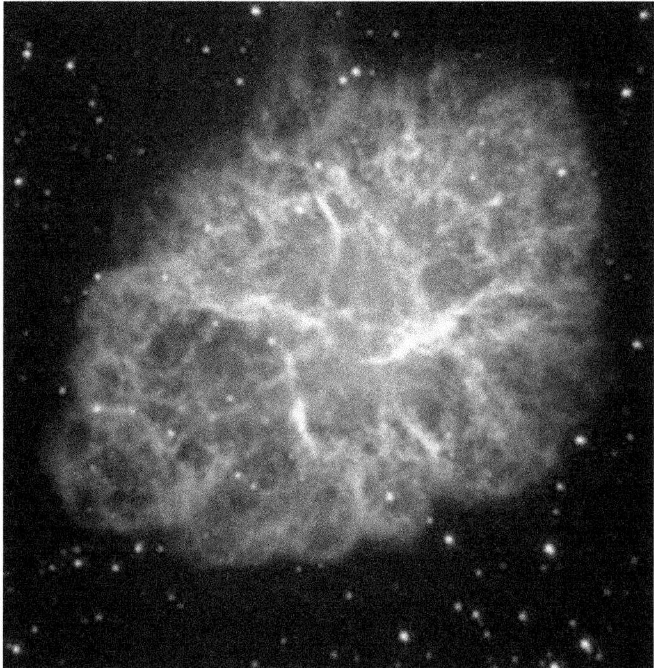

FIGURE 1.3 The Crab Nebula. The remnant of a supernova explosion whose light reached Earth in AD 1054. The nebula is powered by a rotating relativistic neutron star at its core.

relativity predicts that the paths of light rays will be bent when they pass near the Sun and that their time of passage is increased over that predicted by Newtonian theory—both tiny effects that are today routinely incorporated in precision astronomical observations (Chapter 10).

Relativistic Stars

Most stars support themselves against the ever present attractive forces of gravity by the pressure of gas heated by thermonuclear reactions at their cores. When a star runs out of thermonuclear fuel, gravitational collapse ensues. The cores of some collapsing stars wind up supported by nonthermal sources of pressure leading to highly compact white dwarf and neutron stars. With masses on the order of a solar mass and radii of order 10 km, neutron stars are relativistic objects, $GM/c^2R \sim 0.1$, whose properties are discussed in Chapter 24. There is a maximum mass for neutron stars and white dwarfs of a few solar masses. The ongoing collapse of more massive cores leads to black holes.

Black Holes

General relativity predicts that a black hole is created whenever mass is compressed into a volume small enough that the gravitational pull at the surface is too large for anything to escape, even light (Chapters 12 and 15). In Newtonian mechanics, a particle of mass m starting at radius R with velocity V escapes the gravitational attraction of a mass M when its initial velocity is greater than the escape velocity, V_{escape}, at which its kinetic energy balances its negative gravitational potential energy, namely,

$$\frac{1}{2}mV_{\text{escape}}^2 = \frac{GmM}{R}. \tag{1.4}$$

The escape velocity exceeds the velocity of light when

$$\frac{2GM}{c^2R} > 1. \tag{1.5}$$

Although Newtonian analysis is not applicable to a relativistic situation, (1.5) turns out to be the correct relativistic criterion for a spherical mass to be a black hole with R properly interpreted.

The surface that defines a black hole is called its *event horizon*. Mass, information, and observers can fall through it, but, in classical physics, nothing can emerge from it. Although created in nature through often messy gravitational collapse, general relativity predicts that black holes are remarkably simple objects characterized by just a few numbers. As S. Chandrasekhar put it, "The black holes of nature are the most perfect macroscopic objects there are in the universe: the only elements in their construction are our concepts of space and time. And

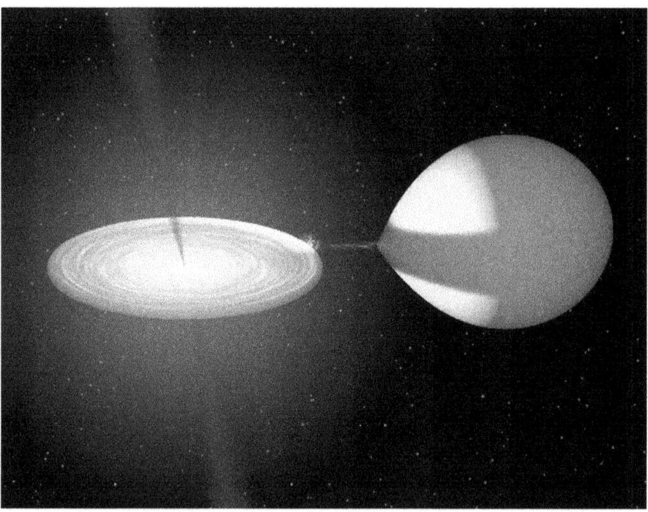

FIGURE 1.4 Simulated image of the X-ray binary GRO J1655-40. A massive star at right is orbiting a black hole (not visible) and shedding mass that falls toward the black hole and forms a disk about it that is so hot it emits X-rays.

since general relativity provides only a single unique family of solutions for their description, they are the simplest objects as well" (Chandrasekhar 1983).

Black holes of a few solar masses have been detected in orbit around a companion star. Supermassive black holes of up to approximately a billion solar masses have been detected at the centers of galaxies. At the center of our own Milky Way there is an approximately three-million solar-mass black hole. Indeed, at the time of writing, there is growing evidence that *all* sufficiently massive galaxies have black holes at their cores.

Although black holes are dark themselves, the strongly curved spacetime around them is the arena for some of the most dramatic phenomena in contemporary astrophysics. Matter falling towards a black hole goes into orbit about it, creating a hot disk that is the source of the radiation from X-ray sources (Figure 1.4). Matter flowing onto a rotating, magnetized black hole is the powerhouse for quasars. Black holes may well be behind gamma ray bursts, which include the biggest explosions since the big bang. (The detection of black holes and their astrophysical importance are the subjects of Chapter 13.)

Gravitational Waves

General relativity predicts that ripples in spacetime curvature can propagate with the speed of light through otherwise empty space. These ripples are *gravitational waves* (Chapter 16). Any mass in nonspherical, nonrectilinear motion produces gravitational waves (Chapter 23), but gravitational waves are produced most copiously in events such as the coalescence of two compact stars, the merger of massive black holes, or the big bang. Mass is in motion in many places in the universe, and this gravitational analog of charge is unscreened. The universe is, therefore, not especially dim in gravitational radiation. Indeed, coalescing black holes at the heart of pairs of merging galaxies could be the most energetic events in the universe with most energy emitted in gravitational waves. The weak coupling to matter (1.2) makes gravitational radiation difficult to detect. However, that same weak coupling is what makes detecting gravitational radiation so interesting. Once produced, little is absorbed. Therefore, gravitational waves could provide a new window on the universe that would enable us to see to the earliest moments of the big bang and to the heart of the formation of black holes.

Gravitational radiation, never directly received on Earth, has been detected by its effect on the orbits of bodies emitting the radiation. The waves can be detected by precise measurements of the relative motion of masses produced as the ripple of spacetime curvature passes by. But waves from the binary star system that is brightest in gravitational radiation at Earth produce a fractional change in the distance between two test masses that is of order only of 1 part in 10^{20}. That is a change smaller than an atom for the 5,000,000-km size of the largest gravitational wave detectors in space contemplated at the time of writing (Figure 1.5).

As big as the experimental challenge is, detectors are now under construction on the surface of the Earth and under study for space that will make gravitational wave astronomy a realistic possibility in the first decades of the 21st century.

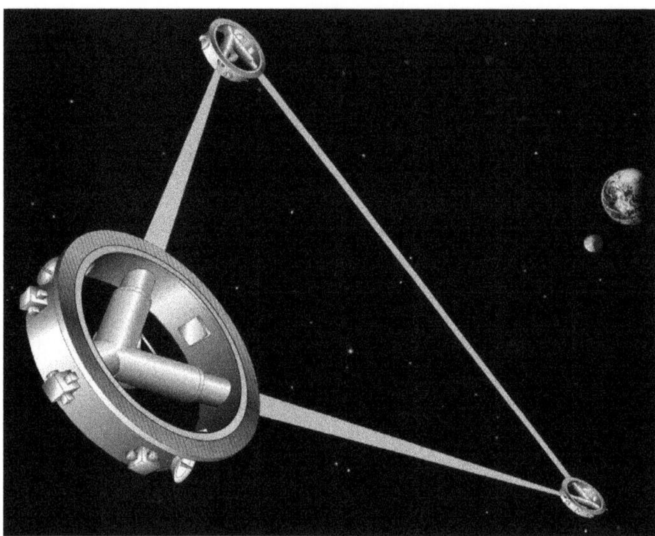

FIGURE 1.5 Artist's conception of the LISA gravitational wave interferometer in space. Laser beams connect three detectors in space separated by 5,000,000 km. Gravitational waves can be detected by observing the small changes they produce in the distances between detectors.

The Universe

As mentioned earlier, gravity governs the structure and evolution of the universe on the largest scales of space and time. These are the scales of *cosmology* (Chapters 17–19).

Observations of the motion of galaxies show our universe is expanding. Observations of their distribution on the largest distance scales show our universe to be remarkably regular today—much the same on average in all places and in all directions. Observations of the cosmic background radiation produced in the big bang show the universe to have been even more regular at the beginning. General relativity predicts how the geometry of space can be curved for such a regular universe. It also governs the evolution of the universe in time, allowing us to understand its origin and history as well as predict its future fate.

General relativity plus present observations imply the universe began in a big bang—a singular moment of infinite density, infinite pressure, and infinite spacetime curvature. Although extreme in these measures, the big bang was remarkably regular in space. Indeed, it is possible that the only deviations from exact uniformity were tiny quantum fluctuations in the density of matter, which condensed under gravitational attraction to eventually become the stars and galaxies we see today. Many properties of the large-scale universe result from the mutual operation of gravitational and particle physics in the earliest moments. Besides planting the seeds of today's large-scale distribution of matter, the earliest moments fixed the abundance of matter to antimatter, matter to electromagnetic, neutrino, and gravitational radiation, and the primordial abundances of the chemical elements.

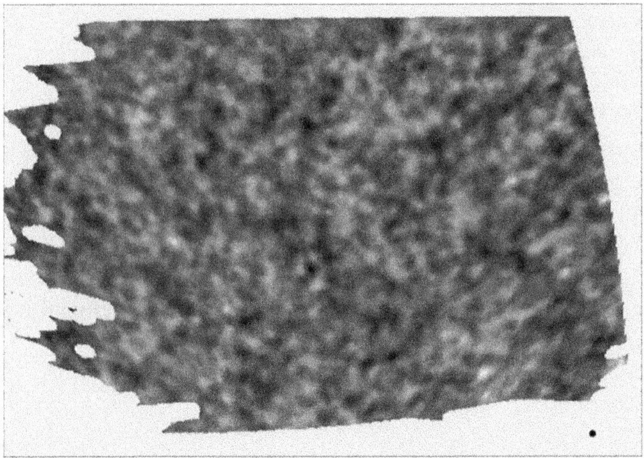

FIGURE 1.6 A picture of the universe some hundreds of thousands of years after the big bang. This map from the Boomerang experiment shows the temperature fluctuations in the microwave background radiation corresponding to irregularities in the universe that later developed into galaxies. The difference in temperature between the lightest and darkest regions is one of order a milliKelvin.

Quantum Gravity

Although this text on classical gravity will touch upon it in only one place (Chapter 13), quantum spacetime deserves to be mentioned in any survey of important phenomena in gravitational physics. Planck's constant $\hbar$ characterizes all quantum phenomena. Quantum gravitational phenomena are characterized by the unique combinations of $\hbar$, G, and c with the dimensions of length, time, energy, and density:

$$\begin{aligned} \ell_{\text{Pl}} &\equiv (G\hbar/c^3)^{1/2} = 1.62 \times 10^{-33}\ \text{cm}, \\ t_{\text{Pl}} &\equiv (G\hbar/c^5)^{1/2} = 5.39 \times 10^{-44}\ \text{s}, \\ E_{\text{Pl}} &\equiv (\hbar c^5/G)^{1/2} = 1.22 \times 10^{19}\ \text{GeV}, \\ \rho_{\text{Pl}} &\equiv c^5/\hbar G^2 = 5.16 \times 10^{93}\ \text{g/cm}^3. \end{aligned} \tag{1.6}$$

These are called the Planck length, the Planck time, the Planck energy, and the Planck density, respectively. Einstein's classical theory of gravity is no longer applicable to phenomena characterized by these scales, because significant quantum fluctuations in the classical geometry of spacetime can be expected. In these regimes, Einstein's theory needs to be replaced by a quantum theory of gravity for which general relativity is the classical limit.

Even a casual glance at the numbers in (1.6) reveals that the domain in which quantum space and time are important is both far from everyday experience and from accessible experiment. As far as we know, there are only two places in the universe where conditions characterized by the Planck scales are realized—the big bang in which the universe started (Chapters 17–19) and the quantum evaporation of black holes (Chapter 13). Yet quantum gravity lies squarely at two frontiers of contemporary physics. The first is the search for a unified theory of the fundamental interactions, including gravity, whose simplicity would emerge at high energies comparable to E_{Pl}. The second is the search for a quantum initial condition of the universe. In the early universe, at the big bang, large and small are one. The largest system is compressed into the smallest size reaching the highest energies. Quantum gravity will not be discussed in this book, but the classical theory of gravity developed here is a prerequisite to understanding this frontier of contemporary physics.

CHAPTER 2

Geometry as Physics

This book is about space, time, and gravity because (as mentioned briefly in Chapter 1) the central idea of general relativity is that gravity arises from the curvature of *spacetime*—the four-dimensional union of space and time. Gravity *is* geometry. This chapter expands a little on the idea that gravity is geometry and then describes how the geometry of space and time is a subject for experiment and theory in physics.

2.1 Gravity Is Geometry

It is an experimental fact that all bodies fall with the same acceleration in a uniform gravitational field—independently of their composition. If Galileo could have dropped a cannonball and a feather from the leaning tower of Pisa in a vacuum, they both would have accelerated towards the ground at 980 $\mathrm{cm/s^2}$. This equality of accelerations is one of the most accurately tested facts in physics. For example, at the time of writing, the accelerations of the Earth and the Moon as they fall toward the Sun are known to be equal to an accuracy of 1.5×10^{-13}. (See Box 2.1 on p. 14; more in Chapter 6.) This experimental fact underlies general relativity.

Figure 2.1 shows a time vs. space plot of the height h of a ball thrown straight upward from the surface of the Earth as a function of time. The ball starts with

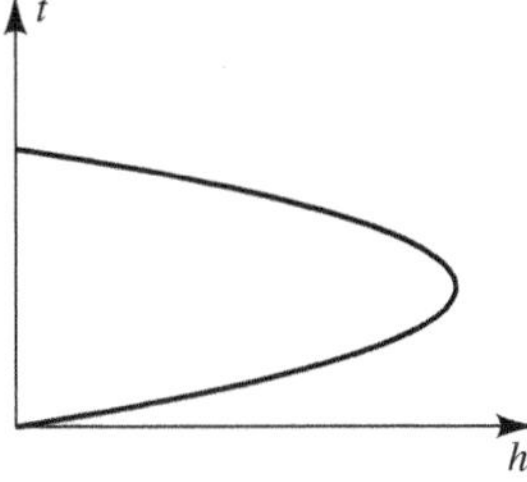

FIGURE 2.1 A ball thrown upward from the surface of the Earth with an initial speed decelerates with the acceleration of gravity, $g = 980\ \mathrm{cm/s^2}$, reaches a maximum height, and returns to Earth. The figure shows the characteristic parabolic curve of time t vs. height h for a particular initial speed plotted with the time axis vertical as is standard in relativity. Any other body thrown upward with the same initial speed would follow the same spacetime curve. In Einstein's general relativity, the bodies are following a straight path in the curved spacetime produced by the Earth's mass.

BOX 2.1 Lunar Laser Ranging Test of the Equality of Accelerations in a Gravitational Field

The most accurate test of the fact that all bodies fall with the same acceleration in a gravitational field to date does not come from a laboratory on Earth but comes from comparing the accelerations of the Earth and the Moon as they fall around the Sun. These match to within a fractional error of less than 1.5×10^{-13} (Williams et al. 1996, Anderson and Williams 2001).

The test is carried out using very precise positions of the Moon relative to the Earth over time determined by measuring the round-trip travel time of a laser pulse from the Earth returned by reflectors on the Moon. This is called *lunar laser ranging*. Currently, the distance to the Moon can be determined to a few centimeters out of a mean Earth–Moon distance of 384,401 km—an accuracy of one part in 10^{10}!

The key to these measurements are corner-cube retroreflectors consisting of three reflecting sides of a cube meeting in one corner. This geometry has the useful property that any incident light ray is reflected back in the direction from whence it came, no matter from what direction it is incident. (See Problem 1.) The Apollo 11, 14, and 15 Moon missions in 1969 and 1971 left behind arrays of from one to three hundred corner reflectors at various locations on the Moon. An additional Russian-French array was left by the Lunakhod II

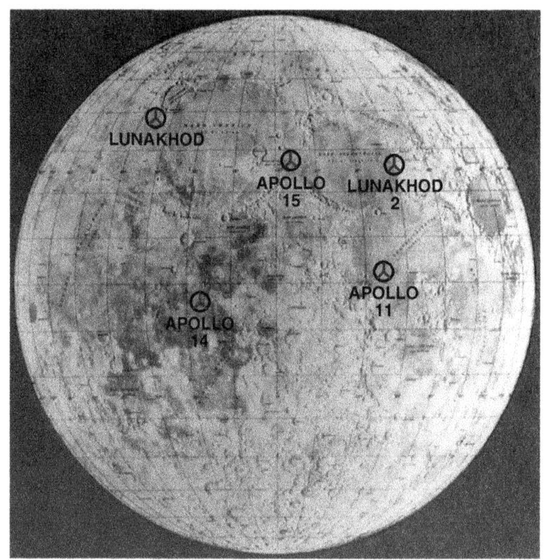

Retroreflector Sites.

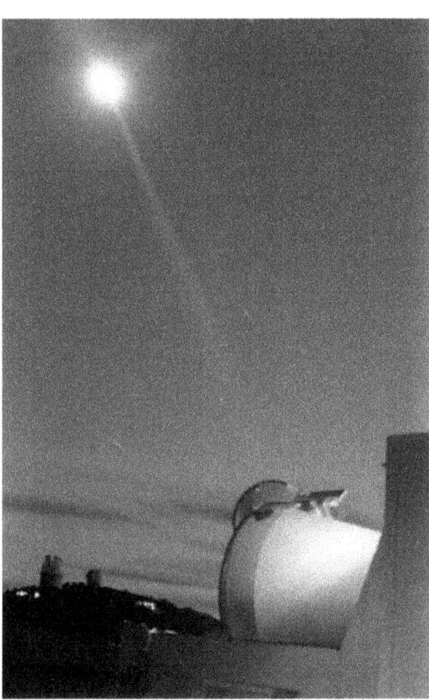

Laser pulse to the Moon at McDonald Observatory.

Retroreflectors on the Moon.

BOX 2.1 (*continued*)

unmanned spacecraft in 1973. Since 1969 a systematic program to determine the Moon's orbit using these devices has been carried out mainly at the McDonald Observatory at Mt. Locke, in Texas, and the Observatoire de Côte d'Azur station in Grasse, France. The lasers currently in use send pulses lasting 200 picoseconds, each containing about 10^{18} photons, about 10 times per second. Diffraction, refraction in the atmosphere, and other effects spread the beam over a 7-km radius on the Moon so that only 10^{-9} of the photons that are sent impinge on the retroreflector. On return, the reflected spot is spread over 20 km so that a 1-m telescope would detect only 10^{-9} of the returning photons. In the end, only one reflected photon is detected every few seconds. Returning photons have been detected for more than thirty years since 1970 at the time of writing.

some initial velocity, decelerates, reaches a maximum height, accelerates downward, and returns to the surface of the Earth. Any other body thrown upward from the same initial position with the same initial velocity would follow *exactly the same curve*.

This uniqueness of trajectory in space and time is a special property of gravity. The motion of a body in a magnetic field depends on what kind of charge it has. Bodies with one sign of charge will be deflected one way, bodies with the opposite charge will be deflected the other, and bodies with no charge will not be deflected at all. Only in a gravitational field do all bodies with the same initial conditions follow the same curve in space and time.

Einstein's idea was that this uniqueness of path in spacetime could be explained in terms of the geometry of the four-dimensional union of space and time called *spacetime*. Specifically, he proposed that the presence of a mass such as Earth curves the geometry of spacetime nearby, and that, in the absence of any other forces, all bodies move on the straight paths in this curved spacetime. Bodies free from forces move on straight lines of three-dimensional Euclidean space in Newtonian mechanics—that's part of Newton's first law. Einstein's idea is that the Earth moves in its orbit around the Sun, not because a force of gravity acts on it, but because it is following the straightest possible path in the slightly non-Euclidean geometry of spacetime produced by the Sun.

2.2 Experiments in Geometry

There is a story that in the late 1820s the great mathematician C. F. Gauss carried out an experiment to verify one of the standard theorems of the Euclidean geometry of space—that the interior angles of a triangle add up to 180°. Using the mountaintops of Hohenhagen, Brocken, and Inselsberg as vertices and assuming that light rays move on straight lines, he is said to have measured the angles, found the sum, and determined 180° to the accuracy with which the angles could be measured. (See Figure 2.2.)

The historical evidence is not conclusive as to whether Gauss actually carried out this experiment. However, he might have done so, and that emphasizes an important point. The sum of the angles was not guaranteed to be 180° from logic alone. Many geometries of physical space are possible that are different from

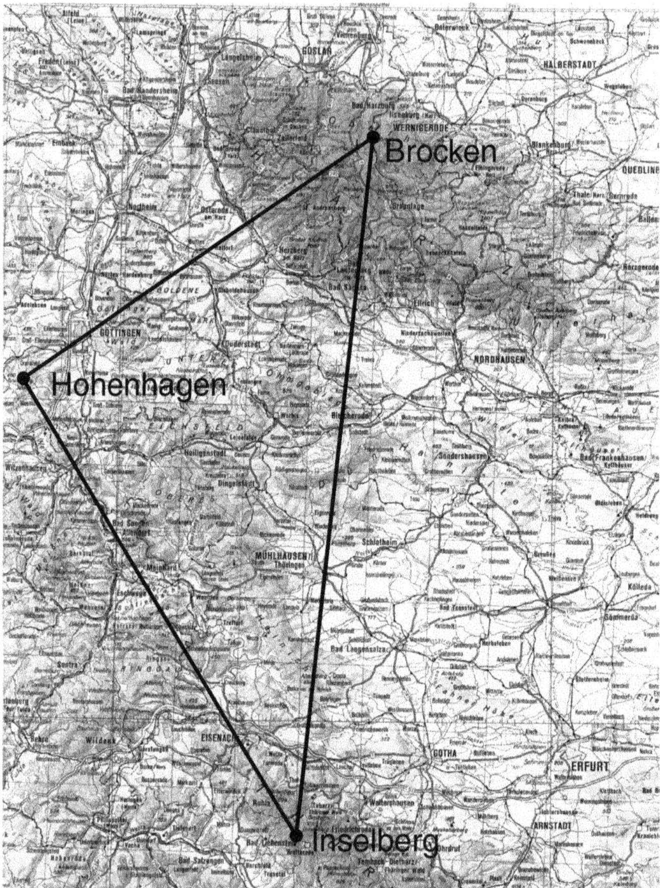

FIGURE 2.2 A modern map of Germany showing the locations of the peaks Hohenhagen, Brocken, and Inselberg, which form the vertices of a triangle that Gauss could have surveyed to check whether the interior angles add up to 180°, as predicted by Euclidean geometry.

Euclid's. These predict different results for the sum of the interior angles of a triangle. The geometry of space is an empirical question. It is a question in physics, subject to measurement, hypothesis, and test. By the end of this book you will know that if Gauss had been able to carry out his experiment with sufficient accuracy, he would have found a small difference in the sum of the angles just due to the mass of the Earth, $M_\oplus$, of order

$$\left|\begin{pmatrix}\text{sum of interior angles}\\ \text{of a triangle in radians}\end{pmatrix}\right| - \pi \sim \frac{(\text{area of triangle})}{R_\oplus^2}\left(\frac{GM_\oplus}{R_\oplus c^2}\right) \tag{2.1}$$

(where $R_\oplus$ is the Earth's radius) together with contributions from the Sun and the other planets. Note the appearance of the ratio GM/Rc^2, which is characteristic of weak relativistic effects, as discussed in Chapter 1. The distances between the mountains are 69 km, 85 km, and 107 km. Using these, this works out to be a difference of order 10^{-15} radians (!). So small a discrepancy could not be detected even with present technology, but modern experiments can detect the deviations from Euclidean geometry produced by the Sun and measure the geometry of space on the very large scales of cosmology. (See Box 2.2.)

BOX 2.2 Determining the Spatial Geometry of the Universe

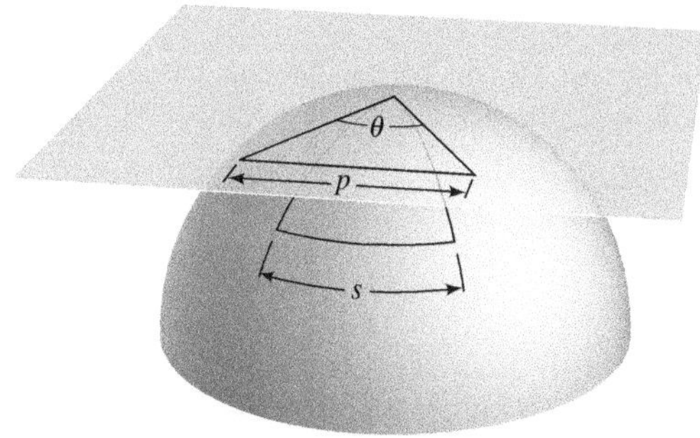

Modern measurements—not so different in character from that attributed to Gauss—determine the curvature of space on the distance scale of the visible universe. General relativity plus observations of the distribution of galaxies and radiation in the universe suggest only a few possibilities for the large-scale geometry of three-dimensional space at a moment of time, as we will see in detail in Chapter 18. The flat geometry of the plane, the positively curved geometry of the surface of a sphere, or the negatively curved geometry of a surface locally like some potato chips[a] are two-dimensional analogs of the possible flat, positively curved, and negatively curved large-scale geometries for three-dimensional space. How can the geometry of space in our universe be measured?

To understand a little of one method, imagine the geometry of space to be fixed in time (in contrast to the geometry of the actual universe, which is expanding). Imagine further that objects of known size p could be identified a known distance, d, away. If the geometry were flat like a plane, the angle θ subtended by these objects would be p/d. But, as illustrated in the accompanying figure, if the geometry were positively curved like the surface of a sphere, an object of smaller size s would subtend the same angle.[b] Alternatively, an object of a given size and distance away will subtend a larger angle on a positively curved surface like a sphere than it will in a flat plane (Problem 6). In a similar way, the angle subtended in a negatively curved space will be smaller. (For details, see Problem 18.12.) This discussion will be corrected to include the expansion of the universe in Chapter 19, but the qualitative result is the same: measuring the angular size of features of known size and distance is one way of determining whether the geometry of intervening space is flat, positively curved, or negatively curved. The cosmic background radiation provides such features.

The cosmic microwave background radiation (CMB) is light from the hot big bang that began the universe. The radiation started from the moment the universe had

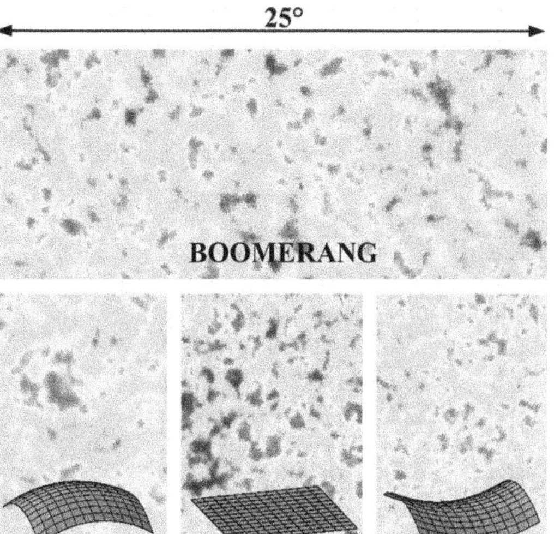

[a]"Crisps" in the UK and elsewhere.

[b]If that is not clear from thinking about the figure, imagine the sphere is made of rubber and could be flattened out on a plane tangent to the North Pole. The angles between lines of longitude at the North Pole do not change, but the equator and other lines of latitude have to be stretched. Thus the angle subtended by an object spanning a range of longitude on a sphere is the same as that subtended by a larger object in a plane.

BOX 2.2 (*continued*)

expanded and cooled enough for the matter to be transparent to radiation. It has been propagating freely to us over the intervening approximately 14 billion years. Were the universe unchanging in time, it would have traveled approximately 14 billion light-years. The temperature of the radiation has cooled to 2.73°K and is very nearly the same in all directions but not quite. Tiny temperature fluctuations of only a few tens of millionths of a degree are observed. The theory of the origin of these fluctuations predicts a spectrum of sizes that is characterized by a known length scale. The fluctuations are, therefore, features with a known spectrum of sizes a known distance away. Observations of their angular size can thus measure the spatial geometry of the universe. The right hand figure on the previous page shows a map of the temperature fluctuations in a 25°-wide region of the sky that were observed by the Boomerang experiment (de-Bernardis et al. 2000). The three figures on the bottom show simulations of what the map would look like based on the theoretical spectrum of original sizes if the geometry were positively curved (left), flat (middle), or negatively curved (right). Quantitative comparisons of the spectrum of angular sizes show that the geometry is very close to flat. (In the near future there will be a more accurate result, but the idea will be the same.) The geometry of space is a measurable, physical question.

2.3 Different Geometries

The idea of different geometries is easily illustrated in two dimensions. In your studies of the Euclidean geometry of the plane, you met the notions of point, straight line, distance, angle, parallel, triangle, circle, chord, etc. Familiar theorems include the one just discussed for a triangle:

$$\sum_{\text{vertices}} \begin{pmatrix}\text{interior}\\ \text{angle}\end{pmatrix} = \pi. \tag{2.2}$$

Another relates the ratio of the circumference to the radius of a circle:

$$\frac{(\text{circumference})}{(\text{radius})} = \frac{C}{r} = 2\pi. \tag{2.3}$$

The surface of a sphere provides an example of a *different* two-dimensional geometry in which such results of plane geometry are replaced by different theorems. Straight lines can be defined on a sphere as the shortest distance between two points, that is, as segments of great circles. Triangles are made up of three intersecting great circles. A circle is the locus of points equidistant (as measured on the surface) from a point which is its center, etc. For a spherical triangle of area A,

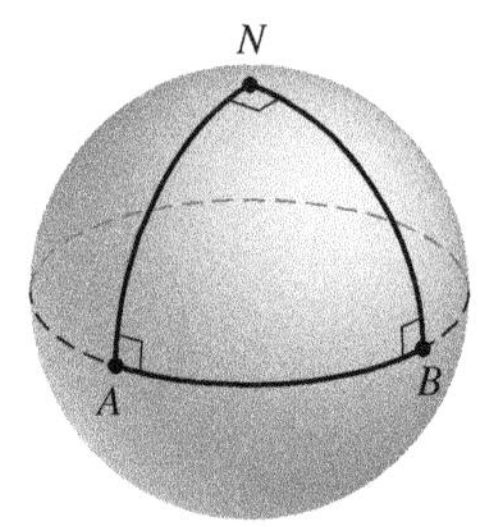

FIGURE 2.3 A spherical triangle NAB where the sum of the interior angles is 270°. The triangle consists of the parts of two lines of longitude 90° apart from the North Pole to the equator and the part of the equator between them. These are all segments of great circles and, therefore, straight lines in the geometry of the sphere.

$$\sum_{\text{vertices}} \begin{pmatrix}\text{interior}\\ \text{angle}\end{pmatrix} = \pi + \frac{A}{a^2} \tag{2.4}$$

where a is the radius of the sphere.[1]

Equation 2.4 shows that the sum of the interior angles of a spherical triangle is always *greater* than π. An example is shown in Figure 2.3. As the size of the triangle becomes smaller and smaller compared with the radius of curvature a, it

[1]Don't get (2.4) mixed up with (2.1). Eq. (2.1) is for triangles formed by light rays in the four-dimensional curved spacetime that surrounds the Earth according to general relativity. Eq. (2.4) is for triangles made up of segments of great circles on the surface of a two-dimensional sphere.

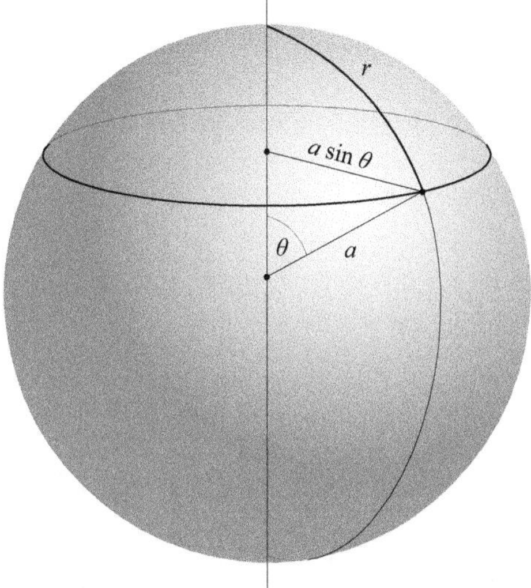

FIGURE 2.4 The relation of the circumference to the radius of a circle in the geometry of the surface of a sphere. A circle is the locus of points on the surface that are equidistant (as measured on the surface) from its center point. In this figure the North Pole has been chosen to coincide with the center of a circle that is a line of constant latitude labeled by θ. The radius r is the distance from the North Pole to this latitude measured along any line of constant longitude.

becomes increasingly difficult to tell the difference between a flat plane and the curved surface of the sphere. For triangles with very small areas ($A/a^2 \ll 1$), the result (2.4) is well approximated by the flat space result (2.2).

With the bit of geometry shown in Figure 2.4, the ratio of the circumference to the radius of a circle on a sphere can be calculated to be

$$\frac{\text{(circumference)}}{\text{(radius)}} = \frac{C}{r} = 2\pi \frac{\sin(r/a)}{(r/a)}. \tag{2.5}$$

Again, if $r \ll a$, the right-hand side reduces to the flat-space result (2.3).

It is not necessary to leave the surface of the earth to determine its geometry. Surveyors (such as Gauss) working on the surface of the earth can measure such things as the interior angles of a triangle and the circumference and radius of circles. By fitting to formulas such as (2.4) and (2.5), they could, in principle, tell if the geometry of the surface was spherical and determine the radius of curvature a. Similarly, by surveying in three dimensions we can, in principle, determine the geometry of space without needing any extra dimensions.

Visualization of three-dimensional curved geometries is not as easy as for two-dimensional curved geometries, which can often be represented as surfaces in Euclidean three-dimensional space. However some simple three-dimensional geometries can be thought of as curved surfaces in a hypothetical four-dimensional Euclidean space. For example, the three-dimensional geometry analogous to the two-dimensional sphere discussed before is the three-dimensional surface of a sphere in four dimensions—a *three-sphere*. If space had a fixed three-sphere geometry, a journey in a straight line in any direction would eventually bring one back to the starting point. However, more detailed information about the geometry of space can be determined locally. For example, it turns out[2] that the volume inside a two-dimensional sphere of radius r in such a spatial geometry is given by

$$V = 4\pi a^3 \left\{ \frac{1}{2} \sin^{-1}\left(\frac{r}{a}\right) - \frac{r}{2a}\left[1 - \left(\frac{r}{a}\right)^2\right]^{1/2} \right\} \tag{2.6a}$$

$$\approx \frac{4\pi r^3}{3}\left[1 + \begin{pmatrix}\text{corrections} \\ \text{of order } (r/a)^2\end{pmatrix}\right], \qquad \text{for small } \frac{r}{a}, \tag{2.6b}$$

where a is the characteristic radius of curvature of the three-sphere geometry. For a two-sphere whose radius is much smaller than a, the volume-radius relation approaches the Euclidean flat-space result, as (2.6b) shows. If three-dimensional space had such a three-sphere geometry, the characteristic radius of curvature a could be determined by careful measurements of the radii and volume of two-spheres. As we will discover in Chapter 18, Einstein's theory predicts this three-sphere geometry as one possibility for the spatial geometry of a uniform universe on very large distance scales. Box 2.2 on p. 17 describes one effort to survey space on these scales.

2.4 Specifying Geometry

In addition to the geometry of the plane and the geometry of the sphere, there are an infinite number of other two-dimensional geometries. For example, there is the geometry on the surface of an egg or the geometry of the surface of a plane with a few hills on it. In three dimensions there are a similarly infinite number of geometries. How are these different geometries described and compared mathematically?

One way to describe a geometry is to embed it as a surface in a higher-dimensional space whose geometry is Euclidean. We have made use of this method in describing two-dimensional geometries as surfaces embedded in three-dimensional Euclidean geometry—planes, spheres, eggs, etc. However, it becomes almost impossible to think of any but the simplest three- and four-dimensional geometries as surfaces in four and five dimensions. Further, the extra dimension is physically superfluous. An *intrinsic* description of geometry that makes use of just the physical dimensions that can be measured is what is needed.

Another idea is to specify a geometry by giving a small number of *axioms*, or postulates, from which the other results of geometry can be derived as theorems. For the geometry of the flat plane, for example, there are Euclid's five axioms: Two points determine a unique line, parallel lines never intersect, etc. Some other

[2]This result will be derived explicitly in Example 7.6.

simple geometries can be characterized in this way with different axioms. For example, the geometry on the surface of a sphere can be summarized by a set of axioms like Euclid's in which the parallel postulate is replaced by the axiom that two parallel lines *always* meet in two points. However, this method also is limited. What axioms describe the geometry on the surface of a potato? We need a more local and detailed description.

The key to a general description of geometry is to use differential and integral calculus to reduce all geometry to a specification of the distance between each pair of nearby (infinitesimally separated) points. From the distance between nearby points, the distances along curves can be built up by integration. Straight lines are the curves of the shortest distance between two points. Angles are ratios of the lengths of arcs to their radii when those radii are small. Areas, volumes, etc., can be constructed by multiple integrals over area and volume elements, themselves specified by the distances between nearby points. By specifying the distances between nearby points and using differential and integral calculus, the most general geometry may be speci-fied. This area of mathematics is called *differential geometry*. We will explore just a few ideas of this subject in the next section.

2.5 Coordinates and Line Element

The Euclidean Geometry of a Plane

A systematic way of labeling points is a prerequisite to a specification of the distance between nearby ones. A system of coordinates assigns unique labels to each point, and there are many systems that do so. In two dimensions, for instance, there are Cartesian coordinates (x, y), polar coordinates (r, ϕ) about some origin, etc. (Figure 2.5).

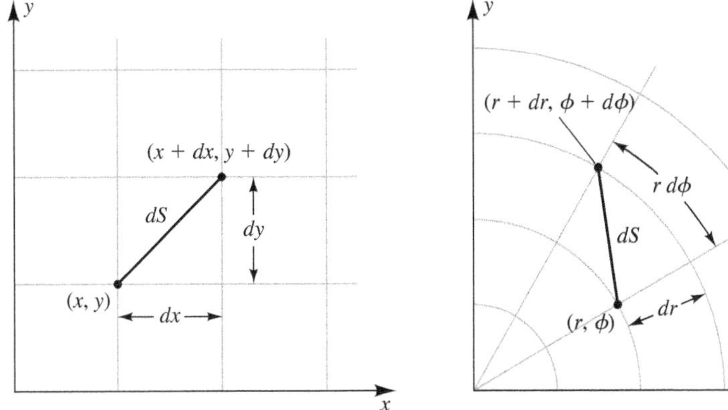

FIGURE 2.5 Cartesian and polar coordinates. Cartesian and polar coordinates are both systematic ways of labeling points in the plane, and the distance between nearby points can be expressed in terms of either.

Nearby points have nearby values of their coordinates. For example, the points (x, y) and $(x+dx, y+dy)$ are nearby when dx and dy are infinitesimal. Similarly, (r, ϕ) and $(r + dr, \phi + d\phi)$ are nearby.

In Cartesian coordinates (x, y), the distance dS between the points (x, y) and $(x + dx, y + dy)$ is (see Figure 2.5)

$$dS = \left[(dx)^2 + (dy)^2\right]^{1/2}. \tag{2.7}$$

The same rule can be expressed in polar coordinates where the distance between the nearby points (r, ϕ) and $(r + dr, \phi + d\phi)$ is (see Figure 2.5)

$$dS = \left[(dr)^2 + (r\,d\phi)^2\right]^{1/2}. \tag{2.8}$$

Expression (2.8) and others like it are valid only if dr and $d\phi$ are small. However, large distances can be built up from these infinitesimal relations by integration. Let's, for example, calculate the ratio of the circumference to the diameter of a circle of radius R. Choosing the origin at the center, the equation for such a circle in Cartesian coordinates is

$$x^2 + y^2 = R^2. \tag{2.9}$$

The circumference C is the integral of dS around the circle. Using (2.7) this is

$$C = \oint dS = \oint \left[(dx)^2 + (dy)^2\right]^{1/2} \tag{2.10a}$$

$$= 2\int_{-R}^{+R} dx \left[1 + \left(\frac{dy}{dx}\right)^2\right]^{1/2}_{x^2+y^2=R^2} \tag{2.10b}$$

$$= 2\int_{-R}^{+R} dx \sqrt{\frac{R^2}{R^2 - x^2}}. \tag{2.10c}$$

Changing variables by writing $x = R\xi$, we have

$$C = 2R\int_{-1}^{1} \frac{d\xi}{\sqrt{1-\xi^2}} = 2\pi R. \tag{2.11}$$

This is the correct answer. The integral could even be taken to define π; by doing it numerically, one could discover that $\pi = 3.1415926535\ldots$.

Deriving the relation between radius and circumference is even easier in polar coordinates, where the equation of the circle is just $r = R$. Evaluating (2.8) on the circle and integrating the resulting dS over it gives

$$C = \oint dS = \int_0^{2\pi} R\,d\phi = 2\pi R. \tag{2.12}$$

The ease of using polar coordinates to arrive at (2.12) shows that, for a given problem, some coordinates are better than others.

By proceeding in this way we could derive all the theorems of Euclidean plane geometry. The angle between two intersecting lines, for example, can be defined as the ratio of the length ΔC of the part of a circle centered on their intersection that lies between the lines to the circle's radius R.

$$\theta \equiv \frac{\Delta C}{R} \qquad \text{(radians)}. \tag{2.13}$$

With this definition we could prove that the sum of the interior angles of a triangle is π. Indeed, we could verify the *axioms* of Euclidean plane geometry from (2.7) or (2.8). All geometry can be reduced to relations between distances; all distances can be reduced to integrals of distances between nearby points; all Euclidean plane geometry is contained in (2.7) or (2.8).

Line Element

To summarize, a geometry is specified by the *line element*, such as (2.7) or (2.8), which gives the distance between nearby points in terms of the coordinate intervals between them in some coordinate system. Conventionally, a line element is written as a quadratic relation for dS^2, e.g.,

$$dS^2 = dx^2 + dy^2 \tag{2.14}$$

with no brackets around the differentials. The form of the line element for a geometry varies from coordinate system to coordinate system [e.g., (2.7) and (2.8)], but the geometry remains the same.

The Non-Euclidean Geometry of a Sphere

An example of a non-Euclidean geometry is provided by the surface of a two-dimensional sphere of radius a. We can use the angles (θ, ϕ) of three-dimensional polar coordinates to label points on the sphere. The distance between points (θ, ϕ) and $(\theta + d\theta, \phi + d\phi)$ can be seen after a little work (Figure 2.6) to be

Line Element for a Sphere

$$dS^2 = a^2\left(d\theta^2 + \sin^2\theta \, d\phi^2\right). \tag{2.15}$$

This is the line element of the surface of a sphere.

Let's use the line element (2.15) to calculate the ratio of the circumference to the radius of a circle on the sphere. By circle we mean the locus of points *on the surface* that are a constant distance (the radius) *along the surface* from a fixed point (the center) *in the surface*. Since no one point is distinguished geometrically from any other on the sphere, we may conveniently orient our polar coordinate system so that the polar axis is at the center of the circle. A circle is then a curve of constant θ. Consider the circle defined by the equation

$$\theta = \Theta \tag{2.16}$$

for constant Θ. The circumference is the distance around this curve. Nearby points along the curve are separated by $d\phi$ but have $d\theta = 0$. Thus, (2.15) gives $dS =$

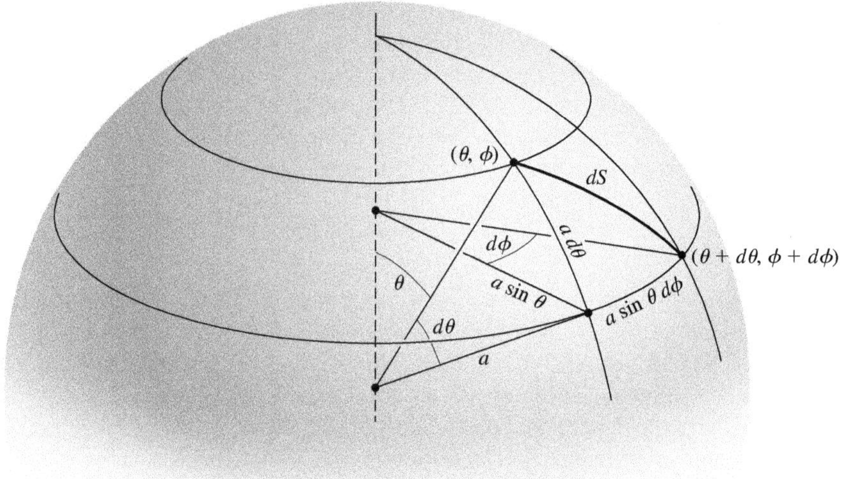

FIGURE 2.6 Deriving the line element on the sphere. The derivation makes use of the fact that the two-dimensional sphere is a surface in three-dimensional Euclidean space. Two infinitesimally separated points at locations (θ, ϕ) and $(\theta + d\theta, \phi + d\phi)$ are indicated. The construction shows that the distance between ϕ and $\phi + d\phi$ along a line of constant latitude θ is $a \sin\theta\, d\phi$. The distance between θ and $\theta + d\theta$ along a line of constant longitude is $a\, d\theta$. Because the θ and ϕ coordinate lines are orthogonal, the sum of the squares of these two differentials gives the square of the distance dS between the two points when $d\theta$ and $d\phi$ are infinitesimally small. This gives (2.15).

$a \sin\Theta\, d\phi$ along the circle, and the circumference is

$$C = \oint dS = \int_0^{2\pi} a \sin\Theta\, d\phi = 2\pi a \sin\Theta. \tag{2.17}$$

The radius is the distance from the center to the circle along a curve for which θ varies but $d\phi = 0$. Along this curve, (2.15) gives $dS = a\, d\theta$, and the radius is

$$r = \int_{\text{center}}^{\text{circle}} dS = \int_0^{\Theta} a\, d\theta = a\Theta. \tag{2.18}$$

Using (2.18) to eliminate Θ in (2.17), the relation between the circumference and radius of a circle in the non-Euclidean geometry of a sphere becomes

$$C = 2\pi a \sin\left(\frac{r}{a}\right). \tag{2.19}$$

In this expression a is a fixed number characterizing the geometry. It measures the scale on which the geometry is curved. When the radius of the circle is much smaller than the radius of the sphere, $r \ll a$, then we have approximately

$$C \approx 2\pi r, \tag{2.20}$$

which is the familiar result in Euclidean geometry. The geometry of the surface of the Earth is the same as a sphere to a good approximation.

The many different projections used to make maps of its surface are just different coordinate systems for expressing the geometry of a sphere as described in Box 2.3.

BOX 2.3 Map Projections

The various projections used to make planar maps of the Earth's surface are examples of a familiar geometry expressed in different systems of coordinates. The geometry is that of the two-dimensional surface of the sphere to an excellent approximation. In usual polar coordinates the line element is given by (2.15), with a being the radius of the Earth. On the surface of the Earth, the angle ϕ is the longitude (measured in radians rather than degrees). The latitude λ is $\pi/2 - \theta$. Expressed in terms of latitude and longitude, the line element is

$$dS^2 = a^2(d\lambda^2 + \cos^2\lambda\, d\phi^2). \tag{a}$$

To make a map we introduce new coordinates x and y on the sphere, defined by relations of the form

$$x = x(\lambda, \phi), \qquad y = y(\lambda, \phi), \tag{b}$$

and use these as rectangular coordinates in the plane to plot the outlines of the continents, locations of cities, etc. Different projections correspond to different choices for the functions $x(\lambda, \phi)$ and $y(\lambda, \phi)$. One can think of these functions as providing a map in the mathematical sense from the sphere into the plane.

There are as many projections as there are different functions. The simplest example is

$$x = (L\phi)/2\pi, \qquad y = (L\lambda)/\pi, \tag{c}$$

where L is the width of the map. This just plots ϕ and λ as x and y on rectangular axes. The result, shown in the accompanying figure, is called an *equirectangular* projection. However, there are more useful projections which preserve some properties of the geometry of the sphere on a plane. Not all properties can be preserved because the geometry of a sphere is different from that of the plane!

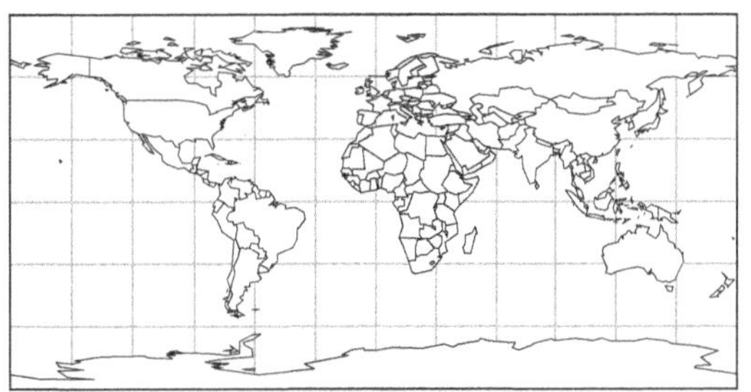

Equirectangular projection.

A wide class of useful projections send longitude linearly into x:

$$x = \frac{L\phi}{2\pi}, \qquad y = y(\lambda). \tag{d}$$

For projections of this kind, the true distances are given by the line element

$$dS^2 = a^2\left[\left(\frac{2\pi}{L}\cos[\lambda(y)]\right)^2 dx^2 + \left(\frac{d\lambda}{dy}\right)^2 dy^2\right]. \tag{e}$$

A simple example of a projection of this kind is the Mercator projection, invented by G. Kramer in 1569 and illustrated below. Kramer's idea was that angles on the map should equal compass bearings on the sphere. That is, the map from the sphere to the plane should preserve angles between different directions from a point. A mariner wishing to sail between Caracas and Lisbon would draw the straight line on the map connecting these two ports. The angle between that line and the y-axis would be the bearing from north that when held constant during the voyage would bring the ship from Caracas

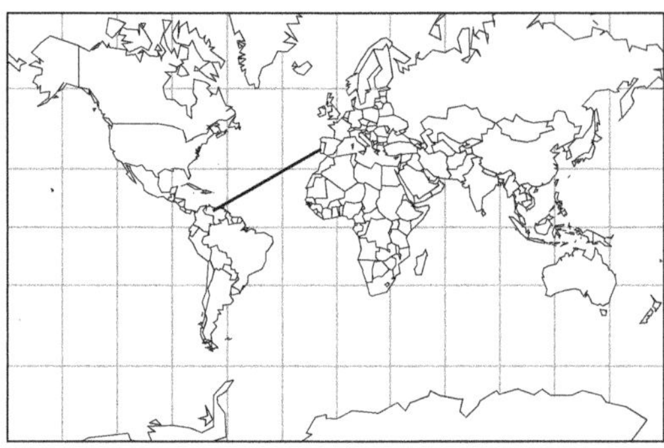

Mercator projection.

to Lisbon. What choice of function $y(\lambda)$ or $\lambda(y)$ would make a map like this?

Angles are *ratios* of distances, as we saw in (2.13). The angle between two directions on a sphere will equal the angle between the corresponding directions on the plane if the line element on the sphere is *proportional* to the line element on the flat plane, $dS^2_{\text{flat}} = dx^2 + dy^2$. Thus, to implement Kramer's idea we seek a function $\lambda(y)$ such that (e) can be written

$$dS^2 = \Omega^2(x, y)(dx^2 + dy^2) \tag{f}$$

for some function $\Omega(x, y)$. Clearly we need

$$\frac{d\lambda}{dy} = \frac{2\pi}{L} \cos \lambda. \tag{g}$$

Choosing $y = 0$ to coincide with $\lambda = 0$ gives

$$y(\lambda) = \frac{L}{2\pi} \int_0^\lambda \frac{d\lambda'}{\cos \lambda'} = \frac{L}{2\pi} \log\left[\tan\left(\frac{\pi}{4} + \frac{\lambda}{2}\right)\right]. \tag{h}$$

Equations (h) and (d) define the Mercator projection. The equator $\lambda = 0$ is mapped to the line $y = 0$. The poles $\lambda = \pm\pi/2$ are mapped to $y = \pm\infty$, respectively.

The proportionality factor between the spherical metric and the flat metric on the plane $\Omega(x, y)$ that was defined in (f) is

$$\Omega(y) = \frac{2\pi a}{L} \cos \lambda(y) = \frac{4\pi a}{L} \frac{e^{2\pi y/L}}{1 + e^{4\pi y/L}}. \tag{i}$$

Most of the familiar properties of the Mercator projection follow from this factor. For example, consider two points at the same latitude separated by a difference in longitude, Δx. The physical distance between these points, ΔS, is

$$\Delta S = \Omega(y)\Delta x \tag{j}$$

and depends on latitude. As $y \to \infty$, the North Pole, this distance shrinks to zero, as it should. True distances in x at higher latitudes are smaller than coordinate distances because of the factor $\Omega(y)$.

The same holds true for areas. A small rectangle on the map of coordinate dimensions Δx and Δy has area

$$\Delta A = [\Omega(y)\Delta x][\Omega(y)\Delta y] = \Omega^2(y)\Delta x \Delta y. \tag{k}$$

Thus, although Greenland looms large on the Mercator projection in coordinate area when compared with South America, for example, its actual area is much smaller.

The Geometry of Some More General Surfaces

The line element of the plane and the sphere were worked out before, starting from a clear picture of these geometries as surfaces in Euclidean space. However, in general relativity it is more usual to be confronted with a line element and have to figure out the properties of the geometry it represents.

Consider as an example the line element

$$dS^2 = a^2(d\theta^2 + f^2(\theta)\, d\phi^2) \tag{2.21}$$

for various possible choices of the function $f(\theta)$. The choice $f(\theta) = \sin\theta$ gives the geometry of the surface of a sphere (2.15). But what surfaces in three-dimensional Euclidean space have intrinsic geometries represented by the line element (2.21) for other choices of $f(\theta)$? There are several clues.

1. Since the line element is the same for all ϕ, it corresponds to a surface that is axisymmetric about an axis.
2. The circumference $C(\theta)$ of a circle of constant θ is (from (2.21))

$$C(\theta) = \int_0^{2\pi} af(\theta)\, d\phi = 2\pi a f(\theta). \tag{2.22}$$

3. The distance from pole to pole is

$$d_{\text{pole-to-pole}} = a \int_0^{\pi} d\theta = \pi a. \tag{2.23}$$

By working out these various metrical properties, a picture can be built up of the surface, as Example 2.1 shows.

Example 2.1. A Peanut Geometry. Consider the surface specified by

$$f(\theta) = \sin\theta(1 - \tfrac{3}{4}\sin^2\theta). \tag{2.24}$$

The surface is symmetric under reflection in the equatorial plane $\theta = \pi/2$. Starting at $\theta = 0$, the circumference of the lines of constant θ (2.22) first increases and then decreases with $f(\theta)$; then it increases and decreases again. At any one θ the circumference is smaller than the corresponding value on a sphere. At the equator, for instance,

$$C\left(\frac{\pi}{2}\right) = 2\pi a\left(1 - \frac{3}{4}\right) = \frac{\pi a}{2}. \tag{2.25}$$

The maximum circumference is $(8\pi/9)a$ at $\theta = \sin^{-1}(\frac{2}{3}) = .73$ radians. Since the distance from pole to pole is πa from (2.23), this surface has the elongated "peanut" shape shown in Figure 2.7.

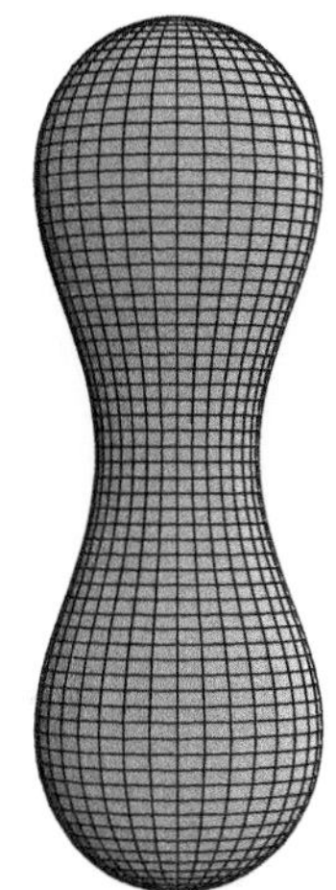

FIGURE 2.7 A surface in flat three-dimensional space with the geometry specified by the line element (2.21) for $f(\theta) = \sin\theta(1 - \frac{3}{4}\sin^2\theta)$. The horizontal rulings are lines of constant θ. The circumference of these varies with θ according to (2.22). The vertical lines are lines of constant ϕ spaced equally around the axis of symmetry. The example looks like the surface of a very symmetric peanut. (You will learn how to construct such surfaces in Section 7.7.)

2.6 Coordinates and Invariance

In the preceding calculation of the ratio of the circumference to the radius for a circle in the plane, the same answer was obtained whether the calculations were done in Cartesian or polar coordinates. It is obvious that the answers *should* be the same. The distance around a circle and the distance from it to its center are defined and meaningful quantities *independent* of the choice of coordinates that are used to label the points in a plane. Presented with a physical disk, we could check whether its edge is a circle by using a tape measure to see whether points on the edge are equidistant from the center. We could then use the tape measure to find the circumference and compute the ratio of circumference to radius. No coordinates are involved in these operations. Coordinates are just a convenient and systematic way of labeling the points in a geometry. They have no meaning in themselves. We could have labeled the points by names Joe, Alice, Fred, ... On maps we do this—New York, Beijing, etc. But such a system of labels is not very systematic and not very convenient for the application of the methods of calculus to problems of geometry. Coordinates *are* a systematic set of labels, but there are an infinite number of different coordinate systems, which are all equivalent. Some may be more convenient for one computation or another, as the calculations of the circumference in polar coordinates in (2.11) and (2.12) show, or more useful for one purpose or another, as the maps in Box 2.3 on p. 25 show, but equivalent answers can be obtained using any of them.

The equivalence of Cartesian and polar coordinates in the plane can be seen generally. Since the two coordinate systems are different ways of labeling the points in a plane there must be a connection between them. A point can be labeled either by coordinates (x, y) or (r, ϕ). The translation between these different labels is called a *coordinate transformation*. In this case it is

$$x = r\cos\phi, \qquad y = r\sin\phi. \tag{2.26}$$

With the aid of the coordinate transformation (2.26), the equivalence between the two line elements (2.7) and (2.8), each expressing the geometry of the plane, may be demonstrated mechanically. Start from (2.7) and compute dx and dy from (2.26)

$$dx = (dr)\cos\phi - r\sin\phi(d\phi), \tag{2.27}$$

$$dy = (dr)\sin\phi + r\cos\phi(d\phi). \tag{2.28}$$

Substitute these into (2.7) and simplify to find the line element (2.8) for dS^2 in polar coordinates. The point here is that the distance between nearby points dS is an *invariant quantity*—independent of the coordinates used to compute it.

The coordinates used in a computation are arbitrary; the answers must be expressed in physically invariant terms. We shall see many more examples of this in the following chapters.

Problems

1. [B] **(a)** In a plane, show that a light ray incident from any angle on a right-angle corner reflector returns in the same direction from whence it came.

(b) Show the same thing in three dimensions with a cubical corner reflector.

2. [S] The center of the Sun is much further away from a terrestrial measurement of angles than the center of the Earth is. But the Sun is also much more massive than the Earth. Using (2.1), *estimate* which would have the greatest effect on a measurement of angles such as is attributed to Gauss.

3. [C] **(a)** Verify the relation (2.4) between the sum of the interior angles of a spherical triangle and its area when two of the angles are right angles.

(b) Prove the relation generally.

4. Draw examples of a triangle on the surface of a sphere for which:
   (a) The sum of interior angles is just slightly greater than π.
   (b) The sum of angles is equal to 2π.
   (c) What is the maximum the sum of angles of a triangle on a sphere can be according to (2.4)? Can you exhibit a triangle where the sum achieves this value?

5. Calculate the area of a circle of radius r (distance from center to circumference) in the two-dimensional geometry that is the surface of a sphere of radius a. Show that this reduces to πr^2 when $r \ll a$.

6. [B] Consider a sphere of radius a and on it a segment of length s of a line of latitude that is a distance d from the North Pole measured on the sphere. What is the angle between the lines of longitude that this segment spans? Is this angle greater or smaller than the angle the segment would subtend at the same distance on a flat plane?

7. Consider the following coordinate transformation from familiar rectangular coordinates (x, y), labeling points in the plane to a new set of coordinates (μ, ν):

$$x = \mu\nu, \qquad y = \frac{1}{2}\left(\mu^2 - \nu^2\right).$$

   (a) Sketch the curves of constant μ and constant ν in the xy plane.
   (b) Transform the line element $dS^2 = dx^2 + dy^2$ into (μ, ν) coordinates.
   (c) Do the curves of constant μ and constant ν intersect at right angles?
   (d) Find the equation of a circle of radius r centered at the origin in terms of μ and ν.
   (e) Calculate the ratio of the circumference to the diameter of a circle using (μ, ν) coordinates. Do you get the correct answer?

8. [A] The surface of an egg is an axisymmetric geometry to a good approximation. In the line element for two-dimensional axisymmetric geometries (2.21), pick an $f(\theta)$ such that the resulting surface would resemble that of an egg. Calculate the ratio of the biggest circle around the axis to the distance from pole to pole.

9. The surface of the Earth is not a perfect sphere. One quarter of the circumference around a great circle passing through the poles is 9985.16 km. This is slightly less than one quarter of the equatorial circumference, 10,0018.75 km, meaning the Earth is slightly squashed. Suppose the surface of the Earth is modeled by an axisymmetric surface with a line element of the kind in (2.21) with

$$f(\theta) = \sin\theta(1 + \epsilon \sin^2\theta)$$

   for some small ϵ. What values of a and ϵ would best reproduce the known polar and equatorial radii?
   Comment: It is not an accident that one quarter of the polar circumference is almost exactly ten million meters. That was the original definition of the meter.

10. [B] *Equal-Area Projections* An equal-area map projection is one for which there is a constant proportionality between areas on the map and areas on the surface of the globe. Given $x = L\phi/2\pi$, what function $y(\lambda)$ would make an equal-area map? (*Hint*: If an infinitesimal area $dxdy$ has the same constant of proportionality to the corresponding infinitesimal area on the sphere wherever it is located, bigger areas will be also proportional.)

11. [B] *Conical Projections* Conical projections map points on the globe into polar coordinates (r,ψ) in the plane of the map. (We use ψ to avoid confusion with the coordinate φ on the sphere.) Thus, in general, $r = r(\lambda, \varphi)$ and $\psi = \psi(\lambda, \varphi)$. A particularly simple class of conical projections uses the North Pole as the origin of the polar coordinates and has $r = r(\lambda)$ and $\psi = \phi$.

(a) For this simple class, express the line element on the sphere in terms of r and ψ.

(b) Find the function $r(\lambda)$ that makes this an equal-area projection in which there is a constant proportionality between each area on map and the corresponding area on the sphere. (*Hint*: See the hint for Problem 10.)

12. [B, N] *Your Personal World Map* The maps in Box 2.3 were made with the *Mathematica* program `WorldPlot`. Make your own projection, centered on your home city, that uses a radial coordinate that represents your view of the importance of the rest of the world.

CHAPTER 3

Space, Time, and Gravity in Newtonian Physics

Chapter 2 introduced the idea of a geometry and how one is described. This chapter discusses the geometry of space and the notion of time assumed in Newtonian mechanics. This discussion will also serve to review aspects of mechanics and special relativity that will be important for later developments.

3.1 Inertial Frames

Newton's First Law

Newtonian mechanics assumes a geometry for space and a particular idea for time. Nowhere is that clearer than in Newton's first law, specifying the motion of *free particles*—particles on which no forces are acting. According to Newton's first law, a free particle moves on a straight line at constant speed. But what geometry defines a "straight line"? What idea of time is used to define "constant speed"?

The straight line of Newton's first law is the shortest distance between two points in three-dimensional Euclidean space. The geometry of space is specified in Cartesian coordinates by the line element

$$dS^2 = dx^2 + dy^2 + dz^2 \tag{3.1}$$

giving the distance dS between points separated by infinitesimal coordinate intervals dx, dy, and dz. This geometry is the natural extension to three dimensions of the geometry of a flat plane. It is, therefore, called *flat* space. Flat, Euclidean geometry is assumed for space in Newtonian mechanics.

To understand how motion is described in the flat space of Newtonian mechanics, imagine a world containing free particles moving this way and that. An observer in a laboratory seeks to describe and understand the motions of the particles that move through it (see Figure 3.1). To describe the motions, the observer can pick a corner of the laboratory as the origin of Cartesian coordinates (x, y, z) oriented along the intersections of the walls and floor that meet at this corner. These coordinates can be used to label the points in space through which a particle moves. The system of coordinates is said to provide a *reference frame*, or *frame* for short.[1]

[1] This book uses the term *frame* as a synonym for a system of coordinates. Although it is not necessary to define the usage of this term very precisely, frames are typically (as here) associated with the laboratory of an observer, and in general cover or are useful for only a limited region of space and time. The inertial frames of Newtonian mechanics and special relativity are exceptions in covering the whole of space and time.

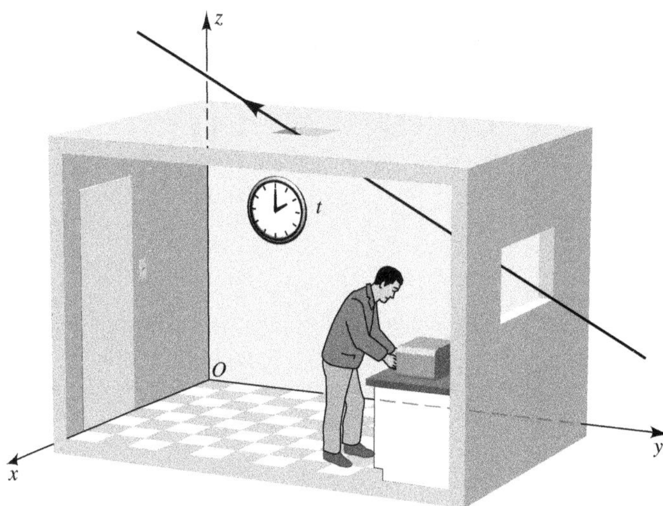

FIGURE 3.1 A laboratory defines a reference frame. An observer in an idealized laboratory can choose one corner as the origin of three Cartesian coordinates (x, y, z) that coincide with the intersections of the walls and floor that meet in that corner. These three coordinates define a reference frame that, together with the time measured by a clock, can be used to describe the motion of particles moving through the laboratory and state Newton's laws of motion.

There are many possible laboratories, which can be moving uniformly, accelerating, rotating with respect to each other, or some combination of these three (see Figure 3.2). Not all these reference frames are equally useful for expressing the laws of mechanics. A particularly useful type of reference frame can be constructed as follows: Pick a free particle to serve as the origin of a Cartesian coordinate system (see Figure 3.3) at all times. At one moment choose three perpendicular Cartesian coordinates (x, y, z) with this origin pointing along the directions set by the axes of three perpendicular gyroscopes. At later moments continue to define (x, y, z) by the directions of these gyroscopes. Equivalently, and more geometrically, propagate the initial axes parallel to themselves (no rotation) as the origin moves along its straight line path. The resulting coordinate system is called an *inertial frame*.[2]

Inertial Frame

The laws of Newtonian mechanics take their standard and simplest forms in inertial frames. An observer in an inertial frame can discover a parameter t with respect to which the positions of all free particles are changing at constant rates. This is *time*. Explicitly, the motion of any one particle can be described by giving its coordinates as a function of time $(x(t), y(t), z(t))$ and its acceleration as zero:

$$\frac{d^2x}{dt^2} = 0, \qquad \frac{d^2y}{dt^2} = 0, \qquad \frac{d^2z}{dt^2} = 0. \tag{3.2}$$

[2]The synonyms used for inertial frames are legion, typically some contraction of *inertial Cartesian coordinate reference frame.*

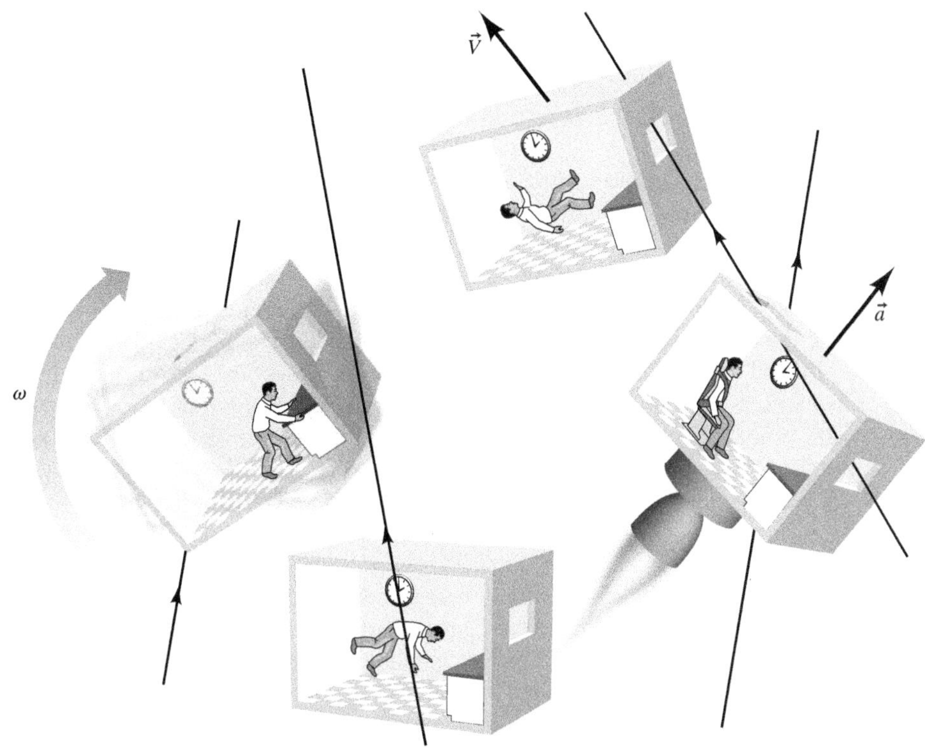

FIGURE 3.2 Not all reference frames are inertial frames. The figure shows four idealized laboratories moving through a world of free particles. Each laboratory defines a reference frame, as illustrated in Figure 3.1. Suppose the bottom laboratory is an inertial frame. A laboratory moving uniformly with respect to the first (top) defines another inertial frame. However, laboratories rotating with respect to the first (left) or accelerating with respect to it (right) do not correspond to inertial frames.

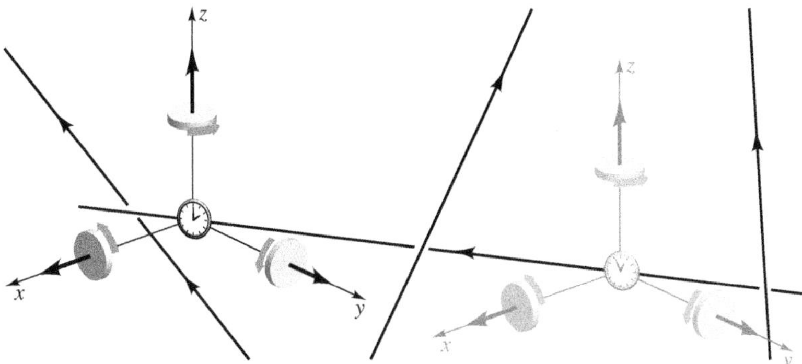

FIGURE 3.3 The construction of an inertial frame. The position of one particle has been chosen as the origin of the frame. Three axes are defined by perpendicular gyroscopes as that particle moves. The resulting system of three Cartesian coordinates (x, y, z) is an inertial frame for describing the motion of the other particles, shown here at two different times.

Equation (3.2) is the expression of Newton's first law. Indeed, inertial frames could be *defined* as Cartesian reference frames for which Newton's first law holds in the form (3.2).

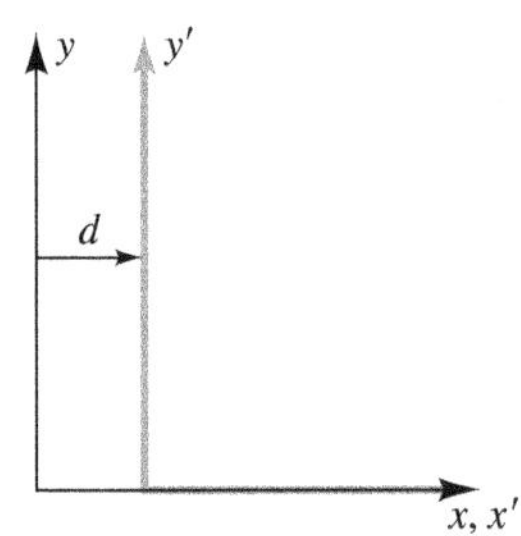

FIGURE 3.4 Two Cartesian coordinate systems related by a displacement d along the x-axis.

Using the laws of mechanics, an observer in an inertial frame can construct a clock that measures the time t. For instance, the position of one free particle could be used to measure t, since its position changes at a constant rate in t.

Not every Cartesian coordinate system is an inertial frame. For instance, the reference frame of a laboratory on the surface of the Earth is not exactly an inertial frame. The equations of motion of a free particle are not (3.2) but include centrifugal and Coriolis terms resulting from the rotation of the Earth as well. The slow precession of a Foucault pendulum is a sure sign that a frame fixed on the Earth is not an inertial frame, but rather it is rotating with respect to them. (See Box 3.1 for another such measurement.)

There are many inertial frames, not just one. In the construction given, three different perpendicular directions could have been chosen for the three axes, defining a new frame (x', y', z') that is *rotated* with respect to the first. A different free particle could have been chosen as the origin defining a frame that is *displaced* with respect to the first and generally moving at a *constant velocity* with respect to it. Rotations, displacements, and uniform motions (or combinations of these) turn out to be the *only* ways inertial frames can differ.

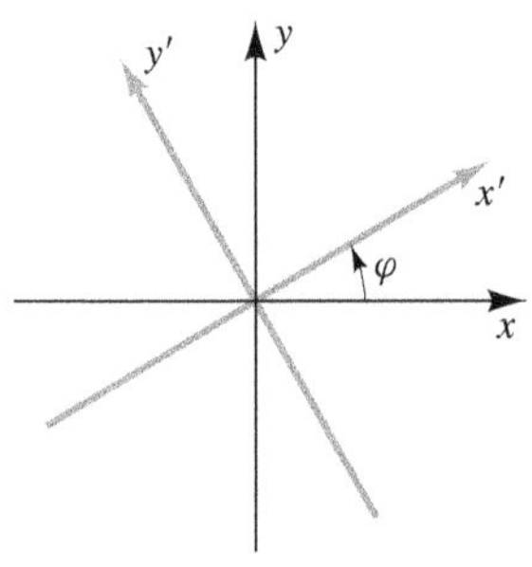

FIGURE 3.5 Two Cartesian coordinate systems related by a rotation through an angle φ about the z-axis.

Any two sets of Cartesian coordinates (x, y, z) and (x', y', z') from different inertial frames are just different ways of labeling the points of three-dimensional flat space. Therefore, there must be a connection between these two different systems of labels—a coordinate transformation. Simple examples of coordinate transformations corresponding to displacements, rotations, and uniform motions are as follows.

1. *Displacement* by a distance d along the x-axis (see Figure 3.4):

$$x' = x - d,\quad y' = y,\quad z' = z. \tag{3.3}$$

2. *Rotation* by an angle φ about the z-axis (see Figure 3.5):

$$x' = (\cos\varphi)x + (\sin\varphi)y,\quad y' = -(\sin\varphi)x + (\cos\varphi)y,\quad z' = z. \tag{3.4}$$

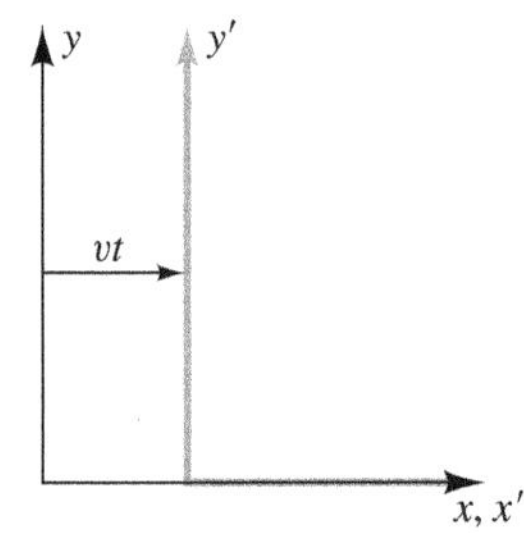

FIGURE 3.6 Two Cartesian coordinate systems related by a uniform velocity v along the x-axis.

3. *Uniform motion* by a velocity v along the x-axis (see Figure 3.6):

$$x' = x - vt,\quad y' = y,\quad z' = z. \tag{3.5}$$

BOX 3.1 Measuring the Rotation of the Earth with a Ring Interferometric Gyro

A laboratory on the surface of the Earth does not define an inertial frame because the Earth is rotating. The Earth's rotation rate can be measured by experiments done entirely inside a closed laboratory on its surface that make no reference to astronomical phenomena such as the rising and setting of the Sun. Observing the precession of a gyroscope or a Foucault pendulum would be one way to make such a measurement. But precise measurements can be made with ring interferometric gyroscopes. The idea behind these devices is illustrated schematically in the figure below in the frame of the gyro. Waves are emitted in phase from one point on a ring to travel in opposite directions around its circumference to detection at their starting point on the ring. If the ring is not rotating, the waves received at any one time have traveled equal distances, are in phase, and constructively interfere. We can use an inertial frame in which the center of the ring is at rest to analyze what happens if the ring is rotating with an angular velocity Ω in that frame. While either wave is moving around the ring, the detector will have rotated by an angle $\Omega \times$ (passage time) from its position at the time of emission. The counter-propagating wave meets the detector after a time interval $\Delta t_{\text{counter}}$ that is shorter than that for the copropagating wave. The distance traveled is $v\Delta t_{\text{counter}}$ where v is the velocity of the wave. This distance is also $(2\pi - \Omega\Delta t_{\text{counter}})R$, where R is the radius of the ring. Equating these two determines $\Delta t_{\text{counter}}$ and shows that the distance is $(2\pi R)/(1+(\Omega R/v))$. A similar expression gives the distance traveled by the copropagating wave, which is the same except that the sign in the denominator is reversed. The difference in distances is

$$(4\pi R^2\Omega/v)[1 - (\Omega R/v)^2]^{-1}.$$

When this distance is an integer number of wavelengths, the two waves will interfere constructively, and when it is an odd half-integer number of wavelengths, they will interfere destructively. This is called the *Sagnac effect.*

The rotation of the Earth has been detected in this way with electromagnetic waves. But remarkably, at the time of writing, the most accurate results employ the quantum de Broglie waves associated with atoms in atom interferometers [e.g., Gustavson, Bouyer, and Kasevich (1997)]. The de Broglie wavelength of a matter wave of a particle with mass m is $h/(mv)$, which for the velocities of the atoms in these experiments is very much smaller than the wavelength of visible light. Since the difference between constructive and destructive interference is half a wavelength, matter wave interferometers could, in principle, yield very precise measurements. Precision measurements of rotation are important, because general relativity predicts that the rotation of matter can influence the rotation of nearby inertial frames as you will learn in Chapter 14.

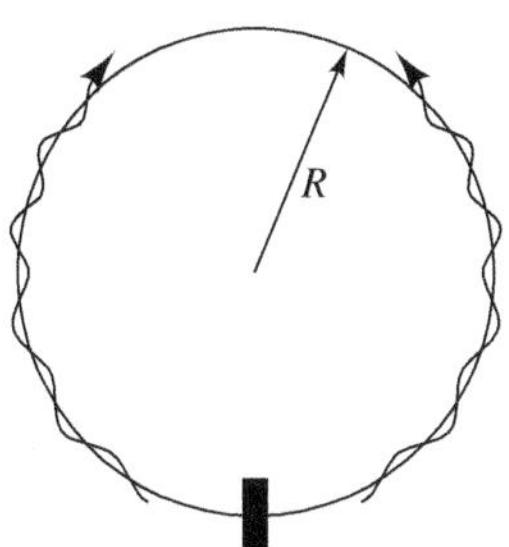

The coordinate transformations (3.3), (3.4), and (3.5) show how the coordinates labeling *position* are connected in different inertial frames. But what about the relationship between the times? We discussed earlier how an observer in one inertial frame could find a time t that led to a simple law of motion for free particles. But will a similarly constructed time t' in a different inertial frame be the same? More specifically, will two events that are simultaneous in one inertial frame be simultaneous in other inertial frames? Newtonian mechanics answers an unequivocal yes to these questions. It is a central *assumption* of Newtonian mechanics that there is a single notion of time for all inertial observers. This is the "absolute," "universal" time that enters in the same way into the laws of motion in any inertial frame. Thus $t' = t$ and the transformation law (3.5) between two inertial frames moving at a uniform speed v with respect to one another can be completed to give

Galilean Transformation

$$\boxed{\begin{aligned} x' &= x - vt, \\ y' &= y, \\ z' &= z, \\ t' &= t. \end{aligned}} \tag{3.6}$$

This is called a *Galilean Transformation*. This Newtonian idea of absolute time is abandoned in special relativity, where notions of time are different in different inertial frames.

3.2 The Principle of Relativity

Newton's first law is not all of mechanics. Newton's second law relates a body's deviations from constant velocity—accelerations—to forces acting on it. However, all Newtonian mechanics, including Newton's second law, is consistent with the following *Principle of Relativity*:

Principle of Relativity

Identical experiments carried out in different inertial frames give identical results.

Suppose you are in a closed laboratory. An experiment checking Newton's first law will determine whether the frame of the laboratory is an inertial frame. But the principle of relativity tells us that there is no experiment of *any* kind that can be carried out inside the laboratory to determine *which* of the infinitely many possible inertial frames the laboratory represents. Put differently, there is no notion of absolute displacement, absolute rotation, or absolute velocity. Contrast this situation with accelerated frames. Blindfolded in a car on an ideally smooth track, it is not possible to tell whether the car is at rest or moving with uniform speed. But it is possible to tell whether it is accelerating. This principle of relativity played an important role in Einstein's discovery of special relativity, as we will see in the next chapter.

"When learning about the laws of physics you find that there are a large number of complicated and detailed laws, laws of gravitation, of electricity and magnetism, nuclear interactions, and so on, but across the variety of these detailed laws there sweep great general principles which all the laws seem to follow." That's how Richard Feynman (1965) characterized principles such as the principle of relativity. You shouldn't expect such principles that pertain to many laws to be too mathematically precise. (What exactly is meant by "identical" in this state-

BOX 3.2 Mach's Principle: What Are the Inertial Frames?

Newtonian mechanics specifies the way inertial frames are related to each other, but it does not specify the way inertial frames are related to the physical properties of the universe. Yet there is a simple physical characterization of the inertial frames. We can approach this by considering the following thought experiment.

Imagine that you are at the North Pole of the Earth under a cloud-covered sky. A Foucault pendulum is suspended vertically. The plane of the pendulum precesses slowly with respect to the surface of the Earth. This shows that the inertial frame in which the plane of the pendulum is stationary is rotating with respect to the surface of the Earth. If the clouds now part, you will find that this inertial frame is at rest with respect to the distant stars or moving uniformly with respect to them. Empirically, the inertial frames of mechanics are at rest with respect to the distant matter in the universe or moving uniformly with respect to it. It was the idea of Bishop Berkeley (1685–1753) and the physicist Ernst Mach (1838–1916) that this connection between the local inertial frames and the distant matter is a necessary one. The connection is therefore sometimes called *Mach's principle*. However, in general relativity, this connection is not necessary. Rather, if the frame where the plane of the pendulum was stationary was rotating with respect to the distant stars, we would say that the universe is rotating. The uniformity of another kind of distant matter—the cosmic background radiation (CMB)—puts stringent upper limits on the rotation of the whole universe. However, the rotation motion of matter does influence inertial frames in general relativity, as we will see in Chapter 14.

ment of the principle of relativity?) But that should not obscure the fact that there is a common property that the detailed laws share. For example, the principle of relativity is sometimes stated as the laws of mechanics take the same form in every inertial frame. It proves to be difficult to give a precise meaning to *form*, but the idea can be illustrated just by Newton's first law. Suppose equations (3.2) hold in one inertial frame. Rotations, displacements, and uniform motions preserve the form of (3.2). To show that, just differentiate (3.3), (3.4), and (3.5) twice with respect to time and use $t = t'$. Since d, θ, and v are constant in time, one finds, in each of the three cases, that (3.2) implies

$$\frac{d^2x'}{dt'^2} = 0, \qquad \frac{d^2y'}{dt'^2} = 0, \qquad \frac{d^2z'}{dt'^2} = 0, \tag{3.7}$$

the same form as in (3.2). The form of the equation of motion for free particles is the same in all inertial frames. In particular, its form is *invariant* under Galilean transformations.

A principle of relativity relating the form of the laws of physics in inertial frames differing by displacements and rotations is possible only because the geometry of Euclidean space shares those symmetries. The laws of physics would not be invariant under displacements and rotations if the geometry of space were curved like the surface of a potato is in two dimensions.

One can verify these symmetries of Euclidean geometry mechanically by examining how its line element

$$dS^2 = dx^2 + dy^2 + dz^2 \tag{3.8}$$

changes under displacements and rotations. The formulas for these transformations are given in (3.3) and (3.4), respectively. Consider, for example, the rotation in (3.4), which can be written

$$\begin{aligned} x &= (\cos\varphi)x' - (\sin\varphi)y', \\ y &= (\sin\varphi)x' + (\cos\varphi)y', \\ z &= z'. \end{aligned} \tag{3.9}$$

Plugging this into (3.8) gives:

$$\begin{aligned} dS^2 &= (\cos\varphi dx' - \sin\varphi dy')^2 + (\sin\varphi dx' + \cos\varphi dy')^2 + dz'^2 \\ &= dx'^2 + dy'^2 + dz'^2. \end{aligned} \tag{3.10}$$

Thus, the form of the line-element is *invariant* under rotations; so, therefore, is Euclidean geometry. The same is true for displacements.

3.3 Newtonian Gravity

Newton's law of gravity specifies the gravitational force $\vec{F}$ that a point mass A with mass M exerts on another point mass B with mass m a distance r away. The force is attractive, directed along the line between the masses, and inversely proportional to r^2:

Newton's Law of Gravity

$$\vec{F}_{\text{grav}} = -\frac{GmM}{r^2}\vec{e}_r. \tag{3.11}$$

Here, $\vec{e}_r$ is the unit vector pointing from A to B, and G is Newton's gravitational constant, 6.67×10^{-8} dyn · cm^2/g^2. This gravitational force on B can be written

$$\vec{F}_{\text{grav}} = -m\vec{\nabla}\Phi(\vec{x}_B), \tag{3.12}$$

where m is B's mass, $\vec{x}_B$ is B's position, and $\Phi(\vec{x})$ is the *gravitational potential* produced by A:

Newtonian Gravitational Potential

$$\Phi(\vec{x}) = -\frac{GM}{r} \equiv -\frac{GM}{|\vec{x} - \vec{x}_A|}. \tag{3.13}$$

If B is attracted by many point masses M_A, $A = 1, 2, \ldots$, at various positions $\vec{x}_A$, the gravitational potential giving the force in (3.12) is the sum of the gravitational potential from each:

$$\Phi(\vec{x}) = -\sum_A \frac{GM_A}{|\vec{x} - \vec{x}_A|}. \tag{3.14}$$

For a continuous distribution of *mass density* $\mu(\vec{x})$ the sum in (3.14) becomes an integral over the mass $\mu(\vec{x})d^3x$ in volume element d^3x, namely,

$$\Phi(\vec{x}) = -\int d^3x' \frac{G\mu(\vec{x}')}{|\vec{x} - \vec{x}'|}. \tag{3.15}$$

Readers familiar with electromagnetism will immediately recognize the similarities between the gravitational potential (3.15) and the electrostatic potential and similarly between the gravitational force law (3.12) and the electrostatic one. The analogy is made explicit in Table 3.1. The origin of these similarities is that both are forces between bodies that vary inversely as the square of the distance between them. Mass is the gravitational analog of electric charge. However, since mass is always positive, the gravitational force is always attractive, unlike the electrostatic force, which is sometimes repulsive.

The analogy between gravity and electrostatics can be pushed further. Introduce the Newtonian gravitational field $\vec{g}$,

Newtonian Gravitational Field

$$\vec{g}(\vec{x}) \equiv -\vec{\nabla}\Phi(\vec{x}), \tag{3.16}$$

which is the gravitational analog of the electric field. The differential form of the law for the gravitational potential (3.15) is

$$\vec{\nabla} \cdot \vec{g}(\vec{x}) = -4\pi G\mu(\vec{x}), \tag{3.17}$$

TABLE 3.1 Newtonian Gravity and Electrostatics

	Newtonian Gravity	Electrostatics
Force between two sources	$\vec{F}_{\text{grav}} = -\frac{GmM}{r^2}\vec{e}_r$	$\vec{F}_{\text{elec}} = +\frac{qQ}{4\pi\epsilon_0 r^2}\vec{e}_r$
Force derived from potential	$\vec{F}_{\text{grav}} = -m\vec{\nabla}\Phi(\vec{x}_B)$	$\vec{F}_{\text{elec}} = -q\vec{\nabla}\Phi_{\text{elec}}(\vec{x}_B)$
Potential outside a spherical source	$\Phi = -\frac{GM}{r}$	$\Phi_{\text{elec}} = \frac{Q}{4\pi\epsilon_0 r}$
Field equation for potential	$\nabla^2\Phi = 4\pi G\mu$	$\nabla^2\Phi_{\text{elec}} = -\rho_{\text{elec}}/\epsilon_0$

Here, $\vec{x}_A$ and $\vec{x}_B$ are the positions of masses M and m in the gravitational case and charges Q and q in the electrostatic case. The distance between them is $r = |\vec{x}_A - \vec{x}_B|$ and $\vec{e}_r = (\vec{x}_B - \vec{x}_A)/r$. $\vec{F}_{\text{grav}}$ is the gravitational force exerted by M on m and $\vec{F}_{\text{elec}}$ is the electric force exerted by Q on q. Φ_{elec} is the electrostatic potential, and ρ_{elec} is electric charge density.

or

Newtonian Field Equation

$$\boxed{\nabla^2\Phi(\vec{x}) = 4\pi G\mu(\vec{x}),} \tag{3.18}$$

where ∇^2 is the Laplacian $\partial^2/\partial x^2 + \partial^2/\partial y^2 + \partial^2/\partial z^2$. This analog of Poisson's equation in electrostatics is the field equation for Newtonian gravity.

Example 3.1. Newton's Theorem. The gravitational field outside a spherically symmetric mass distribution depends only on its total mass. That result is called Newton's theorem. To prove it, integrate both sides of (3.17) over the volume $\mathcal{V}(r)$ inside a sphere of radius r about the center of symmetry whose surface contains all of the mass. One finds

$$\int_{\mathcal{V}(r)} d^3x \vec{\nabla}\cdot\vec{g} = -4\pi G \int_{\mathcal{V}(r)} d^3x\mu(r) = -4\pi GM, \tag{3.19}$$

where M is the total mass. Then use the divergence theorem (also called Gauss' theorem) to express the left-hand side as a surface integral over the sphere of radius r giving

$$\int_r d\vec{A}\cdot\vec{g} = -4\pi GM. \tag{3.20}$$

Because of the spherical symmetry $\vec{g}$ can depend only on r and point only in a radial direction. The surface integral is, therefore, $4\pi r^2|\vec{g}(r)|$, where $|\vec{g}|$ is the magnitude of $\vec{g}$. Thus, if $\vec{e}_r$ is a unit vector in the radial direction,

$$\vec{g}(r) = -\frac{GM}{r^2}\vec{e}_r, \tag{3.21}$$

and depends only on M. Similarly, the gravitational potential outside any spherically symmetric mass distribution also depends only on M when it is normalized to vanish at infinity:

$$\Phi(r) = -\frac{GM}{r}. \tag{3.22}$$

It doesn't matter whether the mass M is concentrated at the center, concentrated in a thin shell, distributed uniformly, or otherwise spherically symmetrically. Nor does it matter whether the mass inside is moving or not as long as it is moving only in radial directions. The field and potential outside a spherically symmetric distribution of mass are given by (3.21) and (3.22) and are always constant in time, since total mass is conserved. In general relativity the curved spacetime outside any spherically symmetric mass distribution also depends only on its total mass.

Example 3.2. Kepler's Law. For a satellite in orbit around a center of gravitational attraction, Kepler's law relates the period of the orbit to its size. Consider by way of example a circular orbit of radius R and period P about a spherically symmetric center of attraction of mass M. The relationship can be derived by equating the centripetal acceleration V^2/R (where V is the linear orbital speed) to the gravitational acceleration, giving

$$\frac{V^2}{R} = \left(\frac{2\pi R}{P}\right)^2 \frac{1}{R} = \frac{GM}{R^2}. \tag{3.23}$$

The resulting relationship is

$$P^2 = \frac{4\pi^2}{GM} R^3. \tag{3.24}$$

This is a special case of the square of the period being proportional to the cube of the semi-major axis.

In Chapter 6 we will see how this Newtonian gravity of forces and accelerations can be reformulated geometrically as a theory of free particles moving in a curved spacetime.

3.4 Gravitational and Inertial Mass

Inserting the gravitational force law (3.12) into Newton's law of motion, $\vec{F} = m\vec{a}$, gives

$$m\vec{a} = -m\vec{\nabla}\Phi \tag{3.25}$$

or

$$\vec{a} = -\vec{\nabla}\Phi. \tag{3.26}$$

This is the statement that all bodies fall with the same acceleration in a gravitational field independently of their mass or composition. As was briefly described in Section 2.1, this universality of free-fall acceleration is at the heart of the geometric understanding of gravity in general relativity.

Greater insight into this universality of free-fall acceleration can be found by distinguishing two roles played by mass in (3.25). Mass on the left-hand side of the equation governs the inertial properties of the body, and in this role it is called the *inertial mass* m_I of the body. This is the mass that occurs generally in Newton's law of motion

Inertial Mass

$$\vec{F} = m_I\vec{a}, \tag{3.27}$$

whatever the origin of the force (gravitational, electromagnetic, elastic, etc.) on the left-hand side of the equation.

The mass on the right-hand side of (3.25) measures the strength of the *gravitational* force between bodies and is therefore called the body's *gravitational mass*, m_G. This is the mass that occurs in the inverse square law [cf. (3.11)]

Gravitational Mass

$$\vec{F}_{\text{grav}} = -\frac{G m_G M_G}{r^2}\vec{e}_r, \tag{3.28}$$

and is analogous to electric charge. Gravitational mass enters the gravitational force law (3.12),

$$\vec{F}_{\text{grav}} = -m_G \vec{\nabla}\Phi(\vec{x}_B), \tag{3.29}$$

and gravitational mass density is the source of the gravitational potential (3.18),

$$\nabla^2\Phi(\vec{x}) = 4\pi G\mu_G(\vec{x}). \tag{3.30}$$

In familiar terms, the gravitational mass gives the *weight* of a body in a given gravitational field $\vec{g}$,

$$\vec{F}_{\text{grav}} = m_G \vec{g}. \tag{3.31}$$

All the masses or mass densities in Table 3.1 are gravitational.

Experiment shows that all bodies fall with the same acceleration in a gravitational field. Inertial mass and gravitational mass must, therefore, be proportional with a proportionality constant that is the same for all bodies. Gravitational mass can be *defined* to be equal to inertial mass for one body, say, the standard kilogram in Sèvres, France. The equality of accelerations then implies it is equal for all bodies:

Gravitational Mass and Inertial Mass Are Equal

$$\boxed{m_I = m_G} \tag{3.32}$$

As Box 2.1 on p. 14 showed, this is one of the most accurately tested relations in physics (more on this in Chapter 6).

This equality between a number m_I, which controls inertia in the general dynamical law for all forces, and a number m_G that measures the coupling strength to a particular force—gravity—is truly remarkable. In Newtonian theory, it appears as an isolated unexplained experimental fact. However, it is this experimental fact that allows a geometric theory of gravity and underlies general relativity. If all bodies with the same initial conditions fall along the same curve independent of their composition, then that curve can be a property of the geometry of spacetime and not of a force acting on the body.

3.5 Variational Principle for Newtonian Mechanics

Physics—where the action is.

(Anon.)

The laws of Newtonian mechanics can be formulated in terms of a variational principle called the principle of extremal action.[3] Extensions of this principle will

[3] Variational principles are sometimes called extremum principles, or action principles.

be the route to formulating the equations of motion of particles in curved spacetime. We review it beginning with the simple case of a particle of mass m moving in one dimension in a potential $V(x)$, whose equations of motion are summarized by the Lagrangian:

$$L(\dot{x}, x) = \frac{1}{2}m\dot{x}^2 - V(x), \tag{3.33}$$

where the dot denotes a time derivative. Newton's law $m\ddot{x} = -dV/dx$ can be expressed as Lagrange's equation

$$-\frac{d}{dt}\left(\frac{\partial L}{\partial \dot{x}}\right) + \frac{\partial L}{\partial x} = 0. \tag{3.34}$$

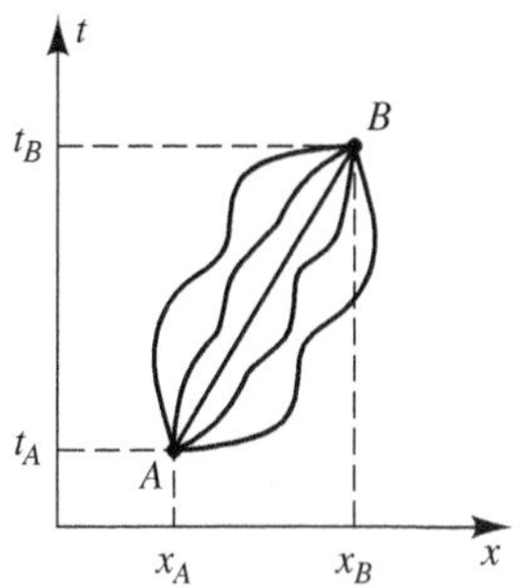

FIGURE 3.7 Many different paths between a position x_A at time t_A and a position x_B at time t_B can be described, but a particle moves on the one obeying Newton's law of motion. That path extremizes the action.

Consider the possible paths between a point x_A at time t_A and a point x_B at time t_B illustrated in Figure 3.7. For each path construct a real number called its *action*:

$$S[x(t)] = \int_{t_A}^{t_B} dt L(\dot{x}(t), x(t)). \tag{3.35}$$

The action is an example of a *functional*—a map from functions (in this case $x(t)$'s) to real numbers.

Among all the curves connecting x_A at t_A with x_B at t_B, those that extremize the action satisfy Lagrange's equation (3.34). That is the variational principle for Newtonian mechanics.

Variational Principle for Newtonian Mechanics

A particle moves between a point in space at one time and another point in space at a later time so as to extremize the action in between.

Put differently, a particle obeying Newton's laws of motion follows a path of extremal action. We now explain what *extremize* means and demonstrate the principle.

The extrema of a function of one variable $f(x)$ are the points where its first derivative vanishes—local maxima, local minima, or saddle points. At any extremum, a small change δx in x produces no first order change δf in the value of the function. That is because, to first order in δx,

$$\delta f = \frac{df}{dx}\delta x, \tag{3.36}$$

and at an extremum, $df/dx = 0$.

The extrema of a function $f(x^1, \ldots, x^n)$ of n variables $x^1, \ldots, x^n$ occur where all the partial derivatives $\partial f/\partial x^a$ vanish, for $a = 1, \ldots, n$. Such an extremum can be characterized as the place where the first variation of the function vanishes,

$$\delta f = \sum_{a=1}^{n} \frac{\partial f}{\partial x^a} \delta x^a = 0, \tag{3.37}$$

for *arbitrary* variations δx^a, $a = 1, \ldots, n$. In many dimensions an extremum does not have to be a maximum or a minimum of the function. It can be a maximum in some directions and a minimum in others.

The extrema of the action functional $S[x(t)]$ are defined by the vanishing of its first-order variation $\delta S[x(t)]$ for arbitrary variations $\delta x(t)$ of the path connecting (x_A, t_A) to (x_B, t_B). To compute $\delta S[x(t)]$ just substitute $x(t) + \delta x(t)$ for $x(t)$ in the definition of the action (3.35), expand to first order in $\delta x(t)$, and integrate once by parts to find:

$$\delta S[x(t)] = \int_{t_A}^{t_B} dt \left[\frac{\partial L}{\partial \dot{x}(t)} \delta \dot{x}(t) + \frac{\partial L}{\partial x(t)} \delta x(t) \right] \tag{3.38a}$$

$$= \frac{\partial L}{\partial \dot{x}(t)} \delta x(t) \bigg|_{t_A}^{t_B} + \int_{t_A}^{t_B} dt \left[-\frac{d}{dt} \left(\frac{\partial L}{\partial \dot{x}(t)} \right) + \frac{\partial L}{\partial x(t)} \right] \delta x(t). \tag{3.38b}$$

Variations of the path that connects x_A at t_A to x_B at t_B necessarily vanish at the endpoints—$\delta x(t_A) = \delta x(t_B) = 0$. The first term in (3.38b) therefore vanishes. The remaining term has to vanish for arbitrary $\delta x(t)$ that meet these conditions for $\delta S[x(t)]$ to vanish. This can happen only if the integrand of the integral in (3.38b) vanishes identically, giving

$$-\frac{d}{dt} \left(\frac{\partial L}{\partial \dot{x}} \right) + \frac{\partial L}{\partial x} = 0. \tag{3.39}$$

The action is extremized by paths that satisfy Lagrange's equation.

This result is not restricted to motion in one dimension. If the Lagrangian is a function of n coordinates $x^a(t)$ and their time derivatives, its extrema satisfy the n equations

Lagrange's Equations

$$-\frac{d}{dt} \left(\frac{\partial L}{\partial \dot{x}^a} \right) + \frac{\partial L}{\partial x^a} = 0, \qquad a = 1, \ldots, n. \tag{3.40}$$

Example 3.3. A Particular Variation. If the action is an extremum with respect to *any* variation away from the path obeying the equations of motion, then it must also be an extremum for any *particular* variation. Consider a free particle $(V(x) = 0)$ moving between x_A at t_A and x_B at t_B. Newton's laws dictate that the free particle travels between these points with a constant velocity, which is $(x_B - x_A)/T$, where $T \equiv t_B - t_A$ is the elapsed time. This is the straight-line path shown in Figure 3.8. When half of the time T has elapsed, the particle is at the po-

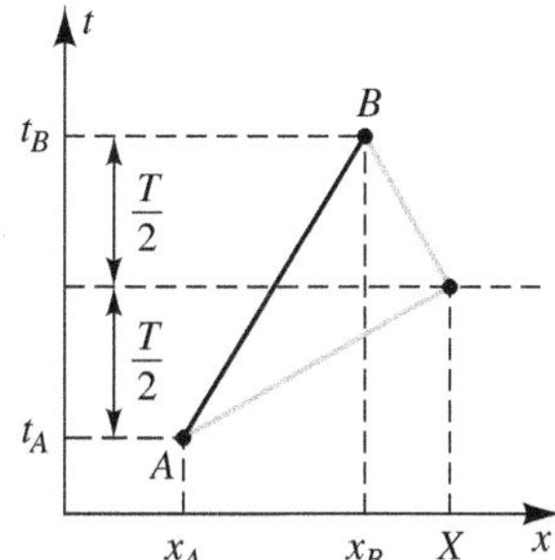

FIGURE 3.8 This figure shows a particular family of particle paths connecting position x_A at time t_A with position x_B at time t_B. Each shaded path consists of two straight segments parametrized by the position X reached in half the time interval between t_A and t_B. The path that extremizes the action is the unshaded straight line connecting the two points. That is the path obeying Newton's laws.

sition $(x_B + x_A)/2$. We compare the action of this path satisfying Newton's laws with paths that move from x_A with a constant velocity to some *different* position X in total time $T/2$ and then with a different constant velocity to get to x_B in time T. Examples are shown in Figure 3.8. The action $S(X)$ for these paths is a function of X, which is easy to calculate from (3.35) because the velocity is constant on each leg, namely, $(X - x_A)/(T/2)$ on the first leg and $(x_B - X)/(T/2)$ on the second. The action along any leg in which the particle is moving with constant velocity V for a time t is $mV^2t/2$. The sum for both legs is

$$S(X) = m\left[(x_B - X)^2 + (X - x_A)^2\right]/T. \tag{3.41}$$

Paths of extremal action occur where $dS/dX = 0$. There is only one solution at

$$X = (x_B + x_A)/2, \tag{3.42}$$

which is the path obeying Newton's laws.

Problems

1. A free particle is moving in an inertial frame (x, y, z) in the xy plane on a trajectory $x = d$, $y = vt$, where d and v are constants in time. Consider a rectangular frame (x', y', z') rotating with respect to the inertial frame with an angular velocity ω about a common z-axis $(z' = z)$. What are the equations of motion obeyed by $x'(t)$, $y'(t)$ and $z'(t)$ in the rotating frame? Sketch the trajectory of the particle in the $x'y'$-plane and show explicitly that it satisfies these equations of motion.

2. Show that Newton's laws of motion are *not* invariant under a transformation to a frame that is uniformly accelerated with respect to an inertial frame of Newtonian mechanics. What are the equations of motion in the accelerated frame?

3. [B, S] How many degrees per hour does the Foucault pendulum described in Box 3.2 on p. 37 precess?

4. Find the gravitational potential inside and outside a sphere of uniform mass density having a radius R and a total mass M. Normalize the potential so that it vanishes at infinity.

5. Consider the functional

$$S[x(t)] = \int_0^T \left[\left(\frac{dx(t)}{dt} \right)^2 + x^2(t) \right] dt.$$

Find the curve $x(t)$ satisfying the conditions

$$x(0) = 0, \qquad x(T) = 1,$$

which makes $S[x(t)]$ an extremum. What is the extremum value of $S[x(t)]$? Is it a maximum or minimum?

6. [B, E, C] Estimate the gravitational self-energy of the Moon as a fraction of the Moon's rest mass energy. Is this ratio larger or smaller than the few parts in 10^{15} accuracy of the Lunar laser ranging test of the equality of gravitational and inertial mass?

CHAPTER

4

Principles of Special Relativity

Einstein's 1905 special theory of relativity requires a profound revision of the Newtonian ideas of space and time that were reviewed in the previous chapter. In special relativity the Newtonian ideas of Euclidean space and a separate absolute time are subsumed into a single four-dimensional union of space and time called *spacetime*. This chapter reviews the basic principles of special relativity, starting from the non-Euclidean geometry of its spacetime.

4.1 The Addition of Velocities and the Michelson–Morley Experiment

Not much needs be known about Maxwell's equations governing electromagnetic fields to conclude that they do not take the same form in every inertial frame of Newtonian mechanics. Maxwell's equations imply that light travels with the speed c that enters as a basic parameter of the equations.[1] But the Galilean transformation (3.6) between inertial frames implies that light should travel with different speeds in different inertial frames moving with respect to each other.

More specifically, suppose (V^x, V^y, V^z) are components of the velocity of a particle[2] measured in one inertial frame, and $(V^{x'}, V^{y'}, V^{z'})$ the components of the velocity measured in a frame moving with respect to the first along its x-axis with velocity v. Then, from (3.6),

$$V^{x'} = \frac{dx'}{dt'} = \frac{dx'}{dt} = \frac{dx}{dt} - v = V^x - v, \tag{4.1}$$

so that together with the trivial transformations of the y and z components one has

Newtonian Addition of Velocities

$$\begin{aligned} V^{x'} &= V^x - v, \\ V^{y'} &= V^y, \\ V^{z'} &= V^z. \end{aligned} \tag{4.2}$$

This is called the Newtonian addition of velocities rule.

[1] You might be used to thinking that quantities called ϵ_0 and μ_0 are the basic parameters in Maxwell's equations, but $\mu_0 \equiv 4\pi \times 10^{-7}$ is a pure number, and $\epsilon_0 = 1/(c^2\mu_0)$.

[2] For the most part, uppercase letters such as $\vec{V}$ are used for the velocities of particles as measured in one inertial frame, and lowercase letters such as $\vec{v}$ are used for the the velocity of one inertial frame with respect to another, but occasionally it's necessary to compromise this convention.

The transformation (4.2) implies that Maxwell's equations can be valid only in one inertial frame because they predict one velocity for light. In the nineteenth century this frame was thought to be the rest frame of the physical medium through which light propagated—the "ether." The velocity of light in any inertial frame moving with respect to the ether rest frame would be given by (4.2).

Michelson–Morley Experiment

In an experiment whose results were published in 1887, Albert Michelson and Edward Morley tested the Newtonian addition of velocities law (4.2) for light. A modern version of their experiment is described in more detail in Box 4.1. Michelson and Morley compared the velocity of light in an Earth-based laboratory in directions along the Earth's orbital motion and perpendicular to it at two different points on the Earth's orbit. (See Figure 4.1.) The motion of the Earth around the Sun means that at most points on its orbit it will be moving with respect to the ether. If it happens to be at rest with respect to the ether at one point in its orbit, then six months later it will be moving with respect to the ether with double its orbital speed. Suppose for simplicity that the Sun is at rest with respect to the ether. If $V_\oplus$ is the Earth's orbital velocity, the Newtonian law for addition of velocities (4.2) implies that the velocity of light perpendicular to the Earth's motion is c, whereas the velocity in directions parallel to it should be $c \pm V_\oplus$. Michelson and Morley detected no difference. Evidently the Newtonian law of

BOX 4.1 A Modern Michelson–Morley Experiment

In 1978 Brillet and Hall set new limits on the isotropy of space with respect to the propagation of light. A He-Ne laser ($\lambda = 3.39\ \mu$m) fed radiation into a Fabry-Perot interferometer—essentially an optical cavity bounded by two mirrors a fixed distance apart. The frequency of this laser was continuously adjusted to keep a standing wave in the cavity. Any variation in the velocity of light would cause a shift in the frequency f of the laser because $f = c/\lambda$. Laser and cavity were mounted on a massive granite table that could be rotated to compare different directions in space. The frequency of the laser was determined by splitting the beam, running one part up the rotation axis, and comparing the result with a stationary reference laser. Were the velocity of light different in two perpendicular directions a $\cos(2\phi)$ dependence of the frequency would result, where ϕ is the rotation angle of the platform. Brillet and Hall found

$$\Delta f/f = (1.5 \pm 2.5) \times 10^{-15}$$

consistent with no variation in the frequency at all. The Newtonian addition of velocities would predict a frequency shift of order $(V_\oplus/c)^2 \approx 10^{-8}$, where $V_\oplus$ is the velocity of the Earth in its orbit. Brillet and Hall's experiment gives a null result on a scale ten million times smaller than the classical prediction.

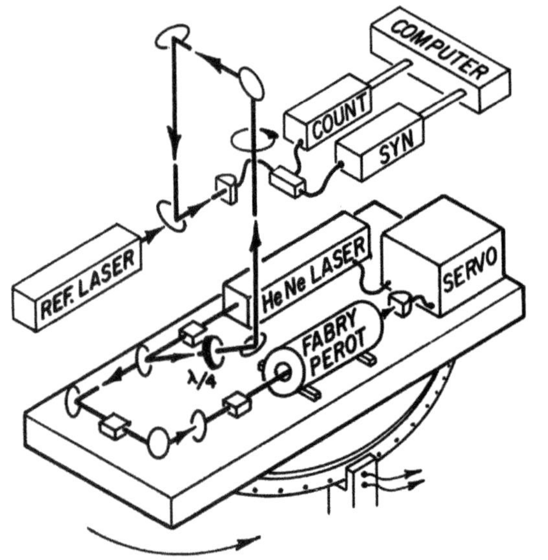

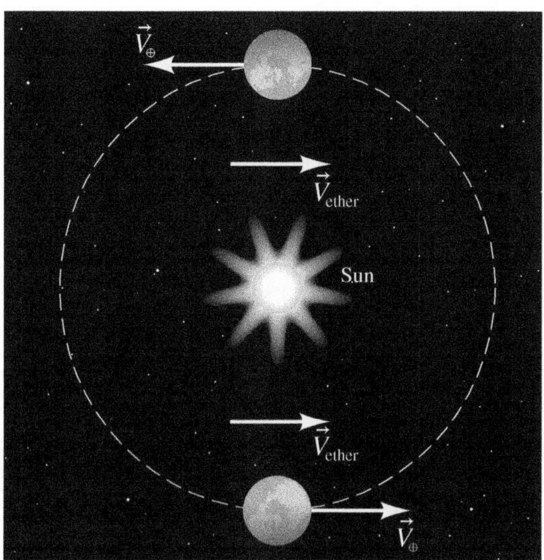

FIGURE 4.1 The Michelson–Morley experiment. Suppose the uniform ether is moving with a velocity $\vec{V}_{ether}$ with respect to the Sun or, equivalently, that the Sun is moving with a velocity $-\vec{V}_{ether}$ with respect to the ether. Let $\vec{V}_\oplus$ be the velocity of the Earth with respect to the Sun at one point in its orbit. At that point, the velocity of the Earth with respect to the ether is $\vec{V}_\oplus - \vec{V}_{ether}$. Six months later the velocity of the Earth is approximately (neglecting the ellipticity of the Earth's orbit) $-\vec{V}_\oplus$, and its velocity with respect to the ether is $-\vec{V}_\oplus - \vec{V}_{ether}$. That is a difference in velocity of $2\vec{V}_\oplus$ no matter what $\vec{V}_{ether}$ is.

addition of velocities was not correct. Either Newtonian mechanics or Maxwell's equations had to be modified. It turned out to be mechanics.

4.2 Einstein's Resolution and Its Consequences

Einstein's 1905 successful modification of Newtonian mechanics is called the *special theory of relativity*, or *special relativity* for short. To formulate it, Einstein assumed that the principle of relativity described in Section 3.2 holds for electromagnetic phenomena as described by Maxwell's equations. In particular, he assumed that the velocity of light had the same value c in all inertial frames—an assumption that from the present perspective is clearly motivated by the Michelson–Morley experiment.[3] But, in accepting the principle of relativity, Einstein did not adopt the Galilean transformation, which implements it in Newtonian mechanics, since this implies the Newtonian velocity addition law. Rather, he found a new connection between inertial frames that is consistent with the same value of the velocity of light in all of them.

The assumption that the velocity of light is the same in every inertial frame requires a reexamination and ultimately the abandonment of the Newtonian idea

[3]The true history is, as usual, more complex (Miller 1981, Pais 1982).

of absolute time. One can see this most clearly by examining the idea of simultaneity. Two events are simultaneous if they occur at the same time. In Newtonian theory two events that are simultaneous in one inertial frame are simultaneous in every other inertial frame because there is a single absolute time. To see the impact on this idea of assuming the constancy of the velocity of light, consider the thought experiment illustrated in Figures 4.2 and 4.3.

Three observers A, B, and O are riding a rocket of length L. O is midway between A and B. A and B each emit light signals directed along the rocket toward O. O receives the signals simultaneously. Which signal was emitted first?

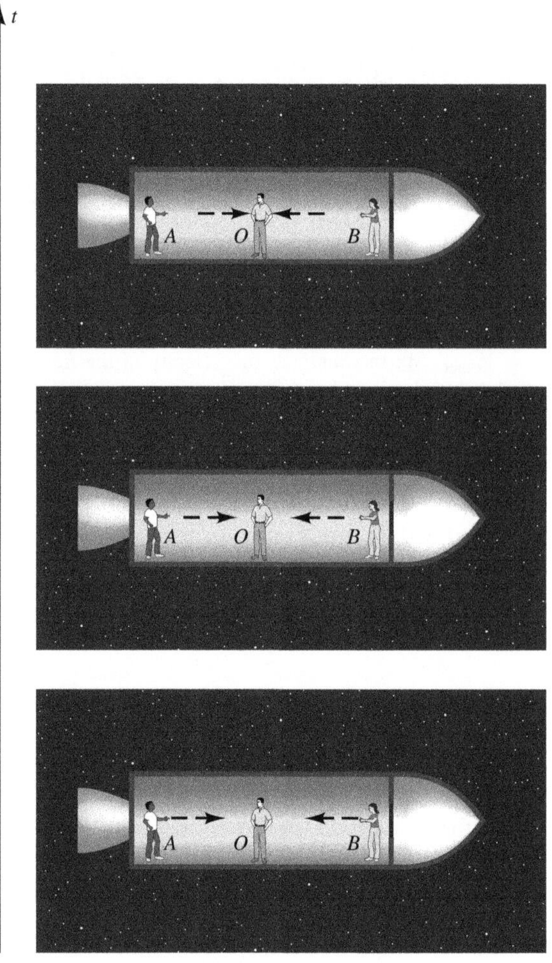

FIGURE 4.2 Three observers, A, B, and O, are riding on a rocket at rest in an inertial frame. Observers A and B are equally distant from O. A and B emit light signals that are received simultaneously by O. Moving upward from the bottom, the figure shows views of the rocket and signals at three equally spaced instants ending with the simultaneous arrival of the signals at O. Since the signals from A and B arrived simultaneously, traveled with speed c, and came from equal distances away, they must have been emitted simultaneously.

The answer depends on the inertial frame if the velocity of light is the same in all of them.

Figure 4.2 shows the inertial frame where the rocket is at rest. An observer at rest in this frame reasons as follows: "The rocket is at rest and the two observers A and B are equal distances away from O. It therefore takes the same length of time for a light signal to propagate from A to O as it does from B to O. Since the signals reached O at the same instant, they were emitted simultaneously."

A different result is obtained in an inertial frame in which the rocket is moving, such as that shown in Figure 4.3. An observer at rest in this frame reasons as

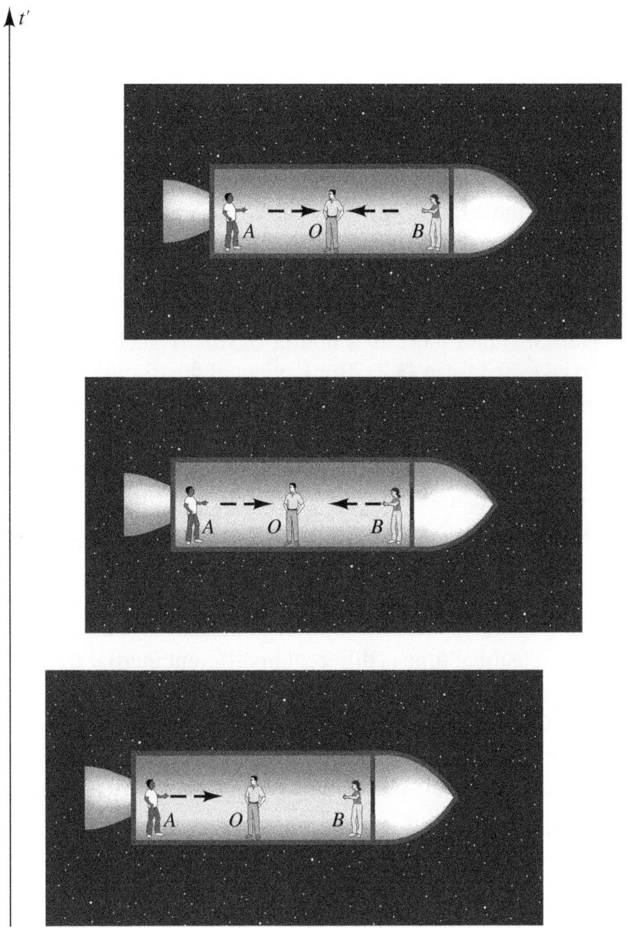

FIGURE 4.3 The same rocket and observers as in Figure 4.2 in an inertial frame in which the rocket is moving to the right with speed V. Moving upwards from the bottom, the figure shows three views equally spaced in time. At the top the signals from A and B are received simultaneously by O. The two bottom views show the emission of the signals from A and B. For the signals to arrive simultaneously at O, the one from A must have been emitted *earlier* than the one from B becaue it has a longer distance to travel. The two signals are not emitted simultaneously.

follows: "The signals are received simultaneously by O. At earlier times when the signals were emitted B was always closer to O's position at reception than A. Since *both signals travel with speed c*, the one from A was emitted earlier than the one from B because it has a longer distance to travel to reach O at the same instant as the one from B."

Thus, two event simultaneous in one inertial frame are not simultaneous in one moving with respect to the first if the velocity of light is the same in both. (Contrast Problem 7.) The Newtonian idea of time must be abandoned. We will next see how.

4.3 Spacetime

Newton's first law—free particles move at constant speed on straight lines—is unchanged in special relativity. The construction of inertial frames described in Section 3.1 and illustrated in Figure 3.3 is therefore also unaltered: start with an origin following the straight-line trajectory of a free particle. At one moment choose three Cartesian coordinates (x, y, z) with this origin. Propagate these axes parallel to themselves as the origin moves to define (x, y, z) at later times. The result is an inertial frame.[4]

For each inertial frame there is a notion of time t such that the law of free particle motion takes the form (3.2). But in view of the discussion of simultaneity in the previous section, there is no reason to accept the assumption of Newtonian physics that the times of different inertial frames will agree. Rather, there is generally a *different* notion of time and simultaneity for each inertial frame. Inertial frames are, therefore, spanned by *four* Cartesian coordinates (t, x, y, z), and a *different* inertial frame has a *different* set of four coordinates (t', x', y', z'). The correct geometric arena for physics is, therefore, not a separate space and absolute time but rather a *four*-dimensional unification of space *and* time called *spacetime*.[5] The separation of spacetime into separate notions of three-dimensional space and one-dimensional time is different in different inertial frames. The transformations between inertial frames moving with respect to each other that are analogous to the Galilean transformations (3.6) will mix space and time, as we will see in Section 4.5.

The defining assumption of special relativity is a geometry for four-dimensional spacetime to which we now turn.

Spacetime Diagrams

To describe four-dimensional spacetime we first introduce a tool, which is so simple it appears trivial, but so powerful it is indispensable. This is the idea of a *spacetime diagram*. A spacetime diagram is a plot of two of the coordinate axes of an inertial frame—two coordinate axes of spacetime. Since there are four axes and

[4]Inertial frames in special relativity are sometimes called Lorentz frames.

[5]Relativists write *spacetime* as one word instead of, for example, space-time to indicate that it is one unified idea.

only two dimensions on a piece of paper, two or at most three of these axes can be drawn. Spacetime diagrams are *slices* or sections of spacetime in much the same way as an x-y plot is a two-dimensional slice of three-dimensional space. A typical example is shown in Figure 4.4. It is convenient to use ct rather than t as an axis, because then both have the same dimension.

A point P in spacetime can be called an *event* because an event occurs at a particular place at a particular time, that is, at a point in spacetime. For example, a supernova explosion happened at the event in spacetime that occurred in A.D. 1054 at the location of the Crab nebula. An event P can be located in spacetime by giving its coordinates (t_P, x_P, y_P, z_P) in an inertial frame, as shown in Figure 4.4.

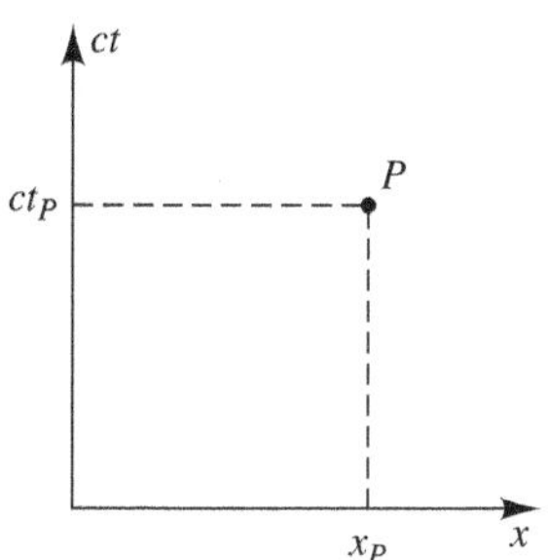

FIGURE 4.4 A spacetime diagram showing a two-dimensional slice of four-dimensional spacetime in the coordinates of a particular inertial frame. An event is a point P in spacetime located at a particular place in space (x_P) at a particular time (t_P).

A particle describes a curve in spacetime called a *world line*. It is the curve of positions of the particle at different instants, i.e., $x(t)$. Figure 4.5 shows a spacetime diagram with two sample world lines. The slope of the world line gives the ratio c/V^x since $d(ct)/dx = c\,dt/dx = c/V^x$. Zero velocity corresponds to infinite slope. A velocity of c corresponds to a slope of unity. Light rays therefore move along the 45° lines in a spacetime diagram. Box 4.2 on p. 55 shows an early example of a spacetime diagram with world lines.

The Geometry of Flat Spacetime

The central assumption of special relativity is a geometry for spacetime. As we learned in Chapter 2, a geometry is specified by a line element that gives the distance between nearby points. It would be appropriate to begin a discussion of special relativity by positing this line element. However, before doing that, consider a simple thought experiment that motivates the form of the line element and connects that with Einstein's assumption that the velocity of light is c in all inertial frames.

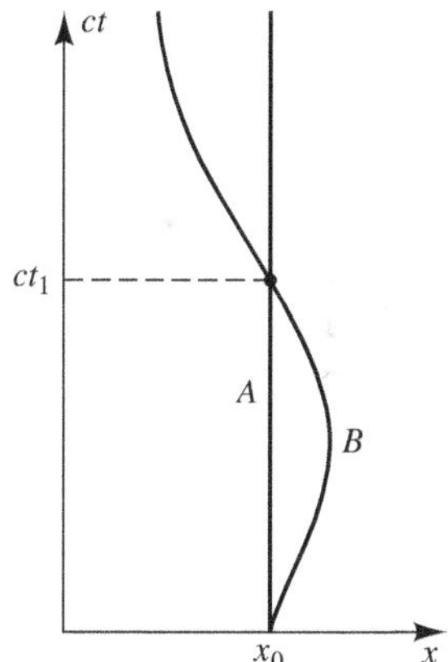

FIGURE 4.5 World lines in spacetime. A is the world line of a particle that sits at rest at x_0 for all time in the inertial frame (ct, x). World line B represents an observer who accelerates away from x_0 at time $t = 0$, decelerates, reverses direction, crosses x_0 at $t = t_1$, and heads off toward negative x.

The thought experiment is illustrated in Figure 4.6. Two parallel mirrors separated by a distance L are at rest in an inertial frame in which events are described by coordinates (t, x, y, z). We take y to be the vertical direction between the mirrors and x the direction parallel to them. A light signal bounces back and forth between the mirrors; the right hand part of Figure 4.6 shows its world line in a spacetime diagram. A clock measures the time interval Δt between the event A of the departure of the light ray and the event C of its return to the same point in space. These two events are separated by coordinate intervals

$$\Delta t = 2L/c, \qquad \Delta x = \Delta y = \Delta z = 0 \tag{4.3}$$

in the inertial frame where the mirrors are at rest.

Analyze the same thought experiment in an inertial frame that is moving with speed V with respect to the (t, x, y, z) inertial frame along the negative x-direction parallel to the mirrors. Locate events in this frame by coordinates (t', x', y', z') with x' parallel to x. In this frame the mirrors are moving with speed V along the positive x'-direction, as illustrated in Figure 4.7. What is the time interval $\Delta t'$ in this frame between the departure of a pulse and its return? Analyze this question as follows: the light ray travels a distance $\Delta x' = V\Delta t'$ in the x'-

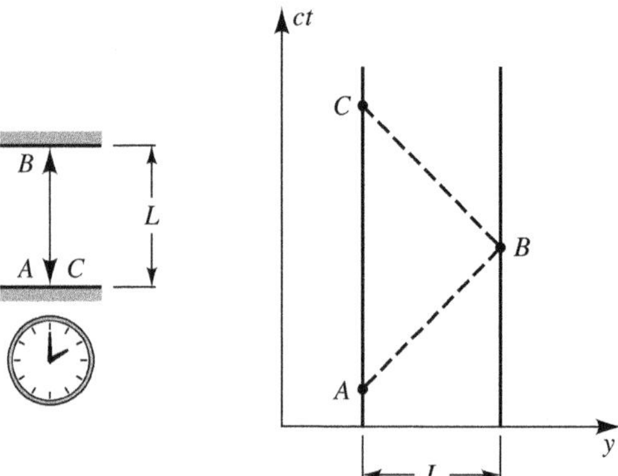

FIGURE 4.6 The left-hand figure shows two parallel mirrors at rest in an inertial frame spanned by coordinates (t, x, y, z). A light pulse bounces back and forth between the mirrors, and a clock measures the time interval $\Delta t = 2L/c$ between the departure of the pulse from the lower mirror at A and its return at C. The world line of the pulse is shown in the spacetime diagram of the right-hand figure, where the y-axis is the vertical direction along which the light ray travels. The events of departure A and return C are separated by the time interval Δt but are at the same spatial point $\Delta x = \Delta y = \Delta z = 0$. The same setup can be regarded as a model of a clock that advances every time the pulse returns to the lower mirror at intervals of $2L/c$ per advance.

direction between emission at A and return at C. The distance traveled in the y'-direction is L, assuming that the transverse distances are the same in both inertial frames. (Work Problem 16 for more support of this.) The total distance traveled between departure and return is, therefore, $2[L^2 + (\Delta x'/2)^2]^{1/2}$. Assuming with Einstein that the velocity of light is c in this inertial frame, the time of travel, $\Delta t'$, is this distance divided by c. Thus the coordinate intervals between A and C in this

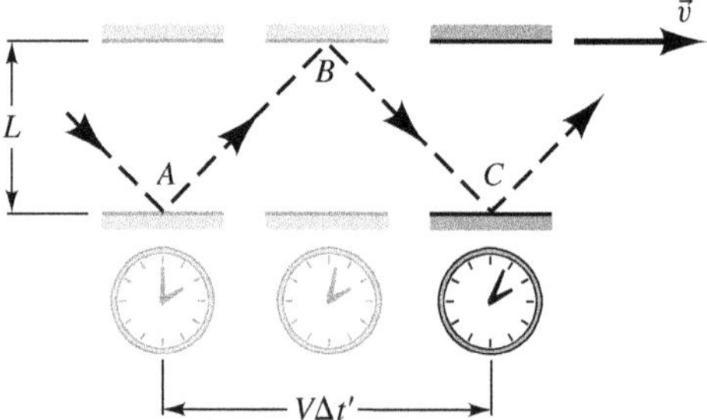

FIGURE 4.7 The thought experiment described in Figure 4.6 is shown here in an inertial frame spanned by coordinates (t', x', y', z'), in which the mirrors are moving with speed V in the x'-direction along their lengths. The path the light pulse travels in a time $\Delta t'$ between departure and return to the lower mirror is shown. The events of departure and return are separated in space by $\Delta x' = V\Delta t'$, $\Delta y' = \Delta z' = 0$. The length of the path traveled is $2[L^2 + (\Delta x'/2)^2]^{1/2}$, and $\Delta t'$ is this length divided by c.

BOX 4.2 Railway Trains in Spacetime

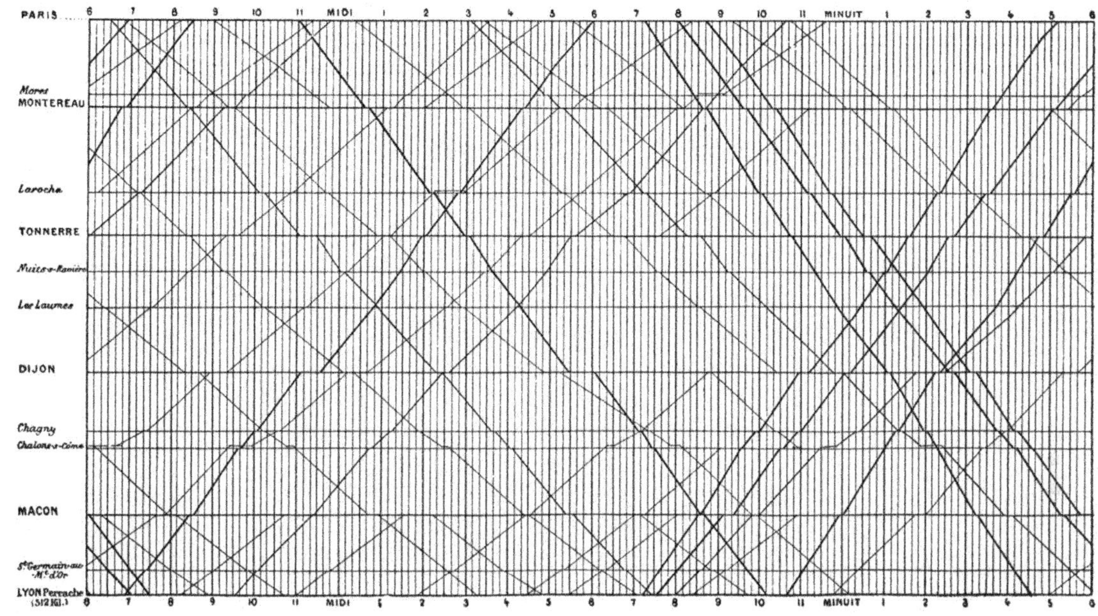

Spacetime diagrams were in use before the advent of special relativity, as this timetable for the railway trains on the Paris–Lyon line reproduced from Marey (1885) shows. Unfortunately the designer of the timetable did not anticipate the convention of relativity and plotted time horizontally. The world lines of the stations (at rest) are horizontal lines. The slanting lines are trains of various speeds moving in between stations and halting at them. Faster trains have steeper slopes, but the time axis is measured in hours, so the 45° lines are not at the speed of light. Rotate the diagram by 90° to view it with the conventions of special relativity.

frame are

$$\Delta t' = \frac{2}{c}\sqrt{L^2 + \left(\frac{\Delta x'}{2}\right)^2}, \qquad \Delta x' = V\,\Delta t', \qquad \Delta y' = 0, \qquad \Delta z' = 0. \tag{4.4}$$

(The right-hand sides of these relations could easily be expressed entirely in terms of V, c, and L, but that isn't necessary at present.)

From (4.3) and (4.4) it is straightforward to derive

$$-(c\Delta t')^2 + (\Delta x')^2 = -4[L^2 + (\Delta x'/2)^2] + (\Delta x')^2 = -4L^2 = -(c\Delta t)^2. \tag{4.5}$$

This mathematical identity is the key to identifying an *invariant*—a quantity which is the same in both frames—and to finding the line element that describes the geometry of spacetime. Since $\Delta x = 0$ and since the Δy's and Δz's are zero in both frames, we can judiciously add them back into the two sides of (4.5) to find that the combination

$$(\Delta s)^2 \equiv -(c\Delta t)^2 + (\Delta x)^2 + (\Delta y)^2 + (\Delta z)^2 \tag{4.6}$$

is the same in both frames. Specifically,

$$(\Delta s)^2 \equiv -(c\Delta t)^2 + (\Delta x)^2 + (\Delta y)^2 + (\Delta z)^2, \tag{4.7a}$$

$$= -(c\Delta t')^2 + (\Delta x')^2 + (\Delta y')^2 + (\Delta z')^2. \tag{4.7b}$$

Although derived from a simple thought experiment, this relation turns out to hold generally in any other thought experiment that involves the time and space separations between two events viewed from two inertial frames. The quantity $(\Delta s)^2$ is *invariant* under the change in inertial frames.

The distance between points defining spacetime geometry must be the same in all systems of coordinates used to label the points. The principle of relativity requires that the line element that defines the distance should have the same form in all inertial frames. The invariance exhibited in (4.7) therefore motivates taking $(\Delta s)^2$ as the squared distance between points in spacetime. More precisely, we will posit the line element[6]

Line Element of Flat Spacetime

$$ds^2 = -(c\,dt)^2 + dx^2 + dy^2 + dz^2 \tag{4.8}$$

(the infinitesimal version of (4.6)) as defining the geometry of four-dimensional spacetime and the starting point for special relativity. By requiring it take the same form in every inertial frame, we will derive the Lorentz transformations that connect inertial frames in Section 4.5.[7] The geometry specified by (4.8) is non-Euclidean (because of the minus sign) but is also flat in a sense we shall make precise in Chapter 21. It is therefore referred to as *flat spacetime*. Sometimes it is called *Minkowski space* after the mathematician H. Minkowski, who proposed it shortly after Einstein introduced special relativity.

Example 4.1. Spacetime Diagrams as Maps of Spacetime. No one would think of confusing the relationships between lengths on a Mercator map of the world with the relationships between true distances on the surface of the Earth. A Mercator map is a projection of the geometry of the globe on a sheet of paper,

[6]There are two possible conventions for the sign of the line element defining the squared distance in others. Also, for the most part, we denote spacetime distances by lowercase letters such as ds^2 and $d\tau^2$ and spatial distances by uppercase letters such as dS^2 and $d\Sigma^2$.

[7]Historically the transformations were derived by Einstein from the assumptions mentioned on p. 49. The idea of spacetime was introduced shortly thereafter by H. Minkowski. This historical sequence is still followed in many elementary texts today.

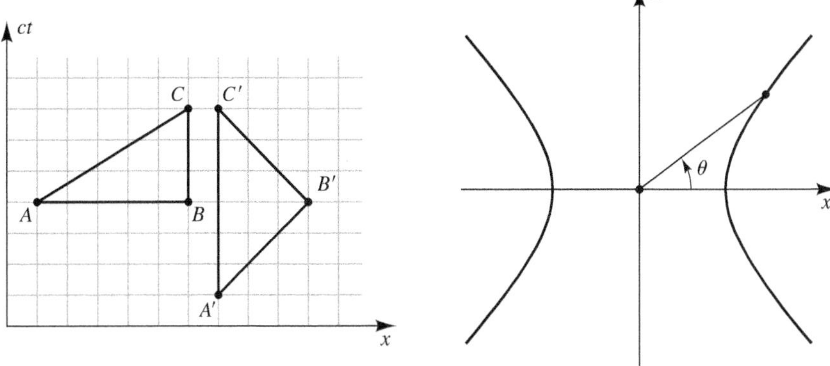

FIGURE 4.8 A little spacetime geometry. The left-hand figure shows a spacetime diagram with two triangles whose properties are discussed in the text. The right-hand figure shows a spacetime analog of a circle—a hyperbola that is a constant spacetime distance from the origin. The hyperbolic angle θ shown is the ratio of the spacetime distance along the hyperbola from the x-axis to the spacetime distance of the hyperbola from the origin.

which has a different geometry. (See Box 2.3 on p. 25). Similarly, a spacetime diagram is a projection of a two-dimensional section of spacetime with a geometry summarized by [cf. (4.6)]

$$(\Delta s)^2 = -(c\Delta t)^2 + \Delta x^2 \tag{4.9}$$

on the plane of a sheet of paper whose geometry is summarized by $(\Delta S)^2 = (\Delta x)^2 + (\Delta y)^2$. Don't get distances on a page displaying a spacetime diagram mixed up with the true distance in spacetime! Test your understanding of this by answering the following questions about the lengths between points in the figures in the spacetime diagram in Figure 4.8. Take length to be the square root of the absolute value of the right-hand side of (4.9), and check your answers with those at the bottom of the page.

(a) Which of the sides of triangle ABC is the longest? Which is the shortest? What are the lengths in the units of the grid?

(b) Which is the shorter path between points A and C—the straight-line path between A and C or the path through the other sides of ABC?

Then for (c) and (d), answer the same questions for triangle $A'B'C'$.

(a) The lengths of the sides are $|AB| = 5$, $|BC| = 3$, and $|AC| = (-3^2 + 5^2)^{1/2} = 4$. The longest side is AB, the shortest, BC. (b) The straight-line path between A and C is the shortest path between those points. (c) The lengths of the sides are $|A'B'| = |B'C'| = (-3^2 + 3^2)^{1/2} = 0$ and $|A'C'| = 6$. The longest side is $A'C'$ and the others are the shortest. (d) The shortest path is $A'B'C'$, which has zero length. In fact, the straight line from A' to C' has the *longest* distance when compared with any nearby path, as we will see in Section 4.4.

There are analogies between the elements of plane geometry and the geometry of a spacetime diagram. One is illustrated in Figure 4.8. The analog of a circle of radius R centered on the origin is the locus of points a constant *spacetime* distance from the origin. This consists of the hyperbolas $x^2 - (ct)^2 = R^2$. The ratios of arcs along a hyperbola to R define *hyperbolic angles*, as shown in the figure, with the relation

$$ct = R \sinh\theta, \qquad x = R \cosh\theta. \tag{4.10}$$

It's useful to be able to understand these analogies, but it does not prove useful in relativity to pursue them too far.

Light Cones

The minus sign in front of the $(c\Delta t)^2$ term is a novel feature of the line element (4.8). The geometry of spacetime is *not* four-dimensional Euclidean geometry. In particular, two points can be separated by distances whose *square* is positive, negative, or zero. When $(\Delta s)^2$ is *positive*, the points are said to be *spacelike separated.* That is the case, for example, when $\Delta t = 0$ and $\Delta x \neq 0$. When $(\Delta s)^2$ is *negative*, the points are said to be *timelike separated*. For instance, that happens when two points are at the same place $\Delta x = \Delta y = \Delta z = 0$ but at different times $\Delta t \neq 0$. When $(\Delta s)^2 = 0$, the two points are said to be *null separated.* For example, there is zero distance between two points with $\Delta y = \Delta z = 0$ but $\Delta x = c\Delta t$. Null separated points can be connected by light rays that move with speed c, so *lightlike separated* is used as a synonym for *null separated.* In summary, there are three kinds of separation:

$$(\Delta s)^2 > 0 \quad \text{spacelike separated,} \tag{4.11a}$$

$$(\Delta s)^2 = 0 \quad \text{null separated,} \tag{4.11b}$$

$$(\Delta s)^2 < 0 \quad \text{timelike separated.} \tag{4.11c}$$

Light Cone

The locus of points that are null separated from a point P in spacetime is its *light cone.*[8] The light cone of P is a three-dimensional surface in four-dimensional spacetime. Part of it (the *future light cone of* P) is generated by light rays that move outward from P. Two of these dimensions correspond to the direction a light ray can go; the third is along the rays. The other part (the *past light cone of* P) is generated by light rays that converge on P. You can think of the future light cone as the surface swept out in spacetime by a spherical pulse of light emitted from the location of P at the time of P. The past light cone is the surface swept out by a spherical pulse converging on P.

Needless to say, nothing in this definition depends on a particular inertial frame; only distances in the geometry of spacetime were used. However, intuition

[8]Some authors prefer the name *null cone* to emphasize that not just light travels at speed c, but also gravitons and possibly some neutrinos, etc. However, the name light cone has a long tradition, and we continue it.

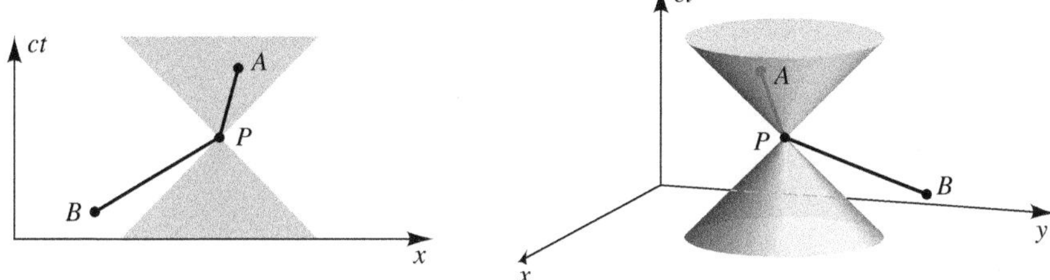

FIGURE 4.9 At left is a spacetime diagram showing a two-dimensional (ct, x) slice of four-dimensional flat space. The 45° lines from point P are the set of points that are null separated from it. They are the intersection of P's light cone with this slice. Point A is timelike separated from P, as are all points in the shaded wedges. The upper shaded wedge is the *inside* of the future light cone; the lower shaded wedge is the *inside* of the past light cone. The unshaded area is the *outside* of the light cone. Point B is spacelike separated from P as are all the points in the unshaded wedges. The figure at right shows the same point P but with one more spatial dimension. The light cone is the locus of points that would be traced out by a pulse of light emitted at P or converging on it. The surface of the pulse would be an expanding or contracting sphere in three spatial dimensions. In this reduced number of dimensions, it appears as the increasing circular cross section of the cone.

about light cones can be built up using the spacetime diagrams of a particular inertial frame. Two examples are shown in Figure 4.9.

Each point P in spacetime has a light cone. Light cones are an important feature of the geometry of spacetime. The points that are timelike separated from P lie *inside* the light cone (like the point A in Figure 4.9). Points that are spacelike separated from P lie *outside* the light cone (like the point B in Figure 4.9).

The paths of light rays are straight lines in spacetime with constant slope corresponding to the speed of light, that is, along null world lines. At every point P along the world line of a light ray, the straight line is *tangent* to the light cone of that point (see Figure 4.10). The distance between two points along a light ray is zero!

Timelike World Lines

Particles with nonzero rest mass move along *timelike* world lines that are always *inside* the light cone of any point along their trajectory (see Figure 4.10). That way their velocity is always less than the speed of light at that point.

It would be consistent with the principles of special relativity discussed so far to have entities with *spacelike* world lines. Such hypothetical entities are called *tachyons* and would never move with a speed *less* than the velocity of light (Problem 15). The existence of tachyons would conflict with other principles of physics such as causality and positive energy (see Problem 5.23). None have ever been observed. We will ignore them from now on and assume that in special relativity particles move at or less than the speed of light.

Light cones therefore define the causal relationships between points in spacetime. An event at P can signal or influence points inside or on its future light cone but not outside it. Information can be received at P only from events inside or on its past light cone but not from events outside it. The relativity of simultaneity

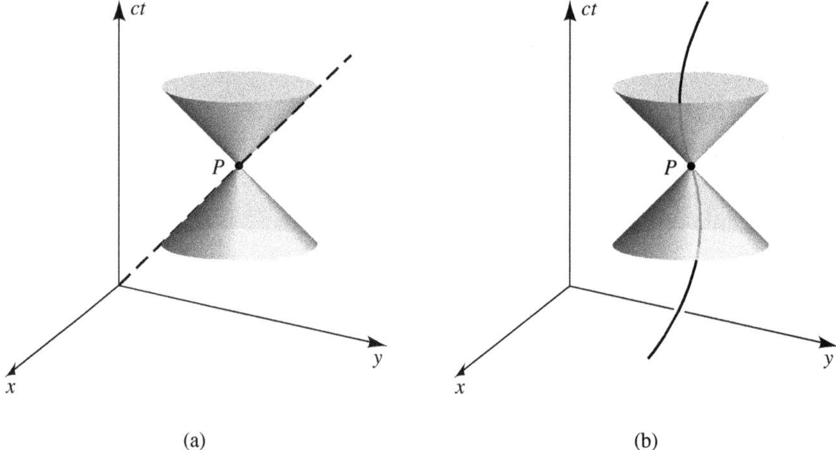

FIGURE 4.10 (a) The path of a light ray must be tangent to the light cone at every point along its trajectory. (b) The timelike path of a particle must lie inside the light cone at every point along its trajectory. These are the invariant ways of saying that light rays move at speed c and particles move with speed less than c in every inertial frame.

means that it does not make sense in general to say that one event is later than another. An event can be later than another spacelike separated event in one inertial frame and earlier in another. But it does make sense to say which is the earlier of two timelike separated events. That's because events to the future of P are inside its future light cone, and the inside and outside of a light cone are properties of the geometry of spacetime—the same in all frames.

The geometric distinction between timelike and spacelike distances is mirrored in the devices used to measure them. A *clock* is a device that measures timelike distances; a *ruler* is a device for measuring spacelike ones. Two nearby points on a timelike world line are timelike separated, $ds^2 < 0$. To measure the distance along a particle's world line, it is convenient to introduce

$$d\tau^2 \equiv -ds^2/c^2. \tag{4.12}$$

Then $d\tau$ is real with units of time. Thus a clock moving along a timelike curve measures the distance τ along it. An alternative name for this distance is the *proper time*—the time that would be measured by a clock carried along the world line.

4.4 Time Dilation and the Twin Paradox

Time Dilation

Just the few facts about the geometry of the spacetime of special relativity can be put to work to derive some its most famous consequences. First is the phenomenon of *time dilation*. The proper time, τ_{AB}, between any two points A and B on

BOX 4.3 Superluminal Motion?

Astronomers observe clouds in radio galaxies moving with velocities apparently exceeding the velocity of light. The radio source 3C345 provides an example. The figure shows a time sequence of maps of the angular positions of clouds, tens of light years across, emerging from the nucleus of this source from Biretta, Moore, and Cohen (1996). The cloud marked C2 is moving outward at an angular rate of approximately .5 mas/yr. (1 mas ≡ 1 milliarcsecond is about the angle a hair in London would subtend if viewed from Paris.)

The linear velocities obtained from this angular velocity and distance using (angular velocity) × (distance) are more than 10 times the velocity of light. (Problem 5.)

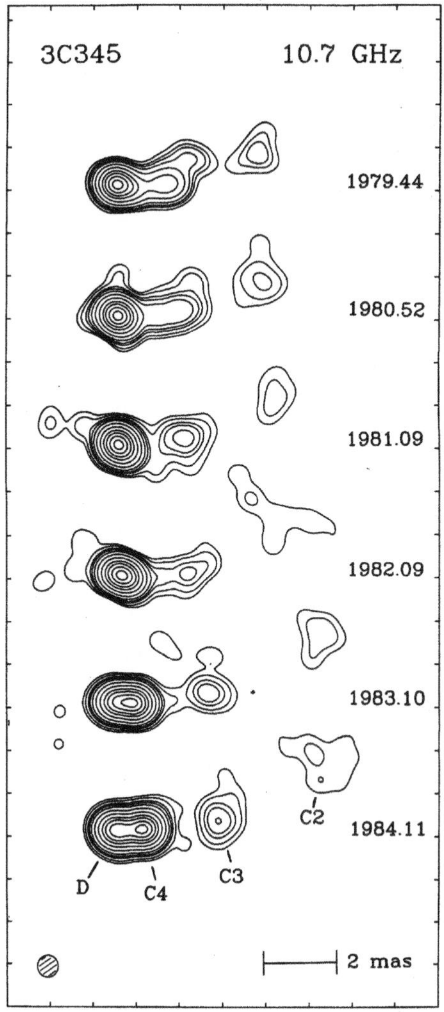

However, this naive calculation is not correct. The clouds are, in fact, moving almost straight toward us with a velocity just below c. As the cloud rapidly approaches, the distance light has to travel to us gets shorter, and the light arrives sooner than it would if the cloud were moving in a transverse direction. This accounts for the apparent superluminal velocities.

This effect can be understood quantitatively with the help of the second diagram. The cloud starts at the nucleus of 3C345 at time $t = 0$ and moves outward at speed V in a direction making an angle θ with the line of sight. Let t_{obs} be the time the observer receives the light emitted from the cloud at time t. The distance traveled can be computed in two ways, which must be equal:

$$c(t_{\text{obs}} - t) = \sqrt{(L - Vt\cos\theta)^2 + (Vt\sin\theta)^2}$$
$$\approx L - Vt\cos\theta, \qquad Vt \ll L.$$

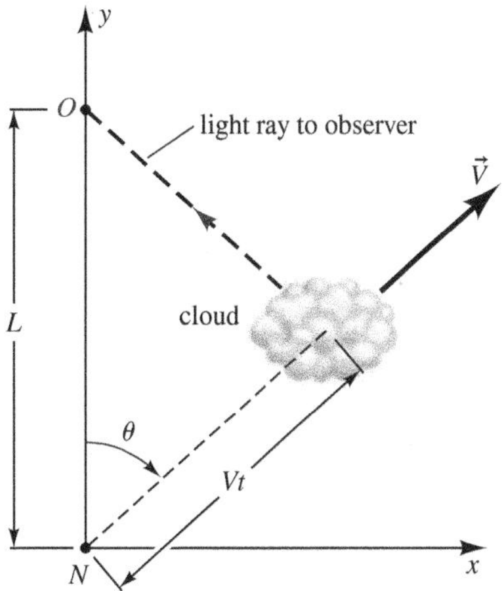

Solve this to find the connection between t and t_{obs}:

$$t_{\text{obs}} = t[1 - (V/c)\cos\theta] + (L/c).$$

The transverse speed, V_T, seen by the observer is

$$V_T = \frac{dx}{dt_{\text{obs}}} = \frac{dx}{dt}\frac{dt}{dt_{\text{obs}}} = \frac{V\sin\theta}{1 - (V/c)\cos\theta}.$$

When θ is small and V is close to c, this can be much larger than c and still consistent with special relativity.

a timelike world line can be computed from the line element (4.8) and (4.12) as

$$\tau_{AB} = \int_A^B d\tau = \int_A^B \left[dt^2 - (dx^2 + dy^2 + dz^2)/c^2\right]^{1/2}, \tag{4.13a}$$

$$= \int_{t_A}^{t_B} dt \left\{1 - \frac{1}{c^2}\left[\left(\frac{dx}{dt}\right)^2 + \left(\frac{dy}{dt}\right)^2 + \left(\frac{dz}{dt}\right)^2\right]\right\}^{1/2}. \tag{4.13b}$$

More compactly,

Time Dilation

$$\boxed{\tau_{AB} = \int_{t_A}^{t_B} dt' \left[1 - \vec{V}^2(t')/c^2\right]^{1/2}.} \tag{4.14}$$

The proper time τ_{AB} is *shorter* than the interval $t_B - t_A$ because $\sqrt{1 - \vec{V}^2/c^2}$ is *less* than unity. That is the phenomenon of *time dilation* summarized imperfectly by the slogan "moving clocks run slow." For time intervals Δt short enough that the velocity $\vec{V}$ is approximately constant over them, it will frequently be useful to make use of the differential form of (4.14),

$$d\tau = dt\sqrt{1 - \vec{V}^2/c^2}. \tag{4.15}$$

Figure 4.11 illustrates the connection.

It should be emphasized that (4.14) or (4.15) hold even for accelerating clocks, i.e., when the velocity is dependent on time.[9] A famous test of this relation for an accelerating clock is described in Box 4.4 on p. 64.

Example 4.2. A Model Clock. The preceding discussion of time dilation did not refer to the workings of any particular clock. Time dilation is consequence of the geometry of spacetime, and all one needs to know about a clock is that it is a device for measuring the distance along timelike curves. Nevertheless, it is instructive to see how time dilation emerges from the workings of a clock, and the model illustrated in Figures 4.6 and 4.7 provides a simple example. The clock mechanism is the bouncing light pulse. The successive returns of the light pulse to the lower mirror are the events defining the successive intervals along the clock's world line (cf. Figure 4.11). The proper time interval $\Delta\tau$ between these events is the time interval between them in the rest frame of the clock, which is $\Delta\tau \equiv \Delta t = 2L/c$ [cf. (4.3)]. The time interval $\Delta t'$ in the frame where that clock

[9] Occasionally one encounters the misconception that special relativity can deal only with motion at constant velocity. Nothing could be further from the truth. This mistaken idea possibly stems from the fact that inertial frames can differ by uniform motion but not accelerated motion. But this is equally true in Newtonian mechanics, which is mainly concerned with explaining accelerated motion. The high-speed motion of particles in high-energy accelerators is an everyday example of accelerated motion described by the principles of special relativity, as Box 4.4 on p. 64 illustrates.

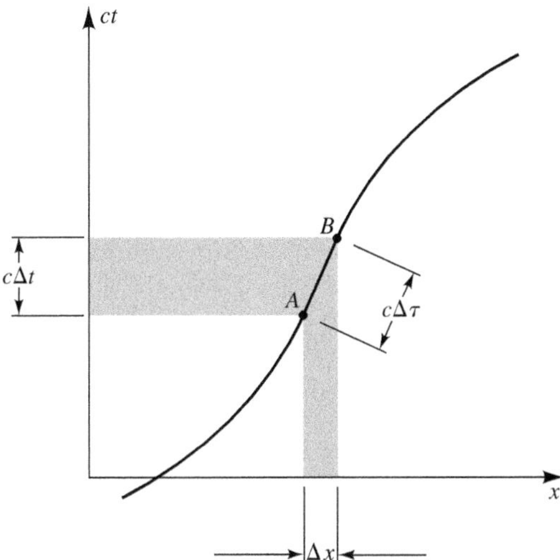

FIGURE 4.11 Proper time and coordinate time. The curve in this figure is the world line of a particle moving in the x-direction. A clock carried with the particle measures the proper time along the world line, which is the spacetime distance along the world line in time units. The proper time, $\Delta\tau$, between two points in spacetime A and B separated by small coordinate intervals Δt and Δx is given by the line element of flat spacetime, (4.8) and (4.12). The interval Δt is longer than $\Delta\tau$; that is time dilation. Judged by the Euclidean geometry of the plane, Δt appears shorter than $\Delta\tau$. But the geometry of a (ct, x) slice of flat spacetime is not Euclidean.

is moving with speed V can be found by eliminating $\Delta x'$ between the first two relations in (4.4). The result is

$$\Delta\tau = \Delta t'(1 - V^2/c^2)^{1/2}. \tag{4.16}$$

This is just the differential relation (4.15) for a clock with speed V. This model clock exhibits time dilation explicitly.

The Twin Paradox

Equation (4.14) shows that the time registered by a clock moving between two points in space depends on the route traveled even if it returns to the same point it started from. This is the source of the famous twin paradox.

Two twins, Alice and Bob, start from rest at one point in space at time t_1 in an inertial frame, as illustrated in Figure 4.12. Alice moves away from the starting point but later returns to rest at the same point at time t_2. Bob remains at rest at the starting point. The time elapsed on Bob's clock is $t_2 - t_1$. The time elapsed on Alice's clock is always *less* than this because $(1 - \vec{V}^2/c^2)^{1/2}$ is always less than 1 in (4.14). The moving twin ages less than the stationary twin.

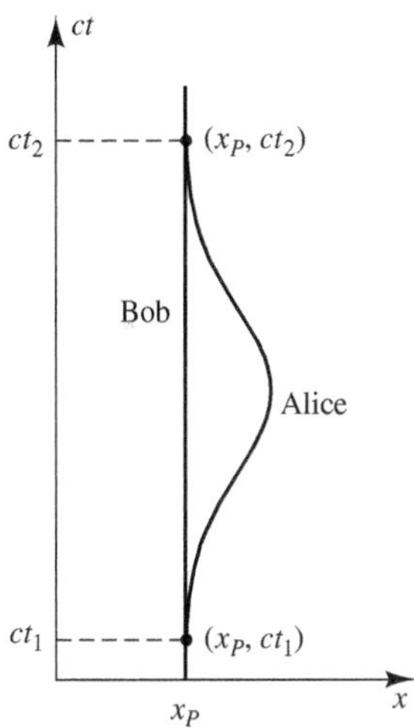

FIGURE 4.12 The twin paradox from a spacetime point of view. Alice and Bob follow two different world lines between the same two spacetime points. The lengths of these curves are different, and consequently the proper time registered by clocks carried along each is different.

BOX 4.4 The CERN Muon Lifetime Experiment

Elementary particles that decay into other particles can serve as a kind of clock. The probability of decay is typically an exponential decay law. The time of decay of any particular particle is uncertain. But in a collection of many particles, a fraction $\exp(-t/\tau_P)$ will remain undecayed after a time t, where the lifetime τ_P is a property of the kind of elementary particle.

Elementary particles can reach velocities close to the velocity of light in particle accelerators. Special relativity predicts that the lifetime of rapidly moving particles should be longer than their lifetime at rest by a factor of $\gamma \equiv (1 - \vec{V}^2/c^2)^{-1/2}$ [cf. (4.14)]. If $\tau_P(\gamma)$ is the lifetime of a particle moving with a speed corresponding to the value of γ, then

$$\tau_P(\gamma) = \gamma\tau_P(1),$$

where $\tau_P(1)$ is the lifetime at rest. Observations of particle decays in accelerators can thus test time dilation.

A particularly accurate test was carried out at CERN in the late 1970s using a special muon storage ring (Bailey et al. 1977). Muons ($\mu^{\pm}$) are elementary particles having either positive or negative charge. They decay into neutrinos and electrons or positrons (depending on their charge) with a lifetime of about 2.2 μs.

Muons circulated in the storage ring in 14-m circular orbits with a measured γ of 29.3 corresponding to $V/c = .9994$. The lifetime of circulating muons of both charges was measured by detecting the electron or positron decay products in counters surrounding the ring. The number of electrons or positrons was monitored as a function of time and fit to a decay law parametrized by the lifetimes $\tau_\mu^{\pm}$ and a number of other parameters affecting the decay, most importantly the muon magnetic moment. The results were $\tau_\mu^+ = 64.419 \pm .058$ μs and $\tau_\mu^- = 64.368 \pm .029$ μs. The lifetimes at rest that would be inferred from time dilation, $\tau_\mu^{\pm}(\gamma)/\gamma$, were compared with independent measurements of the muon lifetime at rest, $\tau_\mu^{\pm}(1)$. The best results were for the μ^+'s:

$$[\tau_\mu^+(1) - \tau_\mu^+(\gamma)/\gamma]/\tau_\mu^+(1) = (2 \pm 9) \times 10^{-4}.$$

This is in excellent agreement with the predictions of special relativity. Even an estimate on the basis of the Newtonian formula V^2/R shows that the centripetal acceleration of the muons is large ($\sim 10^{18}$ cm/s^2), giving good evidence that there is no dependence of time dilation on acceleration.

Example 4.3. Alice accelerates instantaneously to a uniform speed $\frac{4}{5}c$, travels in a straight line away from Bob, eventually instantaneously reverses direction, returns to Bob with the same speed, and decelerates instantaneously to rest. Bob has aged by 50 yr. By how much has Alice aged?

The ages of each are the proper times along their respective world lines between departure and return calculated with (4.14). To understand the contribution of an instantaneous acceleration, first suppose acceleration is uniform over a small time interval of length 2ϵ and take the limit as ϵ vanishes. Then $V = \frac{4}{5}c(t_{\text{mid}} - t)/\epsilon$ between a time ϵ before the midpoint time, t_{mid}, and a time ϵ afterwards. The contribution of this interval to the integral in (4.14) will be proportional to ϵ and negligible as ϵ approaches zero. The same is true for the other two of Alice's accelerations. Thus, the moving clock is running at a rate $[1 - (4/5)^2]^{1/2} = 3/5$ times slower than the stationary clock for 50 yr. Alice will have aged by 30 yr.

The American Heritage dictionary defines *paradox* as "a seemingly contradictory statement that may nonetheless be true." We obtain a paradox by describing the situation from Alice's point of view. Bob moves away at uniform speed, reverses direction, and returns at uniform speed. That seems to be exactly the same as the situation from Bob's point of view, who sees Alice move away at uniform speed, reverse direction, and return at uniform speed. The result is not symmetric; Alice is younger than Bob.[10] However, their situations are not symmetric. Alice and Bob travel two *different* world lines in spacetime with *different* distances between their starting and ending points. Their clocks measure these distances and so read differently.

Straight Lines and Longest Distances

The twin paradox example illustrates an important property of the non-Euclidean geometry of flat spacetime. As (4.14) shows, *every* distinct timelike world line that Alice could follow between points A and B has a shorter length than the straight line curve followed by Bob. (Other curves may look longer in a figure like Figure 4.12, but are in fact shorter because the geometry is non-Euclidean. Recall Example 4.1.) The straight line path is the *longest* distance between any two timelike separated points in flat four-dimensional spacetime.[11] To see this, pick any two timelike separated points A and B. The straight-line path between them is a world line moving with some constant velocity $\vec{V}$. Use that velocity to transform to another inertial frame where the two events occur at the same *place*. That frame is like Bob's discussed above. Any observer like Alice moving on a non-straight path measures a *shorter* spacetime distance between the events than Bob does. Spacetime distances don't depend on the inertial frame used to calculate them in. In three-dimensional space a straight line is the shortest distance between any two points, but in flat spacetime a straight line is the longest distance between two timelike separated points.

4.5 Lorentz Boosts

The Connection Between Inertial Frames

The discussion of the construction of inertial frames in both Newtonian mechanics, Section 3.1, and special relativity, Section 4.3, shows that two inertial frames can differ from one another by rotations, displacements, and uniform motions (or combinations thereof). Rotations and displacements work in the same way as in Newtonian mechanics, but let's now find the transformation associated with uniform motion that generalizes the Galilean transformation (3.6) to special relativity.

The line element (4.8) specifies the geometry of special relativistic spacetime in terms of four rectangular coordinates (t, x, y, z) defining an inertial frame—

[10]For a direct experimental test of the twin paradox in the slightly curved spacetime of the Earth, see Box 6.2 on p. 130.

[11]A straight-line path in curved spacetime is not always a path of longest proper time, but it is a path of extremal proper time. That is discussed on p. 131.

one in which Newton's first law takes the simple form (3.2). The principle of relativity implies that the line element must take the same form in the rectangular coordinates (t', x', y', z') of any other inertial frame. The transformation laws that connect different inertial coordinate frames must therefore be among those that preserve the form of (4.8), and, in fact, are determined by this requirement. They are called *Lorentz transformations.*

We saw in Section 3.2 that the line element of Euclidean space,

$$dS^2 = dx^2 + dy^2 + dz^2, \tag{4.17}$$

is left unchanged by translations and rotations of the rectangular coordinates (x, y, z). Spatial translations and rotations will also preserve the line element (4.8) of special relativistic spacetime because it could be written $-(cdt)^2 + dS^2$. But what *new* transformations preserve the non-Euclidean line element of *four-dimensional* flat spacetime? The most important examples of new transformations are the analogs of rotations between time and space. These are called *Lorentz boosts* and correspond to the uniform motion of one frame with respect to another.

To be definite, consider the analogs of rotations in the (ct, x) plane. These are transformations between (t, x, y, z) and (t', x', y', z') that leave y and z unchanged but mix ct and x. The transformations of this character that leave (4.8) unchanged are the analogs of rotations such as (3.9) but with trigonometric functions replaced by hyperbolic functions because of the non-Euclidean character of spacetime. Specifically:

$$ct' = (\cosh\theta)(ct) - (\sinh\theta)x, \tag{4.18a}$$

$$x' = (-\sinh\theta)(ct) + (\cosh\theta)x, \tag{4.18b}$$

$$y' = y, \tag{4.18c}$$

$$z' = z, \tag{4.18d}$$

where the parameter θ can vary from $-\infty$ to $+\infty$. (In fact, θ is a hyperbolic angle in the sense briefly alluded to in Example 4.1.) It's straightforward to verify by direct calculation that transformation (4.18) preserves line element (4.8).

$$\begin{aligned}(ds)^2 &= -(c\,dt')^2 + (dx')^2 + (dy')^2 + (dz')^2,\\ &= -[\cosh\theta(c\,dt) - \sinh\theta(dx)]^2\\ &\quad + [-\sinh\theta(c\,dt) + \cosh\theta(dx)]^2 + (dy)^2 + (dz)^2,\\ &= -(c\,dt)^2 + (dx)^2 + (dy)^2 + (dz)^2.\end{aligned} \tag{4.19}$$

The coordinates (t', x', y', z') thus span a new inertial frame.

Figure 4.13 shows the new (ct', x') coordinates plotted on the old (ct, x) axes. The similarity to a rotation is apparent, but there are also important differences. A particle at rest at the origin $(x' = 0)$ in the (ct', x') coordinates has the ct' axis as

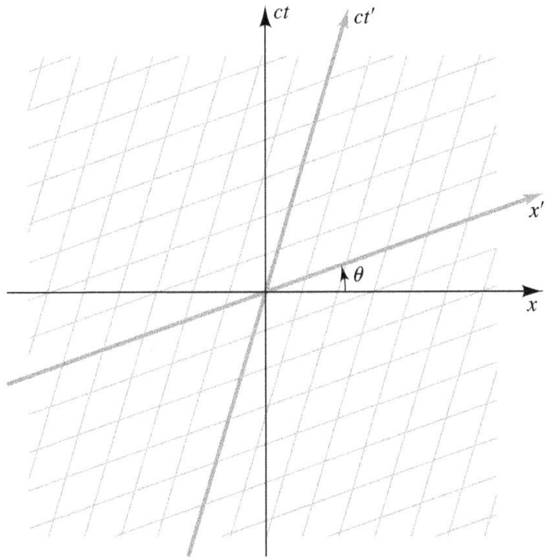

FIGURE 4.13 A Lorentz boost as a change of coordinates on a spacetime diagram. The figure shows the grid of (ct', x') coordinates defined by (4.18) plotted on a (ct, x) spacetime diagram. The (ct', x') coordinates are not orthogonal to each other in the Euclidean geometry of the printed page. But they are orthogonal in the geometry of spacetime. (Recall the analogies between spacetime diagrams and maps discussed in Example 4.1.) The (ct', x') axes have to be as orthogonal as the (ct, x) axes because there is no physical distinction between one inertial frame and another. The orthogonality is explicitly verified in Example 5.2. The hyperbolic angle θ is a measure of the velocity between the two frames.

its world line. In (ct, x) coordinates, that particle is moving with a constant speed along the x-axis. The speed v can be found by putting $x' = 0$ in (4.18b), with the result

$$v = c \tanh \theta. \tag{4.20}$$

A particle at rest at any other value of x' in the (ct', x') coordinates moves in the x-direction with the same speed in the (ct, x) coordinates. The transformation from (t, x, y, z) to (t', x', y', z') is, therefore, from one inertial frame to another moving uniformly with respect to it along the x-axis with speed v. Such transformations are called *Lorentz boosts*.[12]

Lorentz Boost

The identification of (4.18) as a Lorentz boost is made explicit by using (4.20) to eliminate θ in terms of v. After a little algebra in which the identity $\cosh^2\theta - \sinh^2\theta = 1$ plays a useful role, one finds

[12]Especially in elementary treatments, Lorentz boosts are sometimes called Lorentz transformations. As the latter term is used here, a Lorentz transformation is any transformation in coordinates that preserves the line element of spacetime, including rotations and displacements along with Lorentz boosts. Lorentz boosts are a special case of Lorentz transformations.

$$t' = \gamma(t - vx/c^2), \tag{4.21a}$$
$$x' = \gamma(x - vt), \tag{4.21b}$$
$$y' = y, \tag{4.21c}$$
$$z' = z, \tag{4.21d}$$

where we have introduced the standard abbreviation

$$\gamma = (1 - v^2/c^2)^{-1/2}. \tag{4.22}$$

The inverse transformation is obtained just by changing v into $-v$:

$$t = \gamma(t' + vx'/c^2), \tag{4.23a}$$
$$x = \gamma(x' + vt'), \tag{4.23b}$$
$$y = y', \tag{4.23c}$$
$$z = z'. \tag{4.23d}$$

When $v/c \ll 1$, (4.21) reduces to the Galilean transformation (3.6), as it must.

The Relativity of Simultaneity

A number of special relativistic effects can be seen directly from the spacetime diagram of two inertial frames shown in Figure 4.13. For example, consider two events A and B, which are simultaneous for an observer in the (ct', x') frame. They will lie on a line of constant t', as shown in Figure 4.14. However, there will be a difference in time, Δt, between the events in an inertial frame moving with speed v with respect to the first in the negative x' direction. That is the relativity of simultaneity for which we argued in Section 4.2. The quantitative value of the time difference $\Delta t = t_B - t_A$ can be computed from the Lorentz boost connecting the two frames, in particular (4.23a). If $\Delta x' = x'_B - x'_A$ is the distance between the simultaneous $(\Delta t' = 0)$ events in the (ct', x') frame, then

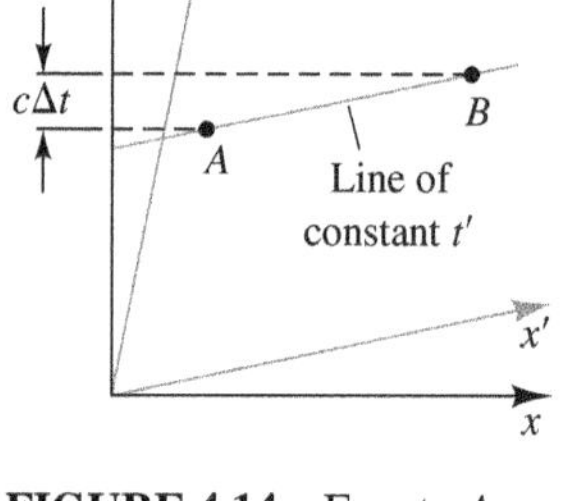

FIGURE 4.14 Events A and B are simultaneous in the (ct', x') frame because they occur at the same value of t'. They are not simultaneous in the (ct, x) frame, where A occurs before B.

$$\Delta t = \gamma(v/c^2)\Delta x'. \tag{4.24}$$

As we argued in Section 4.2, event B is later than event A. Equation (4.24) shows by how much.

Example 4.4. A Toy Model Satellite Location System. Restrict attention for simplicity to two space dimensions: a horizontal one (x) and a vertical one (y). You are lost on the ground at $y = 0$. Overhead, at a height h, a constellation of satellites is moving by with speed V separated from each other by a uniform dis-

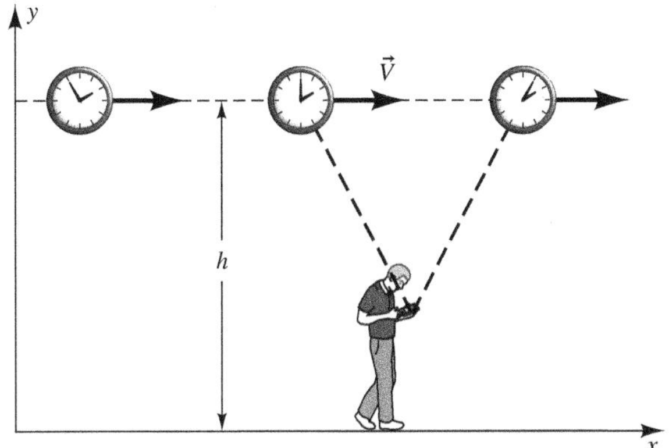

FIGURE 4.15 A toy model satellite location system. Satellites are moving overhead, each broadcasting the time in their rest frame and their location in x at the time of broadcast. From the information received simultaneously from two different satellites, the location on the ground can be determined by taking time dilation, Lorentz contraction, and the relativity of simultaneity into account.

tance L_* in their rest frame. (See Figure 4.15.) The satellites carry clocks, which are all synchronized to read the same time in *their* rest frame. At regular intervals the satellites broadcast the time on their clocks and their horizontal location in x. Simultaneously you receive signals from two neighboring clocks located on either side of you, each reporting the same time at broadcast. Does this mean that you are midway between the two clocks? No, for that to be the case the signals would have to have been emitted simultaneously in *your* rest frame. Because of the relativity of simultaneity, the signal from the clock on the right was emitted a time $\Delta t = \gamma(V/c^2)L_*$ [cf. (4.24)] later in your frame than the signal from the clock on the left. You are, therefore, located closer to that clock than the other one, and with this information you can figure out how much. More generally, you can figure out your location in x from the reported time difference in emission of two signals received simultaneously by taking account of time dilation, Lorentz contraction, and the relativity of simultaneity (Problem 14).

This example was inspired by the Global Positioning System (GPS), which will be described in Chapter 6, but it is a simplification in a number of respects, most importantly, the neglect of gravity. Is the relativity of simultaneity important for GPS? To get an idea let's plug in some GPS numbers in this model, even though a more sophisticated analysis is required. There are 24 GPS satellites, each in a 12-hr orbit. This means that they are moving with speeds $V \sim 4$ km/s in an inertial frame in which the Earth is at rest in orbits a distance $R_s \approx 2.7 \times 10^4$ km from its center. The distance between the satellites is, therefore, approximately $2\pi R_s/24 \sim 7 \times 10^3$ km, and $\Delta t \sim 3 \times 10^{-7}$ s. That is a small error in time, but to achieve a location accuracy of 10 m, the GPS system must have accurate timing to no worse than the light travel time across this distance, which is about 3×10^{-8} s. The relativity of simultaneity is important for the GPS.

Lorentz Contraction

Consider a rod whose length is L_* when measured in its own rest frame. What is its length when measured in an inertial frame in which it is moving with speed V? The spacetime diagram in Figure 4.16 shows graphically why L is different from L_*. The length of the rod is the distance between two *simultaneous* events at its ends. But the notion of simultaneity is different in different inertial frames. The measured length of the rod is, therefore, also different. (See Problem 17 for an explicit example of such a measurement.) The length L in the frame where the rod is moving is the spacetime distance between the ends of the rod at $t' = 0$—the points labeled by •'s in Figure 4.16. This distance can also be computed from (4.6) in the rest frame as:

$$L^2 = L_*^2 - (c\Delta t)^2. \tag{4.25}$$

From (4.21a), the line $t' = 0$ is the line $t = (V/c^2)x$, so $\Delta t = (V/c^2)L_*$. Thus,

Lorentz Contraction

$$\boxed{L = L_*\sqrt{1 - V^2/c^2}.} \tag{4.26}$$

This is Lorentz contraction.

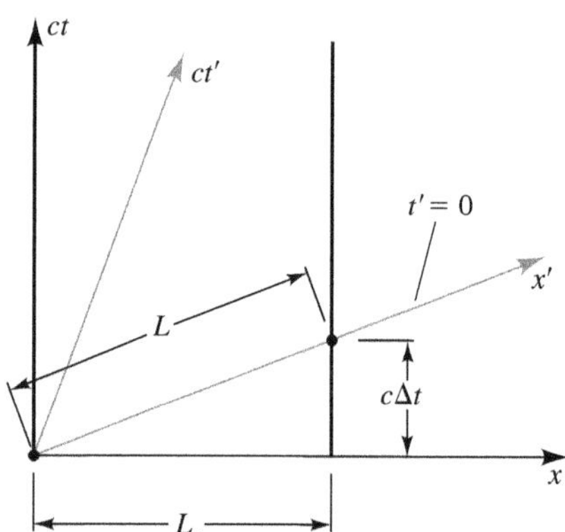

FIGURE 4.16 Lorentz contraction of length. The figure shows the world lines of the ends of a rod oriented along the x-axis in its own rest frame spanned by coordinates (ct, x). The distance L_* between the world lines is the rest length of the rod. Also shown on the same plot are the axes (ct', x') of an inertial frame moving with speed V with respect to the rest frame. In this frame the rod is moving with velocity $-V$ along the x'-axis. The length of the rod L in this frame is the distance between its ends at a single moment of time, t'. The events at the ends at time $t' = 0$ are indicated by •'s in the figure. Although the length L looks longer than L_* in the figure, it is actually shorter because of the non-Euclidean geometry of spacetime.

Addition of Velocities

Having studied the Lorentz boosts connecting different inertial frames, we can now find the relativistic law for addition of velocities that replaces the Newtonian (4.2). Consider a particle whose motion is described by coordinates $x(t), y(t), z(t)$ in one frame and $x'(t'), y'(t'), z'(t')$ in a second frame moving along the x-axis of the first with velocity v. From (4.21) we can compute the relation between the velocity of the particle $\vec{V} = d\vec{x}/dt$ in one frame and the velocity $\vec{V}' = d\vec{x}'/dt'$ in the other, namely,

$$V^{x'} = \frac{dx'}{dt'} = \frac{\gamma(dx - v\,dt)}{\gamma(dt - v/c^2\,dx)}. \tag{4.27}$$

Dividing top and bottom by dt, one finds

$$V^{x'} = \frac{V^x - v}{1 - vV^x/c^2}. \tag{4.28a}$$

Similarly,

$$V^{y'} = \frac{V^y}{1 - vV^x/c^2}\sqrt{1 - v^2/c^2}, \tag{4.28b}$$

$$V^{z'} = \frac{V^z}{1 - vV^x/c^2}\sqrt{1 - v^2/c^2}. \tag{4.28c}$$

This is the *relativistic* rule for the addition of velocities generalizing the Newtonian (4.2) and reducing to it when $v/c \ll 1$.

Example 4.5. The Velocity of Light Is the Same in All Inertial Frames. A particle is moving with speed c along the x-axis in one inertial frame. What velocity does it have in an inertial frame moving with speed v with respect to the first frame along the x-axis?

The answer to this question has to be c, but one can see it directly from (4.28a) with $V^x = c$:

$$V^{x'} = \frac{c - v}{1 - v/c} = c. \tag{4.29}$$

4.6 Units

The attentive reader cannot have failed to notice the symmetry that has been achieved in our formulas by using ct instead of t. The reason can be seen in the line element (4.8). There the constant c emerges as a conversion factor between space units and time units—approximately 3×10^{10} centimeters in every second. From the spacetime point of view, the value of c is a historical accident. It's

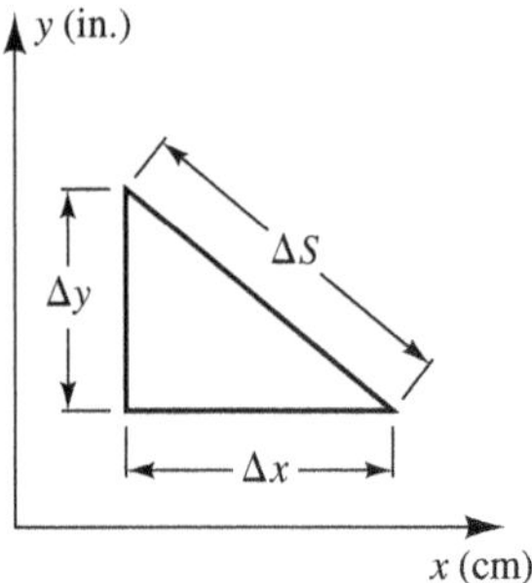

FIGURE 4.17 If distances in the y-direction were measured in inches and distances in the x-direction were measured in centimeters, then the Pythagorean theorem for a right triangle with two sides aligned with these axes would read $\Delta S^2 = C^2\Delta y^2 + \Delta x^2$, where $C = 2.54$ cm/in.

as though in dealing with spatial geometry it had become traditional to measure the y-direction in inches and the x- and z-directions in centimeters. The distance between two nearby points in space would then have been given by

$$dS^2 = C^2dy^2 + dx^2 + dz^2. \tag{4.30}$$

where $C = 2.54$ cm/in. Since space and time are but different directions in a single spacetime continuum, it is desirable to measure them in the same units, either centimeters or seconds. The constant c then gives the conversion factor between these two units. Today the velocity of light is not measured, it is *defined* to be exactly the conversion factor[13]

$$c = 299792458.0000\ldots \text{ m/s}. \tag{4.31}$$

(Zeros all the way out!)

Measuring time in units of length means changing from the mass-length-time ($\mathcal{MLT}$) system of units traditional in mechanics to a mass-length ($\mathcal{ML}$) system. Appendix A gives some discussion of different unit systems and rules for transforming between the ones used in this text.

Measuring both space and time in length units has the effect of putting $c = 1$ everywhere in our formulas. For example, in units where time is measured in centimeters,

$$ds^2 = -dt^2 + dx^2 + dy^2 + dz^2, \tag{4.32}$$

$d\tau^2 = -ds^2$, and velocities are dimensionless. Equation (4.21) for a Lorentz boost becomes

$$t' = \gamma(t - vx), \tag{4.33a}$$

$$x' = \gamma(x - vt), \tag{4.33b}$$

$$y' = y, \tag{4.33c}$$

$$z' = z, \tag{4.33d}$$

where $\gamma \equiv (1 - v^2)^{-1/2}$. For this reason the $\mathcal{ML}$ system of units is informally called $c = 1$ units. Units with $c = 1$ will be used in almost all the rest of this book.

For many practical purposes it is convenient to maintain different units for space and time in a given inertial frame. For example, it is easier to say a lecture is 50 min long than to say it is 899 billion meters long. The c's can always be returned to any expression by identifying those quantities that should be measured in units of time and those that should be measured in units of space. A prescription for doing this is given in Appendix A, but the following example illustrates how it works.

[13]At the time of writing the second is defined to be 9,192,631,770 cycles of the transition radiation between the two lowest energy states of a cesium atom and the meter is defined in terms of the second by (4.31).

Example 4.6. Putting Back the c's. Expressions in units where time is measured in centimeters can be written in units where time is measured in seconds by inserting the conversion factor c in the right places. Consider by way of example the part of a Lorentz boost (4.33a). In $\mathcal{MLT}$ units velocity has dimensions $\mathcal{L}/\mathcal{T}$. The dimensionless v's in (4.33a) must therefore be replaced by v/c's, including in the definition of γ. To get all the terms on the right-hand side of (4.33a) to have the units $\mathcal{T}$ that the left-hand side has, x must be replaced by x/c. The result is (4.21a).

Problems

1. [B,S] Today a TGV train (train à grande vitesse) leaves Paris (Gare de Lyon) at 8:00 and arrives at Lyon (Part Dieu) at 10:04 (using a 24-hr clock). Assuming the train makes no intermediate stops, plot the world line of the train on a copy of the railway spacetime diagram on p. 55. If the distance between Paris and Lyon is 472 km, how fast is the train traveling on average?

2. A rocket ship of proper length L leaves the Earth vertically at speed $\frac{4}{5}c$. A light signal is sent vertically after it which arrives at the rocket's tail at $t = 0$ according to both rocket and Earth-based clocks. When does the signal reach the nose of the rocket according to (a) the rocket clocks; (b) the Earth clocks?

3. A 20-m pole is carried so fast in the direction of its length that it appears to be only 10 m long in the laboratory frame. The runner carries the pole trough the front door of a barn 10 m long. Just at the instant the head of the pole reaches the closed rear door, the front door can be closed, enclosing pole within the 10-m barn for an instant. The rear door opens and the runner goes through. From the runner's point of view, however, the pole is 20 m long and the barn is only 5 m! Thus the pole can never be enclosed in the barn. Explain, quantitatively and by means of spacetime diagrams, the apparent paradox.

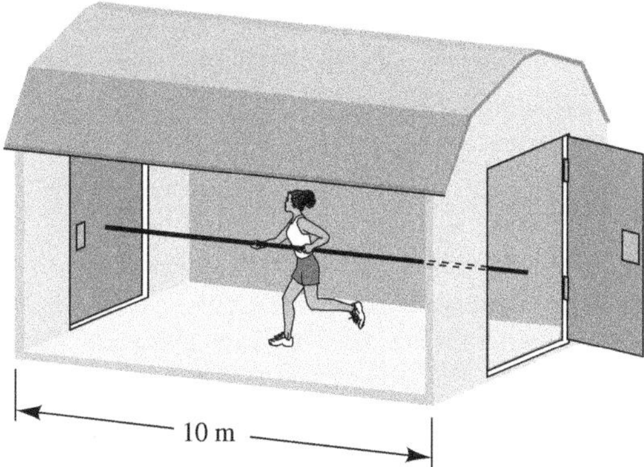

4. A satellite orbits the Earth in the same direction it rotates in a circular orbit above the equator a distance of 200 km from the surface. By how many seconds per day will a clock on such a satellite run slow compared to a clock on the Earth? (Compute just the special relativistic effects.)

5. [B, E] The radio source 3C345 is participating in the expansion of the universe, and its distance can be determined from the redshift arising from its recession velocity and assumptions about our universe. (Work Problem 1 in Chapter 19 when you have studied a little cosmology.) However, a rough idea of the distance can be obtained from Hubble's law relating distance d to observed recession velocity V:

$$V = H_0 d,$$

where $H_0 \approx 72$ (km/s)/Mpc is the Hubble constant. (Look at the endpapers for astronomical units such as the megaparsec (Mpc).) V for 3C345 is about $.6c$. Use these facts together with the data in Box 4.3 on p. 61 to roughly estimate the velocity of the cloud C2 assuming (contrary to fact) that it is moving transverse to the line of sight.

6. Example 4.2 showed how time dilation in a moving clock could be understood in terms of the working of a model clock consisting of two mirrors oriented along the direction of motion. Show that the same result can be derived using a similar clock oriented perpendicular to the direction of motion.

7. [S, P] In (4.4) we deduced a travel time $\Delta t'$ for a pulse of light traveling between two mirrors that were moving with a speed V. This time was different from the travel time Δt in the frame in which the mirrors are at rest, (4.3). In Newtonian physics, with its absolute time, these times would necessarily agree. Carry out the analysis that led to $\Delta t'$ in (4.4) using the principles of Newtonian physics and show that this is the case, assuming that the rest frame of the mirrors is the rest frame of the ether.

8. [S] Calculate the hyperbolic angle between the sides AC and AB of triangle ABC illustrated in Figure 4.8.

9. Consider twins, Joe and Ed. Joe goes off in a straight line traveling at a speed of $\frac{24}{25}c$ for 7 years as measured on *his* clock, then reverses and returns at half the speed. Ed remains at home. Make a spacetime diagram showing the motion of Joe and Ed from Ed's point of view. When they return, what is the difference in ages between Joe and Ed?

10. In the novel *Return from the Stars* by S. Lem, which is concerned with the problems a returning twin in the twin paradox situation might face, there is the following passage:

> "Her eyes were shining and attentive: ... I was thirty then. The expedition ... I was a pilot on the expedition to Fomalhaut. That's twenty-three light years away. We flew there and back in a hundred and twenty years ship time. Four days ago we returned ... The *Prometheus*—my ship—remained on Luna. I came from there today. That's all."[14]

Assuming that all accelerations are instantaneous and the velocity of the *Prometheus* was constant in between, with what speed did it travel from the Earth to Fomalhaut?

11. [C] Alice and Bob are moving in opposite directions around a circular ring of radius R, which is at rest in an inertial frame. Both move with constant speeds V as measured in that frame. Each carries a clock, which they synchronize to zero time at a moment when they are at the same position on the ring. Bob predicts that when next they meet, Alice's clock will read less than his because of the time dilation arising because she has been moving with respect to him. Alice predicts that Bob's clock will

[14] S. Lem, *Return from the Stars*, Harcourt Brace Jovanovich, San Diego, 1989.

read less with the same reasoning. They both can't be right. What's wrong with their arguments? What will the clocks really read?

12. **(a)** Show explicitly that the straight line path between any two points in flat three-dimensional space ($dS^2 = dx^2 + dy^2 + dx^2$) is the shortest distance between them.

(b) Is the straight line path between two spacelike separated points in flat spacetime the shortest distance between them?

13. In an inertial frame two events occur simultaneously at a distance of 3 m apart. In a frame moving with respect to the laboratory frame, one event occurs later than the other by 10^{-8} s. By what spatial distance are the two events separated in the moving frame? Solve this problem in two ways: first by finding the Lorentz boost that connects the two frames and second by making use of the invariance of the *spacetime* distance between the two events.

14. [C] This problem concerns the toy model satellite location system discussed in Example 4.4. Suppose you simultaneously receive broadcasts from two neighboring satellites, A and B that report their locations, x'_A and x'_B, as well as their times of broadcast, t'_A and t'_B, which are equal: $t'_A = t'_B$. The times and positions are in the rest frame of the satellites to which their clocks are all synchronized. Derive a condition that determines your position in x. Evaluate it to find your deviation from the midpoint between the satellites to first order in V/c, where V is the speed of the satellites.

15. Show that the addition of velocities (4.28) implies that (a) if $|\vec{V}| < c$ in one inertial frame, then $|\vec{V}| < c$ in any other inertial frame, (b) if $|\vec{V}| = c$ in one inertial frame, then $|\vec{V}| = c$ in any other inertial frame, and that (c) if $|\vec{V}| > c$ in any inertial frame, then $|\vec{V}| > c$ in any other inertial frame.

16. *Lengths perpendicular to relative motion are unchanged.*

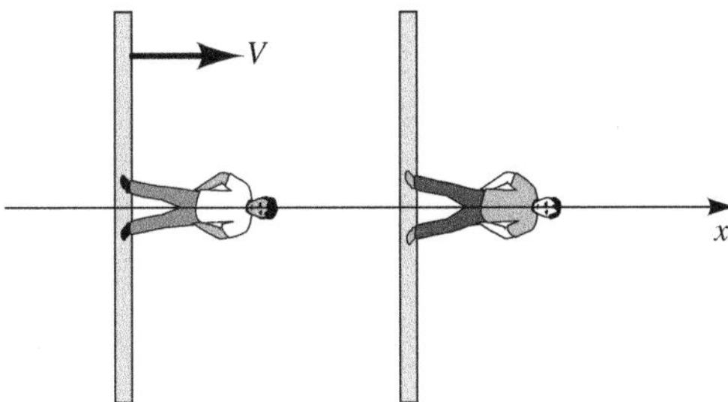

Imagine two meter sticks, one at rest and the other moving along an axis perpendicular to the first and perpendicular to its own length, as shown here. There is an observer riding at the center of each meter stick.

(a) Argue that the symmetry about the x-axis implies that both observers will see the ends of the meter sticks cross simultaneously and that both observers will therefore agree if one meter stick is longer than the other.

(b) Argue that the lengths cannot be different without violating the principle of relativity.

17. *Another derivation of Lorentz contraction.* Example 4.2 showed how the operation of a model clock was consistent with time dilation. This problem aims at showing how Lorentz contraction is consistent with ideal ways of measuring lengths.

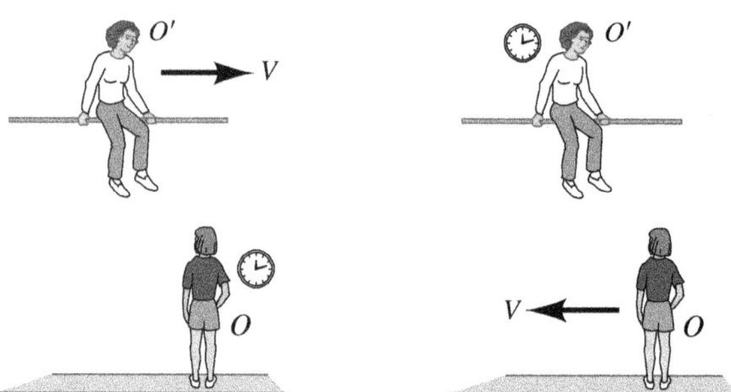

The length of a rod moving with speed V can be determined from the time it takes to move at speed V past a fixed point (left-hand figure). The length of a stationary rod can also be determined by measuring the time it takes a fixed object to move from end to end at speed V (right-hand figure). Taking account of the time dilation between the two times, show that the length of the moving rod determined in this way is Lorentz contracted from its stationary length.

18. [S] Show that for two timelike separated events, there is some inertial frame in which $\Delta t \neq 0$, $\Delta\vec{x} = 0$. Show that for two spacelike separated events there is an inertial frame where $\Delta t = 0$, $\Delta\vec{x} \neq 0$.

19. [C] If a photograph is taken of an object moving uniformly with a speed approaching the speed of light parallel to the plane of the film, it appears rotated rather than contracted in the photograph. Explain why. (Assume the object subtends a small angle from the camera lens.)

CHAPTER 5

Special Relativistic Mechanics

The laws of Newtonian mechanics have to be changed to be consistent with the principles of special relativity introduced in the previous chapter. This chapter describes special relativistic mechanics from a four-dimensional, spacetime point of view. Newtonian mechanics is an approximation to this mechanics of special relativity that is appropriate when motion is at speeds much less than the velocity of light in a particular inertial frame. We begin with the central idea of four-vector.

5.1 Four-Vectors

A *four-vector* is defined as a directed line segment in four-dimensional flat spacetime in the same way as a three-dimensional vector (to be called a three-vector in this chapter) can be defined as a directed line segment in three-dimensional Euclidean space. Boldface letters will denote four-vectors—e.g., **a**—to distinguish them from three-vectors, e.g., $\vec{a}$. The careful terminology *four*-vector and *three*-vector will be kept for this chapter, but succeeding chapters usually refer only to *vectors* and rely on the context to distinguish the two.

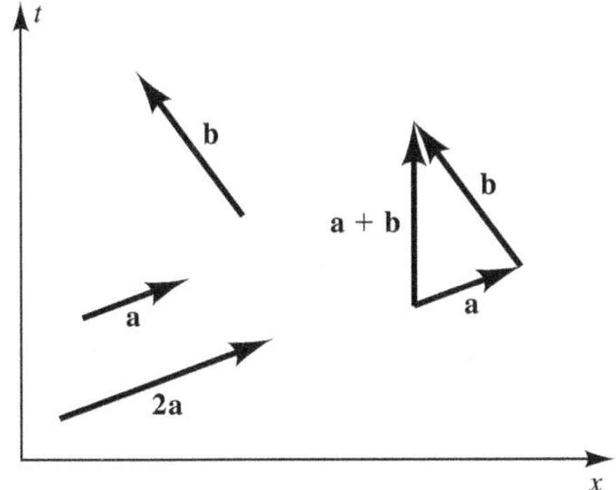

FIGURE 5.1 The addition of four-vectors and their multiplication by numbers. To add two four-vectors **a** and **b**, transport them parallel to themselves until they make a triangle as at right. The sum **a** + **b** is the directed line segment from the tail of the first to the tip of the second. A number α times a four-vector is a four-vector in the same direction with its length α times longer.[1]

[1]For null four-vectors of zero length, first write them as the sum of two four-vectors of nonzero length, multiply those by α, and then add the results.

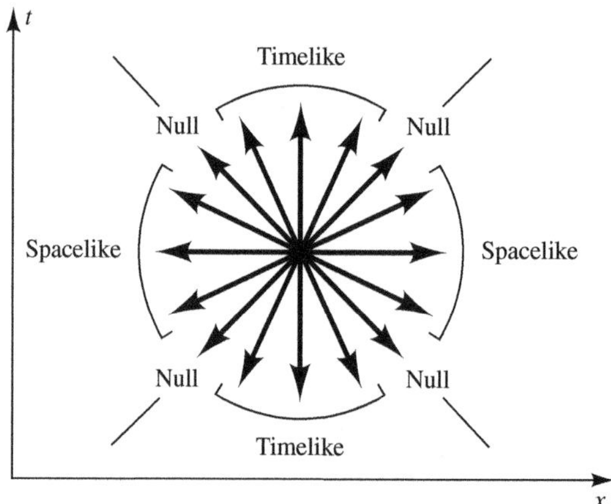

FIGURE 5.2 Timelike, spacelike, and null four-vectors. The three kinds of four-vectors point along timelike, spacelike, and null directions in spacetime, respectively; cf. Figure 4.9. Note that null four-vectors have *zero* length in the non-Euclidean geometry of spacetime.

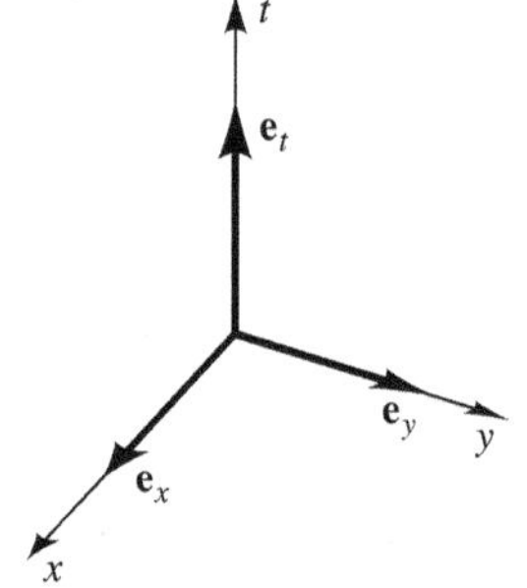

FIGURE 5.3 Basis four-vectors along the coordinate axes.

Four-vectors can be multiplied by numbers, added, and subtracted according to the usual rules for vectors (see Figure 5.1). The *length* of a four-vector is the absolute value of the spacetime distance between its tail and its tip. Four-vectors whose tail and tip have a spacelike separation are called *spacelike*; those whose tail and tip have a timelike separation are called *timelike*; and those whose tail and tip have a null separation are called *null*. Null four-vectors have zero length. Examples of the different types are illustrated in Figure 5.2.

Neither the definition of four-vector just given, nor the rules for addition, multiplication by numbers, and calculating length refer to any particular inertial frame. They are *invariant*—the same in all inertial frames. When the laws of mechanics are formulated in terms of four-vectors, they will necessarily take the same form in every inertial frame, and their predictions will be consistent with the principle of relativity. Therein lies the utility and importance of four-vectors.

Basis Four-Vectors and Components

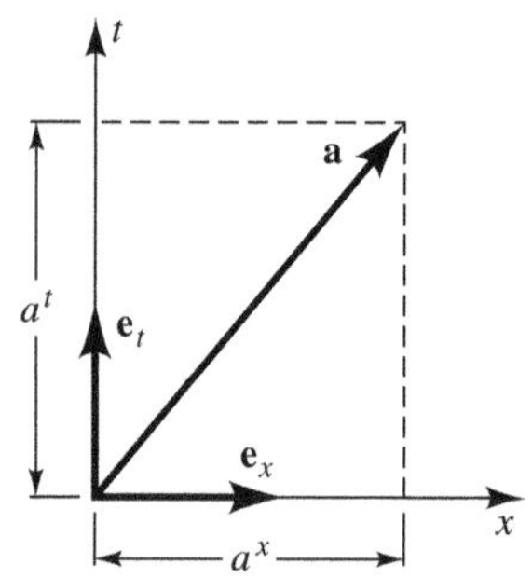

FIGURE 5.4 A four-vector **a** may be specified by its components (a^t, a^x, a^y, a^z) along the coordinate axes.

In a particular inertial frame, *basis four-vectors* can be introduced of unit length pointing along the t, x, y, and z coordinate axes, as shown in Figure 5.3. We call these basis four-vectors $\mathbf{e}_t$, $\mathbf{e}_x$, $\mathbf{e}_y$, and $\mathbf{e}_z$, or, equivalently, $\mathbf{e}_0$, $\mathbf{e}_1$, $\mathbf{e}_2$, $\mathbf{e}_3$, where 0 stands for t, 1 for x, etc. Taken together these four-vectors are called a *basis* for four-vectors because any four-vector can be represented as a linear combination of them as illustrated in Figure 5.4:

$$\mathbf{a} = a^t\mathbf{e}_t + a^x\mathbf{e}_x + a^y\mathbf{e}_y + a^z\mathbf{e}_z. \tag{5.1}$$

The numbers (a^t, a^x, a^y, a^z)—or, equivalently, (a^0, a^1, a^2, a^3) are called the *components* of the four-vector.[2] Components are always written with the component label as a superscript.[3]

Components of a Four Vector

There are some other useful ways of writing (5.1), such as

$$\mathbf{a} = a^0\mathbf{e}_0 + a^1\mathbf{e}_1 + a^2\mathbf{e}_2 + a^3\mathbf{e}_3, \tag{5.2}$$

or, equivalently,

$$\mathbf{a} = \sum_{\alpha=0}^{3} a^\alpha \mathbf{e}_\alpha. \tag{5.3}$$

Summation Convention

Equation (5.3) can be written even more compactly if we introduce the *summation convention* that repeated upper and lower indices are understood to be summed over in any expression. Greek indices are summed from 0 to 3; Roman indices from 1 to 3. Thus,

$$\mathbf{a} = a^\alpha \mathbf{e}_\alpha, \tag{5.4}$$

is the same as (5.3). Similarly for three-vectors,

$$\vec{a} = a^i \vec{e}_i, \tag{5.5}$$

where the $\vec{e}_i$, $i = 1, 2, 3$ coincide with $(\mathbf{e}_1, \mathbf{e}_2, \mathbf{e}_3)$. Any repeated index indicates summation, so (5.4) could also be written

$$\mathbf{a} = a^\beta \mathbf{e}_\beta = a^\gamma \mathbf{e}_\gamma =, \ldots . \tag{5.6}$$

Repeated indices are therefore called *dummy indices* or *summation indices*. We have more to say about the rules of the summation convention in Section 7.3.

Specifying the components of a four-vector and the basis four-vectors is equivalent to specifying the four-vector itself. It is useful to have a number of different ways of listing the components, namely,

$$a^\alpha = (a^t, a^x, a^y, a^z), \qquad a^\alpha = (a^t, a^i), \qquad a^\alpha = (a^t, \vec{a}). \tag{5.7}$$

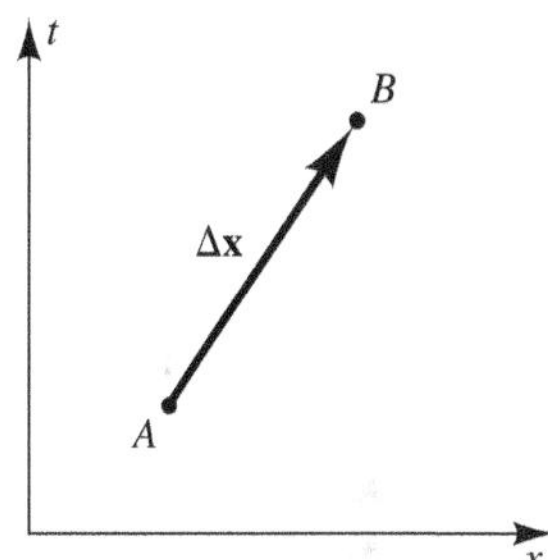

FIGURE 5.5 The displacement four-vector $\Delta\mathbf{x}$ between two points A and B in spacetime.

Example 5.1. Displacement Four-Vectors. A simple example of a four-vector is the displacement four-vector $\Delta\mathbf{x}$ between two events A and B such as those shown in Figure 5.5.

If (t_A, x_A, y_A, z_A) are the coordinates locating event A in a particular inertial frame and (t_B, x_B, y_B, z_B) are the coordinates locating event B, then the components of the displacement four-vector $\Delta\mathbf{x}$ between them are $(t_B - t_A, x_B - x_A,$

[2] Readers with a little mathematical background may know that it is possible to distinguish different *kinds* of components of a four-vector. This distinction will not be necessary until Chapter 20, and until then we refer only to components as defined here.

[3] By now you may be wondering how to write four-vector equations in handwriting since boldface is not easy to reproduce. You can use a wiggly underscore because that is how a printer was instructed to use boldface before electronic typesetting. Thus (5.1) would be

$$\underset{\sim}{a} = a^t \underset{\sim}{e}_t + a^x \underset{\sim}{e}_x + a^y \underset{\sim}{e}_y + a^z \underset{\sim}{e}_z$$

$y_B - y_A, z_B - z_A$). This can be written in a more compact form as

$$\Delta x^\alpha = x_B^\alpha - x_A^\alpha. \tag{5.8}$$

This expression is a shorthand for four equations, one for each value of α, $\alpha = 0, 1, 2, 3$. The index α is called a *free index*—"free" to take on any value from 0 to 3, each value yielding a different equation.

The components of a four-vector are different in different inertial frames because the coordinate basis four-vectors are different. The components of a four-vector—a directed line segment—transform between inertial frames just like the components of a displacement four-vector. For example, for two inertial frames related by a uniform motion v along the x-axis, as in (4.21), the components of a four-vector $\mathbf{a}$ transform as

Lorentz Boost of a Vector

$$a^{t'} = \gamma(a^t - va^x), \tag{5.9a}$$

$$a^{x'} = \gamma(a^x - va^t), \tag{5.9b}$$

$$a^{y'} = a^y, \tag{5.9c}$$

$$a^{z'} = a^z. \tag{5.9d}$$

(If you are wondering where the factors of c in (4.21) went, remember at the end of the previous chapter we said that we would use $c = 1$ units from now on.)

Scalar Product

The *scalar product* is an important idea in the calculus of four-vectors, as it is for three-vectors. The scalar product of two four-vectors $\mathbf{a}$ and $\mathbf{b}$ is denoted by $\mathbf{a} \cdot \mathbf{b}$. It satisfies the usual mathematical rules for scalar products:

$$\mathbf{a} \cdot \mathbf{b} = \mathbf{b} \cdot \mathbf{a}, \tag{5.10a}$$

$$\mathbf{a} \cdot (\mathbf{b} + \mathbf{c}) = \mathbf{a} \cdot \mathbf{b} + \mathbf{a} \cdot \mathbf{c}, \tag{5.10b}$$

$$(\alpha\mathbf{a}) \cdot \mathbf{b} = \alpha(\mathbf{a} \cdot \mathbf{b}), \tag{5.10c}$$

where $\mathbf{a}$, $\mathbf{b}$, $\mathbf{c}$ are any three four-vectors and α is any number.

Calculating scalar products of four-vectors is simple if the scalar products of all pairs of basis four-vectors are known, because if $\mathbf{a} = a^\alpha \mathbf{e}_\alpha$ and $\mathbf{b} = b^\beta \mathbf{e}_\beta$, then

$$\begin{aligned} \mathbf{a} \cdot \mathbf{b} &= (a^\alpha \mathbf{e}_\alpha) \cdot (b^\beta \mathbf{e}_\beta), \\ &= (\mathbf{e}_\alpha \cdot \mathbf{e}_\beta) a^\alpha b^\beta. \end{aligned} \tag{5.11}$$

(There is a double sum in this expression, one sum over α, the other over β.) A special notation is used for the scalar products $\mathbf{e}_\alpha \cdot \mathbf{e}_\beta$ of the basis four-vectors that point along the orthogonal coordinate axes (t, x, y, z) of an inertial frame:

Metric of Flat Spacetime

$$\eta_{\alpha\beta} \equiv \mathbf{e}_\alpha \cdot \mathbf{e}_\beta, \tag{5.12}$$

so that $\mathbf{a} \cdot \mathbf{b}$ can be written

$$\boxed{\mathbf{a} \cdot \mathbf{b} = \eta_{\alpha\beta} a^\alpha b^\beta,} \tag{5.13}$$

Vector Scalar Product

a double sum over α and β implied. The scalar product of all vectors is fixed once the $\eta_{\alpha\beta}$ are known.

The $\eta_{\alpha\beta}$ are determined by the requirement that the scalar product of the displacement four-vector with itself give the square of the distance between the two points it connects:

$$(\Delta s)^2 = \Delta \mathbf{x} \cdot \Delta \mathbf{x}. \tag{5.14}$$

The length of a four-vector defined by the scalar product thus coincides with the length defined as the distance from tail to tip. Comparing this with (4.6) for $(\Delta s)^2$ and noting that $\eta_{\alpha\beta} = \eta_{\beta\alpha}$ as a consequence of (5.12) and (5.10a), we find

$$\eta_{\alpha\beta} = \begin{array}{c} \\ 0 \\ 1 \\ 2 \\ 3 \end{array} \begin{array}{c} \begin{array}{cccc} 0 & 1 & 2 & 3 \end{array} \\ \begin{pmatrix} -1 & 0 & 0 & 0 \\ 0 & 1 & 0 & 0 \\ 0 & 0 & 1 & 0 \\ 0 & 0 & 0 & 1 \end{pmatrix} \end{array}. \tag{5.15}$$

Here, $\eta_{\alpha\beta}$ has been displayed as a diagonal, symmetric matrix. In view of (5.14), the matrix $\eta_{\alpha\beta}$ can be used with the summation convention to express the line element of flat spacetime (4.8) in an especially compact form,

$$\boxed{ds^2 = \eta_{\alpha\beta} dx^\alpha dx^\beta.} \tag{5.16}$$

Line Element of Flat Spacetime

In this role $\eta_{\alpha\beta}$ is called the *metric* of flat spacetime.

Inserting (5.15) in (5.13) gives the following fully equivalent explicit forms for the scalar product of two four-vectors $\mathbf{a}$ and $\mathbf{b}$ in terms of their components in an inertial frame:

$$\mathbf{a} \cdot \mathbf{b} = -a^t b^t + a^x b^x + a^y b^y + a^z b^z, \tag{5.17a}$$

$$\mathbf{a} \cdot \mathbf{b} = -a^0 b^0 + a^1 b^1 + a^2 b^2 + a^3 b^3, \tag{5.17b}$$

$$\mathbf{a} \cdot \mathbf{b} = -a^t b^t + \vec{a} \cdot \vec{b}. \tag{5.17c}$$

As a consequence of a definition that makes no reference to a particular frame, the scalar product is the same in all inertial frames. In a different inertial frame where the components of $\mathbf{a}$ are $(a^{t'}, a^{x'}, a^{y'}, a^{z'})$ and the components of $\mathbf{b}$ are

$(b^{t'}, b^{x'}, b^{y'}, b^{z'})$, the same number $\mathbf{a} \cdot \mathbf{b}$ is given by

$$\mathbf{a} \cdot \mathbf{b} = -a^{t'} b^{t'} + a^{x'} b^{x'} + a^{y'} b^{y'} + a^{z'} b^{z'}. \tag{5.18}$$

This follows from the definition but can be verified explicitly from (5.9).

Example 5.2. Lorentz Boosts Preserve the Orthogonality of Coordinate Axes. The (t', x') axes in Figure 4.13 (now with $c = 1$) don't appear to be orthogonal in the geometry of the printed page, but they are orthogonal in the geometry of spacetime. To see this explicitly in the (t, x) frame, consider a unit displacement four-vector $\mathbf{a}$ along the t' axis and a unit displacement four-vector $\mathbf{b}$ along the x' axis. The (t', x', y', z') components of these four-vectors are $a^{\alpha'} = (1, 0, 0, 0)$ and $b^{\alpha'} = (0, 1, 0, 0)$. These four-vectors are therefore orthogonal because, from (5.17), $\mathbf{a} \cdot \mathbf{b} = 0$ when evaluated in the (t', x', y', z') frame. This means that they are orthogonal in any other inertial frame, but it is instructive to do the calculation explicitly in the (t, x, y, z) frame. From (5.9), the (t, x, y, z) components are

$$a^{\alpha} = (\gamma, v\gamma, 0, 0), \qquad b^{\alpha} = (v\gamma, \gamma, 0, 0). \tag{5.19}$$

From (5.17) again,

$$\mathbf{a} \cdot \mathbf{b} = -\gamma(v\gamma) + (v\gamma)\gamma + 0 + 0 = 0. \tag{5.20}$$

5.2 Special Relativistic Kinematics

Having introduced the idea of four-vectors, let's now turn to their use for describing the motion of a particle in spacetime terms. This is the subject of special relativistic kinematics.

A particle follows a timelike world line through spacetime. This curve can be specified by giving the three spatial coordinates x^i as a function of t in a particular inertial frame. But a more four-dimensional way of describing a world line is to give all four coordinates of the particle x^α as a single-valued function of a parameter σ, which varies along the world line. (See Figure 5.6.) For each value of σ, the four functions $x^\alpha(\sigma)$ determine a point along the curve. Many parameters are possible, but a natural one is the proper time that gives the spacetime distance τ along the world line measured both positively and negatively from some arbitrary starting point. Thus, a world line is described by the equations

$$x^\alpha = x^\alpha(\tau). \tag{5.21}$$

As we discussed in Section 4.3, clocks are devices that measure distance along timelike world lines. The distance τ could be measured by a clock carried along the world line and is called the *proper time* along it.

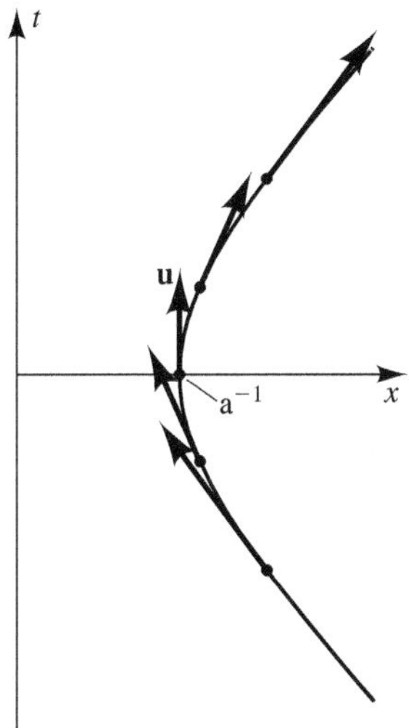

FIGURE 5.6 A simple accelerated world line. This spacetime diagram shows the world line specified parametrically in terms of proper time τ by (5.24). The points label values of $a\tau$ from -1 to 1 in steps of $\frac{1}{2}$. Four-velocity vectors **u** are shown for these points at half-size. The next values of $a\tau$ of 1.5 and -1.5 are off the graph. The points are equidistant along the curve in the geometry of spacetime and the four-vectors are all of equal length. Can you explain why the points appear to increase in separation and the vectors appear to get longer with increasing $|\tau|$ in the geometry of the paper page?

Example 5.3. A Simple Accelerated World Line. A particle moves on the x-axis along a world line described parametrically by

$$t(\sigma) = a^{-1} \sinh \sigma, \qquad x(\sigma) = a^{-1} \cosh \sigma \tag{5.22}$$

where a is a constant with the dimension of inverse length. The parameter σ ranges from $-\infty$ to $+\infty$. For each value of σ, equations (5.22) determine a point (t, x) in spacetime. (The y- and z-dimensions are unimportant for this example and will be suppressed in what follows.) As σ varies, the world line is swept out.

Figure 5.6 shows the world line on a spacetime diagram. It is the hyperbola $x^2 - t^2 = a^{-2}$. The world line could, therefore, alternatively be specified by giving $x(t) = (t^2 + a^{-2})^{1/2}$, but the parametric specification (5.22) is more even-handed between x and t. The world line is accelerated because it is not straight. Proper time τ along the world line is related to σ by [cf. (4.12), (4.8)]

$$d\tau^2 = dt^2 - dx^2 = (a^{-1} \cosh \sigma \, d\sigma)^2 - (a^{-1} \sinh \sigma \, d\sigma)^2 = (a^{-1} d\sigma)^2. \tag{5.23}$$

Fixing τ to be zero when σ is zero, $\tau = a^{-1}\sigma$, and the world line can be expressed with proper time as the parameter in the form (5.21) as

$$t(\tau) = a^{-1} \sinh(a\tau), \qquad x(\tau) = a^{-1} \cosh(a\tau). \tag{5.24}$$

The *four-velocity* is the four-vector $\mathbf{u}$ whose components u^α are the derivatives of the position along the world line with respect to the proper time parameter τ:

Four-Velocity

$$u^\alpha = \frac{dx^\alpha}{d\tau}. \tag{5.25}$$

The four-velocity $\mathbf{u}$ is thus tangent to the world line at each point because a displacement is given by $\Delta x^\alpha = u^\alpha \Delta\tau$. (See Figure 5.7.)

The four components of the four-velocity can be expressed in terms of the three-velocity $\vec{V} = d\vec{x}/dt$ in a particular inertial frame by using the relation (4.15) between t and proper time τ as follows:

$$u^t = \frac{dt}{d\tau} = \frac{1}{\sqrt{1-\vec{V}^2}}, \tag{5.26}$$

and, for example,

$$u^x = \frac{dx}{d\tau} = \frac{dx}{dt}\frac{dt}{d\tau} = \frac{V^x}{\sqrt{1-\vec{V}^2}}. \tag{5.27}$$

In summary, recalling the abbreviation $\gamma \equiv (1-\vec{V}^2)^{-1/2}$ from (4.22),

Four-Velocity in Terms of Three-Velocity

$$u^\alpha = (\gamma, \gamma\vec{V}). \tag{5.28}$$

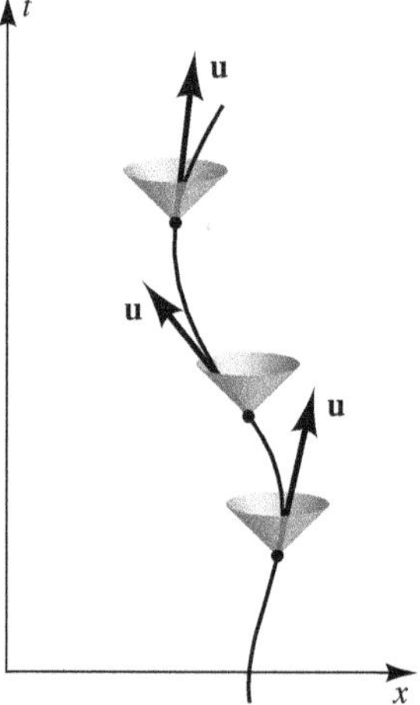

FIGURE 5.7 The four-velocity $\mathbf{u}(\tau)$ at any point along a particle's world line is the unit, timelike tangent four-vector at that point. It lies inside the light cone of that point. This is a two-dimensional plot. But it is evocative and conventional to draw three-dimensional light cones even though there is only two-dimensional information.

An immediate consequence of this result is that the scalar product of $\mathbf{u}$ with itself is [cf. (5.17)]

$$\boxed{\mathbf{u} \cdot \mathbf{u} = -1,} \tag{5.29}$$

Normalization of the Four-Velocity

so that the four-velocity is always a *unit* timelike four-vector. Indeed, this follows directly if (5.13) is used to write the scalar product in the form

$$\mathbf{u} \cdot \mathbf{u} = \eta_{\alpha\beta} \frac{dx^\alpha}{d\tau} \frac{dx^\beta}{d\tau} = -1, \tag{5.30}$$

where the last equality follows from the line element in the form (5.16) and the connection $ds^2 = -d\tau^2$.

Example 5.4. Four-Velocity of a Simple World Line. The four-velocity $\mathbf{u}$ of the world line discussed in Example 5.3 has the components [cf. (5.25)]

$$u^t \equiv dt/d\tau = \cosh(a\tau), \qquad u^x \equiv dx/d\tau = \sinh(a\tau). \tag{5.31}$$

This is correctly normalized:

$$\mathbf{u} \cdot \mathbf{u} = -(u^t)^2 + (u^x)^2 = -\cosh^2(a\tau) + \sinh^2(a\tau) = -1. \tag{5.32}$$

A few examples are shown in Figure 5.6.

The particle's three-velocity is

$$V^x = \frac{dx}{dt} = \frac{dx/d\tau}{dt/d\tau} = \tanh(a\tau). \tag{5.33}$$

This never exceeds the speed of light ($|V^x| = 1$) but approaches it at $\tau = \pm\infty$.

5.3 Special Relativistic Dynamics

Equation of Motion

Newton's first law of motion holds in special relativistic mechanics as well as nonrelativistic mechanics. In the absence of forces, a body is at rest or moves in a straight line at constant speed. This is summarized by

$$\frac{d\mathbf{u}}{d\tau} = 0, \tag{5.34}$$

Newton's First Law

since, in view of (5.28), this equation implies $\vec{V}$ is constant in any inertial frame.

The objective of relativistic mechanics is to introduce the analog of Newton's second law $\vec{F} = m\vec{a}$. There is nothing from which this law can be *derived*, but plausibly it must satisfy certain properties: (1) It must satisfy the principle of relativity, i.e., take the same form in every inertial frame; (2) it must reduce to (5.34) when the force is zero; and (3) it must reduce to $\vec{F} = m\vec{a}$ in any inertial frame when the speed of the particle is much less than the speed of light. The choice

Newton's Second Law

$$\boxed{m\frac{d\mathbf{u}}{d\tau} = \mathbf{f}} \tag{5.35}$$

Rest Mass

naturally suggests itself. The constant m, which characterizes the particle's inertial properties, is called the *rest mass*, and $\mathbf{f}$ is called the *four-force*. Requirement (1) is satisfied because this is a four-vector equation, (2) is evident, and (3) is satisfied with a proper choice of $\mathbf{f}$. This is the correct law of motion for special relativistic mechanics and the special relativistic generalization of Newton's second law. By introducing the *four-acceleration* four-vector $\mathbf{a}$,

Four-acceleration

$$\mathbf{a} \equiv \frac{d\mathbf{u}}{d\tau}, \tag{5.36}$$

the equation of motion (5.35) can be written in the evocative form

$$\boxed{\mathbf{f} = m\mathbf{a}.} \tag{5.37}$$

Although (5.35) represents four equations, they are not all independent. The normalization of the four-velocity (5.29) means

$$m\frac{d(\mathbf{u}\cdot\mathbf{u})}{d\tau} = 0, \tag{5.38}$$

which from (5.36) implies $\mathbf{u}\cdot\mathbf{a} = 0$, and, using (5.37),

$$\mathbf{f}\cdot\mathbf{u} = 0. \tag{5.39}$$

This relation shows that there are only three independent equations of motion—the same number as in Newtonian mechanics. The connection is discussed in more detail soon, and Newton's third law will be discussed as well.

Example 5.5. Required Four-Force. The four-acceleration $\mathbf{a}$ for the world line described in Examples 5.3 and 5.4 has components

$$a^t \equiv du^t/d\tau = a\sinh(a\tau), \qquad a^x \equiv du^x/d\tau = a\cosh(a\tau). \tag{5.40}$$

The magnitude of this acceleration is $(\mathbf{a}\cdot\mathbf{a})^{1/2} = a$, so the constant a is aptly named. The four-force required to accelerate the particle along this world line is $\mathbf{f} = m\mathbf{a}$, where m is the particle's rest mass.

Energy-Momentum

The equation of motion (5.35) leads naturally to the relativistic ideas of energy and momentum. If the *four-momentum* is defined by

$$\mathbf{p} = m\mathbf{u}, \tag{5.41}$$

Four Momentum

then the equation of motion (5.35) can be written

$$\frac{d\mathbf{p}}{d\tau} = \mathbf{f}. \tag{5.42}$$

An important property of the four-momentum follows from its definition (5.41) and the normalization of the four-velocity (5.29)

$$\boxed{\mathbf{p}^2 \equiv \mathbf{p} \cdot \mathbf{p} = -m^2.} \tag{5.43}$$

In view of (5.28), the components of the four-momentum are related to the three-velocity $\vec{V}$ in an inertial frame by

$$p^t = \frac{m}{\sqrt{1 - \vec{V}^2}}, \qquad \vec{p} = \frac{m\vec{V}}{\sqrt{1 - \vec{V}^2}}. \tag{5.44}$$

For small speeds $V \ll 1$,

$$p^t = m + \frac{1}{2}m\vec{V}^2 + \cdots, \qquad \vec{p} = m\vec{V} + \cdots. \tag{5.45}$$

Thus, at small velocities $\vec{p}$ reduces to the usual momentum, and p^t reduces to the kinetic energy plus the rest mass. For this reason $\mathbf{p}$ is also called the *energy-momentum four-vector*, and its components in an inertial frame are written

$$p^\alpha = (E, \vec{p}) = (m\gamma, m\gamma\vec{V}), \tag{5.46}$$

where $E \equiv p^t$ is the *energy* and $\vec{p}$ is the *three-momentum*. Equation (5.43) can be solved for the energy in terms of the three-momentum to give

$$\boxed{E = (m^2 + \vec{p}^2)^{1/2},} \tag{5.47}$$

which shows how rest energy is a part of the energy of a relativistic particle. Indeed, for a particle at rest, (5.47) reduces to $E = mc^2$ in more usual units. This must be the most famous equation in relativity if not one of the most famous ones in all of physics.

An important application of special relativistic kinematics occurs in particle reactions, where the total four-momentum is conserved in particle collisions, corresponding to the law of energy conservation and the conservation of total three-momentum. An example important for astrophysics is given in Box 5.1 on p. 94.

In a particular inertial frame the connection between the relativistic equation of motion (5.35) and Newton's laws can be made more explicit by defining the *three-force* $\vec{F}$ as

$$\frac{d\vec{p}}{dt} \equiv \vec{F}. \tag{5.48}$$

This has the same form as Newton's law but with the relativistic expression for the three-momentum (5.44). Solving problems in the mechanics of special relativity is, therefore, essentially the same as solving Newton's equation of motion. The only difference arises from the different relation of momentum to velocity (5.44). Newton's third law applies to the force $\vec{F}$ just as it does in Newtonian mechanics because, through (5.48), it implies that the total three-momentum of a system of particles is conserved in all inertial frames. Evidently $\vec{f} = d\vec{p}/d\tau = (d\vec{p}/dt)(dt/d\tau) = \gamma\vec{F}$. Using (5.39) and (5.28), the four-force can be written in terms of the three-force as

Newton's Third Law

Four-Force in Terms of Three-Force

$$\boxed{\mathbf{f} = (\gamma\vec{F}\cdot\vec{V}, \gamma\vec{F}),} \tag{5.49}$$

where $\vec{V}$ is the particle's three-velocity. The time component of the equation of motion (5.42) is

$$\frac{dE}{dt} = \vec{F}\cdot\vec{V}, \tag{5.50}$$

which is a familiar relation from Newtonian mechanics. This time component of the equation of motion (5.42) is a consequence of the other three. Thus, in terms of the three-force, the equations of motion take the same form as they do in usual Newtonian mechanics but with the relativistic expressions for energy and momentum. When the velocity is small (5.25) shows that the special relativistic version of Newton's second law (5.48) reduces to the familiar nonrelativistic form. Newtonian mechanics is the low-velocity approximation to special relativistic mechanics.

Example 5.6. A Relativistic Charged Particle in a Magnetic Field. A particle with charge q and rest mass m moves in a *uniform* magnetic field $\vec{B}$ with total energy E. What is the radius of its circular orbit? What are the components of the electromagnetic four-force acting on the particle?

As we have already mentioned, electromagnetism is unchanged in special relativity so that the three-force on a charged particle in a magnetic field is

$$\vec{F} = q(\vec{V}\times\vec{B}), \tag{5.51}$$

where $\vec{V}$ is the velocity of the charge. The particle moves in a circular orbit of radius R at constant speed, obeying the familiar equation of motion (5.48). Therefore,

$$\frac{d\vec{p}}{dt} = \frac{d}{dt}\left(\frac{m\vec{V}}{\sqrt{1-V^2}}\right) = \frac{m}{\sqrt{1-V^2}}\frac{d\vec{V}}{dt}. \tag{5.52}$$

The centripetal acceleration $d\vec{V}/dt$ is given by the usual, purely kinematic relation V^2/R. Therefore,

$$\frac{m\gamma V^2}{R} = qVB. \tag{5.53}$$

Thus,

$$R = \frac{mV\gamma}{qB} = \frac{|\vec{p}|}{qB} = \frac{\sqrt{E^2 - m^2}}{qB}, \tag{5.54}$$

which relates the radius to the total energy. The components of the four-force are $f^t = \gamma \vec{F} \cdot \vec{V} = 0$ (the magnetic field does no work) and a radial component

$$f^r = \gamma F^r = qVB\gamma = \frac{qB}{m}\sqrt{E^2 - m^2}. \tag{5.55}$$

5.4 Variational Principle for Free Particle Motion

Newtonian mechanics can be summarized by a principle of extremal action as reviewed in Section 3.5. The motion of a free particle in special relativity can be summarized by a similar variational principle—the principle of extremal proper time. That principle is already evident from the twin paradox discussion in Section 4.4. The straight lines along which free particles move in spacetime are paths of *longest* proper time between two events. In this section we will demonstrate that this fact constitutes a variational principle that *implies* the free particle equation of motion (5.34). That is important because in Chapter 8 we will turn this argument around. We will *posit* the principle of extremal proper time for a free particle in *curved* spacetime and use it to *derive* the free particle equation of motion.

The variational principle of extremal proper time can be stated as follows:

Variational Principle for Free Particle Motion

The world line of a free particle between two timelike separated points extremizes the proper time between them.

Consider two timelike separated points A and B in spacetime, and all timelike world lines going between them (Figure 5.8). Each curve will have a value of the

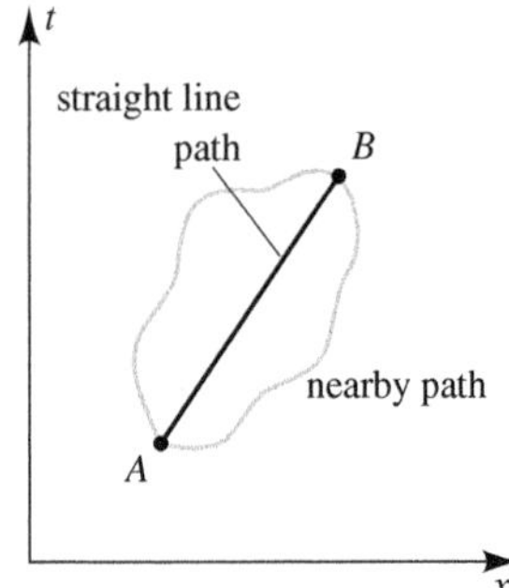

FIGURE 5.8 A straight line between two points is an extremum of the distance between the points when compared with nearby curves (shaded) connecting the two points.

proper time

$$\tau_{AB} = \int_A^B d\tau = \int_A^B \left[dt^2 - dx^2 - dy^2 - dz^2\right]^{1/2}. \tag{5.56}$$

Suppose the world line is described parametrically with parameter σ chosen so that it takes the value $\sigma = 0$ at point A and $\sigma = 1$ at point B for all curves we want to consider. (This would not be the case for the parameter τ.) The world line is then specified by giving the coordinates as a function of σ, namely, $x^\alpha = x^\alpha(\sigma)$. Equation (5.56) can then be written

$$\tau_{AB} = \int_0^1 d\sigma \left[\left(\frac{dt}{d\sigma}\right)^2 - \left(\frac{dx}{d\sigma}\right)^2 - \left(\frac{dy}{d\sigma}\right)^2 - \left(\frac{dz}{d\sigma}\right)^2\right]^{1/2}. \tag{5.57}$$

We seek the world line (or world lines) that extremize τ_{AB}, that is, the curve for which a small variation $\delta x^\alpha(\sigma)$ produces a vanishing variation in the elapsed proper time. This is a familiar type of problem from Newtonian mechanics that was reviewed in Section 3.5. Think of the integrand in (5.57) as the Lagrangian, x^α as the dynamical variables, and σ as the time. Then (5.57) has the same form as action for Newtonian mechanics (3.35). Lagrange's equations are the necessary condition for an extremum both there and here. Specifically,

$$-\frac{d}{d\sigma}\left(\frac{\partial L}{\partial(dx^\alpha/d\sigma)}\right) + \frac{\partial L}{\partial x^\alpha} = 0, \tag{5.58}$$

with

$$L = \left[\left(\frac{dt}{d\sigma}\right)^2 - \left(\frac{dx}{d\sigma}\right)^2 - \left(\frac{dy}{d\sigma}\right)^2 - \left(\frac{dz}{d\sigma}\right)^2\right]^{1/2} = \left[-\eta_{\alpha\beta}\frac{dx^\alpha}{d\sigma}\frac{dx^\beta}{d\sigma}\right]^{1/2}. \tag{5.59}$$

To see what happens, let's write out the Lagrange equation (5.58) for $x^1 \equiv x$:

$$\frac{d}{d\sigma}\left[\frac{1}{L}\frac{dx^1}{d\sigma}\right] = 0. \tag{5.60}$$

However, $L = [-\eta_{\alpha\beta}(dx^\alpha/d\sigma)(dx^\beta/d\sigma)]^{1/2}$ is just $d\tau/d\sigma$, so multiplying by $d\sigma/d\tau$, (5.60) becomes

$$\frac{d^2x^1}{d\tau^2} = 0. \tag{5.61}$$

It is exactly the same for the other coordinates. All four Lagrange equations imply

$$\frac{d^2x^\alpha}{d\tau^2} = 0. \tag{5.62}$$

This is the correct equation of motion for a free particle (5.34). Its solution is the straight world line connecting A and B. The world line of a free particle in flat spacetime is a curve of extremal proper time.

5.5 Light Rays

Zero Rest Mass Particles

The discussion so far has concerned particles with nonzero rest mass, which move at speeds less than the speed of light. Let's now consider particles that move at the speed of light $V = 1$ along null world lines. Examples are the quanta of light and gravity—photons and gravitons—and possibly some kinds of neutrinos.[4] We focus almost exclusively on photons, which are also called *light rays* in their non-quantum aspects, but our treatment would cover any other particle that moves with the speed of light.

Evidently the proper time can no longer be used as a parameter along the world line of a light ray—the proper time interval between any two points on it is zero. However, there are many other parameters that could be used. For example, the curve

$$x = t, \tag{5.63}$$

which has $V = 1$, could be written parametrically as

$$x^\alpha = u^\alpha \lambda, \tag{5.64}$$

where λ is the parameter and $u^\alpha = (1, 1, 0, 0)$. The four-vector $\mathbf{u}$ is a tangent four-vector $u^\alpha = dx^\alpha/d\lambda$ using the parameter λ as τ was used in (5.25). However, here $\mathbf{u}$ is a null vector. Therefore, in contrast to (5.29),

$$\mathbf{u} \cdot \mathbf{u} = 0. \tag{5.65}$$

Different choices of parametrization will give different tangent four-vectors, but all have zero length.

With this choice of parametrization,

$$\frac{d\mathbf{u}}{d\lambda} = 0, \tag{5.66}$$

so that the equation of motion of a light ray is the same as for a particle (5.34). There are many other choices of parametrization for which this is not true. For example, we could have replaced λ by σ^3 in (5.64). As σ varies between $-\infty$ and $+\infty$, the same straight line, $x = t$, would have been described. Equation (5.65) would continue to be true, but (5.66) would not. Parameters for which the equation of motion for a light ray (5.66) has the same form as for particles are called *affine parameters*. There is not a unique affine parameter. For example, if λ is an affine parameter, then a constant times λ is also an affine parameter. Affine parameters

Affine Parameters

[4] There is currently evidence that at least some kinds of neutrinos have small rest masses.

are the most convenient ones to use for light rays because of the simple form of (5.66).

Energy, Momentum, Frequency, and Wave Vector

Photons and neutrinos carry energy and three-momentum. In any inertial frame, the energy of a photon E is connected to its frequency ω by another of Einstein's famous relations,

$$E = \hbar\omega. \tag{5.67}$$

For the three-momentum, note from (5.44) that the three-velocity is given by $\vec{V} = \vec{p}/E$. Since $|\vec{V}| = 1$, this implies that $|\vec{p}| = E$ for a photon, so the three-momentum can be written

$$\vec{p} = \hbar\vec{k}, \tag{5.68}$$

where $\vec{k}$ points in the direction of propagation, has magnitude $|\vec{k}| = \omega$, and is called the *wave three-vector*. In any inertial frame the components of the four-momentum of a photon $\mathbf{p}$ can therefore be written

Four-Momentum of a Photon

$$p^\alpha = \left(E, \vec{p}\right) = \left(\hbar\omega, \hbar\vec{k}\right) = \hbar k^\alpha. \tag{5.69}$$

The four-vector $\mathbf{k}$ is called the *wave four-vector*. Evidently,

$$\mathbf{p} \cdot \mathbf{p} = \mathbf{k} \cdot \mathbf{k} = 0. \tag{5.70}$$

Comparing this with (5.43), we see that photons have zero rest mass, like all particles moving at the speed of light. Both $\mathbf{p}$ and $\mathbf{k}$ are tangent to the world line of a photon. The tangent vector $\mathbf{u}$ could be chosen to coincide with either $\mathbf{p}$ or $\mathbf{k}$ by adjusting the normalization of the affine parameter λ. The equation of motion (5.66) can be written in terms of $\mathbf{p}$ or $\mathbf{k}$ as

$$\frac{d\mathbf{p}}{d\lambda} = 0, \quad \text{or} \quad \frac{d\mathbf{k}}{d\lambda} = 0, \tag{5.71}$$

where λ is an affine parameter.

Doppler Shift and Relativistic Beaming

The relativistic Doppler shift is a simple application of these ideas. Consider a source that emits photons of frequency ω in all directions in the source's rest frame. Suppose in another frame the source is moving with speed V along the x'-axis. What frequency will be observed for a photon that makes an angle α' with the direction of motion? This question is answered by using a Lorentz boost to connect the components of the wave four-vector **k** in the rest frame to those in the observer's frame where the source is moving. Let $k^\alpha = (\omega, \vec{k})$ be components of the wave four-vector **k** of the photon in the frame of the source and $k'^\alpha = (\omega', \vec{k}')$ the components in the frame of the observer. From (5.9),

$$\omega = \gamma(\omega' - Vk^{x'}). \tag{5.72}$$

But $k'^x = \omega' \cos\alpha'$, where α' is the angle between the x'-axis and the direction of the photon in the observer's frame. Thus,

$$\boxed{\omega' = \omega \frac{\sqrt{1-V^2}}{1 - V\cos\alpha'},} \tag{5.73}$$

Doppler Shift

which is the formula for the relativistic Doppler shift. For small V, this is approximately

$$\omega' \approx \omega\left(1 + V\cos\alpha'\right). \tag{5.74}$$

When $\alpha' = 0$, the photon is emitted in the same direction that the source is moving and there is a blue shift of $\Delta\omega' = +V\omega$ in the frequency of the photon. When $\alpha' = \pi$, the photon is moving opposite to the source and there is a red shift of $\Delta\omega' = -V\omega$.

Even photons emitted transverse to the direction of motion of the source ($\alpha' = \pi/2$) are redshifted, although the leading order of this effect is V^2. This is called the transverse Doppler shift, and formula (5.73) shows it is just time dilation.

The phenomenon of relativistic beaming (Figure 5.9) follows from the transformation of the spatial momentum of the photon. Suppose a photon makes an angle α with the x-axis in the source frame where $\cos\alpha = k^x/\omega$. In the observer's frame, the angle it makes with the x'-axis is defined by $\cos\alpha' = k'^x/\omega'$. The Lorentz transformation (5.9) between the two frames connecting (ω, k^x) to (ω', k'^x) shows that these two angles are related by

$$\cos\alpha' = \frac{\cos\alpha + V}{1 + V\cos\alpha}. \tag{5.75}$$

Relativistic Beaming

Thus the half of the photons that are emitted in the forward hemisphere in the source frame ($|\alpha| < \pi/2$) are seen by the observer to be emitted in a smaller cone $|\alpha'| < \alpha'_{1/2}$, where $\cos\,\alpha'_{1/2} = V$. For V close to 1 this opening angle will be small. Photons are thus beamed along the direction of the source by its motion. The Doppler shift implies that the energy of the photons in the forward direction is greater than that in the backward direction, meaning that the *intensity* of the radiation is even more concentrated along the direction of motion (Problem 17). A uniformly radiating body moving toward you is brighter than if it is moving away. That is the phenomenon of relativistic beaming.

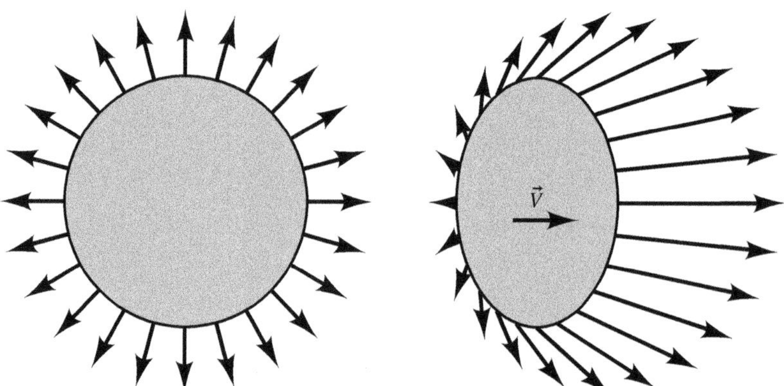

FIGURE 5.9 Relativistic beaming. The figure at left shows a body which radiates equally in all directions in its rest frame. Wave vectors of 24 photons emitted normally to the surface are shown. The figure at right shows the Lorentz contracted body in a frame where it is moving to the right with a speed $V = .75$ together with the wave vectors of the same 24 photons. The photons are increasingly directed along the direction of motion as V approaches the speed of light and their wave vectors have larger components in that direction because of the Doppler effect. The intensity of the radiation is therefore increasingly concentrated along the direction of motion.

BOX 5.1 Cosmic Background Cutoff on Cosmic Ray Energies

The fastest particles in the universe moving below the speed of light with respect to the Earth are the highest-energy cosmic rays. *Cosmic ray* is a general term for an elementary particle or an atomic nucleus propagating through the interstellar medium. Protons are an abundant example. Cosmic rays are detected through the showers of particles they produce when they enter our atmosphere, and energies of up to 3×10^{20} eV have been observed. A collision with a proton in the atmosphere is 100,000 times more energetic than collisions in the most powerful accelerators on Earth. For a proton this corresponds to $\gamma \sim 10^{11}$ and a velocity of only a few parts in 10^{22} less than the velocity of light.

Acceleration mechanisms for cosmic rays are imperfectly understood, but some clues about their origin can be found by understanding their interaction with the photons of the cosmic microwave background radiation (CMB). The CMB is an all-pervasive, blackbody, background of light from the big bang that has cooled to a present temperature of 2.73 K. We study the CMB in detail in Chapter 17, but only a few facts are needed to consider its impact on cosmic rays. The radiation is isotropic with a blackbody spectrum in a frame called the CMB frame. The galaxies are moving only slowly compared to the speed of light with respect to this frame. At a temperature of 2.73 K the characteristic energy of a CMB photon is 2×10^{-4} eV, and there are an average of 400 CMB photons per cm^3.

What happens when a high-energy cosmic ray proton collides with a CMB photon? If the proton is moving fast enough, the collision can initiate reactions like the photoproduction of pions,

$$\gamma + p \rightarrow n + \pi^+ \quad \text{or} \quad \gamma + p \rightarrow p + \pi^0,$$

that will degrade the proton's energy. (Despite the possibility of confusion we persist in using γ both for photon and the factor in Lorentz transformations.) We would not expect to see cosmic ray protons above this energy if their source is distant enough that they would almost surely have collided with a CMB photon in their trip to us. This limit is called the GZK cutoff after the initials of the authors (Greisen-Zatsepin-Kuz'min) who first called attention to the effect.

Evaluating the GZK cutoff energy is an instructive exercise in special relativity. For definiteness consider the first of the processes quoted earlier. The total four-momentum is conserved:

$$\mathbf{p}_\gamma + \mathbf{p}_p = \mathbf{p}_n + \mathbf{p}_\pi . \tag{a}$$

The threshold energy is found most easily in the center-of-mass (CM) frame, where the momenta of the colliding particles are equal and opposite. The threshold occurs when the initial energies are just enough to lead to a neutron and pion both at rest.

At threshold in the CM frame the total energy is $E_n^{CM} + E_\pi^{CM} = m_n + m_\pi$. The total three-momentum is zero by definition, $\vec{p}_n^{\,CM} + \vec{p}_\pi^{\,CM} = 0$. To what energy E_p^{CMB} does this correspond in the CMB frame where photons have a typical energy $E_\gamma^{CMB} \approx 6 \times 10^{-4}$ eV? That threshold proton energy is the GZK cutoff.

This question can be efficiently answered by utilizing the fact that the length of a four-vector is the same in all frames. Evaluating $(\mathbf{p}_n + \mathbf{p}_\pi)^2$ at threshold in the CM frame gives $-(m_n + m_\pi)^2$. The conservation of four-momentum (a) means that this is the same as $(\mathbf{p}_\gamma + \mathbf{p}_p)^2$. Computing that square using $\mathbf{p}_p^2 = -m_p^2$ and $\mathbf{p}_\gamma^2 = 0$ (photons have zero rest mass) gives

$$2\mathbf{p}_\gamma \cdot \mathbf{p}_p - m_p^2 = -(m_n + m_\pi)^2. \tag{b}$$

This relation does not depend on the frame but can be evaluated in terms of the components of the four-momenta in the CMB frame. Suppose the proton with energy $E_p^{CMB} \gg m_p$ is traveling along the x-axis to collide with a photon of energy E_γ^{CMB} traveling in the opposite direction. The CMB frame (t, x) components are

$$\begin{aligned} (p_\gamma^{CMB})^\alpha &= (E_\gamma^{CMB}, -E_\gamma^{CMB}), \\ (p_p^{CMB})^\alpha &\approx (E_p^{CMB}, E_p^{CMB}) \end{aligned} \tag{c}$$

where three-momenta have been expressed in terms of energies using (5.47) and the approximation $E_p^{CMB} \gg m_p$. The scalar product in (b) can be computed in terms of these components and the resulting relation solved for E_p^{CMB}. The result simplifies using the approximation $m_n \approx m_p$ (more than adequate for present purposes) to give

$$E_p^{CMB} \approx \frac{m_p m_\pi}{2E_\gamma^{CMB}}\left(1 + \frac{m_\pi}{2m_p}\right) \approx 3 \times 10^{20} \text{ eV}. \tag{d}$$

This is the GZK cutoff energy. These protons are traveling at a speed V only 5×10^{-24} less than the velocity of light corresponding to a Lorentz gamma factor $\gamma \sim 10^{11}$!

The mean free path λ_{CMB} for a 10^{20} eV proton before a collision with a CMB photon is $\lambda_{CMB} = 1/(\sigma N_\gamma)$, where σ is the cross section for the photo-production process—about 2×10^{-28} cm^2—and N_γ is the number density of CMB photons—about 400 cm^{-3}. These numbers give $\lambda_{CMB} \approx 10^{25}$ cm $\approx$ 10 million light years. This is only a few times the size of the local group of galaxies. It takes a small number of mean free paths for the proton energy to degrade, but protons of that energy can't be coming from too far away. Cosmic rays at very high energies are rare but have been detected at 3×10^{20} eV, and there is no sign of a sharp decrease in numbers that would be expected from the GZK cutoff. One explanation for the high-energy particles is that they were produced close to home.

5.6 Observers and Observations

The predictions of special relativistic mechanics are typically most easily calculated and most easily understood in inertial frames. Observations of observers at rest in an inertial frame are referred to the axes of that frame. For example, the energy of a particle measured by an observer at rest in an inertial frame is the component of the particle's four-momentum along the time axis of that frame. But not every observer is at rest in an inertial frame—observers on the surface of

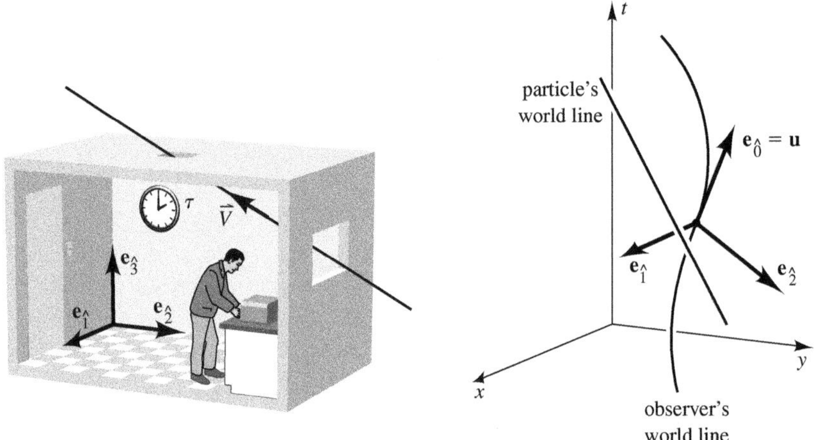

FIGURE 5.10 An observer moving through spacetime may be thought of as inhabiting a local laboratory, as shown at left, that is moving through spacetime on a world line, as shown at right. Three orthogonal space directions inside the laboratory define three spacelike unit four-vectors $\mathbf{e}_{\hat{1}}, \mathbf{e}_{\hat{2}}, \mathbf{e}_{\hat{3}}$. The observer's clocks, at rest in the laboratory, define a time direction $\mathbf{e}_{\hat{0}}$ coinciding with the observer's four-velocity. Observations made by the observer are referred to the basis four-vectors $\mathbf{e}_{\hat{\alpha}}$, $\hat{\alpha} = 0, 1, 2, 3$.

the Earth, for instance, are not. How are the predictions for the observations of accelerated observers calculated?

This question is especially important for general relativity. There are generally no inertial frames in the curved spacetimes of general relativity that extend over all spacetime. As we will see, there are *local* inertial frames in the neighborhood of each point and the neighborhood of the world lines of freely falling observers but no global ones. Therefore, it is crucial to have a systematic way of extracting the predictions for observers who are not associated with global inertial frames. This section describes how to do that in the context of special relativity.

The path of an observer through spacetime is a timelike world line. An observer may be thought of as carrying a laboratory along the world line. At least for astrophysical problems, this laboratory, even if it's the Hubble Space Telescope, will be very small compared to the distances over which physical phenomena take place. We therefore idealize it as being arbitrarily small. Inside the laboratory the observer makes measurements by means of clocks and rulers. (See Figure 3.1.) For example, an observer might measure the velocity of a particle passing through the laboratory by noting that the particle's path made a certain angle with one of the laboratory's walls—this gives the direction—and measuring the time it takes to go the distance across the laboratory, which gives the speed.

Mathematically, this idea of a local laboratory may be idealized as shown in Figure 5.10. An observer carries along four *orthogonal unit* four-vectors $\mathbf{e}_{\hat{0}}, \mathbf{e}_{\hat{1}}, \mathbf{e}_{\hat{2}}, \mathbf{e}_{\hat{3}}$, which define a time direction and three spatial directions, respectively, to which the observer will refer all measurements. Indices with a hat over them are used to emphasize that we are dealing with an *orthonormal* basis—each four-vector with unit length and all four-vectors mutually orthogonal. The timelike unit four-vector $\mathbf{e}_{\hat{0}}$ will be tangent to the observer's world line since that is the direction a clock at rest in the laboratory is moving in spacetime. Since the observer's four-velocity $\mathbf{u}_{\text{obs}}$ is a *unit* tangent vector [cf. (5.29)],

$$\mathbf{e}_{\hat{0}} = \mathbf{u}_{\text{obs}}. \tag{5.76}$$

The observer is free to pick the three spatial basis vectors $\mathbf{e}_{\hat{\imath}}$ as long as they are orthogonal to $\mathbf{e}_{\hat{0}}$ and to each other. Only if the laboratory is at rest in an inertial frame will the $\mathbf{e}_{\hat{\alpha}}$ point along the axes of an inertial frame.

Example 5.7. An Orthonormal Basis for a Simple Accelerating Observer. Consider the observer moving along the accelerated world line described in Examples 5.3 and 5.4. What are the components of a set of orthonormal basis four-vectors for this observer in the inertial frame? These four-vectors will vary with the observer's proper time. The four-vector $\mathbf{e}_{\hat{0}}(\tau)$ is the observer's four velocity $\mathbf{u}_{\text{obs}}(\tau)$, which has components [cf. (5.31)]

$$(\mathbf{e}_{\hat{0}}(\tau))^\alpha = u^\alpha_{\text{obs}}(\tau) = (\cosh(a\tau), \sinh(a\tau), 0, 0). \tag{5.77}$$

The only conditions on the other three four-vectors $\mathbf{e}_{\hat{\imath}}(\tau)$ are that they be orthogonal to $\mathbf{e}_{\hat{0}}(\tau)$, orthogonal to each other, and of unit length. There are many possibilities corresponding to the observer's freedom to orient the spatial axes of the orthonormal frame. The easiest way to satisfy the conditions is to pick $\mathbf{e}_{\hat{2}}(\tau)$ and $\mathbf{e}_{\hat{3}}(\tau)$ to be unit four-vectors in the y- and z-directions, respectively. The remaining four-vector $\mathbf{e}_{\hat{1}}(\tau)$ then has the form $(f(\tau), g(\tau), 0, 0)$ for some functions f and g. Orthogonality with $\mathbf{e}_{\hat{0}}(\tau)$ means

$$\mathbf{e}_{\hat{0}}(\tau) \cdot \mathbf{e}_{\hat{1}}(\tau) = -\cosh(a\tau) f(\tau) + \sinh(a\tau) g(\tau) = 0. \tag{5.78}$$

Unit length means

$$\mathbf{e}_{\hat{1}}(\tau) \cdot \mathbf{e}_{\hat{1}}(\tau) = -f^2(\tau) + g^2(\tau) = 1. \tag{5.79}$$

These two conditions determine f and g. The four-vectors $\mathbf{e}_{\hat{\imath}}(\tau)$ that together with (5.77) make up orthonormal basis four-vectors for the observer are

$$(\mathbf{e}_{\hat{1}}(\tau))^\alpha = (\sinh(a\tau), \cosh(a\tau), 0, 0), \tag{5.80a}$$

$$(\mathbf{e}_{\hat{2}}(\tau))^\alpha = (0, 0, 1, 0), \tag{5.80b}$$

$$(\mathbf{e}_{\hat{3}}(\tau))^\alpha = (0, 0, 0, 1). \tag{5.80c}$$

As discussed before, observers refer observations to the axes of their laboratories and the clocks within them. This means that they measure the components of four-vectors along the basis four-vectors $\{\mathbf{e}_{\hat{\alpha}}\}$ associated with their laboratory. (The notation { } means "set of".) For instance, the energy of a particle measured by an accelerating observer is the component of the particle's four-momentum $\mathbf{p}$ along the basis four-vector $\mathbf{e}_{\hat{0}}$. The three-momentum measured in direction 1 is the component of $\mathbf{p}$ along $\mathbf{e}_{\hat{1}}$, etc. These components are defined by the decomposition [cf. (5.4)]

$$\mathbf{p} = p^{\hat{\alpha}}\mathbf{e}_{\hat{\alpha}}. \tag{5.81}$$

They can be computed as scalar products with the orthonormal basis four-vectors of the observer, with the result:

$$p^{\hat{0}} = -\mathbf{p}\cdot\mathbf{e}_{\hat{0}}, \qquad p^{\hat{1}} = \mathbf{p}\cdot\mathbf{e}_{\hat{1}}, \qquad p^{\hat{2}} = \mathbf{p}\cdot\mathbf{e}_{\hat{2}}, \qquad p^{\hat{3}} = \mathbf{p}\cdot\mathbf{e}_{\hat{3}}. \tag{5.82}$$

To verify these relations, just compute the right-hand sides using (5.81), taking into account that the basis four-vectors are orthonormal. In particular, the energy of the particle E measured by an observer with four-velocity $\mathbf{u}_{\text{obs}}$ is the first of these, or

$$E = -\mathbf{p}\cdot\mathbf{u}_{\text{obs}}. \tag{5.83}$$

The following examples illustrate how this works.

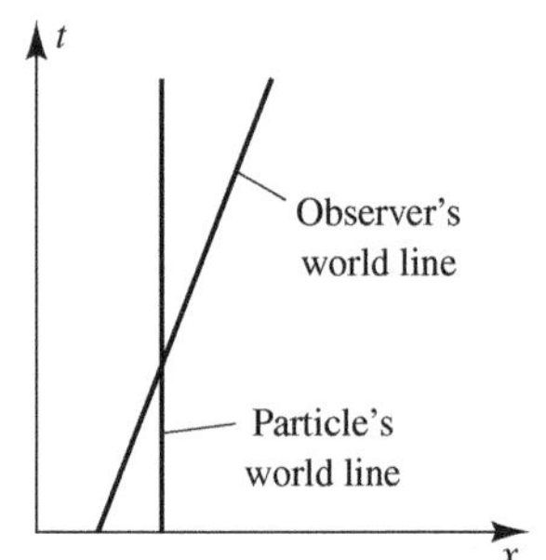

FIGURE 5.11 An observer moving past a stationary particle measures the particle's energy as the component of the four-momentum along the observer's four-velocity.

Example 5.8. Energy of a Stationary Particle Measured by an Observer Moving with Speed V. Consider a particle is at rest in a certain inertial frame (Figure 5.11). An observer is moving with velocity $\vec{V}$ in this frame so that the observer's world line intersects the particle's. From the observer's point of view the particle moves through the observer's laboratory. What energy of the particle would be measured by the observer? We already know the answer. The particle will move through the laboratory with speed V and so the measured energy will be

$$E = m\gamma, \tag{5.84}$$

where m is the particle's rest mass.

Let's see how this comes about by scalar products with the observer's orthonormal basis. In the inertial frame where the particle is at rest, the particle's momentum four-vector has components

$$\mathbf{p} = (m, 0, 0, 0). \tag{5.85}$$

In the same frame the four-velocity of the observer is [cf. (5.28)]

$$\mathbf{e}_{\hat{0}} = \mathbf{u}_{\text{obs}} = (\gamma,\ V\gamma,\ 0,\ 0). \tag{5.86}$$

The energy measured by the observer according to (5.83) is

Energy Measured by an Observer

$$E = -\mathbf{p}\cdot\mathbf{e}_{\hat{0}} = -\mathbf{p}\cdot\mathbf{u}_{\text{obs}} = m\gamma, \tag{5.87}$$

which is the same as (5.84). The energy measured by the observer is just the component of the particle's energy-momentum four-vector along the observer's time direction $\mathbf{e}_{\hat{0}}$.

---

[5]This violates temporarily the rules for balancing indices on both sides of an equation that we have been following implicitly and will be described explicitly in Section 7.3. More pedantically the relations could be written so the indices do balance as:

$$\eta_{\hat{\alpha}\hat{\beta}}p^{\hat{\beta}} = \mathbf{e}_{\hat{\alpha}}\cdot\mathbf{p}.$$

For this simple example the computation (5.87) is excessively complicated. The point is, however, that expression (5.87) is written in an invariant form and can be computed in *any* reference frame. To see the advantage of this consider the following example.

Example 5.9. Frequency Measured by an Accelerating Observer. An observer following the world line of Examples 5.3 and 5.4 observes the light from a star that remains stationary at the origin of the inertial frame, emitting light steadily. Assume for simplicity that the light is emitted at a single optical frequency, ω_*, in the rest frame of the star. What frequency $\omega(\tau)$ will the observer measure as a function of proper time along his or her world line?

In the inertial frame in which the star is stationary, the wave four-vector $\mathbf{k}$ of a photon reaching the observer has components $k^\alpha = (\omega_*, \omega_*, 0, 0)$. The observed frequency $\omega(\tau)$ could be worked out by transforming these components into the instantaneous rest frame of the observer at proper time τ. That is not so very difficult (Problem 18), but it is easier to note that $E = \hbar\omega$ for photons and use (5.87):

$$\omega(\tau) = -\mathbf{k} \cdot \mathbf{u}_{\text{obs}}, \tag{5.88}$$

where $\mathbf{u}_{\text{obs}}$ is the four-velocity (5.31). Explicitly, this gives

$$\omega(\tau) = k^t u^t - k^x u^x = \omega_*[\cosh(a\tau) - \sinh(a\tau)] = \omega_* \exp(-a\tau). \tag{5.89}$$

At early proper times the observer is moving rapidly toward the source and the light is blue-shifted; at late proper times the observer is moving rapidly away from the source and the light is red-shifted.

An observer on the bridge of a starship following the world line (5.31) that is looking at a field of stars will see them for only a limited period of proper time of order $1/a$. Can you explain why?

Problems

1. [S] Consider two four-vectors $\mathbf{a}$ and $\mathbf{b}$ whose components are given by

$$a^\alpha = (-2, 0, 0, 1),$$
$$b^\alpha = (5, 0, 3, 4).$$

   (a) Is $\mathbf{a}$ timelike, spacelike, or null? Is $\mathbf{b}$ timelike, spacelike, or null?

   (b) Compute $\mathbf{a} - 5\mathbf{b}$.

   (c) Compute $\mathbf{a} \cdot \mathbf{b}$.

2. The scalar product between two three-vectors can be written as

$$\vec{a} \cdot \vec{b} = ab \cos\theta_{ab}$$

where a and b are the lengths of $\vec{a}$ and $\vec{b}$, respectively, and θ_{ab} is the angle between them. Show that an analogous formula holds for two timelike four-vectors **a** and **b**:

$$\mathbf{a} \cdot \mathbf{b} = -ab \cosh \theta_{ab},$$

where $a = (-\mathbf{a} \cdot \mathbf{a})^{1/2}$, $b = (-\mathbf{b} \cdot \mathbf{b})^{1/2}$, and θ_{ab} is the parameter defined in (4.18) that describes the Lorentz boost between the frame where an observer whose world line points along **a** is at rest and the frame where an observer whose world line points along **b** is at rest.

3. [S] A free particle is moving along the x-axis of an inertial frame with speed $dx/dt = V$ passing through the origin at $t = 0$. Express the particles's world line parametrically in terms of V using the proper time τ as the parameter.

4. Work out the components of the four-acceleration vector $\mathbf{a} \equiv d\mathbf{u}/d\tau$ in terms of the three-velocity $\vec{V}$ and the three-acceleration $\vec{a} = d\vec{V}/dt$ to obtain expressions analogous to (5.28). Using this expression and (5.28), verify explicitly that $\mathbf{a} \cdot \mathbf{u} = 0$.

5. Make a copy of Figure 5.6 and draw on it the acceleration four-vectors **a** at half-scale. Are these vectors orthogonal to **u**?

6. Consider a particle moving along the x-axis whose velocity as a function of time is

$$\frac{dx}{dt} = \frac{gt}{\sqrt{1 + g^2t^2}},$$

where g is a constant.

(a) Does the particle's speed ever exceed the speed of light?

(b) Calculate the components of the particle's four-velocity.

(c) Express x and t as a function of the proper time along the trajectory.

(d) What are the components of the four-force and the three-force acting on the particle?

7. [C] A particle is moving along the x-axis. It is uniformly accelerated in the sense that the acceleration measured in its instantaneous rest frame is always g, a constant. Find x and t as functions of the proper time τ assuming the particle passes through x_0 at time $t = 0$ with zero velocity. Draw the world line of the particle on a spacetime diagram.

8. [S] A π^0 meson (rest mass 135 MeV) is moving with a speed (magnitude of the three-velocity) $V = c/\sqrt{2}$ in a direction 45° to the x-axis.

(a) Find the components of the four-velocity of the particle.

(b) Find the components of the energy momentum four-vector.

9. [S] In the now-decommissioned Stanford Linear Collider, electrons and positrons were accelerated to energies of approximately 40 GeV in a beam pipe 2 mi long but only a few centimeters in diameter. Steering an electron through such a narrowly defined path over such a distance sounds like a daunting task. But how long is the accelerator in the rest frame of the electron when it has this energy?

10. In the LEP particle accelerator at CERN, electrons and positrons travel in opposite directions around a circular ring approximately 10 km in radius at an energy of 100 GeV apiece.

(a) How close are these particles to moving at the velocity of light?

(b) Electrons and positrons can be stored for 2 h. How many turns will an electron or positron make around the ring in this time?

11. Express the law of addition of parallel velocities in terms of the parameter θ used to describe Lorentz boosts in (4.18). Can you give a geometric interpretation to your result?

12. The 2-mi-long Stanford linear accelerator accelerates electrons to an energy of 40 GeV as measured in the frame of the accelerator. Idealize the acceleration mechanism as a constant electric field $\vec{E}$ along the accelerator and assume that the equation of motion is

$$\frac{d\vec{p}}{dt} = e\vec{E},$$

where $\vec{p}$ is the spatial part of the relativistic momentum $\mathbf{p}$.

(a) Assuming that the electron starts from rest, find its position along the accelerator as a function of time in terms of its rest mass m and $F \equiv e|\vec{E}|$.

(b) What value of $|\vec{E}|$ would be necessary to accelerate the particle to its final energy?

13. [B, S] One reaction for photoproducing pions is

$$\gamma + p \to n + \pi^+.$$

Find the minimum energy (the threshold energy) a photon would have to have to produce a pion in this way in the frame in which the proton is at rest. Is this energy within reach of contemporary accelerators?

14. [B] Compare the energy of the highest energy cosmic rays with the energy of a rock thrown energetically by yourself.

15. [C] A source and detector are spaced a certain angle ϕ apart on the edge of a rotating disk. The source emits radiation at a frequency ω_* in its instantaneous rest frame. What frequency is the radiation detected at? *Hint:* Little information is given in this problem because little is needed.

16. *Aberration* Consider a star, which happens to be directly overhead (the zenith) at midnight in a direction that lies in the plane of the Earth's orbit. To observe the star through a telescope, the telescope axis must be tilted with respect to the zenith direction by a small angle in the direction the Earth is moving in its orbit. Explain why and calculate the angle. To simplify the situation you may assume that the Earth's orbit is approximately circular and, if necessary, that the rotation axis is perpendicular to the orbital plane.

17. [C] *Relativistic Beaming* A body emits photons of frequency ω_* at equal rates in all directions in its rest frame. A detector at rest in this frame a large distance away (compared to the size of the body) receives photons at a rate per unit solid angle $(dN/dtd\Omega)_*$[photons/(s · sr)] that is independent of direction. In an inertial frame (t', x', y', z') in which an observer is at rest the body is moving with speed V along the x'-axis.

(a) Derive (5.75) relating a photon's direction of propagation in the rest frame to the direction of propagation in the observer's frame.

(b) Find the rate at which photons are received per unit solid angle $dN/dt'd\Omega'$ a large distance away in the observer's frame as a function of angle α' from the x'-axis. (*Hint*: Remember that the time interval between the reception of two photons by a stationary observer is not the same as the time interval between their emission if the source is moving.)

(c) Find the luminosity per unit solid angle $dL'/d\Omega'$[erg/(s · sr)] a large distance away as a function of the angle α' in the observer's frame.

(d) Discuss the beaming of number and energy in the observer's frame as the velocity of the source approaches the velocity of light.

18. Work out the frequency as a function of proper time seen by the observer in Example 5.9 by transforming the components of the wave vector of the photons into the instantaneous rest frame of the observer at proper time τ.

19. [S] An observer moves with a constant speed V along the x-axis of an inertial frame. Find the components in that frame of orthonormal basis four-vectors $\{\mathbf{e}_\alpha\}$ to which the observer can refer observations.

20. Consider a particle with four-momentum $\mathbf{p}$ and an observer with four-velocity $\mathbf{u}$. Show that if the particle goes through the observer's laboratory, the magnitude of the three-momentum measured is

$$|\vec{p}| = \left[(\mathbf{p}\cdot\mathbf{u})^2 + (\mathbf{p}\cdot\mathbf{p})\right]^{1/2}.$$

21. [P, A] Assume that in all inertial frames the force on a charged particle is given by the usual Lorentz force law:

$$\vec{F} \equiv \frac{d\vec{p}}{dt} = q(\vec{E} + \vec{V} \times \vec{B}),$$

where q is the charge on the particle, $\vec{V} \equiv d\vec{x}/dt$ is its three-velocity, and $\vec{E}$ and $\vec{B}$ are the electric and magnetic fields as measured in the Lorentz frame. Consider a different inertial frame moving with speed v along the x-axis with respect to the first.

(a) Find the components of the four-force $\mathbf{f}$ in terms of $\vec{E}$ and $\vec{B}$ and the components of the particle's four-velocity $\mathbf{u}$.

(b) Use the transformation law for the components of $\mathbf{f}$ and $\mathbf{u}$ to find the transformation rules that give the electric and magnetic fields in the new inertial frame for the following special fields in the original inertial frame:

(i) An electric field in the x-direction.

(ii) A magnetic field in the x-direction.

(iii) An electric field in the y-direction.

(iv) A magnetic field in the y-direction.

22. [C] *The Relativistic Rocket* A rocket accelerates by ejecting part of its rest mass as exhaust. The speed of the exhaust is a constant value u in the rocket's rest frame. Use the conservation of energy and momentum to find the ratio of final to initial rest mass for a rocket that accelerates from rest to a speed V. *Hint*: Rest mass is not conserved—energy and momentum are conserved. You might want to start by working the same problem in Newtonian mechanics.

23. [C] *Tachyons*

(a) Argue that a kind of particle that always moves *faster* than the velocity of light would be consistent with Lorentz invariance in the sense that if its speed is greater than light in one frame, it will be greater than light in all frames. (Such hypothetical particles are called *tachyons*.)

(b) Show that the tangent vector to the trajectory of a tachyon is spacelike and can be written $u^\alpha = dx^\alpha/ds$, where s is the spacelike interval along the trajectory. Show that $\mathbf{u} \cdot \mathbf{u} = 1$.

(c) Evaluate the components of a tachyons four-velocity $\mathbf{u}$ in terms of the three-velocity $\vec{V} = d\vec{x}/dt$.

(d) Define the four-momentum by $\mathbf{p} = m\mathbf{u}$ and find the relation between energy and momentum for a tachyon.

(e) Show that there is an inertial frame where the energy of any tachyon is negative.

(f) Show that if tacyhons interact with normal particles, a normal particle could emit a tachyon with total energy and three-momentum being conserved.

Comment: The result in (f) suggests that a world containing tachyons would be unstable, and there is no evidence for tachyons in nature.

PART II

The Curved Spacetimes of General Relativity

The idea that gravity *is* the geometry of curved spacetime is introduced. The tools for describing curved spacetimes and the motion of test particles and light rays that probe these curved geometries are developed. The geometries of the exterior of spherical stars, spherical black holes, gravitational waves, and cosmology are explored. The basic tests of general relativity are described.

CHAPTER 6

Gravity as Geometry

With the success of special relativity, it became apparent that the Newtonian theory of gravity, which had been so successfully applied to the mechanics of the solar system for almost 300 years, could no longer be exactly correct. The Newtonian gravitational interaction is instantaneous. The gravitational force $\vec{F}_{12}$ on a mass m_1 at time t due to a second mass m_2 is given in magnitude by [cf. (3.11)]

$$F_{12} = \frac{Gm_1m_2}{|\vec{r}_1(t) - \vec{r}_2(t)|^2}, \tag{6.1}$$

where $\vec{r}_1(t)$ and $\vec{r}_2(t)$ are the positions of the masses at the *same instant of time.* But in special relativity the notion of simultaneity is different in different inertial frames. The Newtonian law (6.1) could be true in only one frame, and it would then single out that frame from all others. The Newtonian law of gravity is thus inconsistent with the principle of relativity.

We will trace out some parts of the path that led Einstein to a new theory of gravity that *is* consistent with the principle of relativity. The result will be general relativity, a theory that is qualitatively different from Newtonian gravity. In general relativity gravitational phenomena arise not from forces and fields, but from the curvature of four-dimensional spacetime. The starting point for these considerations is the equality of gravitational and inertial mass.

6.1 Testing the Equality of Gravitational and Inertial Mass

As discussed briefly in Chapter 2, the equality of gravitational and inertial mass has been tested to extraordinary accuracy. Because of the central importance of this equality for general relativity, it is worthwhile to describe something more of these tests, even if only in a schematic way.

Experiments testing the equality of gravitational and inertial mass seek to compare the accelerations of bodies of different composition falling freely in a gravitational field. The accelerations of the Earth and the Moon falling in the gravitational field of the Sun were compared in the lunar laser ranging experiment described in Box 2.1 on p. 14. The accelerations agree to an accuracy of 1.5×10^{-13}—the most accurate current test to date. Experiments done on the surface of the Earth with torsion pendulums attain a similar accuracy. Such experiments are called Eötvös experiments after R. von Eötvös (1848–1919), who carried out the first modern version. We describe their basic features here.

Imagine two masses of different material at the ends of a rod that is suspended from a fiber in a laboratory on the surface of the Earth, as sketched in Figure 6.1. That is a schematic picture of a torsion pendulum. Because the laboratory is rotating with the Earth, the hanging fiber is not exactly aligned with the local force of gravity. Rather, the fiber hangs at a small angle to that direction so that a small component of the gravitational force can balance the centripetal acceleration arising from the Earth's rotation, as shown in the right-hand member of Figure 6.1 (Problem 1).

The masses are free to move in the direction perpendicular to both the fiber and the rod. Gravity is the only force acting in this "twisting direction" along which the masses are effectively freely falling. Any difference between the acceleration of the two masses would cause the pendulum to twist. Thus, a difference in the equality of their gravitational and inertial masses could be detected.

To understand how this kind of experiment works more quantitatively, denote the two bodies by A and B, their gravitational masses by $m_{A,G}$ and $m_{B,G}$, and

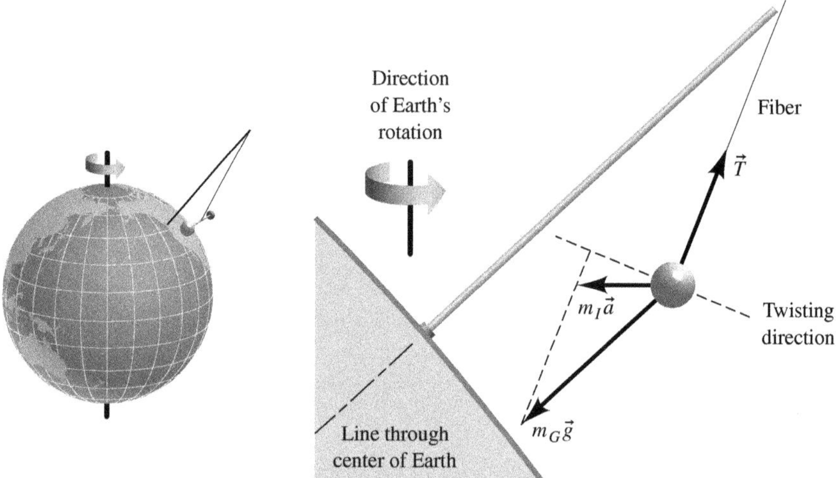

FIGURE 6.1 The left-hand figure is an idealized torsion pendulum for testing the equivalence principle. A rod with two masses of different compositions on its ends is suspended by a fiber from a rigid support so that it is balanced but can twist. The forces acting on a mass in a torsion pendulum are shown at right. This schematic diagram shows a fraction of the Earth's surface together with an end view of the torsion pendulum and the forces acting on one of the masses. The figure's vertical direction is along the Earth's rotation axis off to the left. The mass is rotating with the Earth and therefore has a centripetal acceleration, $\vec{a}$. As a consequence the suspension fiber makes a small angle with respect to the line through the center of the Earth as shown. The force of gravity, $m_G\vec{g}$ and the force from the suspension $\vec{T}$ are also indicated. The dotted line is the "twisting direction" perpendicular to both the balance bar and fiber along which the mass is effectively freely falling. The component of the gravitational force $m_G\vec{g}$ must equal the component of $m_I\vec{a}$ along this direction if the pendulum is not to twist in the frame of the Earth. That can happen only if m_I/m_G is the same for both masses. Small differences in the ratio can, therefore, be detected by the twisting of the balance.

their inertial masses by $m_{A,I}$ and $m_{B,I}$, respectively. Assume that the gravitational field $\vec{g}$, which gives the gravitational force $m_G\vec{g}$ on a mass [cf. (3.31)], is constant over the dimensions of the pendulum. Sources such as the Earth, Sun, and Milky Way will satisfy this easily, but smaller sources closer to the experiment are a significant problem. Denote the component of $\vec{g}$ in the twisting direction by g^t and the components of the accelerations in this direction by a^t_A and a^t_B. Then,

$$m_{A,I}a^t_A = m_{A,G}g^t, \tag{6.2a}$$

$$m_{B,I}a^t_B = m_{B,G}g^t. \tag{6.2b}$$

If the ratio of gravitational to inertial mass is the same for all bodies, the pendulum can be at rest with both bodies having the same centripetal acceleration due to the rotation of the Earth. Any difference of the ratio of gravitational to inertial mass between bodies of different composition would show up as a difference in their accelerations and a twist in the pendulum. From (6.2) the difference in the accelerations as a fraction of their average is

$$\frac{a^t_A - a^t_B}{\frac{1}{2}(a^t_A + a^t_B)} = \frac{\left(\dfrac{m_{A,G}}{m_{A,I}} - \dfrac{m_{B,G}}{m_{B,I}}\right)}{\dfrac{1}{2}\left(\dfrac{m_{A,G}}{m_{A,I}} + \dfrac{m_{B,G}}{m_{B,I}}\right)} \equiv \eta. \tag{6.3}$$

An upper limit on the twist of the pendulum gives an upper limit on η and an upper limit on deviations from equality of gravitational and inertial mass.

The preceding discussion is little more than a cartoon idealization of the actual modern experiments that have been carried out by Roll, Krotkov, and Dicke (1964), Braginsky and Panov (1971), and Su et al. (1994). The pendulum used in the latter experiment is shown in Figure 6.2. A few features can be mentioned. First, four masses, rather than two, are used. This is to minimize the effect of gradients in $\vec{g}$ across the pendulum that would lead to a torque on the pendulum even if gravitational and inertial mass were equal. Clever design is needed to shield the pendulum from such gra-dients and from magnetic, thermal, and other sources of noise. However, the key to achieving great experimental accuracy is to rotate the pendulum slowly with a known period. In the frame of the pendulum the magnitude and sign of a twist-ing torque arising from a difference in gravitational and inertial mass would vary harmonically with precisely this period. By focusing on the Fourier component of the angular position of the pendulum with this period, the signal measuring any deviations from the equality of gravitational and inertial mass can be separated from noise with high accuracy. The result, for example, of the experiments of Su et al. (1994) using masses of beryllium and copper for the quantity η defined in (6.3) is

$$\eta = (-0.2 \pm 2.8) \times 10^{-12}. \tag{6.4}$$

The equality of gravitational and inertial mass is one of the most accurately tested principles in all physics.

FIGURE 6.2 Torsion pendulum used in the experiment of Su, et al. (1994) to test the equality of accelerations for test bodies attracted by the Earth, the Sun, and the matter in our galaxy. The pendulum is small (its overall diameter is about 3 in.) to minimize disturbing effects from local variations in the gravitational force. It hangs from a tungsten fiber, which is so thin that it cannot be seen in this photograph. The circular plate holds four cylindrical test bodies (two of copper and two of beryllium) along with four right-angle mirrors that are part of a sensitive optical system for detecting pendulum twists. The pendulum is suspended in a vacuum and the entire instrument is rotated continuously at about one revolution per hour. A violation of the equality of gravitational and inertial mass would show up as a pendulum twist that varied at this rotation frequency.

6.2 The Equivalence Principle

"There then occurred to me the 'glückischste Gedanke meines Lebens,' the happiest thought of my life, in the following form. The gravitational field has only a relative existence.... *Because for an observer falling freely from the roof of a house there exists*—at least in his immediate surroundings—*no gravitational field.* Indeed, if the observer drops some bodies then these remain relative to him in a state of rest or uniform motion, independent of their particular chemical or physical nature (in this consideration air resistance is, of course, ignored). The observer has the right to interpret his state as 'at rest.'"[1] Thus, Einstein later recalled the origin of his *equivalence principle*, which led him to the discovery of general relativity. Today the equivalence principle is regarded as a heuristic idea whose central content is incorporated automatically and precisely in general relativity where appropriate. However, the idea remains a useful starting point for motivating general relativity, and it is for this purpose that it is described here. The discussion is entirely in the context of Newtonian gravity, where the idea of a gravitational field makes sense. The equivalence principle in general relativity is discussed in Section 7.4.

The modern version of Einstein's observer falling from the roof might be astronauts freely falling around the Earth in the space shuttle (see Figure 6.3). The astronauts are "weightless." Cups, saucers, cannon balls, feathers, and any other objects moving freely within the shuttle remain at rest or in uniform motion with respect to them (neglecting air resistance etc., as Einstein did). From a study of the motion of such objects over a short period of time the astronauts cannot tell whether they are falling freely in the gravitational field of the Earth or are at rest in empty space far from any source of gravitation. In effect, the gravitational field has vanished in the freely falling frame of the space shuttle.

The equality of gravitational and inertial mass is essential to reach this conclusion. If a cannonball and feather fell toward the Earth with *different accelerations*, they would not remain at rest or in uniform motion with respect to each other in

FIGURE 6.3 Astronauts freely falling around the Earth in the space shuttle are weightless and cannot detect the gravitational field of the Earth by experiments done inside the shuttle over a short period of time.

[1] As quoted in Pais (1982), p. 178.

the shuttle's interior. The detection of a small difference in acceleration would suffice to distinguish the presence of a gravitational attraction.

The equality of gravitational and inertial mass not only implies that a gravitational field can be eliminated by falling freely, but also that one can be created by acceleration. Consider an experimenter in a small, closed laboratory at rest on the surface of the Moon or other source of gravitational force as illustrated in Figure 6.4. The laboratory is small enough that the gravitational field is uniform to any accuracy the experimenter can test. The experimenter can carry out experiments with various objects. For example, if a cannonball and a feather are dropped, they will fall to the floor of the laboratory with the same acceleration—the local acceleration of gravity g—because of the equality of gravitational and inertial mass.

Consider the same laboratory in empty space, far from any source of gravitational force, not at rest, but accelerated upward with an acceleration g, as

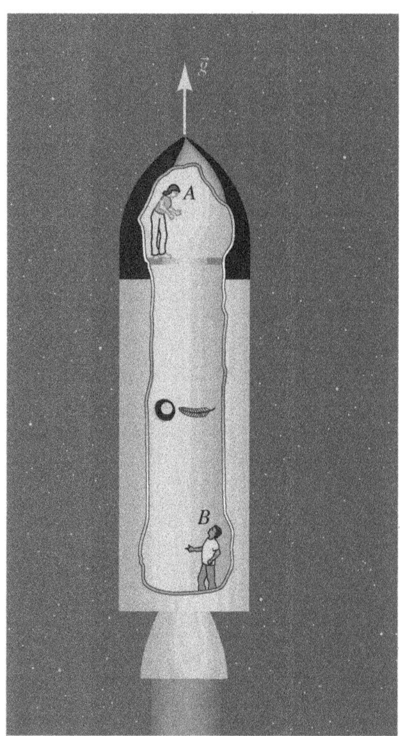

FIGURE 6.4 The equivalence of a uniform acceleration and a uniform gravitational field. On the left is a laboratory at rest on the surface of the Earth. An observer inside lets go of a cannonball and feather. If the gravitational and inertial masses are equal, both fall to the floor with an acceleration g. On the right is a closed laboratory deep in space, far from any sources of gravitational force. The laboratory is being accelerated upward with an acceleration g. An observer inside the laboratory lets go of a cannonball and feather at the same time. Both drop to the floor with acceleration g. An observer inside a closed laboratory cannot distinguish whether they are in one situation or the other.

also illustrated in Figure 6.4. An experimenter inside who drops a cannonball and a feather will observe that they fall to the floor of the laboratory with equal accelerations—the same result as for the laboratory at rest in a gravitational field. By this, or any other mechanical experiment with particles, the experimenter inside cannot tell whether the laboratory is unaccelerated in a uniform gravitational field or accelerated in empty space. The two laboratories are equivalent as far as these experiments are concerned.

The absence of local experiments that distinguish between uniform acceleration and uniform gravitational fields follows immediately from the equality of gravitational and inertial mass as long as those experiments concern the motion of bodies such as cannonballs and feathers. But what about experiments with photons or neutrinos? What about electromagnetic fields or the fields of quantum chromodynamics? Could the two laboratories be distinguished by these effects? Einstein's equivalence principle is the idea that *there is no experiment that can distinguish a uniform acceleration from a uniform gravitational field*. The two are fully "equivalent."

The power of the equivalence principle derives from its assertion that it applies to *all* laws of physics. As an example, if we accept the equivalence principle, we must also accept that light falls in a gravitational field with the same acceleration as material bodies. It is not obvious otherwise how to calculate the effect of gravity on light. There is no Newtonian equation of motion—no $F = m_I a$. Here is how the equivalence principle forces one to the conclusion that light falls in a gravitational field.

In empty space, a light ray will move on a straight line in an inertial frame. Suppose a light ray is observed from a laboratory accelerating transversely to its direction of propagation (Figure 6.5). In the laboratory frame, the light ray will exit at a position below where it entered because the laboratory has accelerated upward in the time the light ray crosses. Thus, in the laboratory frame the light ray will accelerate downward with the acceleration of the laboratory. From the equivalence principle, one can deduce that the same behavior occurs in a uniform gravitational field; i.e., the light ray accelerates downward with the local acceleration of gravity.

6.3 Clocks in a Gravitational Field

Consider the thought experiment illustrated in Figure 6.6. Observer Alice is located a height h above observer Bob in a uniform gravitational field where bodies fall with acceleration g. Two observers at the top and bottom of a tower on the surface of the Earth are in this situation to a good approximation (See, for example, Box 6.1 on p. 118.) Or, for the purposes of the following discussion, we can imagine Alice to be in the nose and Bob to be in the tail of a rocket ship of length h at rest on the surface of the Earth, as shown in Figure 6.6. Alice emits light signals at equal intervals $\Delta\tau_A$ as measured on a clock[2] located at the same

[2]For example, an atomic clock where the unit of time is a defined number of cycles in the transition between the two lowest energy states in a cesium atom. See Appendix A.

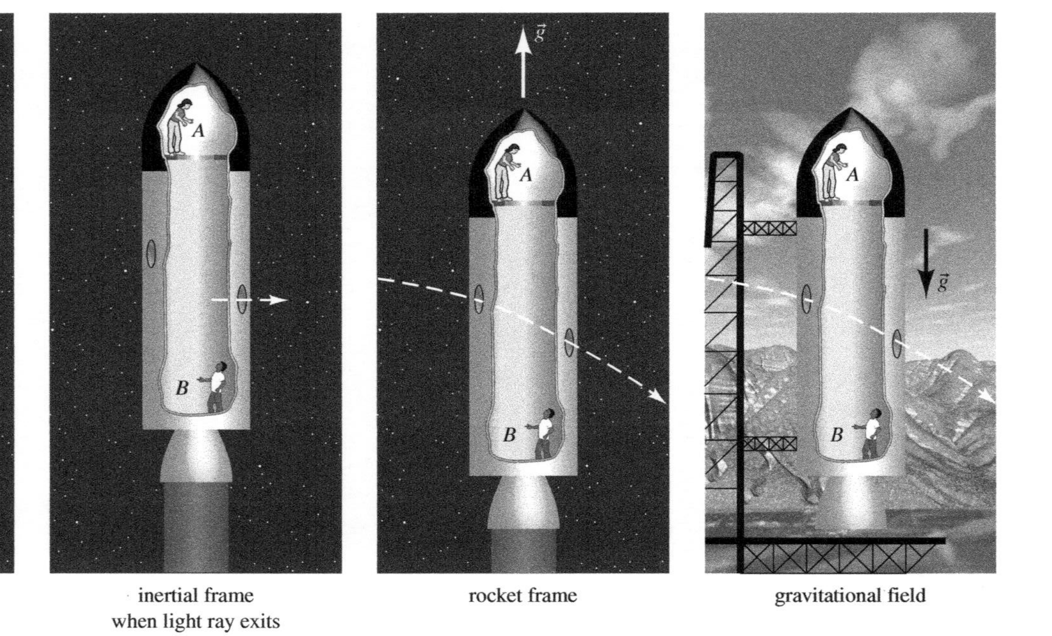

FIGURE 6.5 A light ray traverses a laboratory accelerating upward in empty space. The first two pictures are views from the inertial frame. The path of the light ray is straight. However, the laboratory moves upward in the time the light ray takes to cross. The exit point of the light ray is, therefore, lower in the laboratory than the entry point. Thus, for an observer in the accelerating laboratory the light ray falls with an acceleration g. The equivalence principle implies that the same observation is made in a laboratory at rest in a uniform gravitational field. A light ray in a gravitational field must fall with the same acceleration as other objects. Gravity attracts light.

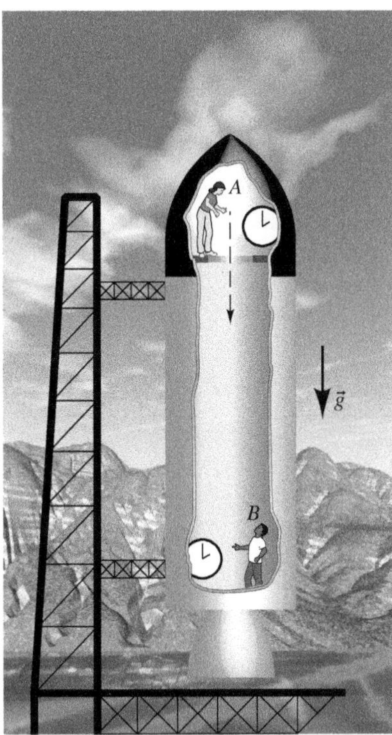

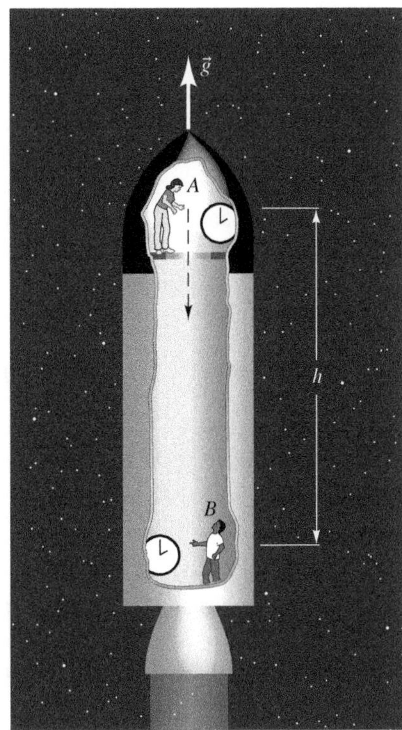

FIGURE 6.6 On the left is a rocket at rest in a uniform gravitational field. Alice, in the nose of the rocket, emits signals at equal intervals on a clock at her height. Bob, in the tail, measures the time interval between receipt of the signals on an identical clock at his location. The equivalence principle implies that the relation between the intervals of emission and reception must be the same as if the rocket ship were accelerating vertically upward far from any source of gravitational attraction, as shown at right. These signals are received at shorter intervals than they are emitted because the accelerating tail is catching up with the signals. The equivalence principle implies that in the gravitational field, the signals are received at a faster rate than they are emitted.

height. At what intervals $\Delta\tau_B$ does Bob receive the signals as measured by an identically constructed clock at his height?

The equivalence principle implies that Bob receives the signals at a faster rate than they are emitted. To see this, imagine that Alice and Bob are in a rocket ship in empty space, far from any source of gravitation, and accelerating with acceleration g. Because of the acceleration Bob catches up with the signals faster and faster and thus receives them at a faster rate than they were emitted. The equivalence principle implies that the same relationship between rates will be observed in the rocket at rest in a uniform gravitational field.

To get a quantitative result for this effect, analyze the accelerating rocket ship in an inertial frame in which, over the time of interest, $(V/c)^2$ is negligible and $(gh/c^2)^2$ is negligible, but in which V/c and gh/c^2 may be important. (For this analysis and the rest of the chapter $c \neq 1$ units will be used.) These two conditions are not essential but greatly simplify the analysis while getting at the central result.[3] When $(V/c)^2$ is negligible, Newtonian mechanics can be used and Lorentz contraction and time dilation neglected. Also, since we will just be comparing time intervals, issues of simultaneity can be neglected. Assuming that $(gh/c^2)^2$ is negligible means that the rocket does not accelerate to relativistic velocities in the time it takes for light to travel from nose to tail.[4] With these assumptions the Newtonian mechanics is adequate for the analysis, and the result for the difference in rates of the clocks will be correct to leading order in gh/c^2.

Suppose the rocket is accelerating along the z-axis. Bob's position in the tail of the rocket is given as a function of time by

$$z_B(t) = \tfrac{1}{2}gt^2 \tag{6.5}$$

if the origin of z is chosen to coincide with Bob's position at $t = 0$. Alice's position in the nose of the rocket is given by

$$z_A(t) = h + \tfrac{1}{2}gt^2. \tag{6.6}$$

Consider the emission of two successive light pulses by Alice and their reception by Bob. Suppose that $t = 0$ is the time the first pulse is emitted, t_1 is the time it is received, $\Delta\tau_A$ is the time the second pulse is emitted, and $t_1 + \Delta\tau_B$ is the time the second pulse is received. The sequence of events is illustrated in Figure 6.7. The distance traveled by the first pulse is

$$z_A(0) - z_B(t_1) = ct_1. \tag{6.7}$$

The distance traveled by the second pulse is shorter and given by

$$z_A(\Delta\tau_A) - z_B(t_1 + \Delta\tau_B) = c(t_1 + \Delta\tau_B - \Delta\tau_A). \tag{6.8}$$

Inserting (6.5) and (6.6) and assuming $\Delta\tau_A$ is *small* so that only *linear* terms in $\Delta\tau_A$ and $\Delta\tau_B$ need be kept, one finds

$$h - \tfrac{1}{2}g{t_1}^2 = ct_1, \tag{6.9a}$$

$$h - \tfrac{1}{2}g{t_1}^2 - gt_1\Delta\tau_B = c(t_1 + \Delta\tau_B - \Delta\tau_A). \tag{6.9b}$$

Subtract (6.9b) from (6.9a) and use (6.9a) again to eliminate t_1. According to the ground rules announced at the start of the calculation, terms such as $(gh/c^2)^2$ can be neglected and only a first approximation to t_1 is needed, namely, $t_1 = h/c$. The result is

$$\Delta\tau_B = \Delta\tau_A\left(1 - \frac{gh}{c^2}\right). \tag{6.10}$$

[3]For a full analysis in special relativity, you can work through Problems 6 and 7 at the end of this chapter.

[4]The same condition allows the neglect of the small differences in acceleration of order $g(gh/c^2)$ between the nose and the tail that are necessary in special relativity for the rocket to accelerate rigidly. See Problems 6 and 7.

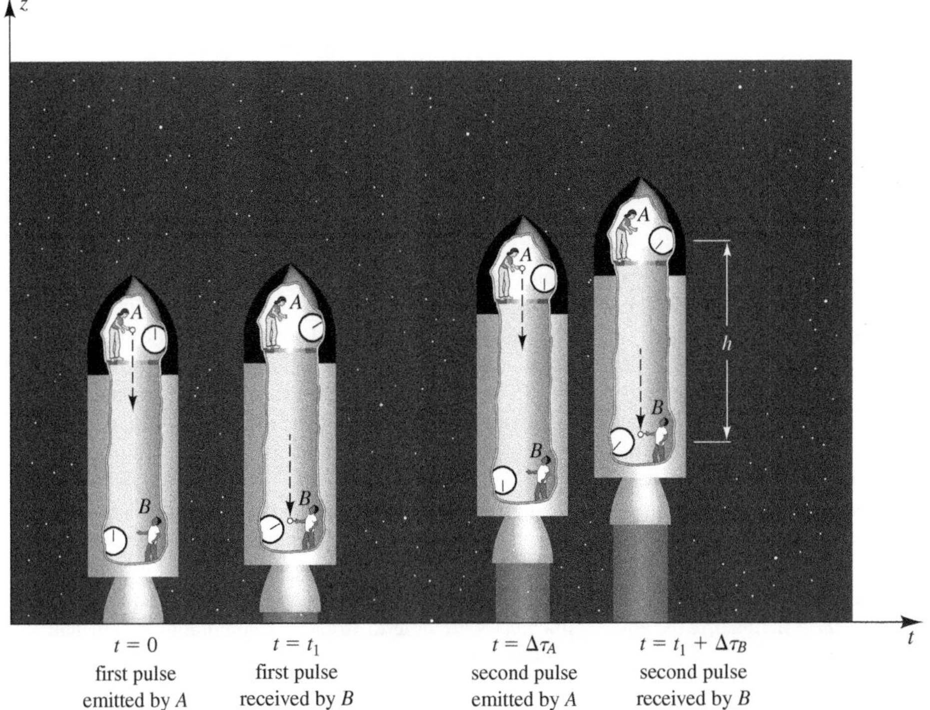

FIGURE 6.7 Alice and Bob are in a rocket accelerating upward in empty space. Alice, in the nose, emits signals at equal intervals on a clock there. The acceleration means that Bob, in the tail, measures a smaller interval between the received signals as discussed in the text. The figure shows the position of the rocket for the emission and reception of two successive signals for the calculation of the quantitative connection between the rates of emission and reception in the text.

The interval at which the pulses are received is smaller by a factor of $(1 - gh/c^2)$ than the interval at which they are emitted.

The equivalence principle tells us that the same effect must occur in a uniform gravitational field (Figure 6.6). Since the *rates* of emission and reception are just $1/\Delta\tau_A$ and $1/\Delta\tau_B$, respectively, and since gh is the gravitational potential difference between A and B,

$$\Phi_A - \Phi_B = gh, \tag{6.11}$$

(6.10) can be expressed in terms of rates as

$$\begin{pmatrix}\text{rate signals}\\ \text{received at } B\end{pmatrix} = \left(1 + \frac{\Phi_A - \Phi_B}{c^2}\right)\begin{pmatrix}\text{rate signals}\\ \text{emitted at A}\end{pmatrix} \tag{6.12}$$

Rates of Emission and Reception in a Gravitational Field

BOX 6.1 A Test of the Gravitational Redshift

The first accurate tests of the prediction of the gravitational redshift by Pound and Rebka (1960) and Pound and Snider (1964) were a realization of the thought experiment in the text in which Alice and Bob compare the rates of emission and reception of signals in a uniform gravitational field. The test used the 22.5-m-high tower of the Jefferson Physical Laboratory at Harvard University. For signals it employed the 14.4-keV gamma rays emitted by the unstable nucleus Fe^{57} when it decays. The frequency of the gamma rays, ω, related to their energy E by $E = \hbar\omega$, can be thought of as the rate of emission at A in (6.12). The gamma rays fell to the bottom of the tower where a similar sample of Fe^{57} acted as a receiver. If a gamma ray's frequency were still that of the emitter, it would be detected through the inverse of the reaction by which it was emitted. But, (6.11) predicts that their frequency should be larger—blueshifted—by a fractional amount $gh/c^2 \sim 10^{-15}$ for $h = 22.5$ m, making the absorption less efficient. By varying the vertical velocity of the source at the top of the tower, the experimenters could produce a Doppler redshift that would compensate for the gravitational blueshift. The velocity that gave maximum absorption was then a direct measure of the gravitational blueshift.

That's a cartoon version of the experiment; the reality was more challenging. The decay of a nucleus does not always produce an exactly 14.4-keV gamma ray. That's just the average over a span of energies called the line width. Futher, when a nucleus emits a gamma ray, it is typically moving with some velocity inside the sample, and in the decay the residual nucleus recoils in an uncontrollable way. Both of these effects lead to a spread in emitted frequency of the gamma ray, which, in ordinary circumstances, would dwarf the tiny gravitational shift. However, by utilizing the then-recently discovered Mössbauer effect, the nuclei could be effectively locked into a crystal lattice, making their recoil velocities much smaller but still leaving a line width about 1000 times greater than the predicted gravitational frequency shift. By filtering the amount of absorption at the frequency of the imposed variation in velocity of the source, the experimenters were able to isolate the gravitational shift and confirm the prediction of (6.12) to an accuracy of about 1%. A more accurate experiment is described in Chapter 10.

to the $1/c^2$ accuracy that (6.10) is valid. In this form the relation holds whatever the relative sizes of Φ_A and Φ_B. When the receiver is at a higher gravitational potential than the emitter, the signals will be received more slowly than they were emitted. When the receiver is at a lower potential than the emitter, the signals will be received more quickly. An experiment confirming this prediction is described in Box 6.1.

Example 6.1. Theorists Age More Quickly at UCSB. These effects are extraordinarily small in ordinary laboratory circumstances. At the author's institution the theorists occupy the top floor of the physics building. The heart of a theorist is a kind of clock. As measured by a clock on the ground floor, a heart will beat more times on the top floor in a given interval of time than the heart of a similar physicist on the ground floor by a factor of $(1 + gh/c^2)$, where $h \approx 30$ m. This is only

$$1 + \frac{(9.8 \text{ m/s}^2)(30 \text{ m})}{(3 \times 10^8 \text{ m/s})^2}, \tag{6.13}$$

which differs from unity only by a few parts in 10^{15}! The theorists whose offices are at the top of the building are older by a few microseconds in 100 yr—a small price to pay for a view.

It seems natural to suppose that result (6.12), although derived for uniform gravitational fields, holds for nonuniform ones as well. This extension would mean that (6.12) holds when $\Phi_A = \Phi(\vec{x}_A)$ and $\Phi_B = \Phi(\vec{x}_B)$. The test of this extension, like the equivalence principle itself, ultimately rests in experiment. This extension leads to the gravitational redshift derived in Example 6.2 below, and to a practical application to the Global Positioning System described in the next section.

Example 6.2. The Gravitational Redshift. The crests of a light wave of definite frequency can be thought of as a series of signals emitted at a rate that is the frequency of the wave. Relation (6.12) between the rates of emission and reception can, therefore, be applied to light. For example, light emitted from the surface of a star with frequency ω_* will arrive at a receiver far from the star with a frequency ω_∞, which is less than ω_*. That is the gravitational redshift. The gravitational potential at the surface of a star of mass M and radius R is $\Phi = -GM/R$; the gravitational potential far away is zero. Equation (6.12) becomes

$$\omega_\infty = \left(1 - \frac{GM}{Rc^2}\right)\omega_* . \tag{6.14}$$

This expression is accurate for small values of GM/Rc^2; the general relation is derived in Section 9.2. The gravitational redshift has been detected in the spectra of white dwarf stars where $M \sim M_\odot$ and $R \sim 10^3$ km and the fractional change in frequency is only $\sim 10^{-3}$.

When the gravitational field is nonuniform the equivalence principle holds only for experiments in laboratories that are small enough and that take place over a short enough period of time that no nonuniformities the gravitational field can be detected.[5]

> **Equivalence Principle**
>
> Experiments in a sufficiently small freely falling laboratory, over a sufficiently short time, give results that are indistinguishable from those of the same experiments in an inertial frame in empty space.

[5] Does the equivalence principle sound mathematically imprecise to you? It is. Principles like this and the principle of relativity that make statements about the laws of physics in advance of their mathematical formulation are generally so. That does not mean they have no content. See the remarks on the principle of relativity on p. 36.

In this form the equivalence principle will also have a meaning in general relativity, as we'll see in Section 7.4. Example 6.3 shows more clearly than any abstract argument how small the laboratory has to be and how short a duration of the experiments is required.

As mentioned in Section 6.2, the equivalence principle in this form doesn't have to hold for a consistent theory of gravitation. But it does hold for many phenomena and is a useful guide for guessing how to generalize known flat spacetime laws to curved spacetime. We'll see examples later.

Example 6.3. Detecting the Earth's Gravitational Field Inside the Space Shuttle. Even inside the freely falling laboratory of the space shuttle there is enough room to detect the gravitational field of the Earth with experiments carried out over a sufficiently long time. Suppose the astronauts release two ping-pong balls at rest in the instantaneous rest frame of the space shuttle and observe the subsequent separation of the two balls. Were the space shuttle in empty space, far from any source of gravitational attraction, the distance between the balls would not change (assuming ideal circumstances, neglecting air resistance, electrostatic forces, mutual gravitational attraction, etc.) However, in the Earth's gravitational field, the ball nearer the Earth will have a slightly greater acceleration toward the Earth's center than the one further away. The distance between them will therefore change, and by measuring this change the astronauts can detect the gravitational field of the Earth.

To estimate the time for a significant change in separation, analyze the balls' motion in an inertial frame in which the center of Earth is at rest (neglecting its motion around the Sun.) For a discussion of their relative motion, the motion of the shuttle itself is irrelevant—the balls are freely falling and we are concerned only with the separation between them. For simplicity suppose that initially the balls are a distance s apart along a radial line from the center of the Earth, as shown in Figure 6.8, and that the nearer ball is released with the right velocity $\vec{V}$ to execute a circular orbit around the Earth of radius R. The farther ball, released with the same velocity $\vec{V}$, will execute a slightly elliptical orbit (Problem 5). The acceleration of the nearer ball toward the Earth's center is $V^2/R = GM_\oplus/R^2 \equiv g$. The farther ball's acceleration is $GM_\oplus/(R+s)^2$. The difference—the relative acceleration a_{rel}—is initially $2g(s/R)$ for $s \ll R$. In a time δt the distance between the balls will change by an amount $\delta s \sim (1/2)(a_{\text{rel}}\delta t^2)$. This can be expressed in terms of the period of the orbit $P = 2\pi R/V$, where $V^2/R = g$, to find the rough result

$$(\delta s/s) \sim (2\pi\delta t/P)^2. \tag{6.15}$$

Thus, very quickly, on the time scale of one orbit, the astronauts will find a significant change in the distance between the balls and detect that they are close to a source of gravitation. However, for a fixed accuracy δs with which positions can be measured, the effect becomes harder and harder to detect the smaller the laboratory (and hence s) and the shorter the times over which experiments can be

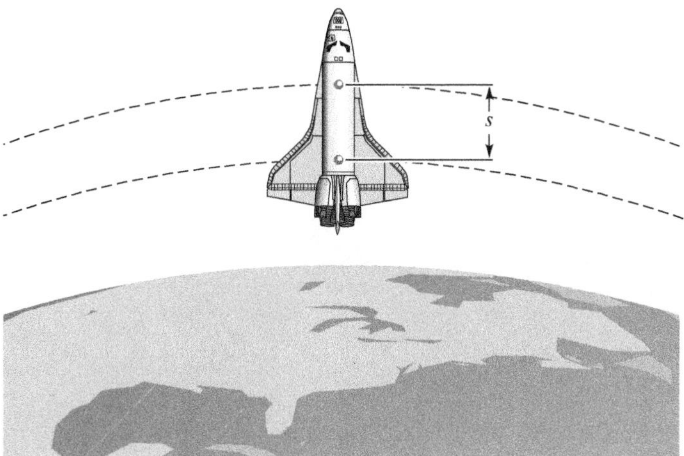

FIGURE 6.8 Astronauts in the space shuttle release two ping-pong balls separated by a distance s along a radial line through the Earth's center. The balls are released with equal velocities as measured in an inertial frame in which the center of the Earth is approximately at rest. In an idealized situation the balls will fall freely around the Earth. However, they execute different orbits, and by measuring the change in their relative separation, the astronauts can detect the presence of a gravitational field in a fraction of an orbit.

carried out. In Chapter 21 we will use the idea of this experiment to find a local measure of the curvature of spacetime.

6.4 The Global Positioning System

The difference (6.12) between rates at which signals are emitted and received at two locations with different gravitational potentials is minute in laboratory circumstances, as (6.13) shows. Yet taking these differences into account is crucial for the operation of the Global Positioning System (GPS) used every day. If the relativistic effects of time dilation discussed in Section 4.4 and the gravitational effects of the present chapter were not properly taken into account, the system would fail after only a fraction of an hour.

The GPS consists of a constellation of 24 satellites, each in a 12-h orbit about the Earth in a total of six orbital planes (see Figure 6.9). Each satellite carries accurate atomic clocks that keep proper time on a satellite to accuracies of a few parts in 10^{13} over a few weeks. Corrections uploaded several times a day from the ground enable accurate time to be kept over longer periods. The details of the operation of the system are complex,[6] but the basic idea is easily explained in an idealization of the real situation.[7]

[6] See, for example, the nearly 800 pages of detail in Parkinson and Spilker (1996).

[7] Another toy model in one dimension related to the GPS but including only the effects of special relativity was discussed in Example 4.4 on p. 68.

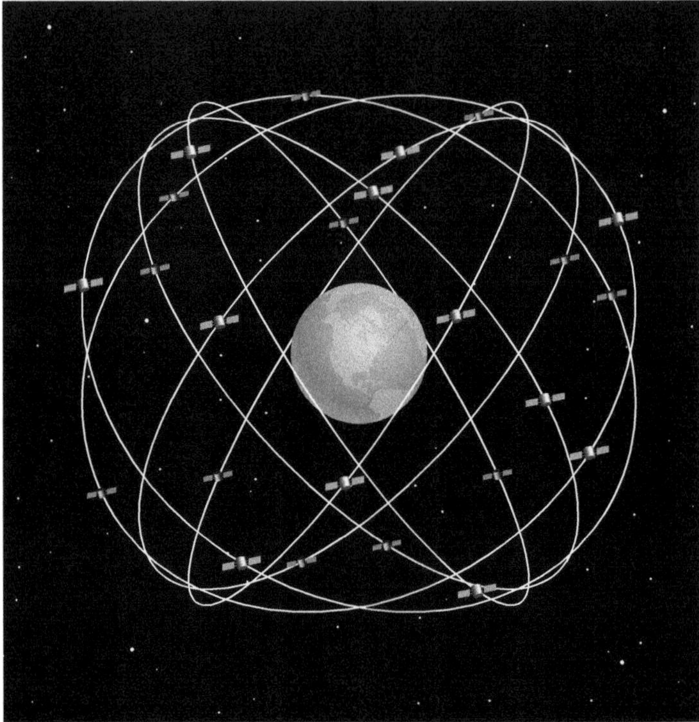

FIGURE 6.9 GPS satellite constellation: The GPS constellation of 24 satellites is arranged in 6 equally spaced orbit planes.

Imagine an inertial frame in which the center of the Earth is approximately at rest for the time it takes a signal to propagate from a satellite to the ground. Periodically each satellite sends out microwave signals encoded with the time and spatial location of emission in the coordinates of this inertial frame. An observer that receives a signal an interval of time later can calculate his or her distance from the satellite by multiplying that time interval by the speed of light c (see Figure 6.10). By using the signals from three satellites the observer's position in space can be narrowed down to the possible intersection points of three spheres. By using four satellites, the observer's position in both space and time can be fixed, even without the observer possessing an accurate clock, giving a complete location in spacetime as illustrated in Figure 6.11. Signals from further satellites reduce any uncertainty further.

Proper time on the satellite clocks has to be corrected to give the time of the inertial frame for at least two reasons: time dilation of special relativity and the effects of the Earth's gravitational field discussed in this chapter. To understand this, suppose a GPS satellite emits signals at a constant rate as measured by its clock. Suppose further that these are monitored by a distant observer at rest in the inertial frame. A clock of this observer, at rest and far from any source of gravitational effects, measures the time of the inertial frame. The signals will be

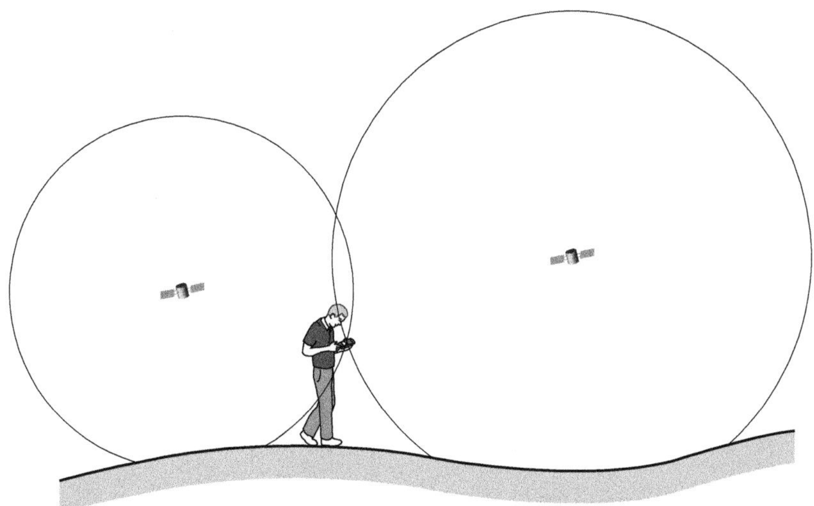

FIGURE 6.10 A GPS satellite emits a signal encoded with its time of emission, t_e, and the location of the satellite. An observer who receives the signal at a time t_r that is an interval $\Delta t = t_r - t_e$ later knows that he or she is located somewhere on a sphere of radius $c\Delta t$ centered on the satellite. Signals from two satellites narrow the location down to the intersection of two spheres.

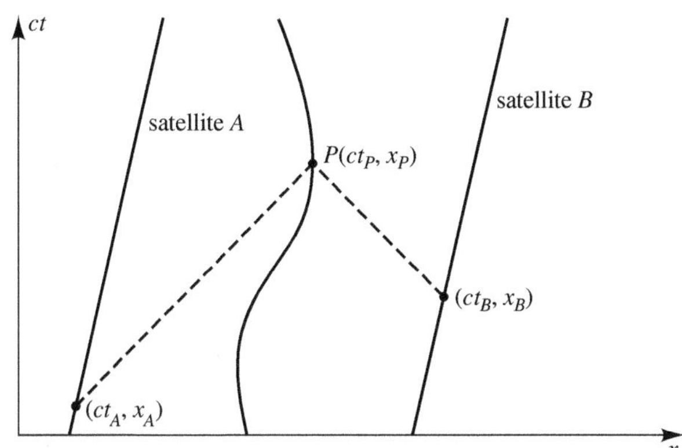

FIGURE 6.11 In one space dimension the signals from just two satellites are sufficient to locate a point P in spacetime where they are received simultaneously. The figure shows the world lines of two satellites in an inertial frame, each sending signals encoded with the coordinates (ct, x) of their emission. These signals move at the speed of light along the 45° lines shown in the diagram. If signals from (ct_A, x_A) and (ct_B, x_B) are received simultaneously at P then the coordinates of P are given by

$$ct_P = \frac{1}{2}\left[c\left(t_A + t_B\right) + \left(x_B - x_A\right)\right],$$
$$x_P = \frac{1}{2}\left[c\left(t_B - t_A\right) + \left(x_B + x_A\right)\right].$$

In a four-dimensional spacetime, a spacetime point can be similarly located with the signals from four satellites.

received at a slower rate than they were emitted. Time dilation of the moving satellite clock is one reason. But another is the difference between the rates of emission and reception (6.12) because the satellite is lower in the gravitational potential of the Earth than the distant observer. Two corrections must therefore be applied to the rate of satellite time to get the time in the inertial frame.

To estimate the magnitude of these corrections, suppose for simplicity that a GPS satellite is in a 12-h circular equatorial orbit of radius R_s from the Earth's center. The parameters of the orbit can all be calculated from Newtonian mechanics to an accuracy sufficient to estimate the magnitude of the special relativistic and gravitational effects. Thus, the satellite's speed, V_s , in the inertial frame is determined by the relation between velocity and period, $V_s = 2\pi R_s/(12 \text{ hr})$, and

$$\frac{V_s^2}{R_s} = \frac{GM_\oplus}{R_s^2}. \tag{6.16}$$

A little calculation from data in the endpapers yields

$$R_s \approx 2.7 \times 10^4 \text{ km} \approx 4.2 R_\oplus, \tag{6.17a}$$

$$V_s \approx 3.9 \text{ km/s}, \qquad V_s/c \approx 1.3 \times 10^{-5}, \tag{6.17b}$$

where $R_\oplus = 6.4 \times 10^3$ km is the radius of the Earth.

With these basic parameters we can estimate the upward corrections to the rate of the satellite clock necessary for it to keep the time of the inertial frame. We write the factor by which the rate must be multiplied as 1 plus a fractional correction. From (4.15), the fractional correction needed to compensate for time dilation is

$$\begin{pmatrix} \text{fractional correction in} \\ \text{rate for time dilation} \end{pmatrix} \approx \frac{1}{2}\left(\frac{V_s}{c}\right)^2 \approx .84 \times 10^{-10} \tag{6.18}$$

to leading order in $1/c^2$. From (6.12), the fractional correction to the rate to compensate for the effect of the gravitational potential is to leading order in $1/c^2$

$$\begin{pmatrix} \text{fractional correction in rate} \\ \text{for the gravitational potential} \end{pmatrix} \approx \frac{GM_\oplus}{R_s c^2} \approx 1.6 \times 10^{-10} \tag{6.19}$$

for the parameters in (6.17). The gravitational correction is bigger than the correction for time dilation.

These corrections are tiny by everyday standards, but a nanosecond is a significant time in GPS operation. A signal from a satellite travels 30 cm in a nanosecond. To meet the announced 2-m accuracy for the military applications of the GPS, times and time differences must be known to accuracies of approximately 6 ns. Keeping time to that accuracy is not a problem for contemporary atomic clocks, but at these accuracies, both time dilation and the gravitational redshift become important for GPS operation. Were they not accounted for, it would take less than a minute to accumulate an error which exceeds the few nanosecond accuracy required. The GPS is a practical application of both special and general relativity.

The actual GPS does not employ an inertial frame whose time is defined by clocks at infinity; rather it uses a frame rotating with the Earth whose time is defined by clocks on its surface. The rates of the satellite clocks must be corrected downward to keep the time of that frame (Problem 14). Further corrections are needed for the relativistic Doppler effect, the relativity of simultaneity (see Example 4.4), the Earth's rotation, the asphericity of the Earth's gravitational potential, the time delays from the index of refraction of the Earth's ionosphere, satellite clock errors, etc.

6.5 Spacetime Is Curved

What is the explanation of the difference between the rates at which signals are emitted and received at two different gravitational potentials?

One explanation is that gravity affects the rates at which clocks run. This would go as follows: in the absence of any gravitational field, two clocks at rest in an inertial frame of flat spacetime both keep track of the time of that frame. In the presence of a gravitational field, spacetime remains flat, but clocks run at a rate that is a factor $(1 + \Phi/c^2)$ different from their rates in empty spacetime, where Φ is the gravitational potential at the location of the clock. Clocks run faster where Φ is positive and slower where Φ is negative. All clocks are affected in exactly the same way. Clocks higher up in a gravitational potential run faster than clocks lower down, and this explains the difference between the rates of emission and reception in (6.12). The discussion of GPS operation in the previous section implicitly took this point of view.

This kind of explanation is not so very different than one that might be proposed by someone who believes that the surface of the Earth is flat and only appears to be curved. The surface is really flat, but as one moves further north the rulers by which distances are measured all become longer. The fact, long known to airline pilots, that the distance between Paris and Montreal appears shorter than the distance between Lagos and Bogota is explained by saying that the true distance is the same, but because rulers in the north are longer, the distance appears to be shorter (see Figure 6.12). A complete theory of this is worked out, including a special "field" that changes the lengths of rulers. For consistency it is soon found that this field must affect *all* lengths in the same way so that the more northerly distances *always* come out shorter. The field has to make not only rulers longer, but also airplanes, pilots, and passengers longer in their east-west directions. Furthermore, it has to change the fundamental atomic constants in such a way that there are fewer air molecules encountered and less fuel used in traveling between Paris and Montreal than between Lagos and Bogota.

The flat spacetime explanation of the time intervals measured by clocks in a gravitational field and the flat-earth explanation of the distances measured by rulers on Earth have one thing in common: they both posit an underlying geometry which is impossible to measure directly because all measuring instruments are

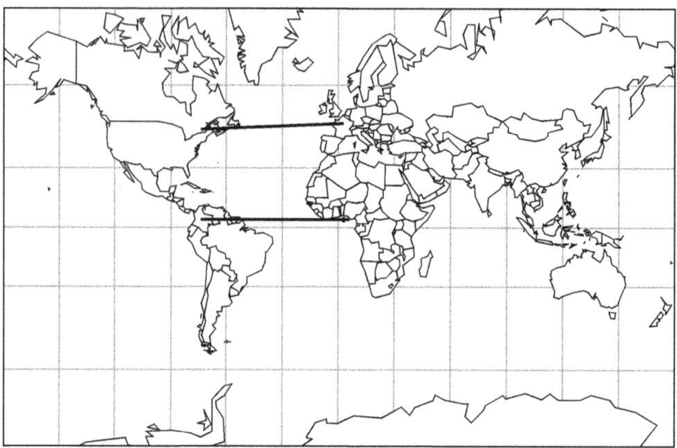

FIGURE 6.12 The flat-earth theory. Flat-earth theorists say that the distance between Montreal and Paris is approximately the *same* as the distance from Bogota to Lagos. The distance only appears to be shorter because of a special field that couples to all matter and lengthens all rulers and other measures of distance in the east-west direction increasingly strongly as one moves to higher latitudes.

affected in the same way. It is simpler, more economical, and ultimately more powerful to recognize that distances on Earth are correctly measured by rulers and that its surface is curved. In the same way it is simpler, more economical, and ultimately more powerful to recognize that clocks correctly measure timelike distances in spacetime and that its geometry is curved. That is the route to general relativity.

6.6 Newtonian Gravity in Spacetime Terms

To gain insight into what a geometric theory of gravity could be like, we first consider a simple model. In this model the flat spacetime geometry of special relativity is modified to introduce a slight curvature that will explain *geometrically* the behavior of clocks we have been discussing. Further, the world lines of extremal proper time in this modified geometry will reproduce the predictions of Newtonian mechanics for motion in a gravitational potential for nonrelativistic velocities.

The model spacetime geometry is specified by the line element ($c \neq 1$ units)

Static Weak Field Metric

$$ds^2 = -\left(1 + \frac{2\Phi(x^i)}{c^2}\right)(cdt)^2 + \left(1 - \frac{2\Phi(x^i)}{c^2}\right)(dx^2 + dy^2 + dz^2), \tag{6.20}$$

where the gravitational potential $\Phi(x^i)$ is a function of position satisfying the Newtonian field equation (3.18) and assumed to vanish at infinity. For example, outside Earth $\Phi(r) = -GM_\oplus/r$ [cf. (3.13)]. This line element is in fact predicted by general relativity for small curvatures produced by time-independent weak sources. That is why it is called *static* and *weak field*. It is a good approximation to the curved spacetime geometry produced by the Sun, for example.

Rates of Emission and Reception

The difference between the rates at which signals are emitted and received is explained from (6.20) in the following way: consider signals propagating along the x-axis emitted at one location, x_A, and received at another, x_B. Figure 6.13 is a (ct, x) spacetime diagram showing the world lines of emitter, receiver, and two light signals propagating between them that are separated on emission at A by an interval Δt in the coordinate t. The world line of a light signal won't be a 45° straight line, as in flat spacetime. But the world lines of both signals will have the same shape because the geometry is independent of t. The world line of the second light signal will be the same as the first but displaced upward by Δt. The signals are, therefore, received at B with the same coordinate separation Δt as they were emitted with at A. But a coordinate separation Δt corresponds to two different *proper* time intervals at the two locations. The coordinate separations between the two emissions at location x_A are Δt and $\Delta x = \Delta y = \Delta z = 0$. The proper time separation $\Delta\tau_A$ between these events is, from (6.20) and $d\tau^2 = -ds^2/c^2$,

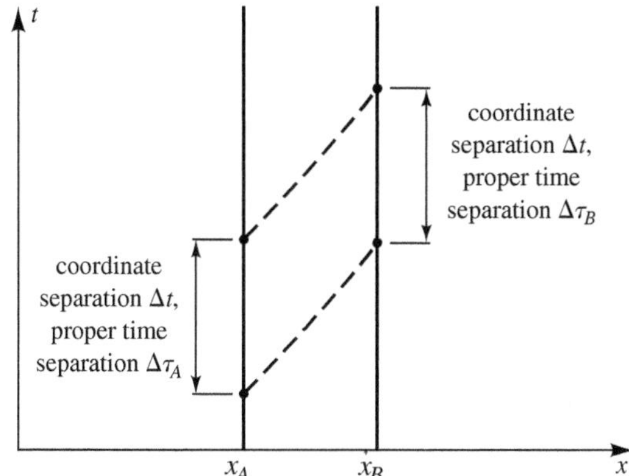

FIGURE 6.13 Emission and reception of light signals in the model curved spacetime (6.20). This spacetime diagram (where $c = 1$) shows the world lines of two stationary observers A and B. Signals are emitted at A with a proper time interval $\Delta\tau_A$ related to a coordinate time interval Δt by (6.21). Since the line element (6.20) is independent of t, the coordinate interval between the reception of the signals is also Δt, but the proper time interval $\Delta\tau_B$ between these events is different from $\Delta\tau_A$. The rate of reception is different from the rate of emission.

$$\Delta\tau_A = \left(1 + \frac{\Phi_A}{c^2}\right)\Delta t, \tag{6.21}$$

accurate to order $1/c^2$, where $\Phi_A \equiv \Phi(x_A, 0, 0)$. (The relation $(1+x)^{1/2} \approx 1 + (1/2)x$, valid for small x, has been used.) Similarly, on reception

$$\Delta\tau_B = \left(1 + \frac{\Phi_B}{c^2}\right)\Delta t. \tag{6.22}$$

Eliminating Δt between these two relations gives

$$\Delta\tau_B = \left(1 + \frac{\Phi_B - \Phi_A}{c^2}\right)\Delta\tau_A. \tag{6.23}$$

This is exactly (6.10) given (6.11); the relation (6.12) then follows. The difference in rates has been explained by the geometry of spacetime.

Newtonian Motion in Spacetime Terms

The Newtonian laws of motion for a particle in a gravitational field can be expressed in geometric terms using the geometry specified by (6.20). Section 5.4 showed that a free particle in flat spacetime follows a path of extremal proper time between any two points. The same principle also gives the motion of a particle in a gravitational potential Φ in the spacetime geometry summarized by (6.20). The argument is the same as in Section 5.4, but with the line element (6.20) instead of that of flat spacetime. The proper time between two points A and B in spacetime depends on the world line between them and is given by

$$\begin{aligned}\tau_{AB} &= \int_A^B d\tau = \int_A^B \left(-\frac{ds^2}{c^2}\right)^{1/2} \\ &= \int_A^B \left[\left(1 + \frac{2\Phi}{c^2}\right)dt^2 - \frac{1}{c^2}\left(1 - \frac{2\Phi}{c^2}\right)(dx^2 + dy^2 + dz^2)\right]^{1/2}\end{aligned} \tag{6.24}$$

integrated along the world line connecting A and B. Using t as a parameter along the world line, the elapsed proper time can be rewritten as

$$\tau_{AB} = \int_A^B dt \left\{\left(1 + \frac{2\Phi}{c^2}\right) - \frac{1}{c^2}\left(1 - \frac{2\Phi}{c^2}\right)\left[\left(\frac{dx}{dt}\right)^2 + \left(\frac{dy}{dt}\right)^2 + \left(\frac{dz}{dt}\right)^2\right]\right\}^{1/2}. \tag{6.25}$$

The quantity in square brackets is just the square of the nonrelativistic velocity $\vec{V}^2$. All our considerations have been accurate only to first order[8] in $1/c^2$, and to

[8] By first order in $1/c^2$ we mean strictly speaking first order in an expansion in the dimensionless comparable small quantities $(V/c)^2$ and Φ/c^2. That has meaning even in units where $c = 1$. We'll use this informal way of referring to such expansions elsewhere.

that order (6.25) is

$$\tau_{AB} \approx \int_A^B dt \left[1 - \frac{1}{c^2}\left(\frac{1}{2}\vec{V}^2 - \Phi\right)\right]. \tag{6.26}$$

You may recognize this as the combination of the effects of time dilation and gravitational potential discussed in connection with the Global Positioning system in Section 6.4 but here emerging in a unified way from spacetime geometry to first order in $1/c^2$. An interesting test of this formula is described in Box 6.2 on p. 130.

The world line that extremizes the proper time between A and B will extremize the combination

$$\int_A^B dt \left(\frac{1}{2}\vec{V}^2 - \Phi\right) \tag{6.27}$$

since the first term in (6.26) doesn't depend on which world line is traveled. The conditions for an extremum are Lagrange's equations, following from the Lagrangian [cf. (3.33), (3.35)]

$$L\left(\frac{d\vec{x}}{dt}, \vec{x}\right) = \frac{1}{2}\left(\frac{d\vec{x}}{dt}\right)^2 - \Phi(\vec{x}, t). \tag{6.28}$$

If multiplied by the mass, (6.28) is just the Lagrangian for a nonrelativistic particle moving in the gravitational potential Φ. Lagrange's equations imply

$$\frac{d^2\vec{x}}{dt^2} = -\nabla\Phi, \tag{6.29}$$

which, when both sides are multiplied by m, is just $\vec{F} = m\vec{a}$.

Newtonian gravity can be expressed completely in geometric terms in the curved spacetime (6.20). (See Table 6.1.) Rather than say the presence of mass produces a gravitational potential Φ, which determines particle motion through

TABLE 6.1 Newtonian and Geometric Formulations of Gravity Compared

	Newtonian	Geometric Newtonian	General Relativity
What a mass does	Produces a field Φ causing a force on other masses $\vec{F} = -m\vec{\nabla}\Phi$	Curves spacetime $ds^2 = -\left(1 + \frac{2\Phi}{c^2}\right)(c\,dt)^2 + \left(1 - \frac{2\Phi}{c^2}\right)(dx^2 + dy^2 + dz^2)$	Curves spacetime
Motion of a particle	$m\vec{a} = \vec{F}$	Curve of extremal proper time (first order in $1/c^2$)	Curve of extremal proper time
Field equation	$\nabla^2\Phi = +4\pi G\mu$	$\nabla^2\Phi = +4\pi G\mu$	Einstein's equation

$m\vec{a} = -m\nabla\Phi$, one can say the presence of mass produces spacetime curvature described by (6.20), and particles move in this geometry along paths of extremal proper time. Concepts of force and effects on clocks have been replaced by geometric ideas. In a sense, the equality of gravitational and inertial mass has been explained because the idea of mass never enters into the description of motion

BOX 6.2 The Twin Paradox Tested

The length of a timelike curve is measured by the proper time of a clock moving along it, and clocks traversing different curves between two spacetime points show different elapsed proper times. That was the geometric resolution of the twin paradox discussed in Section 4.4. It is just as true in the static weak field metric (6.20) as it was in the spacetime of special relativity. In 1971 J. C. Hafele and R. E. Keating carried out an experiment that combined a test of both time dilation and the relative rates of clocks—in effect checking the metric (6.20) (Hafele and Keating 1972). They transported cesium-beam atomic clocks around the Earth on scheduled commercial flights and compared their reading on return to that of a standard clock at rest on the Earth's surface. The experiment was carried out twice—once flying eastward around the world and once westward.

The flying clocks are higher up in the Earth's gravitational potential and—were this the only effect—would seem to run faster compared to surface clocks. However, the flying clocks are also moving relative to the surface clocks and, due to time dilation, would run slower. Thus, there is a competition between these two effects, which is neatly summarized by (6.26) to the $1/c^2$ accuracy sufficient for analyzing this experiment. The t in this formula is not the time that would be registered by either the flying or the surface clocks, but rather the time on a clock at rest in an inertial frame. To compare the flying and surface clocks to each other, first compare them to this standard and thus to each other.

Define $V_g(t)$ to be the speed of the plane with respect to the ground, $h(t)$ to be its altitude, and $V_\oplus = 2\pi R_\oplus/(24\text{ h})$ to be the surface speed of the Earth. Assuming, for simplicity, that the flights were all along the equator, the predicted difference in elapsed proper time between the flying clocks and the surface clock is (Problem 15)

$$\Delta\tau = \frac{1}{c^2}\int dt\left\{gh(t) - \frac{1}{2}V_g(t)[V_g(t) + 2V_\oplus]\right\},$$

where t is the time in an inertial frame to a good approximation at rest with respect to the center of the Earth. There is a significant difference in the size and sign of the second term between eastbound flights, where V_g is positive, and westbound flights, where it is negative.

By keeping careful logs of $h(t)$ and $V_g(t)$ the experimenters could evaluate this formula and compare with the observed readings on their clocks. For the eastbound flight they predicted -40 ± 23 ns (more time elapsed on ground than flying clock) and observed -59 ± 10 ns. For the westbound flight they predicted 275 ± 21 ns and observed 273 ± 7 ns. These were out of total flying times of 41 and 49 h, respectively—timing accuracies of a few parts in 10^{13}. Both observations are in good agreement with the predictions of time dilation and the equivalence principle.

Hafele and Keating on board with their clocks.

of a particle moving under the influence of a curvature-producing mass. The law of motion is the same as that of a free particle, but in a curved spacetime.

In flat spacetime the straight-line path between two points is also a curve of longest proper time, as discussed in Section 4.4. That is also true in curved spacetime if there is just one curve of extremal proper time connecting the two points. But if there is more than one, the path may not be of longest or shortest proper time between two points. It may be just extremal.[9]

You may have noticed that the factor $(1 - 2\Phi/c^2)$ in the *spatial* part of the line element (6.20) played no role to leading order in $1/c^2$ in reproducing either the relativistic relation (6.23) between time intervals on clocks or the Newtonian equation of motion (6.29). Any factor there that is unity to leading order in $1/c^2$ would have worked, including 1. There are, therefore, many curved spacetimes that will reproduce the predictions of Newtonian gravity for low velocities. The particular static, weak field metric (6.20) is the prediction of general relativity. It will give different predictions than other choices for the orbits of light rays. We'll see that in Chapter 10.

What's the matter with the ingredients listed in the second column in Table 6.1 as a geometric theory of gravity? As we have seen, it correctly reproduces the motions of Newtonian theory in the first column for nonrelativistic velocities. The answer is that such a theory is not consistent with special relativity. As we stressed at the beginning of this chapter, the Newtonian gravitational law, whether expressed as (6.1) or the equivalent (3.14), is inconsistent with the principles of special relativity because it specifies an instantaneous interaction between bodies. The asymmetry between space and time in (6.20) shows this in another way. Even in a geometric formulation Newtonian gravity is inconsistent with special relativity. A fully relativistic, geometric theory of gravity would treat space and time on a symmetric footing. This is the case for Einstein's 1915 general theory of relativity. Einstein's theory deals with general geometries not restricted to the form (6.20) and a field equation that these geometries must satisfy generalizing that of Newtonian gravity (3.18). This field equation is called the *Einstein equation* or sometimes *Einstein's equation*. We won't meet up with the Einstein equation until Chapter 21, but in the meantime we will explore many of its consequences. We first need to discuss the mathematical description of curved spacetimes. We do this in the next two chapters.

Problems

1. What angle does the fiber of the torsion balance described in Figure 6.1 make with the direction of the local gravitational field $\vec{g}$? What is the value of g^t in (6.2)? Assume that the experiment is carried out at latitude 47°. (This is the latitude of Seattle, where the experiment of Su et al. described in the text was carried out.)

2. Suppose any twisting of the torsion balance in the modern versions of the Eötvös experiment was measured by bouncing a light off a mirror attached to the bar and

[9] For more insight on this question, work Problem 12 and/or Problem 14.

measuring the time dependence of the angle θ as before. What angular accuracy is needed to test the principle of equivalence to 1 part in 10^{12}? Assume the bar is 4 cm long and the masses are about 10 g each, that the torsion constant of the fiber (analogous to the spring constant for linear motion) is 2×10^{-8} N · m/rad, and that the acceleration of gravity in the twisting direction is as determined in Problem 1.

3. [S] Assuming the acceleration of gravity at the surface of the Earth, how wide does the elevator in Figure 6.5 have to be for the light ray to fall by 1 mm over the course of its transit? Is this a thought experiment that could be realized on the surface of the Earth?

4. Starting from the equivalence principle in the form stated on p. 119, i.e., using only freely falling frames and inertial frames, argue that light must fall in the gravitational field of the Earth.

5. In Example 6.3 concerning freely falling ping-pong balls, assume that the inner ball is released with just the tangential velocity necessary for a circular orbit about the Earth. The outer ball released with the same velocity will, therefore, execute an elliptical orbit. What is the eccentricity of this orbit as a function of s? Sketch the two orbits. Does your picture support the conclusion of the example that there is significant change in the separation of the particles in one period? *Hint*: Look up the details of elliptical orbits in your Newtonian mechanics text.

6. **(a)** Transform the line element of special relativity from the usual (t, x, y, z) rectangular coordinates to new coordinates (t', x', y', z') related by

$$t = \left(\frac{c}{g} + \frac{x'}{c}\right) \sinh\left(\frac{gt'}{c}\right)$$

$$x = c\left(\frac{c}{g} + \frac{x'}{c}\right) \cosh\left(\frac{gt'}{c}\right) - \frac{c^2}{g}$$

$$y = y', \qquad z = z'.$$

for a constant g with the dimensions of acceleration.

(b) For $gt'/c \ll 1$, show that this corresponds to a transformation to a uniformly accelerated frame in Newtonian mechanics.

(c) Show that a clock at rest in this frame at $x' = h$ runs fast compared to a clock at rest at $x' = 0$ by a factor $(1 + gh/c^2)$. How is this related to the equivalence principle idea?

7. **(a)** An accelerated laboratory has a bottom at $x' = 0$ and a top at $x' = h$, both with extent in the y'- and z'-direction. Use the line element derived in part (a) of Problem 6 to show that the height of the laboratory remains constant in time, i.e., the laboratory moves rigidly.

(b) Compute the invariant acceleration $a \equiv (\mathbf{a} \cdot \mathbf{a})^{1/2}$, where $a^\alpha = d^2x^\alpha/d\tau^2$, and show that it is different for the top and bottom of the laboratory.

8. [S] It is not legitimate to mix relativistic with nonrelativistic concepts, but imagine that a photon with frequency ω_* is like a particle with gravitational mass $\hbar\omega_*/c^2$ and kinetic energy $K = \hbar\omega$. Using Newtonian ideas, calculate the "kinetic" energy loss to a photon that is emitted from the surface of a spherical star of radius R and mass M and escapes to infinity. From this calculate the frequency of the photon at infinity. How does this compare with the gravitational redshift in (6.14) to first order in $1/c^2$?

9. A GPS satellite emits signals at a constant rate as measured by an onboard clock. Calculate the fractional difference in the rate at which these are received by an identical clock on the surface of the Earth. Take both the effects of special relativity and gravitation into account to leading order in $1/c^2$. For simplicity assume the satellite is in a circular equatorial orbit, the ground-based clock is on the equator, and that the angle between the propagation of the signal and the velocity of the satellite is 90° in the instantaneous rest frame of the receiver.

10. [C, P] The Earth is approximately 5 billion years old. How much younger are the rocks at the center of the Earth than at the surface? If equal abundances of a radioactive element like ^{238}U with an exponential decay time of 6.5 billion years were present to start, how much more of that element would be present at the center than the surface? Assume the density of the Earth is constant.

11. [E] Aging goes on at a slower rate at the center of a spherical mass than on its surface. Estimate how much mass would need to be assembled in a radius of 10 km such that if you lived at the center for 1 year you would emerge 1 day younger than those who had stayed outside and far away.

12. [S] In the two-dimensional flat plane, a straight-line path of extremal distance is the shortest distance between two points. On a two-dimensional round sphere, extremal paths are segments of great circles. Show that between any two points on the sphere there is an extremal path that provides the shortest distance between them when compared with nearby paths. Show there is another path between the two points which is extremal, but neither the longest or shortest distance between the points when compared with nearby paths. Show that there is no one path that provides the longest distance between the points.

13. Three observers are standing near each other on the surface of the Earth. Each holds an accurate atomic clock. At time $t = 0$ all the clocks are synchronized. At $t = 0$ the first observer throws his clock straight up so that it returns at time T as measured by the clock of the second observer, who holds her clock in her hand for the entire time interval. The third observer carries his clock up to the maximum height the thrown clock reaches and back down, moving with constant speed on each leg of the trip and returning in time T.
Calculate the total elapsed time measured on each clock assuming that the maximum height is much smaller than the radius of the Earth. Include gravitational effects but calculate to order $1/c^2$ only using nonrelativistic trajectories. Which clock registers the *longest* time? Why is this?

14. [C] Consider a particle moving in a circular orbit about the Earth of radius R. Suppose the geometry of spacetime outside the Earth is given by the static weak field metric (6.20) with $\Phi = -GM_{\oplus}/r$. Let P be the period of the orbit measured in the time t. Consider two events A and B located at the same spatial position on the orbit but separated in t by the period P. The particle's world line is a curve of extremal proper time between A and B. As discussed in Section 3.5, that means the proper time around the orbit is a maximum, minimum, or saddle point with respect to *nearby* paths. But we can also ask whether the proper time is longer and shorter than *any* other world line, nearby or not. Analyze this question for the circular orbit by calculating to first order in $1/c^2$ the proper time along the following world lines connecting points A and B in spacetime:

(a) The orbit of the particle itself.

(b) The world line of an observer who remains fixed in space between A and B.

(c) The world line of a photon that moves radially away from A and reverses direction in time to return to B in a time P.

Can you find another curve of extremal proper time that connects A and B?

15. [B] *Twin Paradox Test*

(a) Derive the formula for the elapsed difference in proper time between the flying clocks and the surface clock given in Box 6.2 on p. 130.

(b) Using typical altitudes and speeds for commercial aircraft, *estimate* the value of $\Delta\tau$ for both eastward and westward flights around the world.

CHAPTER 7

The Description of Curved Spacetime

This chapter and the next one cover some basic mathematics needed to describe four-dimensional curved spacetime geometry. Much of this is a generalization of the concepts introduced in Chapter 5 for flat spacetime.

7.1 Coordinates

As discussed in Chapter 2 and as illustrated by flat spacetime in Chapters 4 and 5, a spacetime geometry is summarized by a line element giving the spacetime distance between any two nearby points. Coordinates are a systematic way of labeling the points of spacetime. The choice of coordinates is arbitrary as long as they supply a unique set of labels for each point in the region they cover. For a particular problem one coordinate system may be more useful than another. For example, to solve central force problems in mechanics, it is usually easier to use polar rather than Cartesian coordinates. The laws of motion, however, *can* be expressed in either set of coordinates, and the content is the same.

The arbitrariness of the coordinates can be a difficult point for students to grasp because in almost all elementary parts of physics there are a few coordinate systems that are preferred because they make the laws look simpler. For example, there is the class of inertial frames, in which the general laws of special relativistic mechanics take a simple form. The special symmetries of flat spacetime, expressed by Lorentz transformations, are the reason why inertial frames are so useful. But in general relativity, where spacetime is curved, generally without special symmetries, there will be no class of coordinate systems which simplifies the *general laws*. Particular coordinate systems may simplify particular problems, but no one set of coordinates simplifies all problems. Therefore, experience is needed in formulating general laws in arbitrary coordinates. That is the subject of this chapter.

A line element specifies a geometry, but *many different line elements describe the same spacetime geometry* because different coordinate systems can be used. For example, the flat spacetime geometry of special relativity can be summarized in Cartesian coordinates by [cf. (4.8)]

$$ds^2 = -dt^2 + dx^2 + dy^2 + dz^2 \tag{7.1}$$

in the $c = 1$ units that will be used throughout this and following chapters. The spatial part of the metric can be transformed to spherical polar coordinates by writing

$$x = r\sin\theta\cos\phi, \quad y = r\sin\theta\sin\phi, \quad z = r\cos\theta, \tag{7.2}$$

working out the differentials, e.g.,

$$dz = dr\cos\theta - r\sin\theta\, d\theta, \tag{7.3}$$

and substituting the results into (7.1). The transformed line element is

$$ds^2 = -dt^2 + dr^2 + r^2 d\theta^2 + r^2\sin^2\theta\, d\phi^2. \tag{7.4}$$

This expression for ds^2 *looks* different than (7.1), but it represents the *same* flat spacetime geometry with the points labeled in a different way. An example of another interesting set of coordinates for flat spacetime is given in Box 7.1 on p. 137.

Because the coordinates are arbitrary, you should be careful not to read too much into the names used for any one of them. For example, the line element

$$ds^2 = -dx^2 + dy^2 + y^2 dz^2 + y^2\sin^2 z\, dt^2 \tag{7.5}$$

describes flat spacetime in the same coordinate system as (7.4). Only the *names* of the coordinates have been changed. Despite their names, the coordinate t is an angle, and the direction along x is timelike.

A good coordinate system provides unique labels for each point in spacetime. However, most coordinate systems fail to provide unique labels *somewhere*. For example, in polar coordinates (r, θ, ϕ), the points on the axis $(\theta = 0)$ are labeled by *more* than one set of coordinate values—different ϕ at each r correspond to the same point on the axis. This is a mild example of a *coordinate singularity*. A simple example of a more serious looking singularity is provided by writing the line element of the two-dimensional plane in polar coordinates,

$$dS^2 = dr^2 + r^2 d\phi^2, \tag{7.6}$$

and making the transformation $r = a^2/r'$ for some constant a. The result is

$$dS^2 = \frac{a^4}{r'^4}(dr'^2 + r'^2 d\phi^2). \tag{7.7}$$

This line element blows up at $r' = 0$. But nothing physically interesting happens there; the geometry is a flat plane still! The singularity arises because the coordinate transformation $r' = a^2/r$ has mapped all the points at infinity into $r' = 0$ and thus failed to provide them unique labels. In fact, (7.7) correctly gives an infinite distance between $r' = 0$ and any point with $r' \neq 0$ (Problem 1). The singularities in most coordinate systems mean that different overlapping coordinate patches must be used to cover spacetime so that every point is labeled by a nonsingular set of coordinates. We will see more important examples of this later.

BOX 7.1 The Penrose Diagram for Flat Space

Another example of a useful coordinate system for flat spacetime is the one used to construct its *Penrose diagram*. Begin with the line element for flat spacetime in spheri-cal polar coordinates (7.4). Replace t and r by two new coordinates u and v defined by

$$u \equiv t - r, \qquad v \equiv t + r \qquad \text{(a)}$$

so that the line element becomes

$$ds^2 = -du\,dv + \frac{1}{4}(u - v)^2(d\theta^2 + \sin^2\theta\, d\phi^2). \quad \text{(b)}$$

The (u, v) axes are rotated with respect to the (t, r) axes by 45°, as shown in the (t, r) spacetime diagram. Radial light rays travel on lines of constant u or constant v. That is evident either from the definitions of these coordinates in (a) or because (b) shows that lines of constant θ, ϕ, and either u or v have $ds^2 = 0$.

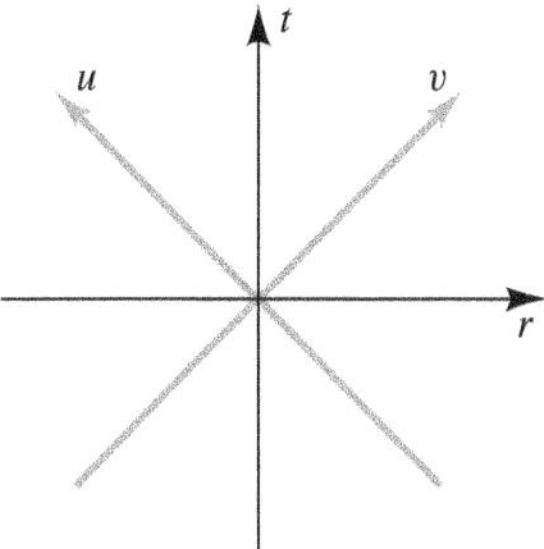

Make a further transformation of u and v to new coordinates u' and v' and corresponding new coordinates t' and r' with the relations:

$$u' \equiv \tan^{-1} u \equiv (t' - r')/2, \quad v' \equiv \tan^{-1} v \equiv (t' + r')/2. \qquad \text{(c)}$$

The t and r coordinates for flat spacetime have the infinite ranges $-\infty < t < +\infty$, $0 < r < +\infty$. But $\tan^{-1} x$ lies between $-\pi/2$ and $+\pi/2$, so the ranges for (u', v') or (t', r') are finite. In fact, all the (t, r) plane of flat spacetime is mapped into the finite region $r' > 0$, $v' < \pi/2$, $u' > -\pi/2$ shown lightly shaded in the (t', r') diagram at top right. This is the Penrose diagram for flat spacetime.

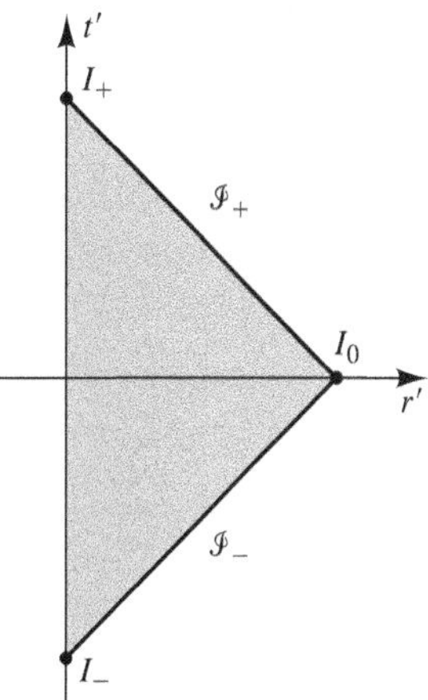

By this mapping of infinity to finite coordinate values, it is possible to distinguish different *kinds* of infinity. Out-going radial light rays—with $t = r +$ constant—are lines of constant u'. They wind up on the boundary $v' = \pi/2$. This is called *future null infinity* and is denoted by $\mathscr{I}_+$ (pronounced "scri plus"). Ingoing radial light rays follow lines of constant v' starting at the boundary $u' = -\pi/2$, called *past null infinity* and denoted by $\mathscr{I}_-$. Particle trajectories that lie *within* the local light cone start from the point $(t' = -\pi,\ r' = 0)$, called *past timelike infinity*, I_-, and wind up at the point $(t' = +\pi,\ r' = 0)$, called *future timelike infinity*, I_+. (Problem 4). Similarly, infinite spacelike curves wind up at the point I_0, which labels a sphere called *spacelike infinity*.

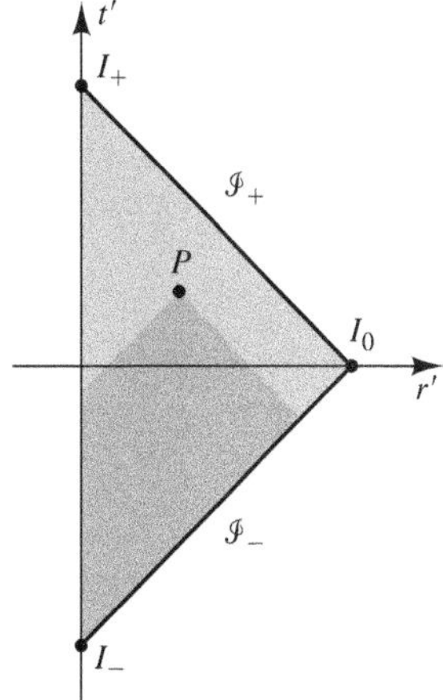

BOX 7.1 (*continued*)

Among other things, Penrose diagrams are useful for describing graphically from which events in spacetime an observer at a given point can receive information. For example, in the final diagram on the previous page, an observer at point P can receive information from events in the heavily shaded area and not from events outside that area. That kind of analysis can be useful for discussing black holes, whose Penrose diagrams can be considerably more complex.

7.2 Metric

To describe a general geometry use a system of four coordinates, x^α, to label the points and specify the line element giving the distance, ds^2, between nearby points separated by coordinate intervals dx^α. That line element will have the form

Metric Defined

$$\boxed{ds^2 = g_{\alpha\beta}(x)dx^\alpha dx^\beta,} \tag{7.8}$$

where $g_{\alpha\beta}(x)$ is a symmetric, position-dependent[1] matrix called the *metric*. For example, the metric for flat spacetime in polar coordinates (7.4) is

$$g_{\alpha\beta}(x) = \begin{matrix} & 0 & 1 & 2 & 3 \\ 0 & -1 & 0 & 0 & 0 \\ 1 & 0 & 1 & 0 & 0 \\ 2 & 0 & 0 & r^2 & 0 \\ 3 & 0 & 0 & 0 & r^2\sin^2\theta \end{matrix} . \tag{7.9}$$

Diagonal metrics such as this can be specified more compactly by writing $g_{\alpha\beta}(x) = \text{diag}(-1, 1, r^2, r^2\sin^2\theta)$.

As a symmetric 4×4 matrix, $g_{\alpha\beta}$ has 10 independent components. The form of $g_{\alpha\beta}$ will be different in different coordinate systems for the same geometry. Since there are 4 arbitrary functions involved in transforming 4 coordinates, there are really only $10 - 4 = 6$ independent functions associated with a metric.

7.3 The Summation Convention

By this point you will have noticed that we have been careful with the placement of indices in expressions. Our conventions in this regard are part of a larger set commonly employed in relativity, and we have used them so that you will have

[1] When dealing with functions of the coordinates, we routinely use the abbreviations $f(x^\alpha)$, or $f(x)$, for $f(x^0, x^1, x^2, x^3)$ where there is no danger of confusion.

as little difficulty as possible in making the transition to more advanced texts. We set out a few rules to help codify the conventions and keep them consistent.

1. The location of the indices must be respected: superscripts (upper indices) for coordinates and vector components to be discussed in Section 7.8 and subscripts (lower indices) for the metric. (In expressions such as the chain rule, $dx^\alpha = (\partial x^\alpha/\partial x'^\beta)dx'^\beta$, the superscript β in the *denominator* acts as a subscript.)
2. Repeated indices always occur in superscript-subscript *pairs* and imply summation. For that reason they are called *summation indices*. One index is as good as any other for indicating a summation, and for this reason summation indices are also called *dummy indices*. Thus, $g_{\alpha\beta}a^\alpha b^\beta$ means the same thing as $g_{\gamma\delta}a^\gamma b^\delta$. Expressions with three or more repeated indices, such as $g_{\alpha\alpha}a^\alpha b^\alpha$, or repeated indices that are not in superscript-subscript pairs, such as $g_{\alpha\beta}g_{\beta\gamma}$, will never occur. If they do, it signals a mistake!
3. Indices that are not summed are called *free indices*. They must balance on both sides of an equation. The value of a free index can be changed if it is changed on both sides of an equation at the same time. The equation

$$g_{\alpha\beta} = g_{\beta\alpha} \tag{7.10}$$

expresses the symmetry of the metric. The indices balance because there is one lower index, α and β, on each side of the equation. An equation such as this can be thought of as a shorthand for an array of equations—one for each of the four possible values of the free indices α and β. Equation (7.10) stands for the 16 equations

$$\begin{array}{cccc} g_{00} = g_{00}, & g_{01} = g_{10}, & g_{02} = g_{20} & \cdots \\ g_{10} = g_{01}, & g_{11} = g_{11}, & g_{12} = g_{21} & \cdots. \\ \cdots & \cdots & \cdots & \cdots \end{array} \tag{7.11}$$

For this reason, a free index can be changed to another free index (not already tied up in a summation) provided it is changed on both sides of an equation at the same time. Changing β to γ in (7.10) gives $g_{\alpha\gamma} = g_{\gamma\alpha}$, which represents the *same* set of 16 relations (7.11). An expression such as $g_{\alpha\beta} = g_{\alpha\gamma}$, in which the indices *don't* balance, is meaningless.

Example 7.1. A Little Test. From the following list of expressions, try to pick out those that are consistent with the summation convention and those that are not, in each case explaining why. Don't worry about what the symbols mean (we will encounter them soon); just try and decide if the summation convention rules are obeyed or not. The answers are at the bottom of the page.

(a) $g_{\alpha\beta}dx^{\alpha}dx^{\beta} = g_{\alpha\beta}dx^{\alpha}dx^{\gamma}$

(b) $g_{\alpha\beta}a^{\alpha}b^{\beta} = g_{\beta\gamma}a^{\beta}b^{\gamma}$

(c) $g_{\alpha\beta}a^{\alpha}b^{\beta} = g_{\alpha\beta}a^{\alpha}c^{\beta}$

(d) $\Gamma^{\alpha}_{\alpha\gamma}a^{\gamma} = g_{\alpha\beta}a^{\alpha}b^{\beta}$

(e) $\Gamma^{\alpha}_{\beta\gamma}a^{\alpha}c^{\beta}c^{\gamma} = b^{\alpha}$

(f) $\partial x^{\alpha}/\partial x^{\beta} = \delta^{\alpha}_{\beta}$

(g) $\partial g_{\alpha\beta}/\partial x^{\gamma} = 0$

(h) $g_{\alpha\beta}\dfrac{\partial x^{\alpha}}{\partial x'^{\gamma}}\dfrac{\partial x^{\beta}}{\partial x'^{\delta}} = g_{\gamma\delta}\dfrac{\partial x^{\gamma}}{\partial x'^{\alpha}}\dfrac{\partial x^{\delta}}{\partial x'^{\beta}}$

(i) $g'_{\alpha\beta}a'^{\alpha}b'^{\beta} = g_{\alpha\beta}a^{\alpha}b^{\beta}$

(j) $a^{\alpha}(g_{\beta\gamma}b^{\beta}b^{\gamma}) = b^{\gamma}$

(k) $\Gamma^{\alpha}_{\alpha\beta} = \Gamma^{\beta}_{\beta\beta}$

(l) $g_{\alpha\beta} = \eta_{\beta\alpha}$

7.4 Local Inertial Frames

The equivalence principle (p. 119) suggests that the local properties of curved spacetime should be indistinguishable from those of the flat spacetime of special relativity. A concrete expression of this physical idea is the requirement that, given a metric $g_{\alpha\beta}(x)$ in one system of coordinates, at each point P of spacetime it is possible to introduce new coordinates x'^{α} such that

$$g'_{\alpha\beta}(x'_P) = \eta_{\alpha\beta}, \tag{7.12}$$

where $\eta_{\alpha\beta} = \text{diag}(-1, 1, 1, 1)$ is the Minkowski metric of flat spacetime and x'^{α}_P are the coordinates locating the point P. This requirement is one of the assumptions of general relativity. It means that at every point there are three space dimensions and one time dimension.

It is not difficult to find new coordinates in which $g'_{\alpha\beta}(x'_P)$ is diagonal at one point P because $g'_{\alpha\beta}(x'_P)$ is a symmetric 4×4 matrix that can always be diagonalized. Once diagonal, the coordinates can be rescaled by constant factors one by one so that the diagonal values of $g'_{\alpha\beta}(x'_P)$ are ± 1. (Work through Problem 8 if you have doubts about this.) However, no coordinate transformation can change the number of $+1$s and the number of -1s in the resulting metric at P. (Try it!) It is an *assumption* that at every point P there are three $+1$s and one -1, as in (7.12). That is just the physical assumption that there are three space dimensions and one time dimension.

How much further can one go in using coordinate transformations to make the metric coincide with that of flat spacetime? Evidently it is not possible to find

In the following OK means consistent with the summation convention and the absence of an OK means its not consistent. (a) Free indices don't balance. (b) OK, all indices are summation indices and the same double sum is indicated on both sides. (c) OK; it could be true if $b^{\alpha} = c^{\alpha}$. (d) OK, all summations are on both sides. (e) Repeated indices are not in upper-lower pairs. (f) OK and true; an upper index in a denominator counts as a lower index. (g) OK, this is 40 equations expressing the vanishing of all first derivatives of all metric coefficients. (h) α and β are free indices on one side but repeated indices on the other. (i) OK. (j) Free indices don't balance. (k) Repeated indices are not in upper-lower pairs. (l) OK.

coordinates in which $g_{\alpha\beta} = \eta_{\alpha\beta}$ over the whole of a curved spacetime. If one could, the spacetime would be flat! But one can find coordinates x'^{α} such that, at a point P, the first derivatives of the metric vanish in addition to (7.12):

$$\boxed{g'_{\alpha\beta}(x'_P) = \eta_{\alpha\beta}, \qquad \left.\frac{\partial g'_{\alpha\beta}}{\partial x'^{\gamma}}\right|_{x=x_P} = 0.} \tag{7.13}$$

Local Inertial Frame

A coordinate system that satisfies these two conditions at a point P is called a *local inertial frame* at the point P. It is like an inertial frame of flat space—but only in an infinitesimal neighborhood of a single point P. That is why it is called a *local* inertial frame. Equations (7.13) can be satisfied at any other point but in a *different* set of coordinates. We postpone a demonstration that it is possible to find a local inertial frame at each point in spacetime until Section 8.4, but a supporting counting argument can be had by working through Problem 9.

Example 7.2. The Metric of a Sphere at the North Pole. The line element of the geometry of a sphere of circumference $2\pi a$ has the form [cf. (2.15)]

$$dS^2 = a^2(d\theta^2 + \sin^2\theta\, d\phi^2) \tag{7.14}$$

in familiar polar angular coordinates (θ, ϕ). At the north pole, $\theta = 0$, the metric doesn't look like the metric of a flat plane, $dS^2 = dx^2 + dy^2$, but we can find coordinates such that it does and, further, such that the first derivatives of the metric vanish in analogy with (7.13). Consider

$$x = a\theta\cos\phi, \qquad y = a\theta\sin\phi. \tag{7.15}$$

Inverting this transformation to find

$$\theta = \sqrt{x^2 + y^2}\Big/a, \qquad \phi = \tan^{-1}(y/x) \tag{7.16}$$

and substituting in (7.14) gives a new form of the line element for the geometry of the sphere. The north pole, where $\theta = 0$, is located at $x = y = 0$. In its neighborhood, where x and y are small, the metric coefficients can be expanded in powers of x and y to find ($x^1 \equiv x, x^2 \equiv y$):

$$g_{AB}(x, y) = \begin{pmatrix} 1 - y^2/(3a^2) & xy/(3a^2) \\ xy/(3a^2) & 1 - x^2/(3a^2) \end{pmatrix} + \begin{pmatrix} \text{terms of third} \\ \text{and higher} \\ \text{order in } x \text{ and } y \end{pmatrix}. \tag{7.17}$$

At the north pole $x = y = 0$, $g_{AB} = \text{diag}(1, 1)$, and $\partial g_{AB}/\partial x^C = 0$ where indices $A, B, \ldots$ range over 1 and 2. How did we find these coordinates? They are examples of Riemann normal coordinates to be discussed in Section 8.4.

It is not possible to find coordinates that make all the *second* derivatives of the metric vanish at a point for a general curved spacetime. (See Problem 9.) As we will see in Chapter 21, when properly organized, those second derivatives are the measure of spacetime curvature at a point.

As mentioned before, local inertial frames give a precise expression to the equivalence principle idea that the geometry of a curved spacetime is locally indistinguishable from that of flat spacetime. Beyond geometry, the same principle suggests that other laws of physics (those of particle motion for instance) take the same form in a local inertial frame as they do in flat space. As discussed in Section 6.2, that is not a requirement for a consistent theory in curved space, but it can be a useful starting point for guessing how known flat spacetime laws can be generalized to work in curved spacetime. We'll see several examples of this later.

7.5 Light Cones and World Lines

The spacetime distance between a point P at x^α and neighboring points can be calculated either in the coordinates of (7.8) or in those of a local inertial frame. The assumption (7.12) therefore means that general relativity inherits the local light cone structure of special relativity described in Section 4.3 and illustrated in Figure 4.9. Points separated from P by infinitesimal coordinate intervals dx^α can be timelike separated, spacelike separated, or null separated as the square of their distance away defined by (7.8) satisfies

Timelike, Null and Spacelike

$$ds^2 < 0 \quad \text{timelike separation,} \tag{7.18a}$$

$$ds^2 = 0 \quad \text{null separation,} \tag{7.18b}$$

$$ds^2 > 0 \quad \text{spacelike separation.} \tag{7.18c}$$

Light rays move along null curves in spacetime along which $ds^2 = 0$. The family of null directions emerging from, or converging on, a point P spans the local future and past light cones at P exactly as described in Section 4.3.

Particles move on timelike world lines which can be specified parametrically by four functions $x^\alpha(\tau)$ of the distance τ along them, just as they can in special relativity (Section 5.2). In curved spacetime the distance between a point A and a point B along a timelike world line is given by the curved spacetime generalization of (4.13),

$$\tau_{AB} = \int_A^B \left[-g_{\alpha\beta}(x)dx^\alpha dx^\beta\right]^{1/2}, \tag{7.19}$$

where the integral is along the world line. A clock carried along this curve measures the spacetime distance τ, which, therefore, is also called the *proper time*. A timelike world line with $ds^2 < 0$ or $d\tau^2 \equiv -ds^2 > 0$ [cf. (4.12)] lies *within* the local light cone at every point along its trajectory as illustrated in Figure 4.10. That is the coordinate invariant statement that the particle is moving less than the velocity of light at that point.

Example 7.3. A World Line and Light Cones in Two Dimensions. Consider the two-dimensional metric

$$ds^2 = -X^2 dT^2 + dX^2, \tag{7.20}$$

and the world line

$$X(T) = A\cosh(T), \tag{7.21}$$

where A is a constant with the dimensions of length. The light cones are the curves with $ds^2 = 0$ that have slopes $dT/dX = \pm 1/X$. A few are shown in Figure 7.1 along with the world line (7.21). A particle's world line is timelike if the size of its slope $|dT/dX|$ is bigger than $1/X$ or, alternatively, if $|dX/dT|$ is *less* than X. Then it is moving at less than the velocity of light *locally*. The world line (7.21) is timelike since $\sinh T < \cosh T$. The proper time along the world line is

$$d\tau^2 \equiv -ds^2 = A^2(\cosh^2 T dT^2 - \sinh^2 T dT^2) = A^2 dT^2. \tag{7.22}$$

Choosing $\tau = 0$ when $T = 0$, $\tau = AT$ and the world line (7.21) may be expressed parametrically as

$$T = \tau/A, \qquad X(\tau) = A\cosh(\tau/A). \tag{7.23}$$

(*Confession*: The metric (7.20) is really just flat space in a different system of coordinates. Can you find the coordinate transformation that puts it in the form $ds^2 = -dt^2 + dx^2$?)

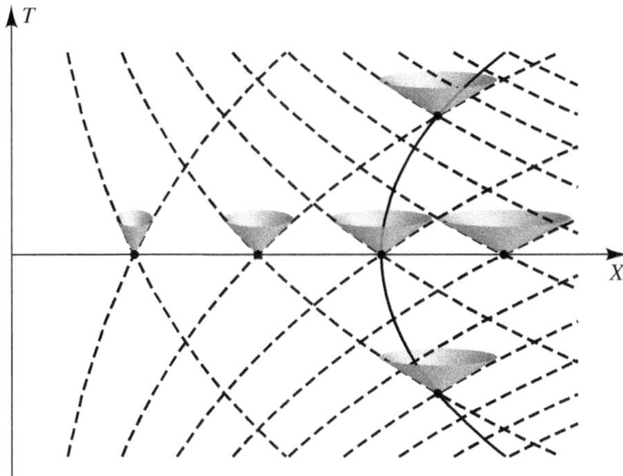

FIGURE 7.1 A spacetime diagram of the two-dimensional spacetime with metric (7.20) with $A = 1$ showing ingoing and outgoing light rays that intersect the $T = 0$ axis at $X = .5, 1, 1.5, 2, \ldots$, and the timelike world line (7.21). A few future light cones are shown. At each point along it, the tangent to the timelike world line lies in the interior of the light cone.

In short, the *local* light cone structure of general relativity is the same as that of flat spacetime. However, the global arrangement of light cones (called the spacetime's *causal structure*) can have interesting properties. Black-hole spacetimes, to be discussed in Chapters 12 and 15, are perhaps the most important examples, but the following unrealistic example of spacetime illustrates the point.

Example 7.4. Warp-Drive Spacetime. This example, due to Alcubierre (1994), uses coordinates (t, x, y, z) and a curve $x = x_s(t)$, $y = 0$, $z = 0$, lying in the t-x plane passing through the origin. The line element specifying the metric is

$$ds^2 = -dt^2 + [dx - V_s(t) f(r_s) dt]^2 + dy^2 + dz^2, \tag{7.24}$$

where $V_s(t) \equiv dx_s(t)/dt$ is the velocity associated with the curve and $r_s \equiv [(x - x_s(t))^2 + y^2 + z^2]$. The function $f(r_s)$ is any smooth positive function that satisfies $f(0) = 1$ and decreases away from the origin to vanish for $r_s > R$ for some R. Evaluating (7.24) on a $t =$ constant slice of spacetime gives $dS^2 = dx^2 + dy^2 + dz^2$. The geometry of each spatial slice is flat and r_s is just the usual Euclidean distance from the curve $x_s(t)$. Spacetime is flat where $f(r_s)$ vanishes, but curved where it does not. Figure 7.2 is a spacetime diagram of the t-x plane. The shaded region is where spacetime is curved.

The light cones at a point in the t-x plane are the curves emerging from the point with $ds^2 = 0$, that is, with

$$ds^2 = -dt^2 + [dx - V_s(t) f(r_s) dt]^2 = 0, \tag{7.25}$$

or, equivalently,

$$\frac{dx}{dt} = \pm 1 + V_s(t) f(r_s). \tag{7.26}$$

The $\pm$ corresponds to the two directions a light ray in the t-x plane can emerge from a point. Figure 7.2 shows the resulting light cones. Where spacetime is flat, the light cones are the usual 45° lines. Inside the region where spacetime is curved, the light cones are tipped over.

To see what is interesting about this arrangement of light cones, consider two stationary space stations whose world lines are shown in Figure 7.2. Imagine a spaceship moving along a curve $x_s(t)$ that connects the two stations in an elapsed coordinate time $T < D$, as shown. That looks like the spaceship has traveled faster than the speed of light. Indeed, such a curve necessarily has to have $V_s(t) > 1$ somewhere, as in the example shown. Were the spacetime flat in between the observers, at those points the spaceship would be moving at a speed greater than light. But the spacetime in between is not flat. Because the light cones are tipped over, the curve is inside the local light cone at every point along it (Problem 11). The spaceship is always moving at *less* than

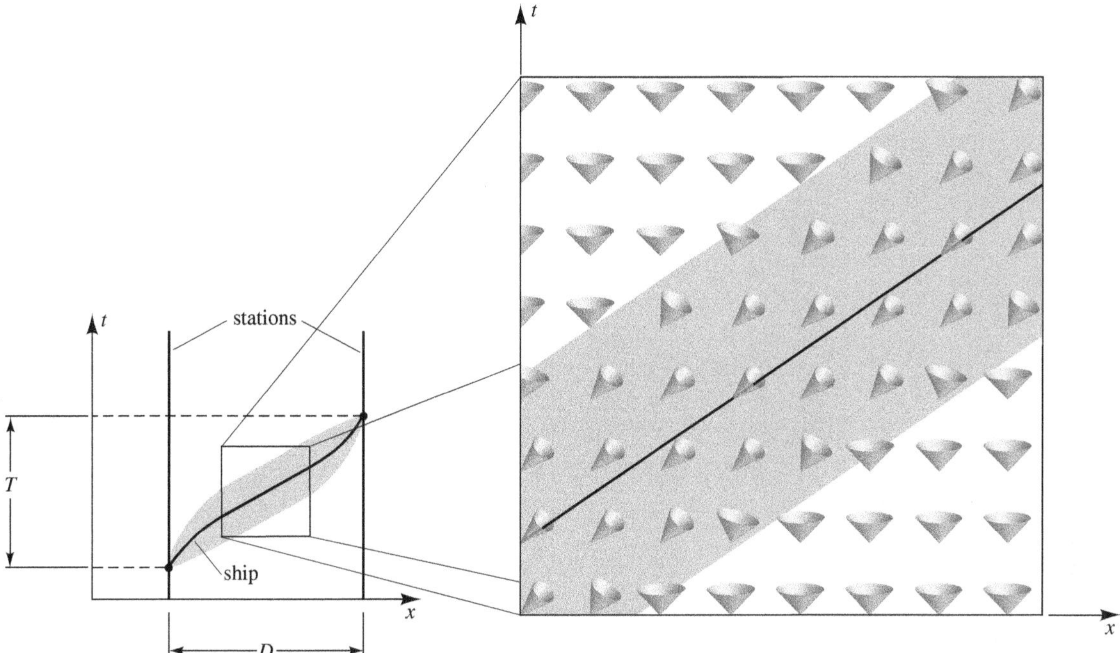

FIGURE 7.2 Light cones in warp-drive spacetime. A spaceship travelling between two space stations along the world line in the figure on the left would be sometimes moving at a speed greater than that of light (as the blowup on the right shows) if these were spacetime diagrams of flat space. But, as described by the warp-drive metric (7.24), there is a bubble of spacetime curvature surrounding the spaceship whose location in spacetime is shaded in these figures. Inside the future light cones are "tipped" as described by (7.26) and as shown in the blowup. At every point, the ship's world line lies within the light cone. The ship is, therefore, always moving locally at *less* than the velocity of light. However, for an observer in the flat space outside who knew nothing of this curvature bubble, the ship would have traversed the distance between the station world lines in a time T that was less than the flat space distance D between them. (The particular light cone structure illustrated assumes $f(r_s) = 1 - (r_s/R)^4$ for $r_s < R$ and zero outside that range.)

the local velocity of light, even if some coordinate velocity such as $V_s = dx_s/dt$ or some coordinate ratio such as D/T is sometimes greater than 1.

Could an advanced civilization build a spaceship that would *create* a region of spacetime curvature surrounding it, such as that represented by this metric? That would be one way of implementing the "warp-drive" of science fiction, enabling travel across the galaxy in times much less than the approximately 100,000-yr minimum needed if spacetime is approximately flat. Alas, spacetimes such as the Alcubierre warp-drive spacetime are excluded in known classical physics. As we will see in Chapter 22, Problem 14, they require matter or fields with *negative* local energy densities. All the classical fields we know about, for example, the electromagnetic fields, have *positive* energy density. Quantum mechanics allows negative energy densities, but physics is far from understanding whether they could be harnessed in this way.

7.6 Length, Area, Volume, and Four-Volume for Diagonal Metrics

For a given metric it is useful to know how to compute lengths of curves, areas, three-volumes, and four-volumes. We already know how to compute the lengths of curves as integrals of ds. For the rest we will consider only the special case of diagonal metrics in which

$$ds^2 = g_{00}(dx^0)^2 + g_{11}(dx^1)^2 + g_{22}(dx^2)^2 + g_{33}(dx^3)^2, \tag{7.27}$$

because almost all our examples will be of this form. In diagonal metrics the coordinates are all orthogonal, so ideas of area and volume can be built up simply. Consider, for example, an element of area shown in Figure 7.3 in the x^1-x^2 surface defined by x^0 = const. and x^3 = const. and suppose the area is defined by *coordinate* lengths dx^1 and dx^2.

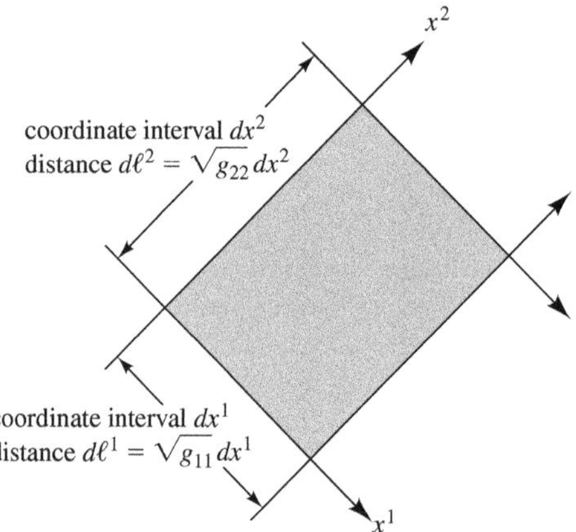

FIGURE 7.3 An element of area is defined by coordinate intervals dx^1 and dx^2. The lengths $d\ell^1$ and $d\ell^2$ of these intervals are related to dx^1 and dx^2 by the metric. If the coordinate lines are orthogonal, the area is $d\ell^1 d\ell^2$.

The proper lengths of two segments will be $d\ell^1 = \sqrt{g_{11}}\,dx^1$ and $d\ell^2 = \sqrt{g_{22}}\,dx^2$, respectively. Since the coordinates are orthogonal, the element of area is then

$$dA = d\ell^1 d\ell^2 = \sqrt{g_{11}g_{22}}\,dx^1 dx^2. \tag{7.28}$$

For three-volume,[2]

$$d\mathcal{V} = \sqrt{g_{11}g_{22}g_{33}}\,dx^1 dx^2 dx^3; \tag{7.29}$$

[2]We use $\mathcal{V}$ for three-volume to distinguish it from speed V.

a similar expression can be constructed for four-volume:

$$dv = \sqrt{-g_{00}g_{11}g_{22}g_{33}}\,dx^0 dx^1 dx^2 dx^3. \tag{7.30}$$

The latter expression has a minus sign so that it is real when applied to flat space. If we define g to be the determinant of $g_{\alpha\beta}$ considered as a matrix, the four-volume element is $dv = \sqrt{-g}\,d^4x$. This is, in fact, the general expression even when the metric is not diagonal. The following examples show how to use these expressions.

Example 7.5. Area and Volume Elements of a Sphere. As a simple example, consider flat spacetime in polar coordinates

$$ds^2 = -dt^2 + dr^2 + r^2(d\theta^2 + \sin^2\theta d\phi^2). \tag{7.31}$$

Using (7.28) and (7.29) we get familiar expressions for an element of area on the surface of a sphere,

$$dA = r^2 \sin\theta\, d\theta\, d\phi, \tag{7.32}$$

and three-volume,

$$d\mathcal{V} = r^2 \sin\theta\, dr\, d\theta\, d\phi. \tag{7.33}$$

Example 7.6. Distance, Area, and Volume in the Curved Space of a Constant Density Spherical Star or a Homogeneous Closed Universe. The spatial metric for these situations turns out to be

$$dS^2 = \frac{dr^2}{1-(r/a)^2} + r^2\left(d\theta^2 + \sin^2\theta\, d\phi^2\right), \tag{7.34}$$

where a is a constant related to the density of matter. (We will see in Section 18.6 that this is one way of expressing the geometry on the three-dimensional surface of a sphere in a fictitious four-dimensional flat space.) Let's calculate the circumference around the equator, area, volume, and distance from center to surface of a sphere of coordinate radius R centred on $r = 0$ in this space.

The equator of the sphere is the curve $r = R, \theta = \pi/2$. Its circumference is

$$C = \oint dS = \int_0^{2\pi} r d\phi = 2\pi R. \tag{7.35}$$

The distance S from center to surface along a line $\theta = \text{const.}, \phi = \text{const.}$, is

$$S = \int dS = \int_0^R \frac{dr}{\sqrt{1-(r/a)^2}} = a \sin^{-1}\left(\frac{R}{a}\right). \tag{7.36}$$

The area of the two-surface $r = R$ is

$$A = \int dA = \int_0^{\pi} d\theta \int_0^{2\pi} d\phi R^2 \sin\theta = 4\pi R^2. \tag{7.37}$$

The volume inside $r = R$ is

$$\begin{aligned} \mathcal{V} = \int d\mathcal{V} &= \int_0^R dr \int_0^{\pi} d\theta \int_0^{2\pi} d\phi \frac{r^2 \sin\theta}{\sqrt{1-(r/a)^2}} \\ &= 4\pi a^3 \left\{ \frac{1}{2} \sin^{-1}\left(\frac{R}{a}\right) - \frac{R}{2a}\left[1 - \left(\frac{R}{a}\right)^2\right]^{1/2} \right\}. \end{aligned} \tag{7.38}$$

Of course, since the space is curved these expressions are different from those of a sphere in flat space. But it is not difficult to see that the familiar results are recovered when $R/a \ll 1$. For a neutron star, where $R \sim 10$ km, $a \sim 15$ km, and $R/a \sim .7$, the deviations from flat space results can be significant.

7.7 Embedding Diagrams and Wormholes

In Chapter 2 we used pictures of curved two-dimensional surfaces embedded in three-dimensional flat space to illustrate such curved two-dimensional geometries as the sphere (Figure 2.6) and a geometry shaped like a peanut (Figure 2.7). These figures are examples of the general idea of *embedding diagrams.* Not every curved two-dimensional geometry can be represented as a curved surface in three-dimensional flat space, but, for the many that can, the resulting embedding diagram is a useful way of visualizing their geometric properties.[3]

At least five dimensions would be required to represent a four-geometry as a surface in a flat space. The result would not be very helpful because it could not be readily pictured. However, it is sometimes possible to embed a *two-dimensional slice* of a four-dimensional geometry in three-dimensional flat space and learn something useful about its properties. Example 7.7 shows more clearly than any general explanation how this works.

Example 7.7. Embedding a Slice of a Wormhole Spacetime. Consider the metric

$$ds^2 = -dt^2 + dr^2 + (b^2 + r^2)(d\theta^2 + \sin^2\theta \, d\phi^2) \tag{7.39}$$

for some constant b with dimensions of length. This metric does not represent a physically realistic spacetime as far as is known but is an easy way to introduce embedding diagrams. The metric (7.39) is similar to the metric of flat spacetime

[3]Not every surface that is curved in flat three-dimensional space has a curved two-dimensional geometry. The surface of a cylinder has a flat geometry, for instance.

written in polar coordinates [cf. (7.4)] and shares a number of properties with it. It is independent of time t. It is spherically symmetric because a surface of constant r and t has the geometry of a sphere. At very large r the spacetime is approximately flat because the metric becomes close to (7.4). However, except for the value $b = 0$, the geometry is not flat but curved in an interesting way, as we will now see.

A t = const. slice of the geometry in (7.39) is a three-dimensional spatial geometry with metric

$$dS^2 = dr^2 + (b^2 + r^2)(d\theta^2 + \sin^2\theta d\phi^2). \tag{7.40}$$

All t = constant slices have the same geometry because the metric is independent of time. Because the spatial metric is spherically symmetric, a picture of it can be built up by looking at two-dimensional slices at a constant angle. For instance, the $\theta = \pi/2$ "equatorial" slice has a geometry described by

$$d\Sigma^2 = dr^2 + (b^2 + r^2)d\phi^2. \tag{7.41}$$

Spherical symmetry implies that any other constant-angle slice has the same geometry. This geometry *can* be visualized as a two-dimensional surface embedded in three-dimensional flat space. Let's find that surface.

The metric (7.41) of the two-dimensional r-ϕ slice has a rotational symmetry inherited from the spherical symmetry of the spacetime (7.39). Send ϕ into ϕ + const. and (7.41) remains unchanged. This suggests that it should be possible to embed the slice as an *axisymmetric* surface in three-dimensional flat space. To investigate this possibility it is convenient to locate points in flat space using cylindrical coordinates (ρ, ψ, z) based on the z-axis. The coordinate ρ is the distance from the axis, ψ is a polar angle around the axis, and z is the distance along the axis. The metric for flat space in these coordinates is

$$dS^2 = d\rho^2 + \rho^2 d\psi^2 + dz^2. \tag{7.42}$$

A surface in flat space can be specified by giving height above the $z = 0$ plane of each point in it, $z(r, \phi)$. We seek a function $z(r, \phi)$ specifying a surface that has the same geometry as (7.41). But to find that, we also have to specify the connection between the coordinates (ρ, ψ) that label a point on the surface in flat space and the coordinates (r, ϕ) that label points in (7.41). In short, to specify an embedding of the surface (7.41) we have to give three functions:

$$z = z(r, \phi), \qquad \rho = \rho(r, \phi), \qquad \psi = \psi(r, \phi). \tag{7.43}$$

Finding the functions in (7.43) is considerably simplified in the case of an axisymmetric surface when we can take $\psi = \phi$ and the functions z and ρ to be independent of these angles, namely,

$$z = z(r), \qquad \rho = \rho(r), \qquad \psi = \phi, \qquad \text{(axisymmetry)}. \tag{7.44}$$

Inserting (7.44) into (7.42) and working out the differentials, we find the following for the line element on the embedded surface:

$$d\Sigma^2 = \left[\left(\frac{dz}{dr}\right)^2 + \left(\frac{d\rho}{dr}\right)^2\right] dr^2 + \rho^2 d\phi^2. \tag{7.45}$$

This will agree with the metric on the slice (7.41) if

$$\rho^2 = r^2 + b^2 \tag{7.46a}$$

and

$$\left(\frac{dz}{dr}\right)^2 + \left(\frac{d\rho}{dr}\right)^2 = 1. \tag{7.46b}$$

Using (7.46a) for ρ, (7.46b) becomes a differential equation for $z(r)$, which can be integrated to give $z(r) = b \sinh^{-1}(r/b)$, with the integration constant chosen so that z vanishes when r does. Eliminating r in favor of ρ yields the equation of the curve in the ρ-z plane:

$$\rho(z) = b \cosh(z/b). \tag{7.47}$$

Figure 7.4 shows a graph of the curve (7.47) in the z-ρ plane. The full axisymmetric surface is generated by rotating this curve around the z-axis. (See Figure 7.5.) The range $0 < r < \infty$ that one might have been tempted to assume by analogy with flat space in fact covers only the half of the surface with $z > 0$. The value $r = 0$ does not label a point, but rather a circle at $\rho = b$ or $z = 0$. The bottom half of the surface with $z < 0$ can be covered by letting r range from $-\infty$ to 0. This surface in three-dimensional flat space has the same geometry as the constant time equatorial slice of the wormhole geometry.

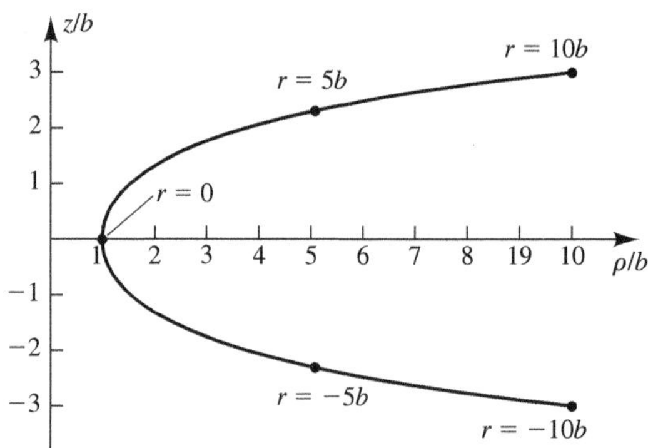

FIGURE 7.4 The curve $\rho = b \cosh(z/b)$, which when rotated around the z-axis, generates the two-dimensional surface shown in Figure 7.5, which has the same intrinsic geometry as (7.41), and is thus an embedding of an (r, ϕ) slice of the wormhole geometry (7.39) in three-dimensional flat space.

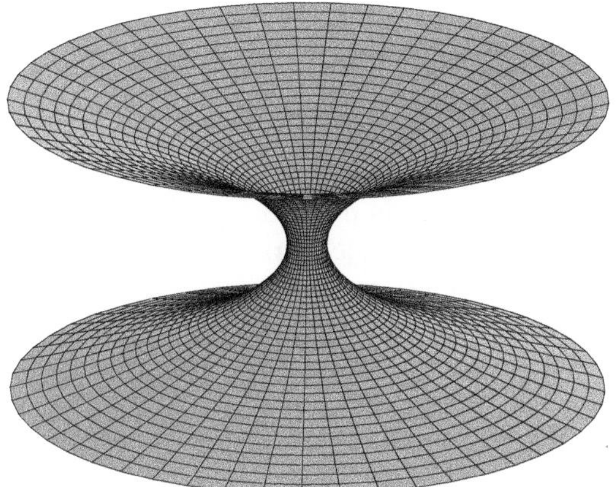

FIGURE 7.5 An embedding of the (r, ϕ) slice of the wormhole geometry (7.39) as a two-dimensional surface in flat three-dimensional space. This surface has *two* asymptotically flat regions connected by a "throat" of circumference $2\pi b$. It is therefore called a "wormhole" geometry.

BOX 7.2 Wormholes in Spacetime

The wormhole in the simple geometry of (7.39) illustrated in Figure 7.5 connects two different asymptotically flat regions of spacetime—two "universes" in the language of science fiction. Even more interesting might be a wormhole connecting two places in our own asymptotically flat region of spacetime, as qualitatively illustrated

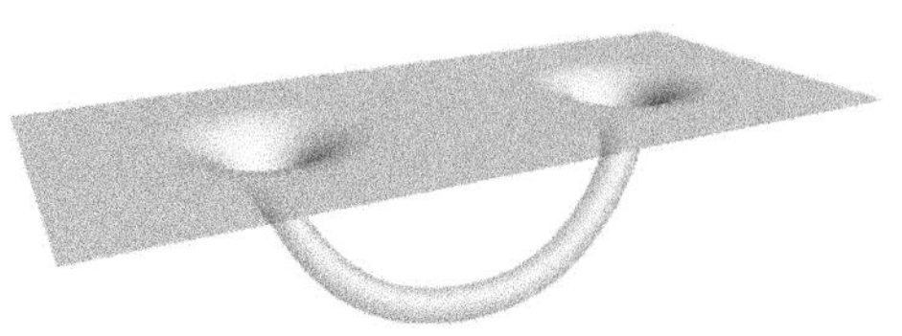

here. The figure shows an embedding diagram of a two-dimensional slice of spacetime at one instant of time in the approximate inertial frame of the asymptotic region. The wormhole "mouths" might appear as roughly spherical regions in space. By crawling through one mouth, one could emerge from the other in a different place, as the distinguished relativist Kip Thorne is shown doing in the figure from his book (Thorne 1994). The distance through the wormhole throat could be much shorter than the distance between the mouths in the region outside, enabling rapid travel between the two places. Indeed, one could imagine arranging the wormhole to connect events in spacetime so that one emerged at an earlier value of time in the approximate inertial frame than the value at which one went in! If the time was early enough, one could walk back in the outside region and meet oneself before one went through the wormhole. That is one way of imagining a machine for going backward in time. (For one that goes forward in time, see Box 9.1 on p. 192.)

There is no need to analyze causal paradoxes that would arise from such a time machine spacetime. The sober truth is that the classical Einstein equation implies that wormholes require matter with negative energy densities, and the energy densities of all known classical fields are positive. Short of invoking quantum fluctuations in spacetime geometry, the future domain of application for wormholes is probably entirely fictional.

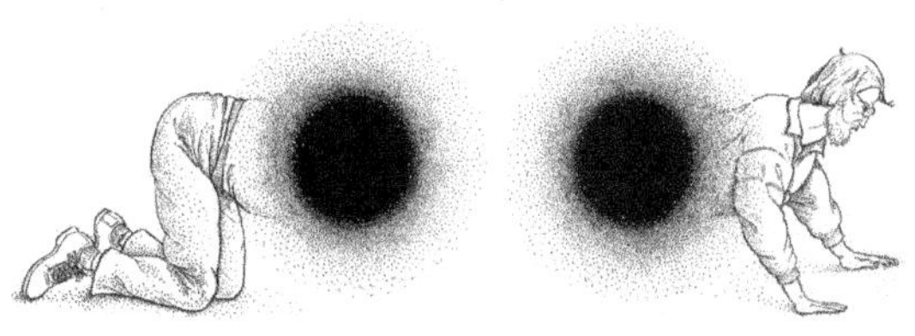

At large ρ (or equivalently large r) we know from (7.41) that the geometry of the surface becomes flat. But there is not just one asymptotically flat region, as in flat space, but two! They are connected by a curved throat of minimum circumference $2\pi b$. This kind of geometry is called a *wormhole*. In the language of science fiction, the wormhole connects two different "universes." One could imagine, for example, two different rockets—in different asymptotic regions—each orbiting the wormhole. In the next chapter the journey between them is described more quantitatively. Other kinds of wormholes are described in Box 7.2.

The surface specified by (7.41) could not be produced from a flat plane by smooth distortions. The geometry has not only a different metric from the flat plane but also a different *topology*.

7.8 Vectors in Curved Spacetime

The definition of vector as a directed line segment introduced in Section 5.1 has to be modified in curved spacetime.[4] Think of defining directed line segments on the surface of a potato! The key to defining vectors in curved spacetime is to recognize that vectorial quantities—momentum, velocity, current density, etc.—are all *local*. They can be measured by an observer in a laboratory located in a small region of spacetime. The way to define vectors in curved spacetime is, therefore, to separate the notions of magnitude and direction and to define direction *locally* by means of small vectors, exactly as a physicist working in a local laboratory would. Larger vectors can be built up algebraically by multiplying them by numbers and adding and subtracting according to the usual flat spacetime rules. A mathematician would call this procedure (described pretty crudely here[5]) defining vectors in a tangent space. Figure 7.6 shows a pictorial representation of the idea.

Vectors are thus defined at a point and there they obey all the usual flat spacetime rules of vector algebra. An assignment of a vector to each point in spacetime in a smooth way, $\mathbf{a} = \mathbf{a}(x)$, is called a *vector field*. Vectors defined at *different* points, however, are in *different* tangent spaces, and there is no way of adding vectors at different points, as there is in flat spacetime. Position vector is another notion that must be abandoned because it is not a local idea. Similarly, displacement vectors must be abandoned, except for the displacement vector between infinitesimally separated points, which is a local quantity.

Let's now review some of the machinery of vector algebra as it applies in curved spacetimes and add a little more to it. At every point, x^α, we can give a basis of four vectors, $\mathbf{e}_\alpha(x)$, in terms of which any other vector can be expressed

[4]What's meant here is that the notion of *four-vector* has to be modified, but recall in Chapter 5 that we warned that we would generally use the word *vector* in future chapters for both four-vectors in spacetime and three-vectors in three-dimensional space and rely on context to distinguish them. In this case *spacetime* is the giveaway.

[5]If immediate mathematical precision is needed, read Chapter 20.

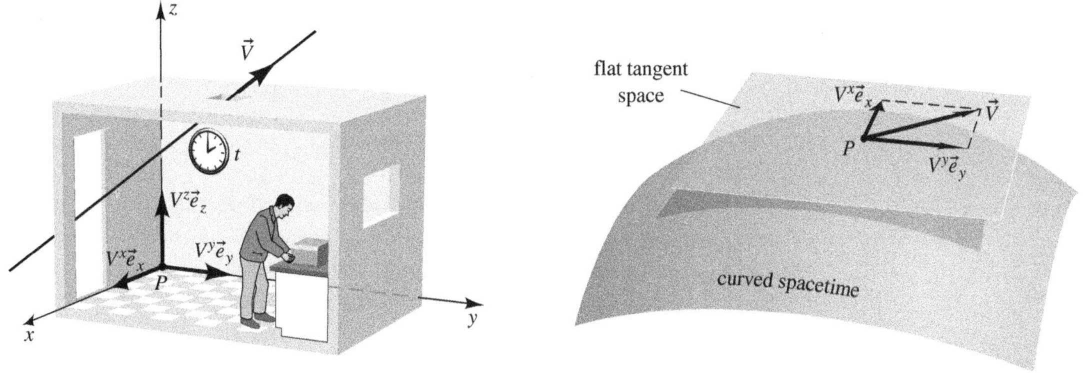

FIGURE 7.6 In physics quantities with magnitude and direction are typically defined locally and can be measured by an observer in a small laboratory located at a point in spacetime. The example of the velocity $\vec{V}$ is shown in this diagram—measured by an observer in a laboratory at left idealized as being at a point P and as a directed line segment in the corresponding tangent space at right. In that tangent space vectors can be added, subtracted, and multiplied by scalars as in flat space, as illustrated by $\vec{V} = V^x\vec{e}_x + V^y\vec{e}_y$.

as a linear combination:

$$\mathbf{a}(x) = a^\alpha(x)\mathbf{e}_\alpha(x). \tag{7.48}$$

The numbers $a^\alpha(x)$ are called the *components* of the vector $\mathbf{a}$ in the basis $\mathbf{e}_\alpha$.

The idea of *scalar product* can be introduced as in flat space. The scalar product between any two vectors $\mathbf{a}$ and $\mathbf{b}$ at the same point can be computed in terms of the components if the scalar products of the basis vectors are known:

$$\begin{aligned}\mathbf{a}\cdot\mathbf{b} &= (a^\alpha\mathbf{e}_\alpha)\cdot(b^\beta\mathbf{e}_\beta)\\ &= (\mathbf{e}_\alpha\cdot\mathbf{e}_\beta)a^\alpha b^\beta.\end{aligned} \tag{7.49}$$

We can pick a basis in which the scalar products are anything we like, but two types of bases are of particular importance.

Orthonormal Bases

An *orthonormal basis* consists of four mutually orthogonal vectors of unit length $\mathbf{e}_{\hat{\alpha}}$, $\hat{\alpha} = 0, 1, 2, 3$. As in Section 5.6, a hat on the index is used to distinguish orthonormal bases and components from other kinds. In spacetime three of the orthogonal unit vectors may be spacelike but one must be timelike. The requirements for an orthonormal basis are, therefore, conveniently summarized by

$$\mathbf{e}_{\hat{\alpha}}(x)\cdot\mathbf{e}_{\hat{\beta}}(x) = \eta_{\hat{\alpha}\hat{\beta}}, \tag{7.50}$$

Orthonormal Basis

where $\eta_{\hat{\alpha}\hat{\beta}} = \mathrm{diag}(-1, 1, 1, 1)$. In terms of orthonormal basis components, the scalar product between vectors is then, from (7.49),

Scalar Product in an Orthonormal Basis

$$\mathbf{a}\cdot\mathbf{b} = \eta_{\hat{\alpha}\hat{\beta}}a^{\hat{\alpha}}b^{\hat{\beta}}. \tag{7.51}$$

Figure 7.7 shows an orthonormal basis oriented along polar coordinates for the flat plane.

As described in Section 5.6, an observer's laboratory may be thought of as defining an orthonormal basis. The timelike vector $\mathbf{e}_{\hat{0}}$ is the observer's four-velocity $\mathbf{u}_{\text{obs}}$, and $\mathbf{e}_{\hat{\imath}}$ are three unit vectors that define the axes of the observer's laboratory. *This type of basis is important because the components in an observer's basis define measurable physical quantities.* Thus, if $\mathbf{e}_{\hat{\alpha}}$ is an orthonormal basis appropriate to a particular observer, $\mathbf{p}$ is the momentum of a particle being observed, and

$$\mathbf{p} = p^{\hat{\alpha}}\mathbf{e}_{\hat{\alpha}}, \tag{7.52}$$

then $E = p^{\hat{t}}$ is the observed energy and $p^{\hat{\imath}}$ are the components of the three-momentum. Exactly as in (5.82), these components can be computed by taking scalar products of $\mathbf{p}$ with the basis vectors. For instance the observed energy is [cf. (5.83)]

Energy Measured by an Observer

$$E = -\mathbf{p}\cdot\mathbf{u}_{\text{obs}}. \tag{7.53}$$

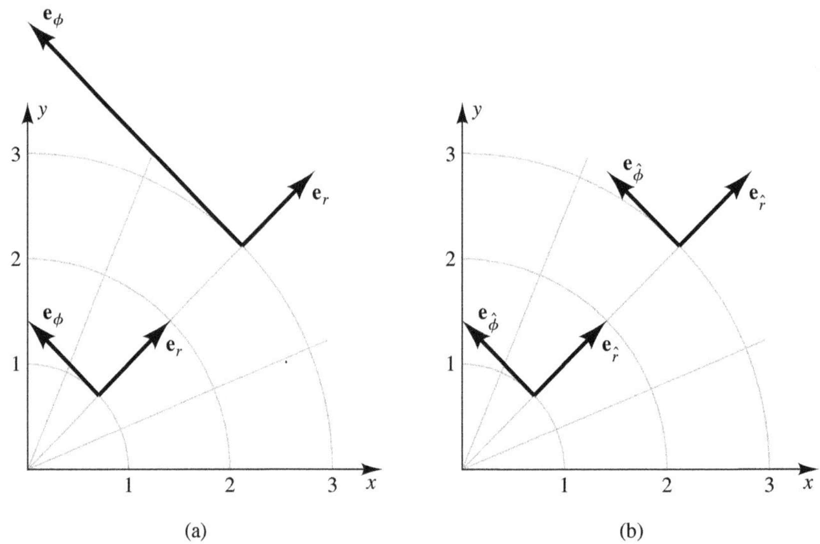

FIGURE 7.7 Coordinate and orthonormal basis vectors for polar coordinates in the plane. At left, the coordinate basis vectors point along the coordinate lines and have lengths $|\mathbf{e}_r| = 1$, $|\mathbf{e}_\phi| = r$. At right, the orthonormal basis vectors shown also point along the same coordinate lines but have unit length.

Coordinate Bases

The four-velocity $\mathbf{u}$ is a familiar example of a vector. Given a world line $x^\alpha(\tau)$, the components of the four-velocity might be expected to be [cf. (5.25)]

$$u^\alpha = \frac{dx^\alpha}{d\tau}. \tag{7.54}$$

But what basis are these components in? To find out note from (7.8) and $d\tau^2 = -ds^2$ that

$$g_{\alpha\beta}u^\alpha u^\beta = g_{\alpha\beta}\frac{dx^\alpha}{d\tau}\frac{dx^\beta}{d\tau} = -1. \tag{7.55}$$

The left-hand side defines $\mathbf{u} \cdot \mathbf{u}$, but not in an orthonormal basis where (7.50) holds. Rather, (7.54) are the components of the four-velocity in a different kind of basis, where

$$\mathbf{e}_\alpha(x) \cdot \mathbf{e}_\beta(x) = g_{\alpha\beta}(x). \tag{7.56}$$

Coordinate Basis

These are the defining relations of a *coordinate basis* where generally

$$\boxed{\mathbf{a} \cdot \mathbf{b} = g_{\alpha\beta}a^\alpha b^\beta.} \tag{7.57}$$

Scalar Product in a Coordinate Basis

Example 7.8. Polar Coordinates in the Plane. Consider polar coordinates in the two-dimensional flat plane. A coordinate basis consists of two vectors $\mathbf{e}_r$ and $\mathbf{e}_\phi$ pointing along the coordinate lines, as shown in Figure 7.7. The metric is

$$dS^2 = dr^2 + r^2\, d\phi^2. \tag{7.58}$$

From (7.56) these vectors are orthogonal because the off-diagonal components of the metric are zero. The lengths of the vectors are given by the square roots of the diagonal components. Although $\mathbf{e}_r$ is a unit vector because $|\mathbf{e}_r| = \sqrt{g_{rr}} = 1$, the length of $\mathbf{e}_\phi$ is $\sqrt{g_{\phi\phi}} = r$.

The vectors of a coordinate basis are in general not unit vectors, as the preceding example shows, nor are they generally mutually orthogonal. Nevertheless, as we'll see in the next chapter, coordinate bases are useful for computation, and we will use them frequently.

Actually we've been using coordinate bases all along in special relativity. Equation (5.12) is the same as (7.56). It just happens that for an inertial frame in flat space, the metric $g_{\alpha\beta}$ is $\eta_{\alpha\beta}$, so the coordinate basis for an inertial frame is also an orthonormal basis. The same is true for the coordinate basis vectors of a local inertial frame [cf. (7.12)]. That won't be true in general in curved space, and therefore it's important to keep the two ideas distinct. The convention of using hats over indices for orthonormal bases and no hats for coordinate bases helps to do that. If a coordinate basis is also orthonormal, it doesn't matter which notation is used.

Working with Coordinate and Orthonormal Bases

Curved spacetimes are explored through the study of test particles and light rays that move in them—both theoretically and experimentally. As we see in the next chapter, the motion of test particles can be directly calculated from equations of motion for the *coordinate basis components* of vectors like the four-velocity. But the coordinate basis components generally cannot be interpreted as predictions for observations.[6] Observers measure components of vectors in their associated *orthonormal* basis. It is, therefore, necessary to be able to deal with both kinds of components. Indeed it would be only a modest oversimplification to say that we will *calculate* in coordinate bases and *interpret* the results in orthonormal bases. Box 7.3 on p. 157 is an exotic illustration of that.

To see how to move back and forth between different bases, let's consider just one coordinate basis, $\{\mathbf{e}_\alpha\}$, and one orthonormal basis, $\{\mathbf{e}_{\hat{\alpha}}\}$. (The notation { } means *set of.*) Despite the similarity in notation, these are *different* sets of vectors, with different lengths and directions. A vector **a** can be expanded in either basis,

$$\mathbf{a} = a^\alpha \mathbf{e}_\alpha = a^{\hat{\beta}} \mathbf{e}_{\hat{\beta}}, \tag{7.59}$$

thus defining the coordinate components a^α and the orthonormal components $a^{\hat{\beta}}$. These components can be connected if the coordinate components $(\mathbf{e}_{\hat{\beta}})^\alpha$ of the orthonormal basis vectors and the orthonormal components $(\mathbf{e}_\alpha)^{\hat{\beta}}$ of the coordinate basis vectors are both known.[7] Then

$$a^\alpha = a^{\hat{\beta}} (\mathbf{e}_{\hat{\beta}})^\alpha, \qquad a^{\hat{\beta}} = a^\alpha (\mathbf{e}_\alpha)^{\hat{\beta}}. \tag{7.60}$$

The notation used here is intended to keep distinct the two kinds of indices in play—one labeling components and the other labeling vectors. For instance, $(\mathbf{e}_{\hat{2}})^1$ is the 1 coordinate component of the vector $\mathbf{e}_{\hat{2}}$, whereas $(\mathbf{e}_3)^{\hat{2}}$ is the $\hat{2}$ orthonormal component of the vector $\mathbf{e}_3$. The following examples illustrate the connection.

Example 7.9. Orthonormal Basis Vectors along Orthogonal Coordinate Directions. Suppose the metric happens to be diagonal in a certain coordinate system having the form (7.27). Any set of four vectors pointing along the four coordinate directions will be mutually orthogonal, so that six of the relations defining an orthonormal basis (7.50) are already satisfied. Making these vectors unit vectors satisfies the rest. One example of an orthonormal basis is, therefore,

[6] Indeed, where a coordinate system becomes singular, as discussed on p. 136, coordinate components can diverge when there is no physical singularity.

[7] If you are lecturing, practice saying this quickly.

BOX 7.3 Extra Dimensions?

The idea that spacetime has more than the four familiar dimensions has a long history in the search for unified theories of the fundamental forces. But how could we be unaware of extra dimensions? One answer is that they could be curled up ("compactified") on microscopic length scales. The simplest case is a five-dimensional spacetime in which the fifth dimension runs around a circle with a very small radius. An example of a line element describing such a spacetime is

$$\begin{aligned} ds^2 &= g_{\mathrm{AB}} dx^{\mathrm{A}} dx^{\mathrm{B}} \\ &= -dt^2 + dx^2 + dy^2 + dz^2 + R^2 d\psi^2, \end{aligned} \tag{a}$$

where $0 \leq \psi < 2\pi$ and A, B, ... range over 0 to 4. Note that R is a constant fixing the size of the circle, not a radial coordinate.

To see how it might be difficult to detect such a fifth dimension, imagine a plane wave of some zero rest mass field $\Phi(x^{\mathrm{A}})$ (like a component of the electromagnetic field) propagating in the spacetime (a). We'll see that if the frequency of the wave is sufficiently low, its propagation is little affected by the extra dimension. Accept that the field for such a wave could have the form

$$\Phi_{\mathbf{k}}(x^{\mathrm{A}}) \propto \cos(\mathbf{k} \cdot \mathbf{x}) \equiv \cos(g_{\mathrm{AB}} k^{\mathrm{A}} x^{\mathrm{B}}) \tag{b}$$

for $x^{\mathrm{A}} = (t, \vec{x}, \psi)$ and a five-dimensional wave vector $\mathbf{k}$ with components $k^{\mathrm{A}} = (\omega, \vec{k}, k^4)$. These are the *coordinate basis components* of $\mathbf{k}$ because in (b) they enter into an expression for the scalar product of the form (7.57). Here, ω is the frequency of the wave, $\vec{k}$ is the three-dimensional wave vector, and k^4 is the component of the wave vector in the fifth dimension [cf. (5.69)]. For a zero rest mass field (recall (5.70)),

$$\mathbf{k} \cdot \mathbf{k} \equiv g_{\mathrm{AB}} k^{\mathrm{A}} k^{\mathrm{B}} = 0. \tag{c}$$

If the fifth dimension runs around a circle, then the field (b) must be periodic in ψ with period 2π. That can happen only at the discrete values of k^4 at which the value of $\mathbf{k} \cdot \mathbf{x}$ when $\psi = 2\pi$ differs from its value at $\psi = 0$ by a multiple of 2π, i.e.,

$$\begin{aligned} g_{44} k^4 (2\pi) &= R^2 k^4 (2\pi) \\ &= 2\pi n, \quad n = 0, 1, 2, \ldots . \end{aligned} \tag{d}$$

The consequence of this periodicity is that k^4 is restricted to the values

$$k^4 = n/R^2, \quad n = 0, 1, 2, \ldots . \tag{e}$$

Condition (c) can then be solved to give the frequency of the wave as follows:

$$\omega^2 = \vec{k}^2 + (n/R)^2. \tag{f}$$

For $n = 0$ this gives the relation $\omega = |\vec{k}|$, as if the wave were propagating in four-dimensional spacetime. Deviations from this relation occur for higher values of n, but these require field quanta with energies

$$E = \hbar\omega \geq \hbar/R. \tag{g}$$

If R is of the order of the Planck length $\ell_{Pl} = (G\hbar/c^3)^{1/2}$ characteristic of quantum gravity described on p. 11, then in $c \neq 1$ units,

$$E \geq (\hbar c/\ell_{Pl}) \sim 10^{19} \text{ GeV}. \tag{h}$$

This is many orders of magnitude above the highest energies available in contemporary accelerators. Were there extra dimensions curled up on such a small scale, we might well not have noticed them yet.

Equation (e) turns out to be the condition that there are an integer number of wavelengths going around the circle in the fifth dimension. However, that is not so very evident from the coordinate basis components of $\mathbf{k}$; k^4 doesn't even have the correct dimension to be inversely related to wavelength. The components of $\mathbf{k}$ in an orthonormal basis are so related. A unit-length basis vector pointing along the fifth dimension has coordinate basis components [cf. (7.61)]

$$(e_{\hat{4}})^{\mathrm{A}} = (0, 0, 0, 0, 1/R), \tag{i}$$

so that the corresponding orthonormal basis component of $\mathbf{k}$ is, from (7.60):

$$k^{\hat{4}} = n/R. \tag{j}$$

Defining the wavelength along the fifth dimension as $2\pi/k^{\hat{4}}$, (j) does mean that there are an integer number of wavelengths in the circle of circumference $2\pi R$.

$$(\mathbf{e}_{\hat{0}})^\alpha = [(-g_{00})^{-1/2}, 0, 0, 0], \tag{7.61a}$$

$$(\mathbf{e}_{\hat{1}})^\alpha = [0, (g_{11})^{-1/2}, 0, 0], \ldots, \text{etc.} \tag{7.61b}$$

Using (7.57) it is easy to check that (7.50) is satisfied.

Example 7.10. Different Bases in Two-Dimensional Polar Coordinates. In the two-dimensional polar coordinate example shown in Figure 7.7, the orthonormal basis vectors $\mathbf{e}_{\hat{r}}$ and $\mathbf{e}_{\hat{\phi}}$ point in the same directions as the corresponding coordinate basis vectors $\mathbf{e}_r$ and $\mathbf{e}_\phi$ but have unit length everywhere. The components of the coordinate basis vectors in the coordinate basis are, by definition,

$$(\mathbf{e}_r)^A = (1, 0), \qquad (\mathbf{e}_\phi)^A = (0, 1) \tag{7.62}$$

and, similarly,

$$(\mathbf{e}_{\hat{r}})^{\hat{A}} = (1, 0), \qquad (\mathbf{e}_{\hat{\phi}})^{\hat{A}} = (0, 1), \tag{7.63}$$

where the indices A and $\hat{A}$ range over 1 and 2. But what about the coordinate components of the orthonormal basis vectors and vice versa? We have

$$(\mathbf{e}_{\hat{r}})^A = (1, 0), \qquad (\mathbf{e}_{\hat{\phi}})^A = (0, 1/r). \tag{7.64}$$

The defining relations for an orthonormal basis (7.50) are easily checked using the metric (7.58). Similarly the orthonormal basis components of the unit vectors of the coordinate basis vectors are

$$(\mathbf{e}_r)^{\hat{A}} = (1, 0), \qquad (\mathbf{e}_\phi)^{\hat{A}} = (0, r). \tag{7.65}$$

The defining relations of a coordinate basis (7.56) are easily checked using (7.51) as are the connections (7.60).

7.9 Three-Dimensional Surfaces in Four-Dimensional Spacetime

Just as there are two-dimensional surfaces in three-dimensional space, there are three-dimensional surfaces in four-dimensional spacetime. They are called *three-surfaces*. *Hypersurface* is another frequently used term. A three-surface can be specified by giving one coordinate as a function of the other three, e.g.,

$$x^0 = h(x^1, x^2, x^3). \tag{7.66}$$

The function h gives the position in x^0 of the point in the surface located by (x^1, x^2, x^3). More symmetrically, a three-surface can be specified by a function $f(x^\alpha)$:

$$f\left(x^{\alpha}\right)=0. \tag{7.67}$$

For the surface specified by (7.66), the difference between its left- and right-hand sides could be the function $f(x^{\alpha})$.

Normal and Tangent Vectors

At each point on a three-surface, there are directions in spacetime that lie *in* the surface, that is, directions that are tangent to it. Tangent vectors $\mathbf{t}$ point in these directions, and there are three linearly independent ones. The normal direction lies along a vector $\mathbf{n}$ at the point that is orthogonal to every tangent vector. That is,

$$\mathbf{n}\cdot\mathbf{t}=0 \tag{7.68}$$

for all tangent vectors $\mathbf{t}$. The vector $\mathbf{n}$ is a *normal* to the surface. (See Figure 7.8.)

A three-surface has its own three-dimensional geometry. The line element defining the intrinsic geometry of the surface is found by using a defining relation such as (7.66) to eliminate one of the coordinates from the line element defining the geometry of spacetime. Some important classes of these surface geometries are discussed in the following.

Spacelike Surfaces

Spacelike surfaces are best introduced by a simple example:

Example 7.11. Constant Time Three-Surfaces in Flat Spacetime. Consider flat-spacetime in the rectangular coordinates (t, x, y, z) of a Lorentz frame. The line element is the now-familiar (4.8) (with $c = 1$). Any constant value of t

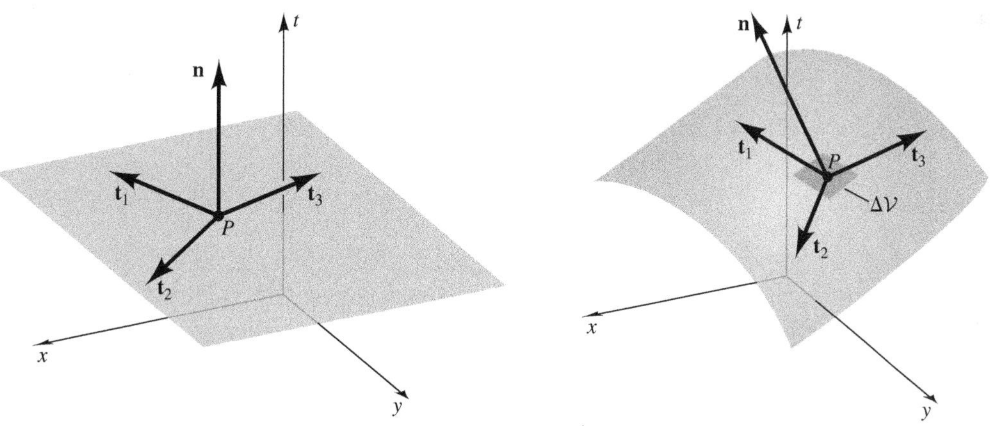

FIGURE 7.8 Spacelike Surfaces. At left is a spacelike surface $t = t_*$ in flat spacetime. At right is a more general example specified by $t = h(x, y, z)$ for some function h. Spacelike tangent vectors such as $\mathbf{t}_1$, $\mathbf{t}_2$, $\mathbf{t}_3$ lie in the surfaces, and timelike normal vectors $\mathbf{n}$ are orthogonal to all tangent directions. The orientation of an element of three-volume $\Delta\mathcal{V}$ in spacetime is specified by its normal four-vector.

specifies a three-surface in flat spacetime, as illustrated in Figure 7.8:

$$t = \text{const.} \tag{7.69}$$

A point in the surface is located by (x, y, z), and the metric obtained by substituting (7.69) into (4.8) is

$$dS^2 = dx^2 + dy^2 + dz^2, \tag{7.70}$$

defining the geometry of flat three-dimensional space. Any vector with a zero time component is a tangent vector **t** to the surface

$$t^\alpha = \left(0, \vec{t}\right). \tag{7.71}$$

A normal vector **n** satisfying (7.68) is

$$n^\alpha = (1, 0, 0, 0). \tag{7.72}$$

This is a unit normal vector because $\mathbf{n} \cdot \mathbf{n} = -1$.

Example 7.11 is a simple case of a *spacelike surface*—one for which each tangent vector is spacelike. As the example also illustrates, spacelike surfaces have timelike normals

$$\mathbf{n} \cdot \mathbf{n} < 0 \qquad \text{(spacelike surface)}. \tag{7.73}$$

Just as the orientation of an element of area ΔA in three-dimensional space is specified by its normal $\vec{n}$, so also the orientation of an element of *volume* $\Delta\mathcal{V}$ in *spacetime* is specified by its normal **n** in spacetime, as illustrated in Figure 7.8.

Spacelike surfaces provide the general notion of "space" in spacetime. Spacetime can be divided into space and time by finding a family of spacelike surfaces such that each point lies on one and only one member. The family of $t =$ const. spacelike surfaces in flat spacetime is a simple example illustrated in Figure 7.9. Another is the family of surfaces with a constant value of the time $t' = \gamma(t - vx)$ of a different inertial frame. In a (t, x) spacetime diagram, the $t =$ const. surfaces are horizontal, and the $t' =$ constant surfaces have a slope v. There are just as many ways of dividing spacetime into space and time as there are such families of spacelike surfaces. Example 7.12 is less trivial.

Example 7.12. A Lorentz Hyperboloid. To see another interesting example of a spacelike three-surface in four-dimensional flat spacetime, start with the line element in usual polar coordinates, (t, r, θ, ϕ), as in (7.4) and consider the surface defined by a constant a through

$$-t^2 + r^2 = -a^2. \tag{7.74}$$

A cross section is the hyperbola illustrated in the t-r spacetime diagram in Figure 7.10. This is called a *Lorentz hyperboloid.* Points on this surface can be labeled

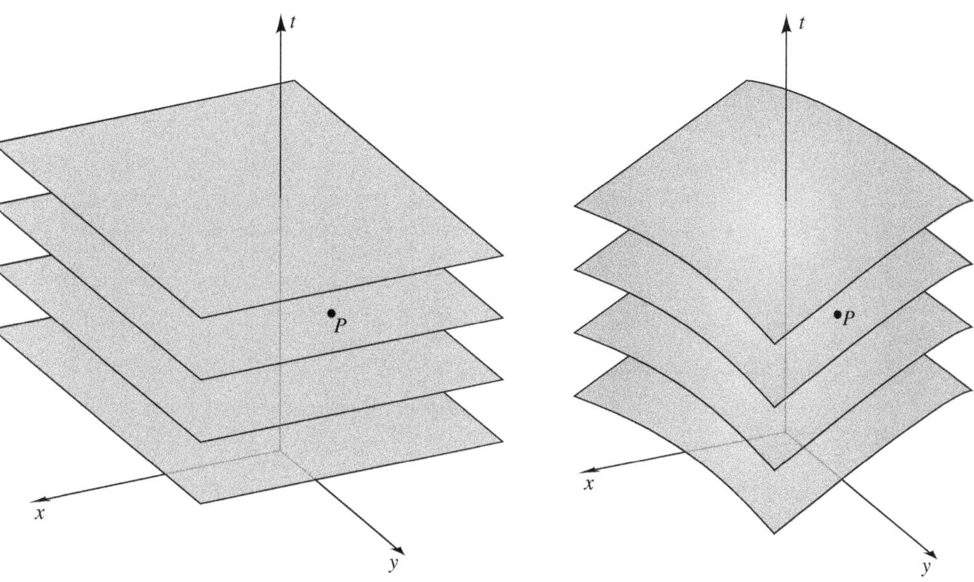

FIGURE 7.9 Space and Time. Families of spacelike surfaces divide spacetime up into space and time. At left is a family of $t = t_* =$ constant surfaces—one surface for each value of t_*. Each point P in spacetime lies on one such surface. The value of t_* can be said to be its time, and the position in the surface gives its location in space. But there are many different families of spacelike surfaces, such as the one illustrated at right, and correspondingly many different ways of dividing spacetime up into space and time.

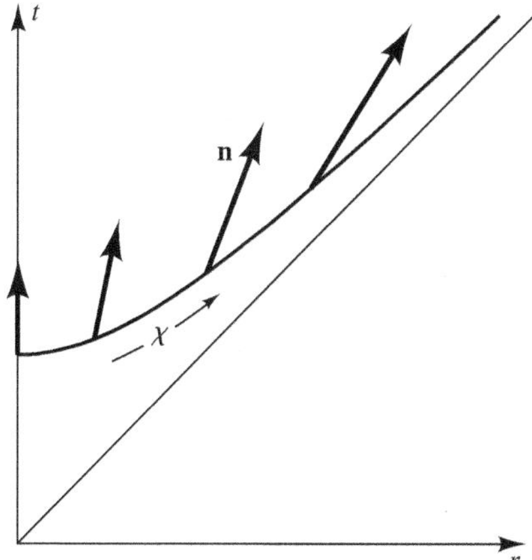

FIGURE 7.10 A Lorentz hyperboloid. This t-r spacetime diagram shows a cross section of the surface defined by (7.74). Points along the curve can by labeled by a coordinate χ, as defined in (7.75). Each point on the curve corresponds to a two-sphere containing the other two directions in the surface—those along θ and ϕ. A sequence of equal-length timelike normal vectors is shown at equally spaced values of χ. These are not normal to the surface nor of equal length in the geometry of the plane. But they are in the geometry of spacetime! At large χ the surface asymptotically approaches the light cone $t = r$, and the normal vector asymptotically lies in the surface.

elegantly by θ, ϕ, and a radial coordinate χ related to t and r by

$$t = a \cosh \chi, \qquad r = a \sinh \chi \tag{7.75}$$

so that (7.74) is satisfied for any $0 < \chi < \infty$. The line element describing the geometry in this surface is the spatial part of the line element found by substituting (7.75) into (7.4). Explicitly,

$$dS^2 = a^2[d\chi^2 + \sinh^2 \chi (d\theta^2 + \sin^2 \theta \, d\phi^2)], \tag{7.76}$$

showing that the surface is indeed spacelike.

A displacement in the surface by a small change $\Delta\chi$ is along the tangent vector [cf. (7.75)],[8]

$$t^\alpha = (a \sinh \chi, \; a \cosh \chi, 0, 0). \tag{7.77}$$

A unit normal vector orthogonal to this direction and the θ- and ϕ-directions in the surface is then

$$n^\alpha = (\cosh \chi, \; \sinh \chi, \; 0, \; 0). \tag{7.78}$$

Note that $\mathbf{n} \cdot \mathbf{n} = -1$, as required for a unit normal to a spacelike surface.

This example is not as abstract as it might seem. The geometry (7.76) is one possibility for the geometry of space in an important class of cosmological models, as we will see in Chapter 18. The family of spacelike hyperboloids obtained by varying a is another way of dividing the spacetime inside the forward light cone of the origin up into space and time (Problem 26).

Null Surfaces

Surfaces generated by light rays are another important class of three-surfaces called *null surfaces*. At each point in a null surface, there is one tangent direction $\boldsymbol{\ell}$ that points along a light ray and is null,

$$\boldsymbol{\ell} \cdot \boldsymbol{\ell} = 0, \tag{7.79}$$

and two orthogonal spacelike directions. The null direction $\boldsymbol{\ell}$ is a normal to the null surface because it is orthogonal to the spacelike directions and also to itself by virtue of (7.79). A normal to a null surface is a null vector that lies in it.

Example 7.13. The Light Cone as a Null Surface. In flat spacetime, the future light cone of the origin illustrated in Figure 7.11 is an example of a null surface. Using time and spatial polar coordinates (t, r, θ, ϕ), an equation for the surface is

$$t = r. \tag{7.80}$$

[8]Don't get the tangent vector t^α mixed up with the coordinate t.

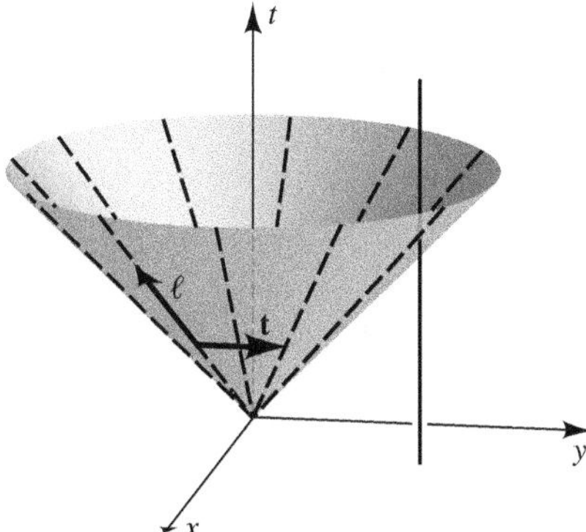

FIGURE 7.11 The future light cone of the origin of an inertial frame. This null surface is generated by radial light rays (the straight lines in the surface) that move outward from a single event. The normal to the surface $\boldsymbol{\ell}$ lies in the surface and along the generating light rays. A tangent vector **t** is also shown. Null surfaces like this one have a "one-way" property: once a timelike world line crosses the surface, it cannot cross it again.

A point in the surface is labeled by (r, θ, ϕ), and its location in spacetime then given by (7.80).

This three-surface is generated by a sphere of light rays moving radially outward from the origin with speed 1. The components $(\ell^t, \ell^r, \ell^\theta, \ell^\phi)$ of the vector $\boldsymbol{\ell}$ along any of these light rays is

$$\ell^\alpha = (1, 1, 0, 0), \tag{7.81}$$

and this is a normal vector to the surface. Two other linearly independent spacelike tangent vectors are $(0, 0, r^{-1}, 0)$ and $(0, 0, 0, (r \sin\theta)^{-1})$, chosen here to be of unit length.

Like the future light cone in flat space, many null surfaces that we will meet are one-way surfaces in the following sense: the world line of a particle can pass through a null surface, as illustrated in Figure 7.11, but it cannot pass through the same null surface again. Think of remaining stationary while the outward-moving sphere of light passes by. At the moment the sphere passes, your world line has crossed the null surface. But you cannot turn around and catch up with any part of it—all parts are moving away at the speed of light. As we will see in Chapter 12 and Chapter 15, the surfaces defining black holes are null surfaces with this one-way property: you can fall through one but you can never get back out.

Problems

1. **(a)** In the singular line element for the plane (7.7), show that the distance between $r' = 0$ and a point with any finite value of r' is infinite.

(b) Find the distance between $r' = 5$ and $r' = \infty$ along the line $\phi = 0$.

2. The following line element corresponds to flat spacetime:

$$ds^2 = -dt^2 + 2dx\,dt + dy^2 + dz^2.$$

Find a coordinate transformation that puts the line element in the usual flat space form (7.1).

3. [C, P] (*The Sagnac Effect*) The Sagnac effect was worked out in an inertial frame in Box 3.1 on p. 35. Two light waves propogate in opposite directions around a rotating ring. The phase of a wave with frequency ω at time t a distance S around the ring is $\Psi \equiv -\omega(t - S) + \text{const}$. (The speed v of a light wave is 1.) When there is a difference in phase of a multiple of 2π the waves constructively interfere.

   It is also possible to work out the Sagnac effect in a frame rotating with the interferometer. The line element of flat spacetime in that frame can be found by defining a new coordinate $\phi = \phi' + \Omega t$. Derive the condition for constructive interference in this frame.

4. [B] In the Penrose diagram for flat space spanned by the coordinates (t', r'), make a rough sketch of the following (a) a curve of constant r and (b) a curve of constant t.

5. Consider the two-dimensional spacetime spanned by coordinates (v, x) with the line element

$$ds^2 = -x\,dv^2 + 2\,dv\,dx.$$

   (a) Calculate the light cone at a point (v, x).

   (b) Draw a (v, x) spacetime diagram showing how the light cones change with x.

   (c) Show that a particle can cross from positive x to negative x but cannot cross from negative x to positive x.

   (*Comment*: The light cone structure of this model spacetime is in many ways analogous to that of black-hole spacetimes to be considered in Chapter 12, in particular in having a surface such as $x = 0$, out from which you cannot get.)

6. [B] Express the line element for flat spacetime in terms of the coordinates (t', r', θ, ϕ) used to construct the Penrose diagram and defined in (a) and (c) in Box 7.1 on p. 137.

7. [S] *Transformation Law for the Metric* A general coordinate transformation is specified by four functions $x'^{\alpha} = x'^{\alpha}(x^{\beta})$.

   (a) Show that the chain rule can be expressed by

$$dx^{\alpha} = \frac{\partial x^{\alpha}}{\partial x'^{\gamma}} dx'^{\gamma}.$$

   (b) Substitute this into the line element (7.8) to show that the transformed metric $g'_{\gamma\delta}$ is given by

$$g'_{\gamma\delta} = g_{\alpha\beta} \frac{\partial x^{\alpha}}{\partial x'^{\gamma}} \frac{\partial x^{\beta}}{\partial x'^{\delta}}.$$

   Make sure your answers are consistent with the summation convention.

8. **(a)** Use the mathematical fact that any real symmetric matrix can be diagonalized by an orthogonal matrix to show that any metric can be diagonalized at one point P

by a linear transformation of the form

$$x'^{\alpha} = M^{\alpha}_{\beta} x^{\beta}.$$

In particular, make clear the connection between orthogonal matrix of the theorem and $g_{\alpha\beta}(x_P)$, and between M^{α}_{β} and the components of the orthogonal diagonalizing matrix.

(b) Find the linear transformation that will diagonalize the warp-drive metric (7.25) at any one point along the trajectory $x_s(t)$.

9. [C] The argument in Section 7.4 shows that at a point P there are coordinates in which the value of the metric takes its flat space form $\eta_{\alpha\beta}$. But are there coordinates in which the first derivatives of the metric vanish at P as they do in flat space? What about the second derivatives? The following counting argument, although not conclusive, shows how far one can go.

The rule for transforming the metric between one coordinate system and another was worked out in Problem 7. This can be expanded as a power (Taylor) series about x_P:

$$x^{\alpha}(x'^{\beta}) = x^{\alpha}(x'^{\beta}_P) + \left(\frac{\partial x^{\alpha}}{\partial x'^{\beta}}\right)_{x_P} (x'^{\beta} - x'^{\beta}_P)$$
$$+ \frac{1}{2}\left(\frac{\partial^2 x^{\alpha}}{\partial x'^{\beta}\partial x'^{\gamma}}\right)_{x_P} (x'^{\beta} - x'^{\beta}_P)(x'^{\gamma} - x'^{\gamma}_P)$$
$$+ \frac{1}{6}\left(\frac{\partial^3 x^{\alpha}}{\partial x'^{\beta}\partial x'^{\gamma}\partial x'^{\delta}}\right)_{x_P} (x'^{\beta} - x'^{\beta}_P)(x'^{\gamma} - x'^{\gamma}_P)(x'^{\delta} - x'^{\delta}_P) + \cdots .$$

At the point x^{α}_P there are 16 numbers $(\partial x^{\alpha}/\partial x'^{\beta})_{x_P}$ to adjust to make the transformed values of the metric $g'_{\alpha\beta}$ equal to $\eta_{\alpha\beta}$. Since there are only 10 $g'_{\alpha\beta}$, we can do this and still have 6 numbers to spare! These 6 degrees of freedom correspond exactly to the 3 rotations and 3 Lorentz boosts, which leave $\eta_{\alpha\beta}$ unchanged. Following this line of reasoning, fill in the rest of the spaces in the following table to show that there is enough freedom in coordinate transformations to make the first derivatives of the metric vanish in addition to (7.12) but not the second derivatives:

	Conditions	Numbers
$g'_{\alpha\beta} = \eta_{\alpha\beta}$	10	16
$\partial g'_{\alpha\beta}/\partial x'^{\gamma} = 0$	?	?
$\partial^2 g'_{\alpha\beta}/\partial x'^{\gamma}\partial x'^{\delta} = 0$	?	?

When properly organized, the second derivatives that cannot be transformed away are the measure of spacetime curvature, as we shall see in Chapter 22. How many of them are there?

10. An observer moves on a curve $X = 2T$ for $T > 1$ in the two-dimensional geometry with metric (7.20).

(a) What are the components of the four-velocity of this observer? Is the curve a timelike one?

(b) Find the components of an orthonormal basis $\mathbf{e}_{\hat{0}}$, $\mathbf{e}_{\hat{1}}$ for this observer.

11. [S] For the warp-drive spacetime in Example 7.4, show that, at every point along the curve $x_s(t)$, the four-velocity of the ship lies *inside* the forward light cone.

12. In the warp-drive spacetime in Example 7.4, how much ship time elapses on a trip between stations that takes coordinate time T?

13. [S] Consider two vector fields $\mathbf{a}(x)$ and $\mathbf{b}(x)$ and a world line $x^\alpha(\tau)$ in a spacetime with metric $g_{\alpha\beta}$. Derive an expression for $d(\mathbf{a} \cdot \mathbf{b})/d\tau$ in terms of partial derivatives of the coordinate basis components of $\mathbf{a}$ and $\mathbf{b}$, the partial derivatives of $g_{\alpha\beta}$, and the components of the four-velocity $\mathbf{u}$.

14. In a certain spacetime geometry the metric is

$$ds^2 = -(1 - Ar^2)^2 \, dt^2 + (1 - Ar^2)^2 dr^2 + r^2(d\theta^2 + \sin^2\theta \, d\phi^2).$$

(a) Calculate the proper distance along a radial line at constant t from the center $r = 0$ to a coordinate radius $r = R$.

(b) Calculate the area of a sphere of coordinate radius $r = R$.

(c) Calculate the three-volume of a sphere of coordinate radius $r = R$.

(d) Calculate the four-volume of a four-dimensional tube bounded by a sphere of coordinate radius R and two $t =$ constant planes separated by a time T.

15. [S] Calculate the area of the peanut illustrated in Figure 2.7.

16. [B] Suppose that you have a map of the world in the Mercator projection as described in Box 2.3 on p. 25. The map is 1m wide. You use the Cartesian coordinates (x, y) described in the box to locate points on the map. Greenland is approximated by a rectangle extending from $x = -5$ cm to $x = -14$ cm and $y = 21$ cm to $y = 38$ cm. The United States is approximated by a rectangle extending from $x = -21$ cm to $x = -34$ cm and $y = 8$ cm to $y = 12$ cm. On the map, therefore, Greenland has an area about 3 times that of the U.S. Use the line element specified in these coordinates by equations (f) and (i) in the box find the true ratio of areas of these rectangles. Caution: These rectangles do not represent the actual areas of Greenland and the U.S. very accurately.

17. [S] Calculate the three-dimensional volume on a $t =$ const. slice of the wormhole geometry (7.39) bounded by two spheres of coordinate radius R on each side of the throat.

18. Consider the three-dimensional space with the line element

$$dS^2 = \frac{dr^2}{(1 - 2M/r)} + r^2(d\theta^2 + \sin^2\theta \, d\phi^2).$$

(a) Calculate the radial distance between the sphere $r = 2M$ and the sphere $r = 3M$.

(b) Calculate the spatial volume between the two spheres in part (a).

19. The surface of a sphere of radius R in four flat Euclidean dimensions is given by

$$X^2 + Y^2 + Z^2 + W^2 = R^2.$$

(a) Show that points on the sphere may be located by coordinates (χ, θ, ϕ), where

$$X = R \sin \chi \sin \theta \cos \phi, \qquad Z = R \sin \chi \cos \theta,$$
$$Y = R \sin \chi \sin \theta \sin \phi, \qquad W = R \cos \chi .$$

(b) Find the metric describing the geometry on the surface of the sphere in these coordinates.

20. *Make the cover* Consider the two-dimensional geometry with the line element

$$d\Sigma^2 = \frac{dr^2}{(1 - 2M/r)} + r^2 d\phi^2 .$$

Find a two-dimensional surface in three-dimensional flat space that has the *same* intrinsic geometry as this slice. Sketch a picture of your surface. (*Comment*: This is a slice of the Schwarzschild black-hole geometry to be discussed in Chapter 12. It is also the surface on the cover of this book.)

21. Consider a two-dimensional flat space with a skew coordinate system, the x^1, x^2 axes making an angle of 45° with each other.

(a) Reproduce the accompanying coordinate grid and draw on it the basis vectors $\mathbf{e}_1, \mathbf{e}_2$ of a coordinate basis associated with x^1, x^2.

(b) Calculate the components of the metric g_{AB} (A, B range over 1, 2) from the scalar product of the basis vectors.

(c) On the coordinate grid, draw a vector $\mathbf{V}$ of length 2 making an angle of 30° with the x^1-axis. Calculate the components V^A for this vector. Can you give a geometric construction for finding V^A?

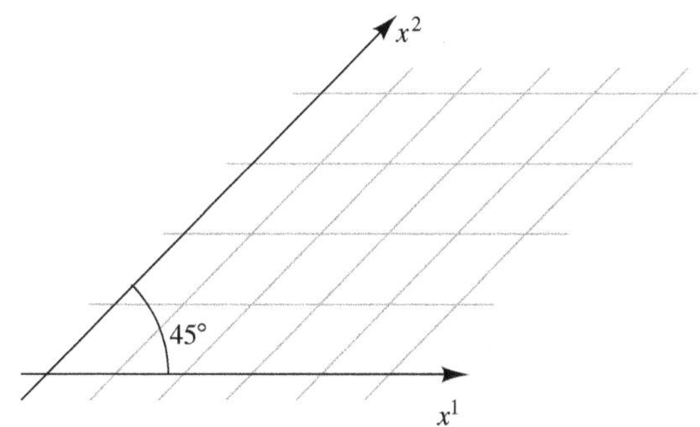

22. [S] **(a)** Find the coordinate basis components of an orthonormal basis for the wormhole metric (7.39) that is oriented along the coordinate lines.

(b) Find the components of the coordinate basis vectors in this orthonormal basis.

23. Show that any two orthonormal bases are related by a Lorentz transformation. More precisely, show that the vectors in one basis are linear combinations of the vectors in another with a matrix of coefficients that define a Lorentz transformation.

24. In an inertial frame (t, x, y, z) consider the spacelike hypersurfaces of constant time t' of another frame moving along the x-axis with a velocity v with respect to the first.

(a) Make a rough graph in a (t, x) spacetime diagram of the family of surfaces separated by equal values of t'. Does every point in flat spacetime lie on one of these surfaces?

(b) Find the (t, x, y, z) coordinate components of a unit normal vector to these spacelike surfaces.

25. [C] *A Toy Model of a Wormhole Connecting Two Regions of Space* Take a plane and delete two disks of equal radius R whose centers are separated by a distance d. Identify points on the edges of one disk with points on the edge of the other as shown, so that all points labeled 1 are identified, all points labeled 2, etc. A free particle or light ray whose straight-line path intersects a point on the left-hand disk would emerge from the identified point on the right-hand disk, as shown, making the same angle with the normal as it went in with.

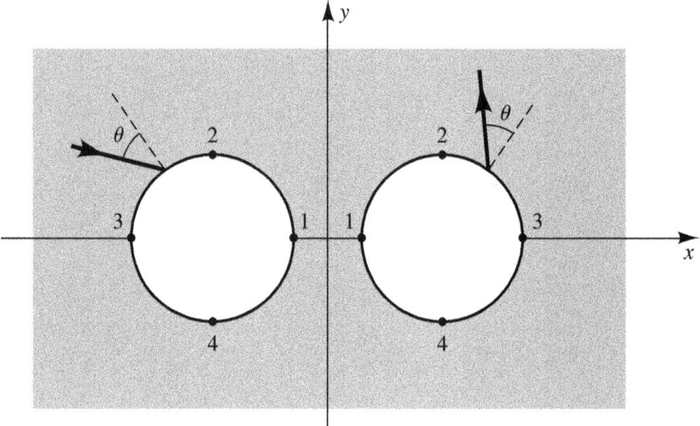

(a) Provide an argument based on the identification that straight-line particle trajectories behave as shown.

(b) Two points lie on the x-axis at locations $x = +L$ and $x = -L$, $L > R + d/2$. A particle starts moving along the x-axis from one point toward the other. What distance has it traveled when it reaches the other point?

(c) Find a closed orbit for a free particle in this geometry. Is your orbit stable against small perturbations?

(d) Suppose two spheres were deleted from three-dimensional flat space and identified in an analogous way. What kind of scene would an observer some distance out along the x-axis see when looking back towards the wormhole mouth?

26. *Another Division into Space and Time* Show that each point inside the forward light cone of the origin $(-t^2 + r^2 < 0)$ lies on some Lorentz hyperboloid of the form (7.74) for some value of a. Points inside can be labeled using a as a time coordinate and (χ, θ, ϕ) as spatial coordinates as in (7.75). Find the line element of flat spacetime in these new coordinates. Sketch the family of spacelike surfaces in a (t, r) spacetime diagram.

CHAPTER 8

Geodesics

Both experimentally and theoretically, the curved spacetimes of general relativity are explored by studying how test particles and light rays move through them. A "test" body has a mass so small that it produces no significant spacetime curvature by itself. Rather, it moves in response to the curvature produced by other bodies with significant masses. A satellite in orbit around the Earth is following a path determined by the slight curvature of spacetime produced by the Earth. However, its own mass is so much smaller than the Earth's that the curvature produced by the satellite can be neglected. It's a test mass.

The equations governing the motion of test particles and light rays in a general curved spacetime are derived and analyzed in this chapter. Only test particles free from any influences other than the curvature of spacetime (electric forces, for instance) are considered. Such particles are called *free* or *freely falling* in general relativity. There is thus a subtle change in how the term *free* is used in general relativity from its usage in Newtonian mechanics. In Newtonian mechanics a free particle is uninfluenced by any force—gravitation included. In general relativity gravitation is not a force but a property of spacetime geometry. In general relativity free means free from any influences *besides* the curvature of spacetime. In both cases a free particle moves in response to just the geometry of spacetime. We begin with the equations of motion for test particles with nonvanishing rest mass moving on timelike world lines, and return to the equations of motion for light rays in Section 8.3.

8.1 The Geodesic Equation

The general principle for the motion of free test particles in curved spacetime is the same as that for flat spacetime discussed in Section 5.4:

Variational Principle for Free Test Particle Motion

The world line of a free test particle between two timelike separated points extremizes the proper time between them.

There are only two differences from the flat-space variational principle for free particle motion in Section 5.4: (1) The word *test* has been added to the statement to make clear that it applies to the motion of bodies that are not a significant source of curvature. (2) The proper time is determined by a general metric $g_{\alpha\beta}(x)$ through (7.19) rather than with the flat metric $\eta_{\alpha\beta}$. In previous chapters, this variational principle was a convenient summary of equations of motion already known. In general relativity, we will deduce the equations of motion *from* the variational principle. Extremal proper time world lines are called *geodesics*, and the equations of motion that determine them comprise the *geodesic equation*.

In previous chapters the principle of extremal proper time was used to derive the equations for geodesics for test particles in particular spacetime geometries. Section 5.4 showed that the principle of extremal proper time implied the equations of motion,

$$\frac{d^2x^\alpha}{d\tau^2} = 0, \tag{8.1}$$

for the coordinates of a test particle in an inertial frame in the *flat* spacetime of special relativity. Section 6.6 showed that the Newtonian equations of motion for a nonrelativistic particle follow from the principle of extremal proper time to leading order in $1/c^2$ in the geometry (6.20). This chapter studies a test particle moving in a general spacetime geometry described by a metric $g_{\alpha\beta}(x)$ and a line element (7.8). The analogies between these cases are exhibited in Table 8.1.

Although we aim at generality, it's appropriate to begin with a simple example —the geodesics of the flat two-dimensional plane viewed as curves of extremal distance. These are spacelike geodesics in space rather than timelike ones in spacetime, but the analogy is close. Of course, a curve of extremal distance between two points in a flat plane is a straight line. But it is instructive to see how this familiar result emerges from first finding the equations that govern geodesics in the plane and then solving them. We'll find the equations in Example 8.1 and solve them in the next section. The equations are simplest in Cartesian coordinates, but the simplest problems don't always make the best examples (Problem 1). We study the geodesics in the plane using polar coordinates, illustrated in Figure 2.5.

TABLE 8.1 Extremal Proper time $\delta \int d\tau = 0$ and Equations of Motion

	Variational Principle	Equation of Motion
Particle in flat spacetime	$\delta \int (-\eta_{\alpha\beta}dx^\alpha dx^\beta)^{1/2} = 0$	$\dfrac{d^2x^\alpha}{d\tau^2} = 0$
Geometric Newtonian	$\delta \int \left[(1+2\Phi/c^2)(cdt)^2 - (1-2\Phi/c^2)(dx^2+dy^2+dz^2)\right]^{1/2} = 0$ (to leading order in$1/c^2$)	$\dfrac{d^2x^i}{dt^2} = -\dfrac{\partial\Phi}{\partial x^i}$ (to leading order in $1/c^2$)
General metric	$\delta \int (-g_{\alpha\beta}\, dx^\alpha dx^\beta)^{1/2} = 0$	$\dfrac{d^2x^\alpha}{d\tau^2} = -\Gamma^\alpha_{\beta\gamma}\dfrac{dx^\beta}{d\tau}\dfrac{dx^\gamma}{d\tau}$

Example 8.1. Equations for Geodesics of the Plane in Polar Coordinates. The metric of the plane in polar coordinates r and ϕ is [cf. (2.8])

$$dS^2 = dr^2 + r^2 d\phi^2. \tag{8.2}$$

A curve between two points A and B can be described parametrically by giving r and ϕ as a function of a parameter σ, which varies between the value $\sigma = 0$ at point A and $\sigma = 1$ at point B. There are many choices of parameter with these properties; it won't matter which one is used. For any particular parameter, a curve is described by two functions $r(\sigma)$ and $\phi(\sigma)$. The distance between A and B is

$$\begin{aligned} S_{AB} &= \int_A^B dS = \int_A^B (dr^2 + r^2 d\phi^2)^{1/2} \\ &= \int_0^1 d\sigma \left[\left(\frac{dr}{d\sigma}\right)^2 + r^2 \left(\frac{d\phi}{d\sigma}\right)^2 \right]^{1/2}. \end{aligned} \tag{8.3}$$

The necessary conditions for an extremum of this distance are Lagrange's equations for the Lagrangian,

$$L\left(\frac{dr}{d\sigma}, \frac{d\phi}{d\sigma}, r\right) = \left[\left(\frac{dr}{d\sigma}\right)^2 + r^2 \left(\frac{d\phi}{d\sigma}\right)^2 \right]^{1/2}. \tag{8.4}$$

These are

$$\frac{d}{d\sigma}\left(\frac{1}{L}\frac{dr}{d\sigma}\right) = \frac{r}{L}\left(\frac{d\phi}{d\sigma}\right)^2, \qquad \frac{d}{d\sigma}\left(\frac{1}{L} r^2 \frac{d\phi}{d\sigma}\right) = 0. \tag{8.5}$$

But as (8.3) shows, the value of L is just $dS/d\sigma$. Therefore, multiplying (8.5) by $d\sigma/dS$, the equations for geodesics using the distance S as the parameter along the curve take the simple form

$$\frac{d^2 r}{dS^2} = r\left(\frac{d\phi}{dS}\right)^2, \tag{8.6a}$$

$$\frac{d}{dS}\left(r^2 \frac{d\phi}{dS}\right) = 0. \tag{8.6b}$$

We solve these equations in the next section.

The procedure for finding the equations for timelike geodesics in spacetime is a straightforward generalization of Example 8.1. The proper time along a timelike world line between two points A and B in spacetime is, from (7.19),

$$\tau_{AB} = \int_A^B d\tau = \int_A^B \left[-g_{\alpha\beta}(x)\, dx^\alpha\, dx^\beta\right]^{1/2}. \tag{8.7}$$

The world line can be described parametrically by giving the four coordinates x^α as a function of a parameter σ that varies between $\sigma = 0$ at endpoint A and $\sigma = 1$

at endpoint B. The proper time between A and B is then

$$\tau_{AB} = \int_0^1 d\sigma \left(-g_{\alpha\beta}(x) \frac{dx^\alpha}{d\sigma} \frac{dx^\beta}{d\sigma} \right)^{1/2}. \tag{8.8}$$

The world lines that extremize the proper time between A and B are those that satisfy Lagrange's equations,

$$-\frac{d}{d\sigma}\left(\frac{\partial L}{\partial(dx^\alpha/d\sigma)} \right) + \frac{\partial L}{\partial x^\alpha} = 0, \tag{8.9}$$

for the Lagrangian

$$L\left(\frac{dx^\alpha}{d\sigma}, x^\alpha \right) = \left(-g_{\alpha\beta}(x) \frac{dx^\alpha}{d\sigma} \frac{dx^\beta}{d\sigma} \right)^{1/2}. \tag{8.10}$$

These are the equations for geodesics in the spacetime with the metric $g_{\alpha\beta}$. We illustrate their construction with the wormhole metric discussed in Example 7.7 on p. 148.

Example 8.2. Equations for Geodesics in a Wormhole Geometry. The line element of the wormhole geometry (7.39) is

$$ds^2 = -dt^2 + dr^2 + (b^2 + r^2)(d\theta^2 + \sin^2\theta \, d\phi^2) \tag{8.11}$$

and the Lagrangian for geodesics (8.10) is

$$L\left(\frac{dx^\alpha}{d\sigma}, x^\alpha \right) = \left\{ \left(\frac{dt}{d\sigma} \right)^2 - \left(\frac{dr}{d\sigma} \right)^2 - (b^2 + r^2) \left[\left(\frac{d\theta}{d\sigma} \right)^2 + \sin^2\theta \left(\frac{d\phi}{d\sigma} \right)^2 \right] \right\}^{1/2}. \tag{8.12}$$

In writing out Lagrange's equations, differentiating the square root in (8.12) produces a factor of $1/L$. However, from (8.8) the value of L is $d\tau/d\sigma$. The inverse factors of L can, therefore, be used to trade derivatives with respect to σ for derivatives with respect to τ. The result is four equations:

$$\frac{d^2t}{d\tau^2} = 0, \tag{8.13a}$$

$$\frac{d^2r}{d\tau^2} = r \left[\left(\frac{d\theta}{d\tau} \right)^2 + \sin^2\theta \left(\frac{d\phi}{d\tau} \right)^2 \right], \tag{8.13b}$$

$$\frac{d}{d\tau}\left[(b^2 + r^2) \frac{d\theta}{d\tau} \right] = (b^2 + r^2) \sin\theta \cos\theta \left(\frac{d\phi}{d\tau} \right)^2, \tag{8.13c}$$

$$\frac{d}{d\tau}\left[(b^2 + r^2) \sin^2\theta \frac{d\phi}{d\tau} \right] = 0. \tag{8.13d}$$

We will apply these equations to understand a property of geodesics in the wormhole geometry in the next section.

A little thought and the preceding examples make clear that the general form of the equations for geodesics in an arbitrary curved spacetime is

$$\boxed{\frac{d^2x^\alpha}{d\tau^2} = -\Gamma^\alpha_{\beta\gamma}\frac{dx^\beta}{d\tau}\frac{dx^\gamma}{d\tau}.} \tag{8.14}$$

Geodesic Equation for Timelike Geodesics

There are four equations—one for each value of the free index α. The coefficients $\Gamma^\alpha_{\beta\gamma}$, called the *Christoffel symbols*, are constructed from the metric and its first derivatives. Taken together these four equations (8.14) are called the *geodesic equation*.[1] The geodesic equation is the basic equation of motion for test particles in a curved spacetime. Equivalently, it could be written in terms of the *coordinate basis* components of the four-velocity $u^\alpha = dx^\alpha/d\tau$ as

$$\boxed{\frac{du^\alpha}{d\tau} = -\Gamma^\alpha_{\beta\gamma}u^\beta u^\gamma.} \tag{8.15}$$

Geodesic Equation for Timelike Geodesics

The Christoffel symbols may be taken to be symmetric in the lower two indices

$$\Gamma^\alpha_{\beta\gamma} = \Gamma^\alpha_{\gamma\beta} \tag{8.16}$$

because an antisymmetric part would not contribute anything to the symmetric sum over β and γ in (8.14). For the simple examples used in this book, it is usually easiest to find the Christoffel symbols by working out the equations for geodesics from the line element, as illustrated in the preceding example, and then reading the Christoffel symbols from them. Even easier is using the *Mathematica* program on the book website. The results of such computations for some important metrics we will study can be found in Appendix B.

Example 8.3. Finding the Christoffel Symbols from the Geodesic Equation. A comparison of the general geodesic equation (8.14) with specific form of equations (8.6) shows that the only nonvanishing Christoffel symbols for the metric of the plane in polar coordinates (8.2) are

$$\Gamma^r_{\phi\phi} = -r, \qquad \Gamma^\phi_{r\phi} = \Gamma^\phi_{\phi r} = 1/r. \tag{8.17}$$

[1]By now the reader may wonder why we call (8.14) the geodesic equatio*n* rather than the geodesic equatio*ns* when four differential equations are involved. It is the same reason it's usual to call $\vec{F} = m\vec{a}$ Newton's equatio*n* of motion rather than Newton's equatio*ns* of motion. Viewed as a vector relation, $\vec{F} = m\vec{a}$ is one equation, even though it comprises three component differential equations. In a similar way (8.15) can be thought of as one equation for the vector four-velocity **u** that comprises four component equations. Notation that makes this clearer is introduced in Chapter 20. A similar distinction arises for the Einstein equation, which comprises 10 component differential equations.

Similarly, from (8.13) the only nonvanishing Christoffel symbols for the wormhole metric (8.11) are

$$\begin{aligned} \Gamma^r_{\theta\theta} &= -r, & \Gamma^r_{\phi\phi} &= -r\sin^2\theta, \\ \Gamma^\theta_{r\theta} = \Gamma^\theta_{\theta r} &= \frac{r}{b^2+r^2}, & \Gamma^\theta_{\phi\phi} &= -\sin\theta\cos\theta, \\ \Gamma^\phi_{r\phi} = \Gamma^\phi_{\phi r} &= \frac{r}{b^2+r^2}, & \Gamma^\phi_{\phi\theta} = \Gamma^\phi_{\theta\phi} &= \cot\theta. \end{aligned} \tag{8.18}$$

Both these answers are displayed using the convention mentioned in (5.7) of using coordinate names to replace specific labels. For instance, in (8.18), $\Gamma^\phi_{r\phi} = \Gamma^3_{13}$, where $x^1 = r$ and $x^3 = \phi$ in (8.11). The repeated ϕ does not indicate summation in this case! Legally this is a violation of the summation convention, but it is also a standard and convenient practice.

By working through Lagrange's equations for the general form of the Lagrangian (8.10), a general expression can be found for the Christoffel symbols in terms of the metric and its derivatives, although we will hardly ever need it. This is sufficiently involved that we defer the calculation to a supplement on the book website, but the answer is

Christoffel Symbols

$$\boxed{g_{\alpha\delta}\Gamma^\delta_{\beta\gamma} = \frac{1}{2}\left(\frac{\partial g_{\alpha\beta}}{\partial x^\gamma} + \frac{\partial g_{\alpha\gamma}}{\partial x^\beta} - \frac{\partial g_{\beta\gamma}}{\partial x^\alpha}\right).} \tag{8.19}$$

If the metric happens to be diagonal in the coordinate system being used, then the calculation of the Γ's from (8.19) is straightforward because there is only one term in the sum on the left-hand side, as illustrated in Example 8.4. If it is not diagonal, then the matrix inverse of $g_{\alpha\beta}$ has to be computed to solve the linear equation (8.19) for the Γ's.

Example 8.4. Finding Christoffel Symbols from the General Formula. To show how the general formula (8.19) works, let's calculate $\Gamma^r_{\phi\phi}$ for the metric (8.2) of a flat, two-dimensional plane in polar coordinates. We'll use indices A, B that run over $x^1 = r$ and $x^2 = \phi$ so that the metric is $g_{AB} = \mathrm{diag}(1, r^2)$. Putting $\alpha = r$, $\beta = \gamma = \phi$ in (8.19) and noting that only one term contributes to the sum on the left because the metric is diagonal gives

$$g_{rr}\Gamma^r_{\phi\phi} = \frac{1}{2}\left(\frac{\partial g_{r\phi}}{\partial\phi} + \frac{\partial g_{r\phi}}{\partial\phi} - \frac{\partial g_{\phi\phi}}{\partial r}\right) = -r. \tag{8.20}$$

Since $g_{rr} = 1$, that gives $\Gamma^r_{\phi\phi} = -r$, as in (8.17).

8.2 Solving the Geodesic Equation—Symmetries and Conservation Laws

The geodesic equation (8.14) is a set of four coupled, second-order, ordinary differential equations for the four coordinates locating a test particle in spacetime as a function of proper time. Given an initial location in spacetime and an initial four-velocity, standard techniques could be used to integrate these equations numerically to find location and four-velocity at later moments of proper time. In very simple cases this can sometimes be done analytically, as Example 8.5 shows.

Example 8.5. Travel Time through a Wormhole. Consider the wormhole geometry described in Example 7.7 on p. 148 and illustrated in Figure 7.5. A traveler starts at a coordinate radius $r = R$ and falls freely and radially through the wormhole throat. For a given initial radial four-velocity $u^r \equiv U$, how much time does it take on the traveler's own clock to fall through the wormhole throat and reach the corresponding point $r = -R$ on the other sheet?

The freely falling traveler is moving on a radial geodesic in the geometry specified by the line element (8.11). Initially the four-velocity is radial:

$$u^\alpha = [(1 + U^2)^{1/2}, U, 0, 0], \tag{8.21}$$

where we have taken the coordinates of (8.11) in the order (t, r, θ, ϕ) and determined u^t so that the normalization condition $\mathbf{u} \cdot \mathbf{u} = -1$ [cf. (7.55)] is satisfied. Spherical symmetry implies that, once moving radially, the traveler stays moving radially. The four-velocity components $u^\theta(\tau)$ and $u^\phi(\tau)$ thus vanish all along the world line. The radial component of the four-velocity changes according to the equation for $d^2r/d\tau^2$ in (8.13). When evaluated at constant θ and ϕ, this is

$$\frac{du^r}{d\tau} = 0. \tag{8.22}$$

Thus, $u^r(\tau)$ is constant along the world line and equal to its initial value U. Integrating $u^r \equiv dr/d\tau = U$ gives

$$r(\tau) = U\tau, \tag{8.23}$$

where the zero of proper time has been chosen to be when the traveler is at $r = 0$ (the throat). The elapsed proper time $\Delta\tau$ between $r = -R$ and $r = +R$ is, thus,

$$\Delta\tau = 2R/U. \tag{8.24}$$

Example 8.5 is exceptional in its tractability. In more general situations, conservation laws, such as those for energy and angular momentum, lead to tractable problems as in Newtonian mechanics. Conservation laws give first integrals[2] of the equations of motion that can reduce the order and number of the equations that have to be solved.

One first integral that is always available comes from the normalization of the four-velocity. In the coordinate basis this reads [cf. (7.55)]

$$\mathbf{u}\cdot\mathbf{u} = g_{\alpha\beta}\frac{dx^{\alpha}}{d\tau}\frac{dx^{\beta}}{d\tau} = -1, \tag{8.25}$$

and for a completely general metric, that will be the only first integral. Further conservation laws arise from symmetries.

In Newtonian mechanics conservation laws are connected to symmetries. To conserve energy, for example, the force must be conservative—derivable from a potential—and that potential must be time independent. To conserve linear momentum along a particular direction, the potential must be constant along that direction. To conserve angular momentum the potential must be spherically symmetric. In short, energy is conserved when there is a symmetry under displacements in time, linear momentum is conserved when there is a symmetry under displacements in space, and angular momentum is conserved when there is a symmetry under rotations.

Conserved quantities for the motion of test particles cannot be expected in a general spacetime that has no special symmetries. A general spacetime metric is time dependent, angle dependent, position dependent, etc. However, when the spacetime has a symmetry, then there is an associated conservation law. For example, if spacetime geometry is independent of time, there is a conserved energy for test particles.

How does one tell if a spacetime geometry has a symmetry? One simple case is if the metric is independent of one of the coordinates, say x^1. Then the transformation

$$x^1 \rightarrow x^1 + \text{const.} \tag{8.26}$$

leaves the metric unchanged. The vector $\boldsymbol{\xi}$ with components

$$\xi^{\alpha} = (0, 1, 0, 0) \tag{8.27}$$

lies along a direction in which the metric doesn't change. The vector with components (8.27) is called the *Killing vector* associated with the symmetry (8.26), after the German mathematician Wilhelm Killing (1847–1923) (not because it's an especially difficult concept!). A Killing vector is a general way of characterizing symmetry in any coordinate system, as Example 8.6 below helps to show.

Killing Vectors

[2]In the usual terminology of Newtonian mechanics, a *first integral* is a function of the coordinates and their first time derivatives, which is constant by virtue of equations of motion, which are second-order differential equations. The conservation laws for energy and angular momentum are examples. A first integral is also called a *constant of the motion*.

Example 8.6. The Killing Vectors of Flat Space. When the metric of flat three-dimensional space is written in usual Cartesian coordinates

$$dS^2 = dx^2 + dy^2 + dz^2, \tag{8.28}$$

there are three evident Killing vectors, $(1, 0, 0)$, $(0, 1, 0)$ and $(0, 0, 1)$, corresponding to the three translational symmetries of flat space. But when polar coordinates are used,

$$dS^2 = dr^2 + r^2 d\theta^2 + r^2 \sin^2\theta \, d\phi^2, \tag{8.29}$$

another Killing vector emerges because the metric is independent of ϕ corresponding to rotational symmetry about the z-axis. This Killing vector has components $(0, 0, 1)$ in polar coordinates and components $(-y, x, 0)$ in Cartesian coordinates. There are two other Killing vectors for flat space corresponding to rotational symmetry about the other two axes. Can you guess their components in Cartesian coordinates?

A symmetry implies a conserved quantity along a geodesic. To see this, recall that the equations for geodesics follow from the principle of extremal proper time and Lagrange's equations (8.9). If the metric—and, therefore, L—is independent of the coordinate x^1, then $\partial L/\partial x^1 = 0$. Equation (8.9) for $\alpha = 1$ then reads

$$\frac{d}{d\sigma}\left[\frac{\partial L}{\partial\left(dx^1/d\sigma\right)}\right] = 0, \tag{8.30}$$

which implies that

$$\frac{\partial L}{\partial(dx^1/d\sigma)} = -g_{1\beta}\frac{1}{L}\frac{dx^\beta}{d\sigma} = -g_{1\beta}\frac{dx^\beta}{d\tau} = -g_{\alpha\beta}\xi^\alpha u^\beta = -\boldsymbol{\xi}\cdot\mathbf{u} \tag{8.31}$$

is conserved along the geodesic. In an arbitrary coordinate system, a conserved quantity along a geodesic is, therefore,

$$\boxed{\boldsymbol{\xi}\cdot\mathbf{u} = \text{const.}} \qquad (\boldsymbol{\xi} \text{ a Killing vector}). \tag{8.32}$$

Conserved Quantities Along a Geodesic

Equally well, one could say that $\boldsymbol{\xi}\cdot\mathbf{p}$ is conserved, where $\mathbf{p}$ is the particle's momentum. A simple example illustrates how to use these conservation laws:

Example 8.7. Geodesics in the Plane Using Polar Coordinates. Integrals of the motion make it straightforward to solve equations (8.6) for all the geodesics in the plane using the polar coordinates of (8.2). In this two-dimensional example, let indices $A, B, \ldots$ run over the values 1 and 2, and label the two polar coordinates as $x^1 = r$, $x^2 = \phi$. The components of the tangent vector $\vec{u}$ are $u^A = dx^A/dS$.

The first integral corresponding to $\vec{u} \cdot \vec{u} = 1$ is provided by dividing both sides of the line element (8.2) by dS^2:

$$\left(\frac{dr}{dS}\right)^2 + r^2\left(\frac{d\phi}{dS}\right)^2 = 1. \tag{8.33}$$

Another first integral arises because the metric (8.2) is independent of ϕ. The associated Killing vector, $\vec{\xi}$, has coordinate basis components $\xi^r = 0$ and $\xi^\phi = 1$. A conserved quantity is, therefore,

$$\ell \equiv \vec{\xi} \cdot \vec{u} = g_{AB}\xi^A u^B = r^2 \frac{d\phi}{dS}, \tag{8.34}$$

whose conservation also follows directly from the geodesic equation (8.6b).

Inserting this result into (8.33) gives

$$\frac{dr}{dS} = \left(1 - \frac{\ell^2}{r^2}\right)^{1/2}. \tag{8.35}$$

This equation could be easily integrated to find r as a function of S, but it is really the shape of the geodesic we are interested in—r as a function of ϕ or, alternatively, ϕ as a function of r. Dividing (8.34) by (8.35) gives

$$\frac{d\phi}{dr} = \frac{d\phi/dS}{dr/dS} = \frac{\ell}{r^2}\left(1 - \frac{\ell^2}{r^2}\right)^{-1/2}. \tag{8.36}$$

This can be integrated to give

$$\phi = \phi_* + \cos^{-1}\left(\frac{\ell}{r}\right), \tag{8.37}$$

where ϕ_* is an integration constant. Thus, the shape of the geodesic is given by

$$r\cos(\phi - \phi_*) = \ell. \tag{8.38}$$

Expanding the cosine in (8.38) and using $x = r\cos\phi$, $y = r\sin\phi$ gives

$$x\cos\phi_* + y\sin\phi_* = \ell, \tag{8.39}$$

which is the general equation of a straight line. Thus, we recover the familiar straight lines as curves of extremal distance in the flat plane—straight lines the hard way!

8.3 Null Geodesics

The previous sections of this chapter have explored the paths followed by free particles through curved spacetime. These are the timelike geodesics. Light rays are also important for exploring spacetime geometry. Light rays move along null world lines for which $ds^2 = 0$. More concretely, if $x^\alpha(\lambda)$ is the path of a light ray through spacetime parametrized by some parameter λ and $u^\alpha \equiv dx^\alpha/d\lambda$ is the tangent vector, then

$$\mathbf{u} \cdot \mathbf{u} = g_{\alpha\beta}(x)\frac{dx^\alpha}{d\lambda}\frac{dx^\beta}{d\lambda} = 0. \tag{8.40}$$

This equation is not enough, however, to determine the trajectory completely—it is one equation for four unknowns. We need the analog of the geodesic equation, (8.14). Ultimately this would have to be derived from the laws of electromagnetism generalized to curved spacetime, but we can argue for its form using the equivalence principle.

The flat spacetime equation of motion for a light ray (5.66) can be written

$$\frac{d^2x^\alpha}{d\lambda^2} = 0, \tag{8.41}$$

where λ is an affine parameter. We seek a generalization of this law to curved spacetime that (1) reduces to this form in a local inertial frame and (2) takes the same form in every coordinate system. It must satisfy the latter requirement because the coordinates are arbitrary. We already have a law that does this—the geodesic equation (8.14). The natural generalization of (8.41) that satisfies requirements (1) and (2) is

$$\frac{d^2x^\alpha}{d\lambda^2} = -\Gamma^\alpha_{\beta\gamma}\frac{dx^\beta}{d\lambda}\frac{dx^\gamma}{d\lambda}. \tag{8.42}$$

Geodesic Equation for Null Geodesics

Null curves that satisfy (8.42) are called *null geodesics*. Light rays move on null geodesics. The affine parameter λ is not a spacetime distance—the distance along a light ray is zero! Rather, it is a parameter chosen so that (8.42) takes the form of the geodesic equation.

8.4 Local Inertial Frames and Freely Falling Frames

Riemann Normal Coordinates

Section 7.4 introduced the idea of a local inertial frame—coordinates centered on a point P in spacetime in which $g_{\alpha\beta} = \eta_{\alpha\beta}$ at the point and the first derivatives of the metric vanish [cf. (7.13)]. In these coordinates the Christoffel symbols vanish at P and the geodesic equation (8.14) takes the same form as for a free particle in flat space [cf. (5.62)]:

$$\left.\frac{d^2x^\alpha}{d\tau^2}\right|_P = 0. \tag{8.43}$$

Therefore, to an approximation that can be made better and better as the extent of the frame becomes small, free particles move for a moment on straight lines. These coordinates are thus the analogs of inertial frames in Newtonian mechanics but only locally near a point.

The understanding of geodesics achieved in this chapter allows us to redeem the pledge made in Chapter 7 to explicitly construct at least one system of coordinates defining a local inertial frame.

Pick a point P in spacetime to serve as the origin of the coordinate system. Pick a basis of four orthonormal vectors $\{\mathbf{e}_\alpha\}$ at that point. (We drop the hat on indices that distinguish orthonormal bases from coordinate ones because it will turn out that the coordinate basis of the constructed coordinates coincides with this orthonormal basis.) These might be the orthonormal basis vectors of the laboratory of an observer at P, for example (cf. Section 5.6). Pick a direction from P defined by a unit vector $\mathbf{n}$ and send out a geodesic in that direction. The point reached after a distance s (if the geodesic is spacelike) can be labeled by the coordinates

$$x^\alpha \equiv s n^\alpha, \tag{8.44}$$

where the n^α are the components of $\mathbf{n}$ in the basis $\{\mathbf{e}_\alpha\}$. (See Figure 8.1.) Repeat this procedure for all different directions $\mathbf{n}$ (using proper time τ instead of s if the direction is timelike and filling in by continuity if it is null). The result is a coordinate system that uniquely labels points close enough to P that spacetime curvature has not caused the geodesics to cross. *Riemann normal coordinates* is the name given to this system of coordinates. We now show they constitute a local inertial frame.

Riemann Normal Coordinates

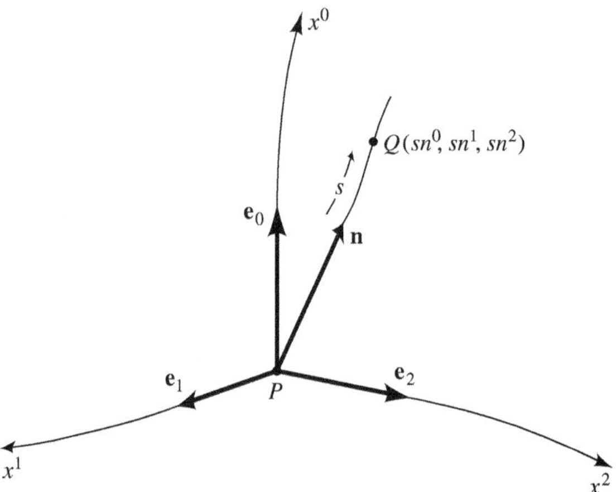

FIGURE 8.1 Riemann normal coordinates define a local inertial frame (LIF). A choice of four orthonormal vectors $\{\mathbf{e}_\alpha\}$ at a point P starts the construction of a local inertial frame there. A point Q a distance s along the geodesic starting in a direction $\mathbf{n}$ is assigned the coordinates $x^\alpha = sn^\alpha$. The four-coordinate axes of the LIF are along the geodesics starting in the four orthogonal directions. Eventually the curvature of spacetime may lead geodesics to cross and the coordinate system to become singular.

The orthonormal vectors $\{\mathbf{e}_\alpha\}$ are the coordinate basis vectors of the local inertial frame at P and, therefore [cf. (7.56)],

$$g_{\alpha\beta}(x_P) = \eta_{\alpha\beta}. \tag{8.45}$$

This is the first of the requirements (7.13) for a local inertial frame. The second, that the derivatives of the metric vanish at P, can be seen as follows:

Every geodesic through P is labeled by some fixed direction $\mathbf{n}$ and obeys the geodesic equation (8.14) if timelike, and the same equation with s replacing τ if spacelike. Evaluating (8.14) with (8.44) at P, one finds

$$\Gamma^\alpha_{\beta\gamma}\Big|_P n^\beta n^\gamma = 0. \tag{8.46}$$

But this equation has to hold for all unit vectors $\mathbf{n}$, which implies

$$\Gamma^\alpha_{\beta\gamma}\Big|_P = 0. \tag{8.47}$$

All the Christoffel symbols can vanish only if all the derivatives of the metric vanish [cf. (8.19)]. Riemann normal coordinates therefore define a local inertial frame.

Example 8.8. Riemann Normal Coordinates at the North Pole of a Sphere. The line element of the geometry of a sphere of circumference $2\pi a$ has the form [cf. (2.15)]

$$dS^2 = a^2(d\theta^2 + \sin^2\theta\, d\phi^2) \tag{8.48}$$

in familiar angular coordinates (θ, ϕ). The procedure for constructing Riemann normal coordinates at the north pole can be implemented as follows: The unit vectors $\vec{e}_1$ and $\vec{e}_2$, pointing in the $\phi = 0$ and $\phi = \pi/2$ directions respectively, constitute a convenient orthonormal basis. A unit vector $\vec{n}$ pointing in the ϕ direction has components $n^A = (\cos\phi, \sin\phi)$ in this basis. Consider the point (θ, ϕ). The geodesic connecting it to the north pole is part of the great circle whose longitude equals ϕ. The geodesic distance between (θ, ϕ) and the north pole is then $s = a\theta$. The Riemann normal coordinates of the point (θ, ϕ) are, therefore,

$$x^A = (sn^1, sn^2) = (a\theta\cos\phi,\ a\theta\sin\phi). \tag{8.49}$$

Example 7.2 showed that in these coordinates the metric takes the form $g_{AB} = \mathrm{diag}(1, 1)$ and that its first derivatives vanish there.

Freely Falling Frames

Riemann normal coordinates are not the only way of defining a local inertial frame. Indeed, the equivalence principle suggests that one can go much further than making the Christoffel symbols vanish just at a point. It suggests that the

BOX 8.1 Drag-free Satellites

Realizing a freely falling frame is easy in principle. Launch a satellite into empty space, release it in a non-rotating state, and voilà, the frame of the satellite's interior is a freely falling frame. But in reality space is not so empty. Residual atmospheric drag, radiation pressure, and other forces can cause deviations of a small satellite ($\sim$1000 kg) that are significant for precision experiments in gravitational physics. For the GP-B experiment (Box 14.1 on p. 305), testing the predictions of general relativity for the motion of gyroscopes, nongravitational accelerations must be less than about $\sim 10^{-13}$ m/s^2. For space tests of the equality of gravitational and inertial mass contemplated for the next decade, they must be less than $\sim 10^{-14}$ m/s^2, and for gravitational wave detectors in space even less. Residual atmospheric drag in a near-Earth orbit can be $\sim 10^{-6}$ m/s^2.

Drag-free satellites are a realistic way of realizing a freely falling frame. The idea is illustrated in the accompanying figure. The experimental platform floats freely inside the satellite, which protects it from perturbing forces such as those described above. The sheltered experimental platform therefore follows a geodesic in spacetime. Accurate sensors detect the location of the experimental platform relative to the protective frame of the satellite. The satellite uses thrusters to steer itself so it remains centered about the experiment. In effect, the thrusters cancel the accelerations produced by perturbing forces. Evidently the sensors must be able to detect the accelerations of the satellite to the tiny accuracies mentioned here, and the satellite itself must not significantly perturb the motion of the platform. However, this is not the place to review the ingenious solutions to these technological challenges. Drag-free satellites provide a realistic approximation to a freely falling frame.

geodesic equation should also reduce to (8.43) in the frame of a sufficiently small freely falling laboratory *over some period of time*. The laboratory of an orbiting space shuttle described in Example 6.3 on p. 120 is one example of an approximate freely falling laboratory. Drag-free satellites described in the Box 8.1 are another.

The mathematical idealization of a freely falling laboratory is a system of coordinates in which the Christoffel symbols vanish all along a geodesic, not just at one point on it. We will call such a coordinate system a *freely falling frame*.[3] A freely falling frame is a local inertial frame all along a geodesic.

The construction of a freely falling frame parallels the construction of inertial frames in Newtonian mechanics (Section 3.1) and special relativity (Section 4.3).

[3]The more usual names are *Fermi normal coordinates* or *proper reference frame of a freely falling observer*. We depart from the usual terms because freely falling frame is a shorter way of capturing the essential idea. Note, however, that any local inertial frame can be said to be "freely falling," since the acceleration of its origin vanishes at the spacetime point P at which it is defined [cf. (8.43)]. In some texts, therefore, a local inertial frame defined at one point in spacetime is called a freely falling frame. Here we mean a frame defined along a geodesic.

(Recall Figure 3.3.) Pick a free test particle moving on a geodesic. The proper time τ along the geodesic will serve as the time coordinate, with the position of the test particle as the origin of spatial coordinates. At one moment of proper time, orient gyroscopes along three orthogonal directions. At later moments use the directions set by these gyroscopes to construct spatial coordinates x^i by a similar procedure to the one used to construct Riemann normal coordinates. The resulting coordinates (τ, x^i) constitute a freely falling frame in which the Christoffel symbols vanish along the geodesic at the origin, $x^i = 0$. We will not demonstrate this here because we lack the laws of how gyroscopes move in curved spacetime. These are provided in Chapter 14, and we return again to freely falling frames in Chapter 20.

Freely falling frames are as close as one can come in curved spacetime to the inertial frames of Newtonian mechanics and special relativity. But as Example 6.2 showed, astronauts in their freely falling space shuttle can detect the effects of spacetime curvature with experiments done over a long enough time over sufficient spatial distance. Correspondingly in a freely falling frame, the Christoffel symbols vanish only on the defining geodesic, not at every point labeled by the coordinates.

Problems

1. [S] Use Cartesian coordinates to write out and solve the geodesic equations for a two-dimensional flat plane and show that the solutions are the straight lines.

2. In usual spherical coordinates the metric on a two-dimensional sphere is [cf. (2.15)]

$$dS^2 = a^2\left(d\theta^2 + \sin^2\theta\, d\phi^2\right),$$

where a is a constant.

(a) Calculate the Christoffel symbols "by hand".

(b) Show that a great circle is a solution of the geodesic equation. (*Hint*: Make use of the freedom to orient the coordinates so the equation of a great circle is simple.)

3. A three-dimensional spacetime has the line element

$$ds^2 = -\left(1 - \frac{2M}{r}\right)dt^2 + \left(1 - \frac{2M}{r}\right)^{-1} dr^2 + r^2 d\phi^2.$$

(a) Find the explicit Lagrangian for the variational principle for geodesics in this spacetime in these coordinates.

(b) Using the results of (a) write out the components of the geodesic equation by computing them from the Lagrangian.

(c) Read off the nonzero Christoffel symbols for this metric from your results in (b).

4. [A] *Rotating Frames* The line element of *flat spacetime* in a frame (t, x, y, z) that is rotating with an angular velocity Ω about the z-axis of an inertial frame is

$$ds^2 = -[1 - \Omega^2(x^2 + y^2)]\,dt^2 + 2\Omega(y\,dx - x\,dy)\,dt + dx^2 + dy^2 + dz^2.$$

(a) Verify this by transforming to polar coordinates and checking that the line element is (7.4) with the substitution $\phi \to \phi - \Omega t$.

(b) Find the geodesic equations for x, y, and z in the rotating frame.

(c) Show that in the nonrelativistic limit these reduce to the usual equations of Newtonian mechanics for a free particle in a rotating frame exhibiting the centrifugal force and the Coriolis force.

5. Derive the Christoffel symbols $\Gamma^{\phi}_{r\phi}$ and $\Gamma^{\theta}_{\phi\phi}$ for the wormhole metric (7.39) directly from the general formula (8.19) and not starting from variational principle of extremal proper time.

6. Show by direct calculation from the geodesic equation (8.15) that the norm of the four-velocity $\mathbf{u} \cdot \mathbf{u}$ is a constant along a geodesic.

7. [S] Consider a particle of mass m moving in a central potential $V(r)$ in nonrelativistic Newtonian mechanics. Write down the Lagrangian for this system in polar coordinates. Using the method of Section 8.2, show that invariance under rotations about the z-axis implies the conservation of the z component of the angular momentum.

8. Verify the claim in Example 8.6 that the Killing vector corresponding to the rotational symmetry of flat space about the z-axis has components $(-y, x, 0)$ in Cartesian coordinates. In the same coordinates find the components of the Killing vectors corresponding to the rotational symmetry of flat space around the y- and x-axes.

9. Consider the two-dimensional spacetime with the line element

$$ds^2 = -X^2 dT^2 + dX^2.$$

Find the shapes $X(T)$ of all the timelike geodesics in this spacetime.

10. Show that any one of the four rectangular coordinates of an inertial frame is an affine parameter for a light ray in flat spacetime.

11. Solve for the *null* geodesics in three-dimensional flat spacetime using polar coordinates so the line element is $ds^2 = -dt^2 + dr^2 + r^2 d\phi^2$. Do light rays move on straight lines?

12. *The Hyperbolic Plane* The hyperbolic plane defined by the metric

$$dS^2 = y^{-2}(dx^2 + dy^2),\ y \geq 0$$

is a classic example of a curved two-dimensional surface.

(a) Show that points on the x-axis are an infinite distance from any point (x, y) in the upper half-plane.

(b) Write out the geodesic equations.

(c) Show that the geodesics are semicircles centered on the x-axis or vertical lines, as illustrated.

(d) Solve the geodesic equations to find x and y as functions of the length S along these curves.

Remark: This example was important in the history of geometry. Euclid's fifth postulate for Euclidean geometry states that for a straight line L and a point P, there is only one straight line (a geodesic) through P that does not intersect L. (That straight

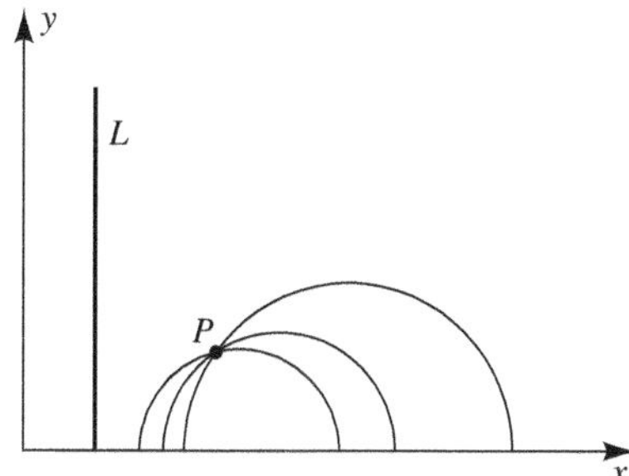

line is the one *parallel* to L.) The sphere is an example for which there are no such straight lines through P (all great circles intersect.) The hyperbolic plane is a constant *negative* curvature example (see Chapter 21), where there are an *infinite* number of straight lines through P that do not intersect L (see the example in the accompanying figure).

13. [S] Construct Riemann normal coordinates for flat space by the procedure discussed in Section 8.4 using the origin of an inertial frame as the point P and four unit vectors pointing along its axes. Do the resulting coordinates coincide with the inertial frame coordinates?

14. [C] *Fermat's Principle of Least Time* Consider a medium with an index of refraction $n(x^i)$ that is a function of position. The velocity of light in the medium varies with position and is $c/n(x^i)$. Fermat's principle states that light rays follow paths between two points in *space* (not spacetime!) that take the least travel time.

(a) Show that the paths of least time are geodesics in three-dimensional space with the line element

$$dS^2_{\text{fermat}} = n^2(x^i)\, dS^2,$$

where dS^2 is the usual line element for flat three-dimensional space, e.g.,

$$dS^2 = dx^2 + dy^2 + dz^2.$$

(b) Write out the geodesic equations for the extremal paths in (x, y, z) rectangular coordinates.

15. [C] *The Lunenberg Lens* A sphere of radius R with an index of refraction that varies with radius as

$$n(r) = \left[2 - \left(\frac{r}{R}\right)^2\right]^{1/2}$$

is called a Lunenberg lens. Use the results of Problem 14 to show that it has the property that any bundle of parallel rays incident from one direction is focused on one point on the surface of the sphere.

CHAPTER 9

The Geometry Outside a Spherical Star

The simplest curved spacetimes of general relativity are the ones with the most symmetry, and the most useful of these is the geometry of empty space outside a spherically symmetric source of curvature, for example, a spherical star. This is called the *Schwarzschild geometry* after Karl Schwarzschild (1873–1916), who solved the Einstein equation to find it in 1916. To an excellent approximation this is the curved spacetime outside the Sun and therefore leads to the predictions of Einstein's theory most accessible to experimental test. We show in Chapter 21 that the Schwarzschild geometry is a solution of the vacuum Einstein equation—the Einstein equation for curved spacetime devoid of matter. In this chapter we explore the geometry of Schwarzschild's solution, assuming it's given. We will concentrate on predicting the orbits of test particles and light rays in the curved spacetime of a spherical star that exhibit some of the famous effects of general relativity—the gravitational redshift, the precession of the perihelion of a planet, the gravitational bending of light, and the time delay of light. The next chapter describes experiments and observations that check these predictions and test Einstein's theory.

9.1 Schwarzschild Geometry

In a particularly suitable set of coordinates, the line element summarizing the Schwarzschild geometry is given by ($c \neq 1$ units)

Schwarzschild Metric

$$ds^2 = -\left(1 - \frac{2GM}{c^2 r}\right)(c\,dt)^2 + \left(1 - \frac{2GM}{c^2 r}\right)^{-1} dr^2 + r^2\left(d\theta^2 + \sin^2\theta\, d\phi^2\right). \tag{9.1}$$

The coordinates are called *Schwarzschild coordinates* and the corresponding metric $g_{\alpha\beta}(x)$ is called the *Schwarzschild metric*. It has the following important properties:

- *Time Independent* The metric is independent of t. There is a Killing vector $\boldsymbol{\xi}$ associated with this symmetry under displacements in the coordinate time t, which has the components [cf. (8.27)]

$$\xi^\alpha = (1, 0, 0, 0) \tag{9.2}$$

(listed in the order (t, r, θ, ϕ)) in the coordinate basis associated with (9.1).

- *Spherically Symmetric* The geometry of a two-dimensional surface of constant t and constant r in the four-dimensional geometry (9.1) is summarized by the line element

$$d\Sigma^2 = r^2(d\theta^2 + \sin^2\theta\, d\phi^2). \tag{9.3}$$

This describes the geometry of a sphere of radius r in flat three-dimensional space [cf. (2.15)]. The Schwarzschild geometry thus has the symmetries of a sphere with regard to changes in the angles θ and ϕ. In (9.1) or (9.3) this is evident for the ϕ-direction because the metric is independent of ϕ—invariant under rotations about the z-axis. The Killing vector associated with this symmetry is [cf. (8.27)]

$$\eta^\alpha = (0, 0, 0, 1). \tag{9.4}$$

There are Killing vectors associated with the other rotational symmetries but we won't need them.

The Schwarzschild coordinate r has a simple geometric interpretation arising from spherical symmetry. It is *not* the distance from any "center." Rather, it is related to the area A of the two-dimensional spheres of fixed r and t by the standard formula

$$r = (A/4\pi)^{1/2}. \tag{9.5}$$

This follows from (9.3), (7.28), and (7.37).

- *Mass M* If GM/c^2r is small, the coefficient of dr^2 in the line element (9.1) can be expanded to give

$$ds^2 \approx -\left(1 - \frac{2GM}{c^2r}\right)(c\,dt)^2 + \left(1 + \frac{2GM}{c^2r}\right)dr^2 + r^2\left(d\theta^2 + \sin^2\theta\, d\phi^2\right). \tag{9.6}$$

This is exactly the form of the static, weak field metric (6.20) with a Newtonian gravitational potential Φ given by

$$\Phi = -\frac{GM}{r}. \tag{9.7}$$

This leads to the identification of the constant M in the Schwarzschild metric (9.1) with the *total mass* of the source of curvature.

In Newtonian physics the Sun's mass is determined by measuring the period and size of the orbit of a test body (the Earth) and using Kepler's law [cf. (3.24)] to relate these to the mass of the source of gravitational attraction. In general relativity the mass of a stationary source of spacetime curvature is *defined* by this kind of experiment. Any form of energy is a source of spacetime curvature, including the energy in electromagnetic fields, nuclear interaction energy, etc., and, in a rough sense that will be clearer later, the energy in spacetime curvature itself. The limit of very large orbits should, therefore, be taken to define a *total mass*, that includes all of these. The larger the orbit, the more accurately its properties are determined by the Newtonian approximation as (9.6) and the discussion in Section 6.6 show. The total mass of a stationary body can, therefore, be defined by Kepler's law for a very large orbit, and, since that is determined by the Newtonian potential (9.7), the constant M in the Schwarzschild metric (9.1) is the total mass.

The geometry outside a spherically symmetric source is thus characterized by a single number—the total mass M—and not on how that mass is radially distributed inside the source. That's the relativistic version of Newton's theorem for the Newtonian gravitational potential discussed in Example 3.1.

- *Schwarzschild Radius* There is obviously something interesting happening to the metric at the radii $r = 0$ and $r = 2GM/c^2$. The latter is called the *Schwarzschild radius* and is the characteristic length scale for curvature in the Schwarzschild geometry. It turns out, however, that the surface of a static star is always outside these radii. The Schwarzschild radius of the Sun, for instance, is $2GM_\odot/c^2 = 2.95$ km—much smaller than the radius of the solar surface 6.96×10^5 km. At the surface the Schwarzschild geometry joins a different geometry inside the star. As long as one sticks to the outsides of static stars, one doesn't have to worry about the radii $r = 2GM/c^2$ and $r = 0$. However, we will have to face up to these radii in Chapter 12 when we consider the gravitational collapse of a star to zero radius and the formation of a black hole.

Equation (9.1) exhibits the Schwarzschild geometry in mass-length-time ($\mathcal{MLT}$) units. The expression is a little simpler in the $\mathcal{ML}$ units that are convenient for special relativity, where $c = 1$ and both space and time have the same dimension of length. A system of units convenient for general relativity also puts $G = 1$ by measuring mass in units of length through the conversion

$$M(\text{in cm}) = \frac{G}{c^2}M(\text{in g}) = .742 \times 10^{-28}\left(\frac{\text{cm}}{\text{g}}\right)M(\text{in g}). \tag{9.8}$$

Geometrized Units

In these units, for example, the mass of the Sun is $M_\odot = 1.47$ km and the mass of the Earth is $M_\oplus = .44$ cm. These $\mathcal{L}$ units are called *geometrized units*, or $c = G = 1$ units. To convert an expression in geometrized units back to $\mathcal{MLT}$ ones, it is necessary only to insert the correct factors of G and c, replacing, for example, M by GM/c^2, τ by $c\tau$, $dx^i/d\tau$ by $(1/c)(dx^i/d\tau)$, etc. Appendix A gives a list of such transformation rules as well as a brief general discussion of units.[1]

[1]Does this discussion mean that the value of Newton's constant can be defined like the value of c is? Not at present because the unit of gravitational mass is defined in terms of inertial mass—the standard kilogram—whose gravitational properties are determined by measurement. See the discussion in Appendix A.

In geometrized units the Schwarzschild line element has the form

$$ds^2 = -\left(1 - \frac{2M}{r}\right) dt^2 + \left(1 - \frac{2M}{r}\right)^{-1} dr^2 + r^2 \left(d\theta^2 + \sin^2\theta \, d\phi^2\right). \tag{9.9}$$

Schwarzschild Metric (geometrical units)

Explicitly the metric $g_{\alpha\beta}$ is

$$g_{\alpha\beta} = \begin{array}{c} \\ t \\ r \\ \theta \\ \phi \end{array} \begin{array}{c} \begin{array}{cccc} \quad t \quad & \quad r \quad & \theta & \phi \end{array} \\ \left(\begin{array}{cccc} -(1-2M/r) & 0 & 0 & 0 \\ 0 & (1-2M/r)^{-1} & 0 & 0 \\ 0 & 0 & r^2 & 0 \\ 0 & 0 & 0 & r^2\sin^2\theta \end{array} \right) \end{array} . \tag{9.10}$$

Both theoretically and experimentally the Schwarzschild geometry can be studied through the orbits of test particles and light rays. Observations of the small effects predicted by general relativity on the orbits of planets and trajectories of light rays in the solar system are important tests of the theory. The following discussion concentrates on the effects that lead to experimental tests beginning with the gravitational redshift.

9.2 The Gravitational Redshift

Consider an observer stationed at a fixed Schwarzschild coordinate radius R who emits a light signal. When emitted, the signal has frequency ω_* as measured by this stationary observer. The light signal propagates out to infinity, not necessarily along a radial path, where its frequency is measured by another stationary observer (see Figure 9.1). The frequency ω_∞ received by an observer at infinity is less than ω_*. That is the gravitational redshift worked out from the equivalence principle to first order in $1/c^2$ in Example 6.2. The following discussion derives it exactly in the Schwarzschild geometry.

The change in frequency is related to the change in energy of an emitted photon because for any observer, $E = \hbar\omega$. In Newtonian physics the change in kinetic energy of a particle moving in a time-independent potential can be easily calculated from the conservation of energy arising from time-displacement invariance. This suggests that the efficient way to calculate the change in frequency of a photon moving in the time-independent Schwarzschild geometry is to make use of the conserved quantity that arises because of its time-displacement invariance. This conserved quantity is $\boldsymbol{\xi} \cdot \mathbf{p}$ [cf. (8.32)], where $\mathbf{p}$ is the photon's four-momentum and $\boldsymbol{\xi}$ is the Killing vector (9.2) associated with time-displacement symmetry. Let's see how to do that.

The energy of the photon measured by an observer with four-velocity $\mathbf{u}_{\text{obs}}$ is

$$E = -\mathbf{p} \cdot \mathbf{u}_{\text{obs}}, \tag{9.11}$$

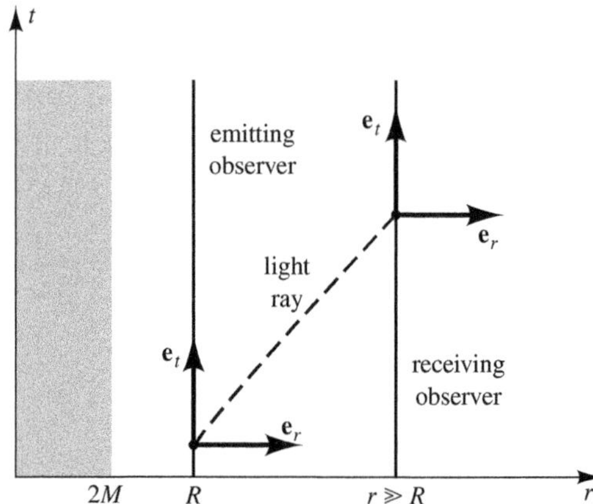

FIGURE 9.1 A spacetime diagram showing the world lines of two stationary observers outside of a spherically symmetric mass. One observer hovers at radius R, the other is "at infinity," that is, at a radius $r \gg R$. A photon is emitted from radius R with frequency ω_* as measured in the laboratory of a stationary observer at R. The photon propagates along the dotted world line until it is detected by the observer at infinity with frequency ω. Two of the orthonormal vectors associated with each laboratory are indicated schematically. The frequency ω_∞ is less than ω_* because the photon loses energy climbing out of the gravitational well of the central mass. That is the gravitational redshift.

as described in (7.53) and the discussion following it. Since the energy of a photon is related to its frequency by $E = \hbar\omega$,

$$\hbar\omega = -\mathbf{p} \cdot \mathbf{u}_{\text{obs}}, \tag{9.12}$$

giving the frequency measured by an observer with four-velocity $\mathbf{u}_{\text{obs}}$. The spatial components u^i_{obs} of the four-velocity are zero for a stationary observer. The time component $u^t_{\text{obs}}(r)$ of a stationary observer at radius r is determined by the normalization condition [cf. (8.25)]

$$\mathbf{u}_{\text{obs}}(r) \cdot \mathbf{u}_{\text{obs}}(r) = g_{\alpha\beta} u^\alpha_{\text{obs}}(r) u^\beta_{\text{obs}}(r) = -1. \tag{9.13}$$

Since $u^i_{\text{obs}}(r) = 0$, this implies

$$g_{tt}(r)[u^t_{\text{obs}}(r)]^2 = -1, \tag{9.14}$$

and, using the metric (9.10), this gives

$$u^t_{\text{obs}}(r) = \left(1 - \frac{2M}{r}\right)^{-1/2}. \tag{9.15}$$

Thus,

$$u^\alpha_{\text{obs}}(r) = [(1 - 2M/r)^{-1/2}, 0, 0, 0] = (1 - 2M/r)^{-1/2} \xi^\alpha, \tag{9.16}$$

where $\boldsymbol{\xi}$ is the Killing vector (9.2) associated with the time independence of the Schwarzschild metric. For a stationary observer at radius r, therefore,

$$\mathbf{u}_{\text{obs}}(r) = (1 - 2M/r)^{-1/2}\boldsymbol{\xi}. \tag{9.17}$$

Using (9.17) in (9.12), the frequency of the photon measured by the stationary observer at radius R is,

$$\hbar\omega_* = \left(1 - \frac{2M}{R}\right)^{-1/2}(-\boldsymbol{\xi}\cdot\mathbf{p})_R, \tag{9.18}$$

where the subscript R indicates that the quantities are to be evaluated at the radius $r = R$ in Schwarzschild coordinates. Similarly, at infinite radius

$$\hbar\omega_\infty = (-\boldsymbol{\xi}\cdot\mathbf{p})_\infty. \tag{9.19}$$

But from (8.32) the quantity $\boldsymbol{\xi}\cdot\mathbf{p}$ is *conserved* along the photon's geodesic. It is the same at infinity as it is at radius R. The frequencies are, therefore, related by

$$\boxed{\omega_\infty = \omega_*\left(1 - \frac{2M}{R}\right)^{1/2}.} \tag{9.20}$$

Gravitational Redshift

The frequency at infinity is *less* than the frequency at R by a factor $(1-2M/R)^{1/2}$. The photon has suffered a gravitational redshift.

Equation (9.20) may be expanded in powers of $2M/R$ when that is small, as for the Sun. The first two terms reproduce the approximate result (6.14) derived from the principle of equivalence.

9.3 Particle Orbits—Precession of the Perihelion

Let's now examine the orbits of test particles following timelike geodesics in the Schwarzschild geometry. These test particles might be the planets orbiting our Sun or particles of an accretion disk orbiting a neutron star or black hole.

Conserved Quantities

The study of geodesics in the Schwarzschild geometry is considerably aided by the laws of conservation of energy and angular momentum that hold because the metric is independent of time and spherically symmetric. In particular, since the metric is independent of t and ϕ, the quantities $\boldsymbol{\xi}\cdot\mathbf{u}$ and $\boldsymbol{\eta}\cdot\mathbf{u}$ are conserved [cf. (8.32)], where $\mathbf{u}$ is the four-velocity of the particle and $\boldsymbol{\xi}$ and $\boldsymbol{\eta}$ are given by (9.2) and (9.4). These quantities are so useful it is convenient to give them special

BOX 9.1 Time Machines

In science fiction, time machines transport a traveler forward or backward in time. General relativity—the theory of space and time—supplies the principles for analyzing whether time machines are possible and practical.

Relativity provides several examples of time machines that transport an observer to events in the future faster than other observers. The twin paradox setup discussed on p. 63 is the simplest example. As viewed in an inertial frame in flat spacetime, one twin accelerates away from a stationary twin, reaches speeds close to the velocity of light, and returns. The accelerating twin returns younger than the stationary twin who follows a geodesic—the curve of *longest* proper time. If accelerated to high enough velocities, the returning twin can participate in events far to the future of the lifetime of any stationary human observer. That is transportation forward in time. *Any* spacetime therefore abounds in forward time machines—two points that can be connected by timelike curves with two different lengths.

Curved spacetime provides different kinds of forward time machines. Construct a spherical shell of mass M and radius R and go live inside. The exterior geometry is Schwarzschild. Inside spacetime is flat. (There would be no force in Newtonian gravity because there is no mass inside any sphere of symmetry. This also holds in relativity.) Your clocks inside the shell will run slower than clocks at infinity by the gravitational redshift factor $(1 - 2M/R)^{1/2}$ [cf. (9.20)]. Suppose, for example, you wanted to know by the end of a day the output of a computation that would take a hundred years to carry out on your laptop. Or suppose that you wanted to watch the next hundred years of television in a day. Leave your laptop and television outside the shell and go inside the shell to watch. How big and how massive a shell would you need to construct? You would need an M and R such that $(1 - 2M/R)^{1/2} = 1/(100 \times 365) \approx 3 \times 10^{-5}$. That is, the radius of the shell R could only be very *slightly* bigger than twice its mass M. Assuming that one needs a reasonable-size living room inside of, say, $R \sim 10$ m, the mass required would be $M \approx 5$ m $\approx (1/300)M_{\odot}$ or a shell 4 times the mass of Jupiter. There is no material that would support the resulting stress, and the shell has to be considerably larger and much more massive to have low enough stresses (Problem 4).

In Chapter 12 we will learn that a shell is not really needed to construct a forward time machine. Hovering outside a black hole near $R = 2M$ is equally effective. That, however, requires an expenditure of energy to create the thrust to balance the gravitational attraction of a black hole. The no-cost option is to fall freely into the black hole. But then one can never return, and the maximum time to view the future even for the largest black holes in the known universe is about three hours before destruction in a singularity.

What about traveling backward in time? The world line of an observer can't turn backwards in time because to do so, it would have to be moving faster than the speed of light at some point. The only way to travel backward in time to an earlier point in one's history is if spacetime has closed timelike curves. It's possible to cook up spacetimes with this property. Take flat spacetime in a particular Lorentz frame and identify points along the $t = 0$ surface with points on a $t = T$ surface. Spacetime is then curled up in the t-direction like a cylinder, and closed timelike curves of constant $\vec{x}$ go around it. But there is no evidence that our universe has such an exotic topological structure, and, if energy is positive, general relativity prohibits the evolution of closed timelike curves in a space with a simple topological structure like the one we believe we live in. Thus, although it is possible in principle to go forward into the future, we probably cannot revisit the past, at least in the classical theory of gravity.

names. We'll call them[2] $-e$ and ℓ. Their explicit forms are

$$e = -\boldsymbol{\xi} \cdot \mathbf{u} = \left(1 - \frac{2M}{r}\right) \frac{dt}{d\tau}, \tag{9.21}$$

Conserved Energy per Unit Rest Mass

$$\ell = \boldsymbol{\eta} \cdot \mathbf{u} = r^2 \sin^2\theta \frac{d\phi}{d\tau}. \tag{9.22}$$

Conserved Angular Momentum per Unit Rest Mass

At large r the constant e becomes energy per unit rest mass because in flat space, $E = mu^t = m(dt/d\tau)$ [cf. (5.41)]. Energy per unit rest mass is what we'll call it everywhere. We'll call the conserved quantity ℓ the angular momentum per unit rest mass because that's what it is at low velocities. Thus, there is a conserved energy and angular momentum for particle orbits.

Effective Potential and Radial Equation

The conservation of angular momentum implies that the orbits lie in a "plane," as do the orbits in Newtonian theory. To see this, fix your attention on a particular instant and let $\vec{u}$ denote the spatial components of the particle's four-velocity. Orient the coordinates so $d\phi/d\tau = 0$ at that instant and the particle is at $\phi = 0$, i.e., so that $\vec{u}$ lies in the meridional "plane" $\phi = 0$. According to (9.22) this implies $\ell = 0$, so that $d\phi/d\tau$ is zero everywhere along the geodesic. The particle thus remains in the meridional "plane" $\phi = 0$. Having once established this, it is simpler to reorient the coordinates so that the particle orbits are in the equatorial "plane." Thus for the rest of the discussion we consider $\theta = \pi/2$ and $u^\theta = 0$.

Orbits in a Plane

The normalization of the four-velocity supplies another integral for the geodesic equation in addition to those for energy (9.21) and angular momentum (9.22). Explicitly, this third integral reads

$$\mathbf{u} \cdot \mathbf{u} = g_{\alpha\beta} u^\alpha u^\beta = -1. \tag{9.23}$$

These three integrals can be used to express the three nonzero components of the four-velocity in terms of the constants of the motion e and ℓ. Writing (9.23) out for the Schwarzschild metric (9.10), and taking account of the equatorial plane condition $u^\theta = 0$, $\theta = \pi/2$ gives

$$-\left(1 - \frac{2M}{r}\right)(u^t)^2 + \left(1 - \frac{2M}{r}\right)^{-1}(u^r)^2 + r^2(u^\phi)^2 = -1. \tag{9.24}$$

Writing $u^t = dt/d\tau$, $u^r = dr/d\tau$, and $u^\phi = d\phi/d\tau$ and using (9.21) and (9.22) to eliminate $dt/d\tau$ and $d\phi/d\tau$, (9.24) can be rewritten as

[2]Don't get e mixed up with the eccentricity of an orbit. We'll denote that by ϵ.

$$-\left(1-\frac{2M}{r}\right)^{-1}e^2+\left(1-\frac{2M}{r}\right)^{-1}\left(\frac{dr}{d\tau}\right)^2+\frac{\ell^2}{r^2}=-1. \tag{9.25}$$

With a little further rewriting, this can be put in the form

$$\frac{e^2-1}{2}=\frac{1}{2}\left(\frac{dr}{d\tau}\right)^2+\frac{1}{2}\left[\left(1-\frac{2M}{r}\right)\left(1+\frac{\ell^2}{r^2}\right)-1\right]. \tag{9.26}$$

We have written the expression in this form to show the correspondence with the energy integral of Newtonian mechanics. By defining the constant

$$\mathcal{E}\equiv(e^2-1)/2 \tag{9.27}$$

and the effective potential

Effective Potential for Radial Motion of Particles

$$V_{\text{eff}}(r)\equiv\frac{1}{2}\left[\left(1-\frac{2M}{r}\right)\left(1+\frac{\ell^2}{r^2}\right)-1\right]=-\frac{M}{r}+\frac{\ell^2}{2r^2}-\frac{M\ell^2}{r^3} \tag{9.28}$$

the correspondence becomes exact:

$$\mathcal{E}=\frac{1}{2}\left(\frac{dr}{d\tau}\right)^2+V_{\text{eff}}(r). \tag{9.29}$$

Thus, the techniques for treating orbits by effective potentials in Newtonian mechanics can be applied to the orbits in the Schwarzschild geometry. Indeed the form of the effective potential (9.28) differs from that of a $-M/r$ Newtonian central potential by only the additional $-M\ell^2/r^3$ term. That term, however, will have important consequences for orbits, as we explore shortly.

Greater insights into (9.29) may be obtained by considering its nonrelativistic limit. To do this, first put back the factors of c and G by replacing τ by $c\tau$ and M by GM/c^2. The conserved quantity ℓ is replaced by ℓ/c, where ℓ continues to mean $r^2(d\phi/d\tau)$. The effective potential $V_{\text{eff}}(r)$ becomes

$$V_{\text{eff}}(r)=\frac{1}{c^2}\left(-\frac{GM}{r}+\frac{\ell^2}{2r^2}-\frac{GM\ell^2}{c^2r^3}\right). \tag{9.30}$$

The dimensionless constant e is the total energy per unit rest mass. Anticipating a correspondence with the usual Newtonian energy, let's define E_{Newt} by

$$e\equiv\frac{mc^2+E_{\text{Newt}}}{mc^2}. \tag{9.31}$$

Using (9.30) and (9.31), (9.29) becomes

$$E_{\text{Newt}}\left(1+\frac{E_{\text{Newt}}}{2mc^2}\right)=\frac{m}{2}\left(\frac{dr}{d\tau}\right)^2+\frac{L^2}{2mr^2}-\frac{GMm}{r}-\frac{GML^2}{c^2mr^3}, \quad (9.32)$$

where $L = m\ell$. This has the same form as the energy integral in Newtonian gravity with additional relativistic corrections to the potential on the right proportional to $1/r^3$ and to the energy on the left. The Newtonian limit is recovered when these relativistic derivatives are dropped and the τ-derivative replaced by a t-derivative.

Returning to the analysis of the relativistic orbits, consider the properties of the effective potential $V_{\text{eff}}(r)$. A few simple properties are immediate from its definition (9.28):

$$V_{\text{eff}}(r) \underset{r\to\infty}{\to} -\frac{M}{r}, \qquad V_{\text{eff}}(2M) = -\frac{1}{2}. \quad (9.33)$$

For large values of r the potential is close to the Newtonian effective potential for motion in a $1/r$ potential, as Figure 9.2 illustrates. That is because the first two terms in (9.28) are the same as in Newtonian theory. However, as r decreases, the attractive $1/r^3$ correction from general relativity becomes increasingly important.

The extrema of the effective potential can be found from solving $dV_{\text{eff}}/dr = 0$. There is one local minimum and one local maximum, whose radii r_{min} and r_{max} are

$$r_{\substack{\text{min}\\\text{max}}} = \frac{\ell^2}{2M}\left[1 \pm \sqrt{1-12\left(\frac{M}{\ell}\right)^2}\right]. \quad (9.34)$$

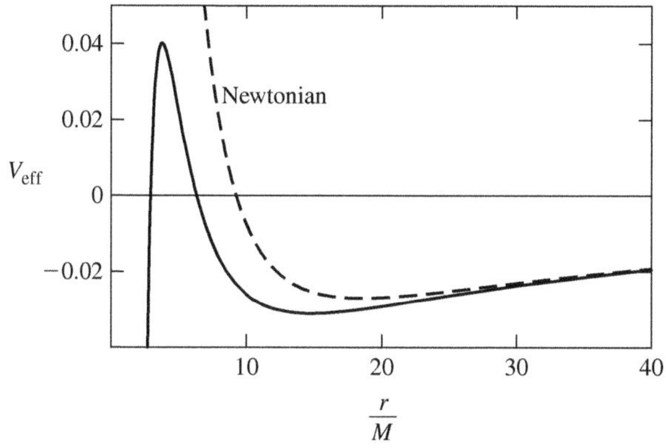

FIGURE 9.2 The relativistic and Newtonian effective potentials for radial motion compared for $\ell/M = 4.3$. The relativistic effective potential $V_{\text{eff}}(r)$ is defined by (9.28) and we take the Newtonian effective potential to be the first two terms of that. The two are close for large r, as shown, but differ significantly for small r, where the $1/r^3$ term in (9.28) becomes important. In particular the infinite centrifugal barrier of Newtonian theory becomes a barrier of finite height. For the Earth in orbit around the Sun, $\ell/M \sim 10^9$ and the differences between the Newtonian and relativistic potential over the orbit of the Earth are tiny but detectable in precise measurements, as we see in Chapter 10.

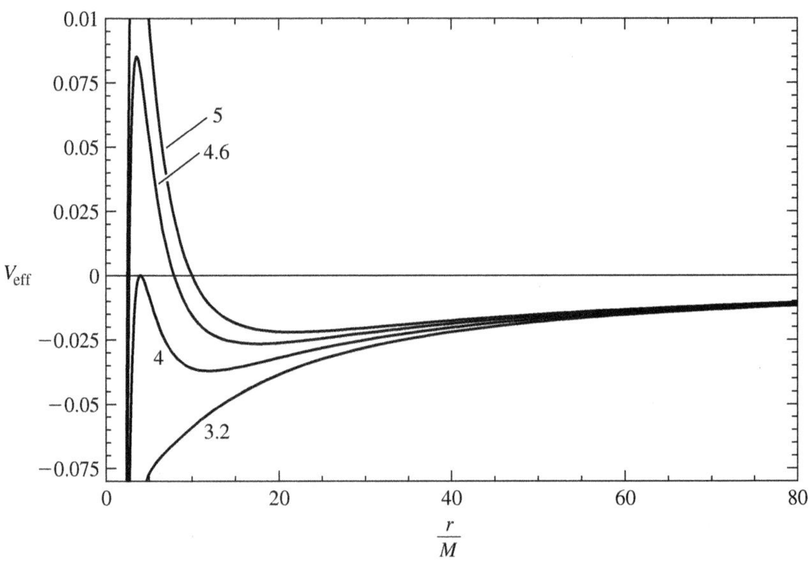

FIGURE 9.3 The effective potential $V_{\text{eff}}(r)$ for radial motion for several different values of ℓ. The values of ℓ/M label the curves.

Figure 9.3 is a plot of V_{eff} for various values of ℓ. If $\ell/M < \sqrt{12} = 3.46$, there are no real extrema and the effective potential is negative for all values of r. If $\ell/M > \sqrt{12}$ the effective potential has one maximum and one minimum. The maximum lies above $V_{\text{eff}} = 0$ if $\ell/M > 4$ and otherwise lies below it. There is a centrifugal barrier, but it has a maximum height, in contrast to the one in Newtonian theory that has infinite height. (See Figure 9.2.)

The qualitative behavior of an orbit depends on the relationship between $\mathcal{E} \equiv (e^2 - 1)/2$ and the effective potential in (9.29), just as in a Newtonian central force problem. Turning points occur at the radii r_{tp}, where $\mathcal{E} = V_{\text{eff}}(r_{\text{tp}})$, because that's where the radial velocity vanishes. If $\ell/M < \sqrt{12}$, there are no turning points for positive values of $\mathcal{E}$. An inwardly directed particle falls all the way to the origin. This is in contrast to Newtonian theory, where as long as $\ell \neq 0$ there is a positive centrifugal barrier that will reflect the particle (see Figure 9.2). Figure 9.4 shows four types of orbits for values of $\ell/M > \sqrt{12}$, along with their qualitative shapes. Circular orbits are possible at the radii (9.34) at which the effective potential has a maximum or a minimum. The orbit at the maximum is unstable because a small increase in $\mathcal{E}$ will lead the particle to escape to infinity or collapse to $r = 0$. The orbit at the minimum is stable. There are bound orbits for $\mathcal{E} < 0$ that oscillate between two turning points. (The planets are moving in bound orbits in the spacetime geometry of the Sun to a good approximation.) Orbits with positive $\mathcal{E}$ but less than the maximum of the effective potential are scattering orbits that come in from infinity, orbit the center of attraction, and then return.

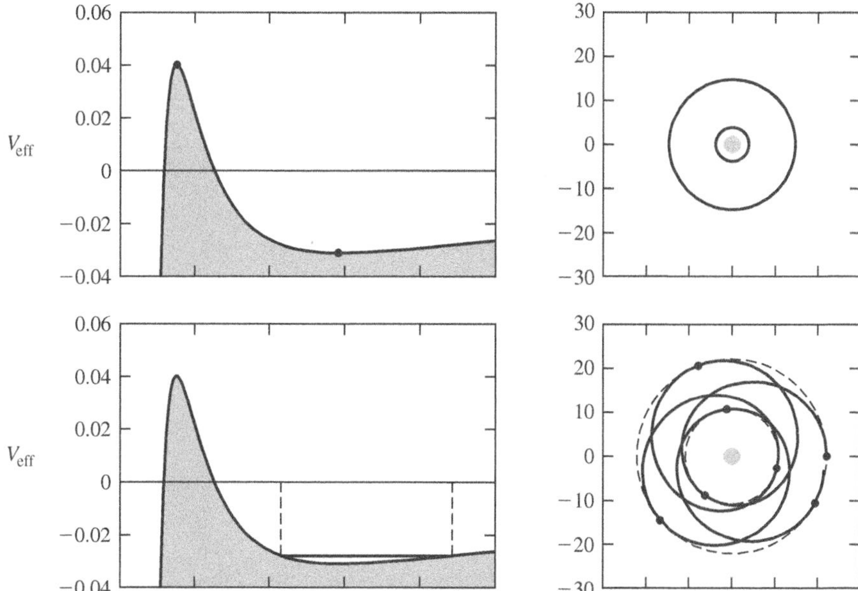

FIGURE 9.4 Four kinds of orbits in the Schwarzschild geometry. The pairs of figures on this page and the next show four orbits corresponding to different values of $\mathcal{E}$ for the illustrative value $\ell/M = 4.3$. The potential and its relationship to $\mathcal{E}$ are shown at left. The horizontal axis in these plots is r/M. The vertical axis is $V_{\rm eff}(r)$. Horizontal lines indicate the values of $\mathcal{E}$. The vertical dashed lines are at turning points. The dots denote the possible locations of circular orbits. The shapes of the corresponding orbits are shown in the figures at right where Schwarzschild r and ϕ are plotted as polar coordinates in the plane. The shaded region at the center of each plot corresponds to $r < 2M$. The top figure on this page shows two circular orbits—the outer one is stable, the inner one is unstable. The next figure shows a bound orbit in which the particle moves between two turning points marked by the dotted circles. The positions of closest approach (perihelion) and furthest excursion (aphelion) are indicated by dots. The precession of the perihelion is large for this relativistic orbit. (*Continued on next page.*)

Those with $\mathcal{E}$ greater than the maximum plunge into the center of attraction. In the following we will calculate the detailed properties of the orbits that are most important for future applications.

Radial Plunge Orbits

The simplest example of an orbit is the radial free fall of a particle from infinity—$\ell = 0$. The particle can start at infinity with various values of its kinetic energy corresponding to different positive values of $\mathcal{E}$, but starting from rest is an especially simple case. Then $dt/d\tau = 1$ at infinity, $e = 1$ from (9.21), or, equivalently, $\mathcal{E} = 0$ from (9.27).

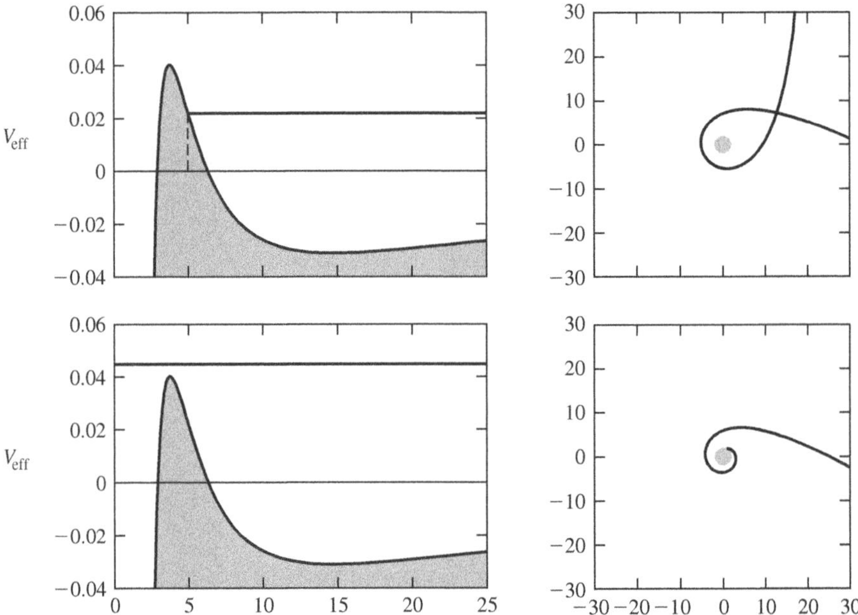

FIGURE 9.4 *continued.* The first figure on this page shows a scattering orbit. The particle comes in from infinity, passes around the center of attraction and moves out to infinity again. This is a highly relativistic orbit which differs significantly from a Newtonian parabola. The last pair of figures shows a plunge orbit in which the particle comes in from infinity, moves part way around the central mass, and then plunges into the center. This kind of orbit is not possible in Newtonian mechanics for a particle moving in a $1/r$ central potential.

From (9.26) with $e = 1$ and $\ell = 0$, we have

$$0 = \frac{1}{2}\left(\frac{dr}{d\tau}\right)^2 - \frac{M}{r}, \tag{9.35}$$

which gives the radial component of the four-velocity $dr/d\tau$. Taken together with the time component $dt/d\tau$ given by (9.21), the four-velocity is

$$u^\alpha = ((1 - 2M/r)^{-1}, -(2M/r)^{1/2}, 0, 0). \tag{9.36}$$

By writing (9.35) in the form

$$r^{1/2}dr = -(2M)^{1/2}d\tau, \tag{9.37}$$

both sides can be integrated to give r as a function of τ. The negative square root is appropriate for a geodesic going inward. The result is

$$r(\tau) = (3/2)^{2/3}(2M)^{1/3}(\tau_* - \tau)^{2/3}, \tag{9.38}$$

where τ_* is an arbitrary integration constant that fixes the proper time when $r = 0$. The Schwarzschild time can be conveniently found by first calculating t as a function of r and then using (9.38) to get it as a function of τ. Computing the derivative dt/dr from (9.21) with $e = 1$ and (9.35), we find

$$\frac{dt}{dr} = -\left(\frac{2M}{r}\right)^{-1/2}\left(1 - \frac{2M}{r}\right)^{-1}, \tag{9.39}$$

which on integration gives

$$t = t_* + 2M\left[-\frac{2}{3}\left(\frac{r}{2M}\right)^{3/2} - 2\left(\frac{r}{2M}\right)^{1/2} + \log\left|\frac{(r/2M)^{1/2}+1}{(r/2M)^{1/2}-1}\right|\right], \tag{9.40}$$

where t_* is another integration constant. There is thus a whole family of freely falling observers that start from rest at infinity. They may be labeled by giving the time they cross a particular radius or by giving their radius at a particular time. Either way this fixes t_*. The relation $t = t(\tau)$ can then be found by substituting (9.38) into (9.40).

Several important features of radial plunge orbits can be seen from (9.38) and (9.40). From (9.40), $r \to \infty$ as $t \to -\infty$, so the particle is falling inward from infinity. From (9.38) we see that from any fixed value of r on the trajectory, it takes only a finite proper time to reach $r = 2M$, even though (9.40) shows it takes an infinite amount of coordinate time t. This is just one indication that the Schwarzschild coordinates are flawed at $r = 2M$. Points are labeled by infinite coordinate values when they are actually only a finite distance away. We learn more about this in Chapter 12.

Example 9.1. Escape Velocity. An observer maintaining a stationary position at Schwarzschild coordinate radius R launches a projectile radially outward with velocity V, as measured in his or her own frame. How large does V have to be for the projectile to reach infinity with zero velocity? This is the escape velocity V_{escape}.

The outward-bound projectile follows a radial geodesic since there are no forces acting on it. At infinity a projectile at rest has $e = 1$. Since e is conserved, the observer must launch the projectile with a minimum value $e = 1$. This requires a four-velocity $\mathbf{u}$, which is the same as (9.36) but with the sign of u^r reversed. The energy E measured by the observer is $-\mathbf{p} \cdot \mathbf{u}_{\text{obs}}$ from (5.87), where $\mathbf{u}_{\text{obs}}$ is the stationary observer's four-velocity and $\mathbf{p} = m\mathbf{u}$ is the projectile's four momentum if m is its rest mass. The four-velocity of a stationary observer at radius R is given by (9.16). The result for the energy required at launch to escape is, therefore,

$$\begin{aligned} E &= -\mathbf{p} \cdot \mathbf{u}_{\text{obs}} = -m\mathbf{u} \cdot \mathbf{u}_{\text{obs}} = -mg_{\alpha\beta}u^\alpha u^\beta_{\text{obs}} \\ &= -mg_{tt}u^t u^t_{\text{obs}} = m\left(1 - \frac{2M}{R}\right)^{-1/2}. \end{aligned} \tag{9.41}$$

The fourth equality is because the four-velocity $\mathbf{u}_{obs}$ of a stationary observer has only a t component and because the Schwarzschild metric is diagonal. The fifth is from substituting the values of the metric (9.10) and the four-velocities from (9.36) and (9.16). In the observer's frame the energy of a particle E is related to its speed V by $E = m/\sqrt{1-V^2}$ [cf. (5.46)]. Thus, the escape velocity is

$$V_{\text{escape}} = \left(\frac{2M}{R}\right)^{1/2}. \tag{9.42}$$

This is, coincidently, the same formula as in Newtonian theory. As R approaches $2M$, the velocity necessary to escape approaches the velocity of light.

Stable Circular Orbits

Stable circular orbits occur at the radii $r = r_{\text{min}}$ of the minima of the effective potential given in (9.34). These radii decrease with decreasing ℓ/M, but stable circular orbits are not possible at arbitrarily small radii. From (9.34), the innermost stable circular orbit (called the ISCO in relativistic astrophysics) in the Schwarzschild geometry occurs when $\ell/M = \sqrt{12}$ at the radius

Innermost Stable Circular Orbit

$$\boxed{r_{\text{ISCO}} = 6M.} \tag{9.43}$$

That fact is important for the structure of X-ray sources, as we will see in Chapter 11.

The angular velocity of a particle in a circular orbit is the rate at which angular position in the orbit changes with time. The rate Ω with respect to the Schwarzschild coordinate time t is the rate measured with respect to a stationary clock at infinity, where t and the proper time of such a clock coincide. It is, for any equatorial orbit,

$$\Omega \equiv \frac{d\phi}{dt} = \frac{d\phi/d\tau}{dt/d\tau} = \frac{1}{r^2}\left(1-\frac{2M}{r}\right)\left(\frac{\ell}{e}\right), \tag{9.44}$$

where the last equality follows from (9.21) and (9.22). Circular orbits of radius r have values of ℓ and e determined by two requirements: First, the potential is a minimum at the radius of the orbit leading to the relation between r and ℓ in (9.34). Second, the value of $\mathcal{E}$ equals the value of the effective potential at that minimum. From (9.29) or (9.26) this gives $e^2 = (1-2M/r)(1+\ell^2/r^2)$. These two requirements can be solved for the ratio ℓ/e of circular orbits:

$$\frac{\ell}{e} = (Mr)^{1/2}\left(1-\frac{2M}{r}\right)^{-1} \quad \text{(circular orbits)}. \tag{9.45}$$

Substituting this in (9.44) gives

Angular Velocity in Stable Circular Orbit

$$\boxed{\Omega^2 = \frac{M}{r^3} \quad \text{(circular orbits).}} \tag{9.46}$$

This has the same form as the nonrelativistic Kepler's law. The period in Schwarzschild coordinate time is $2\pi/\Omega$, and (9.46) says that the square of the period is proportional to the cube of the radius of the orbit. This simple agreement between relativistic and nonrelativistic theory is, however, just a fortuitous consequence of the choice of Schwarzschild coordinate time to measure the angular velocity and Schwarzschild coordinate radius to measure the location of the orbit. The rate of change of angular position with respect to proper time, for example, is given by a more complicated formula (Problem 9).

The components of the four-velocity of a particle in a circular orbit are then

$$u^{\alpha} = u^{t}(1, 0, 0, \Omega) \tag{9.47}$$

with the angular velocity Ω given by (9.46). The component u^t is determined by the normalization condition $\mathbf{u} \cdot \mathbf{u} = -1$ in a way similar to (9.15) for a stationary observer. Now, however, there is a contribution from the angular velocity, and a similar calculation gives

$$u^{t} = \left(1 - \frac{2M}{r} - r^{2}\Omega^{2}\right)^{-1/2} = \left(1 - \frac{3M}{r}\right)^{-1/2} \quad \text{(circular orbits).} \tag{9.48}$$

The Shape of Bound Orbits

To find the shape of an orbit means finding r as a function of ϕ or, equivalently, ϕ as a function of r. To do this solve (9.29) for $dr/d\tau$, solve (9.22) with $\theta = \pi/2$ for $d\phi/d\tau$, and divide the first into the second. One finds

$$\frac{d\phi}{dr} = \pm\frac{\ell}{r^{2}}\frac{1}{[2(\mathcal{E} - V_{\text{eff}}(r))]^{1/2}} = \pm\frac{\ell}{r^{2}}\left[e^{2} - \left(1 - \frac{2M}{r}\right)\left(1 + \frac{\ell^{2}}{r^{2}}\right)\right]^{-1/2}. \tag{9.49}$$

The sign corresponds to the direction in ϕ the particle moves with increasing r. The function $\phi(r)$ can be found simply by integrating the right-hand side. The result can be expressed in terms of elliptic functions but not in a very enlightening way for those not familiar with them. One especially important property is the question whether the orbits close. When we mention one *orbit* we will mean a passage between two successive inner turning points (or equivalently between two successive outer turning points). The orbits are said to *close* if the magnitude of the angle swept out in this passage $\Delta\phi$ is 2π. If it is not 2π, then the inner turning point is said to *precess*, and the amount of precession per orbit is

$$\delta\phi_{\text{prec}} = \Delta\phi - 2\pi \tag{9.50}$$

as illustrated in Figure 9.6.

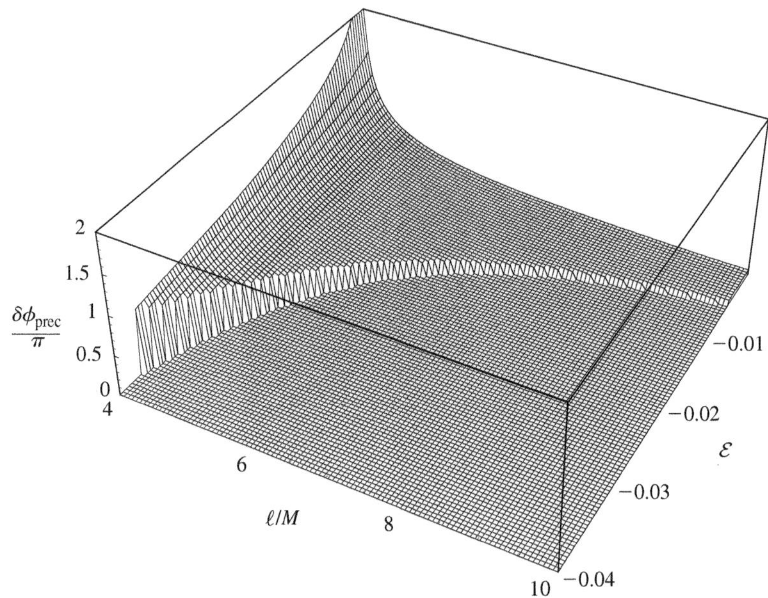

FIGURE 9.5 The precession of the perihelion $\delta\phi_{\text{prec}}$ in the Schwarzschild geometry for bound orbits characterized by the parameters $\mathcal{E} = (e^2 - 1)/2$ and ℓ. This is a plot of $\delta\phi_{\text{prec}}$ as defined by (9.50) and the integral (9.51). There are no bound orbits for the flat region in the foreground where $\delta\phi_{\text{prec}}$ is plotted as zero. The boundary is the curve of ℓ vs. e for circular orbits. Large values of ℓ correspond to orbits that are far from the star where relativistic effects are small. [See (9.45), for example, for the connection between ℓ and the radius of circular orbits.] That is the limit in which (9.57) is a good approximation, and the case important for the planets in the solar system.

The angle $\Delta\phi$ swept out in passing between successive inner turning points at r_1 is just twice the angle swept out between the turning points r_1 and r_2. Thus,

$$\Delta\phi = 2\ell \int_{r_1}^{r_2} \frac{dr}{r^2} \left[e^2 - \left(1 - \frac{2M}{r}\right)\left(1 + \frac{\ell^2}{r^2}\right) \right]^{-1/2}. \tag{9.51}$$

The turning points r_1 and r_2 are the places where $dr/d\tau$ vanishes along the orbit. From (9.26) these are places where the denominator of (9.51) vanishes. Thus, to find $\Delta\phi$ one has only to carry out the integral in (9.51) between the radii where the denominator vanishes. Figure 9.5 shows a plot of a numerical evaluation.

For applications in the solar system, $\Delta\phi$ needs to be evaluated only to the next order in $1/c^2$ after the Newtonian. To accomplish this, first put back in the factors of G and c^2 in (9.51), as described in the discussion leading to (9.32), to give

$$\Delta\phi = 2\ell \int_{r_1}^{r_2} \frac{dr}{r^2} \left[c^2\left(e^2 - 1\right) + \frac{2GM}{r} - \frac{\ell^2}{r^2} + \frac{2GM\ell^2}{c^2 r^3} \right]^{-1/2}. \tag{9.52}$$

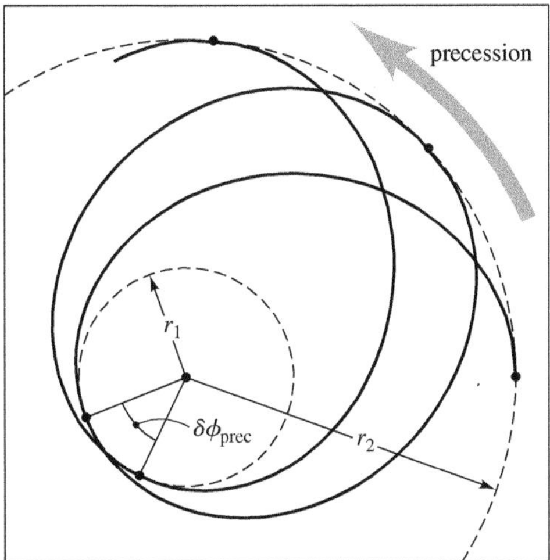

FIGURE 9.6 The shape of a bound orbit outside a spherical star. This is a picture of the orbital plane of the bound orbit whose radial motion is of the kind illustrated in the second pair of plots of Figure 9.4 The planet moves from a minimum radius r_1 out to a maximum radius r_2 and back to the same minimum radius. However, unlike the Keplerian ellipse of Newtonian gravitational theory, the orbit does not close. Rather, the angular position of the closest approach advances slightly on each return by an angle called the precession of the perihelion for a planet around the Sun. The figure shows a little over two orbits of a test mass that starts from the 3 o'clock position. The two positions of closest approach at the inner turning radius are indicated by dots. The angle between them is the precession of the perihelion per orbit.

In the bracket the constant term is not of order c^2, as it appears, but is of order unity, because from (9.31)

$$e^2 = 1 + \frac{2E_{\text{Newt}}}{mc^2} + \cdots \tag{9.53}$$

in an expansion in $1/c^2$. As we saw in (9.30), the first three terms in the bracket thus represent the Newtonian energy, gravitational potential, and centrifugal potential. The last term is of order $1/c^2$ with respect to the first three and represents the relativistic correction. It affects the orbits as a small additional $1/r^3$ term in the Newtonian potential would.

In the Newtonian approximation, in which the last term in the denominator in (9.52) is negligible, it is not difficult to see that $\Delta\phi$ is exactly 2π and that, therefore, the orbits close. Neglecting the last term in the bracket, introducing a new variable $u = 1/r$ the integral in (9.52) can be rewritten in the form

$$\Delta\phi = 2\int_{u_2}^{u_1} \frac{du}{[(u_1 - u)(u - u_2)]^{1/2}}, \tag{9.54}$$

where $u_1 = 1/r_1$, $u_2 = 1/r_2$ ($u_1 > u_2$) are roots at which the quadratic expression in the denominator of (9.52) vanishes. This integral is easily looked up and the result is $\Delta\phi = 2\pi$ for all values of u_1 and u_2.

Expanding the integral (9.52) to find the first-order relativistic correction to the Newtonian result is a little tricky. You can read about one method of proceeding in Problem 15. By working through that problem you can find that after one orbit,

$$\delta\phi_{\text{prec}} = 6\pi\left(\frac{GM}{c\ell}\right)^2, \qquad \text{(first order in } 1/c^2\text{)}. \tag{9.55}$$

To this accuracy we can use the Newtonian orbits to evaluate ℓ in terms of the usual parameters: eccentricity, ϵ, and semimajor axis, a. Recall from your intermediate mechanics text that in Newtonian mechanics,

$$\ell^2 = \left(r^2\frac{d\phi}{d\tau}\right)^2 \approx \left(r^2\frac{d\phi}{dt}\right)^2 = GMa\left(1-\epsilon^2\right). \tag{9.56}$$

Thus,

Precession of the Perihelion

$$\boxed{\delta\phi_{\text{prec}} = \frac{6\pi G}{c^2}\frac{M}{a\left(1-\epsilon^2\right)}} \qquad \begin{pmatrix}\text{small } GM/c^2a\\ \text{per orbit}\end{pmatrix}. \tag{9.57}$$

This is the relativistic precession of the inner turning point of the Keplerian ellipse per orbit. When applied to the Sun, the inner turning point is called the perihelion, and this is the precession of a planet's perihelion.[3] The largest effect is for the smallest a—the planets closest to the Sun. For Mercury the predicted rate of precession is about 43 seconds of arc per century—a tiny number but one detected by precision measurements, as we see in the next chapter.

9.4 Light Ray Orbits—The Deflection and Time Delay of Light

The calculation of light ray orbits in the Schwarzschild geometry parallels the calculations of particle orbits, but with important differences. As discussed in Section 5.5 and Section 8.3, the world lines of light rays can be described by giving the coordinates x^α as functions of any one of a family of affine parameters λ. The null vector $u^\alpha \equiv dx^\alpha/d\lambda$ is tangent to the world line. Because the Schwarzschild metric is independent of t and ϕ, the quantities

$$e \equiv -\boldsymbol{\xi}\cdot\mathbf{u} = \left(1-\frac{2M}{r}\right)\frac{dt}{d\lambda}, \tag{9.58}$$

$$\ell \equiv \boldsymbol{\eta}\cdot\mathbf{u} = r^2\sin^2\theta\frac{d\phi}{d\lambda}, \tag{9.59}$$

[3]If it's a binary star system, the inner turning point is called the *periastron*.

are conserved along light ray orbits. These are the analogs of (9.21) and (9.22) in the particle case. Indeed, if the normalization of λ is chosen so that $\mathbf{u}$ coincides with the momentum $\mathbf{p}$ of a photon moving along the null geodesic, then e and ℓ are the photon's energy and angular momentum at infinity. A third integral is supplied by the requirement that the tangent vector be null [cf. (8.40)]:

$$\mathbf{u} \cdot \mathbf{u} = g_{\alpha\beta} \frac{dx^\alpha}{d\lambda} \frac{dx^\beta}{d\lambda} = 0. \tag{9.60}$$

The 0 rather than the -1 of (9.23) on the right-hand side of this equation is the only real difference between the particle case and the light ray case.

The derivation of an energy integral for the radial motion of light rays parallels the steps leading from (9.23) to (9.29). Writing out (9.60) for the orbit of a light ray in the equatorial plane $\theta = \pi/2$ gives

$$-\left(1 - \frac{2M}{r}\right)\left(\frac{dt}{d\lambda}\right)^2 + \left(1 - \frac{2M}{r}\right)^{-1}\left(\frac{dr}{d\lambda}\right)^2 + r^2\left(\frac{d\phi}{d\lambda}\right)^2 = 0. \tag{9.61}$$

Using (9.58) and (9.59) to eliminate $dt/d\lambda$ and $d\phi/d\lambda$, respectively, we have

$$-\left(1 - \frac{2M}{r}\right)^{-1} e^2 + \left(1 - \frac{2M}{r}\right)^{-1}\left(\frac{dr}{d\lambda}\right)^2 + \frac{\ell^2}{r^2} = 0. \tag{9.62}$$

Multiplying by $(1 - 2M/r)/\ell^2$, this can be put in the form

$$\boxed{\frac{1}{b^2} = \frac{1}{\ell^2}\left(\frac{dr}{d\lambda}\right)^2 + W_{\text{eff}}(r).} \tag{9.63}$$

Here

$$b^2 \equiv \ell^2/e^2, \tag{9.64}$$

and

$$\boxed{W_{\text{eff}}(r) \equiv \frac{1}{r^2}\left(1 - \frac{2M}{r}\right).} \tag{9.65}$$

Effective Potential for Photon Orbits

Equation (9.63) has the form of an energy integral for radial motion with $W_{\text{eff}}(r)$ playing the role of the effective potential and b^{-2} playing the role of the energy. This relation can be used to analyze light ray orbits in much the same way that (9.29) was used to analyze particle orbits. However, unlike the particle case, where distinct values of e and ℓ determined different orbits, the physical properties of light ray orbits can depend only on their ratio, ℓ/e. That is because of the freedom in normalizing the affine parameter, λ. If λ is multiplied by a constant

K, it is just as good an affine parameter because (9.60) and the geodesic equation (8.42) are still satisfied. Physical predictions can't change by changing the affine parameter in this way, but e and ℓ are each divided by K. Therefore, only the ratio ℓ/e has physical significance and determines the properties of light ray orbits. Calculations of physical properties of light ray orbits, such as their shape, should automatically yield a result that depends only on the ratio ℓ/e. If they don't, there is a mistake in the calculation!

The sign of ℓ indicates which way the light ray is going around the center of attraction. We'll define $b \equiv |\ell/e|$ since that is what the shape of the orbits depends on. To see what b is, consider orbits that reach infinity. At infinity space is flat and Cartesian coordinates can be introduced that are related to Schwarzschild polar coordinates in the usual way, e.g., in the equatorial plane

$$x = r\cos\phi, \qquad y = r\sin\phi. \tag{9.66}$$

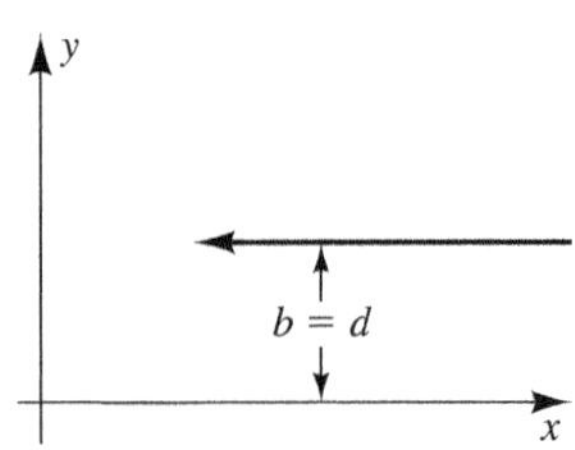

FIGURE 9.7 A segment of the orbit of an inwardly directed light ray far from the source of gravitational attraction is shown in this plot using Cartesian coordinates defined in (9.66). The light ray is moving inward on a straight line with speed 1 a distance d from the x-axis through the center of spherical symmetry. This distance is the *impact parameter* and is $b \equiv |\ell/e|$, as demonstrated in the text.

Consider a light ray moving parallel to the x-axis a distance of d away from it, as shown in Figure 9.7. Far away from the source of curvature, the light ray is moving in a straight line. For $r \gg 2M$, the quantity b is

$$b \equiv \left|\frac{\ell}{e}\right| \approx \frac{r^2 d\phi/d\lambda}{dt/d\lambda} = r^2\frac{d\phi}{dt}. \tag{9.67}$$

For very large r we have $\phi \approx d/r$, and $dr/dt \approx -1$, giving

$$\frac{d\phi}{dt} = \frac{d\phi}{dr}\frac{dr}{dt} = \frac{d}{r^2}. \tag{9.68}$$

Thus,

$$b = d. \tag{9.69}$$

The constant b is thus the *impact parameter* of a light ray that reaches infinity. It is defined to be positive. In geometrized units b has dimensions of length from (9.67). We will define it so it has the dimensions of length in any system of units as is appropriate for an impact parameter. Thus in $c \neq 1$ units $b \equiv |\ell/(ce)|$ if ℓ has the units of angular momentum per unit rest mass.

The plots on the left-hand side of Figure 9.8 show the shape of $W_{\text{eff}}(r)$. It vanishes at large r and has one maximum at $r = 3M$. The height at the maximum is

$$W_{\text{eff}}(3M) = \frac{1}{27M^2}. \quad (\text{maximum of } W_{\text{eff}}) \tag{9.70}$$

Circular orbits of light rays of radius $r = 3M$ are possible at this maximum if $b^2 = 27M^2$. However, these circular orbits are unstable since a small change in b results in an orbit that moves away from the maximum. A circular light ray orbit would not be possible around the Sun because the solar radius is much larger than $3M_\odot \approx 4.5$ km, but, as we'll see in Chapter 12, there can be circular light ray orbits outside a black hole.

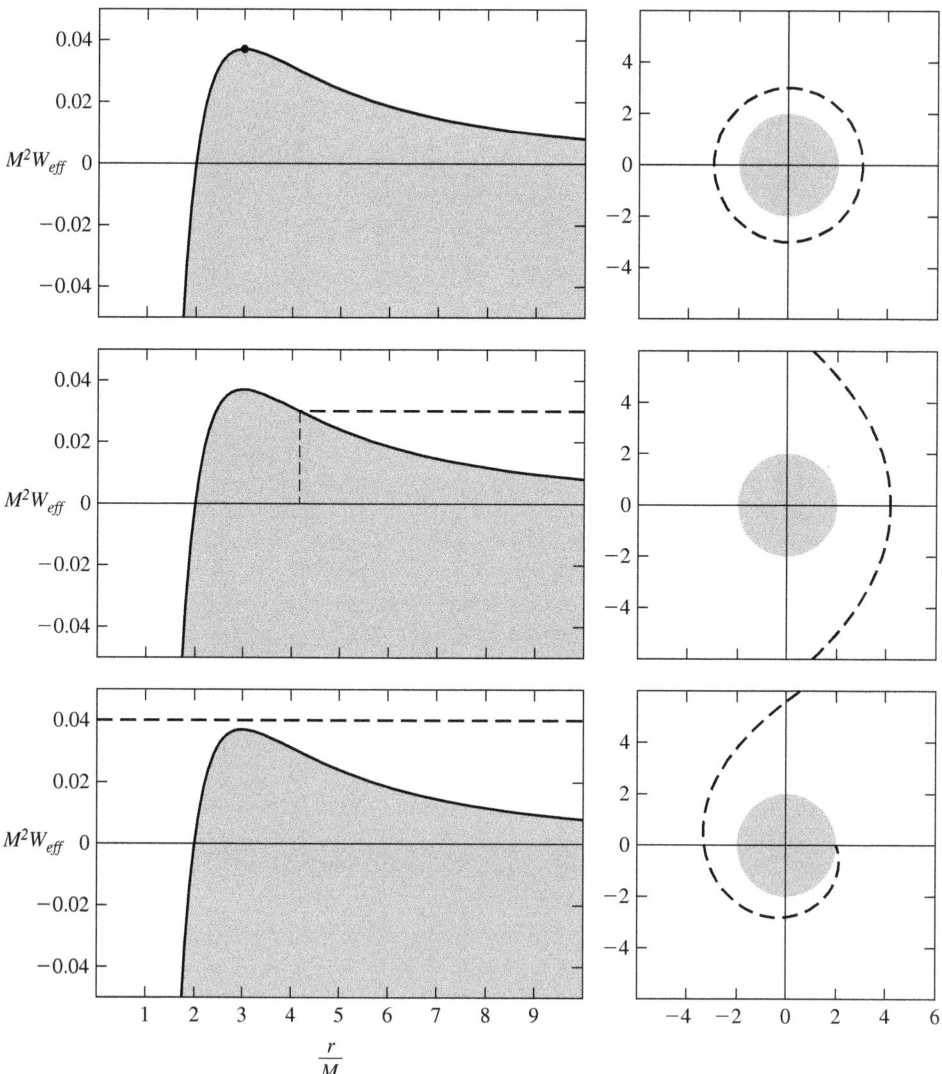

FIGURE 9.8 Three kinds of light ray orbits in the Schwarzschild geometry. The figure shows three orbits corresponding to different values of b. The potential and its relationship to $1/b^2$ are shown at left. The horizontal axis is r/M. The vertical axis is $M^2W_{\text{eff}}(r)$. The heavy dotted lines are the values of $1/b^2$. The shape of the orbit at right. From the top down there are a circular orbit, a scattering orbit, and a plunge orbit.

The qualitative character of other light ray orbits depends on whether $1/b^2$ is greater or less than the maximum height of W_{eff}, as shown in Figure 9.8. First consider orbits that start from infinity. If $1/b^2 < 1/(27M^2)$, then the orbit will have a turning point and again escape to infinity, as in the second of the examples in Figure 9.8. The light from a star being bent around the Sun is following one

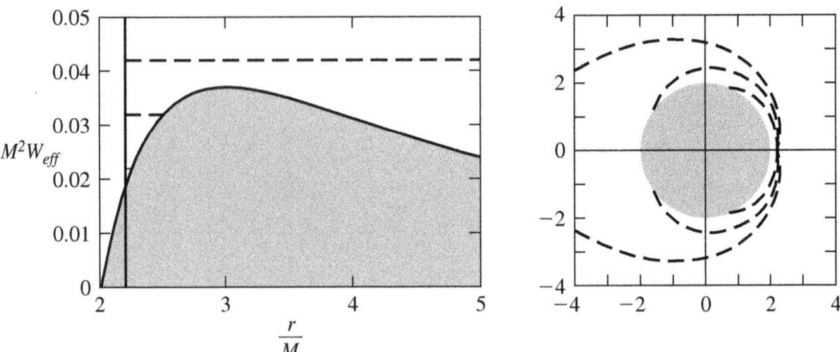

FIGURE 9.9 Light rays emitted between $r = 2M$ and $r = 3M$. A stationary observer at a radius $r = R = 2.2M$ emits light rays in various different outward directions corresponding to different values of b^2. This figure shows what happens to three cases $(M/b)^2 = .022, .032, .042$—values that were chosen to make intelligible plots. The left-hand plot shows a detail of the effective potential $M^2W_{\text{eff}}(r)$ together with a vertical line marking $r = 2.2M$ and horizontal lines marking the various values of b^{-2}. The right-hand plot shows the equatorial plane spanned by the Cartesian (x, y) defined in (9.66), together with the orbits of pairs of light rays with these values of b^{-2} emitted in directions above and below the x-axis. A radial light ray with $b \equiv |\ell/e| = 0$ or infinite b^{-2} (not shown) will escape. Light rays with values of b^{-2} higher than the maximum of the barrier $1/(27M^2) = .037/M^2$, making sufficiently small angles with the radial direction, will also escape like the pair with the value $(M/b)^2 = .042$ illustrated. Light rays with values of b^{-2} less than the height of the barrier will not escape. They move outward for a bit but then fall back through the radius $r = 2M$ like the pairs with the values $(M/b)^2 = .022, .032$. There is thus a critical angle ψ_{crit} with respect to the radial direction such that light rays emitted with less than this angle escape, but those with greater than this angle do not. Its value is given in (9.74). As $R \to 2M$, the opening angle for escaping light rays goes to zero, and essentially no light can escape.

of these scattering orbits, and measurements of the amount of deflection is an important test of general relativity, as we will see shortly. If $1/b^2 > 1/(27M^2)$, then the light ray will plunge all the way into the origin and be captured, as in the last pair in Figure 9.8.

Similar considerations hold for trajectories that start at small radii between $r = 2M$ and $r = 3M$, as shown in Figure 9.9. If $1/b^2 > 1/(27M^2)$, the light ray will escape. If $1/b^2 < 1/(27M^2)$ there is a turning point and the light ray falls back onto the center of attraction. Since $b^2 = \ell^2/e^2$, these criteria mean that if the light ray starts with sufficiently small angular momentum, i.e., is aimed sufficiently near the radial direction, then it will escape. Otherwise it falls back on the source of attraction. The situation is illustrated in Figure 9.9 and discussed quantitatively in Example 9.2.

Example 9.2. How Much Light Escapes to Infinity? A stationary observer stationed at a radius $R < 3M$ sends out light rays in various directions in the equatorial plane $\theta = \pi/2$, making angles ψ with the radial direction. Radial light rays with $\psi = 0$ have $b = 0$ and escape. What is the critical angle ψ_{crit} beyond which the light rays will fall back into the center of attraction, as illustrated in Figure 9.9? The answer depends on the connection between b and ψ, which can be found by analyzing the initial velocity of the light ray in an orthonormal basis $\{\mathbf{e}_{\hat{\alpha}}\}$ associated with the laboratory of the observer. The vector $\mathbf{e}_{\hat{0}}$ is the observer's timelike four-velocity and points along the t-direction. It is simplest to choose the three spacelike basis vectors to be oriented along the orthogonal coordinate axes at the position of the observer. Denote these by $\mathbf{e}_{\hat{r}}$, $\mathbf{e}_{\hat{\theta}}$, and $\mathbf{e}_{\hat{\phi}}$. In this orthonormal basis the angle between the direction of the light ray and the radial direction is given by

$$\tan\psi = \frac{u^{\hat{\phi}}}{u^{\hat{r}}} = \frac{\mathbf{u}\cdot\mathbf{e}_{\hat{\phi}}}{\mathbf{u}\cdot\mathbf{e}_{\hat{r}}}, \tag{9.71}$$

where the connection between orthonormal basis components and inner products with basis vectors in (5.82) has been used. To calculate the scalar products in (9.71) the coordinate basis components of the basis vectors $\mathbf{e}_{\hat{r}}$ and $\mathbf{e}_{\hat{\phi}}$ are needed in the equatorial plane along with the coordinate basis components of $\mathbf{u}$ given by solving (9.59) and (9.63) for $u^r = dr/d\lambda$ and $u^\phi = d\phi/d\lambda$. These components of the basis vectors can be found by following Example 7.9, and are

$$(\mathbf{e}_{\hat{r}})^\alpha = [0, (1-2M/R)^{1/2}, 0, 0], \tag{9.72a}$$

$$(\mathbf{e}_{\hat{\phi}})^\alpha = [0, 0, 0, 1/R], \tag{9.72b}$$

where the components are listed in the order (t, r, θ, ϕ). The scalar products in (9.71) can then be computed utilizing (7.57), (9.10), (9.60), (9.72), and solving (9.63) for $dr/d\lambda$, with the following results:

$$\mathbf{u}\cdot\mathbf{e}_{\hat{\phi}} = g_{\phi\phi}(\mathbf{e}_{\hat{\phi}})^\phi u^\phi = \frac{\ell}{R}, \tag{9.73a}$$

$$\mathbf{u}\cdot\mathbf{e}_{\hat{r}} = g_{rr}(\mathbf{e}_{\hat{r}})^r u^r = \left(1-\frac{2M}{R}\right)^{-1/2} \ell \left[\frac{1}{b^2} - \frac{1}{R^2}\left(1-\frac{2M}{R}\right)\right]^{1/2}. \tag{9.73b}$$

The ratio of these gives $\tan\psi$, according to (9.71). The critical opening angle ψ_{crit} below which light rays escape to infinity occurs when $b^2 = 27M^2$:

$$\tan\psi_{\text{crit}} = \frac{1}{R}\left(1-\frac{2M}{R}\right)^{1/2}\left[\frac{1}{27M^2} - \frac{1}{R^2}\left(1-\frac{2M}{R}\right)\right]^{-1/2}. \tag{9.74}$$

(Recall that $2M < R < 3M$.)

At $R = 3M$ the quantity in the square bracket vanishes because that's the maximum of the effective potential $1/(27M^2)$. There $\psi_{\text{crit}} = \pi/2$. That is just what could be expected from the existence of the circular orbit at that radius. The light ray making the circular orbit is just on the borderline between escaping to infinity and falling into the center of attraction. As R decreases below $3M$, the critical angle gets less and less until finally it vanishes altogether at $R = 2M$. At that point, no light gets out except the exactly radial light ray. Viewed from the exterior, a flashlight held by the stationary observer at radius R and emitting in all directions would appear dimmer and dimmer as R approaches $2M$. This anticipates the black hole phenomenon discussed in Chapter 12.

The Deflection of Light

From the discussion of light rays proceeding from infinity with a large impact parameter, it is evident that all material bodies will bend light trajectories somewhat. This effect is important because the deflection of light by the Sun is one of the most important experimental tests of general relativity, and the deflection of light by galaxies is the mechanism behind gravitational lenses to be discussed in the next chapter. The angle of interest is the deflection angle $\delta\phi_{\text{def}}$, defined as in Figure 9.10. This angle is a property of the shape of the orbit of a light ray. The shape of a light ray orbit can be calculated in the same way as the shape of a particle orbit. Solve (9.59) for $d\phi/d\lambda$, solve (9.63) for $dr/d\lambda$, divide the second into the first, and then simplify using (9.64) and (9.65) to find

$$\frac{d\phi}{dr} = \pm\frac{1}{r^2}\left[\frac{1}{b^2} - W_{\text{eff}}(r)\right]^{-1/2}. \tag{9.75}$$

The sign gives the direction of the orbit; integration gives its shape. In particular, the magnitude of the total angle swept out as the light ray proceeds in from infinity and back out again $\Delta\phi$ is just twice the angle swept out from the turning point $r = r_1$ to infinity. Thus,

$$\Delta\phi = 2\int_{r_1}^{\infty}\frac{dr}{r^2}\left[\frac{1}{b^2} - \frac{1}{r^2}\left(1 - \frac{2M}{r}\right)\right]^{-1/2}. \tag{9.76}$$

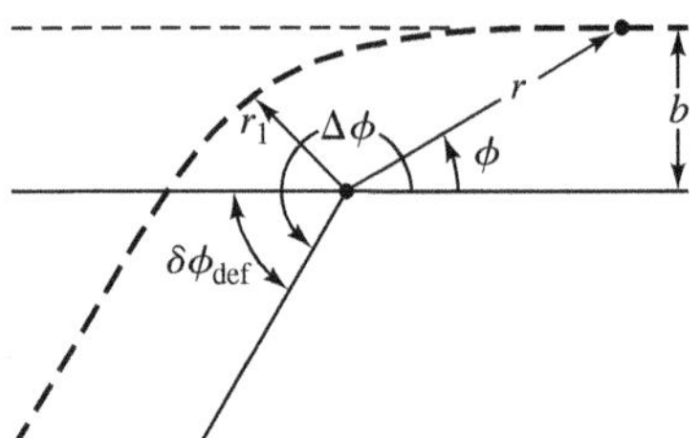

FIGURE 9.10 Quantities needed for calculating the deflection of light $\delta\phi_{\text{def}}$ by a spherical star. In this schematic diagram a light ray enters at right with an impact parameter b corresponding to a scattering orbit as in the second pass of plots in Figure 9.8. It approaches the center of attraction until the turning point at $r = r_1$, after which it moves out to infinity, emerging deflected by an angle $\delta\phi_{\text{def}}$. That deflection angle is the total angle swept out in the orbit $\Delta\phi$ less π.

The turning point r_1 is the radius where $1/b^2 = W_{\text{eff}}(r_1)$, i.e., the radius where the bracket in the preceding expression vanishes. By introducing a new variable w defined by

$$r = (b/w), \tag{9.77}$$

the expression for $\Delta\phi$ becomes

$$\Delta\phi = 2\int_0^{w_1} dw \left[1 - w^2\left(1 - \frac{2M}{b}w\right)\right]^{-1/2}, \tag{9.78}$$

where w_1 is the value of w at which the bracket vanishes. The angle $\Delta\phi$ swept out in one pass thus depends only on the ratio M/b. A plot of its behavior for large values of this ratio is shown in Figure 9.11.

For the bending of light by the Sun, the smallest value for b is approximately the solar radius $R_\odot = 6.96\times10^5$ km, whereas $M_\odot = 1.47$ km. The value of $2M/b$ is $\sim 10^{-6}$. The integral (9.78) can be expanded in powers of $2M/b$ to find an analytic expression for the deflection adequate for such small values. Expanding the integral requires a trick similar to the one needed for the expansion of (9.52), but since the algebra is not as messy we include a few steps to show how it goes. First rewrite (9.78) in the form

$$\Delta\phi = 2\int_0^{w_1} dw \left(1 - \frac{2M}{b}w\right)^{-1/2}\left[\left(1 - \frac{2M}{b}w\right)^{-1} - w^2\right]^{-1/2}. \tag{9.79}$$

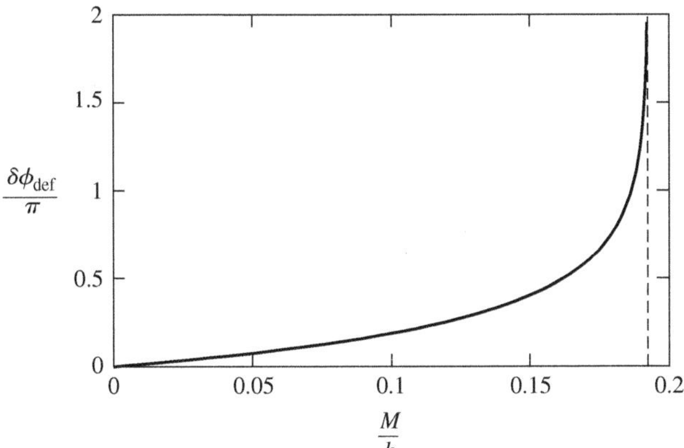

FIGURE 9.11 The deflection of light as a function of impact parameter. This is a rough plot of the angle $\delta\phi_{\text{def}}$ defined by (9.82) and the integral in (9.78) as a function of M/b. For values of $M/b < 2 \times 10^{-6}$ that are relevant for the deflection of light by the Sun, the linear approximation (9.83) is more than adequate. The deflection angle increases with smaller b, becoming infinite at the value $\sqrt{27}M$ at which an incoming photon would be injected into a circular orbit.

Next expand both inverse factors of $1 - (2M/b)w$ in powers of $2M/b$ and keep only the linear terms. The result is

$$\Delta\phi = 2\int_0^{w_1} dw \frac{1 + (M/b)w}{\left[1 + (2M/b)w - w^2\right]^{1/2}}, \tag{9.80}$$

w_1 being all along a root of the denominator. The integral is now in a form where it can be looked up in a table or done using an algebraic integration program. The result is

$$\Delta\phi \approx \pi + \frac{4M}{b} \tag{9.81}$$

for small M/b. From Figure 9.10 we see that the deflection angle $\delta\phi_{\text{def}}$ is related to $\Delta\phi$ by

$$\delta\phi_{\text{def}} = \Delta\phi - \pi. \tag{9.82}$$

Thus,

$$\delta\phi_{\text{def}} = \frac{4M}{b} \quad (\text{small } M/b). \tag{9.83}$$

This is the relativistic deflection of light when M/b is small. Reinserting the factors of G and c, it can also be written (remember b has dimensions of length)

Deflection of light

$$\boxed{\delta\phi_{\text{def}} = \frac{4GM}{c^2 b}} \quad (\text{small } GM/c^2 b). \tag{9.84}$$

For a light ray just grazing the edge of the Sun, the deflection angle is 1.7″ (″ is the standard notation for seconds of arc). We discuss how that's measured in the next chapter.

The Time Delay of Light

Another interesting relativistic effect found in the propagation of light rays is the apparent delay in propagation time for a light signal passing near the Sun. This is important because radar-ranging techniques can measure this delay and give another test of general relativity, and the time delay of light is a relevant correction for other observations. The effect is called the *Shapiro time delay* after Irwin Shapiro (1929–) who predicted it and led the first measurements of it to test general relativity. To see what's involved, imagine the following experiment: a radar signal is sent from the Earth to pass close to the Sun and reflect off another planet or a spacecraft. The time interval between the emission of the first pulse and the reception of the reflected pulse is measured. What does relativity predict for this number? We already have the machinery to answer this question.

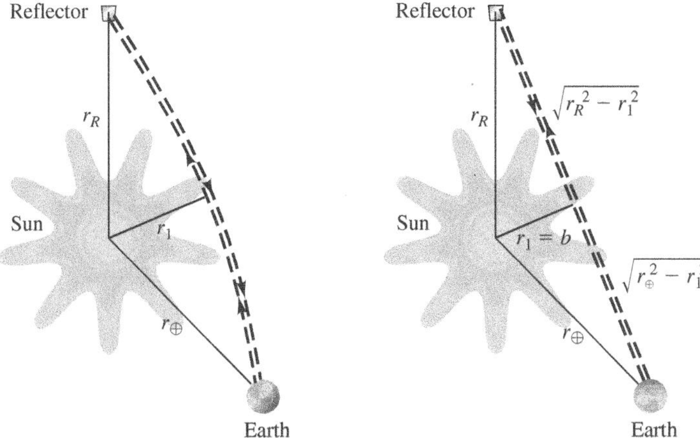

FIGURE 9.12 At left is a schematic diagram of the radar-ranging time delay experiment. Radar waves are sent from the Earth to a distant reflector so that they pass close to the Sun. They are deflected as all electromagnetic radiation is. There is an excess time delay between sending and return above what would be expected were the signals propagating along straight lines in flat spacetime as shown in the right-hand figure. That time delay caused by the curvature of the spacetime in the vicinity of the Sun is an important test of general relativity.

The geometry of the situation is illustrated in Figure 9.12. The path of the radar signals will be curved because they are deflected by the Sun, although we have greatly exaggerated the effect in the figure. The quantities $r_\oplus$ and r_R are radii of the orbits of the Earth and the reflector, respectively, in Schwarzschild coordinates centered on the Sun. These are not enough to specify the orbit because they do not fix the orientation of the planets relative to the Sun. Only one other distance is needed to do this, and we choose it to be the Schwarzschild radius of closest approach, r_1.

The Earth can be thought of as stationary over the round-trip travel time of the pulse (about 41 min). The total time interval between the emission and return of a pulse as measured by a clock on Earth is the Schwarzschild coordinate time interval $(\Delta t)_{\text{total}}$ between these events corrected for the influence of the Earth on spacetime and other effects. To calculate $(\Delta t)_{\text{total}}$ we need t as a function of r along the path of the pulse. This is like finding the shape of the orbit in the t-r plane and can be found in much the same way that we found ϕ as a function of r for the deflection problem. Solve (9.58) for $dt/d\lambda$ and (9.63) for $dr/d\lambda$ and divide the second into the first to find

$$\frac{dt}{dr} = \pm\frac{1}{b}\left(1 - \frac{2M}{r}\right)^{-1}\left[\frac{1}{b^2} - W_{\text{eff}}(r)\right]^{-1/2}, \tag{9.85}$$

where the $+$ sign is appropriate for when the radius is increasing and the $-$ sign applies when it is decreasing. Over the whole of the pulse's trajectory, the radius decreases from $r_\oplus$ to a minimum value r_1 at the turning point—the point of closest approach—and then increases again to r_R. On the return journey the pulse repeats this sequence in inverse order. The total elapsed time is

$$(\Delta t)_{\text{total}} = 2t\,(r_\oplus, r_1) + 2t\,(r_R, r_1)\,, \tag{9.86}$$

where $t(r, r_1)$ is the travel time from the turning point r_1 to a radius r given by

$$t(r, r_1) = \int_{r_1}^{r} dr \frac{1}{b}\left(1 - \frac{2M}{r}\right)^{-1}\left[\frac{1}{b^2} - W_{\text{eff}}(r)\right]^{-1/2}. \tag{9.87}$$

The parameters b and r_1 are related by

$$\frac{1}{b^2} = W_{\text{eff}}\,(r_1)\,. \tag{9.88}$$

For solar system experiments we need to evaluate the integral in (9.87) only to first order in M. The integral can be carried out in that approximation by expanding the integrand similarly to the case of the deflection of light (9.79). Equation (9.88) shows that to first order in M,

$$b = r_1 + M + \cdots\,, \tag{9.89}$$

where the neglected terms are of order $M(M/r_1)$. This result can be used to eliminate b from the answer. The result is

$$t(r, r_1) = \sqrt{r^2 - r_1^2} + 2M \log\left[\frac{r + \sqrt{r^2 - r_1^2}}{r_1}\right] + M\left(\frac{r - r_1}{r + r_1}\right)^{1/2}. \tag{9.90}$$

The first term in this expression is the Newtonian expression for the propagation time, as is seen from right-hand figure in Figure 9.12. The next terms represent the relativistic corrections, which increase the propagation time over the Newtonian value. The total time delay is obtained by substituting (9.90) in (9.86).

This division of the time delay into a Newtonian contribution and relativistic correction depends crucially on the use of the Schwarzschild radial coordinate in (9.90). Make a small change in the radial coordinate by an amount proportional to M and this division would change. Only the total elapsed time that is measured is a physical quantity. Nevertheless, the experimental results are usually quoted in terms of the *excess delay* over that which would be expected in Newtonian theory (see Figure 9.12) using Schwarzschild coordinates and (9.90):

$$(\Delta t)_{\text{excess}} \equiv (\Delta t)_{\text{total}} - 2\sqrt{r_\oplus^2 - r_1^2} - 2\sqrt{r_R^2 - r_1^2}. \tag{9.91}$$

The biggest effect occurs when r_1 is close to the solar radius. For $r_1/r_R \ll 1$ and $r_1/r_\oplus \ll 1$, expression (9.91) simplifies to give to a good approximation:

Time Delay of Light

$$\boxed{(\Delta t)_{\text{excess}} \approx \frac{4GM}{c^3}\left[\log\left(\frac{4r_R r_\oplus}{r_1^2}\right) + 1\right],} \tag{9.92}$$

where the factors of G and c have been reinserted. We describe the comparison of this expression with experiment in the next chapter.

Results like these for the time intervals measured by particular observers for light to travel over large distances do not mean that the velocity of light differs from c in general relativity. If you take 10 days to cross the United States it does not mean that your velocity is the distance traveled divided by 10 days. Velocity is a property of *each point* of a trajectory in Newtonian mechanics, special relativity, and general relativity. As discussed in Section 7.5, the local light cone structure of spacetime guarantees that velocity is always c for light as summarized by the condition that the four-velocity of a light ray is null: $\mathbf{u} \cdot \mathbf{u} = 0$ at each point along its world line.

Problems

1. [S] An advanced civilization living outside a spherical neutron star of mass M constructs a massless shell concentric with the star such that the area of the inner surface is $144\pi M^2$ and the area of the outer surface is $400\pi M^2$. What is the physical thickness of the shell?

2. Positrons are produced in the dense plasma surrounding a neutron star, which is accreting material from a binary companion, and electrons and positrons annihilate to produce γ rays. Assuming the neutron star has a mass of $2.5M_{\odot}$ (solar masses) and a radius of 10 km, at what energy should a distant observer look for the γ rays being emitted from the star by this process? Assume that both electron and positron are nearly at rest with respect to the star when they annihilate.

3. An observer is stationed at fixed radius R in the Schwarzschild geometry produced by a spherical star of mass M. A proton moving radially outward from the star traverses the observer's laboratory. Its energy E and momentum $|\vec{P}|$ are measured.

(**a**) What is the connection between E and $|\vec{P}|$?

(**b**) What are the components of the four-momentum of the proton in the Schwarzschild coordinate basis in terms of E and $|\vec{P}|$?

4. [B, E] Suppose the shell discussed in Box 9.1 on p. 192 is to be designed so the g-forces experienced by an observer falling into the shell are to be less than $20g$, where $g = 9.8\ \text{m/s}^2$. If the observer falls feet first into the shell, these g-forces are the *difference* between the force per unit mass at the observer's head and feet. *Estimate* using Newtonian theory how massive and how big would the shell have to be to meet this design criterion.

5. Sketch the qualitative behavior of a particle orbit that comes in from infinity with a value of $\mathcal{E}$ exactly equal to the maximum of the effective potential, V_{eff}. How does the picture change if the value of $\mathcal{E}$ is a little bit larger than the maximum or a little bit smaller?

6. [S] An observer falls radially inward toward a black hole of mass M whose exterior geometry is the Schwarzschild geometry, starting with zero kinetic energy at infinity. How much time does it take, as measured on the observer's clock, to pass between the radii $6M$ and $2M$?

7. Two particles fall radially in from infinity in the Schwarzschild geometry. One starts with $e = 1$, the other with $e = 2$. A stationary observer at $r = 6M$ measures the speed of each when they pass by. How much faster is the second particle moving at that point?

8. A spaceship is moving without power in a circular orbit about a black hole of mass M. (The exterior geometry is the Schwarzschild geometry.) The Schwarzschild radius of the orbit is $7M$.

(a) What is the period of the orbit as measured by an observer at infinity?

(b) What is the period of the orbit as measured by a clock in the spaceship?

9. Find the relation between the rate of change of angular position of a particle in a circular orbit with respect to proper time and the Schwarzschild radius of the orbit. Compare with (9.46).

10. Find the linear velocity of a particle in a circular orbit of radius R in the Schwarzschild geometry that would be measured by a stationary observer stationed at one point on the orbit. What is its value at the ISCO?

11. Show that a small radial displacement δr from the unstable circular orbit at the maximum of the effective potential V_{eff} will grow as

$$\delta r \propto e^{\tau/\tau_*},$$

where τ is the proper time along the particle's trajectory and τ_* is a constant. Evaluate τ_*. Explain its behavior as the radius of the orbit approaches $6M$.

12. A comet starts at infinity, goes around a relativistic star of mass M and goes out to infinity. The impact parameter at infinity is b. The Schwarzschild radius of closest approach is R. What is the speed of the comet at closest approach as measured by a stationary observer at that point?

13. [N, C] Particle orbits in the Schwarzschild geometry generally do not close after one turn. Explain why there should be a set of values $\mathcal{E}(\ell)$ for which orbits close for a given number of turns greater than one. Using the *Mathematica* program on the book website or otherwise find a value of $\mathcal{E}$ for which the orbit closes after four turns when $\ell = 4.6$ making a kind of clover leaf pattern.

14. In Newtonian mechanics one of Kepler's laws says that equal areas are swept out in equal time as a particle moves around an elliptical orbit in a $1/r$ potential. Consider the area outside a radius $R > 2M$ that is swept out by an orbit in the Schwarzschild geometry that stays outside this radius. Does Kepler's area law hold true using either proper time or Schwarzschild time?

15. [A] *Precession of the Perihelion of a Planet* To find the first order in $1/c^2$ relativistic correction to the angle $\Delta\phi$ swept out in one bound orbit, one might be tempted to expand the integrand in (9.52) in the small quantity $2GM\ell^2/c^2r^3$ and keep only the first two terms. This would be a mistake because the resulting integral would diverge near a turning point such as $\int^{r_2} dr/(r_2 - r)^{3/2}$, whereas the original integral is finite. There are several ways of rewriting the integrand so it can be expanded. One trick is to factor $(1 - 2GM/c^2r)$ out of the denominator so that it can be written

$$\Delta\phi = 2\ell \int_{r_1}^{r_2} \frac{dr}{r^2}\left(1 - \frac{2GM}{c^2r}\right)^{-1/2}\left[c^2e^2\left(1 - \frac{2GM}{c^2r}\right)^{-1} - \left(c^2 + \frac{\ell^2}{r^2}\right)\right]^{-1/2}.$$

The factor in the brackets is then still the square root of a quantity quadratic in $1/r$ to order $1/c^2$. To derive the expression (9.55) evaluate this expression as follows.

(a) Expand the factors of $(1-2GM/c^2r)$ in the preceding equation in powers of $1/c^2$, keeping only the $1/c^2$ corrections to Newtonian quantities and using (9.53).

(b) Introduce the integration variable $u = 1/r$, and show that the integral can be put in the form

$$\Delta\phi = \left[1 + 2\left(\frac{GM}{c\ell}\right)^2\right] 2\int_{u_2}^{u_1} \frac{du}{\left[(u_1 - u)(u - u_2)\right]^{1/2}} + \frac{2GM}{c^2}\int_{u_2}^{u_1} \frac{u\,du}{\left[(u_1 - u)(u - u_2)\right]^{1/2}} + \left(\begin{array}{c}\text{higher}\\ \text{order in } 1/c^2\end{array}\right).$$

(c) The first integral (including the 2) is just the one in (9.54) and equals 2π. Show that the second integral gives $(\pi/2)(u_1 + u_2)$ and that this equals $\pi\, GM/\ell^2$ to lowest order in $1/c^2$.

(d) Combine these results to derive (9.55).

16. A beam of photons with a circular cross section of radius a is aimed toward a black hole of mass M from far away. The center of the beam is aimed at the center of the black hole. What is the largest radius $a = a_{\text{max}}$ of the beam such that all the photons in the beam are captured by the black hole? The capture cross section is πa_{max}^2.

17. Calculate the deflection of light in Newtonian gravitational theory assuming that the photon is a "nonrelativistic" particle that moves with speed c when far from all sources of gravitational attraction. Compare your answer to the general relativistic result.

18. Suppose in another theory of gravity (not Einstein's general relativity) the metric outside a spherical star is given by

$$ds^2 = \left(1 - \frac{2M}{r}\right)\left[-dt^2 + dr^2 + r^2(d\theta^2 + \sin^2\theta\, d\phi^2)\right].$$

Calculate the deflection of light by a spherical star in this theory assuming that photons move along null geodesics in this geometry and following the steps that led to (9.78). When you get the answer see if you can find a simpler way to do the problem.

19. [N] Write a *Mathematica* program for the *null* geodesics in the Schwarzschild geometry analogous to the one on the website for particle geodesics. Use this program to illustrate the orbits with impact parameters a little above and a little below the critical impact parameter for a circular orbit.

20. **(a)** What is the speed of a particle in the smallest possible unstable circular orbit in the Schwarzschild geometry as measured by a stationary observer at that radius?

(b) What is the connection of this orbit to the unstable circular orbit of a photon in the Schwarzschild geometry?

21. [E] Suppose a neutron star were luminous so that features on its surface could be viewed with a telescope. The gravitational bending of light means that not only the hemisphere pointing toward us could be seen but also part of the far hemisphere.

Explain why and *estimate* the angle measured from the extension of the line of sight on the far side above which the surface could be seen. This would be $\pi/2$ if there were no bending, but less than that because of the bending. A typical neutron star has a mass of $\sim M_{\odot}$ and a radius of $\sim$ 10 km.

22. [N, C] *Looking for Black Holes with Lasers* Suppose primordial black holes of mass $\sim 10^{15}$ g were made in the early universe and are now distributed throughout space. If an observer shines a laser on a black hole some of the light is backscattered to the observer. A search for such primordial black holes could in principle be carried out by shining lasers into space and looking for the backscattered radiation.

(a) Explain why some light is backscattered.

(b) Suppose the flux of photons $[(\text{number})/\text{m}^2 \cdot \text{s}]$ in the laser beam is f_*, the mass of the black hole is M, and it is a distance R away. Derive a formula for the number of photons per second that will be returned to a collecting area of radius d at the origin of the beam. Assume that the width of the beam is much larger than the size of the black hole. (*Hint*: A little numerical integration is required to get an accurate answer for this problem.)

(c) Could the lasers described in Box 2.1 on p. 14 hope to detect such a black hole?

CHAPTER

10

Solar System Tests of General Relativity

The previous chapter's analysis of the orbits of test particles and light rays in the Schwarzschild geometry identified four effects of general relativity that can be tested in the solar system: the gravitational redshift, the deflection of light by the Sun, the precession of the perihelion of a planetary orbit, and the time delay of light. This list does not exhaust the tests that can be carried out in the solar system, but they are among the more important. This chapter describes some experiments that measure these effects and confirm the predictions of general relativity in the solar system to a typical accuracy of a fraction of 1%.

The discussion in this chapter is in no sense a review of the experimental situation in general relativity either in the past or at the time of writing. Rather, we present a discussion of *representative* experiments that are currently among the most accurate but are not necessarily *the* most accurate.

The experiments are described only schematically, but they are discussed in enough detail so that the major sources of error are mentioned. For a real appreciation of the ingenuity and effort that goes into these very precise measurements, you should consult the original papers to which references are given.

10.1 Gravitational Redshift

Any theory of gravity consistent with the principle of equivalence will predict a gravitational redshift, as we saw in Chapter 6. To leading order in $1/c^2$, the value of the gravitational redshift depends just on the principle of equivalence and not on the details of the gravitational theory. Tests of the gravitational redshift are, therefore, more of a test of this principle than the details of general relativity.

The obvious place to look for the gravitational redshift is in spectral lines emitted from atoms far down in the gravitational potential of a massive body such as a star. The effect has been seen in the Sun, white dwarfs, and some active galactic nuclei. However, at the time of writing, the most accurate tests of the gravitational redshift are not carried in the deep gravitational potentials of massive bodies but through experiments near the surface of the Earth. The redshifts are much smaller, but the ability to control an experiment is much greater.

In the 1976 experiment of R. Vessot and M. Levine (1979), a rocket carrying an accurate hydrogen-maser atomic clock was launched in an orbit reaching 10^4 km above the Earth's surface. During the experiment, the position of the rocket is monitored from the ground, as is the frequency f_0' of a signal emitted at a fixed

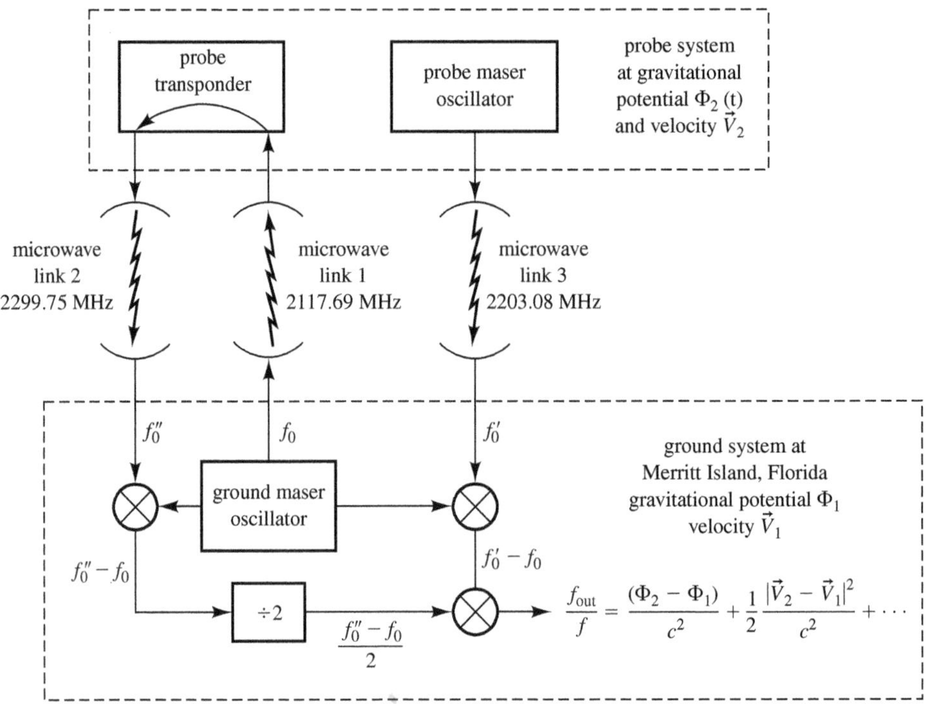

FIGURE 10.1 A schematic diagram of the rocket experiment of Vessot and Levine (1979) measuring the gravitational redshift. The top dotted box shows the package carried on the rocket; the bottom dotted box shows the package on the ground. The signal from the rocket clock at frequency f_0' is shown. Also shown is the uplink signal at frequency f_0, which is transponded into the downlink signal at f_0''. Half the difference between these frequencies is proportional to the first-order Doppler shift due to the velocity of the rocket. When this is subtracted out from the clock signal, the leading terms in $1/c$ are the gravitational redshift and the second-order Doppler shift, whose value is known from $(f_0 - f_0'')/2$.

frequency according to the orbiting clock, thus effectively monitoring its rate. Figure 10.1 is a schematic diagram of how this signal was analyzed. (We'll use f for frequency in this discussion to correspond to that diagram.) To the $1/c^2$ accuracy needed to analyze the experiment, the observed frequency f_0' is shifted from the frequency of the emitting clock by the sum of the special relativistic Doppler shift[1] (5.73) and the general relativistic gravitational redshift (6.12). The Doppler shift (5.73) can be expanded in powers of the velocity of the rocket. Only the first two terms—called the first- and second-order Doppler shifts—are relevant for the experiment. The order of magnitude of the first-order Doppler shift of a signal emitted with frequency f_* is

[1]You might wonder whether the effect of time dilation should be included as well. But (5.73) includes *all* special relativistic effects. Time dilation is essentially the factor in its numerator.

$$\frac{\Delta f_{\text{Doppler}}}{f_*} \approx \frac{V}{c} \sim \left(\frac{gh}{c^2}\right)^{1/2} \sim 10^{-5}, \tag{10.1}$$

where V is the velocity of the rocket and the estimate was made using a fraction of the velocity needed to reach an altitude h of 10^4 km. (We return to $c \neq 1$ units in this chapter on experiment.) The second-order Doppler shift is of order of the square of this. The gravitational redshift is [cf. (6.12)]

$$\frac{\Delta f_{\text{grav}}}{f_*} \approx \frac{gh}{c^2} \sim 10^{-10}. \tag{10.2}$$

A major experimental problem is now clear. The effect to be measured is five orders of magnitude less than the competing first-order Doppler effect.

The ingenious experimental solution (Figure 10.1) is to send a signal of known frequency f_0 to a transponder on the rocket, which then sends it back again at the frequency it was received. The first-order Doppler shifts of these uplink and downlink signals will add—the source is moving away from the receiver in both cases. However, the gravitational and second-order Doppler shifts will cancel because they are the same both on uplink and downlink. The transponded signal thus arrives at the surface with a frequency f_0'', which is shifted from f_0 by the first-order Doppler effect twice and with no gravitational or second-order Doppler shifts. It is thus a direct measure of the velocity of the rocket. The difference $(f_0'' - f_0)/2$ is subtracted from $f_0' - f_0$ automatically when the data are taken. The subtraction cancels the dominant-order $1/c$ Doppler shifts, leaving the $1/c^2$ gravitational redshift and second-order Doppler effects. The latter are known from the velocity of the rocket, determined from $f_0'' - f_0$ and other monitoring of the rocket orbit. The result is an accurate test of the gravitational redshift. The predicted and observed values differ by

$$\left|\frac{(\Delta f_{\text{grav}}/f_*)_{\text{obs}} - (\Delta f_{\text{grav}}/f_*)_{\text{pred}}}{(\Delta f_{\text{grav}}/f_*)_{\text{pred}}}\right| \leq 2 \times 10^{-4}. \tag{10.3}$$

10.2 PPN Parameters

Einstein's general relativity is not the only theory of relativistic gravity that has been proposed over the years, although at present it is essentially the only seriously considered theory consistent with experimental tests in the solar system. In discussing these experimental tests, it is useful to have a framework in which the predictions of different theories are parametrized in a systematic way. The parametrized-Post-Newtonian (PPN) framework has become the standard way of doing this.

To understand the idea behind the PPN framework, imagine another theory of gravity. Suppose that, like general relativity, the theory predicts that mass curves spacetime and that light rays and test particles move on geodesics in that spacetime. The geometry outside the Sun would be spherically symmetric to an excellent approximation but would differ in detail from the Schwarzschild geometry (9.1) predicted by Einstein's theory. The differences relevant for the experimental tests can be summarized in a few PPN parameters.

As we show in detail in Section 21.4, with an appropriate choice of coordinates, the most general, static, spherically symmetric metric can be put in the form

$$ds^2 = -A(r)\,(c\,dt)^2 + B(r)\,dr^2 + r^2(d\theta^2 + \sin^2\theta\,d\phi^2). \tag{10.4}$$

You might wonder why there isn't an arbitrary function $C(r)$ in front of the $d\theta^2 + \sin^2\theta\,d\phi^2$. Were there one, a new radius $r' = [C(r)]^{1/2}$ could be defined such that the new metric takes the form (10.4) with r' replacing r everywhere. Then, just changing the name of r' to r, we'd get to the form (10.4). The Schwarzschild geometry (9.1) has this form for particular functions A and B. Now imagine expanding the metric (10.4) in inverse powers of c, thereby obtaining the Newtonian limit and the *post*-Newtonian corrections. Assuming that the mass M is the only stellar parameter that determines the spherical geometry outside the star, this must be an expansion in powers of GM/c^2r. (That is the only dimensionless combination of G, M, c, and r.)

Any relativistic theory of gravity must agree with the well-tested results of Newtonian theory in the nonrelativistic limit. The discussion in Section 6.6 showed that the predictions for orbits in this limit are determined by the first relativistic correction to the geometry of flat space in $g_{tt}(r)$. Agreement with Newtonian theory therefore requires

$$A(r) = 1 - \frac{2GM}{c^2r} + \cdots, \qquad B(r) = 1 + \cdots. \tag{10.5}$$

Agreement with the static weak field metric (6.20) predicted by general relativity would fix more terms in $B(r)$, but, as mentioned in Section 6.6, those terms don't affect the small-velocity, Newtonian predictions. To get the first *post-Newtonian* corrections, we keep the next terms in both A and B:

$$A(r) = 1 - \frac{2GM}{c^2r} + 2(\beta - \gamma)\left(\frac{GM}{c^2r}\right)^2 + \cdots, \tag{10.6a}$$

$$B(r) = 1 + 2\gamma\left(\frac{GM}{c^2r}\right) + \cdots. \tag{10.6b}$$

The coefficients in front of the post-Newtonian terms are related to the PPN parameters β and γ according to standard usage. These parameters may be different in different theories of gravity. For general relativity the values are those of the Schwarzschild metric (9.1):

$$\text{general relativity:} \quad \gamma = 1, \qquad \beta = 1. \tag{10.7}$$

The bending of light by the Sun, the precession of perihelion of a planet, and the time delay of light can all be worked out for the PPN metric obtained by inserting (10.6a) and (10.6b) in (10.4) (e.g., Problem 4). The results to leading order in $1/c^2$ are as follows:

- For the deflection angle $\delta\phi_{\text{def}}$ of a light ray passing by a mass M at an impact parameter b [cf. (9.84)]:

$$\delta\phi_{\text{def}} = \left(\frac{1+\gamma}{2}\right)\left(\frac{4GM}{c^2 b}\right). \tag{10.8}$$

- For the precession $\delta\phi_{\text{prec}}$ of the perihelion of a planet per orbit:

$$\delta\phi_{\text{prec}} = \frac{1}{3}(2+2\gamma-\beta)\,\frac{6\pi GM}{c^2 a(1-\epsilon^2)}, \tag{10.9}$$

  where M is the mass of the orbited star, a is the orbit's semimajor axis, and ϵ is eccentricity [cf. (9.57)].
- For the "excess" time delay of light, Δt_{excess}, in the approximation that the radii $r_\oplus$ of the emitter at the Earth and responder r_R are much greater than the distance r_1 of closest approach to the gravitating body [cf. (9.92)]:

$$\Delta t_{\text{excess}} = \left(\frac{1+\gamma}{2}\right)\frac{4GM}{c^3}\left[\log\left(\frac{4r_\oplus r_R}{r_1^2}\right)+1\right]. \tag{10.10}$$

These three experimental tests can be used to measure the values of β and γ and compare with the general relativistic values (10.7).

10.3 Measurements of the PPN Parameter γ

The deflection of light by the Sun and the time delay of light are two experiments that directly determine the value of the PPN parameter γ.

Deflection of Light by the Sun

Light rays will bend in the curved spacetime of the Sun as shown in Figure 10.2, by an amount given in (10.8). For a light ray that just grazes the limb of the Sun, general relativity predicts

$$[\delta\phi_{\text{def}}]_{\text{predicted}} = 1.75''. \tag{10.11}$$

A measurement of this deflection for light from stars carried out in 1919 was one of the first tests of general relativity.

The deflection given by (10.8) is greatest for stars closest to the Sun. However, stars close to the Sun can be seen only during a solar eclipse, when the light from the solar disk is blocked by the Moon. A photograph of a region of the sky about an eclipse is compared with a photograph of the same region months later when the Sun has moved from the field. As shown in Figure 10.2, the deflection means that, when the Sun is in the field of view, the angular position of a star is shifted *away* from the center of the solar disk. The predicted shift decreases with the angular distance of a star from the Sun.

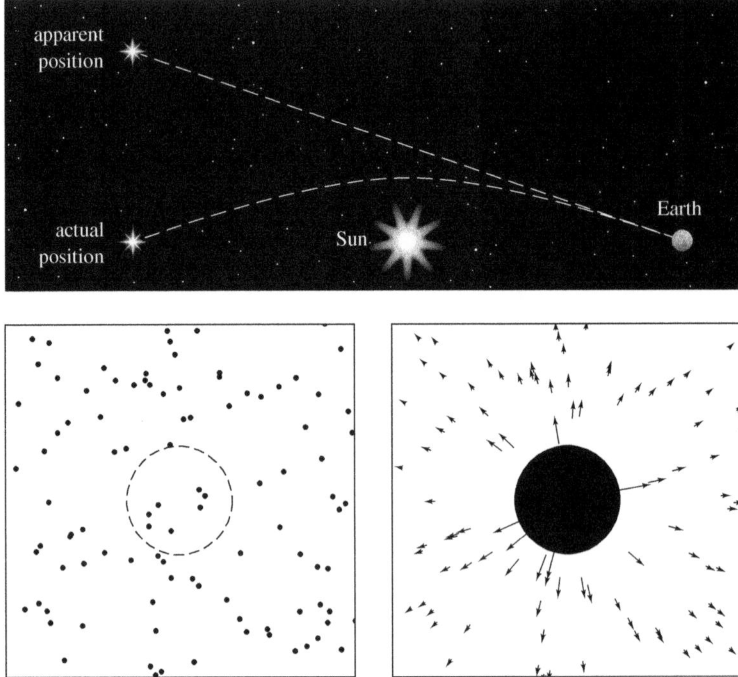

FIGURE 10.2 The top figure shows how a star whose image is deflected by the curved spacetime produced by the Sun appears to be at a greater angular separation from the Sun than it actually is. The bottom figure illustrates the outward deflection for a field of stars when the Sun is in the field of view. See Figure 10.3 for some actual data. The shift becomes smaller with increasing angular distance from the Sun. The effect is greatly exaggerated in these figures. The angular diameter of the Sun viewed from Earth is $959''$ but the deflection of light at the edge of the Sun is only $1.75''$.

Under normal conditions, the fluctuations of stellar positions due to refraction through fluctuations in the Earth's atmosphere are comparable or larger than the predicted deflection. Measurements must, therefore, be carried out on a large number of stars to average out these fluctuations. Useful eclipses are often in remote places, where mechanical and thermal difficulties of temporary observation posts can produce significant systematic errors. Some data from a 1922 eclipse observation are shown in Figure 10.3. One can get some feel for the difficulty of the observations from the scatter in the directions of the displacements. Despite the difficulties, the best observations gave results consistent with $\gamma = 1$ to accuracies such as 5%.

Far better measurements can be made today with radio telescopes and radio sources instead of stars, although the idea is exactly the same. The Sun is not very bright in the radio band, so observations can be made of sources close to the Sun at all times. Further, radio interferometry provides much better angular resolution than optical instruments. Excellent measurements were made by Edward Fomalont and Richard Sramek (Fomalont and Sramek 1975) at the National Ra-

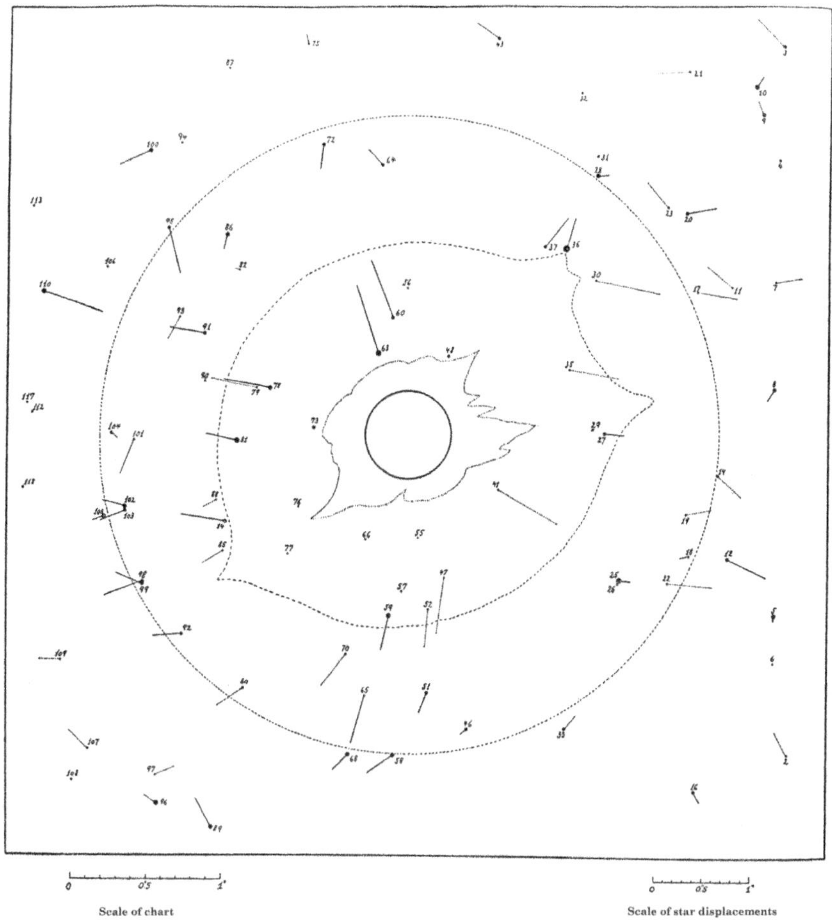

FIGURE 10.3 Star displacements observed by Campbell and Trumpler (1923) in an eclipse in 1922. The solar disk is at the center surrounded by a dotted line indicating the corona. The variety of directions of the displacements is some measure of the difficulty in making the measurements.

dio Astronomy Observatory (NRAO) in Green Bank, West Virginia in 1974 and 1975 using long-baseline interferometry (LBI), as illustrated in Figure 10.4.

Two telescopes separated by a baseline B and operating at a wavelength λ are pointed toward a radio source e.g., a distant quasar, as shown in Figure 10.4. The two signals are carried by cable to a common point, added together, and averaged over some convenient time interval. Since there is a difference $B \sin\theta$ between the distances the two signals travel, they will interfere constructively if this difference is an even multiple of half a wavelength and destructively if the difference is an odd multiple of half a wavelength. The sum of the two signals will be multiplied by a factor (assuming equal intensities)

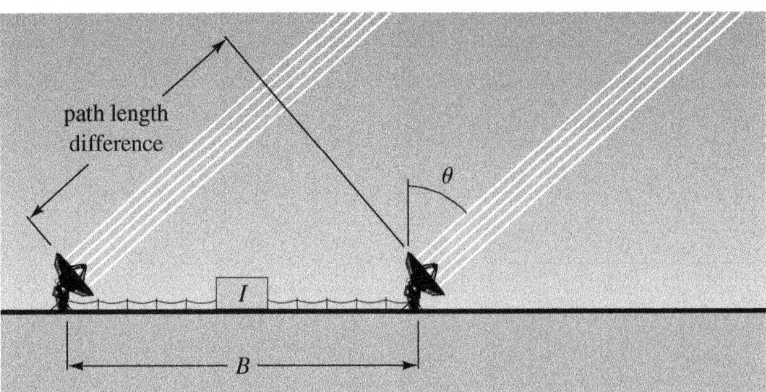

FIGURE 10.4 Radio interferometry. Two radio telescopes, a distance B apart, are pointing toward the same distant object, whose position makes an angle θ with the zenith. The path-length difference means that the two signals will interfere constructively for some angles and destructively for others. That enables a very precise determination of the angular position of the object. In long-baseline interferometry (LBI) ($B \sim 20$ km), the telescopes are close enough that their signals can be combined in real time. In very long baseline interferometry (VLBI) ($B \sim 1000$ km), signals are recorded separately at each location and later combined.

$$1 + \cos\left(\frac{2\pi B \sin\theta}{\lambda}\right). \tag{10.12}$$

As the Earth rotates, θ will change and the sum of the signals will vary proportionally to the preceding function—sometimes interfering constructively, sometimes destructively. From the observation of these patterns of interference and a knowledge of the Earth's rotation speed, $\sin\theta$ can be measured. The accuracy is ultimately determined by the phase stability of the system, which is typically .01 to .1 of the phase in (10.12). An angular accuracy of $.01\lambda/B$ is thus obtained.

In the NRAO experiment, four radio telescopes were used, three of which are shown in Figure 10.5. The effective baseline was $B = 35$ km, so that at frequencies of a few gigahertz the expected angular accuracy would be less than or about $.01''$—more than enough to measure the $1.75''$ bending predicted by general relativity for the deflection of light by the Sun.

The observations proceed as follows (Figure 10.6): Three radio sources, 0111+ 02, 0119 + 11, and 0116 + 08, were used. They are less than 10° apart, nearly collinear, small in angular extent, and reasonably strong. Every April 11 the Sun occults the source 0116 + 08. Its angular position is measured as a function of time using the other two sources as references. The results are compared with those predicted by general relativity (see Figure 10.6).

The major source of error in the experiment arises from the propagation of the signals through the solar corona. The solar corona is a gas of ionized particles above the solar surface, which, like any medium, has an index of refraction $n(r)$ and bends light. The index of refraction can be modeled by (SI units)

$$n(r) = 1 - \frac{e^2 N(r)}{2\epsilon_0 m\omega^2} \tag{10.13}$$

FIGURE 10.5 Three of the four radio telescopes used in the NRAO experiment to measure γ. These three 85-m dishes are separated by a few kilometers and themselves make up an interferometer. The participation of a fourth telescope further away gives the effective 35-km baseline.

where $N(r)$ is the density of particles with mass m and charge e and ω is the frequency of the radiation. The bending due to the corona must be separated out to get at the general relativistic effect. With an adequate model of the solar corona, the two effects can be partially separated by making measurements at several different frequencies. The bending due to the index of refraction is frequency dependent, whereas the general relativistic deflection is not. Measurements at different frequencies can thus determine both γ and some information about $N(r)$. In the NRAO experiment, two frequencies of 8.1 GHz and 2.7 GHz were used.

The average results of the 1974 and 1975 experiments give

$$\gamma = 1.007 \pm 0.009, \tag{10.14}$$

which shows truly impressive agreement with Einstein's theory.

VLBI [e.g., Lebach et al. (1995)] gives a slightly more accurate determination of γ. The principle of VLBI is the same as LBI, except that the two antennas are not connected. The signals are recorded separately and later added. This permits baselines comparable to the diameter of the Earth and a consequent improvement in angular accuracy.

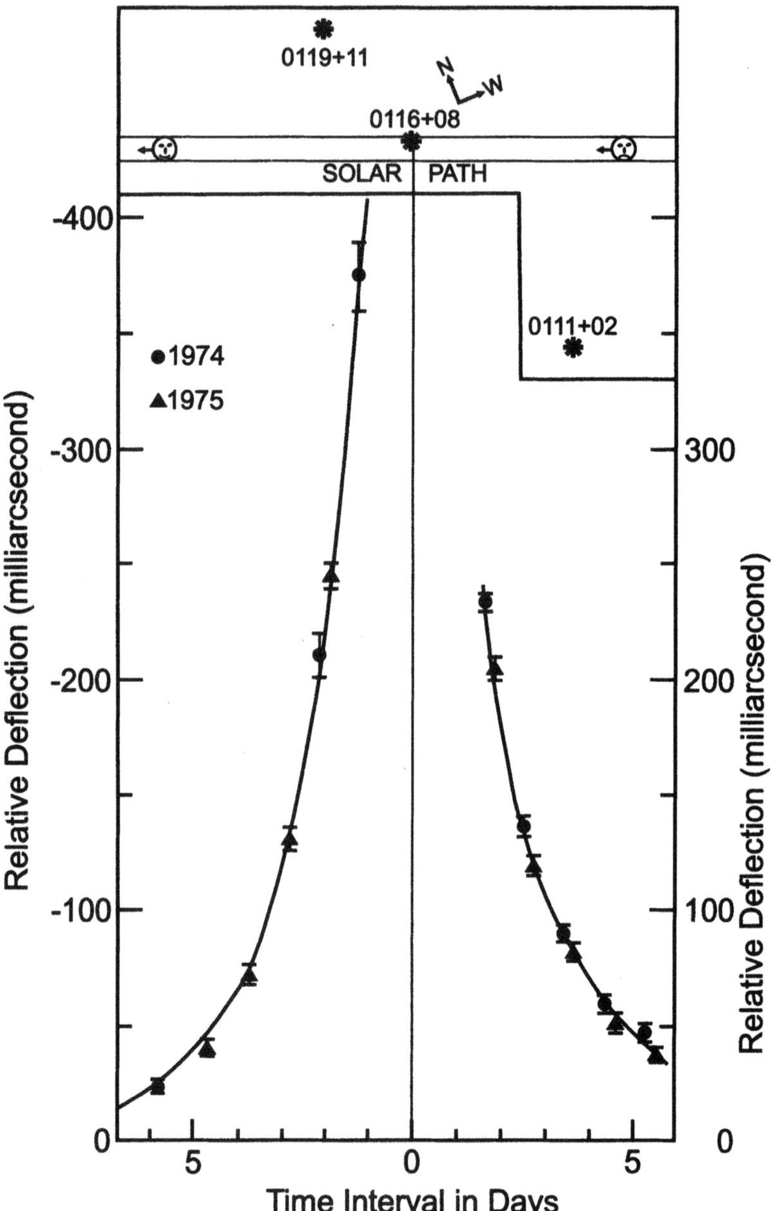

FIGURE 10.6 The deflection of the light from the radio source 0116 + 08 as a function of time as it is occulted by the Sun. The top part of this figure illustrates the path of the Sun on the sky and the relative positions of the three radio sources involved. The bottom part shows the experimental data for the deflection in angular position of 0116 + 08 measured relative to the other two sources. The solid curve is the prediction of general relativity.

Time Delay of Light

A classic measurement of the time delay of light was carried out in conjunction with the Viking mission to Mars in 1976 (Shapiro et al., 1977). All four of the Viking vehicles—two landers and two orbiters—carried radar transponders. Each lander had a transponder that transmitted in S band ($\sim$ 10-cm wavelength) and each orbiter had transponders that transmitted in both S and X band ($\sim$ 3-cm wavelength). The dual-frequency capability is important because, like radio waves, the dispersive effect of the solar corona is important for radar waves. The advantage of the Mars landers for transponders, as opposed to the orbiters, is that the orbit of Mars is predictively determined by gravitational forces and negligibly affected by nongravitational forces such as the buffeting by the solar wind that can be significant for the orbiters. A very accurate theoretical model to fit the data can thus be constructed.

Recall that the experiment's goal is to measure the "excess" delay in the round-trip travel time for a radar signal from Earth to Mars. This is given by (10.10). This Schwarzschild time delay has to be corrected for various additional sources of

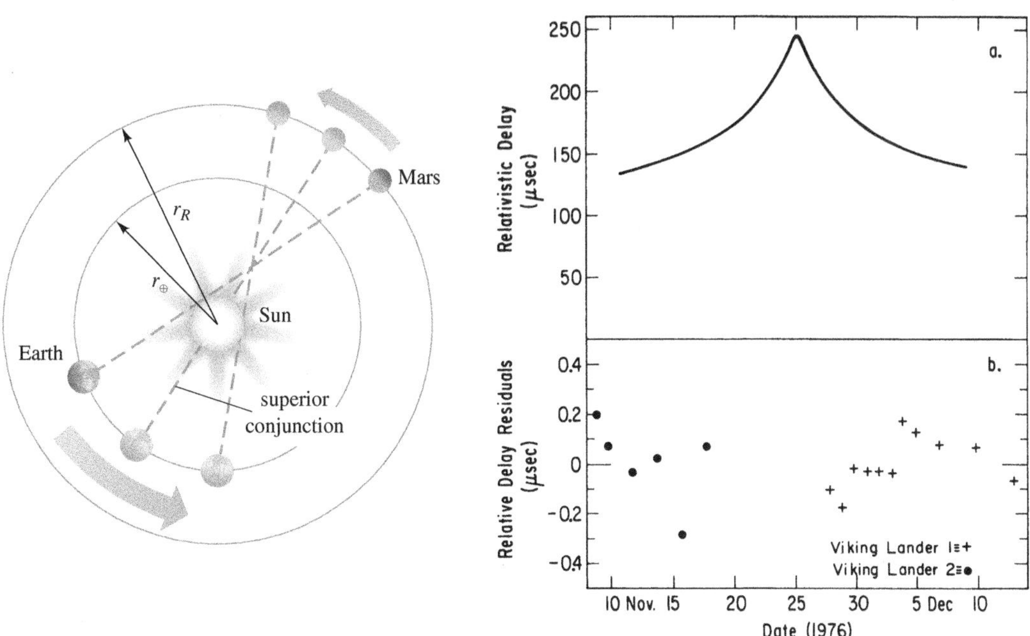

FIGURE 10.7 The measurement of the time delay of light carried out on the Viking Mars mission (Shapiro, et al., 1977). As described in the text, a radar signal sent from Earth is returned from the Viking lander on Mars, and the difference between the time of return and the time of emission is monitored as a function of time. The figure at left is a schematic diagram of the configuration of the two planets during the experiment. Near the time of superior conjunction, November 26, 1976, the signals passed close to the Sun, and general relativistic effects on the time delay could be accurately measured (Problem 10). Signals were not blocked by the Sun at superior conjunction because the orbits of Earth and Mars are not exactly in the same plane. The figure at right shows the measured excess delay vs. time. They are accurately fit by the prediction of general relativity.

delay, such as propagation through the solar corona and the curvature of spacetime produced by the Earth, but for purposes of discussion we focus just on (10.10).

The sequence of positions of the Earth, Mars, and the Sun during the experiment is shown schematically in the left part of Figure 10.7. Because the Earth moves in its orbit with a higher angular velocity than Mars, the distance of closest approach to the Sun r_1 will at first get smaller and then larger. The predicted delay as a function of time will thus look as in the top right part of Figure 10.7.

The excess delay is largest when r_1 is as small as it can be—the radius of the Sun, $R_\odot$. The maximum delay is about $(\Delta t)_{\max} \approx (4GM/c^3)[\log(4r_R r_\oplus/R_\odot^2)+1] \approx 247\mu\text{s}$. This is a delay out of a total round-trip travel time of roughly $2(r_R + r_\oplus)/c \approx 2.51 \times 10^3$ s ≈ 41 min! An accuracy of one part in 10^7 is, therefore, necessary to see the effect, and one part in 10^9 is needed to measure it to 1% accuracy. This is even more remarkable when one realizes that to get the accuracy needed for a 1% measurement, all orbits must be known to an accuracy of about 1 km—which might be the height of a typical mountain on the surface of the planet! Fortunately, atomic clocks keep time accurately—to better than one part in 10^{12}—and the round-trip travel time can be measured to 10 ns. The chief source of the error is not the measurement of the time delay but in the interpretation of the data in terms of a theoretical model, including the corrections for the solar corona and the orbital motions of the bodies involved. The corrections from the corona themselves can be as high as 100 μs.

The result for γ is

$$\gamma = 1.000 \pm 0.002. \tag{10.15}$$

This few tenths of a percent accuracy is one of the most accurate quantitative tests of Einstein's theory to date.

10.4 Measurement of the PPN Parameter β—Precession of Mercury's Perihelion

Mercury is the closest planet to the Sun, and its orbit has the largest precession of its perihelion. However, comparison of the general relativistic prediction for the precession of the perihelion of Mercury with observation is not easy. The predicted precession due to general relativity is, from (9.57),

$$\delta\phi_{\text{prec}} = 42.98''/\text{century}. \tag{10.16}$$

The observed precession from an Earth-based laboratory is

$$\delta\phi = 5599.74'' \pm 0.41''/\text{century}. \tag{10.17}$$

There are various known Newtonian effects to be subtracted from this observation, but the relative size of (10.17) and (10.16) indicates how well these must be known to determine the residual precession due to general relativity. Determining the orbits of the planets is a complex observational problem at the level of accuracy needed to test relativity. Radar ranging has supplied accurate positions of the inner planets as a function of time since 1966. Less accurate optical observations dating back to the eighteenth century also help. Satellite flybys provide another source of data. All these data are fit to a model, which includes as parameters the masses, semimajor axes, eccentricities, etc., of the Newtonian theory of the planetary motion as well as the post-Newtonian relativistic parameters and the solar mass quadrupole moment.

The largest Newtonian subtraction is the precession of the equinoxes. The observed precession (10.17) is referred to an Earth-based reference frame. However, the rotation axis of the Earth is precessing with respect to an inertial frame with a period of about 26,000 yr. This contributes a $\delta\phi$ of $5025.64'' \pm .50''$/century.[2]

The gravitational attractions of the other planets mean that Mercury does not move in an *exactly* $1/r$ Newtonian gravitational potential. The orbit will precess just from these Newtonian perturbations. The total precession from these perturbations can be inferred from Newtonian mechanics and the observations of the planetary orbits. The most accurate determination of the precession of Mercury's perihelion unexplained by Newtonian mechanics is $42.98'' \pm 0.04''$/century (Shapiro 1990)—exactly the prediction of general relativity. When combined with the best observations of the PPN parameter γ discussed previously, this gives for the PPN parameter β:

$$\beta = 1.000 \pm 0.003. \tag{10.18}$$

Thus, provided there are no additional corrections to be made, observations are in excellent agreement with the prediction of general relativity. The chief candidate for an additional correction would be a mass quadrupole moment of the Sun.

The previous chapter's calculation of the precession of the perihelion assumed that the source of curvature is exactly spherical. But the Sun is not exactly spherical. It is rotating, and the resulting centripetal accelerations mean that the Sun is slightly "squashed" along the rotation axis—although still axisymmetric about it to an excellent approximation. Outside an axisymmetric distribution of mass, the Newtonian gravitational potential $\Phi(r, \theta)$ can be expanded in inverse powers of r. Assuming the distribution is symmetric under inverting the axis of symmetry, the first two terms—called the mass monopole and mass quadrupole terms—are

$$\Phi(r, \theta) = -\frac{GM}{r} + J_2 \frac{GM}{r}\left(\frac{R}{r}\right)^2 \left(\frac{3\cos^2\theta - 1}{2}\right) + \cdots . \tag{10.19}$$

Here, θ is the polar angle measured from the rotation axis, R is the mean radius of the body, and J_2 is a dimensionless measure of the mass quadrupole moment. Readers who have had a course in electromagnetism will recognize (10.19) as the standard multipole expansion of the axisymmetric solutions of Laplace's equation

[2]The exact definitions of this number and (10.17) at this level of accuracy are not explained because the relevant facts for this discussion are just that they can be precisely determined and are much larger than (10.16).

(3.18), and the polynomial in the angles as the Legendre polynomial $P_2(\cos\theta)$. If you aren't familiar with any of this, just plug (10.19) into Laplace's equation to verify that it is a solution.

From (10.19), a solar quadrupole moment would mean a Newtonian gravitational potential in the equatorial plane $\theta = \pi/2$ of the form

$$\Phi(r) = -\frac{GM}{r} - \frac{J_2 GMR^2}{2r^3}. \tag{10.20}$$

This extra $1/r^3$ potential will cause a precession of the perihelion just in Newtonian mechanics. Indeed, it makes an additional contribution to the effective potential (9.30), which has exactly the same form as the relativistic term $GM\ell^2/(c^2r^3)$. Thus, observations of the orbits of the planets can determine only a combination of the PPN parameters $(2 + 2\gamma - \beta)/3$ [cf. (10.9)] and J_2.

The Sun has a quadrupole moment because it is rotationally distorted, and the value of J_2 can be determined from a model of the interior and the angular velocity there. The rotational period on the surface is about 27 days at the equator, and angular velocity in the interior can be determined by observing the precise frequencies of modes of oscillations of the Sun—an area of study called *helioseismology*—and understanding the effect of rotation on these frequencies. The result of Brown et al. (1989) is that $J_{2\odot} \sim 10^{-7}$, roughly what would be expected if the Sun were uniformly rotating and too small to make any significant contribution to the precession of the perihelion and the determination of β in (10.18) at the levels of accuracy available.

Problems

1. [E] Estimate the gravitational redshift of light from the surface of the Sun. Discuss the possibility of measuring this effect given that the velocities of matter in convection cells at the surface of the Sun is of order 1 km/s. Is there one part of the surface that is better than another for making the observation?

2. Is the experiment of Vessot and Levine sensitive enough to say anything about the parameters β and γ? Is the third-order Doppler effect important in analyzing the experiment?

3. Evaluate the maximum deflection of light by the Sun predicted by general relativity in seconds of arc.

4. Derive (10.8) for the deflection of light as a function of the parametrized post-Newtonian parameters.

5. Evaluate, in seconds of arc per century, the precession of the perihelion of Mercury, Venus, and Earth as predicted by general relativity.

	Semimajor axis 10^6 (km)	Eccentricity	Mass/$M_\oplus$	Period (yr)
Mercury	57.91	.2056	.054	.241
Venus	108.21	.0068	.815	.615
Earth	149.60	.0167	1.000	1.000

$M_\oplus = 5.977 \times 10^{24}$ kg

6. Evaluate the precession of the perihelion of Mercury caused by a Newtonian quadrupole potential of the form given in (10.20), and show that with the observed value of $J_{2\odot}$ it is too small to correct the determined value of the PPN parameter β.

7. *Solar Oblateness and the Precession of the Perihelion* Measuring the shape of the solar surface is an alternative way of determining the solar quadrupole moment. Optical measurements can determine the solar oblateness, defined by

$$\Delta = \frac{\text{(radius at equator)} - \text{(radius at pole)}}{\text{(mean radius)}}$$

If the surface of the Sun is a surface of equal gravitational potential, this oblateness can be used to determine the solar mass quadrupole moment. Early measurements gave values for Δ as large as 5×10^{-5}. (Later measurements gave a much lower value for Δ.)

(a) Explain why the surface of the Sun is a surface of equal gravitational potential if the centripetal accelerations due to the rotation at the surface are a negligible contribution to the Sun's distortion (contrary to fact).

(b) Calculate the value of J_2 from the oblateness using (10.20) and assuming that Φ is constant on the surface of the Sun.

(c) Calculate the magnitude of the precession of the perihelion of Mercury that would result from $\Delta \sim 10^{-5}$.

8. [P, E] Starting from (10.12), make a rough estimate of the angular accuracy that could be expected in the NRAO experiment to detect the deflection of light. Under ideal circumstances, what size optical telescope above the atmosphere in space would be needed to achieve the same accuracy?

9. [E] *Estimate* the amount by which radio signals used in the quasar bending of light observation would be bent by the solar corona. The corona is reasonably well modeled by a free electron gas whose index of refraction is

$$n(r) = 1 + \frac{2\pi e^2 N(r)}{m\omega^2},$$

where the electron density $N(r)$ may be taken to be 10^8 cm^{-3} out to twice the solar radius. The frequencies used in the NRAO experiment were 8.1 GHz and 2.7 GHz.

10. Assuming that general relativity correctly predicts the excess time delay measured in the Viking experiment, what can you infer from the data in Figure 10.7 about the closest a radar pulse involved in the experiment came to the Sun? Express your answer in solar radii from the center.

CHAPTER 11

Relativistic Gravity in Action

The orbits of test particles and light rays in the Schwarzschild geometry that were worked out in Chapter 9 are not only important for the delicate tests of general relativity in the solar system discussed in Chapter 10. They are also central to a number of astrophysical applications. This chapter introduces three of these applications—gravitational lensing, relativistic frequency shifts from accretion disks, and weighing stars in binary pulsars. Some tests of Einstein's theory were the subject of the previous chapter; some of its applications are the subject of this.

11.1 Gravitational Lensing

The gravitational attraction of mass deflects light, as we saw in Chapter 9. Because of this bending there can be multiple pathways for light to use in traveling from a source to an observer, as illustrated in Figure 11.1. An intervening mass can, therefore, produce multiple images of a distant source. Acting in this way a concentration of mass is called a *gravitational lens*.[1] Gravitational lensing has become an important tool for astronomy. A gravitational lens can give information about the source that is imaged, about the object acting as a lens, and about the intervening large-scale geometry of the universe when source, lens, and observer are at cosmological distances from one another.

Realistic gravitational lenses may be clusters of distant galaxies without any special symmetries. Light may propagate through them as well as around. This book, however, will consider only the simplest case of lensing by a small spherical mass, which is assumed to be the only relevant source of spacetime curvature. For a lens at cosmological distances, the curvature of the universe must be considered as well; conversely, gravitational lenses give information about that curvature. However, the simple example of lensing by a spherical mass in an asymptotically flat spacetime will illustrate the basic physics of gravitational lenses.

[1]The images are not focused in the sense that all the light from one point on the source is brought to one point in an image, as with some idealizations of optical lenses with which you may be familiar. For that reason the observer does not have to be a special distance from the lens in order to see the images. *Lens* is used in a more general sense.

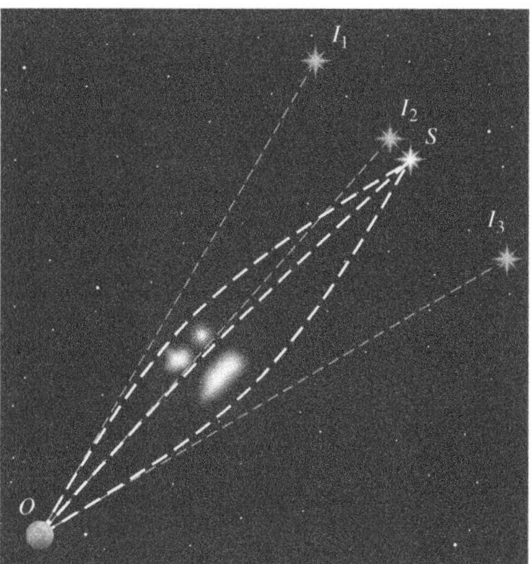

FIGURE 11.1 The idea behind a gravitational lens. Intervening mass can bend light from a distant source S to produce multiple pathways for light to travel from it to an observer O. The observer sees these as multiple images of the source. The diagram illustrates how images of one source S could be produced at angular locations I_1, I_2, and I_3. Almost everything about the diagram is exaggerated for clarity. Realistically, the size of the lens is tiny compared to the distances involved, the bending angles are minute, and the images unlikely to line up in a plane.

Lens Geometry and Image Position

The deflection angle α for a light ray passing by a mass M at an impact parameter $b \gg M$ is given by (9.83) and is

$$\alpha = \frac{4GM}{c^2 b} \equiv \frac{2R_S}{b}. \tag{11.1}$$

Here, shorthand expressions α for the deflection angle $\delta\phi_{\text{def}}$ and R_S for the Schwarzschild radius $2GM/c^2$ have been introduced.

The geometry of a spherical gravitational lens is shown in Figure 11.2. It is important to appreciate the scale of this figure. If the lens is a galaxy at a cosmological distance, bending the light from an even more distant source, then typically[2]

$$M \sim 10^{11} M_\odot, \qquad R_S \sim 10^{11} \text{ km}, \tag{11.2a}$$

$$D_S \sim D_L \sim D_{LS} \sim 1 \text{ Gpc} \sim 3 \times 10^{22} \text{ km}. \tag{11.2b}$$

[2]The *parsec* (pc) is a standard unit in galactic and extragalactic astronomy. One parsec = 3.086×10^{13} km or 3.262 light-years. The units kiloparsecs (kpc), megaparsecs (Mpc), and gigaparsecs (Gpc) are useful. *Very* roughly, distances between neighboring stars in the galaxy are of order pc, the size of the galaxy is of order kpc, distances to nearby galaxies are of order Mpc, and the size of the visible universe is measured in Gpc. (See Figure 17.7 for the origin of the unit.)

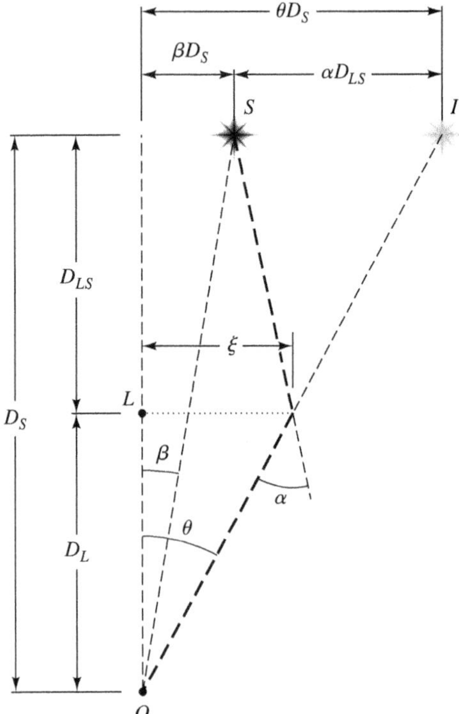

FIGURE 11.2 The geometry of a gravitational lens in the thin lens approximation. O is the observer. L is the location of the lensing mass at a distance D_L from the observer. S is the source located a distance D_S from the observer and D_{LS} from the lens. The figure shows the source-lens-observer plane. The heavy dashed line shows the path of a light ray from source to observer. The ray passes by the lens with an impact parameter that differs negligibly from the distance ξ and is deflected by an angle $\alpha = 4GM/(c^2\xi)$, where M is the mass of the lens. In the thin lens approximation, the lens is treated as a point and all the deflection takes place in a transverse plane at the position of the lens, L. An image of the source, I, appears at an angle θ from the observer-lens axis rather than its true direction, β. The transverse distances in this diagram are all greatly exaggerated. Were they drawn to true scale it would not be possible to distinguish any of the lines in the figure. The relationship between the transverse distances at the top of the figure constitutes the lens equation.

The characteristic radius over which the bending occurs is the Schwarzschild radius R_S, which is much smaller than the distances D_L, D_S, and D_{LS} over which the light propagates. That is typical of realistic lensing situations. Therefore, to an excellent approximation, the light rays propagate as straight lines in flat space when far from the lens, and all the deflection occurs at the lens. That is the *thin lens approximation* assumed in Figure 11.2. Specifically, in the thin lens approximation, source and lens are approximated as points. The deflection angle is assumed to be given by (11.1) for all values of b. All the deflection is assumed to take place in a plane normal to the line of sight at the location of the lens. Of course, these approximations will break down, for example when b is comparable to R_S, but they allow a simple and elegant description of many realistic lensing situations.

In realistic situations, all the angles in Figure 11.2 are very small, so that distances transverse to the line of sight are well approximated by (angle) × (distance). The relationship between the transverse distances at the top of Figure 11.2 is

$$\boxed{\theta D_S = \beta D_S + \alpha D_{LS}} \tag{11.3}$$

Lens Equation

and is called the *lens equation*. Because $b \approx \xi$ and $\xi \approx \theta D_L$ in the small-angle approximation, the lens equation can be written using (11.1) as

$$\theta = \beta + \frac{\theta_E^2}{\theta}, \tag{11.4}$$

where

$$\boxed{\theta_E \equiv \left[2R_S\left(\frac{D_{LS}}{D_S D_L}\right)\right]^{1/2}} \tag{11.5}$$

Einstein Angle

is called the *Einstein angle*. The solutions of (11.4) determine the angular position of the images on the sky.

To understand the significance of the Einstein angle, consider the degenerate case, where the source, lens, and observer are exactly in line. The symmetry about this axis implies that the image of the source is spread out over a circular ring called the *Einstein ring*. The Einstein ring makes an angle $\theta = \theta_E$ with the axis, as easily follows from (11.4) when $\beta = 0$.

The Einstein angle sets the characteristic angular scale for gravitational lensing phenomena. Consider the lensing of a star within the galaxy by a solar mass size object between us and the star. In this case, $M \sim M_\odot$, $R_S \sim 1$ km, and $D_L \sim D_S \sim D_{LS} \sim 10$ kpc $\sim 10^{17}$ km. This implies an Einstein angle $\theta_E \sim 10^{-3}{}''$. This is well below the accuracies achievable by contemporary telescopes. But, as we will see shortly, lensing by stellar mass objects is detectable by observing the change in brightness of the images with time due to relative motion between the lens and source. Because of the small angles involved, this situation is often called *microlensing*. For the lensing by a galaxy and source at cosmological distances with the parameters of (11.2), the Einstein angle is $\theta_E \sim 1''$. That *is* resolvable by optical telescopes. This situation is sometimes called *macrolensing*.

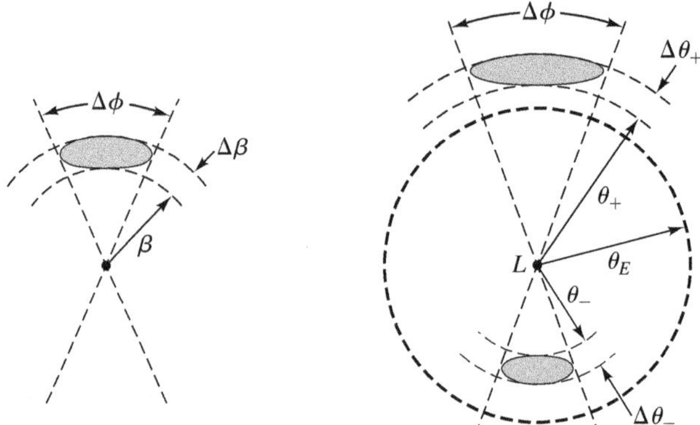

FIGURE 11.3 The images of a distant galaxy created by an intervening spherical "point" gravitational lens. Two figures of angular positions on the sky are shown, illustrating the effect of a spherical lens located at L at the center of each figure. The left hand figure shows the galaxy as it would appear if the lens did not deflect light. The galaxy is located at an angle β from the observer-lens axis and has angular dimensions $\Delta\phi$ and $\Delta\beta$. The right hand figure shows the action of the lens. Two images have been created at angles $\theta_\pm$ from the observer-lens axis. One of these is inside the Einstein angle θ_E, the other is outside. The azimuthal width of the image $\Delta\phi$ is preserved by the lens. The polar angle and width are changed, resulting in a distortion of the images into arcs. When the lens is of finite but small size, there is a third image behind it.

The solutions to (11.4) give generally the location of two images in the source-lens-observer plane:

Image Positions

$$\boxed{\theta_\pm = \frac{1}{2}\left[\beta \pm \left(\beta^2 + 4\theta_E^2\right)^{1/2}\right].} \tag{11.6}$$

The arrangement of these images produced by a spherical mass is shown in Figure 11.3. There are two images on opposite sides of the position of the lens—one at a position greater and one at a position less than the Einstein angle. Lensing by realistic, transparent, extended sources turns out always to produce an *odd* number of images. In the limit of a small but finite-size spherical lens, there is a third image *behind* the lens besides the two at the locations given by (11.6) (Problem 2).

The lens positions (11.6) are independent of the frequency of the light. Unlike optical lenses, gravitational lenses are *achromatic*.

Realistic lenses are more complicated than the simple spherical mass discussed here, but the principles are the same. Figure 11.4 is a beautiful image of a more complex lensing system exhibiting multiple images distorted into arcs.

By measuring the angles $\theta_\pm$ between the position of the lens and the positions of the images, the Einstein angle θ_E can be determined using (11.6). If distances to the lens and source can be estimated, then the mass of the lens can be determined

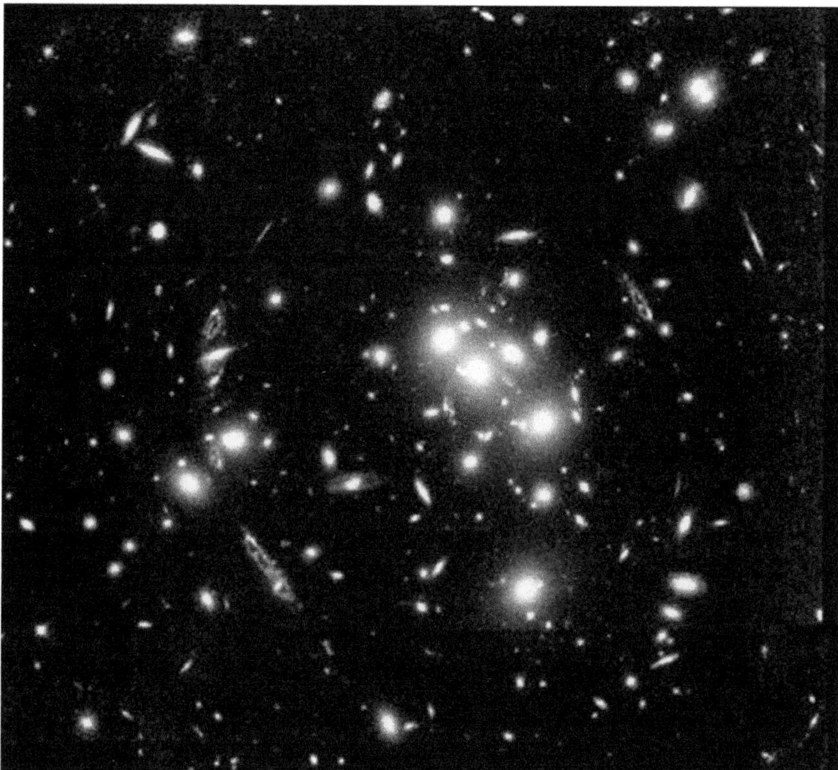

FIGURE 11.4 A Hubble space telescope image of the galaxy cluster 0024+1654 acting as a gravitational lens. The mass in the foreground cluster of galaxies (the bright, diffuse images in the center) acts as a gravitational lens for a more distant galaxy. The geometry of the lens (not a point) is such that multiple images of the distant galaxy are produced close to the radius of the Einstein ring. The images are distorted into arcs.

from (11.5) and (11.1). Gravitational lensing can, therefore, be used to detect mass in the universe whether it is visible or not.

Image Shape and Brightness

Up until now we have tacitly assumed that the source and its images are points. But the change in shape and brightness of a finite-angular-size image are among the most important properties of a gravitational lens. The left diagram in Figure 11.3 shows a finite-size galaxy image as it might appear if the lens at L had no mass and did not deflect light. In the notation of Figure 11.2, the image is located at an angular separation β from the lens (i.e., at the location of the source) and has angular dimensions $\Delta\beta$ and $\Delta\phi$ (assumed small). The right figure shows the action of the lens. Two images have been created at the positions $\theta_{\pm}$ given by (11.6). The symmetry about the observer-lens axis implies that the light ray's value of ϕ is unchanged by the deflection of the lens. The azimuthal angular width of the image $\Delta\phi$ is thus preserved. The polar width $\Delta\theta$ *is* changed by an amount that can be determined by differentiating (11.6) to find

$$\Delta\theta_{\pm} = \frac{1}{2}\left[1 \pm \frac{\beta}{\left(\beta^2 + 4\theta_E^2\right)^{1/2}}\right]\Delta\beta. \tag{11.7}$$

The images of the galaxy are thus elongated and distorted.

Not only is the shape of an image changed by gravitational lensing, but its brightness is also. That change in brightness is the key to the use of gravitational lensing to detect small massive bodies, as will be described shortly.

To understand the brightness of a lensed image, let's start with a simple example. Imagine a plate heated to a high temperature so that it radiates approximately like a black body—each small piece of its surface radiating uniformly in all directions. A detector placed at a distance directly above the plate—so that it is viewed face-on—will record a certain flux (energy/time) of radiation. But the same detector at the same distance along a direction making an angle with the normal to the plate will receive *less* flux, as shown in Figure 11.5. That is because the plate subtends a *smaller* solid angle when viewed obliquely than when viewed face-on. A detector viewing the plate edge-on, for instance, would receive no radiation if the plate has negligible thickness. The factor of proportionality between the flux Δf received from a small piece of the surface and the solid angle $\Delta\Omega$ it subtends is called the plate's *surface brightness*, namely,

$$\Delta f = (\text{surface brightness}) \times \Delta\Omega. \tag{11.8}$$

To see what that means for lensing, let's consider the concrete case of the lensing of a star. Gravitational lensing does not change the surface brightness of a lensed star. That is a property of the star. But it can change the solid angle subtended by the star because that is a property of the trajectories of the light rays between the star and detector. Lensing can, therefore, change the brightness (flux)

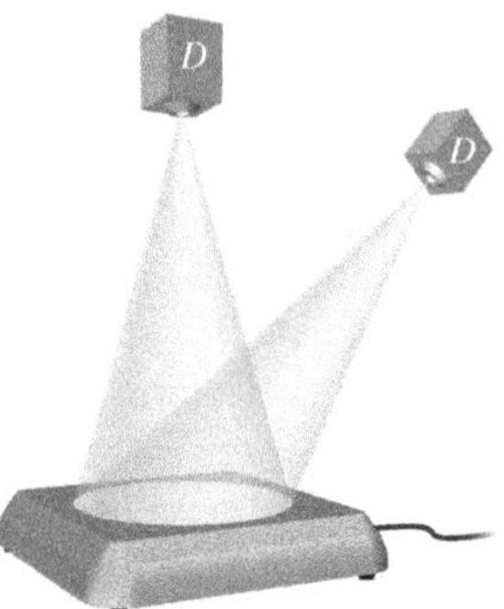

FIGURE 11.5 The surface of a hot plate radiates equally in all outward directions. A detector D viewing the plate face-on measures a different flux than when viewing it obliquely because the plate subtends different solid angles at the different positions of the detector.

of the image from what it would have been were the lens not present, as (11.8) shows.

Another way of looking at this is to think of what happens to the light radiated from a little piece of the surface of the star. Light rays are radiated in all outward directions isotropically. Some rays will intersect a distant detector, but most will miss and not be registered. The bending of light can change which rays intersect the distant detector and how many intersect it. If more rays intersect than would if the lens were not present, the detector receives more light and the image is brighter than without the lens. If fewer rays intersect, then the image is dimmer. The total brightness of all the images seen by a given observer can be greater than without the lens, as we will see. In such situations the light bending by the lens has directed more rays to the distant detector than would have gone there were the lens absent.

From this discussion it follows that the ratio of the brightness of the images $I_\pm$ at the positions $\theta_\pm$ to the unlensed brightness I_* will be the ratio of the solid angles $\Delta\Omega_\pm$ that the images subtend when the lens is present to the value $\Delta\Omega_*$ they would subtend were it not. Using the familiar expression for an element of solid angle in polar coordinates, this is

$$\frac{I_\pm}{I_*} = \frac{\Delta\Omega_\pm}{\Delta\Omega_*} = \left|\frac{\theta_\pm \Delta\theta_\pm \Delta\phi}{\beta\Delta\beta\Delta\phi}\right|. \tag{11.9}$$

Since $\Delta\phi$ is preserved, the magnification is

$$\frac{I_\pm}{I_*} = \left|\left(\frac{\theta_\pm}{\beta}\right)\left(\frac{d\theta_\pm}{d\beta}\right)\right| = \frac{1}{4}\left(\frac{\beta}{\left(\beta^2 + 4\theta_E^2\right)^{1/2}} + \frac{\left(\beta^2 + 4\theta_E^2\right)^{1/2}}{\beta} \pm 2\right) \tag{11.10}$$

from (11.6) and (11.7). Since $x + 1/x \geq 2$ for any x, the expression in brackets is always positive. The image outside the Einstein ring is brighter, and the one inside is dimmer.

For microlensing by stars where the images cannot be resolved, the *total* magnification is of interest:

$$\boxed{\frac{I_{tot}}{I_*} \equiv \frac{I_+ + I_-}{I_*} = \frac{1}{2}\left(\frac{\beta}{\left(\beta^2 + 4\theta_E^2\right)^{1/2}} + \frac{\left(\beta^2 + 4\theta_E^2\right)^{1/2}}{\beta}\right).} \tag{11.11}$$

Total Image Brightness

This function of β is always *greater* than unity. The gravitational lens therefore always enhances total brightness, and if the source is close to the observer-lens axis so that β is small, this enhancement can be substantial. As we will see shortly, this enhancement is the reason that gravitational lenses can be detected and used even when the individual images cannot be resolved.

Timing of Fluctuations

A fluctuation in the brightness of the source will produce a later fluctuation in the image when it arrives at Earth. However, the arrival times of the two images can be different for two reasons: First, the path length traversed by the two images is different because the angles θ_+ and θ_- are different; second, the relativistic time delay discussed in Section 9.4 will be different for the same reason. We will not try to calculate this time delay in any detail, but just a simple estimate of the difference in path length suffices to show that the difference in arrival times can be significant. Take for simplicity the case $D_L = D_{LS} = D_S/2$ and $\beta \ll \theta_E \ll 1$. A little plane geometry and Figure 11.2 shows that, to first order in β, the difference in path length is approximately (Problem 4)

$$\Delta D \approx \beta\theta_E D_S, \qquad \beta \ll \theta_E \ll 1. \tag{11.12}$$

This relation vanishes when $\beta = 0$ reflecting the symmetry between the two paths that holds in that case. The result is proportional to the only length in the problem, and vanishes with θ_E when the mass of the lens goes to zero (as it should). For the lensing by a galaxy of a source at cosmological distances [cf. (11.2b)], we have $\Delta D \approx 4(\beta/\theta_E)R_S$ and the difference in arrival times $\Delta D/c$ can be measured in weeks. The effect is observed and is important in determining cosmological parameters such as the expansion rate of the universe, although we do not discuss that here.

Microlensing

As we will learn in Chapter 17, there is considerable evidence that the matter visible in stars and galaxies is only a small fraction of the total matter in the universe. Even our own galaxy must be surrounded by a halo that is more massive than the stars and dust that we can see. Of what is the undetected matter made? Jupiter-size objects, white dwarf stars, and black holes are examples of one class of candidates called massive compact halo objects (MACHOs). The mass range for such objects might be from a few hundredths to several solar masses. They are dark, so they are difficult to detect by any means other than their gravitational interactions. Gravitational microlensing provides a tool for detecting them.

Suppose our galaxy does have a halo of MACHOs, each moving in the collective gravitational potential of the mass in the galaxy. Imagine examining a star in a nearby galaxy *outside* the halo. The stars in the Large Magellanic Cloud (LMC), a small satellite of our own galaxy, are an important example. If the trajectory of a MACHO in the halo takes it close to the line of sight to a star in the LMC, the MACHO will act briefly as a gravitational lens. The combined brightnesses of the star images will increase and then decrease as the MACHO moves by. The change will be given by (11.11), with θ_E related to the mass of the MACHO by (11.5) and with β changing with time due to the motion of the MACHO. Thus, the MACHO can be detected from the change in brightness of the distant star, even though the angular deflection of the light is far below the resolving power of optical telescopes, as we discussed earlier. The characteristic time scale for the variation can be estimated as the time, $t_{\rm var}$, it takes for a MACHO to move an angular distance equal to the Einstein angle θ_E. Roughly estimating $\theta_E \sim 10^{-3}\,''$ for a solar mass that is a galactic radius away, $D_L \sim 10$ kpc and estimating $V \sim 200$ km/s for the typical velocities of stars in the galaxy, this time for variation $t_{\rm var}$ is

$$t_{\rm var} = \frac{\theta_E D_L}{V} \sim \frac{(10^{-3}\,'')(10\ {\rm kpc})}{200\ {\rm km/s}} \sim .2\ {\rm yr}. \tag{11.13}$$

Conversely, a measurement of the time of variation and estimates of the velocities and distances to the source and lens enable the Einstein angle to be determined from (11.13) and the mass of the lens from (11.5) (Problem 7). That is how microlensing can weigh dark objects in the galaxy.

The chance of a MACHO crossing the line of sight to any particular star is very small. But if a great number of stars are examined, the chance of detecting a MACHO in some of them becomes significant. Several such observing programs are now under way. Dedicated telescopes, electronic imaging, and high-speed software enable astronomers to study hundreds of thousands of stars over periods of hundreds of days. Figure 11.6 shows the light curve of one event from the MACHO collaboration (Alcock et al. 1997). In this way gravity—which couples to all matter—can be used as a tool to probe the dark matter in the universe through gravitational lensing.

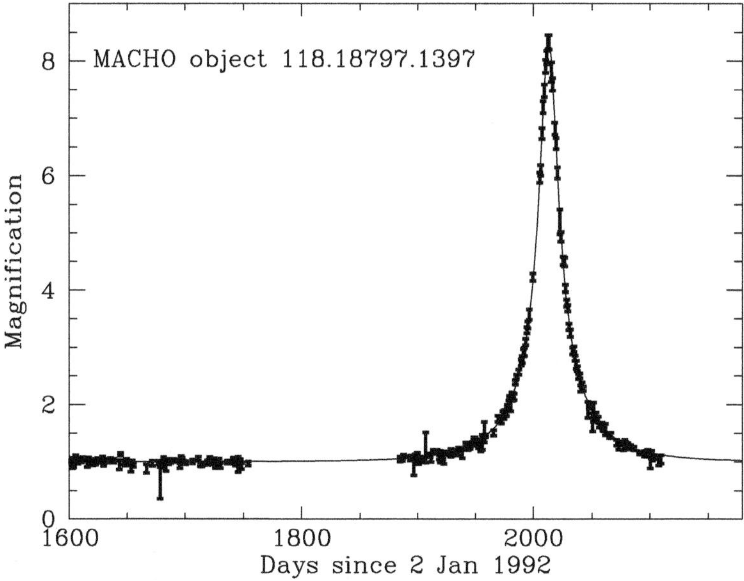

FIGURE 11.6 Light curves for a microlensing event from the MACHO collaboration. The figure shows the light curve of a star in the bulge of the galaxy lensed by an intermediate object. The vertical axis is $I_{\rm tot}/I_*$. Data are plotted along with a fit from (11.11) to the parameters specifying the angular speed of lensing object as it moves across the sky and the closest angular approach of the lens to the source (Problems 6 and 7).

11.2 Accretion Disks Around Compact Objects

Accretion Disks in Astrophysics

The curved spacetime of the Sun is accessible to experimental investigation because we are moving through it. However, the Sun is not a very compact object and as a consequence the curvature outside the Sun is never very large. The ratio M/R, which characterizes relativistic effects in the Sun's geometry, is only of order 10^{-6}. The most compact objects in the universe are two of the endstates of stellar evolution—black holes and neutron stars, to be described in Chapters 12 and 24, respectively. For black holes $M/R \sim .5$, and for typical neutron stars $M/R \sim .2$. We explore these compact objects further in subsequent chapters, but they have one thing in common: their exterior spacetimes are the Schwarzschild geometry if they are not rotating. The motion of matter and light can be used to observe and explore these geometries utilizing the techniques and results of the previous two chapters. Nearby matter, for example from a companion star, can naturally fall onto such objects in a process called *accretion*. This matter is a source of test particles whose motions probe the spacetime geometry.

Consider, for example, a black hole or neutron star in mutual orbit with a more normal companion star—one like the Sun, for instance. The binary pair can lose orbital energy—by gravitational radiation among other mechanisms—decreasing the size of its orbit. The orbit can become small enough that the outermost layer of the companion is more strongly attracted to the compact object than to its own center. In that case the more normal star will shed mass, which will fall (accrete) onto the compact object. Conservation of its initial orbital angular momentum means that the accreting material does not fall directly onto the compact object but rather forms a disk around it called an *accretion disk*. Various dissipative mechanisms associated with interactions between the particles in the disk cause them to slowly lose energy and angular momentum and gradually spiral toward the compact object. They spiral slowly inward on nearly circular orbits until they reach the innermost stable circular orbit [cf. (9.43)], after which they fall rapidly into the compact object. The energy they lose leaves the disk as radiation—characteristically at X-ray wavelengths for compact objects around a solar mass. (See Box 11.1.) That is why accretion disks around solar mass compact objects are the likely explanation of galactic X-ray sources.

Accretion disks also surround the 10^6–$10^9 M_\odot$ supermassive black holes that are possibly at the centers of almost every sufficiently massive galaxy, including our own (Section 13.2). Disks around such supermassive black holes are cooler than those around solar mass–size compact objects, as the estimates in Box 11.1 suggest. But that does not mean their luminosity is negligible. As we will see in Section 13.2 and Box 15.1, accretion disks around black holes at the centers of galaxies play a central role in explaining active galactic nuclei, such as quasars. These include the most energetic steady sources of radiation in the universe.

Evidence for Compact Objects in the Spectra of X-Ray Sources

Information about the geometry of a compact object can be obtained by observing the motion of particles in a surrounding accretion disk and the light rays emitted

BOX 11.1 Accretion Power

Just a little physics is needed to make simple estimates of the luminosity and temperature of accretion disks around compact objects.

In steady state the luminosity (total rate of emitted energy) and temperature of an accretion disk are determined by the rate $\dot{M}$ at which mass is accreting. In steady state, changes in the gravitational potential energy of $\dot{M}$ grams of matter per second are being turned into radiated energy. The higher $\dot{M}$, the greater the luminosity and temperature.

There is an upper limit to the rate at which mass can be accreted by a compact object in a spherical, steady manner. As the rate is increased, the increasing pressure of outgoing photons on infalling matter will eventually exceed the gravitational attraction of the compact object. Consequently, there is an upper limit to the luminosity, called the Eddington limit. The typical luminosities of observed X-ray sources range from a few percent up to nearly the limiting value, even though the accretion is not spherically symmetric.

To estimate the Eddington limit, let M denote the mass of the compact object and L denote the luminosity in radiation. A simple Newtonian analysis is adequate for the kind of crude estimate we are looking for here. The flux of energy across a surface a distance r from the center is $L/(4\pi r^2)$. The flux of momentum is $L/(4\pi r^2 c)$ because (momentum) $=$ (energy)$/c$ for photons. The scattering of outgoing radiation off infalling matter gives rise to an effective outward pressure. To estimate how much of the outward momentum is transferred to infalling matter, we can use the Thomson cross section σ_T for the scattering of low-energy light from an electron (SI units):

$$\sigma_T = \frac{8\pi}{3}\left(\frac{e^2}{4\pi\epsilon_0 m_e c^2}\right)^2 = 0.665 \times 10^{-24}\ \mathrm{cm}^2. \quad \text{(a)}$$

Here, e is the electron's charge and m_e is its mass. (Scattering from protons also contributes to the effective outward force, but the cross section is approximately a million times smaller.) The momentum transferred to one electron at radius r per unit time is, therefore, $\sigma_T L/(4\pi r^2 c)$. A momentum per unit time is a force, and if we equate this to the force of gravity, noting that there is about one nucleon of mass m_p for each electron in the infalling matter, we find the Eddington limit for pure ionized hydrogen:

$$\frac{G m_p M}{r^2} = \frac{\sigma_T L_{\mathrm{Edd}}}{4\pi r^2 c}. \quad \text{(b)}$$

The limiting Eddington luminosity is, therefore,

$$L_{\mathrm{Edd}} = \frac{4\pi G c m_p M}{\sigma_T} = 1.3 \times 10^{38} (M/M_\odot)\ (\mathrm{erg/s}). \quad \text{(c)}$$

(For comparison, the luminosity of the Sun is $L_\odot = 3.8 \times 10^{33}$ erg/ sec.) The luminosities of typical X-ray sources are a modest percentage of L_{Edd}. Converting energy to radiation by accreting into a deep gravitational potential well is thus competitive with the thermonuclear burning that is the source of radiation in stars.

The characteristic energy of radiation from the accretion disk at a radius R from the compact object can be roughly estimated by equating the luminosity to that of a black body of size R and temperature T, although the radiation spectrum is not typically thermal. If the luminosity L is a fraction ε of L_{Edd}, then

$$4\pi R^2 \sigma T^4 = \varepsilon L_{\mathrm{Edd}}. \quad \text{(d)}$$

(Here σ is the Stefan-Boltzmann constant characterizing the radiation from a blackbody, not a cross section —it is standard notation.) Using (c) this gives

$$T \sim 5 \times 10^7 \left(\frac{GM}{c^2 R}\right)^{1/2} \left(\varepsilon \frac{M_\odot}{M}\right)^{1/4} \mathrm{K} \sim 5 \left(\frac{GM}{c^2 R}\right)^{1/2} \left(\varepsilon \frac{M_\odot}{M}\right)^{1/4} \mathrm{keV}. \quad \text{(e)}$$

At a neutron star's surface, $GM/c^2 R \sim .1$. The innermost stable circular orbit of the accretion disk around a spherical black hole has $GM/c^2 R \sim 1/6$ [cf. (9.43)]. In either case, for $M \sim M_\odot$, $\varepsilon \sim .5$, we find $T \sim$ few keV. That explains why the accretion disks around solar mass black holes or neutron stars are X-ray sources. Accretion disks around the massive 10^6–$10^9 M_\odot$ black holes found at the center of almost every sufficiently massive galaxy are correspondingly cooler.

from it. X-ray spectroscopy provides one tool for observing the consequences of this motion.

Consider, for example, the accretion disks surrounding the supermassive black holes at the center of galaxies discussed before. The disk temperatures are cool enough that some heavy nuclei, such as iron, retain bound electrons (Problem 8). Excited by X-ray flares above the disk, even partially ionized iron atoms can de-excite (fluoresce) by emitting a 6.4-keV photon, giving a spectral line in the X-ray spectrum. However, by the time these photons reach the observer at infinity, they have a different energy. Roughly, they will be gravitationally redshifted by an amount that depends on the radius of their emission. Further, they will be Doppler-shifted by an amount that depends on the velocity of the emitting matter and whether it is moving toward or away from the observer. The result of integrating these effects over the contributions from various parts of the disk is a much broadened iron line whose shape contains information about the geometry around the accreting object.

The techniques developed in Chapter 9 allow the calculation of the red shift from any part of the disk. To focus on a simple but quantitative model, let's assume the central black hole is nonrotating, so the geometry outside is described by the Schwarzschild metric (9.9). Let's also assume a thin planar disk. The Schwarzschild coordinates can be oriented so the disk is in the equatorial plane $\theta = \pi/2$. Figure 11.7 shows the geometry when the disk is edge-on to a distant observer. (We'll return later to the case when it is face-on.) Let ω_* be the natural frequency of an emitted photon—6.4 keV$/\hbar$—and ω_∞ the frequency as observed by a distant observer. This depends on the radius r and angular position ϕ from which the photon is emitted. Let $\mathbf{u}_{\text{src}}(r, \phi)$ be the four-velocity of the matter from which the photon is emitted with four-momentum $\mathbf{p}(r, \phi)$, and $\mathbf{u}_{\text{rec}}$ be the four-velocity of

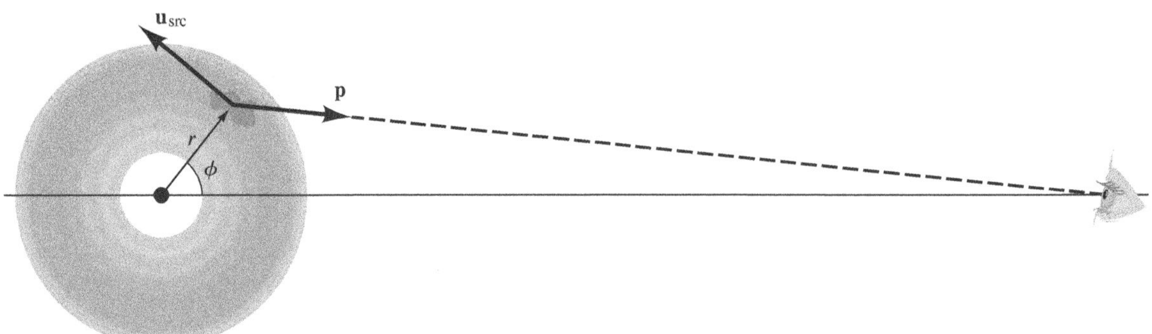

FIGURE 11.7 A schematic view from above of an accretion disk surrounding a compact object such as a black hole at the center of a galaxy. Excited atoms in the heavily shaded region of the disk are emitting a spectral line of frequency ω_* in their rest frame as they rotate around the compact object with an angular velocity appropriate to their radius. The figure shows the emitting source and a photon connecting it to a distant observer. The frequency received by the observer will be modified by relativistic effects arising from the source's motion and the curvature of spacetime produced by the compact object. The integrated effect of the photons from many different regions in the disk will be a spectral line whose broadened shape carries information about the geometry of the compact object.

the stationary observer at infinity who receives the photon with four-momentum $\mathbf{p}(\infty)$. In general, we have, from the discussion of Section 9.2 [cf. (9.12)],

$$\frac{\omega_\infty}{\omega_*} = \frac{\mathbf{u}_{\text{rec}} \cdot \mathbf{p}(\infty)}{\mathbf{u}_{\text{src}}(r, \phi) \cdot \mathbf{p}(r, \phi)}. \tag{11.14}$$

The receiving observer at infinity is stationary, with four-velocity

$$\mathbf{u}^\alpha_{\text{rec}} = (1, 0, 0, 0) = \xi^\alpha, \tag{11.15}$$

where $\boldsymbol{\xi}$ is the Killing vector (9.2) associated with invariance of the Schwarzschild metric under displacements in t. The emitting matter is in a circular orbit about the center. Its angular velocity is $\Omega(r) = d\phi/dt = (M/r^3)^{1/2}$, as given by (9.46), where r is the Schwarzschild coordinate radius of the orbit. The four-velocity of the emitting matter at location (r, ϕ) is, therefore,

$$u^\alpha_{\text{src}}(r, \phi) = [u^t_{\text{src}}(r), 0, 0, u^\phi_{\text{src}}(r)] = u^t_{\text{src}}(r)[\xi^\alpha + \Omega(r)\eta^\alpha], \tag{11.16}$$

where $\boldsymbol{\eta}$ is the Killing vector (9.4) associated with invariance of the Schwarzschild metric under translations in ϕ. The time component $u^t_{\text{src}}(r)$ is determined in terms of the other components by the normalization condition $\mathbf{u} \cdot \mathbf{u} = -1$ [cf. (9.48)]:

$$u^t_{\text{src}}(r) = \left[1 - \frac{2M}{r} - r^2\Omega^2(r)\right]^{-1/2} = \left(1 - \frac{3M}{r}\right)^{-1/2}. \tag{11.17}$$

The frequency shift (11.14) can now be evaluated in terms of the conserved quantities $e \equiv -\mathbf{p} \cdot \boldsymbol{\xi}$ and $\ell \equiv \mathbf{p} \cdot \boldsymbol{\eta}$, defined for photon orbits by (9.58) and (9.59), and their ratio $b \equiv |\ell/e|$. The conserved quantities e, ℓ, and b depend on the location of the source (r, ϕ) but we won't indicate that explicitly. (Recall from Section 9.4 that a photon's four-velocity can be normalized so that it is $\mathbf{p}$.) In terms of e and ℓ the scalar products in (11.14) are

$$\mathbf{u}_{\text{rec}} \cdot \mathbf{p}(\infty) = \boldsymbol{\xi} \cdot \mathbf{p}(\infty) = -e, \tag{11.18a}$$

$$\begin{aligned}\mathbf{u}_{\text{src}}(r, \phi) \cdot \mathbf{p}(r, \phi) &= u^t_{\text{src}}(r)[\boldsymbol{\xi} + \Omega(r)\boldsymbol{\eta}] \cdot \mathbf{p}(r, \phi) \\ &= u^t_{\text{src}}(r)[-e + \Omega(r)\ell]. \end{aligned}\tag{11.18b}$$

The result for the frequency shift is

$$\frac{\omega_\infty}{\omega_*} = \{u^t_{\text{src}}(r)[1 \pm \Omega(r)b]\}^{-1} \qquad \begin{pmatrix}\text{toward} \; - \\ \text{away} \; + \end{pmatrix}, \tag{11.19}$$

with a plus sign if the emitting matter is on the side of the disk moving *away* from the observer and a minus sign on the other side, where it is moving *toward* the observer.

It remains to evaluate b for photons emitted at radius r for various values of ϕ. For simplicity, we won't do this for general ϕ but only for two special cases—(1)

when the photon is emitted from matter moving transverse to the observer, i.e., when $\phi = 0$ or $\phi = \pi$, and (2) at angles $\pm\phi_t$ where a photon emitted tangentially to the disk reaches the observer.

Photons from the transversely moving matter at $\phi = 0$ or $\phi = \pi$ propagate to the observer along an axis through the center of the disk. They therefore have $b = 0$ and $\ell = 0$ so the frequency shift from (11.19) and (11.17) is

$$\frac{\omega_\infty}{\omega_*} = \left(1 - \frac{3M}{r}\right)^{1/2} \qquad \text{(transverse motion).} \tag{11.20}$$

The photon is redshifted from whatever radius in the disk it is emitted.

The second case, when the photon is emitted from matter moving either directly toward or away from the observer, requires a computation of b. Recall from the definitions of e and ℓ, in (9.58) and (9.59) that

$$b \equiv \left|\frac{\ell}{e}\right| = \frac{r^2|p^\phi(r,\phi)|}{(1-2M/r)p^t(r,\phi)}. \tag{11.21}$$

The radial component $p^r(r, \pm\phi_t)$ vanishes because $\pm\phi_t$ are angles where photons emitted tangentially to the disk reach the observer. (The radial component of the four-velocity $\mathbf{u}_{\text{src}}$ always vanishes because we have assumed the orbits are circular, but the radial component of a connecting photon vanishes at only two places on the orbit of the emitting matter.) Then, the condition that the photon four-momentum is null (5.70) is enough to evaluate b:

$$\mathbf{p}\cdot\mathbf{p} = -\left(1 - \frac{2M}{r}\right)[p^t(r, \pm\phi_t)]^2 + r^2[p^\phi(r, \pm\phi_t)]^2 = 0. \tag{11.22}$$

The result for b from (11.21) and (11.22) is

$$b = r\left(1 - \frac{2M}{r}\right)^{-1/2} \qquad \left(\begin{matrix}\text{directly}\\ \text{toward or away}\end{matrix}\right) \tag{11.23}$$

and for the frequency shift from (11.19), when $|\phi| = \pi/2$

$$\frac{\omega_\infty}{\omega_*} = \left(1 - \frac{3M}{r}\right)^{1/2}\left[1 \pm \left(\frac{r}{M} - 2\right)^{-1/2}\right]^{-1} \qquad \left(\begin{matrix}\text{toward} -\\ \text{away} +\end{matrix}\right). \tag{11.24}$$

For small values of M/r, this frequency shift is approximately

$$\begin{aligned}\frac{\omega_\infty}{\omega_*} &= 1 \pm \left(\frac{M}{r}\right)^{1/2} - \frac{M}{2r} + \cdots \\ &= 1 \pm V + \frac{1}{2}V^2 - \frac{M}{r} + \cdots,\end{aligned} \qquad \left(\begin{matrix}\text{toward} +\\ \text{away} -\end{matrix}\right) \tag{11.25}$$

where we have used[3] $(M/r)^{1/2} = \Omega r \equiv V$. The terms involving V in the last of (11.25) are the lowest orders of the Doppler effect [cf. (5.73)] and the remaining term is leading-order gravitational red shift [cf. (9.20)].

[3]Don't mix up the velocity V with potential energy.

The observed spectral line will consist of photons coming from different radii in the disk. The *smallest* radius is that of the innermost stable circular orbit—$r = 6M$ for the Schwarzschild metric [cf. (9.43)]. The smallest frequency for an edge on disk is, therefore, $\omega_\infty/\omega_* = \sqrt{2/3} = .47$. The smallest frequency for a face-on disk would be $\omega_\infty/\omega_* = 1/\sqrt{2} = .71$. If the central object is rotating, the smallest frequency can be even lower than these values. In general the 6.4-keV line will be broadened with a smallest frequency (maximum redshift), which depends on the size and rotation of the central object and the inclination of the disk to the line of sight. Further, although we will not analyze it in detail here, the shape of the line will be influenced by the relativistic beaming discussed in Section 5.5 and possibly other sources of emission. Relativistic beaming will increase the intensity of the blue end of the line over the red end.

Figure 11.8 shows an Fe line observed in the Seyfert I galaxy MCG-6-30-15 by Tanaka et al. (1995) and the fit to that line by assuming a Schwarzschild geometry and a 30° inclination angle for the disk. The line is redshifted to a maximum value comparable to that discussed before, and the intensity increases from red to blue. At the time of writing, the data are not detailed enough to distinguish a rotating from nonrotating central object or to determine much about the inclination of the disk. The fact that the maximum redshift is reached,

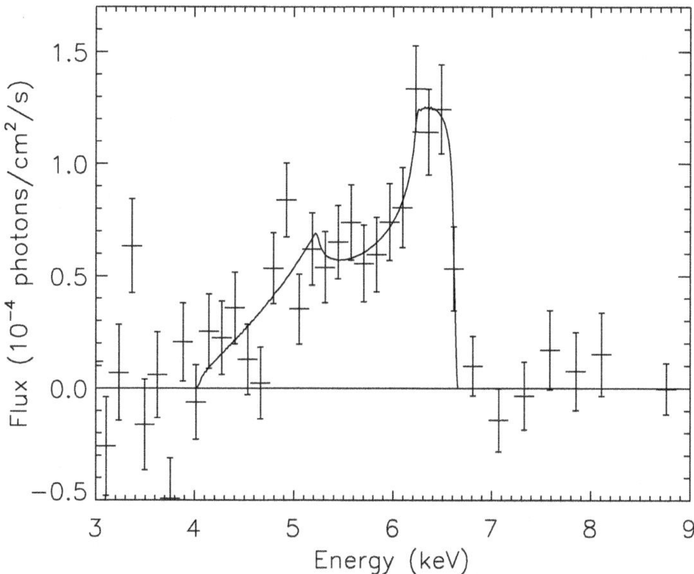

FIGURE 11.8 The broad Fe line observed in the Seyfert I galaxy MCG-6-30-15 by the ASCA X-ray satellite in July 1994 (Tanaka et al., 1995). The continuum X-ray emission has been subtracted to reveal the line. The line corresponds to a 6.4-keV line that has been broadened—mostly to lower (redshifted) energies. The solid line is a fit to the data assuming a model for a disk around a nonrotating (Schwarzschild) black hole inclined at an angle of 30° to the line of sight. Other features of this object suggest that it may be rapidly rotating and more detailed data will lead to more accurate probes of the central geometry.

however, suggests that the object is a black hole; a compact star would have a radius typically larger than the innermost stable circular orbit of the Schwarzschild geometry. Progress in X-ray observations will help us understand more about the innermost regions of such objects.

11.3 Binary Pulsars

As mentioned before, the exterior geometries of neutron stars are some of the best places to see the effects of general relativity. Russell Hulse and Joseph Taylor's 1974 discovery of the binary pulsar PSR B1913+16 has enabled us to do just that with great precision. Observations since with the Arecibo radio telescope in Puerto Rico (Figure 11.9) have been of great importance for general relativity. Hulse and Taylor were awarded the Nobel Prize in 1993 for the discovery of PSR B1913+16.

PSR B1913+16 is a pair of neutron stars in orbit about each other with an approximately 7.75-h period. A neutron star supports itself against the collapsing force of gravity, not by thermal pressure like the Sun, but by the forces arising from the Pauli exclusion principle and nuclear interactions between neutrons. These forces are effective only at nuclear densities and above, which is why neutron stars are so compact. A typical neutron star is slightly more than a solar mass of matter in a radius of 10 km. None of these properties of the stars are impor-

FIGURE 11.9 The Arecibo Radio telescope with which the measurements of the signals from the Hulse-Taylor binary pulsar PSR B1913+16 were carried out.

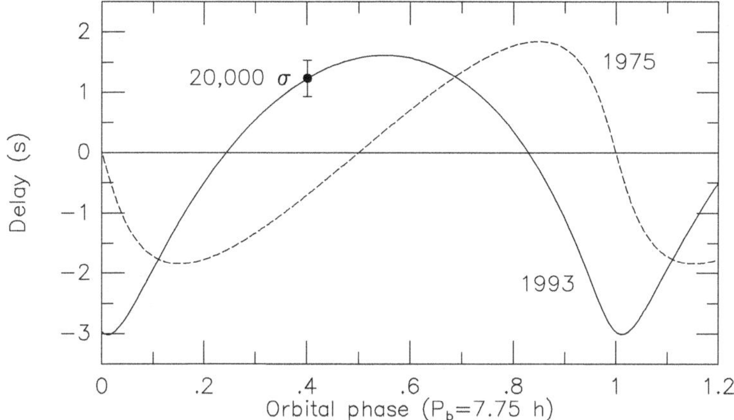

FIGURE 11.10 The delay in the arrival time of the pulses from the binary pulsar PSR B1913+16 as a function of orbital phase. The horizontal axis is time measured as a fraction of the orbital period. The vertical axis is the relative advance or delay from the average arrival time in seconds that is caused by the motion of the pulsar in its orbit about the companion neutron star. The pattern of delays is shown for two different years. There are differences in the shape of the pattern because of the different orientation of the pulsar's orbit with respect to Earth. But there is also an overall shift in the pattern in orbital phase due to the cumulative general relativistic precession of the periastron of the pulsar's orbit. Note the size of the error bar which concisely expresses the remarkable precision of these measurements.

tant for an analysis of their orbit, except for their compact size. This means they can be idealized as point masses and their orbits analyzed by generalizations of the calculations in Chapter 9. A number of such binary neutron star systems are known at the time of writing, but the first discovered—PSR B1913+16—has been studied the longest and in many ways is the most useful for general relativity. Let's consider it as an example.

Relativistic effects on the orbit of PSR B1913+16 are large. The precession of the periastron—the orbital position of closest approach analogous to the perihelion of a planet—is (see Figure 11.10)

$$\delta\phi_{\text{prec}} = 4.22659^\circ \pm .00004^\circ/\text{yr}. \tag{11.26}$$

This is nearly forty thousand times larger than the precession of the perihelion of Mercury (10.16). There is a similar amplification of other effects of relativistic gravity discussed in previous chapters. Both the gravitational redshift and the time delay of light can be observed in this system. How is it done?

One of the neutron stars is a pulsar—a magnetized star whose rapid rotation generates a surrounding plasma that serves as a source of beamed radio waves detectable as periodic pulses at Earth (hence the name *pulsar*).[4] The rotation period of an object as massive and compact as a neutron star is very stable against

[4]Box 24.2 contains a little more detail on pulsars.

external perturbations. The pulsar is, therefore, a remarkably accurate clock. Measurements of the times of arrival of the pulses over an epoch of many years gave for the rotational period

$$P_{\text{rot}} = 0.059029997929613 \pm .000000000000007 \text{ s} \tag{11.27}$$

on July 7, 1984, about 6 h after midnight GMT. The period is not exactly constant but increases slowly, chiefly due to the emission of electromagnetic radiation by the rotating magnetized star. The measured rate of increase, $\dot{P}_{\text{rot}}$, on the same date was 8.62713×10^{-18}.

To appreciate how relativistic gravity can be used to measure the properties of binary pulsar systems, it is first useful to understand how they are analyzed in Newtonian gravity.[5] There, the elliptical orbit of a binary pair of normal stars is characterized by its period[6] P_b, eccentricity ϵ, semimajor axis a (half the maximum distance between the stars), and further parameters that describe the stars' masses and how their mutual orbit is oriented in space and time. What are observed in a typical binary system are the Doppler shifts of the spectral lines of *one* of the stars over time. This shift [cf. (5.73)] measures the component of the observed star's velocity along the line of sight as a function of time. This is called the *radial velocity* curve and contains much information about the mutual orbit of the two stars. Although the details of the analysis will not be given here, both the period P_b and eccentricity ϵ can be inferred from the radial velocity curve. But only the combination $a_1 \sin i$ can be determined, where a_1 is the semimajor axis of the orbit of the observed star about the center of mass and i is the inclination of the orbital plane to the line of sight, defined so that $i = \pi/2$ corresponds to an edge-on orbit. (The semimajor axis a is the sum $a_1 + a_2$ for each star.) This is not enough information in Newtonian mechanics to determine either the masses of the individual stars or their total, but only a combination of masses and i called the mass function.[7] General relativity does better.

What are observed for binary pulsar systems are the arrival times of the radio pulses with the extraordinary precision mentioned before. The pulse arrival times contain all the Doppler shift information used in the Newtonian analysis determining P_b, ϵ, and $a_1 \sin i$, where a_1 is the semimajor axis of the pulsar's orbit. For PSR B1913+16, $P_b = 27906.980895 \pm 0.000002$ s, $\epsilon = 0.617132 \pm 0.000003$, and $a_1 \sin i = 2.34176 \pm 0.00001$ light-seconds. But the arrival times contain more information. In particular, they contain information about the various $1/c^2$ relativistic effects that affect the motion of the binary system and the propagation of the radio signals through it. These $1/c^2$ effects can be used to extract more information about the masses than is possible in the Newtonian approximation.

For example, as already mentioned, a large value is observed for the precession of the periastron (11.26). Although the general relativistic prediction (9.57)

[5] You might want to review your Newtonian mechanics text if any of the terms used here are unfamiliar.

[6] Don't get the orbital period P_b of the mutual orbit of the two stars mixed up with the rotational period P_{rot} of the one star that is a pulsar.

[7] If you want to learn more about this now, look at Example 13.1.

for the precession angle per orbit $\delta\phi_{\text{prec}}$ was derived only for test masses in Chapter 9, the result turns out to hold for binary systems with M replaced by the total mass M_{tot} of the pulsar and its companion. Given the eccentricity ϵ determined in the Newtonian approximation, the periastron precession fixes M_{tot}/a, as (9.57) shows. Kepler's law,[8]

$$P_b^2 = \frac{4\pi^2}{GM_{\text{tot}}}\, a^3, \tag{11.28}$$

gives another relation between M_{tot} and a, which enables both to be determined. The result for M_{tot} is $M_{\text{tot}} = 2.82827 \pm .00004 M_\odot$. (For a see Problem 10.)

The precession of the periastron is not the only $1/c^2$ relativistic effect that can be detected from the pulse arrival times. The contributions to the Doppler effect of order $1/c^2$ [cf. (5.73)] can be measured as well as the Shapiro time delay of the signals as they propagate across the orbit [cf. (9.92)]. Without going into detail, these enable the individual masses of both the pulsar and its companion to be determined—$M_{\text{pulsar}} = 1.442 \pm .003 M_\odot$ and $M_{\text{comp}} = 1.386 \pm .003 M_\odot$. Thus, properties of the binary system that cannot be determined in Newtonian gravity can be measured through relativistic corrections. Further, the determination of the rate of change in the orbital period has yielded the first detection of the effects of gravitational radiation and test of its prediction by general relativity, as we discuss in Section 23.7. Binary pulsars are a laboratory for general relativity and the relativistic corrections to orbits are a tool for astronomy.

Problems

1. At what radius would an observatory have to orbit the Sun in order to use it as a gravitational lens to image more distant objects?

2. *An odd number of gravitational lens images* Realistic gravitational lenses are not point sources, as assumed in the discussion in Section 11.1 but rather are a mass distribution. A lens that is a distribution of mass produces an odd number of images. For a simple model, assume that the gravitational lens is a transparent disk of radius r_* and constant surface mass density σ oriented perpendicularly to the line of sight. Using the thin lens approximation show that, in addition to the two images given by (11.6), there is a third image inside the angle subtended by the disk and find its angular position θ. Assume only the mass inside the deflection radius affects the bending of light.

3. When the line of sight to a star is far from the line of sight to a gravitational lens, the effects of lensing should become negligible. Show that when $\beta \gg \theta_E$, $\theta_+ \approx \beta$, $\theta_- \approx 0$, $I_+/I_* \approx 1$, and $I_- \approx 0$. Explain why these results mean that gravitational lensing is negligible.

4. Derive the path length difference in (11.12).

[8]See a basic mechanics book or (3.24) when one mass is much greater than the other and the orbits are circular.

5. [E] Equation (11.12) estimates the path-length difference traveled by light making up two images in a gravitational lens. The difference in arrival times of the light from the two images due to this effect is $\Delta D/c$. *Estimate* whether the Shapiro time delay discussed in Section 9.4 is a competitive effect.

6. In a typical microlensing event, a moving gravitational lens passes close to the line of sight to a distant source. The magnification I_{tot}/I_* defined by (11.11) increases in time and then decreases. Express the predicted ratio in terms of time measured in units of time t_{var} to cross the Einstein angle θ_E and p, the ratio $\beta_{\text{closest}}/\theta_E$, where β_{closest} is the smallest angular separation between lens and source. Plot the ratio I_{tot}/I_* as a function of time in these units for $p = .1$, $p = .3$, and $p = .7$. Do your curves look like the one in Figure 11.6?

7. **(a)** For the lensing event in Figure 11.6, what is the ratio β/θ_E when the lens comes closest to intersecting the line of sight to the lensed star? (Working Problem 6 may help with this.)

(b) What is the value of t_{var}—the time for the angular position of the lens to move by an amount θ_E? (You can make a rough estimate or fit the data with the results of Problem 6.)

(c) Assuming that the lens is moving with a velocity of $V = 200$ km/s transverse to the line of sight and located halfway between the Earth and the center of the galaxy, estimate the mass of the lens. (The distance between the Sun and the galactic center is approximately 8.5 kpc.)

8. [E, P] *Estimate* the energy in eV necessary to pull the last electron off of an Fe atom. Above what temperatures (in keV and K) will Fe atoms be completely ionized? At a temperature of 2 keV, how many of an Fe atom's electrons would you expect to remain?

9. [B, E] An X-ray source with a luminosity $L = 3 \times 10^{36}$ erg/s is powered by accretion onto a black hole with mass $6M_\odot$. Assuming all the radiation is released at the innermost stable circular orbit, *estimate* the rate $\dot{M}$ at which mass is being accreted by the black hole in $M_\odot$/yr.

10. **(a)** From the data on PSR B1913+16 given in the text, determine the semimajor axis of the orbit a and its angle of inclination i with respect to the line of sight. (Three significant figures is adequate.)

(b) What does this tell you about the companion star? Could it be a normal star like the Sun?

CHAPTER 12

Gravitational Collapse and Black Holes

The life history of a star is the story of the interplay between the contracting force of gravity and the expanding forces of gases heated by reactions that combine nuclei and release energy—a process aptly called *thermonuclear burning*. A star begins its life with the gravitational collapse of a cloud of interstellar gas consisting mostly of hydrogen and helium that is momentarily cooler, denser, or lower in kinetic energy than its surroundings. Compressional heating raises the core temperature high enough to ignite the thermonuclear reactions, which burn hydrogen to make helium and release energy. The star then reaches a steady state in which the energy lost to radiation is balanced by that produced by thermonuclear burning of hydrogen. This is the present state of our Sun.

Eventually, however, a significant fraction of the hydrogen in the star's core is exhausted and there is no longer enough thermonuclear fuel to provide the energy lost to radiation. Gravitational contraction resumes. Again, compressional heating raises the temperature until the reactions which burn helium to make other elements ignite. The star becomes brighter and its surface temperature changes. Eventually a significant amount of the helium will be exhausted, the core will again contract, and a new stage of thermonuclear burning will be initiated.

Where does this evolution end? It cannot continue indefinitely because the element ^{56}Fe has the highest binding energy per nucleon of any nucleus made in significant quantities in stars. (See Figure 12.1.) Nuclei like iron or others near it in the periodic table of elements cannot be burned to release any significant amount of energy leaving more bound nuclei behind. They are already the most bound nuclei. These "iron peak nuclei" are, therefore, the ashes of thermonuclear burning.

What happens to a star when it runs out of thermonuclear fuel? There are two possibilities: Either the end state is an equilibrium star, supported against the force of gravity by a *nonthermal source of pressure*, or the star never reaches equilibrium and the end state is *ongoing gravitational collapse*.

There are several possible nonthermal sources of pressure, discussed in much more detail in Chapter 24. There is pressure because the Pauli exclusion principle forbids two electrons from being in the same quantum state. This is called electron Fermi pressure. There are similar Fermi pressures for neutrons and protons. There are the nonthermal pressures arising from repulsive nuclear forces. Stars supported against the forces of gravitational collapse by the Fermi pressure of electrons are called *white-dwarf stars*, or—more simply and usually—*white dwarfs*. *Neutron stars* are supported by the Fermi pressure of neutrons and by nuclear forces. These two equilibrium end states of stellar evolution are much

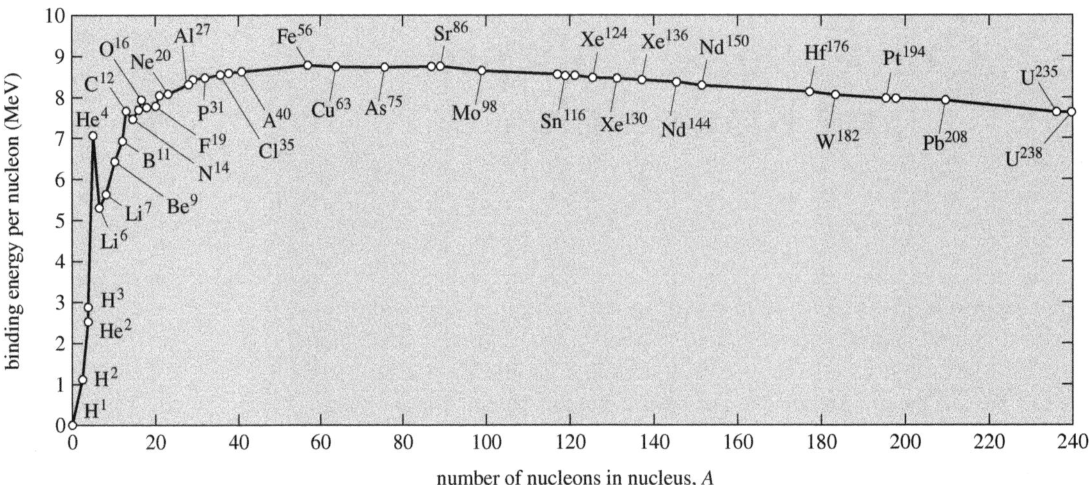

FIGURE 12.1 Atomic nuclei are bound collections of protons and neutrons (nucleons). The binding energy of a nucleus is the difference between its total energy and the energy of its constituent nucleons when they are dispersed. This figure shows the binding energy per nucleon as a function of the total number of nucleons in a nucleus, A.

smaller and denser than ordinary stars. A white dwarf might have a mass of the same order as the Sun ($M_\odot \sim 1.5$ km) but with a radius of only a few thousand kilometers. A neutron star of the same mass might have a radius of 10 km! Spacetime is moderately curved outside a neutron star because $M/R \sim 1/10$. We won't discuss these stars in detail until Chapter 24 because a decent discussion involves the properties of matter at very high densities, which requires physics beyond general relativity.

Rather, in this chapter, we concentrate on the second end state of stellar evolution—the state of ongoing gravitational collapse leading to a black hole. This possibility must exist in nature because there is a maximum amount of nonrotating matter that can be supported against gravitational collapse by Fermi pressure or nuclear forces. (See Box 12.1 on p. 257.) This mass is in the neighborhood of $2M_\odot$. (The exact value is uncertain because our knowledge of the properties of matter above nuclear densities is uncertain.) There are many stars more massive than this upper limit. It is likely that some must wind up in a state of ongoing collapse. It is the properties of this state we now explore.

12.1 The Schwarzschild Black Hole

Eddington–Finkelstein Coordinates

To get at the essential physics of gravitational collapse, let's consider the idealized case where the collapsing body and the spacetime outside it are spherically symmetric. Newton's theorem (see Example 3.1 on p. 40) shows that the Newtonian gravitational potential outside a spherically symmetric body is given by

BOX 12.1 The Maximum Mass of White Dwarfs

White dwarfs support themselves against gravity by the pressure of electrons arising from the Pauli exclusion principle—no two electrons can be in the same quantum state. This pressure is called Fermi pressure; the corresponding compressional energy is called the Fermi energy. A rough *estimate* of the maximum mass that can be supported against gravity by Fermi pressure can be made by studying the competition between the gravitational energy and the Fermi energy of a spherical configuration of radius R consisting of A electrons and A protons (so that it is electrically neutral). This estimate is backed up by detailed calculations in Chapter 24, specifically Figure 24.5.

The heavier protons supply most of the mass and the lighter electrons supply most of the pressure. Since the electrons exclude each other, we can think of each of them as occupying a volume of characteristic size λ such that there is a total of A electrons in the spherical volume of radius R. That is, $\lambda \sim R/A^{1/3}$. From the de Broglie relation $p = 2\pi\hbar/\lambda$, the characteristic momentum of the electrons (called their Fermi momentum), p_F, is

$$p_F \sim \hbar/\lambda \sim A^{1/3}\hbar/R. \tag{a}$$

If the sphere is compressed, R shrinks, p_F rises, the Fermi energy of the electrons rises, and work has to be done to make the compression. For simplicity assume that the compression has been carried out to the point that the electrons are relativistic and their individual energies are $E = [(p_F c)^2 + (m_e c^2)^2]^{1/2} \approx p_F c$. This assumption turns out to be justified for the most massive white dwarfs (Problem 2). The total Fermi energy in this approximation is

$$E_F \sim A(p_F c) \sim A^{4/3}\hbar c/R. \tag{b}$$

The protons supply most of the gravitational energy E_G, which is roughly

$$E_G \sim -G(m_p A)^2/R. \tag{c}$$

Here, m_p is the proton mass, and $m_p A$ is the total mass of the configuration. The gravitational potential energy is negative.

Both the Fermi energy and the gravitational energy vary as $1/R$. If A is sufficiently large, the total energy will be negative and it will be energetically favorable for the configuration to collapse. The critical A at which gravitational collapse becomes favored is

$$A_{\text{crit}} \sim (\hbar c/G m_p^2)^{3/2} \sim 10^{57}. \tag{d}$$

The critical mass is

$$M_{\text{crit}} \sim m_p A_{\text{crit}} \sim M_\odot \tag{e}$$

to an order of magnitude. The exact solution for the maximum mass is called the Chandrasekhar mass and is about $1.4M_\odot$, as discussed in Chapter 24.

$-GM/r$, whether or not the body is changing with time. The potential outside is, therefore, independent of time because the mass is conserved. A similar theorem in general relativity demonstrates[1] that, even though the mass distribution is time dependent, the geometry *outside* a spherically symmetric gravitational collapse is the time-independent Schwarzschild geometry already explored in Chapter 9.

As the collapse proceeds, more and more of the Schwarzschild geometry (9.1) is uncovered. We now have to face up to the singularities in the Schwarzschild metric at the radii $r = 2M$ and $r = 0$ and the significance of the change in sign of g_{tt} and g_{rr} at $r = 2M$. This section discusses the properties of the Schwarzschild geometry all the way down to $r = 0$ without including the collapsing matter. We return to the details of spherical collapse in the next section.

The singularity in the Schwarzschild metric at $r = 2M$ turns out not to be a singularity in the geometry of spacetime, but a singularity in Schwarzschild

[1] You can demonstrate it yourself after reading Chapter 21 by working Problem 18.

coordinates. It is a *coordinate singularity* in the sense discussed on p. 136. To show this, it is only necessary to exhibit one coordinate system in which the metric is not singular at $r = 2M$. There are many, but Eddington–Finkelstein coordinates are an especially simple example. Using these coordinates we will be able to understand why the Schwarzschild geometry is a black hole.

To introduce Eddington–Finkelstein coordinates, begin with Schwarzschild coordinates (t, r, θ, ϕ), in which the metric is summarized by (9.9), and trade the Schwarzschild time coordinate t for a new coordinate v defined by

$$t = v - r - 2M \log\left|\frac{r}{2M} - 1\right|. \tag{12.1}$$

Starting from either $r < 2M$ or $r > 2M$ and transforming t to v in the line element (9.9) gives the same result (Problem 3):

Schwarzschild Geometry in Eddington–Finkelstein Coordinates

$$ds^2 = -\left(1 - \frac{2M}{r}\right) dv^2 + 2dv\, dr + r^2(d\theta^2 + \sin^2\theta\, d\phi^2). \tag{12.2}$$

This is not a new geometry! It's the same time-independent, spherically symmetric geometry represented by the Schwarzschild metric (9.9), but with a different system of coordinates for labeling the points.

The fact that (12.2) was obtained by starting from the Schwarzschild metric starting from either $r < 2M$ or $r > 2M$ shows that these two regions, although separated by a singularity in the Schwarzschild metric, are in fact smoothly connected. Moreover, the absence of any singularity at $r = 2M$ in (12.2) shows that the singularity there in Schwarzschild coordinates is just a coordinate singularity. The line element (12.2) is fit for describing physics outside, at, and inside the Schwarzschild radius. Its nonsingular character shows that observers falling through the radius $r = 2M$ will see nothing special about the local spacetime. Eddington–Finkelstein coordinates are therefore useful for the study of ongoing gravitational collapse.

At large r, the metric (12.2) approaches a flat metric—the usual flat metric (7.4) with t replaced by $v - r$, because the logarithm in (12.1) becomes negligible compared to r. The line element (12.2) therefore bridges both large and small r regions. The metric is off-diagonal with $g_{vr} = g_{rv} = 1$, but that is a small price to pay for its advantages in providing a nonsingular connection between physics at large and small r.

Contrast the situation at $r = 2M$ with that at $r = 0$. There the metric is singular in *both* the Schwarzschild and Eddington–Finkelstein coordinate systems. As we will see quantitatively in Section 21.3, $r = 0$ is a place of infinite spacetime curvature and infinite gravitational forces—a real physical singularity. Observers falling into $r = 0$ will definitely see something special about the local spacetime. They will be destroyed.

Light Cones of the Schwarzschild Geometry

The key to understanding the Schwarzschild geometry as a black hole is the behavior of radial light rays. These move along world lines for which $d\theta = d\phi = 0$ (radial) and $ds^2 = 0$ (null), i.e., from (12.2), those for which

$$-\left(1 - \frac{2M}{r}\right) dv^2 + 2dv\, dr = 0. \tag{12.3}$$

An immediate consequence is that some radial light rays move along the curves

$$\boxed{v = \text{const.} \qquad \text{(ingoing radial light rays).}} \tag{12.4}$$

From (12.1) we see that for $r > 2M$ these are *ingoing* light rays because as t increases, r must decrease to keep v constant. The other possible solution to (12.3) is

$$-\left(1 - \frac{2M}{r}\right) dv + 2dr = 0. \tag{12.5}$$

This can be solved for dv/dr and the result integrated to find that these radial light rays move on the curves

$$\boxed{v - 2\left(r + 2M \log\left|\frac{r}{2M} - 1\right|\right) = \text{const.} \qquad \begin{pmatrix}\text{radial light rays} \\ \text{outgoing } r > 2M \\ \text{ingoing } r < 2M\end{pmatrix}.} \tag{12.6}$$

When one of these light rays is far from the black hole, it is *outgoing* because (12.6) becomes $t = r +$ constant as (12.1) shows. But when $r < 2M$, these light rays are *ingoing* because r decreases as v increases.

There is one special solution to (12.3) in addition to the null curves $v =$ const. and (12.6). The curve $r = 2M$ satisfies (12.3), describing light rays that are neither ingoing nor outgoing but instead are stationary.

Figure 12.2 is a spacetime diagram showing the world lines of the Schwarzschild geometry's radial light rays plotted in Eddington–Finkelstein coordinates. Null lines of constant v have been plotted at a 45° angle as they would usually be in flat space by using $\tilde{t} \equiv v - r$ as the vertical coordinate. The light rays at $r = 2M$ are indicated by the heavy solid line. Future light cones at a few intersections are indicated. These tip further and further toward $r = 0$ as that radius is approached. Radial light rays behave qualitatively differently outside the Schwarzschild radius $r = 2M$ than inside it. At every point with $r > 2M$, one radial ray (the $v =$ const. one) is moving inward to smaller and smaller values of r. The other radial ray is moving outward to larger and larger values of r. In con-

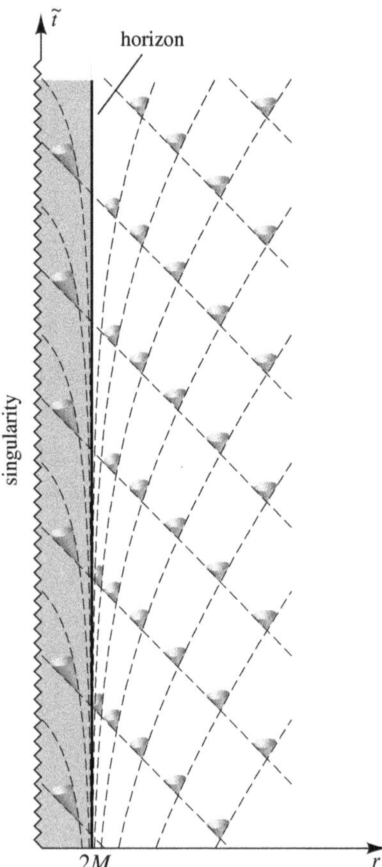

FIGURE 12.2 Radial light rays of the Schwarzschild geometry. Typical radial light rays of the Schwarzschild geometry are plotted in Eddington–Finkelstein coordinates $(\tilde{t} \equiv v - r, r)$. Two radial light rays run through each point in the diagram. There are the ingoing light rays moving along the curves $v =$ const. or, equivalently, along the curves $\tilde{t} = -r +$ const. to eventually reach the singularity at $r = 0$. The other radial light ray through each point is given by (12.6). These propagate outward to infinity if they are in the region $r > 2M$ but collapse inward to the singularity if they are in the region $r < 2M$. Light thus cannot escape the region $r < 2M$. Neither can particles whose timelike world lines must lie *within* the light cone at every point they traverse. The heavy vertical line at $r = 2M$ is the horizon of the Schwarzschild black hole, which divides the region (unshaded) from which a light ray can escape to infinity from each point, from the region where none can (lightly shaded). To see a representation of these features in one more spatial dimension, turn to Figure 12.4.

trast, for $r < 2M$, *both* radial light rays are moving inward to smaller and smaller values of r and eventually to the singularity at $r = 0$. At the boundary $r = 2M$ separating the two regions, one radial ray moves inward while the other remains stationary, hovering at the Schwarzschild radius. The surface $r = 2M$ thus divides

BOX 12.2 A Myth about Black Holes

Black holes irresistibly suck things in. That is a common misconception in science fiction. In fact, a spherical black hole of mass M attracts exterior mass no more strongly than a spherical star of mass M. Their exterior spacetimes are the same Schwarzschild geometry. Were the Sun somehow replaced tomorrow by a spherical black hole of the same mass, our climate would be significantly modified, but the orbit of the Earth would be almost unchanged. ("Almost" because the Sun is not exactly spherical.)

But there *is* a sense in which it is more difficult to escape from close to a black hole (or indeed any spherical mass) than from a Newtonian center of attraction of the same mass. Imagine using the thrust of a rocket to hover at a constant Schwarzschild coordinate radius R outside a spherical black hole of mass M. How much thrust would the rocket of mass m need to exert? The four-force $\mathbf{f}$ required to maintain this orbit can be found from the natural generalization of Newton's second law, $F = ma$, to curved spacetime:

$$f^\alpha = m\left(\frac{d^2x^\alpha}{d\tau^2} + \Gamma^\alpha_{\beta\gamma}\frac{dx^\beta}{d\tau}\frac{dx^\gamma}{d\tau}\right). \qquad \text{(a)}$$

This reduces to the law of motion in special relativity (5.35) when space is flat and to the law of geodesic motion (8.14) when the force vanishes. For a stationary orbit at radius R, $dx^\alpha/d\tau \equiv u^\alpha = [(1-2M/R)^{-1/2}, 0, 0, 0]$ (cf. (9.16)). Using this to evaluate (a), the coordinate radial component of $\mathbf{f}$ is $f^r = mM/R^2$. But it is the radial component $f^{\hat{r}}$ in the orthonormal basis of an observer riding with the rocket that is important for counting the required thrust. This is (Problem 17):

$$f^{\hat{r}} = m\left(1 - \frac{2M}{R}\right)^{-1/2}\frac{M}{R^2}. \qquad \text{(b)}$$

The required thrust is larger than the Newtonian mM/R^2 and infinitely larger as the radius R approaches $2M$.

---

spacetime into two regions: the region outside $r = 2M$ from which light can escape to infinity and the region inside $r = 2M$, where gravity is so strong that not even light can escape. This is the defining feature of a *black hole* geometry. The surface $r = 2M$ is called the *event horizon* (or, often more briefly, just the *horizon*) of the black hole.

Geometry of the Horizon and Singularity

The horizon $r = 2M$ is a three-dimensional null surface in spacetime of the kind discussed generally in Section 7.9. Its normal vector points in the r-direction and is a null vector (Problem 12). Like the future null cone in flat space, the horizon has a one-way property—once crossed it is not possible to cross back. However, unlike the light cones of flat space, the horizon is stationary, not expanding. The horizon is generated by those radial light rays that neither fall into the singularity nor escape to infinity.

A $v =$ const. slice of the horizon is a two-surface with the metric $d\Sigma^2 = (2M)^2(d\theta^2 + \sin^2\theta\, d\phi^2)$. (To see this just put $r = 2M$ and $v =$ const. in (12.2).) This is the geometry of a sphere with area $A = 16\pi M^2$, which is called the *area of the horizon*. The area doesn't change with v in the time-independent Schwarzschild geometry. However, it would change if matter fell into the black hole in a spherically symmetric way. Then the mass would go up and the area would increase. (This situation is considered further on p. 268.)

When polar coordinates are used to label points in flat spacetime [cf. (7.4)], $r = 0$ is a timelike world line that is always inside the local light cone—a place in space at all times. That is not the case in the Schwarzschild geometry. Inside $r = 2M$ surfaces of constant r are *spacelike*. Put $r =$ constant in (12.2) and you get the line element of a surface in which every direction is spacelike because g_{vv} is positive for $r < 2M$. In particular, the singularity $r = 0$ is a spacelike surface. The $r =$ constant spacelike surfaces inside $r = 2M$ define a decomposition of spacetime into space and time in which r is the time of the kind described in Section 7.9. The $r = 0$ singularity in the Schwarzschild geometry is not a place in space; it is a moment in time. (For a different perspective see Box 12.4 on p. 273.)

12.2 Collapse to a Black Hole

Armed with the previous section's picture of the Schwarzschild geometry as a black hole, let's return to the collapse of a spherically symmetric star. The radially moving particles at the surface of the collapsing star follow timelike world lines that lie inside the light cone at each point of spacetime they pass through, just like any other particle (Figure 4.10). The world line of the surface of a collapsing ball of pressureless matter that starts from rest at infinity provides a simple instance that is discussed in Example 12.1 and illustrated in Figures 12.3 and 12.4.

Outside the collapsing surface, the geometry of spherically symmetric collapse is the Schwarzschild geometry, including the horizon after the star has crossed the Schwarzschild radius $r = 2M$ and the singularity after it hits $r = 0$. Inside the surface (the heavily shaded region in Figure 12.3) the geometry is different, dependent on the detailed properties of the matter, but matching Schwarzschild geometry at the surface.[2] For the present discussion we won't need to know anything more about it.

Example 12.1. World Line of the Surface of Collapsing Sphere of Dust. "Dust" means pressureless matter in relativity parlance. Since there are no pressure forces, the outermost particles forming the surface of a collapsing sphere of dust are freely falling and follow radial geodesics in the Schwarzschild geometry. The radial geodesic for a sphere that starts from rest at $t = -\infty$ at $r = \infty$ was calculated in Chapter 9 [cf. (9.38), (9.40)]. The relation between r and the proper time τ along this geodesic was given in (9.38) as

$$r(\tau) = (3/2)^{2/3}(2M)^{1/3}(\tau_* - \tau)^{2/3}, \tag{12.7a}$$

where τ_* is an integration constant determining just which radial geodesic the surface of the star follows. The Schwarzschild time coordinate t at which the surface crosses the radius $r = 2M$ is infinite according to (9.40). However, that infinity is the consequence of the singular nature of Schwarzschild coordinates there. The proper time measured by the falling observer to reach the horizon from any

[2] If you have studied electromagnetism you will be familiar with a similar situation. The electric potential inside a spherical distribution of charge depends on how the charge is distributed, but the potential outside is the $1/r$ potential determined by the total charge.

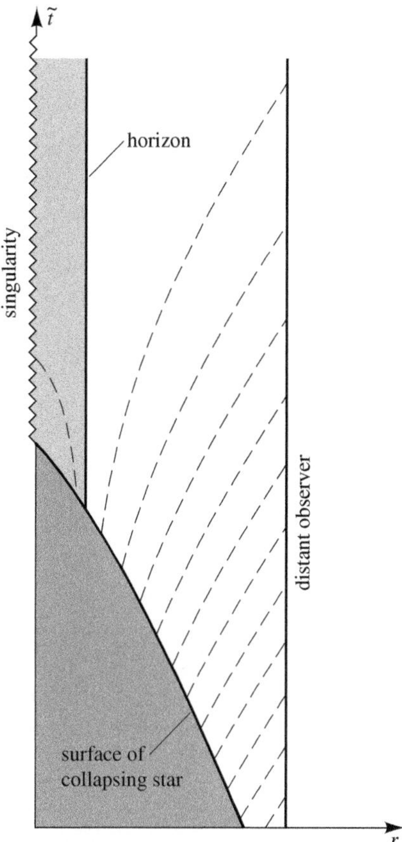

FIGURE 12.3 The story of two observers in the geometry of a collapsing spherical star. One observer stays at a fixed Schwarzschild coordinate radius r_R outside the star. The other follows its surface to smaller and smaller radii, sending out light signals at equal proper time inter-vals according to a clock falling with the surface. These light signals propagate out to the distant observer along the dotted curves shown. Only light rays emitted before the radius $r = 2M$ is crossed reach the distant observer. The distant observer, therefore, never sees the surface of the star cross $r = 2M$. The pulses arrive separated by longer and longer in-tervals as measured by the distant observer's clock. The light from the falling star becomes dimmer and dimmer and increasingly redshifted. A black hole is formed. Only the part of this Eddington–Finkelstein diagram outside the surface of the collapsing star (not heavily shaded) is meaningful. At the surface the geometry matches the geometry inside the star, which is not the Schwarzschild geometry.

starting radius is finite, as (12.7a) shows. Further, the values of the nonsingular coordinates v and $\tilde{t}$ are also finite, since from (9.40) and (12.1) (choosing $\tilde{t} = 0$ when $r = 0$) v as a function of r is given by

$$\frac{v(r)}{2M} = -\frac{2}{3}\left(\frac{r}{2M}\right)^{3/2} + \frac{r}{2M} - 2\left(\frac{r}{2M}\right)^{1/2} + 2\log\left[1 + \left(\frac{r}{2M}\right)^{1/2}\right]. \quad (12.7b)$$

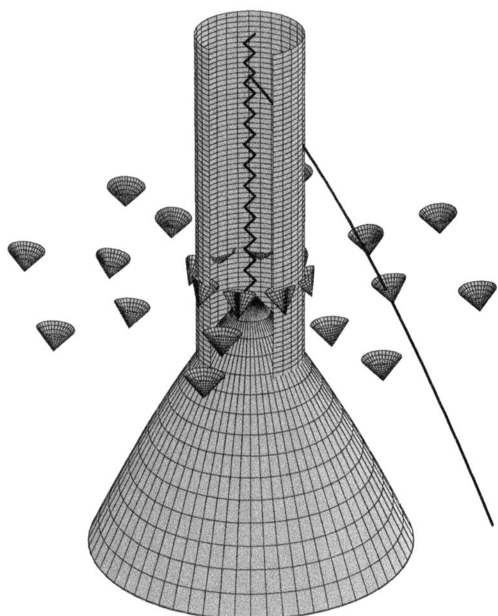

FIGURE 12.4 The formation of a black hole. Some essential features of a spheri-cally symmetric gravitational collapse that forms a black hole are shown in this three-dimensional spacetime diagram. Eddington–Finkelstein coordinates ($\tilde{t} \equiv v - r,\ r, \phi$) are used as cylindrical coordinates to label points in the diagram—$\tilde{t}$ vertically, r as radius from the axis of symmetry, and ϕ as azimuthal angle about that axis. The bottom surface is the world sheet swept out by the surface of the collapsing star as it progresses to smaller and smaller radii and eventually to a singularity at $r = 0$ [cf. (12.7)]. The vertical cylinder is the horizon at the Schwarzschild radius $r = 2M$. The horizon conceals the singularity from any distant observer but has been cut away in the illustration to reveal it. The world line of an observer falling freely from rest at infinity through the horizon and into the singularity is shown. The orientation of the future light cones at different radii on one $\tilde{t} =$ const. surface is shown. These tip more and more toward the center as they get closer to it, as illustrated with radial light rays in Figure 12.2.

Once across the horizon, the sphere hits the singularity at $r = 0$ in a further proper time of $4M/3$. That is of order 10^{-5} s for a solar mass of dust. One of these geodesics is the surface illustrated in Figure 12.3 and also in Figure 12.4.

Two Observers—The Inside Story

To understand the observational consequences of spherical collapse, consider the two observers whose world lines are illustrated by heavy lines in Figure 12.3. One observer rides on the surface of the star down to $r = 0$; the other observer remains outside at a large fixed radius $r = r_R$. The geometry in the unshaded region outside the surface of the star is the Schwarzschild geometry described by the metric (12.2). In the heavily shaded region the geometry is replaced by the geometry inside the star. Suppose the falling observer carries a clock and communicates with the distant one by sending out light signals at equally spaced times according to this clock. The world lines of these light rays are the dotted lines in Figure 12.3.

The pulses emitted after the surface of star has crossed the radius $r = 2M$ do not progress to larger and larger values of r. Rather, as shown in Figure 12.2, they progress to *smaller* and *smaller* values of r and eventually wind up in the singularity at $r = 0$. Once across the surface $r = 2M$, the gravitational attraction is so strong that even light cannot escape to infinity but rather falls back onto the singularity. No particle paths can escape from $r < 2M$ either; their paths must lie within the light cones (Problem 10). Therefore, once inside the Schwarzschild radius $r = 2M$, there is no way in which the falling observer can communicate with the distant one. Conversely, there is no way a distant observer can receive information from anywhere inside $r = 2M$.

Once across the Schwarzschild radius $r = 2M$, gravitational collapse to a singularity is the inevitable fate of the star. No new source of pressure at high densities can save it from collapse to zero size and infinite density. As long as the collapse remains spherical, the surface must travel *some* timelike radial world line, and *all* of these lead to the singularity at $r = 0$, as Figure 12.2 shows. Even if the star becomes nonspherical inside the horizon, it turns out that collapse to a singularity is inevitable. (See Box 12.3 on p. 266.) For the observer riding down with the star, there is also no way to escape destruction in the singularity once across the radius $r = 2M$. Utilizing a rocket the observer could leave the surface, but all possible timelike world lines that the rocket could travel lead to the singularity at $r = 0$ in a finite proper time (Problem 14). (That is evident from Figure 12.2 for radial world lines; for nonradial ones work Problem 10.) The inevitable singularity in geometry remains hidden from any observer outside the black hole. Look at Figure 12.4 for a three-dimensional spacetime diagram of these essential features of gravitational collapse.

Two Observers—The Outside Story

Although the history of the collapse as viewed by the observer who follows the star down is more dramatic, it is the sequence of events seen by the distant observer that is more important for astrophysics because we are (we hope) distant observers of any gravitational collapse. The distant observer never sees the star cross the radius $r = 2M$. The last light signal to reach the distant observer is emitted just before the star crosses this radius. Furthermore, the pulses emitted at equal intervals by the falling observer arrive spaced by increasingly longer intervals at the clock of the distant observer. For large values of r_R, that clock measures time intervals of $\tilde{t}$ to a good approximation [cf. (12.1)] (or intervals in the Schwarzschild time t to a similar good approximation). Figure 12.3 makes clear that the interval between received signals becomes longer and longer at later and later values of $\tilde{t}$. The light from the star is thus increasingly redshifted, with the red shift going to infinity for the light emitted as the star's surface reaches the radius $r = 2M$. (Example 12.2 gives a quantitative discussion.)

BOX 12.3 Trapped Surfaces and Singularities

Consider a spherical star whose surface is outside its Schwarzschild radius and is surrounded by a concentric sphere T of larger radius. Imagine that the sphere T emits a flash of light. One spherical pulse will travel outward to larger Schwarzschild radii, increasing in area. The other will travel to smaller radii, decreasing in area.

Now imagine a similar sphere T' surrounding a collapsing spherical star, where both are inside the horizon of the star's Schwarzschild geometry. As Figure 12.2 makes clear, both inward and outward pulses move to smaller Schwarzschild coordinate radii. The areas of both are decreasing.

The sphere T' is an example of a *closed trapped surface*—a closed spacelike two-surface such that the areas of pulses of light emitted from each little element of surface decrease in both possible directions. Any matter on the sphere T' is "trapped" between the two pulses because it cannot move faster than light. Since both pulses are headed to zero area, the matter caught between must wind up inside a sphere of zero area—a singularity.

That is the rough idea behind one of the *singularity theorems* of general relativity. The preceding discussion assumed spherical symmetry, but the singularity theorems apply in much more general circumstances. Very roughly, if a closed, trapped surface forms in a spacetime, then a singularity is inevitable if matter energy is positive enough that gravity remains attractive. Precise meanings can be given to *singularity* and *positive enough* as well as a mathematically careful statement of the theorem.[a]

Singularity theorems show that singularities are inevitable in many physical situations in general relativity. They are the reason, for example, for our confidence in the big bang singularity at the beginning of the universe.

[a] They can be found in Hawking and Ellis (1973).

Example 12.2. The Redshift of Light Received from a Collapsing Star. Eddington–Finkelstein coordinates can be used to analyze quantitatively how quickly the redshift of the light from the surface of a collapsing star following (12.7) goes to infinity with the time of receipt by a distant observer. Recall the situation illustrated in Figure 12.3. An observer on the surface of a collapsing star sends out radial light rays at equal, small intervals of proper time $\Delta\tau$ or, equivalently, with constant frequency $\omega_* = 2\pi/\Delta\tau$. The light rays emitted when the surface is crossing a sphere labeled by (v_E, r_E) are received by a stationary observer at a distant radius $r = r_R$ at a proper time t_R separated by intervals Δt_R on that observer's clock, i.e., with frequency $\omega_R(t_R) = 2\pi/\Delta t_R$. How does $\omega_R(t_R)$ vary with t_R?

The outgoing light ray connecting the events of emission and reception is one of the curves (12.6). The left-hand side of (12.6) evaluated at the point of emission (v_E, r_E) of the falling observer must be the same as when evaluated at the point of reception (v_R, r_R). For values of r_E close to $2M$, the logarithm in (12.6) dominates all other terms, but it is negligible compared to r_R for large values of r_R. Also, for large r_R the proper time along the stationary observer's world line is approximately the same as Schwarzschild coordinate time. Thus, from (12.1), $v \approx t_R + r_R$. Keeping only the dominant terms in the equality of the left-hand side of (12.6) between emission and reception gives

$$-4M \log\left(\frac{r_E}{2M} - 1\right) \approx t_R - r_R \tag{12.8}$$

or, equivalently,

$$\frac{r_E}{2M} - 1 \approx e^{-(t_R - r_R)/4M}. \tag{12.9}$$

This relation shows how, as t_R becomes large, the radius r_E from which the distant observer is receiving light approaches $2M$ exponentially with a characteristic time $4M$.

To calculate the redshift, think about the intervals Δt_R with which the signals are received by the stationary observer. The intervals in r at which the signals are emitted are $\Delta r_E = u^r \Delta\tau$, where u^r is the (negative) radial component of the four-velocity of the collapsing surface. From (12.9), $\Delta\tau$ and the corresponding Δr_E are connected to the interval in reception Δt_R by

$$-\frac{|u^r|\Delta\tau}{2M} = \frac{\Delta r_E}{2M} \approx -\frac{\Delta t_R}{4M} e^{-(t_R - r_R)/4M}. \tag{12.10}$$

The frequency of reception is $2\pi/\Delta t_R$. Since $|u^r|$ is finite as the surface crosses $r = 2M$ in nonsingular Eddington–Finkelstein coordinates [you can calculate it from (12.7a)], (12.10) implies the following behavior for the received frequency as a function of t_R:

$$\omega_R(t_R) \propto \omega_* e^{-t_R/4M}. \tag{12.11}$$

Equations (12.9) and (12.11) show how both the radius $r = 2M$ and infinite redshift are approached exponentially on a characteristic time scale of $4M$ as viewed by a stationary external observer. Similarly, the luminosity declines to zero exponentially with a time scale of order $4M$. In time units

$$4M = 2.0 \times 10^{-5} \left(\frac{M}{M_\odot}\right) \text{ s}. \tag{12.12}$$

For stellar-size objects this time scale is very small by usual astrophysical standards. For generic spherical gravitational collapse, the approach to a black hole is extremely rapid.

Because the light is redshifted, the energy per photon is less ($E = \hbar\omega$). Radial photons thus arrive both less and less frequently and with lower and lower energy. Sufficiently nonradial photons don't make it to infinity at all, as Example 9.2 shows. All these effects mean that the distant observer sees the luminosity of the collapsing star going to zero. As Example 12.2 shows quantitatively, the time scale for this approach to darkness is very short in realistic situations—of order 10^{-5} s for a freely collapsing solar mass.

In summary, the distant observer very quickly sees a spherical gravitational collapse slow down, grow dark, and become indistinguishable from a *time-independent* Schwarzschild geometry. As long as the collapse is spherical, all records of the star's history and the details of its collapse are erased from the exterior geometry. The second end state of stellar evolution—ongoing gravitational collapse—is thus remarkably simple in the case of spherical symmetry when viewed from outside the Schwarzschild radius. The geometry is time independent and characterized by a single number, M. As is described briefly in Section 12.4 and in more detail in Chapter 15, the outcome of realistic nonspherical gravitational collapse is also believed to be a black hole and similarly simple.

The rapid approach of a spherical collapsing star to a dark Schwarzschild geometry means that black holes cannot be detected by radiation coming from them.[3] But as discussed in Chapter 13, black holes can be detected by observing the properties of matter in orbit around them in much the same way that the orbit of Earth would reveal the presence of the Sun even were it not shining. Indeed, black holes can in principle be distinguished from dark stars by the way they absorb radiation and the way they emit gravitational radiation if perturbed, although these topics are not considered in this text.

The Area of the Horizon Increases

Figure 12.3 is a spacetime diagram showing the behavior of light rays in the exterior Schwarzschild geometry of a spherical collapsing star of mass M that forms a black hole. The horizon is the null three-surface generated by those radial light rays that neither escape to infinity nor fall into the singularity but remain at $r = 2M$. In Figure 12.3 the horizon ends at the star's surface only because the geometry inside is not portrayed. But the horizon continues inside as well. Imagine an observer at the center of the star sending out radial light rays. Those that reach the surface just as it is crossing the radius $r = 2M$ will remain at that radius. They generate the horizon not only outside the star but inside as well, as illustrated *qualitatively* by Figure 12.5a. (For a quantitative example, work Problem 18). Inside the star the horizon grows in radius and area until it reaches the surface, after which it remains stationary if nothing further falls into the black hole.

Figure 12.5b shows what happens if a thin shell of matter of mass M_{shell} falls into the black hole after it has formed. After the shell has fallen in, the horizon will be at the radius $r = 2(M + M_{\text{shell}})$. Before that it will be generated by those radial light rays that start at the center, pass through the surface of the collapsing star at a radius $r > 2M$, and then move outward to reach the shell just as it is crossing the radius $r = 2(M + M_{\text{shell}})$, as sketched qualitatively in Figure 12.5b. Inside the shell, but outside the star, the horizon is not at $r = 2M$ but is expanding through larger radii, its area always increasing. This example illustrates two important properties of event horizons in spherical black-hole spacetimes that are true in more general circumstances: A horizon's location at any one moment depends on the geometry of spacetime to the future of that moment. Horizon area increases provided the energy of matter is sufficiently positive (Problem 19).

[3]In classical physics at least; for quantum physics see the discussion of the Hawking effect in Section 13.3.

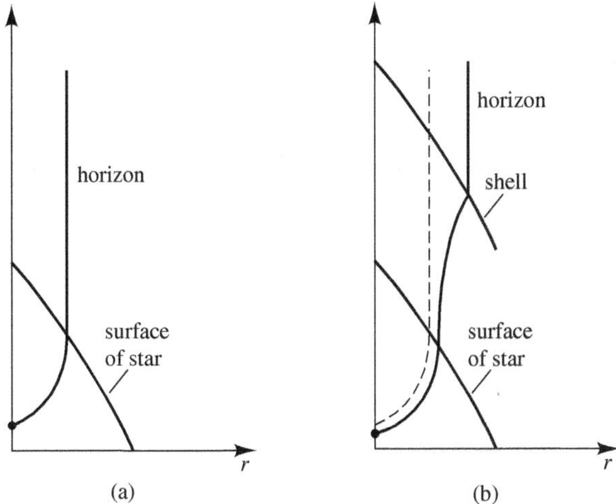

FIGURE 12.5 (a) The horizon inside a collapsing star. This figure is a schematic spacetime diagram showing how the horizon is generated by radial light rays that start at the center and make it to the surface just as the star is crossing its Schwarzschild radius, after which they remain stationary. The horizon grows in area inside the star. (b) This figure is a schematic spacetime diagram showing what happens if a shell of matter falls in after the collapse. The horizon is generated by the radial light rays, which start at the center and reach the shell as it is crossing the Schwarzschild radius corresponding to its mass plus that of the star. (The dotted line shows the location of the horizon without the shell.) The horizon is thus increasing in area between the surface of the star and the shell. (The vertical axes are not labeled in these figures because we haven't specified a set of coordinates inside the star. Outside you can take it to be $\tilde{t}$.)

12.3 Kruskal–Szekeres Coordinates

One quality of "understanding" is being able to describe the same thing from several different points of view. The Schwarzschild geometry is a clear instance. Schwarzschild coordinates (t, r, θ, ϕ) are the most direct way of understanding phenomena far from the center of a collapsing star, such as the approach to the geometry of flat space or the orbits of test particles and light rays. But, because of their singular nature at $r = 2M$, Schwarzschild coordinates are not as useful for understanding the nature of the event horizon of a black hole or the singularity at $r = 0$. Rather, a nonsingular coordinate system like the Eddington–Finkelstein coordinates gives a clear view of these regions. The Kruskal–Szekeres coordinates introduced in this section are an alternative to Eddington–Finkelstein coordinates that give a different perspective on physics near a Schwarzschild black hole. If you feel that your understanding is already sufficient, you can skip this section.

Relation to Schwarzschild Coordinates

Kruskal–Szekeres coordinates are denoted by (V, U, θ, ϕ). The θ, ϕ coordinates are the same as the Schwarzschild polar angles, but Schwarzschild t and r are traded for V and U according to the following coordinate transformations:

$$\left.\begin{aligned} U &= \left(\frac{r}{2M}-1\right)^{1/2} e^{r/4M}\cosh\left(\frac{t}{4M}\right) \\ V &= \left(\frac{r}{2M}-1\right)^{1/2} e^{r/4M}\sinh\left(\frac{t}{4M}\right) \end{aligned}\right\} \quad r > 2M, \tag{12.13a}$$

$$\left.\begin{aligned} U &= \left(1-\frac{r}{2M}\right)^{1/2} e^{r/4M}\sinh\left(\frac{t}{4M}\right) \\ V &= \left(1-\frac{r}{2M}\right)^{1/2} e^{r/4M}\cosh\left(\frac{t}{4M}\right) \end{aligned}\right\} \quad r < 2M. \tag{12.13b}$$

The result of carrying out these transformations on the Schwarzschild metric (9.1) in either the region with $r > 2M$ or that with $r < 2M$ is (Problem 20)

Schwarzschild Geometry in Kruskal–Szekeres Coordinates

$$ds^2 = \frac{32M^3}{r}e^{-r/2M}\left(-dV^2 + dU^2\right) + r^2\left(d\theta^2 + \sin^2\theta\, d\phi^2\right). \tag{12.14}$$

Here, r is considered as a function of V and U, $r = r(V, U)$, defined implicitly by the relation

$$\left(\frac{r}{2M}-1\right)e^{r/2M} = U^2 - V^2 \tag{12.15}$$

which is derivable both from (12.13a) and (12.13b). The Kruskal–Szekeres metric (12.14) is not singular at $r = 2M$, showing again that the singularity there in Schwarzschild coordinates is just a coordinate singularity.

Considerable insight into both this coordinate transformation and the nature of the Schwarzschild geometry itself can be obtained by plotting lines of constant coordinates r and t on a U, V grid. This is called a Kruskal diagram, and one is illustrated in Figure 12.6. From (12.15) we see that lines of constant r are curves of constant $U^2 - V^2$ and therefore hyperbolas in the UV plane. The value $r = 2M$ corresponds to either of the straight lines $V = \pm U$. The value $r = 0$ corresponds to the hyperbola

$$V = +\sqrt{U^2+1}. \tag{12.16}$$

[To see that the positive square root is to be taken, use (12.13b).] In a similar way, the lines of constant t can be also plotted on the Kruskal diagram. From (12.13) one finds

$$\tanh\left(\frac{t}{4M}\right) = \frac{V}{U}, \qquad r > 2M, \tag{12.17a}$$

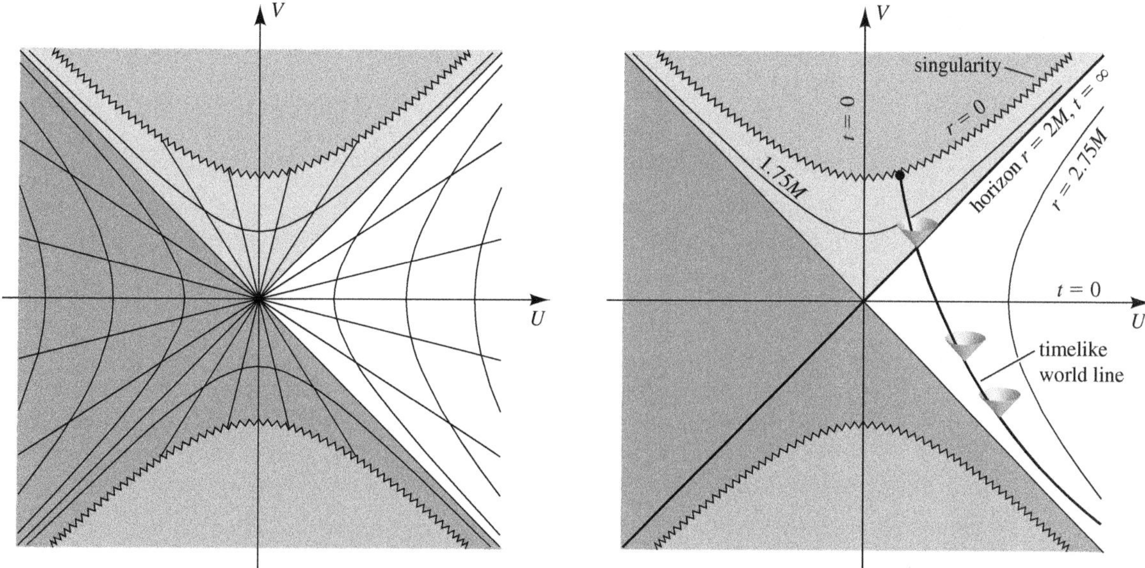

FIGURE 12.6 Kruskal diagrams. Two views are shown of a two-dimensional slice of the Schwarzschild geometry defined by the two Kruskal–Szekeres coordinates (U, V). On the left some lines of constant Schwarzschild coordinates r and t are plotted. Hyperbolas of r-values 0, $1.75M$, $2M$, $2.25M$, $2.75M$, and $3.25M$ are plotted along with straight lines with t-values 0, $\pm M$, $\pm 2.75M$, $\pm 3M$, $\pm 3.25M$ and $\pm\infty$. The shaded regions in these diagrams correspond to the similarly shaded regions in Figure 12.2. The unshaded region outside the black hole is covered by $2M < r < \infty$ and $-\infty < t < +\infty$ or $-\infty < v < +\infty$. We live there. Only the region $V > -U$ is covered by Eddington–Finkelstein coordinates $0 < r < \infty$, $-\infty < v < +\infty$. The heavily shaded region with $V < -U$ is part of the Kruskal extension discussed in Box 12.4. The medium shaded regions above and below the $r = 0$ singularities are not part of the spacetime at all. On the right the singularity at $r = 0$, the horizon at $r = 2M$, and an infalling timelike world line with a few future light cones are indicated. Radial light rays move along 45° lines in the Kruskal diagram.

$$\tanh\left(\frac{t}{4M}\right) = \frac{U}{V}, \qquad r < 2M. \tag{12.17b}$$

Thus, lines of constant t are lines of constant U/V—straight lines through the origin. The value $t = +\infty$ corresponds to $U = V$, whereas $t = -\infty$ corresponds to $U = -V$. The value $t = 0$ corresponds to the line $V = 0$ for $r > 2M$, whereas for $r < 2M$ it is the line $U = 0$. The unshaded quadrant of the Kruskal diagram with $U > 0$, $-U < V < U$ is covered by the Schwarzschild coordinates $-\infty < t < +\infty$, $2M < r < \infty$. The entire region covered by Eddington–Finkelstein coordinates $-\infty < v < +\infty$, $0 < r < \infty$ is mapped into the part of the diagram with $V > -U$. That is the region through which the world line of the collapsing star's surface moves, and, as explained earlier, only that part outside the star's surface is relevant for spherical collapse. For the significance of the region $V < -U$, see Box 12.4 on p. 273.

For the entire range of (V, U, θ, ϕ) the metric component g_{VV} is negative whereas g_{UU}, $g_{\theta\theta}$, $g_{\phi\phi}$ are always positive. The direction along V is thus always timelike, and the direction along U is always spacelike. Contrast this with

Schwarzschild coordinates: Increasing t is a timelike direction for $r > 2M$ but a spacelike one for $r < 2M$; increasing r is a spacelike direction for $r > 2M$ but timelike for $r < 2M$.

Light Cones, Horizon, and Spherical Collapse

Radial light rays are especially easy to analyze in Kruskal coordinates. Radial light rays move along curves for which $d\theta = d\phi = 0$ (radial) and for which $ds^2 = 0$ (null). From (12.14) these are just the curves

$$V = \pm U + \text{const.} \tag{12.18}$$

Radial light rays thus move along the 45° lines in the Kruskal diagram, and the light cones at each point make an angle of 45° with the vertical. Particle world lines are timelike and must lie inside the light cone at every point they pass

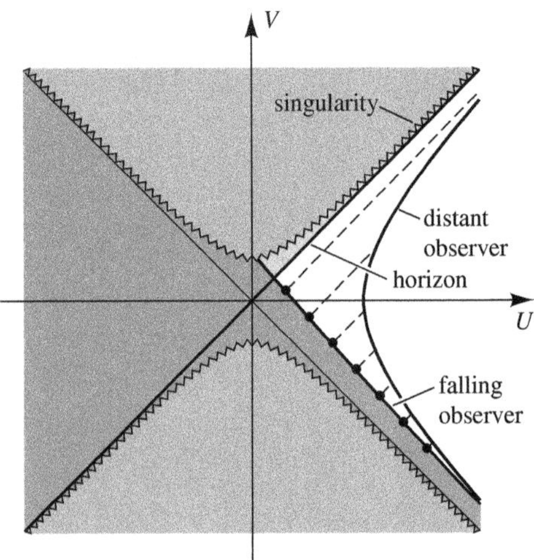

FIGURE 12.7 The world lines of two observers are illustrated in this Kruskal diagram that retells the story in Figure 12.3 but in Kruskal coordinates. The similarly shaded regions correspond in the two diagrams. One observer rides the surface of a collapsing ball of pressureless dust from large r at $t = -\infty$ down to the singularity at $r = 0$ along the world line given by (12.7a) and (12.7b). Only the unshaded and lightly shaded parts of the diagram outside the surface are relevant for the story of spherical collapse. The singularity formed at $r = 0$ is shown, as well as the horizon at $r = 2M$, which is the null curve $U = V$ on this diagram. The region inside the horizon is lightly shaded. The world line of an observer who remains stationary at large r is also illustrated. That is a hyperbola in the region $r > 2M$. The world lines of light rays emitted at equal intervals of proper time by the falling observer are shown. These are received by the distant observer at longer and longer intervals of proper time as the collapse progresses. The last light ray to reach the distant observer follows the 45° line just below the horizon. Once across the horizon, all timelike world lines lead to the singularity at $r = 0$; the star collapses to zero radius and the falling observer is destroyed.

through. Radial particle world lines must, therefore, lie within the 45° lines with slopes that are greater than unity. Indeed, $|dV/dU| > 1$ for a particle world line whether it is moving radially or not (Problem 21). The world line of the surface of a collapsing star discussed in Example 12.1 is shown in a less busy version of the same Kruskal diagram in Figure 12.7. Only the unshaded and lightly shaded regions on that diagram represent the spacetime outside the collapsing star.

Essential features of the Schwarzschild black-hole geometry that we discovered in Eddington–Finkelstein coordinates can be seen from a different perspective in the Kruskal diagram in Figure 12.6. The singularity at $r = 0$ is clearly revealed as a spacelike surface. The horizon at $r = 2M$ is the 45° line $V = U$, showing it to be a null surface generated by the radial light rays that remain stationary at $r = 2M$. Inside $r = 2M$ (above the $V = U$ line in Figure 12.6), all

BOX 12.4 The Kruskal Extension of the Schwarzschild Geometry

Only the region of the Kruskal diagram outside the world line of the surface of the collapsing star (the unshaded region in Figure 12.7) is relevant for spherical collapse. However, purely theoretically it is possible to think of the Schwarzschild geometry as a static, spherically symmetric solution of the Einstein equation with no matter sources. Viewed that way, there is no reason not to think of the whole region of Kruskal coordinates (U, V) bounded by the singularities at $r = 0$ as a spacetime. This is called the Kruskal *extension* of the Schwarzschild geometry because Schwarzschild coordinates ranging over $2M < r < \infty$ and $-\infty < t < +\infty$ cover only the quadrant with $U > 0$, $-U < V < U$. Eddington–Finkelstein coordinates cover more—the half plane above $V = -U$. (See the discussion on p. 271.) Kruskal–Szekeres coordinates cover the whole thing.

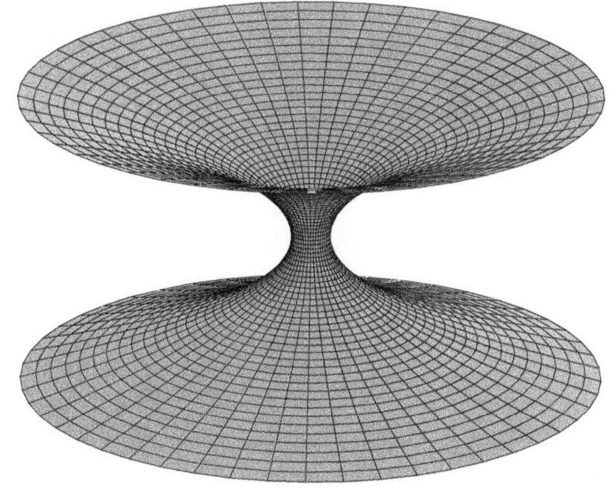

The Kruskal extension possesses *two* spacelike surfaces where $r = 0$ and the geometry is singular—the hyperbolas $V = \pm(U^2 + 1)^{1/2}$. It has *two* asymptotically flat regions—one where $U \to +\infty$ and the other where $U \to -\infty$. These facts alone show that the Kruskal extension is nothing like a spacetime surrounding a point mass. Indeed, on spacelike surfaces such as $V = 0$, there is no singularity at all, just empty curved space! Moving radially on this surface from $U = \infty$ to $U = -\infty$, the function $r(U, 0)$ decreases to a minimum value of $2M$ but then increases to infinity in the second asymptotically flat region. The embedding diagram of the $V = 0$, $\theta = \pi/2$ two-surface at left, constructed according to the methods in Section 7.7, reveals that the Kruskal extension is a wormhole connecting two asymptotically flat regions of spacetime (Problem 24). But it is not a static wormhole like the toy geometry in Example 7.7. Rather, as V moves to larger or smaller values, the minimum radius of the wormhole throat decreases and eventually pinches off in a singularity at $r = 0$. For this reason, if our universe somehow contained one of these wormholes, it would not be possible to move fast enough to get through it while it is evolving from singularity to singularity. Can you see this directly on the Kruskal diagram (Problem 25)?

timelike and null world lines lead to the singularity at $r = 0$, demonstrating its inevitable formation once the star's surface crosses the Schwarzschild radius. No light rays or timelike world lines escape from the inside of the horizon, and events there remain hidden from any observer outside.

The world lines of the two communicating observers discussed in Section 12.2 and illustrated in Figure 12.3 are shown in the Kruskal diagram in Figure 12.7. The distant observer runs along a hyperbola of large, fixed r. The light rays emitted at equal proper time intervals by the falling observer move on the dotted 45° lines shown. They evidently are received less and less frequently at later and later times by the distant observer, leading to the increasing redshift of the light from the collapsing star and the extinction of its luminosity. The last light ray received is emitted just before the star and falling observer plunge through the Schwarzschild radius.

BOX 12.5 The Penrose Diagram for the Schwarzschild Geometry

By a careful choice of new coordinates (U', V'), it is possible to relabel the points of a Kruskal diagram so that light rays continue to propagate along 45° lines in the new coordinates such that points at infinity are labeled by finite coordinate values rather than infinite ones. The resulting picture of the whole slice of the Kruskal extension (Box 12.4) of the Schwarzschild geometry in a finite region of the (U', V') plane is called the Penrose diagram for the Schwarzschild geometry and is a useful tool for picturing its global spacetime structure. The construction of a Penrose diagram for flat spacetime was described in Box 7.1 on p. 137, and the construction for the Schwarzschild geometry is closely parallel. Begin with the Schwarzschild geometry in Kruskal–Szekeres coordinates (12.14) and replace coordinates U and V with

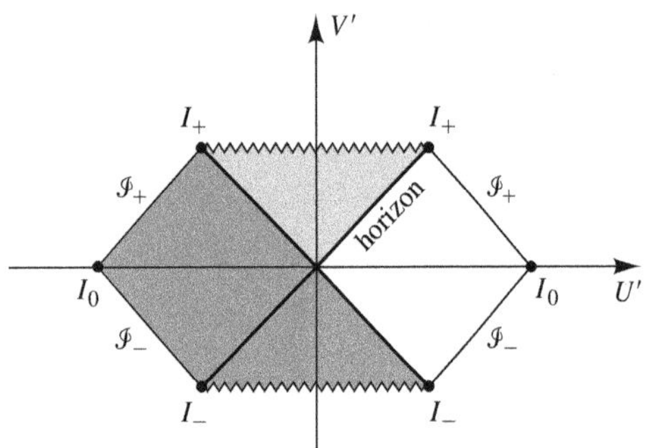

two new coordinates u and v defined by[a]

$$U = (v - u)/2, \qquad V = (v + u)/2. \qquad \text{(a)}$$

The uv axes are just the UV axes rotated by 45° so that light rays move on curves of constant u or v. Introduce other coordinates (u', v') and (U', V') defined by

$$u' \equiv \tan^{-1}(u) \equiv V' - U',$$
$$v' \equiv \tan^{-1}(v) \equiv V' + U'. \qquad \text{(b)}$$

Light rays move on curves of constant u' and v', i.e., the 45° lines in the $U'V'$ plane. The infinite ranges of u and v are each mapped in to the finite range $(-\pi/2, \pi/2)$ for u' and v'. With a little work one sees that the hyperbola $r = 0$, $V > 0$ maps into the line $V' = \pi/4$, $-\pi/4 \leq U' \leq \pi/4$, whereas the one with $V < 0$ maps into the same line at $V' = -\pi/4$. The horizon $V = U$ maps into the same 45° line in the $U'V'$ plane. The resulting Penrose diagram is shown at left. As in flat space, it is possible to identify different kinds of infinity: future and past null infinity $\mathcal{I}_\pm$, where light rays at infinity start and wind up, future and past timelike infinity $I_\pm$, where timelike world lines start and wind up, and spacelike infinity I_0, which all spacelike surfaces at infinity intersect. There are *two* sets of these, one for each asymptotic region. With this diagram we can see that the horizon is the boundary of the region of spacetime that can be connected to future null infinity by a light ray.

[a] Don't mix up this v with the Eddington–Finkelstein v coordinate!

In many respects the causal properties of the Schwarzschild black hole are more directly revealed in Kruskal–Szekeres coordinates than in the Eddington–Finkelstein ones. Further, Kruskal–Szekeres coordinates are the basis for other useful representations of black-hole geometry. (See Boxes 12.4 and 12.5 on p. 273 and p. 274, respectively.) But Kruskal–Szekeres coordinates are not very useful for analyzing the orbits of test particles and light rays at large distances from the black hole. Eddington–Finkelstein coordinates have the virtue of working both near and far from the black hole. To understand phenomena as exotic as a black hole, it is helpful to have several different perspectives, which in relativity means several different coordinate systems.

12.4 Nonspherical Gravitational Collapse

Realistic collapse situations are not exactly spherical. A precollapse star may be distorted by rotation. Realistic supernova explosions by which massive stars end their lives are certainly not spherical. Therefore, the question naturally arises as to how much of this picture of spherical collapse persists in a realistic case. This section describes, without proof, how some features of the simple spherical model are expected to hold in nonspherical gravitational collapse.

- *Formation of a Singularity.* As we have seen, once the surface of a spherical collapsing star crosses the Schwarzschild radius, $r = 2M$, gravitational collapse to a singularity is inevitable. The geometry allows no escape from the region inside the horizon or for the collapse to stop. The formation of a singularity in spherical gravitational collapse is a specific illustration of the *singularity theorems* of general relativity alluded to briefly in Box 12.3 on p. 266. Roughly speaking, these theorems show that any gravitational collapse that proceeds far enough results in a singularity in spacetime geometry. The singularity formed in spherical collapse is thus not an artifact of the special symmetry but a feature of more general collapse situations.
- *Formation of an Event Horizon.* The singularity formed in spherical collapse is inside the horizon, hidden from observers outside. The fact that it is hidden is important, because a singularity is a place where the predictive power of the theory breaks down, but information about this breakdown can never reach observers outside.

  The *cosmic censorship conjecture* that will be discussed in more detail in Section 15.1 holds that the singularities formed in any generic, realistic collapse are hidden inside the horizons of black holes. This has not yet been proven to be a consequence of the Einstein equation; indeed, even a precise formulation of the conjecture is lacking. But no generic exception to the idea has been found to date.

  If the cosmic censorship conjecture is true, then the geometry outside of any realistic complex, nonspherical gravitational collapse, dependent on the detailed properties of realistic matter, is a black-hole geometry with a horizon analogous to that of the Schwarzschild geometry. The Schwarzschild black hole is not the most general black hole allowed by general relativity. But the general case described in Chapter 15 is not much more complex. The general black hole depends on only two parameters—its mass and angular momentum. Therefore, if cosmic censorship is true, the outcome of realistic nonspherical gravitational collapse is as remarkably simple as its spherical idealization.

- *Area Increase.* As described on p. 268, the area of a black hole increases when mass falls into it in a spherically symmetric way. However, even if mass falls in a nonspherically symmetric fashion, the area of the horizon still increases. That is a consequence of the *area increase theorem* for black holes. This behavior of the area of a black hole recalls the increase in entropy in thermodynamics. In Section 13.3 we will see that in quantum mechanics the entropy of a black hole *is* proportional to its area.

Example 12.3. A Lump Falls into a Black Hole. A lump of matter falls radially into a black hole from one direction. That is not a spherically symmetric situation. The resulting black hole will oscillate, emitting gravitational radiation, but eventually it will settle down to another Schwarzschild black hole because its angular momentum has not changed. Could the energy carried away by gravitational radiation be greater than the mass that fell in, leaving a lower mass Schwarzschild black hole than the initial one? The answer is no because the area of the black hole cannot decrease and the areas of Schwarzschild black holes are related to their mass by $A = 4\pi(2M)^2$.

Problems

1. [P] How many protons must combine to make He nuclei every second to provide the luminosity of the Sun? Estimate how long the Sun could go on at this rate before all its protons were used up.

2. [E, B] Follow through the order of magnitude estimate for the maximum mass of white dwarfs in Box 12.1 *without* assuming that the electrons are necessarily relativistic.

   (a) Sketch the behavior of the total energy $E_{TOT}(A, R) = E_F(A, R) + E_G(A, R)$ as a function of R both for values of A greater and less than A_{crit} defined in (d) of the box.

   (b) Find the radius $R_*(A)$ for which $E_{TOT}(A, R)$ has a minimum in R. This is an estimate of the radius of the equilibrium star where gravitational and Fermi pressure forces balance. How does R_* compare to the radii of white dwarf stars quoted in the text?

   (c) Show that there is a value A_{crit} above which no equilibrium is possible. Find its value and compare with A_{crit} estimated in the text.

(d) Are the electrons relativistic at the equilibrium?

3. [S] Carry out the transformation from Schwarzschild to Eddington–Finkelstein coordinates defined by (12.1) to get the line element (12.2).

4. Consider the spacetime specified by the line element

$$ds^2 = -\left(1 - \frac{M}{r}\right)^2 dt^2 + \left(1 - \frac{M}{r}\right)^{-2} dr^2 + r^2(d\theta^2 + \sin^2\theta\, d\phi^2).$$

Except for $r = M$, the coordinate t is always timelike and the coordinate r is spacelike.

(a) Find a transformation to new coordinates (v, r, θ, ϕ) analogous to (12.1) that sets $g_{rr} = 0$ and shows that the geometry is not singular at $r = M$.

(b) Sketch a $(\tilde{t}, r)$ diagram analogous to Figure 12.2 showing the world lines of ingoing and outgoing light rays and the light cones.

(c) Is this the geometry of a black hole?

5. An observer falls radially into a spherical black hole of mass M. The observer starts from rest relative to a stationary observer at a Schwarzschild coordinate radius of $10M$. How much time elapses on the observer's own clock before hitting the singularity?

6. An observer decides to explore the geometry outside a Schwarzschild black hole of mass M by starting with an initial velocity at infinity and then falling freely on an orbit that will come close to the black hole and then move out to infinity again. What is the closest that the observer can come to the black hole on an orbit of this kind? How can the observer arrange to have a long time to study the geometry between crossing the radius $r = 3M$ and crossing it again?

7. [E] A meter stick falls radially into a center of Newtonian gravitational attraction produced by one solar mass located at a point. Using Newtonian physics *estimate* the distance from the point at which the meter stick would break or be crushed.

8. Can an observer who falls into a spherical black hole receive information about events that take place outside? Is there any region of spacetime outside the black hole that an interior observer cannot eventually see? Analyze these questions using a diagram such as the one in Figure 12.2.

9. [S] Darth Vader is pursuing some Jedi knights. The Jedi knights plunge into a large black hole seeking the source of the force. Darth Vader knows that, once inside, any light emitted from his light-ray gun moves to smaller and smaller Schwarzschild radii. He decides to try it by firing in the radial direction. Should he worry that light from his gun will fall back on him before his destruction in the singularity?

10. Show that the slopes of the curves of $\tilde{t}$ vs. r of *nonradial* light rays in an Eddington–Finkelstein diagram like Figure 12.2 must lie *within* the light cones defined by the *radial* light rays.

11. Negative mass does not occur in nature. But just as an exercise, analyze the behavior of radial light rays in a Schwarzschild geometry with a negative value of M. Sketch the Eddington–Finkelstein diagram showing these light rays. Is the negative mass Schwarzschild geometry a black hole?

12. [S] Check that the normal vector to the horizon three-surface of a Schwarzschild black hole is a null vector.

13. [C] **(a)** An observer falls feet first into a Schwarzschild black hole looking down at her feet. Is there ever a moment when she cannot see her feet? For instance, can she see her feet when her head is crossing the horizon? If so, what radius does she see them at? Does she ever see her feet hit the singularity at $r = 0$ assuming she remains intact until her head reaches that radius? Analyze these questions with an Eddington–Finkelstein or Kruskal diagram.

(b) *Is it dark inside a black hole?* An observer outside sees a star collapsing to a black hole become dark. But would it be dark inside a black hole assuming a collapsing star continues to radiate at a steady rate as measured by an observer on its surface?

14. [C] Once across the event horizon of a black hole, what is the *longest* proper time the observer can spend before being destroyed in the singularity?

15. [C] A spaceship whose mission is to study the environment around black holes is hovering at a Schwarzschild coordinate radius R outside a spherical black hole of mass M. To escape back to infinity, crew must eject part of the rest mass of the ship to propel the remaining fraction to escape velocity. What is the largest fraction f of the rest mass that can escape to infinity? What happens to this fraction as R approaches $2M$?

16. In Section 9.2 we derived formula (9.20) for the gravitational redshift from a stationary observer. We started from the conservation law (8.32) arising from the time trans-lation symmetry of the Schwarzschild geometry. Use similar techniques to derive an expression for the redshift of light emitted radially from a star in free fall collapse as a function of the time t_R the radiation is received by a distant observer. Compare your result with (12.11). Equation (9.20) held for nonradial radiation as well. Do you expect that for radiation from the surface of a collapsing star? (Note that in (9.20) R was the radius of the stationary observer emitting the radiation, whereas in the example that led to (12.11), R was the location of the observer receiving the radiation.)

17. [B] Derive the rocket thrust (b) in Box 12.2.

18. [C] *The Horizon Inside a Collapsing Shell* Consider the collapse of a spherical shell of matter of very small thickness and mass M. The shell describes a spherical three-surface in spacetime. Outside this surface the geometry is the Schwarzschild geometry with this mass. Inside make the following assumptions: (1) The world line of the shell is known as a function $r(\tau)$ going to zero at some finite proper time; (2) the geometry inside the shell is flat; (3) the geometry of the three-surface of the collapsing shell is the same inside as outside.

(a) Draw two spacetime diagrams: one like that in Figure 12.2 and another corresponding to the spacetime inside the shell in a suitable set of coordinates. Draw the world line of the shell in both, and indicate how points on the inside and outside correspond. Locate the horizon inside the shell as well as outside.

(b) How does the area of horizon inside the shell change moving along the light rays which generate it?

19. Figure 12.5b illustrates the area of the horizon of a spherical black hole if a shell of mass M_{shell} later fell into it. The discussion assumed that the shell was made of usual matter with $M_{\text{shell}} > 0$. What would happen if it were negative? Would the area of the horizon always increase? Illustrate the qualitative behavior of the horizon as well as that of light rays for the case of a negative mass shell with a diagram like Figure 12.5b. Also show the behavior of the light rays emitted from the center that would have generated the horizon if the shell had not crossed.

20. [S, A] Explicitly carry out the transformation from Schwarzschild to Kruskal coordinates defined in (12.13). Find the metric in Kruskal coordinates for both $r > 2M$ and $r < 2M$.

21. Show that in a Kruskal diagram $|dV/dU|$ must be greater than unity for a timelike particle world line even if it is moving nonradially.

22. Two observers in two rockets are hovering above a Schwarzschild black hole of mass M. They hover at a fixed radius R such that

$$\left(\frac{R}{2M} - 1\right)^{1/2} e^{R/4M} = \frac{1}{2}$$

and fixed angular position. (In fact, $R \approx 2.16M$.) The first observer leaves this position at $t = 0$ and travels into the black hole on *a straight line in a Kruskal diagram* until destroyed in the singularity at the point where the singularity crosses the line $U = 0$. The other observer continues to hover at R.

(a) On a Kruskal diagram, sketch the world lines of the two observers.

(b) Is the observer who goes into the black hole following a timelike world line?

(c) What is the latest Schwarzschild time after the first observer departs that the other observer can send a light signal that will reach the first before being destroyed in the singularity?

23. Formula (12.11) is for the redshift of light from a collapsing star as a function of the time t_R it is received by a distant stationary observer. It could have been worked out in any nonsingular coordinate system for the Schwarzschild geometry. Derive the same result using Kruskal coordinates.

24. [B, N] Construct embedding diagrams for slices of the Kruskal extension of the Schwarzschild geometry for the values $V = .9$ and $V = .999$ that are analogous to that for $V = 0$ in Box 12.4. You may exhibit the cross section of the axisymmetric two-dimensional surface if it is easier. How do these embedding diagrams support the statement that the wormhole of the Kruskal extension is not constant in time? What happens when $V > 1$?

25. [B, S] Suppose that the black hole in the center of our galaxy were really described by the maximal Kruskal extension instead of having been produced by collapsing stars. Using a Kruskal diagram, explain why it would not be possible to traverse from one asymptotic region of the Kruskal extension to the other (the question in Box 12.4). But could an observer see light from stars on the other side of the extension even if they could not travel there? If so what would they look like?

26. [B] Show that the boundaries of the Penrose diagram for the Kruskal extension of the Schwarzschild geometry are as given in Box 12.5.

27. If the area of a black hole must always increase, show that a black hole can never bifurcate into two black holes preserving total mass.

CHAPTER

13

Astrophysical Black Holes

Black holes are the outcome of unhalted gravitational collapse. Gravitational collapse to a black hole occurs on a wide range of mass scales in the universe because gravity is an attractive and universal force (Chapter 1). This chapter describes black holes of three different origins, with three different mass scales, how they have or could be identified, and sketches how they are at the heart of some of the most energetic phenomena in astrophysics.

- *Black Holes in X-Ray Binaries.* The collapse of massive stars in supernova explosions can result in black holes with masses ranging up to of order 10 solar masses. When these are members of binary systems, they can be detected by their influence on the orbit of the companion star and by the radiation from accretion disks that may form around them, as described in Section 11.2.
- *Black Holes in Galaxy Centers.* The deep gravitational potential wells at the centers of galaxies are natural sites of gravitational collapse. Galaxies may undergo collapse of their cores, endure collapse produced by merger with another galaxy, or perhaps even be formed around black holes. The resulting *supermassive black holes* at the center of galaxies range from millions to billions of solar masses.
- *Exploding Primordial Black Holes.* The evidence of the cosmic background radiation is that the distribution of matter in the early universe was remarkably smooth with tiny fluctuations in the density that seeded the formation of today's galaxies. (See Box 2.2 on p. 17 or Chapter 17.) Some early fluctuations might have grown and collapsed under the action of gravitational attraction and produced small black holes. Small *primordial black holes* with masses of order 10^{14} g ($\sim 10^{-19} M_\odot$) would be exploding today via the quantum mechanical Hawking process sketched in Section 13.3.

Black holes have been detected in X-ray binaries and in the centers of galaxies, although the goal of confirming the detailed predictions of general relativity for their geometries is still for the future. However, black holes are not interesting merely because they check general relativity. They also contribute to the explanation of frontier astrophysical phenomena. Their role in X-ray sources was described in Section 11.2; this chapter briefly describes their role in active galactic nuclei such as quasars. No exploding primordial black holes have been detected at this time, but if they are, they will shed light on the union of gravity and quantum theory.

In the short compass of a chapter it is not possible to do justice either to the challenges of the observations or the wealth of physics that underlies black-hole astrophysics. Rather, this chapter aims to introduce some basic facts and sketch some important mechanisms that link the black holes of general relativity to realistic observations.

13.1 Black Holes in X-Ray Binaries

Approximately two-thirds of all stars are members of binary pairs in which one star orbits another. In some of these binaries, a massive star may exhaust its nuclear fuel and undergo gravitational collapse, producing a supernova explosion. The explosion's remnant may be a binary system consisting of a compact end-state of stellar evolution object—a neutron star or a black hole—and a normal star. If the orbit is small enough, the normal star may shed material that falls onto the compact object. This material spirals into the deep gravitational potential of the neutron star or black hole, forming an accretion disk around it as described in Section 11.2 and illustrated in Figure 13.1.

Various dissipation mechanisms cause the orbiting material in the disk to lose energy and slowly spiral deeper into the gravitational potential well of the compact object. The released energy heats the inner regions of the disk to high enough

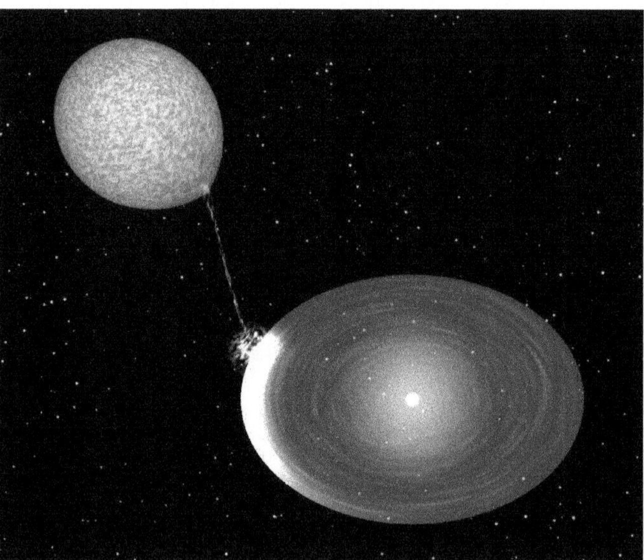

FIGURE 13.1 Computer simulated image of the X-ray binary A0620-00. One member of a binary star system has collapsed to form a black hole. The remaining star is close enough that it sheds mass onto the compact object. Conservation of angular momentum organizes the mass flowing onto the compact object into an accretion disk. The disk is heated by the dissipation of orbital energy to a temperature at which it emits X-rays.

temperatures that X-rays are produced copiously. Gravitational potential energy of the accreting matter is thus converted to X-ray luminosity (see Box 11.1 on p. 245). The result is an X-ray binary.

X-ray binaries are detectable by X-ray telescopes orbiting the Earth and are the brightest X-ray sources in the sky. Their optical counterparts are identified as a star with unusual time variation or spectrum at the location of the X-ray source within the errors of the X-ray observations. Periodic Doppler shifts in the spectral lines from the optical source may reveal it as a star in mutual orbit with a compact object that could possess an X-ray emitting accretion disk. Thus, X-ray binaries are identified.

The mass of the compact object is the chief factor determining its identification as a black hole. As mentioned in Chapter 12, the maximum mass of a neutron star is roughly several solar masses (more on this in Section 24.6). If the mass of the X-ray source is larger than this maximum mass, it is presumed to be a black hole.

Information about the masses can be obtained from the Doppler shift of spectral lines of the companion star. This measures the component of the star's velocity toward or away from us [cf. (5.73)] called its *radial velocity*. A plot of radial velocity versus time is called a *radial velocity curve*; Figure 13.2 is an example. The radial velocity curve contains much information about the mutual orbit of the binary pair, but not enough to determine the individual masses of the stars. What can be determined (see Example 13.1) is the *mass function* $f(M_1, M_2, i)$ defined

Radial Velocity Curve

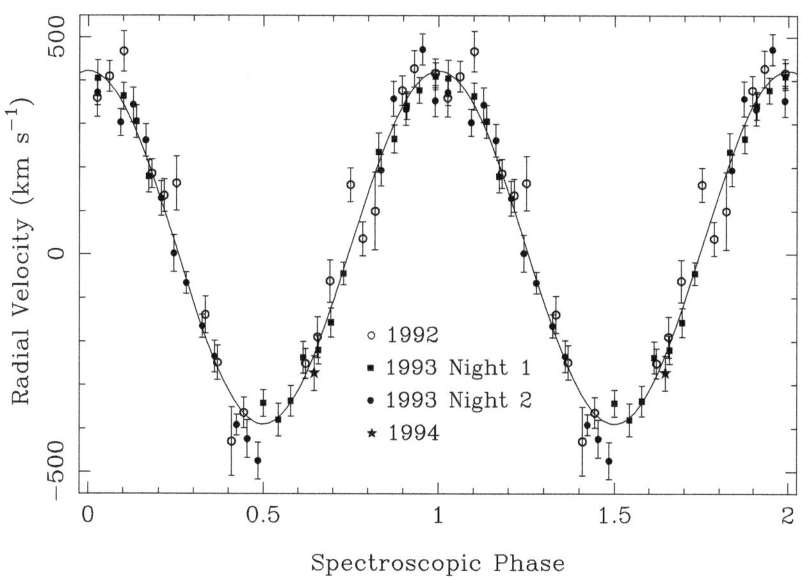

FIGURE 13.2 The radial velocity curve of the normal star orbiting the black hole in the X-ray binary Nova Muscae. The symmetric form of the curve indicates that the mutual orbit is close to circular. The period of the orbit is .423 days. Radial velocity curves like this one determine the mass function of the binary system. Its value in this case is approximately $3M_\odot$.

by

Mass Function

$$f(M_1, M_2, i) \equiv \frac{(M_2 \sin i)^3}{(M_1 + M_2)^2}. \tag{13.1}$$

Here, M_2 is the mass of the observed star, M_1 is the mass of the compact object, and i is the angle of inclination of the orbital plane to the line of sight defined so that $i = \pi/2$ corresponds to an edge-on orbit. The mass M_2 of the observed edge-on star can be estimated from its spectrum if it is a normal star. The mass function then gives a range of values for the mass M_1 of the compact object corresponding to the unknown value of sin i. A number of X-ray binary systems exist whose compact member has a mass above the upper limit for white dwarf and neutron stars. These are presumed to be black holes.

Example 13.1. The Mass Function for a Binary Star System with Circular Orbits. Two stars with masses M_1 and M_2 are in circular orbits about their common center of mass. The orbital plane makes an angle i with the line of sight, as defined before. From observations of the time variation of Doppler shifts in spectral lines of the star with mass M_1, its radial velocity can be inferred as well as the period of the orbit (see Figure 13.2). What can be said about the masses of the stars from these observations assuming Newtonian gravitational physics?

Let r_1 and r_2 be the distances of the stars from the center of mass and $r \equiv r_1 + r_2$ be their distance apart. Since $M_1 r_1 = M_2 r_2$, it follows that

$$r_1 = [M_2/(M_1 + M_2)]r, \qquad r_2 = [M_1/(M_1 + M_2)]r. \tag{13.2}$$

For a circular orbit with period P,

$$\frac{M_1 V_1^2}{r_1} = \frac{M_1}{r_1}\left(\frac{2\pi r_1}{P}\right)^2 = \frac{G M_1 M_2}{r^2}, \tag{13.3}$$

where V_1 is the orbital speed of the star with mass M_1. Using (13.2), this implies

$$(2\pi/P)^2 r^3 = G(M_1 + M_2). \tag{13.4}$$

This is Kepler's (third) law. The maximum radial velocity of the star with mass M_1 is

$$(V_1^r)_{\rm max} = V_1 \sin i = \left(\frac{2\pi G}{P}\right)^{1/3} \frac{M_2 \sin i}{(M_1 + M_2)^{2/3}}. \tag{13.5}$$

Reorganized, this is

$$f(M_1, M_2, i) = (V_1^r)^3_{\rm max}\left(\frac{P}{2\pi G}\right), \tag{13.6}$$

showing how the mass function can be determined from the observed radial velocity and period for circular orbits. For noncircular orbits it can also be determined from more details of the radial velocity curve.

13.2 Black Holes in Galaxy Centers

Observation

The centers of galaxies are natural places to look for concentrations of mass that might have arisen from the infall of gas and stars that managed to dissipate kinetic energy and fall to the bottom of the galaxy's gravitational potential well and form a black hole. There is convincing observational evidence for black holes at the center of a number of galaxies that have been studied carefully, including our own. These *supermassive black holes* range in mass from roughly $10^6 M_\odot$ to $10^9 M_\odot$. There is growing evidence that there may be a black hole at the center of almost every sufficiently massive galaxy.

A black hole in a galaxy center cannot be observed directly. First of all, it is black! Second, its size is much smaller than the characteristic dimensions of the galaxy. A typical spiral galaxy might have a spiral "bulge" with a diameter of a few kiloparsecs.[1] A $10^9 M_\odot$ black hole has a Schwarzschild radius of order 10^9 km (or 10^{-4} pc)—roughly the size of our solar system and 10 million times smaller than the dimensions of the bulge. Black holes can be observed only indirectly by their effects on surrounding matter.

The identification of a supermassive black hole in a galaxy center is the result of a mosaic of consistent observations. It is impossible here to convey even an impression of these detailed analyses. However, the basic idea is to detect a concentration of mass by its gravitational influence on visible nearby matter. The velocity $V(r)$ in a circular orbit of radius r about a spherically symmetric mass distribution is related to the mass $M(r)$ interior to r by the Newtonian relation:

$$\frac{V^2(r)}{r} = \frac{GM(r)}{r^2}. \tag{13.7}$$

By measuring $V(r)$ for a variety of r's, $M(r)$ can be determined. If the results show a large mass in a small region at the center of a galaxy, that's evidence for a black hole at its core. Even when the distribution is not spherically symmetric, (13.7) can be used to estimate masses, and Newtonian mechanics can be used to connect velocities to mass in more general situations.

A clean example is provided by the galaxy[2] NGC4258 (Figure 13.3). Radio interferometry (Section 10.3) allows astronomers to obtain detailed information about the central region of this galaxy. Water vapor is a trace element in the disk of material that surrounds the center. Condensations ("cloudlets") in the disk give rise to water maser emission at a wavelength of 1.35 cm. These masers are bright point sources that serve as test particles that can be tracked with radio interferometry, thus probing the spacetime geometry at the center of the galaxy. Their

[1] A parsec is a much-used unit of distance in galactic and extragalactic astronomy. One parsec = 3.086×10^{13} km, or 3.262 light-years.

[2] Most bright galaxies are known by their numbers in historically important catalogs. NGC refers to the *New General Catalog* published between 1895 and 1908, itself a revision of the *General Catalog* published in 1864. One of the earliest was the Messier catalog from 1781, whose numbers are still used. M31 is the Andromeda galaxy, for instance. Similarly, 3C denotes the *Third Cambridge Catalog* of radio sources. There are many more!

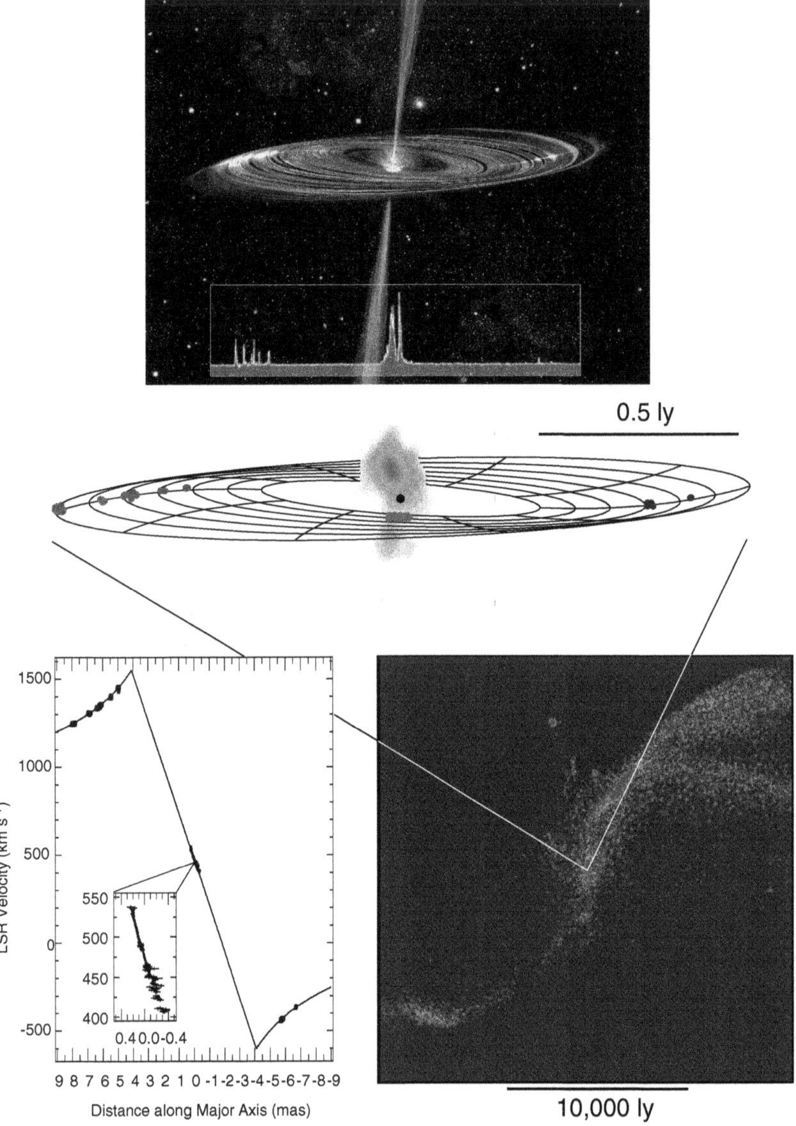

FIGURE 13.3 Evidence for the presence of a supermassive black hole in the galaxy NGC4258. The top panel is an artist's sketch of the molecular accretion disk at the center of this galaxy, which was detected by means of maser emission from trace concentrations of water vapor. Below this is the spectrum of the emission, which shows discrete features corresponding to cloudlets that have different velocities and, hence, different Doppler shifts. The middle picture shows a radio image of the very center of the disk. The small clumps are the images of the maser-emitting water clouds obtained with radio interferometry superposed on a grid representing the unseen portions of the disk. The plot in the lower left shows the radial velocities of the clouds as a function of position along the major axis of the image. The velocities trace a Keplerian profile corresponding to a central mass of $3.5 \times 10^7 M_\odot$. The deviations from Keplerian behavior are so small (less than 0.3%) that the central mass must lie almost entirely inside the inner edge of the disk at a radius of 0.13 pc. A black hole is the only known object which could have so large a mass compressed into so small a volume. The image in the lower right panel shows radio emission from the jets, which emerge along the rotation axis of the molecular accretion disk (the jets are indicated by the cones in the artist's sketch).

velocities fall accurately on the "Keplerian profile" predicted by Newtonian mechanics [cf. (13.7)] for a central mass of $3.5 \times 10^7 M_\odot$, as shown in Figure 13.3. Thus, there is good evidence for a black hole at the center of NGC4258.

There is good evidence for a more modest black hole of $3 \times 10^6 M_\odot$ at the center of our own galaxy. Observing in the infrared using adaptive optics that partially compensate for turbulence in the Earth's atmosphere, the motion of stars near the galactic center can be observed over time with exceptional accuracy. Figure 13.4 shows the observed motion of stars on the sky over a period of years. The existence of a black hole at the center of our galaxy can be inferred from these motions and Newtonian mechanics.

Only in a few galaxies can the rotation of individual objects be measured as in the two preceding examples. Even when the motion of individual objects cannot be detected, the *average* motion of stars can be measured by the consequent

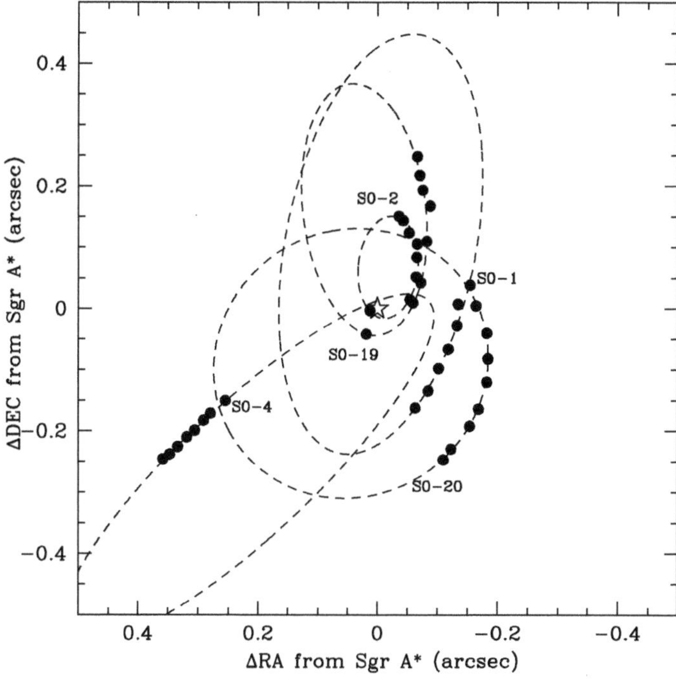

FIGURE 13.4 Evidence for a black hole at the center of our galaxy from Ghez et al (2002). Angular positions vs. time are shown for stars near the radio source Sgr A* at the center of our galaxy located by the ☆. The points are yearly averages of infrared position observations obtained with adaptive optics techniques over a 7-year period. (When a faint star is near a brighter one only one position is shown.) The Earth based polar coordinate system of right ascension (RA) and declination (DEC) is used. The projections on the sky of Keplerian ellipses fitted to the data are shown with the earliest position near the label of the orbit. (The center of attraction is not necessarily at the focus of the projected ellipse if the stars are moving radially as well.) The observed motions are consistent with a $3.7 \pm 0.2 \times 10^6 M_\odot$ black hole at the location of the radio source.

Doppler broadening of the integrated spectral lines from many stars. A mass concentration at the center increases broadening near it. This kind of evidence suggests that there may well be a black hole at the center of every sufficiently massive galaxy.

Black Holes and Active Galactic Nuclei

Certain otherwise normal looking galaxies emit intense radiation from their cores that is sometimes more luminous than all the other stars in the galaxy put together. This radiation is not starlight. There is significant emission over much too broad a spectrum of wavelengths to be the approximately black-body radiation characteristic of stars. These are galaxies with an active galactic nucleus (AGN). AGNs are a class of objects that include quasars and number among its members some of the most energetic persistent sources of radiation in the universe.

Bright AGNs can be detected at distances where the host galaxy is too faint to be seen, and most identified AGNs are in this class. AGNs are characterized by a variety of interesting features, only a few of which are mentioned here.[3]

- *High Luminosity*. The luminosity of observed AGNs ranges from $\sim 10^{42}$ erg/s to $\sim 10^{48}$ erg/s. To put this in perspective, note that a typical galaxy luminosity is $\sim 10^{44}$ erg/s. The brightest AGNs are, therefore, 10,000 times brighter than all the stars in a typical galaxy.
- *Small Size*. The size of the emitting region is estimated as the light travel distance in a time over which the source varies. Sizes of a light-month are not unusual. By contrast, the size of the visible disk of our galaxy is about 60,000 light-years.
- *Broad, Nonthermal Spectrum*. There are AGNs that emit more or less equally in the X-ray, optical, and radio bands of the electromagnetic spectrum. That is nothing like the approximately black-body spectrum of a star or even of any possible collection of stars.
- *Radio Jets*. AGNs are at the heart of radio sources possessing jets of outflowing matter extending over dimensions much larger than any galaxy, such as that shown in Figure 13.5.

What could be the tiny source of the extraordinary power emitted by AGNs and driving the extended jets of radio sources? The "best-buy" answer of contemporary astrophysics is that the source is a rotating supermassive black hole at the core of the AGN with a mass of order millions to billions of solar masses. The class of rotating black holes is discussed in more detail in Chapter 15; the Schwarzschild geometry is the limiting case of zero rotation.

There are two ways in which a rotating black hole can power an AGN. One is gravitational binding of accreting matter. This mechanism is similar to that behind X-ray sources considered in Section 11.2 but operating on much larger mass and length scales. As discussed in Box 13.1 on p. 290, gravitational binding

[3]For more detail, see Krolik (1999), for example.

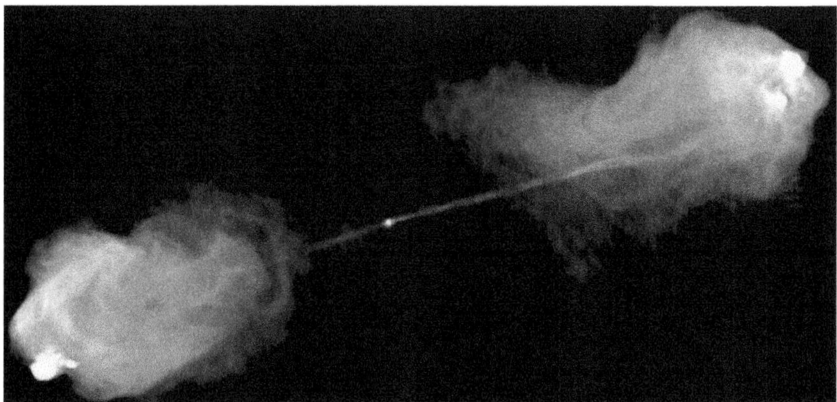

FIGURE 13.5 The radio source Cygnus A. The picture is a map at a wavelength of 22 cm of the radio source Cygnus A (3C 405)—one of the brightest radio sources on the sky and an example of a class of sources that are among the most energetic persistent sources in the universe. The double radio lobes are produced by jets of energetic particles emitted by the core object at the tiny dot between them. The lobes result when these jets are slowed down by thin intergalactic gas. The distance between the lobes is about 450,000 light-years. This is much larger than the size of a typical galaxy. The engine behind this powerful object is plausibly a rotating black hole roughly the diameter of our solar system and that is located in a galaxy midway between the lobes but not visible in the radio.

onto a compact object is in principle more efficient than thermonuclear fusion in converting rest mass to radiated energy. It is therefore a natural source of power for these energetic sources. But the electromagnetic extraction of the *rotational energy* of the black hole described in Box 15.1 on p. 326 is another important mechanism. Probably both contribute to the total luminosity of a radio source, but the extraction of rotational energy provides a natural mechanism for the jets, orienting them along the rotation axis of the black hole.

13.3 Quantum Evaporation of Black Holes—Hawking Radiation

Nothing can escape from the interior of a black hole in classical general relativity. However, when quantum mechanics is taken into account, black holes shine like a blackbody with a temperature inversely related to their mass, as discovered by Stephen Hawking in 1974. This temperature is tiny and negligible for the solar mass size and supermassive black holes discussed earlier in this chapter. But it is important for primordial black holes of much smaller mass that might have formed in the early universe and, in particular, leads to their explosive evaporation. This section presents a brief introduction to the Hawking effect restricting attention to spherical black holes. The discussion is necessarily limited because a full treatment, although not very difficult, requires the tools of quantum field theory.

BOX 13.1 Thermonuclear Fusion vs. Gravitational Binding

Why are black holes and compact relativistic stars at the heart of so many energetic phenomena in the universe? The answer is that gravitational binding is a more efficient mechanism for releasing rest energy than thermonuclear fusion—the process that powers stars and H-bombs.

At the core of the Sun, hydrogen is burning to make helium through a chain of thermonuclear reactions. The bottom line, however, is the transition

$$4\,{}^1\text{H} \rightarrow {}^4\text{He} + (26.731 \text{ MeV of released energy}). \qquad \text{(a)}$$

As powerful as this reaction is when compared to chemical binding, the energy released is only a small fraction of the rest energy involved:

$$\frac{\text{(released energy)}}{\text{(rest energy)}} = \frac{27 \text{ MeV}}{4 \times 938 \text{ MeV}} \sim 1\%. \qquad \text{(b)}$$

This is typical of any thermonuclear process [cf. Figure 12.1].

Contrast thermonuclear burning with gravitational binding. In forming a black hole accretion disk particles make a transition from an approximately free state a large distance away to a central bound orbit with lower energy. In the geometry of a Schwarzschild black hole, the energy per unit rest mass e is generally (9.21)

$$e = (1 - 2M/r)u^t. \qquad \text{(c)}$$

For a circular orbit u^t is $(1 - 3M/r)^{-1/2}$ [cf. (9.48)]. The energy per unit rest mass difference between a free particle and one bound in a circular orbit of radius r that is available for release is, therefore,

$$\frac{\text{(released energy)}}{\text{(rest energy)}} = 1 - (1 - 2M/r)(1 - 3M/r)^{-1/2}. \qquad \text{(d)}$$

For the innermost stable circular orbit at $r = 6M$, this is 6%. Gravitational binding to a Schwarzschild black hole is therefore approximately six times more efficient in releasing energy than thermonuclear fusion.

For rotating black holes, up to 42% of rest energy can be released in principle. That is why black holes and relativistic stars are at the heart of some of the most energetic phenomena in nature.

Particles and antiparticles can annihilate to produce energy in the form of radiation. Conversely, with sufficient energy, particle-antiparticle pairs can be created—in particle accelerators, for instance. Even the zero-energy vacuum of empty space exhibits quantum fluctuations in which particle-antiparticle pairs are created with tiny separations only to annihilate a tiny time later. Energy conservation is the reason such fluctuations can't persist for any significant time in flat spacetime (Problem 6). The energy of the particle and antiparticle must both be positive, but the vacuum has zero energy. The permanent creation of a particle-antiparticle pair from the vacuum in flat spacetime would violate energy conservation.

However, consider vacuum fluctuations that create a particle-antiparticle pair in the vicinity of the horizon of a black hole. There is nothing special about the geometry of a small spacetime region containing part of the horizon. Indeed, if the region is sufficiently small, its geometry is indistinguishable from that of flat spacetime (Section 7.4). In such a region, sometimes a vacuum fluctuation will create a particle outside the horizon and an antiparticle inside, as illustrated in Figure 13.6. (It could be a particle created inside and an antiparticle outside; the following discussion would be the same.) The conserved quantity in the Schwarzschild geometry that is analogous to total energy in flat space is the value of the

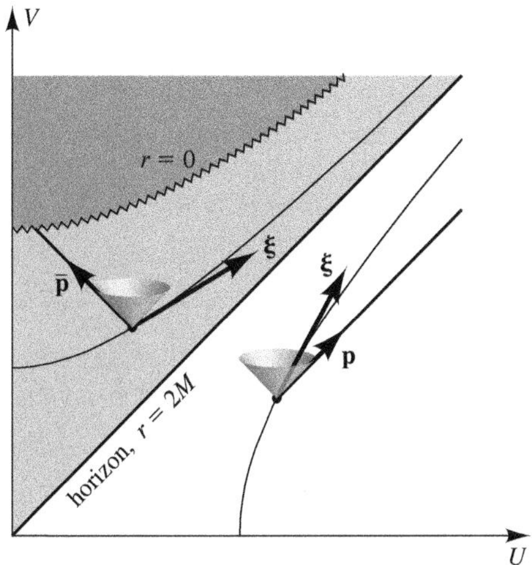

FIGURE 13.6 The Hawking Effect. This Kruskal diagram shows a rest-mass zero particle-antiparticle pair that has been created by a vacuum fluctuation in the vicinity of the horizon of a black hole. The particle and antiparticle happen to have been created on the opposite sides of the horizon with four-momenta $\mathbf{p}$ and $\bar{\mathbf{p}}$ respectively. The Killing vec-tor $\boldsymbol{\xi}$ at the location of the two is shown. The components $\mathbf{p}\cdot\boldsymbol{\xi}$ and $\bar{\mathbf{p}}\cdot\boldsymbol{\xi}$ must be equal and opposite so that the conserved quantity $(\mathbf{p}+\bar{\mathbf{p}})\cdot\boldsymbol{\xi}$ has the value of the vacuum—zero. The directions of $\boldsymbol{\xi}$ along hyperbolas of constant r [cf. (12.15)] are shown at the locations of the two particles at their place of creation. Outside the horizon $\boldsymbol{\xi}$ is timelike and $-\mathbf{p}\cdot\boldsymbol{\xi} > 0$. Inside the horizon $\boldsymbol{\xi}$ is spacelike and $-\bar{\mathbf{p}}\cdot\boldsymbol{\xi}$ *can* be negative to satisfy the conservation law. The outside particle propagates to infinity, where it is seen as Hawking radiation. The mass of the black hole is reduced by the energy lost to the escaping particle, which is the value of $-\bar{\mathbf{p}}\cdot\boldsymbol{\xi}$ for the interior particle. The directed line segments in this figure are only schematic representations of the directions of the vectors at the two points.

inner product of the total four-momentum of the created particles with the Killing vector $\boldsymbol{\xi}$ defined in Section 9.1. If $\mathbf{p}$ is the four-momentum of the particle and $\bar{\mathbf{p}}$ is that of the antiparticle, conservation requires

$$\boldsymbol{\xi}\cdot\mathbf{p}+\boldsymbol{\xi}\cdot\bar{\mathbf{p}}=0 \tag{13.8}$$

for any fluctuation from the vacuum. Since the components of $\boldsymbol{\xi}$ are $(1, 0, 0, 0)$ in Schwarzschild coordinates [cf. (9.2)]

$$\boldsymbol{\xi}\cdot\boldsymbol{\xi}=-(1-2M/r). \tag{13.9}$$

Thus, outside the horizon ($r > 2M$) $\boldsymbol{\xi}$ is timelike, but inside ($r < 2M$) it is spacelike. That difference is the origin of the Hawking radiation.

Outside the horizon $-\boldsymbol{\xi}\cdot\mathbf{p}$ must be positive because it is proportional to the energy that would be measured by an observer whose four-velocity lies along the timelike direction $\boldsymbol{\xi}$ [cf. (7.53)]. Were the antiparticle also outside the horizon, there would be a similar requirement for $-\boldsymbol{\xi}\cdot\bar{\mathbf{p}}$, and the conservation condition (13.8) could not be satisfied. But for the antiparticle inside, where $\boldsymbol{\xi}$ is spacelike, there is no such requirement. There $-\boldsymbol{\xi} \cdot \bar{\mathbf{p}}$ is not an energy for *any* observer. In fact, it is a component of spatial momentum for an observer with an associated spacelike basis vector pointing along $\boldsymbol{\xi}$. Components of spatial momentum do not have to be positive. Thus the process of pair creation from the vacuum is allowed by the conservation laws appropriate to the Schwarzschild geometry, provided the particles are created on opposite sides of the horizon. The process goes best *near* the horizon because the characteristic separations of created pairs are very small.

In the example illustrated in Figure 13.6, the particle can propagate out to infinity, where it will be seen as radiation from the black hole. The result from many such created pairs is Hawking radiation. Overall energy conservation means that the mass of the black hole, as measured at infinity, must decrease by an amount that is just the value of $-\boldsymbol{\xi} \cdot \bar{\mathbf{p}}$ for the antiparticle inside. The process works in the same way if a particle is created inside and an antiparticle outside. A black hole will radiate equal numbers of particles and antiparticles.

The Hawking process could happen anywhere along the horizon created in a gravitational collapse. However, the flux can be expected to be steady well after the collapse because the black hole geometry is time independent. So it proves on detailed analysis.

With a little dimensional analysis we can guess the steady rate dM/dt at which the black hole loses mass by Hawking radiation as determined by a stationary observer at infinity whose proper time is t. We expect the rate to be proportional to Planck's constant $\hbar$ since this is a quantum mechanical process. In geometrized units $\hbar$ is the square of the Planck length (1.6)

$$\hbar(\text{in cm}^2) = G\hbar(\text{in erg} \cdot \text{s})/c^3 \equiv \ell_{Pl}^2 = 2.62 \times 10^{-66}\ \text{cm}^2, \tag{13.10}$$

which governs all quantum gravitational phenomena.

If the particles and antiparticles radiated by the black hole have zero rest mass then the only length scale, other than that provided by $\hbar$, is the mass M of the black hole. The only combination of M and $\hbar$ that is proportional to $\hbar$ and dimensionless like dM/dt is $\hbar/M^2$. Thus, the rate at which the black hole is losing mass by Hawking radiation can be expected to be given by a formula of the form ($G = c = 1$ units):

$$\frac{dM}{dt} = -\nu \frac{\hbar}{M^2} \tag{13.11}$$

with an undetermined dimensionless constant ν. The careful quantum field theory calculation gives a rate of just this form.

The field theory calculation gives much more information beyond the overall rate (13.11). It gives the distribution of emitted particles by energy. Remarkably, all these detailed properties can be summarized in one simple fact: the black hole emits as though it were a black body with temperature

$$k_B T = \frac{\hbar}{8\pi M}, \tag{13.12}$$

where k_B is Boltzmann's constant. This expression is in geometrized ($c = G = 1$) units where $k_B T$—an energy—has dimensions of length. Putting back the G's and

c's, this is

$$k_B T = \frac{c^3 \hbar}{8\pi G M}. \tag{13.13}$$

Hawking Temperature of a Spherical Black Hole

In fact, this relation implies the rate of emission (13.11) that we guessed on purely dimensional grounds. A blackbody emits with a flux σT^4, where σ is the Stefan-Boltzmann constant $\sigma \equiv \pi^2 k_B^4 / (60\hbar^3 c^2)$. The rate at which energy is emitted across a surface with the area of the black hole's horizon, $16\pi M^2$, is just of the form (13.11) with the constant $\nu = 1/(15{,}360\pi)$. This argument neglects the effects of the geometry on the radiation as it propagates, which make the correct value of ν a little different from this, but the form (13.11) follows exactly.

Expressed in terms of a solar mass, the temperature of a black hole is

$$T = 6.2 \times 10^{-8} \left(\frac{M_\odot}{M}\right) \text{ K}. \tag{13.14}$$

The temperature of a solar mass black hole that might have formed by the gravitational collapse of a massive star is thus negligibly small. Solar mass and supermassive black holes are, in fact, gaining more energy by absorbing the 2.73 K cosmic background radiation than they are losing by the Hawking radiation—both processes being completely negligible. But the situation for primordial black holes is different.

As a black hole radiates, it loses mass, becomes hotter, radiates faster, becomes even hotter, radiates even faster, etc. The form of $M(t)$ is easily calculated from (13.11). For a black hole that evaporates completely at time t_*,

$$M(t) = [3\nu\hbar(t_* - t)]^{1/3}. \tag{13.15}$$

Therefore, the rate of emission becomes very large just before the time t_* at which the black hole evaporates completely (see Figure 13.7).

However, as Example 13.2 shows, only black holes with masses less than about 10^{14} g—about the mass of a mountain on Earth—are hot enough that they radiate a significant fraction of their mass over the age of the universe. Some primordial black holes created from collapse of density fluctuations in the early universe could have masses in this range. The detection of the explosion of one of these black holes would be a significant confirmation of quantum black-hole physics as well as information about the early universe where the primordial black holes were formed.

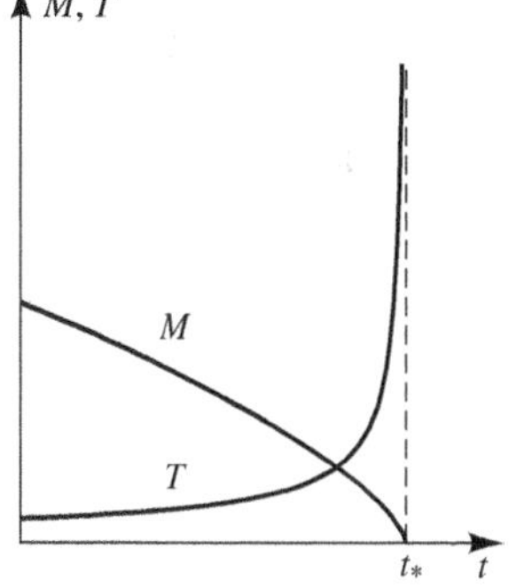

FIGURE 13.7 The evaporation of a black hole. A black hole evaporates from Hawking radiation in a finite time. As the mass decreases, the temperature rises. The black hole therefore radiates at an increasingly rapid rate as it shrinks, resulting in an explosive end.

Example 13.2. Black Hole Lifetimes. Equation (13.15) can be turned around to give an estimate[4] for the lifetime τ_{Hawk} of a spherical black hole of mass M

[4] It's an estimate because (13.11) neglects the effects on the propagation of the Hawking radiation in the geometry of the black hole, and the evolution in its final instants involve uncertain physics at very high energies.

that completely evaporates due to Hawking radiation:

$$\tau_{\text{Hawk}} \approx \frac{1}{3\nu}\frac{M^3}{\hbar} = 8.3 \times 10^{-26}\left(\frac{M}{1\text{ g}}\right)^3 \text{ s.} \tag{13.16}$$

The lifetime of a solar-mass black hole is vastly longer than the approximately 14-billion-year age of the universe. Black holes formed in the early universe with masses of around 10^{14} g would be exploding today. Smaller black holes would have evaporated earlier; larger ones will evaporate later.

Let some mass fall into a black hole and the hole's properties change. Its mass goes up, its area goes up, and its temperature goes down. Remarkably, small changes in the properties of black holes turn out to be governed by thermodynamics, obeying both its first and second laws. Within the limited realm of spherical black holes, we already have enough information to deduce the form of the first law of thermodynamics for black holes. Small changes in a black hole that do no work obey

$$dM = T\,dS_H, \tag{13.17}$$

where S_H is the entropy of the black hole. More generally, for rotating black holes, there will be additional terms in (13.17) representing work done on the hole by external torques (Problem 15.15). The entropy of a black hole can be inferred from (13.17) and the relation (13.12) for the temperature. Assuming the entropy of a zero-mass black hole is zero, we find the remarkable relation

Entropy of a Black Hole

$$\boxed{S_H = \frac{k_B}{4\hbar}A} \tag{13.18}$$

called the *Bekenstein–Hawking formula*. Entropy is area for black holes.

As discussed in Section 12.2, classically the area of a spherical black hole increases in any process that changes its mass. The identification of area with entropy (13.18) from the first law of thermodynamics is thus consistent with the *second law of thermodynamics* as well—entropy increases. Quantum mechanically, the mass of a black hole can decrease by the Hawking process. But it can decrease only at the expense of creating disordered thermal radiation in which the entropy is a maximum for a given mass. The total entropy of black hole plus emitted radiation goes up, consistent with the second law. That is the idea of how the second law of thermodynamics can be generalized to take account of black holes.

Problems

1. Figure 13.2 shows the radial velocity curve of the black-hole X-ray binary Nova Muscae. The symmetric form of the curve indicates that the mutual orbit is close to circular.

Estimate the value of the mass function for this system. The period of the orbit is .423 days.

2. *The Roche Lobe*
   (a) Consider two point masses M_1 and M_2 held at fixed positions in space a distance d apart. Sketch contour lines of constant total Newtonian gravitational potential in a plane through the axis connecting the masses. Find the position between the stars at which the Newtonian gravitational force on a test particle vanishes.
   (b) Suppose the star with mass M_1 is surrounded by a fluid envelope whose mass contributes negligibly to the gravitational potential. Explain why the boundary of the envelope must lie on an equipotential. Sketch the shape of that boundary when material from the envelope is just about to flow onto the second mass. That is the *Roche lobe*. Compare with Figure 13.1.

   Comment: In this problem the masses were imagined to be fixed in space. In a model of a binary star system they would be rotating around one another. For a harder problem work out the shape of the Roche lobe taking proper account of this rotation.

3. In the image of the radio source Cygnus A in Figure 13.5 one jet is much brighter than the other. Rotating black hole models of the source suggest that the two jets emerge in opposite directions along the rotation axis. What famous effect of special relativity could contribute to an explanation of the difference in brightness? Assuming the intensities differ by a factor of 100; and that the axis makes an angle of 45° with respect to the line of sight, what can you say about the velocity of the sources of the visible radiation in the jets?

4. [E] Figure 13.4 shows the orbits of stars around the $3 \times 10^6 M_\odot$ black hole at the center of our galaxy approximately 9 kpc (kpc = kiloparsec) away. Make a rough estimate of the predicted linear orbital velocities as a function of angular separation from the center by assuming that the stars are in circular, Newtonian orbits whose plane is perpendicular to the line of sight. How do your results compare with the velocities that can be estimated from the angular positions that are shown over several years?

5. What is the mass of a black hole formed at the beginning of the universe that would explode by the Hawking process at the time the universe becomes transparent to radiation—approximately 400,000 yr after the big bang?

6. [E] *Estimate* how long an electron-positron pair created in a vacuum fluctuation can last, assuming that the fluctuation can violate energy conservation for a time Δt consistent with the energy-time uncertainty principle $\Delta E \Delta t > \hbar$.

7. [E] *Estimate* the distance at which the energy received at Earth from an exploding primordial black hole in the last one second of its life would be comparable to that received from a nearby star in the same period. (For definiteness take the star to have the luminosity of the Sun and be 10 pc away.)

CHAPTER 14

A Little Rotation

The Schwarzschild geometry that underlies much of the physics in previous chapters is exactly spherically symmetric. It is an excellent approximation to the geometry outside a nonrotating star and is the exact geometry outside a nonrotating black hole. However, no body in nature is *exactly* nonrotating. The Sun, for example, is rotating at the equator with a period of approximately 27 days. As a consequence of the resulting centripetal acceleration, the Sun is not exactly spherically symmetric but is slightly squashed along the rotation axis. But it is not very much out of round; an equatorial diameter is less than a part in a 100,000 longer than a diameter along the rotation axis. The small value of that difference is why the Schwarzschild geometry is an excellent approximation to the curved spacetime geometry outside the Sun.

The curved spacetimes produced by rotating bodies have a richer and more complex structure than the Schwarzschild geometry, as the discussion of rotating black holes in the next chapter illustrates. But there is one limiting case that is accessible. This is the case of *slow rotation*, when the body is rotating sufficiently slowly that only deviations from the spherically symmetric Schwarzschild metric that are *first order* in the angular velocity or angular momentum are of significance. Since centripetal accelerations are *second order* in the angular momentum, the shape of the body is not rotationally distorted to *first order*. It remains spherical. Why then is there any change in the exterior geometry of spacetime? The answer is that general relativity predicts that curvature is produced, not only by the distribution of mass-energy, but also by its motion. When the curvature of spacetime is small and the velocities V of the sources are also small, these effects are typically of order V/c smaller than the GM/Rc^2 effects of the mass distribution itself. This is not unlike electromagnetism, where fields are produced not only by charge distributions but also by currents. Pursuing this analogy, these $(V/c)(GM/Rc^2)$ effects are sometimes called *gravitomagnetic*. In this chapter we explore one simple example of a gravitomagnetic effect—the dragging of inertial frames by a slowly rotating body. In this chapter the dragging is small; in the next chapter on rotating black holes it will be large.

14.1 Rotational Dragging of Inertial Frames

Box 3.2 on p. 37 described the empirical fact that the inertial frames of special relativity are not rotating with respect to the frame in which the distant matter in the universe is at rest. Were all this distant matter somehow to start rotating, the local inertial frames—those in which the plane of the Foucault pendulum described in the box does not precess—might be expected to rotate along with it. If only a small part of the matter in the universe is set into rotation, then the inertial frames might be dragged along slightly. General relativity predicts such rotational dragging of inertial frames.

Even the rotation of the Earth drags the inertial frames in its vicinity slightly. Dimensionally, at the surface of the Earth, the induced angular velocity of the inertial frames with respect to infinity, ω, might be guessed to be related to the Earth's angular velocity $\Omega_\oplus$ by

$$\omega \sim \left(\frac{GM_\oplus}{R_\oplus c^2}\right)\Omega_\oplus, \tag{14.1}$$

where $M_\oplus$ and $R_\oplus$ are the mass and radius of the Earth, respectively. Later in this chapter we will confirm this estimate, which gives

$$\omega \sim .3''/\text{yr}. \tag{14.2}$$

The inertial frames therefore rotate each year by an angle that is roughly that subtended by a football field on the Moon.[1] Even so, at the time of writing, satellite experiments are underway to detect this small effect predicted by general relativity. (See Box 14.1 on p. 305.)

A gyroscope is a natural test body with which to observe the dragging of inertial frames because the spin of a gyro points in a fixed direction in an inertial frame. A discussion of gyroscopes in curved spacetime is, therefore, an appropriate place to begin a discussion of the dragging of inertial frames.

14.2 Gyroscopes in Curved Spacetime

A small test body with spin could be called a *test gyro*, or *test spin*. Studying the behavior of the spin is another way to explore the geometry of spacetime. This section describes the equation of motion for the spin of a freely falling test gyro.[2]

A test gyro moves along a *timelike* geodesic whose four-velocity $\mathbf{u}(\tau)$ obeys the geodesic equation (8.15):

$$\frac{du^\alpha}{d\tau} + \Gamma^\alpha_{\beta\gamma} u^\beta u^\gamma = 0. \tag{14.3}$$

In addition to its four-velocity, the gyro is described by a spacelike *spin four-vector* $\mathbf{s}(\tau)$. In a local inertial frame in which the gyro is at rest, we expect the spin to be a spatial vector $s^\alpha = (0, \vec{s})$. Since the components of the gyro's four-

[1] It doesn't much matter which kind of football.

[2] Only the equation of motion of freely falling gyros moving on geodesics is considered in this chapter, for simplicity. The behavior of the spin for an accelerated gyro is more complicated.

velocity **u** in that frame are $u^\alpha = (1, \vec{0})$, we have

$$\mathbf{s} \cdot \mathbf{u} = 0. \tag{14.4}$$

More generally, the condition (14.4) holds in any frame. The total spin $s_* = (\mathbf{s} \cdot \mathbf{s})^{1/2}$ of the gyro is a constant of its motion, independent of τ (Problem 1). Its units are the units of angular momentum. As in classical mechanics, the angular motion of a constant magnitude angular momentum is called *precession.*

In flat spacetime the equation of motion for a spin simply expresses its constancy in an inertial frame:

$$\frac{ds^\alpha}{d\tau} = 0 \qquad \text{(flat or LIF).} \tag{14.5}$$

Three considerations help guess the form of the curved spacetime generalization of (14.5), which determines how the spin of a test gyro changes as it moves along a geodesic. (1) The equivalence principle suggests that the equation should reduce to (14.5) in a local inertial frame. (2) The equation should be linear in the components of the spin so that a larger spin precesses in the same way as a smaller one. (3) The equation should take the same form in all coordinate systems. The equation of motion that satisfies these three criteria is

Gyroscope Equation

$$\boxed{\frac{ds^\alpha}{d\tau} + \Gamma^\alpha_{\beta\gamma} s^\beta u^\gamma = 0.} \tag{14.6}$$

Criterion 1 is satisfied because the Γs vanish in a local inertial frame. Criterion 2 is evidently satisfied. To check criterion 3, (14.6) can be transformed to a different set of coordinates to see if its form remains the same. This is a straightforward but tedious calculation using the transformation of the metric worked out in Problem 7.7 among other relations. The more energetic or skeptical might want to check this right away. The more patient might want to wait until Chapter 20, where an elegant demonstration is given.

We'll call (14.6) the *gyroscope equation.* It specifies how the components of the spin of a test gyro change as it moves along its geodesic (Figure 14.1.) It is not difficult to check that among its predictions are that $\mathbf{s} \cdot \mathbf{s}$ and $\mathbf{s} \cdot \mathbf{u}$ are constant along the geodesic (Problem 1).

14.3 Geodetic Precession

First consider the behavior of a gyroscope in orbit around a *nonrotating* spherical body of mass M. For simplicity let's consider a circular orbit in the equatorial plane. An observer riding with the gyro will see its spin precess in the equatorial plane. In the observer's frame, where the gyro is at rest, the spin has only spatial components [cf. (14.4)], its magnitude is constant, and the symmetry under re-

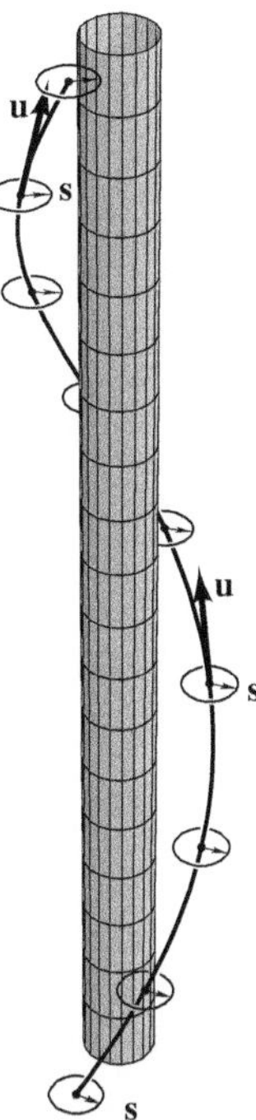

FIGURE 14.1 A gyroscope in orbit about a spherically symmetric, nonrotating body with an orbital velocity small compared to the speed of light. In this spacetime diagram, time points upward and space is horizontal. Vertical (time) distances have been compressed by a factor of five with respect to the horizontal (space) distances to get the diagram to fit on the page. The tube is the world sheet of the surface of the body about which the world line of the gyro twists. The spin **s** is perpendicular to the four-velocity of the gyro **u**, although that relationship is not so evident with the reduced scale of time. The spin remains fixed in a local inertial frame falling with the gyro but precesses with respect to infinity because of the curvature of spacetime produced by the body. This is called the geodetic precession.

flections in the equatorial plane shows that it remains in the equatorial plane if it started in it. Thus limited, precession in the plane is all the gyro can do.

Suppose at the start of an orbit the observer orients the gyro in a direction in the equatorial plane (say in the direction of a distant star). General relativity predicts that on completion of an orbit, the gyro will generally point in a different direction making an angle $\Delta\phi_{\text{geodetic}}$ with the starting one. That change in direction is called *geodetic precession* and is illustrated schematically in Figure 14.2. The value of $\Delta\phi_{\text{geodetic}}$ is straightforwardly derived, not in the orthonormal basis of a rotating observer, but in the coordinate basis where the equations of motion for the gyro are given by (14.6). In Schwarzschild coordinates the line element is the familiar (9.9):

$$ds^2 = -\left(1 - \frac{2M}{r}\right) dt^2 + \left(1 - \frac{2M}{r}\right)^{-1} dr^2 + r^2(d\theta^2 + \sin^2\theta \, d\phi^2). \qquad (14.7)$$

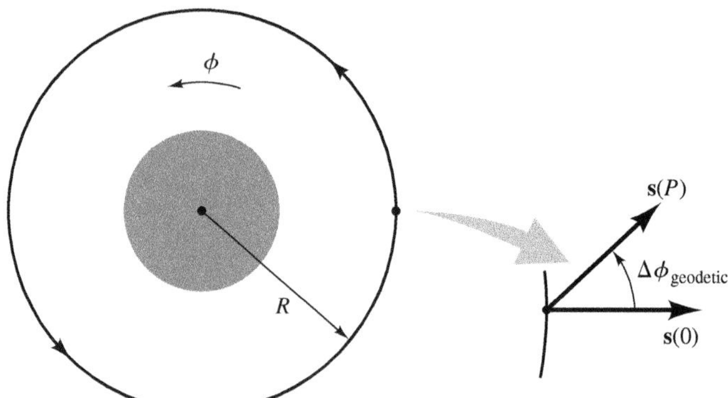

FIGURE 14.2 Geodetic precession. This is a schematic view of the equatorial plane of a *nonrotating* spherical body. A gyroscope orbits in a circle of Schwarzschild coordinate radius R. At the start of one orbit at $t = 0$, its spin is oriented in the radial direction. At the completion of one orbit, its spin has been rotated by an angle $\Delta\phi_{\text{geodetic}}$ in the direction of orbital motion.

Suppose the circular orbit has Schwarzschild coordinate radius R and lies in the equatorial plane $\theta = \pi/2$. Then the only spatial part of the four-velocity $\mathbf{u}$ is in the ϕ-direction and given by

$$u^\phi \equiv \frac{d\phi}{d\tau} = \frac{d\phi}{dt}\frac{dt}{d\tau} = \Omega u^t, \tag{14.8}$$

where $\Omega \equiv d\phi/dt$ is the orbital angular velocity. This is related to the mass M and radius R by (9.46)

$$\Omega^2 = \frac{M}{R^3}. \tag{14.9}$$

The remaining component of the four-velocity, u^t, is determined by the condition $\mathbf{u} \cdot \mathbf{u} = -1$. It isn't needed for the following calculations but is given in (9.48). The components of $\mathbf{u}$ are thus

$$\mathbf{u} = u^t(1, 0, 0, \Omega), \tag{14.10}$$

with u^t given by (9.48) and Ω by (14.9).

The gyroscope equation (14.6) with the Schwarzschild metric (14.7) and the four-velocity (14.10) predicts the evolution of the gyro's spin direction during the orbit. Suppose the spin has magnitude s_* and initially points in the r-direction in the equatorial plane. We now solve (14.6) for the four-components $(s^t, s^r, s^\theta, s^\phi)$ as functions of time given this initial condition. Two components can be disposed of immediately: Initially $s^\theta = 0$ and it remains zero because of the (north pole)–(south pole) symmetry of the Problem (Problem 2). The component s^t is related to the remaining spatial components by (14.4). Written out this is

$$\mathbf{s}\cdot\mathbf{u} = -\left(1-\frac{2M}{R}\right)s^t u^t + R^2 s^\phi u^\phi = 0. \tag{14.11}$$

Solving for s^t using (14.8) yields

$$s^t = R^2\Omega\left(1-\frac{2M}{R}\right)^{-1} s^\phi. \tag{14.12}$$

Only the components s^r and s^ϕ remain to be solved for. The gyro equation (14.6) can be organized to give two *linear* equations for s^r and s^ϕ. The Christoffel symbols for the Schwarzschild metric that are needed to write out r and ϕ components of (14.6) are straightforward to work out but can also be found in Appendix B. Using these, (14.10), and (14.12) to eliminate s^t, the r and ϕ components of the gyro equation (14.6) lead to two coupled equations for s^r and s^ϕ:

$$\frac{ds^r}{dt} - (R-3M)\Omega s^\phi = 0, \tag{14.13a}$$

$$\frac{ds^\phi}{dt} + \frac{\Omega}{R}s^r = 0. \tag{14.13b}$$

Here, τ-derivatives have been converted to t-derivatives using $u^t d\tau = (dt/d\tau)\,d\tau = dt$.

Eliminating s^r from (14.13b) using (14.13a) gives

$$\frac{d^2 s^\phi}{dt^2} + \left(1-\frac{3M}{R}\right)\Omega^2 s^\phi = 0. \tag{14.14}$$

This is the familiar equation for a simple harmonic oscillator with a frequency

$$\Omega' = \left(1-\frac{3M}{R}\right)^{1/2}\Omega. \tag{14.15}$$

The solution for $s^r(t)$ and $s^\phi(t)$ in which the spin starts out at $t=0$ pointing in the r-direction ($s^\phi(0)=0$) is

$$s^r(t) = s_*\left(1-\frac{2M}{R}\right)^{1/2}\cos\left(\Omega' t\right), \tag{14.16a}$$

$$s^\phi(t) = -s_*\left(1-\frac{2M}{R}\right)^{1/2}\left(\frac{\Omega}{\Omega' R}\right)\sin\left(\Omega' t\right), \tag{14.16b}$$

where the normalization of the solution has been chosen so that $(\mathbf{s}\cdot\mathbf{s})^{1/2} = s_*$ (Problem 3).

The spin started out at $t=0$ pointing along a unit vector $\mathbf{e}_{\hat{r}}$ in the r-direction with components $(0, (1-2M/r)^{1/2}, 0, 0)$. Let's see what angle it makes with this vector after one complete orbit in a time $P = 2\pi/\Omega$. The cosine of that angle is given by the scalar product of $\mathbf{e}_{\hat{r}}$ with a unit vector in the spin direction at time $t = P$, namely,

$$\left[\left(\frac{\mathbf{s}(t)}{s_*}\right)\cdot \mathbf{e}_{\hat{r}}\right]_{t=P} = \cos\left(2\pi \frac{\Omega'}{\Omega}\right) = \cos\left[2\pi\left(1-\frac{3M}{R}\right)^{1/2}\right]. \tag{14.17}$$

The spin, therefore, comes back after one orbit rotated by an angle

Geodetic Precession

$$\boxed{\Delta\phi_{\text{geodetic}} = 2\pi\left[1-\left(1-\frac{3M}{R}\right)^{1/2}\right] \qquad \text{(per orbit)}} \tag{14.18}$$

in the direction of motion, as illustrated in Figure 14.2.

To get the angle measured by an observer riding with the spin, we should perform a Lorentz boost to get the components of a radial direction in the observer's frame in the Schwarzschild coordinate basis. But, since the radial direction is transverse to the direction of motion, the components of a vector in the observer's radial direction coincide with those of $\mathbf{e}_{\hat{r}}$ [cf. (5.9)]. Equation (14.18), therefore, also gives the geodetic precession measured by a comoving observer.

For the small values of M/R available in the solar system, the geodetic precession is approximately (putting the Gs and c's back in)

$$\Delta\phi_{\text{geodetic}} \approx \frac{3\pi GM}{c^2R}, \qquad \frac{GM}{Rc^2} \ll 1. \tag{14.19}$$

A gyroscope in a circular orbit of radius R about the Earth comes back, rotated by

$$\Delta\phi_{\text{geodetic}} \approx 6.5\times 10^{-9}\left(\frac{R_\oplus}{R}\right)\ \text{rad}, \tag{14.20}$$

where $R_\oplus = 6378$ km is the radius of the Earth. That corresponds to a precession *rate* of $8.4''(R_\oplus/R)^{5/2}$ per year (Problem 5). That is a very small number but one the GP-B satellite hopes to measure. (See Box 14.1 on p. 305.) Indeed, calculated to leading order in $1/c$ in the PPN metric given in (10.4) and (10.6), the geodetic precession is (Problem 7)

$$\Delta\phi_{\text{geodetic}} \approx \left(\gamma+\frac{1}{2}\right)\frac{2\pi GM}{c^2R}. \tag{14.21}$$

A measurement of $\Delta\phi_{\text{geodetic}}$ is thus a determination of the PPN parameter γ and a test of the value $\gamma = 1$ predicted by general relativity.

14.4 Spacetime Outside a Slowly Rotating Spherical Body

Set a spherical body into slow and uniform rotation about one of its axes and the metric outside the body will change from the spherically symmetric Schwarzschild geometry. As mentioned at the start of this chapter, the changes arising from the rotational distortion of the body will be second order in the body's angular velocity, Ω, or, equivalently, second order in its angular momentum, J. However, in addition to these changes, general relativity also predicts a change in the metric that is *first order* in J and, therefore, more important at small J. We quote the form of this change in a system of coordinates that reduce to Schwarzschild coordinates when $J = 0$, assuming that the polar axis of the coordinate system coincides with the rotation axis:

Metric Outside a Slowly Rotating Body

$$ds^2 = ds^2_{\text{Schwarz}} - \frac{4GJ}{c^3r^2}\sin^2\theta(r\,d\phi)(c\,dt) + \begin{pmatrix}\text{terms of quadratic}\\ \text{and higher order in } J\end{pmatrix}. \tag{14.22}$$

Here, ds^2_{Schwarz} is the line element for the spherically symmetric Schwarzschild geometry in Schwarzschild coordinates (14.7), and the factors of G and c have been restored. We will derive this metric of a slowly rotating body from the Einstein equation in Section 23.3.

The dimensionless ratio GJ/c^3r^2 governs the effects of rotation. For a body rotating with angular velocity Ω, estimates based on Newtonian theory would give

$$J \sim I\Omega \sim MR^2\Omega \sim MRV \tag{14.23}$$

where M and R are the body's mass and radius, I is its moment of inertia, and V is a characteristic rotational velocity. The governing dimensionless ratio is, thus,

$$\frac{GJ}{c^3R^2} \sim \left(\frac{GM}{c^2R}\right)\left(\frac{V}{c}\right). \tag{14.24}$$

Thus, spacetime curvature outside a rotating body depends on its *velocity* as well as its mass and is one order in $1/c$ higher than $1/c^2$ effects such as the gravitational redshift. These are gravitomagnetic effects.

14.5 Gyroscopes in the Spacetime of a Slowly Rotating Body

To illustrate how the effects of rotation on the geometry of spacetime can be studied with gyroscopes, we consider the thought experiment shown schematically in Figure 14.3. A laboratory carrying a gyroscope falls freely down the rotation axis of the slowly rotating Earth. Initially the spin axis of the gyro is oriented perpendicular to the rotation axis pointing in an azimuthal direction ϕ_*.

Were the Earth not rotating, the gyro's spin axis would remain fixed as it falls—always pointing along the same azimuthal angle ϕ_*. This can be verified by solving the gyroscope equation (14.6), but it follows more immediately from the symmetry of the Schwarzschild metric under $\phi \to -\phi$. The gyro could not precess without breaking this symmetry. The geodetic precession is, therefore, zero for

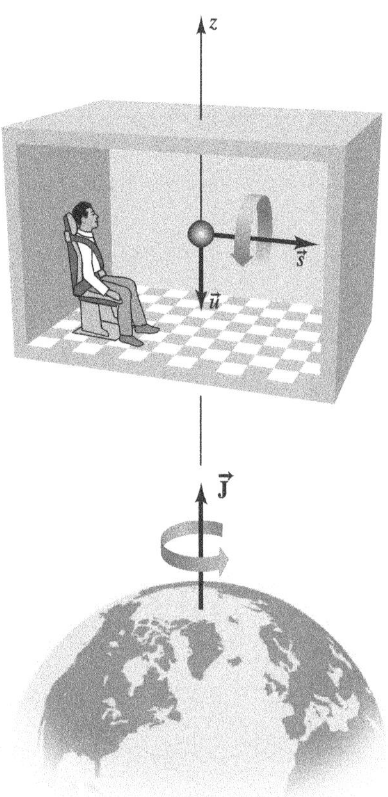

FIGURE 14.3 A thought experiment illustrating the dragging of inertial frames by a slowly rotating body. A gyroscope is freely falling along the rotation axis of the Earth which is rotating with angular momentum J. The spin of the gyro is initially perpendicular to the axis. Were the Earth not rotating ($J = 0$), the spin of the gyroscope would remain fixed with respect to the distant stars (cf. Box 3.2 on p. 37). Because of the rotation, the spin precesses in the same direction the body is rotating with an angular velocity $2GJ/c^2z^3$— the Lense-Thirring precession. The local inertial frames along the axis are thus "dragged" by the body's rotation.

this orbit. But the rotation of the Earth breaks this symmetry [cf. (14.22)] and the gyroscope precesses with time, as we now calculate.

The precession of the gyro on its downward plunge is determined by the gyroscope equation (14.6) in the metric (14.22) because it is following a geodesic. Since we expect the rate of precession to be small for the Earth [cf. (14.2)], let's solve for the precession rate to the leading order in $1/c$, which is $1/c^3$— the order of the rotational correction to the metric in (14.22). Cartesian coordinates (x, y, z) related to Schwarzschild (r, θ, ϕ) by the familiar connections $x = r\sin\theta\cos\phi$, etc. [cf. (7.2)] are convenient for the calculation because polar coordinates are singular along the axis which the gyro falls. Making use of these

BOX 14.1 Gravity Probe B

On April 20, 2004 NASA launched the Gravity Probe B (GP-B) satellite carrying an experiment to measure both the geodetic precession and the dragging of inertial frames by the rotating Earth. The satellite will carry four precision gyroscopes, one of which is shown in the figure below. The gyroscopes are spheres of fused

quartz 1.5 in. in diameter. Each gyroscope is electrostatically suspended by the saucer-shaped electrodes in the two halves of the housing. The gyros are spun up by gas entering the housing. After reaching 150 Hz the gas is pumped out and the sphere spins freely. To ensure that the gyroscopes are operating in as ideal free-fall conditions as possible, one of them is used as "proof" mass. Small thrusters controlled by feedback loops keep the case centered on the freely falling gyro, ensuring a drag-free, free-fall environment for the other three (see Box 8.1 on p. 182). The entire assembly is kept super cooled at liquid helium temperatures, enabling the direction of the gyro to be read out by a superconducting-quantum-interference-device (SQUID).

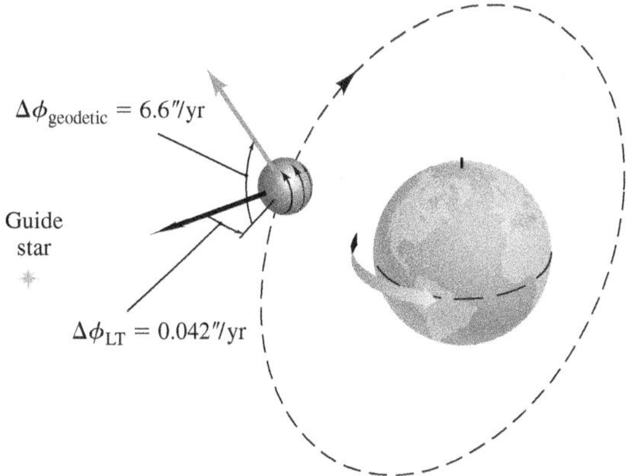

The satellite will be flown in a 640-km polar orbit, as illustrated above. A telescope enables the satellite to be locked onto a guide star. The figure shows the displacement of the gyro spin initially pointing at the guide star that is due to the geodetic and Lense-Thirring precession (both greatly exaggerated). The predicted geodetic precession of $6.6''$/yr should be testable to an accuracy better than .01%, resulting in an accurate determination of the PPN parameter γ [cf. (14.21)]. The predicted dragging of inertial frames of $.042''$/yr should be measurable to an accuracy of 1%—a conclusive demonstration of a "gravitomagnetic" effect.

transformations, the metric (14.22) becomes

$$ds^2 = (ds^2)_{\text{Sch-Cart}} - \frac{4GJ}{c^3r^2}(c\,dt)\left(\frac{x\,dy - y\,dx}{r}\right) + \left(\begin{matrix}\text{terms of quadratic}\\ \text{and higher order in } J\end{matrix}\right). \tag{14.25}$$

where $(ds^2)_{\text{Sch-Cart}}$ is the Schwarzschild metric in Cartesian coordinates. (We don't need its explicit form.) Solving the gyroscope equation to find the precession to the leading order $1/c^3$ is simplified by noting that the GM/c^2r factors in the Schwarzschild geometry can't contribute to the final answer because they would give terms such as $(GM/c^2r) \times (GJ/c^3r^2)$ to the precession rate, which are at best of order $1/c^5$. That means the calculation of the leading-order precession rate can be carried out putting $M = 0$ in (14.25) or, equivalently, replacing $(ds^2)_{\text{Sch-Cart}}$ by the flat space line element in Cartesian coordinates $(ds^2)_{\text{flat}} = -(cdt)^2 + dx^2 + dy^2 + dz^2$. That really helps with the algebra.

Writing out the gyroscope equation is further simplified by the fact that the gyro is moving along the z-axis, where the rotational perturbation to the metric vanishes, although its derivatives with respect to x and y do not. The four-velocity of the laboratory has t and z components in (t, x, y, z) coordinates. The spin we take to lie in the xy plane. Thus,

$$u^\alpha = (u^t, 0, 0, u^z), \tag{14.26}$$

$$s^\alpha = (0, s^x, s^y, 0). \tag{14.27}$$

It is not difficult to show that if the spin starts out in this plane, it remains in it (Problem 8). The condition $\mathbf{s} \cdot \mathbf{u} = 0$ is automatically satisfied on the z-axis.

Calculating in the metric (14.25) with the flat metric replacing $(ds^2)_{\text{Sch-Cart}}$ shows that, on the z-axis, the only nonvanishing Christoffel symbols that occur in the gyroscope equation (14.6) for s^x and s^y are, to leading order in $1/c$,

$$\left(\Gamma^x_{yt}\right)_{z\text{-axis}} = \frac{2GJ}{c^2 z^3}, \qquad \left(\Gamma^y_{xt}\right)_{z\text{-axis}} = -\frac{2GJ}{c^2 z^3}. \tag{14.28}$$

Writing out the equations for s^x and s^y, we find to leading order in $1/c$ (using $u^t \equiv dt/d\tau$ to convert the τ-derivatives to t-derivatives)

$$\frac{ds^x}{dt} = -\frac{2GJ}{c^2 z^3} s^y, \tag{14.29a}$$

$$\frac{ds^y}{dt} = +\frac{2GJ}{c^2 z^3} s^x. \tag{14.29b}$$

These equations describe a gyroscope that precesses with respect to the coordinate axes (x, y, z) in the same direction as the Earth is rotating. This is called the Lense-Thirring precession. At a distance from the center z the instantaneous rate is

$$\Omega_{LT} = \frac{2GJ}{c^2 z^3}. \tag{14.30}$$

This precession is calculated in a frame in which the center of the body is at rest and the gyroscope is falling. But exactly the same precession would be observed by an observer in the freely falling laboratory. That is because the Lorentz boost along the z-axis that connects the two frames does not affect the transverse component of the spin s^x and s^y and because to leading order in $1/c$ there is no effect of time dilation.

We can estimate the magnitude of the Lense-Thirring precession due to the Earth by noting that

$$J_\oplus = I_\oplus \Omega_\oplus \equiv M_\oplus \mathfrak{R}^2_\oplus \Omega_\oplus \tag{14.31}$$

where $I_\oplus$ is the Earth's moment of inertia, conveniently summarized by its radius of gyration $\mathfrak{R}_\oplus$. For the Earth $\mathfrak{R}_\oplus/R_\oplus \approx .576$. Thus,

$$\Omega_{LT} = 2\left(\frac{GM_\oplus}{c^2 R_\oplus}\right)\left(\frac{\mathfrak{R}_\oplus}{R_\oplus}\right)^2\left(\frac{R_\oplus}{z}\right)^3 \Omega_\oplus. \tag{14.32}$$

Plugging in the numbers from the endpapers (plus the obvious $\Omega_\oplus = 2\pi/24$ h), we find

$$\Omega_{LT} = .22''\left(\frac{6378\text{ km}}{z}\right)^3 / \text{ yr}. \tag{14.33}$$

This is a small effect indeed, but one the GP-B satellite experiment (see Box 14.1 on p. 305) expects to measure to an accuracy of 1%.

A gyroscope such as that of the GP-B satellite in a realistic orbit about the Earth would experience both the geodetic and the Lense-Thirring precession. The Lense-Thirring precession depends on the latitude as

$$\boxed{\vec{\Omega}_{LT} = \frac{G}{c^2 r^3}\left[3\left(\vec{J}\cdot\vec{e}_r\right)\vec{e}_r - \vec{J}\right]} \tag{14.34}$$

Lense-Thirring Precession

which agrees with (14.30) when $\vec{r}$ points along $\vec{J}$. For those who have studied electromagnetism, this is a characteristic dipole pattern with $\vec{J}$ playing the role of the dipole moment.

Example 14.1. Measuring the Angular Momentum of a Rotating Body. In Section 9.1 we saw how the mass of a body can be determined from the behavior of the orbit of a test particle a long way away because of the asymptotic properties of the Schwarzschild geometry. In a similar way the angular momentum of a steadily rotating axisymmetric body can be determined from the behavior of gyroscopes a long way away. If the body is slowly rotating, a measurement of the Lense-Thirring precession of the spin of three suitably oriented gyros in suitable orbits give the three components of the angular momentum $\vec{J}$ from (14.34). However, the same measurements give the angular momentum of a *rapidly* rotating body if it is axisymmetric and rotating steadily, provided the gyros are sufficiently far away. That is because rotational effects are a small correction to the metric of flat space for some sufficiently large r no matter how big $\vec{J}$ is. Therefore, the metric (14.22) is an excellent approximation to the asymptotic form a long way away from any steadily rotating axisymmetric body (as will be demonstrated from the Einstein equation in Section 23.3). Indeed, just like mass, angular momentum can be *defined* in terms of the asymptotic properties of spacetime geometry. That is important for understanding the mass and angular momentum of black holes, which are solutions to the Einstein equation with no matter at all.

In a realistic experiment such as GP-B, the precession of a gyroscope is not measured with respect to a system of coordinate axes (x, y, z). Rather, the angle between the gyro spin and the light from a distant "guide" star is monitored as a function of time. (See the figure in Box 14.1 on p. 305) The predicted time dependence of this angle involves a number of other effects besides the Lense-Thirring precession, such as the aberration of the light from the guide star described in Problem 5.16, the bending of light by the Earth, and even—in principle—its bending by the rotation of the Earth (Problem 9). Those are not considered here, but the end result is that the (x, y, z) axes can be thought of as tied to the distant matter of the universe and the spin precesses with respect to that matter at the Lense-Thirring rate.

14.6 Gyros and Freely Falling Frames

An observer in the freely falling laboratory of the thought experiment described previously could use gyroscopes to construct a freely falling frame, as described in Section 8.4. The observer orients the spins of three gyros along three mutually orthogonal spatial directions in the lab. The three spins remain orthogonal as the laboratory falls (Problem 1). Three unit vectors along their directions $\mathbf{e}_{\hat{1}}(\tau)$, $\mathbf{e}_{\hat{2}}(\tau)$, and $\mathbf{e}_{\hat{3}}(\tau)$, together with the four-velocity of the laboratory $\mathbf{u}(\tau) \equiv \mathbf{e}_{\hat{0}}(\tau)$, constitute an orthonormal basis for each point along the geodesic. These vectors are the *coordinate* basis vectors for a system of coordinates in which the Christoffel symbols vanish all along the geodesic. They are thus the coordinate basis vectors of a freely falling frame. A supplement to this chapter on the book website leads you through an explicit construction of this coordinate frame as well as demonstration that the Christoffel symbols vanish. As a consequence of the rotation of the Earth, the three basis vectors along the gyro spins precess at the Lense-Thirring angular velocity (14.34) with respect to the coordinates (x, y, z), which are tied to matter at infinity. It is in this sense that inertial frames are dragged along by a rotating body.

Problems

1. Show that the gyroscope equation (14.6) implies $\mathbf{s} \cdot \mathbf{s}$ and $\mathbf{s} \cdot \mathbf{u}$ are constant along the geodesic followed by a gyro. Show that for any two gyros A and B moving along the same geodesic, $\mathbf{s}_A \cdot \mathbf{s}_B$ is constant.

2. Check explicitly from the gyroscope equation (14.6) that, if the spatial part of the spin initially points in the equatorial plane (so that $s^\theta = 0$), it remains pointing in the equatorial plane for a circular orbit lying in that plane.

3. [S] Check that for the solution to the gyroscope equation given in (14.16), the magnitude of the spin, $(\mathbf{s} \cdot \mathbf{s})^{1/2}$, remains constant in time and equal to the s_* specified by the solution.

4. [A] **(a)** Consider the gyroscope in circular orbit about a nonrotating body discussed in Section 14.3. Find the coordinate components of the orthonormal basis $\mathbf{e}_{\hat{\alpha}}$ of

an observer who is moving with the spin and keeps the spatial parts of the two spacelike basis vectors $\mathbf{e}_{\hat{1}}$ and $\mathbf{e}_{\hat{3}}$ pointing along the r and ϕ directions, respectively.

(b) Project the spin vector of (14.16) onto the orthonormal basis constructed in (a) to obtain

$$s^{\hat{1}} = s_* \cos\left(\Omega' t\right), \qquad s^{\hat{3}} = -s_* \sin\left(\Omega' t\right),$$

showing very explicitly how the spin precesses in a comoving frame.

5. [S] Substitute the numbers in (14.19) to evaluate the total geodetic precession of a gyroscope in orbit around the Earth and the rate of geodetic precession.

6. [S] What is the largest possible geodetic precession for a stable circular orbit in the Schwarzschild geometry?

7. Work through the derivation of geodetic precession again using the PPN metric given in (10.4) and (10.6). Show that

$$\Delta\phi_{\text{geodetic}} \approx \left(\gamma + \frac{1}{2}\right) 2\pi \left(\frac{GM}{Rc^2}\right),$$

so that a measurement of the geodetic precession is another way to determine the PPN parameter γ.

8. Consider the thought experiment described in Section 14.5 concerning a gyro freely falling along the rotation axis of a slowly rotating body. Show from the gyroscope equation (14.6) that if the spatial part of the spin starts out in the x-y plane it remains in the x-y plane to leading $1/c^3$ order.

9. [C] General relativity predicts that, because the Sun is rotating, a light ray passing by will be deflected slightly by an amount additional to the deflection of light in the Schwarzschild geometry considered in Section 9.4. Calculate the amount and direction of this deflection to lowest nonvanishing order in $1/c$ assuming that the orbit is in the equatorial plane perpendicular to the axis of rotation. Estimate the magnitude of this effect for the Sun. Is it an important correction to the results of the observations discussed in Section 10.3? (*Hint*: Before doing any algebra, think about what terms in the metric will contribute to the final answer in leading order in $1/c$.)

10. [B] The figure in Box 14.1 on p. 305 shows schematically the shift of the spin of a gyro due to the geodetic precession and frame dragging effects after one orbit around the rotating Earth. Explain the directions of the shifts of the gyro and calculate the magnitude of the two effects using (14.34) for the Lense-Thirring part of the precession.

CHAPTER

15 Rotating Black Holes

The Schwarzschild black holes discussed in Chapter 12 are not the most general black hole spacetimes predicted by general relativity. They are simple objects—exactly spherically symmetric and characterized by a single parameter, the total mass M. Remarkably, the most general stationary black-hole solutions of the vacuum (no matter) Einstein equation are not much more complicated. They are described by the family of geometries discovered by Roy Kerr in 1963 and called *Kerr black holes*. Members of the family depend on just two parameters—the total mass M and total angular momentum J. Kerr black holes are the rotating generalizations of the Schwarzschild black hole. This chapter gives an elementary introduction to their properties.

15.1 Cosmic Censorship

The treatment of gravitational collapse in Chapter 12 assumed exact spherical symmetry, greatly simplifying the discussion. The Schwarzschild geometry is the unique spherically symmetric solution of the vacuum Einstein equation. Exactly spherically symmetric collapse therefore proceeds by revealing more and more of the Schwarzschild geometry as the radius of the collapsing body contracts, no matter what its internal constitution. Once the Schwarzschild radius is crossed, the horizon is formed, which shields the inevitable singularity from observers at infinity.

Realistic gravitational collapse is not spherically symmetric. The analysis of nonspherical collapse is a complex question that typically can be addressed only by numerical simulation of the Einstein equation. Gravitational radiation from the time-dependent collapsing mass distribution is just one of the issues that has to be addressed (Chapter 23). Yet the evidence of both theoretical investigation and numerical simulation is that the endstate of *any* realistic gravitational collapse that proceeds far enough is remarkably simple, analogous in many ways to the special case of spherical collapse. From the perspective of an observer who collapses with the matter, the end is inevitably a singularity. From the perspective of a distant observer, the endstate is indistinguishable from a time-independent Kerr black hole characterized by just a mass M and angular momentum J, with a horizon that conceals the singularity within it.

At the time of writing, there is no rigorous proof from the Einstein equation that a generic gravitational collapse that proceeds far enough inevitably forms a black hole, concealing singularities from observers outside. Rather, this is a conjecture called the *cosmic censorship conjecture* that was discussed briefly in Section 12.4. The cosmic censorship conjecture is supported by various pieces of theoretical evidence too detailed to go into in this text. We will assume it holds for collapse of the kinds of matter that occur in realistic astrophysical situations.

Cosmic Censorship

To appreciate the predictive power of cosmic censorship, imagine two neutron stars in mutual orbit about one another, spiraling ever closer because of energy lost to gravitational radiation[1] and eventually merging to form a black hole. The initial state is described by a great number of parameters—the masses of the stars, their orbital size, period, and eccentricity, their rotational periods, the compositions of their interiors, the configurations of their atmospheres, the geography of the tiny mountains on their surfaces, etc. The whole range of the classical and quantum physics of matter from ordinary densities to beyond that of nuclear matter is necessary to understand this system in detail, as we will see in Chapter 24. By contrast, the final Kerr black hole that is formed in the merger is characterized by just two parameters—mass and angular momentum—and its external properties can be understood from classical gravitational physics alone. Whatever postcollapse physics transpires near the resulting black hole—the behavior of an accretion disk for instance—it happens in one of the family of Kerr geometries. Kerr black holes thus provide the cleanest connection between fundamental gravitational physics and realistic astrophysics.

15.2 The Kerr Geometry

The spacetime around a rotating black hole with mass M and angular momentum J can be summarized by the line element ($c = G = 1$ units)

$$
\begin{aligned}
ds^2 = &-\left(1 - \frac{2Mr}{\rho^2}\right) dt^2 - \frac{4Mar\sin^2\theta}{\rho^2} d\phi\, dt + \frac{\rho^2}{\Delta} dr^2 + \rho^2\, d\theta^2 \\
&+ \left(r^2 + a^2 + \frac{2Mra^2\sin^2\theta}{\rho^2}\right)\sin^2\theta\, d\phi^2,
\end{aligned}
\tag{15.1}
$$

Kerr Metric of a Rotating Black Hole

where

$$a \equiv J/M, \qquad \rho^2 \equiv r^2 + a^2\cos^2\theta, \qquad \Delta \equiv r^2 - 2Mr + a^2. \tag{15.2}$$

The (t, r, θ, ϕ) coordinates used here are called *Boyer-Lindquist coordinates* and are analogous to the Schwarzschild coordinates for a nonrotating black hole in ways that will become clearer shortly. The parameter a is called the *Kerr parameter*. It has the dimensions of length in geometrized units, just like the mass.

[1] As described in detail in Section 23.7.

The metric (15.1) is a solution of the vacuum Einstein equation. A number of the important properties of the Kerr geometry follow immediately from this line element.

- *Asymptotically Flat.* For $r \gg M$ and $r \gg a$, the line element becomes

$$ds^2 \approx -\left(1 - \frac{2M}{r}\right) dt^2 + \left(1 + \frac{2M}{r}\right) dr^2$$
$$+ r^2(d\theta^2 + \sin^2\theta \, d\phi^2) - \frac{4Ma}{r^2} \sin^2\theta (r \, d\phi) dt + \cdots , \tag{15.3}$$

  where just the leading terms in each metric coefficient as r becomes large and the $1/r$ corrections to that behavior (if any) have been retained. This asymptotic form establishes that the Kerr geometry approaches the geometry of flat spacetime far from the black hole. As discussed in Section 9.1, the total mass causing this spacetime curvature could be determined from the orbit of a distant satellite. Similarly, as discussed in the previous chapter, the angular momentum could be determined from the precession of a distant orbiting gyroscope. Comparison of the metric (15.3) with (14.22) confirms the identification of M with the total mass and J with the total angular momentum.
- *Stationary, Axisymmetric.* The metric (15.1) is independent of t (stationary) and independent of ϕ (axisymmetric). The two Killing vectors that correspond to these symmetries are $\boldsymbol{\xi}$ and $\boldsymbol{\eta}$:

$$\xi^\alpha = (1, 0, 0, 0), \qquad \text{(stationary)}, \tag{15.4}$$
$$\eta^\alpha = (0, 0, 0, 1), \qquad \text{(axisymmetric)}, \tag{15.5}$$

  where the components are given in their usual order (t, r, θ, ϕ). In addition, the Kerr metric is unchanged by a reflection in the equatorial plane $\theta = \pi/2$, which sends θ into $\pi - \theta$. These are all symmetries to be expected of the geometry of a rotating body. However, as is also to be expected, the geometry is not spherically symmetric. The explicit dependence of g_{tt} and g_{rr} on θ is enough to show that.
- *Schwarzschild When Not Rotating.* When $a = 0$, the metric (15.1) reduces to the Schwarzschild metric in Schwarzschild coordinates (9.9). The Kerr family thus includes the Schwarzschild black hole in the special case of zero angular momentum.
- *Coordinate Singularities, Real Singularities and Horizon.* The Kerr metric (15.1) is singular when ρ or Δ vanishes. The singularity at $\rho = 0$, which happens when $r = 0$ and $\theta = \pi/2$, is a real singularity—a place of infinite spacetime curvature. It is the generalization of the real curvature singularity in the Schwarzschild geometry at zero value of the Schwarzschild coordinate r, with which it coincides when $a = 0$.

  The quantity Δ vanishes at the radii

$$r_\pm = M \pm \sqrt{M^2 - a^2}, \tag{15.6}$$

assuming $a \leq M$. Like the singularity in Schwarzschild coordinates at the Schwarzschild radius, the singularities in the metric (15.1) at these two radii turn out to be coordinate singularities. Indeed, the radius r_+ coincides with the radius $2M$ of the coordinate singularity in the Schwarzschild metric when $a = 0$. By working through Problem 3, you can transform (15.1) to new coordinates where the metric is not singular at these radii. The radius r_+ turns out to be the horizon that makes the Kerr metric a black hole. We'll show that in the next section, but one property of the horizon can be noted immediately: The singularities where $\rho = 0$ are safely inside the horizon. The Kerr geometry displays a rich and interesting structure inside the horizon for $r < r_+$ (for example, at $r = r_-$), but our strategy will be to focus exclusively on the properties outside of the horizon, which are the ones important for astrophysics.[2]

Not all values of M and a correspond to a Kerr black hole. The radius of the horizon r_+ exists only for $a \leq M$ [cf. (15.6)]. The angular momentum J of a black hole is, therefore, limited by its mass squared. Ordinary bodies like the Sun are not subject to this limitation (Problem 1). Black holes with the limiting value $a = M$ ($J = M^2$) are called *extreme Kerr black holes*. They are important in astrophysics for the following reason: Matter falling onto a black hole forms an accretion disk, as discussed in Section 11.2. The matter that falls into the black hole after spiraling down through the accretion disk to the innermost stable circular orbit and then plunging into the black hole carries angular momentum with it—increasing total J closer and closer to the extreme limit $J = M^2$. Detailed study of the accretion of hot radiating matter shows a is limited to about $.998M$, but that is very close to the extremal limit. Near extreme rotating black holes thus develop naturally in many astrophysical situations.

The energy released by gravitational binding of the accreting matter to nearly extreme Kerr black holes makes objects such as the X-ray sources and active galactic nuclei discussed in Chapter 13 some of the most powerful energy sources in the universe. (See Box 13.1 on p. 290.) The detectable effects of this released energy is one of the most important ways that black holes can be identified. We will return to this in Section 15.4.

15.3 The Horizon of a Rotating Black Hole

The horizon of a black hole is the null three-surface interior boundary of the spacetime region from which a light ray can escape to infinity from any point. The horizon bounds the region from which a distant observer can receive information in principle. The Schwarzschild black hole, discussed in Chapter 12, provides the simplest example. There the horizon is the null three-surface at $r = 2M$. There is a light ray from any point outside of that radius that will take a signal to a distant observer. No light ray escapes from any point inside $r = 2M$.

[2]For the inside see, for example, Hawking and Ellis (1973).

The boundary of the region from which light can escape is the three-surface generated by those light rays that neither escape to infinity nor fall into the interior. The horizon, therefore, is a *null* three-surface—one in which at each point there is one null tangent direction that is orthogonal to two independent spacelike tangent directions. (Recall the discussion on p. 162.) The horizon of a stationary, axisymmetric black hole can, therefore, be expected to be a stationary, axisymmetric, null, three-surface generated by those light rays that hover between collapse into the interior and escape to infinity.

The three-surface $r = r_+$ is a stationary, axisymmetric, null, three-surface of the Kerr geometry, which is, in fact, the horizon of the Kerr family of black holes. To show that $r = r_+$ is a null surface, consider its tangent vectors $\mathbf{t}$. The t-, θ-, and ϕ-directions are tangent to a surface of constant r. The general tangent vector could have components in any of these three directions but will have no component in the r-direction:

$$t^\alpha = (t^t, 0, t^\theta, t^\phi). \tag{15.7}$$

The surface is null if, at each point, one null tangent vector $\boldsymbol{\ell}$ can be found along with two orthogonal spacelike tangent vectors (Section 7.9). From the form of the Kerr metric in Boyer-Lindquist coordinates (15.1), the condition $\boldsymbol{\ell} \cdot \boldsymbol{\ell} = 0$ for a vector of the form (15.7) reads

$$\boldsymbol{\ell} \cdot \boldsymbol{\ell} = g_{tt}(\ell^t)^2 + 2g_{t\phi}\ell^t\ell^\phi + g_{\phi\phi}(\ell^\phi)^2 + g_{\theta\theta}(\ell^\theta)^2 = 0, \tag{15.8}$$

all evaluated at $r = r_+$. After a little algebra this becomes

$$\left(\frac{2Mr_+ \sin\theta}{\rho_+}\right)^2 \left(\ell^\phi - \frac{a}{2Mr_+}\ell^t\right)^2 + \rho_+^2 \left(\ell^\theta\right)^2 = 0, \tag{15.9}$$

where ρ_+ denotes $\rho(r_+, \theta)$. The only solution to (15.9) is $\ell^\theta = 0$ and $\ell^\phi = (a/2Mr_+)\ell^t$. Up to a multiplicative constant, the unique null vector in the $r = r_+$ three-surface is

$$\ell^\alpha = (1, 0, 0, \Omega_H), \tag{15.10}$$

where Ω_H is defined by

Angular Velocity of a Rotating Black Hole

$$\boxed{\Omega_H = \frac{a}{2Mr_+}.} \tag{15.11}$$

It is not difficult to complete the argument that $r = r_+$ is a null surface by finding two spacelike tangent directions that are orthogonal to $\boldsymbol{\ell}$ and to each other. You can easily check that the directions (0, 0, 1, 0) and (0, 0, 0, 1) will do the job (Problem 6). The reason this is so easy is obvious in retrospect. If $r = r_+$ is a null surface, then $\boldsymbol{\ell}$ is also its normal—automatically orthogonal to every other tangent vector. It is thus necessary only to exhibit two orthogonal vectors of the form (15.7) that are independent of $\boldsymbol{\ell}$ to conclude that $r = r_+$ is a null surface.

The null directions $\boldsymbol{\ell}$ are tangent vectors of the light rays that form the horizon. It is not too difficult to show from the geodesic equation that light rays that start in these directions remain stuck on the horizon at $r = r_+$ (Problem 5). These horizon-generating light rays are rotating with respect to infinity, as the nonzero value of ℓ^ϕ in (15.10) makes clear. Their angular velocity is $d\phi/dt = (d\phi/d\lambda)/(dt/d\lambda) = \ell^\phi/\ell^t$. This is Ω_H given in (15.11). You may have wondered what is rotating in a rotating black hole. After all, it is just empty spacetime. It is the light rays forming the horizon that are rotating with angular velocity $\Omega_H = a/(2Mr_+)$, as shown in Figure 15.1. The angular velocity Ω_H is, therefore, called the angular velocity of the black hole. Further, inertial frames are dragged with the rotation, as discussed in Section 14.6. Indeed, the angular momentum of the black hole could be determined by measuring this dragging.

Even though the Kerr horizon has a constant Boyer-Lindquist coordinate radius r_+, its intrinsic geometry is not spherically symmetric. Putting $r = r_+$ in the Kerr line element (15.1) and taking a $t =$ const. slice yields a two-dimensional surface with the line element

$$d\Sigma^2 = \rho_+^2(\theta)\, d\theta^2 + \left(\frac{2Mr_+}{\rho_+(\theta)}\right)^2 \sin^2\theta\, d\phi^2. \tag{15.12}$$

This is not the geometry on a sphere. For instance, the distance around the equator, $\theta = \pi/2$, is $4\pi M$. But the distance around the poles is less—$7.6M$ in the case

FIGURE 15.1 A spacetime diagram of the equatorial plane ($\theta = \pi/2$) of an extreme ($a = M$) rotating black hole. Boyer-Lindquist coordinates (t, r, ϕ) are used as cylindrical coordinates in this plot, with t running vertically. The horizon is a cylinder at the radius $r = r_+ = M$ that extends in the t-direction. The light rays that generate this null surface (heavier lines) rotate around it with angular velocity $\Omega_H = 1/(2M)$.

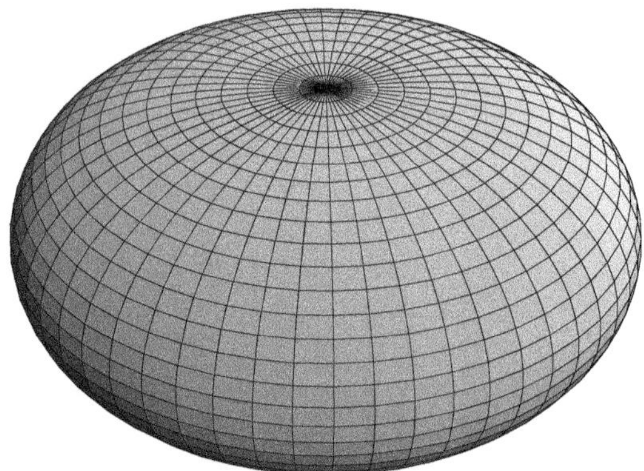

FIGURE 15.2 The horizon of a rotating black hole. The figure shows a surface in three-dimensional flat space that has the same intrinsic geometry as a t = const. slice of the horizon of a Kerr black hole [cf. (15.12)]. The value of a/M is .86—approximately the maximum value for which such a slice is embeddable in flat space as an axisymmetric surface. Lines of constant ϕ (longitude) meeting at a pole are shown, as well as lines of constant θ (colatitude). The surface is characteristically squashed along the rotation axis in a way that is roughly analogous to the distortion of a ball of fluid when it rotates.

of an extreme black hole (Problem 7). Figure 15.2 shows a two-surface in flat space that has the same geometry as a t = constant slice of the horizon. This kind of squashed spherical shape is qualitatively what might be expected for a rotating body. The area of the horizon is easily calculated from (15.12) and is

Horizon Area

$$A = 8\pi M r_+ = 8\pi M \left(M + \sqrt{M^2 - a^2}\right). \tag{15.13}$$

The Kerr horizon at $r = r_+$ is a "one-way" surface like the forward light cone in flat space (Section 7.9) and the horizon of a Schwarzschild black hole (Section 12.1). Particles can cross it once, but never again. As in the Schwarzschild geometry, particles and light rays can cross from the outside in but not from the inside out. Hence, no information about events inside the horizon can reach infinity. All light rays originating there are confined inside. The Kerr geometry is a black hole.

15.4 Orbits in the Equatorial Plane

The orbits of test particles and light rays in the Kerr geometry are remarkable both for the complex behaviors they can exhibit and the extent to which these can be treated by analytical techniques. For instance, a general orbit will not be confined to a "plane." Orbits in the Schwarzschild geometry stay in a plane because of the conservation of test particle angular momentum, itself a consequence of that geometry's spherical symmetry. (Recall the discussion on p. 193.) But the Kerr geometry is not spherically symmetric, only axisymmetric. Only the component of angular momentum along the symmetry axis is conserved. There are orbits confined to the equatorial plane ($\theta = \pi/2$), but the general orbit will not lie in a plane. However, to give a manageable introductory account of Kerr metric orbits, this text will confine attention to the simple case of orbits in the equatorial plane.[3] In particular, the discussion will be directed at calculating the binding energy of the innermost stable circular orbit. This is the central number for understanding why up to 42% of the rest energy of the inflowing matter can be released in radiation during accretion.

The analysis of the orbits in the equatorial plane proceeds in exactly the same way as for the orbits in the Schwarzschild geometry discussed in Chapter 9. Only the algebra is more complicated. First note that the symmetry of the Kerr geometry with respect to reflections in the equatorial plane ($\theta \to \pi - \theta$) implies that there *are* orbits in the equatorial plane $\theta = \pi/2$ with a zero θ-component of their four-velocity $\mathbf{u}$. These are governed by the Kerr metric (15.1) restricted to $\theta = \pi/2$:

$$ds^2 = -\left(1 - \frac{2M}{r}\right)dt^2 - \frac{4aM}{r}dt\,d\phi + \frac{r^2}{\Delta}dr^2 + \left(r^2 + a^2 + \frac{2Ma^2}{r}\right)d\phi^2. \tag{15.14}$$

Orbits are parametrized by the conserved energy per unit mass, e, and the angular momentum per unit mass along the symmetry axis, ℓ, arising from the t-independence and ϕ-independence of the Kerr metric, respectively. In terms of the Killing vectors (15.4) and (15.5) associated with these symmetries, these conserved quantities are[4]

Conserved Energy and Angular Momentum per Unit Rest Mass

$$e \equiv -\boldsymbol{\xi} \cdot \mathbf{u}, \tag{15.15a}$$

$$\ell \equiv \boldsymbol{\eta} \cdot \mathbf{u}. \tag{15.15b}$$

The interpretation of these quantities as the energy and angular momentum per unit rest mass arises from evaluating them at infinity as in the Schwarzschild case in Section 9.3. The conserved angular momentum along the symmetry axis, ℓ, is also the total angular momentum for equatorial orbits. As always, there is one additional general integral of the geodesic equation that follows from the normalization of the four-velocity

$$\mathbf{u} \cdot \mathbf{u} = -1. \tag{15.16}$$

Inspection of the Kerr metric (15.1) shows that e and ℓ are linear combinations of u^t and u^ϕ:

$$-e = g_{tt}u^t + g_{t\phi}u^\phi, \tag{15.17a}$$

$$\ell = g_{\phi t}u^t + g_{\phi\phi}u^\phi. \tag{15.17b}$$

[3] For the general case, see especially Chandrasekhar (1983).

[4] Don't get ℓ the conserved quantity mixed up with $\boldsymbol{\ell}$ the null generator of the horizon. Unfortunately, both notations are standard. It shouldn't be possible to mistake ℓ for the length of $\boldsymbol{\ell}$ because that is zero!

These equations can be solved for u^t and u^ϕ to find

$$\frac{dt}{d\tau} = \frac{1}{\Delta}\left[\left(r^2 + a^2 + \frac{2Ma^2}{r}\right)e - \frac{2Ma}{r}\ell\right], \tag{15.18a}$$

$$\frac{d\phi}{d\tau} = \frac{1}{\Delta}\left[\left(1 - \frac{2M}{r}\right)\ell + \frac{2Ma}{r}e\right]. \tag{15.18b}$$

In turn, these relations can be substituted into $\mathbf{u} \cdot \mathbf{u} = -1$ together with $u^\theta = 0$ to yield a radial equation for $dr/d\tau$. This can be written in the same form as (9.26) for the Schwarzschild metric

$$\frac{e^2 - 1}{2} = \frac{1}{2}\left(\frac{dr}{d\tau}\right)^2 + V_{\text{eff}}(r, e, \ell), \tag{15.19}$$

where the effective potential governing radial motion is

Effective Potential for Equatorial Particle Orbits

$$\boxed{V_{\text{eff}}(r, e, \ell) = -\frac{M}{r} + \frac{\ell^2 - a^2(e^2 - 1)}{2r^2} - \frac{M(\ell - ae)^2}{r^3}.} \tag{15.20}$$

The radial equation for light rays follows similarly from (15.18) and $\mathbf{u} \cdot \mathbf{u} = 0$. Its form depends on the impact parameter $b \equiv |\ell/e|$, as in the discussion of Schwarzschild photon orbits in Section 9.4, but also on whether the orbit is going with the rotation of the black hole (corotating) or against it (counterrotating). That is determined by the sign of ℓ and conveniently summarized by a parameter $\sigma \equiv \text{sign}(\ell)$, which is just that sign. The radial equation is

$$\frac{1}{\ell^2}\left(\frac{dr}{d\lambda}\right)^2 = \frac{1}{b^2} - W_{\text{eff}}(r, b, \sigma), \tag{15.21}$$

where the photon effective potential $W_{\text{eff}}(r, b, \sigma)$ is

Effective Potential for Equatorial Light Ray Orbits

$$W_{\text{eff}}(r, b, \sigma) = \frac{1}{r^2}\left[1 - \left(\frac{a}{b}\right)^2 - \frac{2M}{r}\left(1 - \sigma\frac{a}{b}\right)^2\right]. \tag{15.22}$$

The effective potentials (15.20) and (15.22) have the same three inverse r-dependences as those for the Schwarzschild geometry, (9.28) and (9.65), to which they reduce when $a = 0$. An important difference is that the potentials are energy and angular momentum dependent. For example, particles or light rays that fall from infinity rotating in the same direction as the black hole (positive values of ℓ or σ) move in a different effective potential than initially counterrotating particles (negative values of ℓ or σ). These differences reflect, in part, the rotational frame dragging of the spinning black hole. As the following examples show, particles are dragged around by its rotation.

Example 15.1. The Orbit of a Radially Infalling Particle. A particle falls into a Kerr black hole from infinity, initially moving radially ($\ell = 0$) with zero kinetic energy ($e = 1$). The shape of the orbit $\phi(r)$ can be calculated by integrating $d\phi/dr$, which can be found from (15.18b) and (15.19):

$$\frac{d\phi}{dr} = \frac{d\phi/d\tau}{dr/d\tau} = -\frac{2Ma}{r\Delta}\left[\frac{2M}{r}\left(1+\frac{a^2}{r^2}\right)\right]^{-1/2}. \tag{15.23}$$

(Remember $dr/d\tau$ is negative.) The angle $\Delta\phi$ swept out in falling to radius r can be found by integrating this from ∞ to r.

Example 15.2. Mad Scientist Seeks to Destroy Black Hole and Violate Cosmic Censorship. Kerr black holes are restricted to have values of a less than M. The Kerr metric (15.1) is a solution of the vacuum Einstein equation for $a > M$ but doesn't represent a black hole because there is no horizon [cf. (15.6)]. The Kerr geometry's singularity is visible from infinity for $a > M$—an example of a *naked singularity*. Cosmic censorship would be violated if gravitational collapse produced a Kerr solution with $a > M$.

A mad scientist seeks to destroy a Kerr black hole and violate cosmic censorship by letting a particle with $\ell > 2Me$ fall from infinity into an extremal ($a = M$) Kerr black hole in its equatorial plane. He reasons that, for a particle of rest mass m, the mass M of the hole will increase by $\delta M = me$ and the angular momentum will increase by $\delta J = m\ell$. The change in $a = J/M$ will be $\delta a = (m/M)(\ell - ae)$, which will be greater than the change in the mass if $\ell > 2Me$.

But the particle won't fall into the black hole if the maximum height of the effective potential in (15.20) is greater than $(e^2 - 1)/2$ [cf. (15.19)]. Rather, the particle will execute a scattering orbit and return to infinity in a way analogous to a scattering orbit in the Schwarzschild geometry illustrated in Figure 9.4.

Finding the maximum of the effective potential is a straightforward but messy calculation from (15.20) with $a = M$. The marginal case $\ell = 2Me$ is especially simple. Then the maximum is at $r = M$ and the maximum value of V_{eff} is $(e^2 - 1)/2$—just high enough to prevent the particle from getting inside the horizon and destroying the black hole. Plotting the potential in a few other cases will convince you that no particle with $\ell > 2Me$ will ever fall in.

Examples such as this build confidence in the validity of the cosmic censorship conjecture.

Many interesting properties of the orbits of particles and light rays in the equatorial plane could be explored with the radial equations (15.19) and (15.21) and the equations (15.18) for the other components of the four-velocity. We could calculate the radii of circular orbits, the radii of unstable circular photon orbits, the deflection of light, the shape of bound orbits, etc. These are all different, depending upon whether the particle or light ray is rotating with the black hole (corotating) or in the opposite direction (counterrotating). For instance, in the geometry of an extremal Kerr black hole ($a = M$), there is a corotating unstable circular photon orbit at $r = M$ and a counterrotating unstable circular orbit at $r = 4M$ (Problem 11). However, as we already mentioned, for an introductory discussion it seems appropriate not to catalog all these interesting properties but rather to focus on the one property most important for astrophysics—the binding energy of the innermost stable circular particle orbit.

For a particle to describe a circular orbit at radius $r = R$, its initial radial velocity must vanish. From (15.19) that is the condition

$$\frac{e^2 - 1}{2} = V_{\text{eff}}(R, e, \ell). \tag{15.24a}$$

But to stay on a circular orbit the radial acceleration must also vanish. Differentiating (15.19) with respect to τ leads to the condition

$$\left.\frac{\partial V_{\text{eff}}(r, e, \ell)}{\partial r}\right|_{r=R} = 0. \tag{15.24b}$$

Stable orbits are ones for which small radial displacements away from R oscillate about it rather than accelerate away from it. Just as in Newtonian mechanics, that is the condition that the effective potential must be a minimum:

$$\left.\frac{\partial^2 V_{\text{eff}}(r, e, \ell)}{\partial r^2}\right|_{r=R} > 0. \tag{15.24c}$$

Equations (15.24) determine the ranges of e, ℓ, and R allowed for stable circular orbits in the Kerr geometry. At the *innermost* stable circular orbit (ISCO)—the one just on the verge of being unstable—(15.24c) becomes an equality. The three equations (15.24) can then be solved for the values of e, ℓ, and $R \equiv r_{\text{ISCO}}$ that

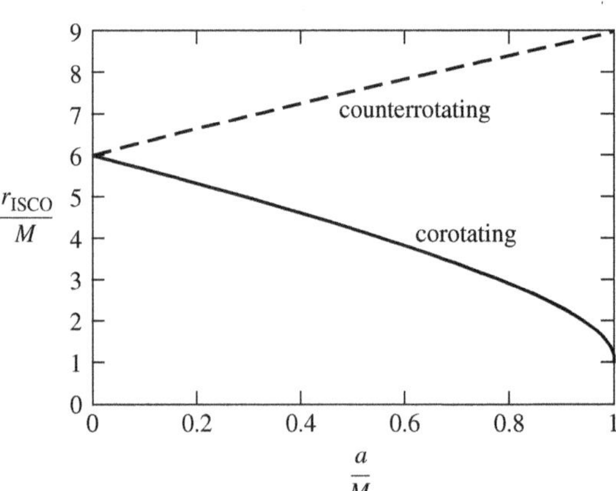

FIGURE 15.3 The Boyer-Lindquist radius r_{ISCO} of the innermost stable circular orbit in the equatorial plane of a rotating black hole. The solid line gives the radius of the innermost corotating orbit; the dashed line gives the radius of the innermost counterrotating orbit. Both coincide with the Schwarzschild value $r_{\text{ISCO}} = 6M$ at zero rotation $a/M = 0$ [cf. (9.43)]. At the extreme limit $a/M = 1$, the radius is $r_{\text{ISCO}} = M$ for corotating orbits and $r_{\text{ISCO}} = 9M$ for counterrotating orbits.

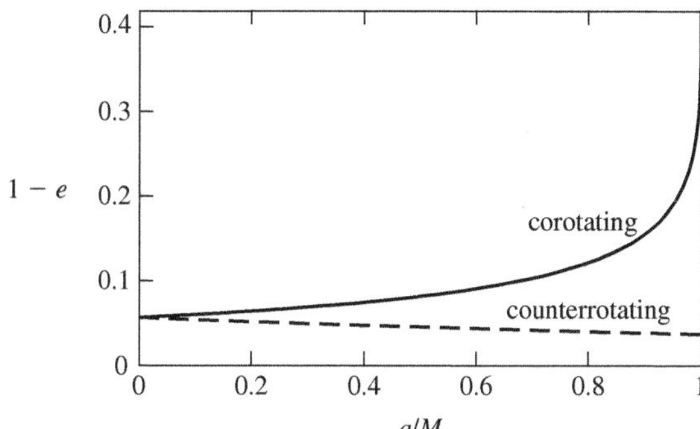

FIGURE 15.4 The binding energy per unit rest mass $1 - e$ of the innermost stable circular orbit in the equatorial plane of a rotating black hole. The solid line corresponds to corotating orbits, the dashed one to counterrotating orbits. For an extreme $a/M = 1$ Kerr black hole, the maximum fractional binding energy rises to 42%. For realistic black holes spun up by accretion to the value $a/M = .998$, the maximum fractional binding energy is approximately 30%. That makes gravitational binding much more efficient than any thermonuclear process for releasing energy.

characterize this orbit. The solution of the three algebraic equations can be carried out analytically. However, it is more instructive to present the results graphically, as in Figure 15.3 for the radius and Figure 15.4 for the binding energy as functions of a/M. The case of an extremal black hole is especially simple. For example, the parameters of the marginally stable corotating circular orbit are

$$e = \frac{1}{\sqrt{3}}, \quad \ell = \frac{2M}{\sqrt{3}}, \quad r_{\text{ISCO}} = M \quad \begin{pmatrix} \text{innermost stable corotating} \\ \text{circular orbit for } a = M \end{pmatrix}. \tag{15.25}$$

The innermost stable counterrotating circular orbit is further out as Figure 15.3 shows.

The binding energy of any orbit is the difference between the energy of a particle at rest at infinity (including rest energy) and the energy of the same particle moving in that orbit as measured from infinity. Since e is the energy measured from infinity per unit rest mass, the binding energy per unit rest mass is $1 - e$. This is the fraction of rest energy that can be released in the process of gravitational binding. Figure 15.4 shows the binding energies for the innermost stable circular orbits in the Kerr geometry found by solving the three equations (15.24) when (15.24c) is an equality. The most bound orbit is the innermost stable corotating orbit whose e was given in (15.25). The fraction of rest energy that can be released in making a transition from an unbound orbit far from an extremal black hole to the most bound innermost stable circular orbit is $(1 - 1/\sqrt{3}) \approx 42\%$. Realistic astrophysical black holes have slightly smaller values of a. This reduces the binding energy significantly because the curve of $1 - e$ vs. a/M is steep near $a = M$, as Figure 15.4 shows. But it is still much more efficient than the typical few percent from thermonuclear burning. (See Box 13.1 on p. 290.)

15.5 The Ergosphere

Stationary Observers

Perhaps nothing illustrates the effect of rotation on spacetime geometry more graphically than the plight of observers who wish to remain stationary with respect to infinity, allowing no spatial coordinate r, θ or ϕ of their world lines to change with time. These observers must use rocket power or some other source of thrust to remain on stationary world lines, because in the absence of such forces they would fall into the black hole. An observer equipped with an arbitrarily large amount of rocket power can hover arbitrarily close to the horizon of a Schwarzschild black hole. But for a rotating black hole there is a limit to how close to the horizon a stationary observer can get, as this section will show.

The four-velocity of a stationary observer $\mathbf{u}_{\text{obs}}$ has only a time component $u^t_{\text{obs}} = dt/d\tau$:

$$u^\alpha_{\text{obs}} = (u^t_{\text{obs}}, 0, 0, 0). \tag{15.26}$$

This four-velocity is a unit *timelike* vector $\mathbf{u}_{\text{obs}} \cdot \mathbf{u}_{\text{obs}} = -1$. Writing this condition out using (15.26) and (15.1) gives

$$\mathbf{u}_{\text{obs}} \cdot \mathbf{u}_{\text{obs}} = g_{tt}(u^t_{\text{obs}})^2 = -\left(1 - \frac{2Mr}{\rho^2}\right)(u^t_{\text{obs}})^2 = -1. \tag{15.27}$$

Where this condition can be satisfied, it determines u^t_{obs}. But sufficiently close to the horizon of a Kerr black hole, it cannot be satisfied at all. That is because g_{tt} vanishes on a surface

Ergosphere Outer Boundary

$$\boxed{r = r_e(\theta) = M + \sqrt{M^2 - a^2 \cos^2\theta}.} \tag{15.28}$$

and is positive inside it. Inside this surface no stationary observers with timelike four-velocities of the form (15.26) are possible.

Evidently $r_e(\theta) > r_+$, so the surface $r_e(\theta)$ lies outside the horizon, as shown in Figure 15.5. The region between this surface and the horizon is called the *ergosphere* of the black hole for reasons that will become clear. When $a = 0$, r_e coincides with r_+. This is the correct result for the Schwarzschild geometry—no stationary observers are possible inside the horizon at $r = 2M$, where $g_{tt} > 0$. Rotation has allowed this region forbidden to stationary observers to extend outside the horizon.

Even if no amount of rocket power will permit an observer to remain at fixed r, θ, and ϕ inside the ergosphere, it is possible to remain at fixed r and θ by rotating with respect to infinity in the same direction as the black hole. Such an observer would have a four-velocity of the form

$$u^\alpha_{\text{obs}} = u^t_{\text{obs}}(1, 0, 0, \Omega_{\text{obs}}), \tag{15.29}$$

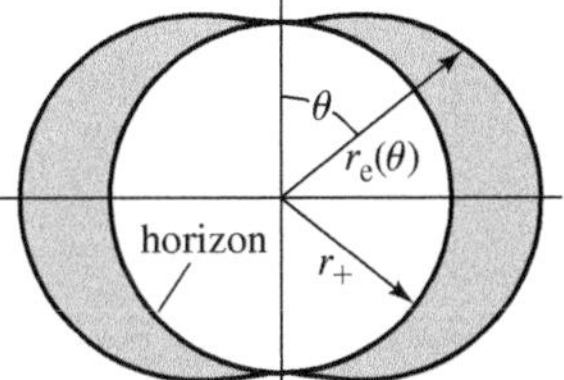

FIGURE 15.5 The ergosphere. This is a plot of the location of the horizon $r = r_+$ and the ergosphere boundary $r = r_e(\theta)$ using r and θ as polar coordinates on a flat plane for $a/M = .95$. The rotation axis of the black hole runs vertically. (This is not an embedding diagram; the horizon, for instance, is not spherical as it appears in this plot [cf. Figure 15.2].) The ergosphere is the shaded region in between these two surfaces. Inside the ergosphere no observer can remain at rest with respect to infinity.

that is, with $\mathbf{u}_{\text{obs}} = u^t_{\text{obs}}(\boldsymbol{\xi} + \Omega_{\text{obs}}\boldsymbol{\eta})$. But for each r and θ inside the ergosphere, there is a minimum as well as a maximum angular velocity for which $\mathbf{u}$'s of the form (15.29) will be timelike (Problem 14).

Extracting Rotational Energy

Classically, it is not possible to get energy out of a Schwarzschild black hole, but it is possible to extract rotational energy from a Kerr black hole. The electromagnetic coupling of a black hole to an exterior environment that was mentioned in Section 13.2 and is described in Box 15.1 on p. 326 is a realistic way of doing this. However, a simple thought experiment shows how the ergosphere could be, in principle, exploited to extract rotational energy in another way. Consider the hypothetical situation shown schematically in Figure 15.6 called a *Penrose process*. A particle (in) starts at infinity and falls into the ergosphere of a Kerr black hole. There it decays into two particles, (out) and (bh). Particle (bh) falls down through the horizon, but particle (out) escapes to infinity. It's possible to arrange the decay so that escaping particle (out) carries more energy away to infinity than particle (in) carried in, thus extracting energy from the black hole.

Energy-momentum must be preserved in the decay. (It is a local process, analyzable in a freely falling frame where physics is locally indistinguishable from flat space.) Thus,

$$\mathbf{p}_{\text{in}} = \mathbf{p}_{\text{out}} + \mathbf{p}_{\text{bh}} \tag{15.30}$$

at the point of decay. The energy of a particle of rest mass m_{out} that reaches infinity is $E_{\text{out}} = -\mathbf{p}_{\text{out}} \cdot \boldsymbol{\xi} = m_{\text{out}}e$, where $\boldsymbol{\xi}$ is the Killing vector (15.4) [cf. (7.53)]. Taking the scalar product of (15.30) with the Killing vector $\boldsymbol{\xi}$ gives, for E_{out},

$$E_{\text{out}} = E_{\text{in}} - E_{\text{bh}}. \tag{15.31}$$

Were particle (bh) to reach infinity, its energy would be E_{bh}—necessarily positive. Equation (15.31) would then require $E_{\text{out}} < E_{\text{in}}$—less energy out than in. But

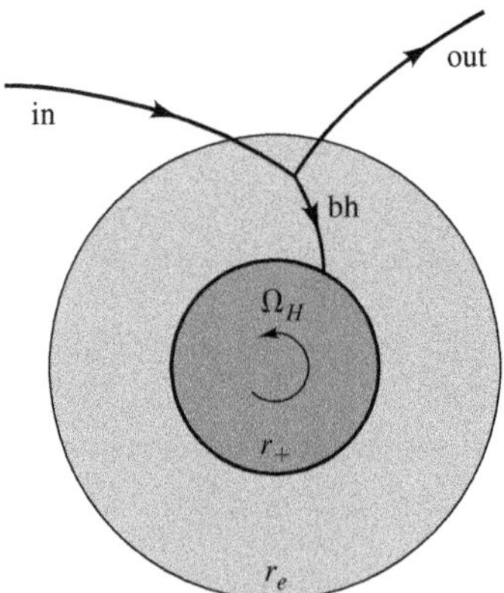

FIGURE 15.6 A Penrose process. This is a schematic view of the equatorial plane of a rotating black hole. The inner circle is the event horizon at $r = r_+$. The outer circle is the boundary of the ergosphere at $r_e = 2M$. The region in between is the ergosphere. In a Penrose process, a particle (in) falls into the ergosphere from infinity and decays into two particles (out) and (bh). Particle (out) escapes to infinity, and particle (bh) plunges through the horizon into the black hole. It's possible to choose the momenta of the three particles so that energy-momentum is conserved in the decay and the particle (out) emerges with more energy than particle (in) carried in. The rotational energy of the black hole is correspondingly reduced.

(bh) never gets outside the ergosphere where $\boldsymbol{\xi}$ is a spacelike vector:

$$\boldsymbol{\xi} \cdot \boldsymbol{\xi} = g_{tt} > 0, \qquad \text{(inside the ergosphere).} \tag{15.32}$$

The quantity $-E_{\text{bh}}$ is, therefore, not an energy locally, but rather a component of spatial momentum—possibly positive, possibly negative. For decays where $E_{\text{bh}} < 0$, $E_{\text{out}} > E_{\text{in}}$, and energy will be extracted from the black hole.[5]

When the Penrose process extracts energy from the black hole, the negative E_{bh} of the infalling particle reduces the total mass of the black hole. But as we will see shortly, the infalling particle also reduces the angular momentum of the black hole. That is the sense in which the Penrose process extracts the rotational energy of the black hole.

To see that the infalling particle reduces the black hole's angular momentum, consider corotating observers inside the ergosphere with four-velocities (15.29) related to the Killing vectors $\boldsymbol{\xi}$ and $\boldsymbol{\eta}$ by

$$\mathbf{u}_{\text{obs}} = u^t_{\text{obs}}(\boldsymbol{\xi} + \Omega_{\text{obs}}\boldsymbol{\eta}). \tag{15.33}$$

[5]Does this analysis sound familiar? It is similar to that used in Section 13.3 to explain the Hawking radiation from a Schwarzschild black hole. There the Killing vector $\boldsymbol{\xi}$ corresponding to t-translation symmetry is timelike outside the horizon and spacelike inside. One member of a pair created in a vacuum fluctuation is inside the horizon with $-\boldsymbol{\xi} \cdot \mathbf{p} < 0$ which is allowed because $\boldsymbol{\xi}$ is spacelike there. The outside partner with $-\boldsymbol{\xi} \cdot \mathbf{p} > 0$ carries positive energy away to infinity. There is Hawking radiation from rotating black holes as well. See Problem 15.

Like all observers these must measure a positive energy of the particle that falls into the black hole. Thus [cf. (7.53)],

$$-(\boldsymbol{\xi} + \Omega_{obs}\boldsymbol{\eta}) \cdot \mathbf{p}_{bh} \geq 0 \tag{15.34}$$

for any of the range of values of Ω allowed by the condition $\mathbf{u} \cdot \mathbf{u} = -1$. Hence,

$$E_{bh} \geq \Omega_{obs} L_{bh}, \tag{15.35}$$

where $L_{bh} = m_{bh}\ell_{bh}$ is the angular momentum of the particle that falls onto the black hole, m_{bh} being its rest mass. The allowed values of Ω_{obs} are positive—the rotating observers rotate in the same direction as the black hole (Problem 14). To extract energy, E_{bh} must be negative, implying L_{bh} is also negative. This negative L_{bh} of the infalling particle thus *reduces* the angular momentum of the black hole. Rotational energy can be extracted in this way until the angular momentum of the black hole is reduced to zero.

The mass and angular momentum of a black hole are reduced in a Penrose process that extracts energy from it. But the area of the black hole's event horizon always increases or remains constant. Example 15.3 shows how this works for Penrose processes, but the result is general. The black-hole area-increase theorem of general relativity shows that classically any interaction between a black hole and physically reasonable matter can only increase its area or leave it unchanged.

Example 15.3. Area Increase in the Penrose Process. The Penrose process can reduce the mass of a rotating black hole. Can it also reduce its area? The particle (bh) that falls into the black hole changes the hole's mass by $\Delta M = E_{bh}$ and its angular momentum by $\Delta J = L_{bh}$. The consequent change in the area ΔA can be worked out from (15.6) and (15.13):

$$\Delta A = \frac{8\pi}{\kappa}(\Delta M - \Omega_H \Delta J), \tag{15.36}$$

where Ω_H is given by (15.11) and κ is

$$\kappa \equiv \frac{(M^2 - a^2)^{1/2}}{2Mr_+}. \tag{15.37}$$

Relation (15.35) shows that $\Delta M > \Omega_{obs}\Delta J$ for any angular velocity Ω_{obs} allowed an observer at constant r inside the ergosphere. Observers near the horizon that almost move like the null generators of the horizon have the largest angular velocities (Problem 14). The limiting value is Ω_H. Therefore, $\Delta M > \Omega_H \Delta J$, and the area of a black hole is always *increased* by any Penrose process.

The area-increase theorem can be used to gain a more precise understanding of the rotational energy of a black hole. Define the *irreducible mass* M_{ir} of a rotating black hole in terms of the area of its horizon by

$$M_{ir} \equiv \left(\frac{A}{16\pi}\right)^{1/2}. \tag{15.38}$$

BOX 15.1 Tapping the Rotational Energy of a Black Hole

The active galactic nuclei (AGNs) described in Section 13.2 include some of the most energetic persistent sources of radiation in the universe. Nearly extreme Kerr black holes are the probable engines driving AGNs. The gravitational binding that accompanies accretion is one source of energy (Box 13.1). The rotational energy of the black hole exhibited in (15.39) is another. This box describes very qualitatively by a few dimensional arguments the *Blandford-Znajek mechanism* by which rotational energy can be extracted electromagnetically from a rotating black hole.[a] We begin with the simple example of the unipolar (homopolar) generator in electromagnetism.

Consider a cylindrical conductor of radius r_C rotating with an angular velocity Ω about its axis and immersed in a uniform magnetic field $\vec{B}$ pointing along that axis, as shown.

The rotation produces a force on the charge carriers in the conductor located at a distance ρ from the axis given by

$$\vec{F}(\rho) = q(\vec{V} \times \vec{B}) = q(\Omega\rho)B\vec{e}_{\hat{\rho}}, \tag{a}$$

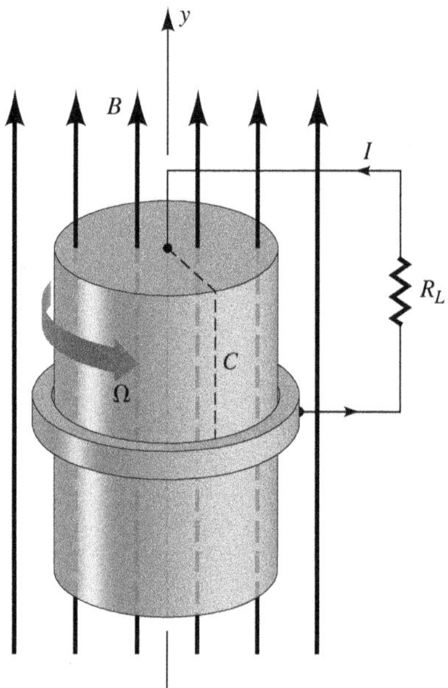

where q is the charge and $\vec{e}_{\hat{\rho}}$ is a unit vector pointing radially away from the axis. A voltage $V = q^{-1} \int_C \vec{F} \cdot d\vec{S}$ will develop across the two stationary contacts shown, where the line integral is over any path C in the conductor connecting them.[b] For the path shown in the figure, the only contribution to the voltage drop is from the axis to the radius r_C:

$$V = \int_0^{r_C} d\rho \rho\Omega B = \frac{1}{2}\Omega B r_C^2 = \frac{\Omega F_B}{2\pi}, \tag{b}$$

where F_B is the magnetic flux threading the conductor. (The magnetic field from currents induced in the conductor is neglected.) The last form for the voltage holds for any axisymmetric shape, such as a sphere (Problem 17).

Suppose the contacts are connected by wires to an external resistance (the load). A current will flow, and power will be dissipated in the load. The rotating conductor thus acts as an electric generator. The current is $I = V/(R_C + R_L)$, where R_L is the resistance of the load and R_C is the resistance of the conductor. The power supplied to the load P_L is maximized when $R_L = R_C$—an example of impedance matching. Under these conditions, the maximum power delivered to the load is

$$(P_L)_{\max} = I^2 R_C = \frac{V^2}{4R_C} = \frac{\Omega^2 F_B^2}{16\pi^2 R_C}. \tag{c}$$

Black holes have electromagnetic properties that are analogous in many ways to ordinary conductors. These follow from Maxwell's equations in black-hole spacetimes. A little qualitative discussion supports the assertion: if an observer drops an electric charge into a black hole, subsequent observers can detect the charge by the long range electric field that develops outside the black hole. At very large r that field is radial and falls off like $1/r^2$, in accord with Coulomb's law. From the point of view of distant observers, the dropped charge never crosses the horizon (Section 12.2) but remains there, forming a surface charge distribution analogous to that of a conductor. *Black holes can therefore carry electric charge.*

Suppose positive charges fall into a black hole at a steady rate from one side and negative charges fall into it at the same rate from the other side. The net charge of

[a] For more details see, for example, Thorne et al. (1986).

[b] In this box $\vec{v}$ is for velocity, V is for voltage, ρ is for distance from the axis, and R is for resistance. Don't mix these up with other uses of the same symbols elsewhere in the text.

the black hole remains zero, but a net charge has been transferred from one side to the other. From the point of view of an observer outside, the black hole has carried current. *Black holes can therefore conduct electric current.* A black hole can dissipate electromagnetic energy, for example, by absorbing electromagnetic waves. Energy is also dissipated when a black hole carries current because it is not a perfect conductor. *Black holes therefore have electrical resistance.* We estimate its value below.

The analogies between black holes and conductors suggest that the rotational energy of a black hole could be tapped if it were immersed in a magnetic field and wired up in a way analogous to the unipolar generator shown earlier. The expected power output can be estimated from (c), but to evaluate that we need to estimate the electrical resistance of a black hole, R_H. Let's try a simple dimensional estimate. The dimensions of resistance follow from Ohm's law, $R = V/I$. In Gaussian units that are convenient for this purpose, the electric potential a distance r away from a point charge q is $\Phi_{\text{elec}} = q/r$. The dimensions of voltage are thus $[q]/\mathcal{L}$, where $[q]$ are the dimensions of charge. The dimensions of current are $[q]/\mathcal{T}$. Hence, the dimensions of resistance in Gaussian units are $[R] = \mathcal{T}/\mathcal{L}$. In units where $c = 1$, resistance is dimensionless! Therefore, $R \sim 1$ is a reasonable guess for the resistance of a black hole in geometrized units. This corresponds to c^{-1}(s/cm) in Gaussian units or about 30 ohms in SI units. This is not so very different from the approximately 376 ohm "impedance of free space" characterizing a wave guide radiating into the vacuum. A black hole *is* empty curved space.

The rate at which rotational energy is extracted from a black hole of mass M and angular velocity Ω_H can be roughly estimated from (c) using this guess for the resistance and estimating the hole's size by $r_H \sim M$. The result is

$$P_L \sim \frac{\Omega_H^2 (B\pi r_H^2)^2}{16\pi^2 R_H} \tag{d}$$

$$\sim \left(10^{45}\frac{\text{erg}}{\text{s}}\right)(\Omega_H M)^2 \left(\frac{M}{10^9 M_\odot}\right)^2 \left(\frac{B}{10^4\ \text{gauss}}\right)^2.$$

If the $10^9 M_\odot$ black holes at the centers of some galaxies are rotating with even a modest value of the dimensionless product $\Omega_H M$ and are immersed in a magnetic field of some thousand gauss, this is more than sufficient to supply the power requirements for active galactic nuclei that were discussed in Section 13.2.

But what supplies the magnetic field and what are the analogs of the wires connecting the unipolar generator to the load that are necessary to make this mechanism work? The answer is that currents in an accretion disk (see the following figure) supply the necessary magnetic field, and the black hole makes its own wires, as we now describe.

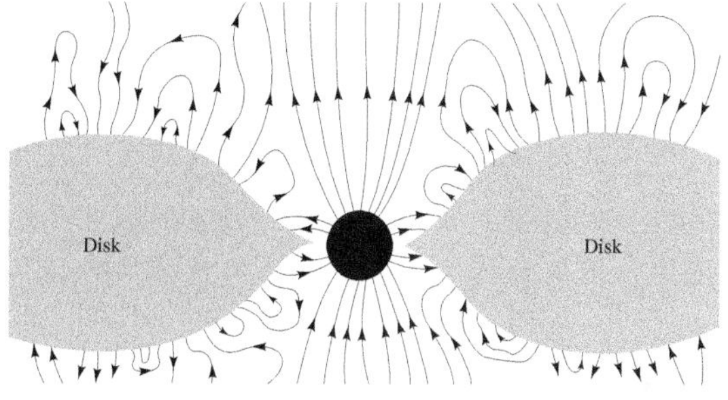

Were there no conducting connection between the accretion disk and the black hole, there would still be a voltage drop between them of order given by (b):

$$V \sim \Omega_H F_B \sim \Omega_H B\pi r_H^2 \tag{e}$$

$$\sim (10^{20}\ \text{volts})(\Omega_H M)\left(\frac{M}{10^9 M_\odot}\right)\left(\frac{B}{10^4\ \text{gauss}}\right).$$

This enormous voltage and the accompanying electric field would quickly accelerate any stray electron to relativistic velocities. The electron would radiate photons, which could produce electron-positron pairs (Problem 18). These would, in turn, accelerate, radiate, and produce more pairs. The resulting cascade would very quickly fill up the neighborhood of the black hole with a conducting plasma of electrons and positrons, electric and magnetic fields. This is the electrically conducting link between the black hole and the outside necessary for the unipolar generation of power.

The question of how all this power is turned into radio jets is complex, and we will not attempt to give even a qualitative discussion here, except to note that the electric fields in the vicinity of a rotating black hole could provide efficient acceleration and the rotation axis provides a natural axis for the jets. In this way, rotating black holes acting as unipolar generators could drive the jets of active galactic nuclei.

Whatever happens classically to a black hole, its irreducible mass must increase or stay constant. Equation (15.13) can then be used to express the total mass in terms of M_{ir} and the angular momentum J, with the result

Irreducible Mass

$$M^2 = M_{\text{ir}}^2 + \frac{J^2}{4M_{\text{ir}}^2}. \tag{15.39}$$

A Penrose process can reduce the value of J to zero but can never lower the value of M_{ir}.

At this time, the Penrose process is not believed to be a significant source of power for active galactic nuclei (Section 13.2) or X-ray sources (Section 11.2). The conditions under which it operates do not occur frequently enough to compete with the energy released in accretion for example. But the Penrose process does illustrate simply how, in principle, the rotational energy of a black hole can be accessed classically provided its area always increases. The *electromagnetic* extraction of a black hole's rotational energy by the Blandford–Znajek mechanism described in Box 15.1 on p. 326, on the other hand, can be an important source of power for active galactic nuclei.

Problems

1. [E] *Estimate* the Kerr parameter a for the Sun and the Earth. Are they bigger or smaller than their rest masses?

2. [S] Reversing the direction of time reverses the angular momentum and the direction of rotation of a rotating body. Show that the action of $t \rightarrow -t$ on (15.1) is the same as sending $J \rightarrow -J$. What happens when $\phi \rightarrow -\phi$?

3. [A] Show that the transformation

$$dt = dv - \frac{r^2 + a^2}{\Delta} dr, \qquad d\phi = d\psi - \frac{a}{\Delta} dr$$

applied to the Kerr metric in Boyer-Lindquist coordinates leads to a coordinate system for the Kerr geometry which is nonsingular at $r = r_{\pm}$. *Comment*: These are the generalization of the Eddington-Finkelstein coordinates for spherical black holes discussed in Section 12.1, as can be seen by comparing the above transformation formulas with (12.1) when $a = 0$.

4. Show that when starting at $r = r_+$, all future-directed light rays in the Kerr geometry move to smaller values of r. Use a nonsingular coordinate system such as that given in Problem 3.

5. [A] The null directions on the horizon of a rotating black hole were identified in (15.10). But does a light ray that starts out in one of these directions remain on the horizon? Use the geodesic equation for light rays in the Kerr geometry to show that it does. Show also that the light ray remains at a fixed value of θ.

6. Show explicitly that the two vectors $(0, 0, 1, 0)$ and $(0, 0, 0, 1)$ on the horizon $r = r_+$ are (a) spacelike and (b) orthogonal to each other and to the null generator $\boldsymbol{\ell}$ (15.10).

7. Show that the distance around the poles in the horizon geometry (15.12) is always less than the distance around the equator.

8. [N] Construct the embedding diagram for a $t =$ const. slice of the horizon of a Kerr black hole for values of a/M equal to 0, .5, and .86. The intrinsic geometry is given by (15.12). Figure 15.2 shows the result for $a/M = .86$. Does your construction explain why there is a maximum value of a/M for which such an embedding is possible?

9. Show that the surface $r = r_-$ is another stationary axisymmetric null surface *inside* the Kerr black hole.

10. *Surrounded by a Horizon!* Consider the metric

$$ds^2 = -\left(1 - \frac{r^2}{R^2}\right) dt^2 + \left(1 - \frac{r^2}{R^2}\right)^{-1} dr^2 + r^2 \left(d\theta^2 + \sin^2\theta \, d\phi^2\right).$$

This metric is not asymptotically flat, but imagine that we were living at the center near $r = 0$. Show that were we to cross the radius $r = R$ we could never return.

11. [A] Show that in the geometry of an extremal Kerr black hole of mass M there are circular light ray orbits in the equatorial plane at Boyer-Lindquist radii $r = M$ rotating with the black hole (corotating) and $r = 4M$ in the opposite direction (counterrotating).

12. The angular velocity $\Omega = d\phi/dt$ of circular orbits of Boyer-Lindquist radius r in the Kerr geometry is given by the simple formula:

$$\Omega = \pm \frac{M^{1/2}}{r^{3/2} \pm a M^{1/2}}.$$

Here the upper sign refers to corotating orbits and the lower one to counterrotating orbits. Explain how to derive this formula, and exhibit the algebraic equations from which it follows. However, don't try and solve the equations unless you really like algebra!

13. [S] Just because the Boyer-Lindquist radii of the corotating innermost stable circular orbits in the Kerr geometry are less than the corresponding radius $r = 6M$ in the Schwarzschild geometry [cf. Figure 15.3] doesn't mean that those orbits are closer to the black hole. After all, these are just coordinate radii in different geometries. The circumference is one invariant measure of the size of the orbit. Use Figure 15.3 to plot the circumference of the innermost stable corotating orbit in the Kerr geometry for the values 0, .2, .4, .6, .8, and 1 of a/M. Is the circumference of an innermost stable corotating circular orbit in the Kerr geometry always bigger or smaller than the innermost stable circular orbit in the Schwarzschild geometry? Can you explain what happens when $a/M = 1$?

14. Work out the range of angular velocities Ω_{obs} allowed an observer inside the ergosphere who remains at a fixed value of r. Show that this range becomes increasingly limited as the observer is located closer and closer to the horizon and is eventually limited to the single value Ω_H.

15. [C] *Temperature of a Rotating Black Hole*

(a) An axisymmetric body is spinning about its symmetry axis with angular velocity Ω and angular momentum J along the axis. Show that, in Newtonian mechanics, the work required to increase the angular momentum by a small amount ΔJ is $\Omega \Delta J$.

(b) Reorganize (15.36) for the change in area A of a rotating black hole given changes in its mass and angular momentum into a form like the first law of thermodynamics, assuming that the entropy of the black hole is $k_B A/4\hbar$, as in the Schwarzschild case [cf. (13.18)]. Find the Hawking temperature of a Kerr black hole.

(c) Show that the temperature of an extreme ($a = M$) black hole is zero and explain this fact from properties of the Kerr geometry.

16. [B, E] An active galactic nucleus with a luminosity of 10^{46} ergs is powered by the rotational energy of an extreme rotating black hole as described in Box 15.1 on p. 326. *Estimate* the maximum time the active galactic nucleus can radiate in this way. Compare your answer to the present age of the universe, approximately 15 billion years.

17. [B, P, S] Show that (b) in Box 15.1 on p. 326 gives the voltage developed across any axisymmetric conductor rotating around its symmetry axis.

18. [B, E, P] Consider a rotating black hole with $M \sim 10^9 M_\odot$, $\Omega_H M \sim 1$ immersed in a magnetic field $B \sim 10^4$ gauss as described in Box 15.1. *Estimate* how far an electron in the vicinity of the black hole has to move in the electric field there before it acquires enough energy to make a further electron-positron pair in a collision with a similar electron or positron.

CHAPTER 16

Gravitational Waves

Mass produces spacetime curvature. That is a central lesson of general relativity. The static spherical mass of the Sun produces the Schwarzschild geometry outside it. Mass in (nonspherical, nonuniform) motion is the source of ripples of curved spacetime, which propagate away at the speed of light. These propagating ripples in spacetime curvature are called *gravitational waves*. Their free propagation will be discussed in this chapter, their form will be derived from the Einstein equation in Chapter 21, and their production will be described in Chapter 23.

There are many important sources of gravitational waves in the universe—binary star systems, supernova explosions, collapse to black holes, and the big bang are all examples. Gravitational waves provide a window for exploring these astronomical phenomena that is qualitatively different from any band of the electromagnetic spectrum—X-rays, visible light, infrared, or radio waves.

The universe is not especially faint in gravitational radiation because of the great variety of possible sources. But the weakness of the gravitational interaction in everyday circumstances that was described in Chapter 1 means that gravitational waves are not easily detected. At the time of writing, the effects of gravitational waves have been observed on the orbit of the binary pulsar described in Sections 11.3 and 23.7, but gravitational waves have not yet been detected on Earth. However, a worldwide network of laser-interferometer detectors sensitive enough to register the gravitational radiation from realistic sources is being constructed. Some principles of their operation are discussed in Section 16.4.

The weak coupling to matter that makes gravitational waves so difficult to detect is also what makes them so interesting astrophysically. Once produced, little is absorbed. Gravitational waves in principle could enable us to see closer to the horizon of a black hole and to earlier moments in the universe than with any form of electromagnetic radiation.

This chapter focuses on weak gravitational waves propagating in nearly flat spacetime empty of matter. This is at once the most useful and most tractable example. Most useful because gravitational waves any distance from realistic sources are very small ripples of curved spacetime. Most tractable because the difficult nonlinear Einstein equation can be solved in a manageable linear approximation, as we will see in Chapters 21 and 23. Solutions in this approximation are called *linearized* gravitational waves. This chapter assumes the form of the linearized solutions that are derived in Chapter 21. We aim at analyzing these to exhibit the following facts.

Linearized gravitational waves

- propagate with the speed of light;
- are transverse;
- have two independent polarizations;
- can be detected by their effect on the relative motion of test masses;
- carry energy.

16.1 A Linearized Gravitational Wave

The simplest example of a gravitational wave spacetime is a small ripple of curvature propagating in one direction and independent of the other two. The direction of propagation is called the *longitudinal* direction; the two perpendicular directions are called *transverse*. Since the wave is the same everywhere in the transverse directions, it is a *plane wave*.

In the (t, x, y, z) coordinates of an inertial frame, the metric of a flat spacetime is $g_{\alpha\beta}(x) = \eta_{\alpha\beta}$, where $\eta_{\alpha\beta} = \text{diag}(-1, 1, 1, 1)$. Metrics of geometries that are *close* to flat can be written

Metric Perturbations

$$g_{\alpha\beta}(x) = \eta_{\alpha\beta} + h_{\alpha\beta}(x), \tag{16.1}$$

where the amplitudes $h_{\alpha\beta}(x)$ are small perturbations to the flat space metric. These *metric perturbations* describe the gravitational wave.

A simple example of a plane gravitational wave spacetime propagating in the z-direction is provided by the following metric perturbations:

$$h_{\alpha\beta}(t, z) = \begin{matrix} \\ t \\ x \\ y \\ z \end{matrix} \begin{matrix} t & x & y & z \\ \begin{pmatrix} 0 & 0 & 0 & 0 \\ 0 & 1 & 0 & 0 \\ 0 & 0 & -1 & 0 \\ 0 & 0 & 0 & 0 \end{pmatrix} \end{matrix} f(t - z). \tag{16.2a}$$

Here, $f(t - z)$ is any function of $t - z$ provided $|f(t - z)| \ll 1$. With these perturbations, the line element for spacetime is

A Plane Gravitational Wave Spacetime

$$\boxed{ds^2 = -dt^2 + [1 + f(t - z)]dx^2 + [1 - f(t - z)]dy^2 + dz^2.} \tag{16.2b}$$

The geometry (16.2) represents a wave of curvature propagating in the positive z-direction with speed 1, that is, with the velocity of light. The size (amplitude) and shape of the propagating ripple in curvature are determined by the function f. The quantities $h_{\alpha\beta}$ and $f(t - z)$ are all dimensionless, so the amplitude of a gravitational wave is a dimensionless number.

For example, the choice $f(t-z) = a\exp[-(t-z)^2/\sigma^2]$ represents a Gaussian wave packet. The wave packet has width σ and a maximum height a. The wave packet propagates along the z-axis at the speed of light without changing its shape. The choice $f(t-z) = a\sin[\omega(t-z)]$ represents a gravitational wave of amplitude a and definite frequency ω. The corresponding wavelength λ is $2\pi/\omega$.

The metric displayed in (16.2) does not solve the Einstein equation exactly. Rather, it solves that equation expanded to first (or linearized) order in the amplitude of the wave, as we will see in Chapter 21. These linearized gravitational waves are excellent approximations to a true solution of the Einstein equation when their amplitude is small. The amplitude of the gravitational waves that might be detectable in the first generation of laser interferometer detectors described in Section 16.4 is of order $\sim 10^{-21}$. The linear approximation is truly excellent for them. Linearized waves have the important property that they can be added to produce other linearized waves that solve the Einstein equation to the same accuracy. That is not true of the full nonlinear Einstein equation.

The gravitational wave metric exhibited in (16.2) is not merely one of a large number of possible forms. As we'll see in Section 21.5, with a suitable choice of coordinates, the general linearized plane gravitational wave propagating in the z-direction can always be written as a sum of a wave of the form (16.2), corresponding to one polarization, and a second, closely related, form corresponding to another polarization. They are both exhibited in (16.17).

16.2 Detecting Gravitational Waves

How could a propagating ripple in spacetime curvature be detected? The answer is the same as for the other spacetimes we have studied: spacetime curvature is detectable through the motion of test bodies—bodies that move along the geodesics of the curved spacetime but whose masses are so small that they produce no significant spacetime curvature on their own.

The motion of a single test body is not enough to detect a gravitational wave. Imagine studying the motion of one test mass in a frame freely falling with it. There it remains at rest, gravitational wave or no. Its motion is indistinguishable from a test body in flat spacetime consistent with the equivalence principle. A study of the relative motion of at least *two* bodies is required to detect a gravitational wave and, indeed, any curvature of spacetime.

To make this idea quantitative, consider a gravitational wave packet of the form (16.2) propagating in the z- direction. Before the wave passes them, two test masses A and B are at rest in the coordinates (t, x, y, z) of (16.2). For simplicity take A to be at the origin $(0, 0, 0)$ and B to lie at spatial position (x_B, y_B, z_B). The initial four-velocities of both test masses are

Linearized Gravitational Wave Spacetime

$$u^\alpha_{(A)} = u^\alpha_{(B)} = (1, \vec{0}). \tag{16.3}$$

(The subscripts (A) and (B) are not vector indices but labels to distinguish the two test masses. Parentheses have been put around them to emphasize that.) Before the wave passes, spacetime is flat and the test masses remain at rest,

$$x^i_{(A)}(\tau) = (0,0,0), \qquad x^i_{(B)}(\tau) = (x_B, y_B, z_B). \tag{16.4}$$

To predict the motion after the wave hits, the geodesic equation must be solved for each particle using the metric (16.2). Since the amplitude of the wave is small, we will solve only for the corrections $\delta x^i_{(A)}(\tau)$ and $\delta x^i_{(B)}(\tau)$ to the motion that are *first order* in the amplitude of the wave.

For either test mass the geodesic equation for the spatial coordinates $x^i(\tau)$ is (8.14)

$$\frac{d^2x^i}{d\tau^2} = -\Gamma^i_{\alpha\beta}\frac{dx^\alpha}{d\tau}\frac{dx^\beta}{d\tau}. \tag{16.5}$$

These equations need be evaluated only to first order in the amplitude of the wave to calculate the first-order changes $\delta x^i(\tau)$. Because $\Gamma^i_{\alpha\beta}$ vanishes in the unperturbed (flat) spacetime, one finds

$$\frac{d^2\delta x^i}{d\tau^2} = -\delta\Gamma^i_{\alpha\beta}\,u^\alpha u^\beta = -\delta\Gamma^i_{tt}, \tag{16.6}$$

where $\delta\Gamma^i_{\alpha\beta}$ are the first-order changes in the Γ's and the u^α are the *unperturbed* four-velocities given by (16.3). The Christoffel symbol Γ^i_{tt} is easily evaluated for the metric (16.2). It vanishes; therefore, so does $\delta\Gamma^i_{tt}$. Thus,

$$\frac{d^2\delta x^i}{d\tau^2} = 0 \tag{16.7}$$

to first order in the amplitude of the wave. Initially, $\delta x^i = 0$ and the test masses are at rest, so $d(\delta x^i)/d\tau = 0$. Equation (16.7) then implies $\delta x^i(\tau) = 0$ for all τ for both test masses:

$$\delta x^i_{(A)}(\tau) = \delta x^i_{(B)}(\tau) = 0. \tag{16.8}$$

Therefore, the *coordinate* positions of the particles remain unchanged as the wave passes to first order in the amplitude of the wave.

The *distance* between the two test masses changes with time even if their *coordinate* separation does not, as illustrated in Figure 16.1. The following example illustrates why.

Example 16.1. The Change $\delta L(t)$ in the Distance Between Two Test Masses in a Plane Orthogonal to the Direction of Propagation. Consider a wave of the form (16.2) traveling in the z-direction and two test masses—one at the origin and the other a coordinate distance L_* away along the x-axis. The distance between them is L_* in the unperturbed flat spacetime. In the gravitational wave spacetime (16.2) the distance between them measured along the x-axis, $L(t)$, is

$$L(t) = \int_0^{L_*} dx[1 + h_{xx}(t,0)]^{1/2} \approx L_*\left[1 + \frac{1}{2}h_{xx}(t,0)\right], \tag{16.9}$$

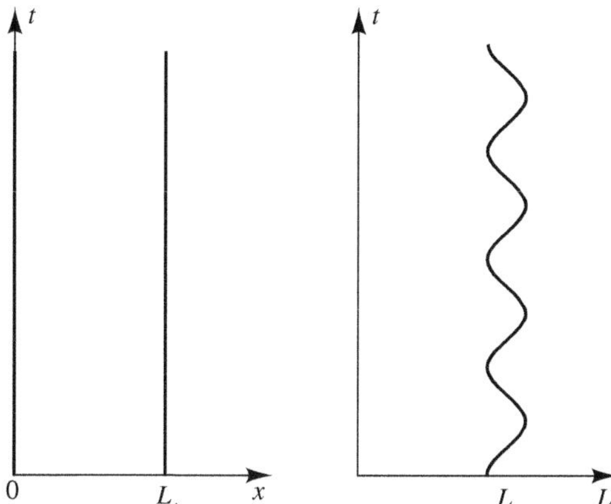

FIGURE 16.1 Test particle motion in a gravitational wave spacetime. This figure shows a t-x spacetime diagram of a spacetime in which a gravitational wave is propagating in the z-direction. Two test particles are located initially at coordinates $x = 0$ and $x = L_*$. As the wave passes, the coordinate separation of the two particles does not change, but the distance between them, $L(t) = L_* + \delta L(t)$, oscillates with the frequency of the wave, as in the right hand figure. The amplitude of oscillation shown here is much larger than that expected in realistic detectors where important sources would contribute $\delta L/L_* \sim 10^{-21}$ at Earth.

giving for the change in distance, $\delta L(t)$,

$$\frac{\delta L(t)}{L_*} = \frac{1}{2} h_{xx}(t, 0) \qquad \begin{pmatrix} \text{test masses on} \\ x\text{-axis at } z = 0 \end{pmatrix}. \tag{16.10}$$

The distance between the test masses thus changes with time according to the time variation of the wave, as shown in Figure 16.1. If the wave has a definite frequency ω, amplitude a, and phase δ so that $f(t - z) = a \sin[\omega(t - z) + \delta]$, we have

$$\frac{\delta L(t)}{L_*} = \frac{1}{2} a \sin(\omega t + \delta). \tag{16.11}$$

The fractional change in distance along the x-axis oscillates periodically with half the amplitude of the gravitational wave.

This example is straightforwardly generalized to the case where one test mass is at the origin, and the other is at an arbitrary location in the plane transverse to the direction of propagation before the wave passes. Suppose the second test mass is distance L_* from the origin in the direction of a unit vector $\vec{n}$ in the plane $z = 0$ perpendicular to the direction of propagation. The distance $L(t)$ calculated along the path that is a straight line in flat space changes as

$$\boxed{\frac{\delta L(t)}{L_*} = \frac{1}{2} h_{ij}(t, 0) n^i n^j \qquad \begin{pmatrix} \text{test masses in} \\ z = 0 \text{ plane} \end{pmatrix}.} \tag{16.12}$$

Fractional Strain Produced by a Gravitational Wave

Borrowing a term from elasticity, the ratio $\delta L/L_*$ is called the fractional *strain* produced by the gravitational wave.

The spatial distance between test masses in (16.12) was calculated along a path that is a straight line (geodesic) in the unperturbed flat spacetime. Just because the gravitational wave doesn't change the coordinates of the test masses doesn't mean that the path is still a geodesic in the curved spacetime of the wave. It isn't. A more realistic choice from the point of view of the laser interferometer detectors, discussed in Section 16.4, would be to calculate separations along the path of a light ray between the test masses. In flat spacetime light rays follow straight-line paths, as they do in any spacetime. But in the curved spacetime of the gravitational wave, their straight paths will deviate by an amount $\delta x^i_{(\ell)}(\lambda)$ from the coordinates of the flat space straight-line path. That change will be first order in the amplitude of the wave (Problem 4). However, there is no corresponding first-order change in the distance between the test masses. Straight lines are paths of extremal distance in flat space from which any small deviation produces only a second-order change in distance. (Recall the discussion of extremal principles in Section 3.5.) Equation (16.12) therefore gives the change in distance along the path of a light ray as well to first order in the amplitude of the wave.

Equation (16.12) shows how gravitational waves can be detected by observing the relative motion of two or more test masses. Indeed, we'll see in Section 21.2 that the relative motion of test particles is one way of *defining* the local curvature of spacetime in general. The gravitational wave detectors discussed in Section 16.4 are based on this principle. Of course, in a realistic detector the test masses can't be arranged ahead of time in a plane perpendicular to the direction of an incoming wave. But the generalization of (16.12) to two masses in an arbitrary orientation is straightforward (Problem 3).

16.3 Gravitational Wave Polarization

The gravitational wave metric (16.2) leads to no change in separation between two test masses lying along the z-axis—the longitudinal direction. The metric perturbation h_{zz} vanishes in the formula analogous to (16.9). Only transverse (x-y) separations change with time as the gravitational wave passes by. Thus, like electromagnetic waves, gravitational waves are *transverse*.

A clearer picture of the characteristic signature of a gravitational wave can be obtained by considering not just two test particles, but many of them. Imagine free test masses initially arranged in a circle in the x-y plane at $z = 0$ with another mass at their center, as shown in Figure 16.2. A plane gravitational wave of the form (16.2) with $f(t - z) = a \sin[\omega(t - z)]$ passes by in the z-direction. The (x, y, z) coordinate positions of the test masses remain unchanged [cf. (16.8)]; in particular, the test masses remain in the x-y plane. The distance in the x-y plane between the central mass and those in the circle—each following its own geodesic—will change with time according to the metric (16.2). To calculate these distances it is convenient to introduce new coordinates (X, Y) for the x-y plane at $z = 0$ defined by

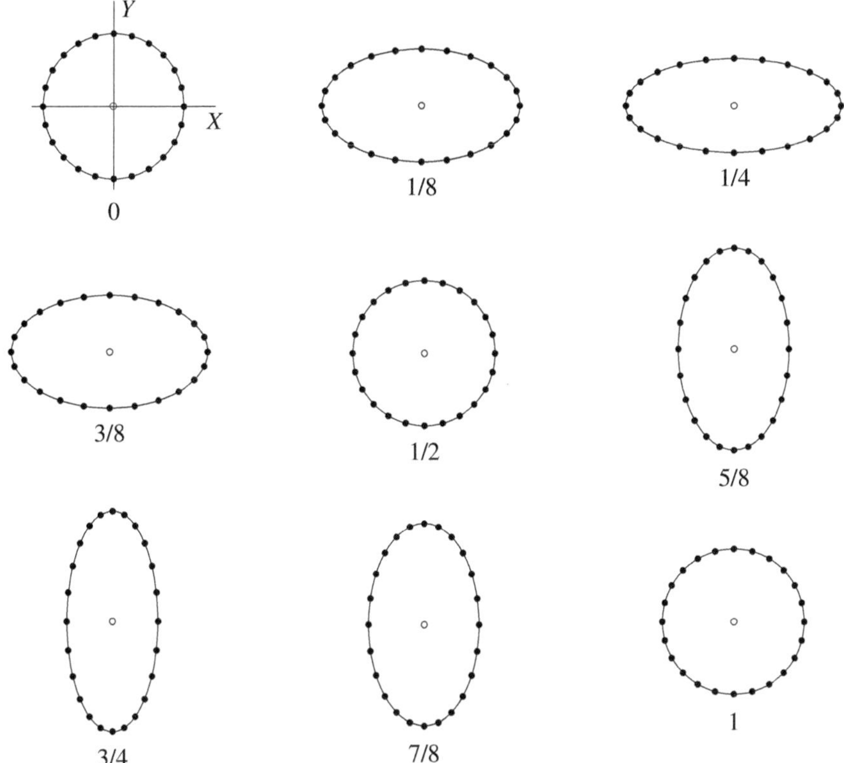

FIGURE 16.2 The signature of a gravitational wave. The figure shows the time behavior of the positions of 24 test masses in the plane transverse to the direction of propagation of the gravitational wave in (16.2) with definite frequency and amplitude $a = .8$. At time $t = 0$, test masses • are at rest in a circle about a central test mass ∘. The subsequent time behavior of their locations in the (X, Y) coordinates defined in (16.13) is shown at the fractions of a period indicated. The circle is first squeezed in the Y-direction and expanded in the X-direction to make an elliptical pattern. After a quarter-period it is the X-direction which is squeezed, and the Y-direction which is expanded, and so on. This pattern is one polarization of a gravitational wave. The other is the same sequence of displacements but rotated by 45°. An amplitude of $a = .8$ is only marginally linear, but an amplitude of $a \sim 10^{-21}$ that might be seen in realistic detectors would not show any effect on the scale of the figure.

$$X = (1 + \tfrac{1}{2}a \sin \omega t)x, \qquad Y = (1 - \tfrac{1}{2}a \sin \omega t)y. \tag{16.13}$$

The line element in the x-y plane in these new coordinates is the familiar $dS^2 = dX^2 + dY^2$ of the flat Euclidean plane plus negligible corrections of order a^2. The x- and y-coordinates of a test mass don't change in time, but X and Y vary with t according to (16.13). Distances between test masses in the x-y plane can be calculated from their $X(t)$ and $Y(t)$ coordinates using Euclidean plane geometry to first order in the amplitude of the wave.

The resulting behavior in time of the ring of test masses in the plane transverse to the wave is shown in Figure 16.2. The general pattern is an ellipse (Problem 6) whose axes oscillate periodically in time but out of phase with each other. In one phase of the oscillation, the ellipse squeezes in the X-direction and expands in the Y-direction. A quarter-period later the X-direction is expanding and the Y-direction is contracting. This pattern of oscillation is characteristic of all gravi-tational radiation and one of the ways it can be identified.

The metric in (16.2) is not the most general possible. It is an example of but one of the two independent *polarizations* of a gravitational wave. To find the other polarization, imagine rotating the x-y axes by an angle θ. Equation (3.4) gives the relation between (x, y) and the rotated coordinates (x', y'). If we choose $\varphi = -45°$ $(= -\pi/4$ rad) for instance, then

$$x = \frac{1}{\sqrt{2}}(x' + y'), \qquad y = \frac{1}{\sqrt{2}}(y' - x'). \tag{16.14}$$

Substituting these transformations into the line element (16.2b) shows how the parts of the metric transform. The flat space part, $\eta_{\alpha\beta}$, is unchanged by any rotation, and it is not difficult to see that

$$h_{x'x'} = 0, \qquad h_{x'y'} = h_{y'x'} = h_{xx} = -h_{yy}, \qquad h_{y'y'} = 0. \tag{16.15}$$

But there is nothing physically to distinguish the new coordinates from the old! If we drop the primes we have another, different solution of the linearized Einstein equation in the (t, x, y, z) coordinates. It has the form

$$h_{\alpha\beta}(t, z) = \begin{pmatrix} 0 & 0 & 0 & 0 \\ 0 & 0 & 1 & 0 \\ 0 & 1 & 0 & 0 \\ 0 & 0 & 0 & 0 \end{pmatrix} f(t - z). \tag{16.16}$$

This is a second polarization linearly independent of the first. The behavior of a ring of test particles is just the same as in Figure 16.2 but rotated by 45°. Waves of the form (16.2) are usually called the $+$ (plus) polarization, while those of the form (16.16) are called the $\times$ (cross) polarization. The most general gravitational wave propagating in the positive z-direction is thus of the form

General Linearized Gravitational Wave in the z-direction

$$h_{\alpha\beta}(t, z) = \begin{pmatrix} 0 & 0 & 0 & 0 \\ 0 & f_+(t - z) & f_\times(t - z) & 0 \\ 0 & f_\times(t - z) & -f_+(t - z) & 0 \\ 0 & 0 & 0 & 0 \end{pmatrix} \tag{16.17}$$

for different functions $f_+(t - z)$ and $f_\times(t - z)$. The most general linearized gravitational wave is a superposition of metric perturbations of the form (16.17), with different directions of propagation and different functions f_+ and $f_\times$ for each direction.

16.4 Gravitational Wave Interferometers

At the time of writing, a number of gravitational wave detectors are under construction around the world based on the principle of the Michelson interferometer illustrated in Figure 16.3.

Imagine three test masses hung from wires and free to swing in horizontal directions. Mirrors are attached to two of the test masses (M). The third (S) supports a beam splitter, as shown. An incident beam of light from a laser (L) splits into beams running along the two perpendicular arms of the interferometer. These are reflected, recombined, and detected. Assume for simplicity that the arms of the interferometer are oriented along the x-and y-axes of a frame falling freely with the beam splitter (S), as shown. Assume further that we are interested in the beams that combine and enter the detector (D). When these beams are recombined, they will interfere constructively if twice the length of the two arms, $L_{(x)}$ and $L_{(y)}$, differ by an integral number of wavelengths and interfere destructively if the lengths differ by an odd number of half-wavelengths:

$$\Delta L \equiv 2(L_{(x)} - L_{(y)}) = n\lambda, \qquad n = 0, 1, 2, \ldots \quad \text{constructive interference,} \tag{16.18a}$$

$$\Delta L \equiv 2(L_{(x)} - L_{(y)}) = (n + \tfrac{1}{2})\lambda, \qquad n = 0, 1, 2, \ldots \quad \text{destructive interference.} \tag{16.18b}$$

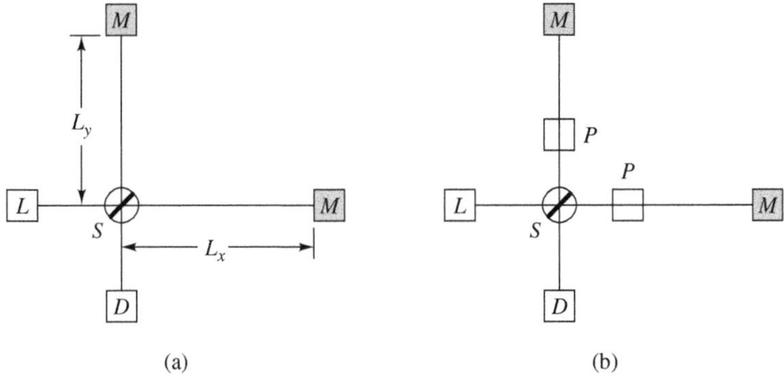

FIGURE 16.3 (a) At left is a schematic diagram of a Michelson interferometer gravitational wave detector. Three test masses are suspended vertically in order to be free to move horizontally in the plane of this top view. One test mass (S) carries a beam splitter and the other two (M) carry mirrors defining the ends of the two arms of the interferometer. A beam from a laser L is split at (S) into two beams running along the perpendicular arms. These are reflected, recombined, and detected in the detector, D. Small differences in the lengths of the two arms, such as would be caused by an incident gravitational wave, are detected as changes in the degree of interference. (b) The figure at right, while still schematic, is somewhat closer to the actual design. Two additional test masses carrying partially reflecting mirrors (P) are introduced that, together with the test masses with mirrors, define the arms of the interferometer. The two mirrors on each arm act as a cavity in which the beam is reflected back and forth many times, thus effectively increasing the length of the arms and the sensitivity of the interferometer.

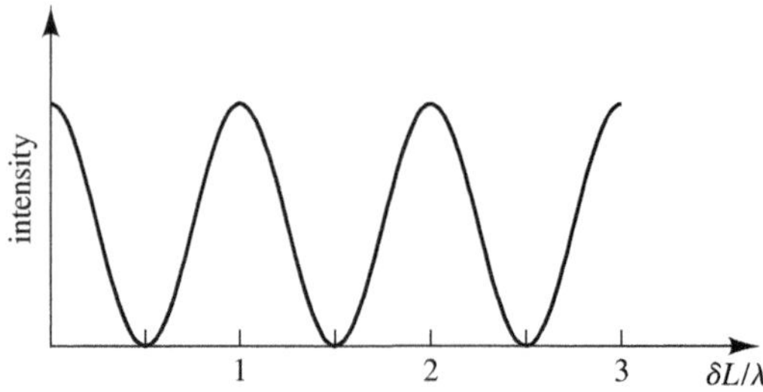

FIGURE 16.4 An idealized interference pattern. This figure shows how the intensity in the detector D in Figure 16.3 would change as the length difference δL between the two arms of the interferometer varies between the conditions for constructive interference and destructive interference given in (16.18). The curve is the $[1 + \cos(2\pi\delta L/\lambda)]$ pattern appropriate for the idealized situation of exactly monochromatic waves, equal intensities in both arms, and no losses (Problem 10). Small changes in δL can be detected from the changes in intensity they produce. The initial LIGO detector can see the intensity changes corresponding to variations in δL of order of a billionth of the wavelength of visible light and a very small change in the degree of interference. This corresponds to sensitivity in $\delta L/L$ of order 10^{-21}.

Suppose the difference in arm lengths $\Delta L = L_{(x)} - L_{(y)}$ varies. As ΔL moves through the conditions for constructive and destructive interference (16.18), the intensity of the combined beams in the detector would look something like the idealized curve shown in Figure 16.4.

An incident gravitational wave will change the lengths of the arms and, therefore, change the way the two beams interfere. One can think of the beam splitter (S) as the central mass in the pattern in Figure 16.2 and the masses with mirrors as two of the test masses in the surrounding pattern. For simplicity assume that the wave is of the form (16.2) normally incident along the z-axis with a definite frequency ω. As the wave passes, $L_{(x)}$ will expand and contract; $L_{(y)}$ contracts and expands out of phase, as illustrated in Figure 16.2, and given quantitatively by (16.13), assuming the x-y axes are oriented along the interferometer arms. Specifically,

$$\frac{\delta L_{(x)}}{L_{(x)}} = +\frac{1}{2}a\sin(\omega t), \qquad \frac{\delta L_{(y)}}{L_{(y)}} = -\frac{1}{2}a\sin(\omega t). \tag{16.19}$$

The amplitude a and frequency ω can be measured by monitoring the interference pattern. More generally, the shape of an incident wave packet could be found. Equations (16.18) and (16.19) show that the longer the interferometer, the greater the change in interference pattern for a given amplitude wave. Other things being equal, therefore, the longer the interferometer, the more sensitive it can be.

Although based on the Michelson interferometer idea, realistic interferometers are much more sophisticated. In particular, instead of the simple configuration of test masses illustrated in Figure 16.3a, they employ two or more cavities formed by one test mass with a highly reflecting mirror (M) and another with a partially reflecting mirror (P). (See Figure 16.3b.) Tuned to resonance, the cavities behave

FIGURE 16.5 LIGO gravitational wave detector at Hanford, Washington. LIGO stands for Laser Interferometer Gravitational (Wave) Observatory. The 4-km-long concrete covers of the beam pipes housing the arms of two interferometers can be seen stretching toward the horizon and to the right. The two ends of the interferometer illustrated in Figure 16.3 are at the ends of these pipes. The beam splitter is in the building housing the end station in the foreground. There is a similar facility in Livingston, Louisiana.

as though there were multiple reflections of the beam from end to end before recombination with the beam from the other arm. This greatly increases the effective length of the interferometer and its sensitivity. Changes of a tiny fraction of an interference fringe shown in Figure 16.4 are expected to be detectable.

The LIGO gravitational wave detector about to go into operation at the time of writing consists of three such interferometers. Two are located at Hanford, Washington, one of which has 4-km-long arms. The third interferometer with 4-km arms is located in Livingston, Louisiana. Figure 16.5 shows a view of the Hanford site. Interferometers in different locations allow for coincidence detection of gravitational wave events, giving some information about the direction from which the waves arrive but also allowing spurious signals arising from local disturbances to be rejected. Seismic rumbling of the Earth is a simple example of a kind of noise that the experiments have to deal with. But the various sources of noise and the ingenious methods used to overcome them is a topic we cannot pursue here.[1]

In its initial data run in 2002–2004, LIGO is expected to achieve a strain sensitivity of 10^{-21}. To appreciate this achievement it is enough to recall that this means, in effect, monitoring the positions of the test masses to a fraction of the dimension of an atomic nucleus. With improvements later in the decade, LIGO

[1]For more information, see Saulson (1994).

should be able to detect gravitational radiation from pairs of neutron stars spiraling toward each other and eventually coalescing at an event rate of perhaps a few per year. We'll discuss more about such sources in Chapter 23.

16.5 The Energy in Gravitational Waves

No Local Gravitational Energy in General Relativity

The energy density in a Newtonian gravitational field is given by

$$\epsilon_{\text{Newt}}(\vec{x}) = -\frac{1}{8\pi G}[\vec{\nabla}\Phi(\vec{x})]^2 = -\frac{1}{8\pi G}[\vec{g}(\vec{x})]^2, \tag{16.20}$$

where $\Phi(\vec{x})$ is the Newtonian gravitational potential and $\vec{g}(\vec{x})$ is the Newtonian gravitational field (3.16). If you have studied electromagnetism, you will recognize these expressions as the analogs of those for the energy density in the electric field. The Newtonian gravitational energy density has the same form, except that the sign is appropriately negative for the attractive force of gravity. Because gravity is attractive, the energy of an assembly of mass is *lower* than it is when dispersed. If you haven't studied electromagnetism, Problem 11 leads you through the standard derivation of (16.20), although its exact form will not be important for us.

What is the energy density corresponding to (16.20) in general relativity? There is none. Pursuing the analogy between Newtonian gravity and general relativity, one might look for an expression constructed from the first derivatives of the metric. But the result should be independent of the choice of coordinates, and the first derivatives of the metric all vanish in a local inertial frame. There is no candidate expression. Its absence is consistent with the principle of equivalence idea that gravitational effects vanish in a freely falling laboratory of sufficiently small size over a sufficiently small length of time.

There is a deeper reason why there is no local notion of gravitational energy density in general relativity, which has to do with the connection between conserved quantities and the symmetries of spacetime (Section 8.2). The conserved energy and angular momentum of particle orbits in the Schwarzschild geometry followed directly from its time displacement and rotational symmetries. The conserved energy and momentum of the electromagnetic field can be seen to be consequences of the symmetries of the flat spacetime that is assumed in Maxwell's theory. But general relativity does not assume a fixed spacetime geometry. It is a theory *of* spacetime geometry, and there are no symmetries that characterize all spacetimes. The absence of a local gravitational energy in general relativity is part of the profound shift in viewpoint from gravity as a force field operating *in* spacetime to gravity *as* curved spacetime.

Energy in Gravitational Waves in the Short Wavelength Approximation

Even though there is no local notion of energy in a gravitational field, the total mass-energy of spacetime turns out to have meaning when spacetime is asymptotically flat. Defining the mass of a black hole by the asymptotic behavior of the Schwarzschild or Kerr geometry is a specific example of this. In between the two extremes of a total energy and no local energy is the *approximate* notion of energy density of a weak gravitational wave whose wavelength, λ, is much shorter than the scale of curvature $\mathcal{R}$ of the background spacetime through which it propagates. This energy is not exactly local. It is an average energy density over spacetime volumes whose dimensions are larger than λ but much smaller than $\mathcal{R}$.

The approximations involved in defining this energy density become increasingly accurate as $\lambda/\mathcal{R}$ becomes small. For example, they can be made arbitrarily accurate for calculating the energy lost by a source to gravitational radiation simply by calculating it very far from the source where space is nearly flat and $\mathcal{R}$ is nearly infinite.

The energy density in short-wavelength gravitational radiation is derived in a web supplement to Chapter 22. However, just a few simple plausibility and dimensional arguments allow us to guess its form for the simple plane waves under discussion in this chapter. Consider the wave in (16.2) with a definite frequency, so $f(t-z) = a\sin[\omega(t-z)]$. Like waves on a string or electromagnetic waves, we would expect the energy density to be proportional to the *square* of the amplitude of the wave a. Energy density has dimensions $(\text{length})^{-2}$ in geometrized units $(G = c = 1)$. [It is $(\text{energy})/(\text{length})^3 \sim \mathcal{M}/(\mathcal{L}\mathcal{T}^2)$ in $\mathcal{MLT}$ units, but mass and time have the dimensions of length in geometrized units.] The only quantity with dimensions of length in the plane gravitational wave metric (16.2) with definite frequency ω is the inverse of that frequency proportional to the wavelength. Therefore, we guess that the energy density must be proportional to $\omega^2 a^2$. In the web supplement we show this to be correct and derive the following for the energy density, ϵ_{GW}, in one polarization of a gravitational wave in the short-wavelength approximation:

$$\epsilon_{GW} = \frac{\omega^2 a^2}{32\pi}. \tag{16.21}$$

You might wonder what happened to the space and time dependence in (16.2). But recall that (16.21) is not a local energy density exactly, but an average over several wavelengths in space and time. A $\sin\omega(t-z)$ dependence for $f(t-z)$ in (16.2) averages to $\frac{1}{2}$ when squared.

Other properties of the wave follow immediately from this expression. Consider, by way of example, the flux of gravitational wave energy f_{GW} through a surface normal to the direction of propagation. This is the energy per unit time crossing a unit area of the surface. Because the waves propagate with the speed of light 1, this is the same as the energy in a cylinder of unit length and unit area behind the surface. In short, in $c = 1$ units the energy flux is the same as the energy density:

$$\boxed{f_{GW} = \frac{\omega^2 a^2}{32\pi}.} \tag{16.22}$$

Energy Flux in a Linearly Polarized Plane Wave

The energy density and flux in a wave that is a superposition of different polarizations is the sum of those in each polarization. Once you have studied Chapter 22 you can read a derivation of these relations from the Einstein equation in a supplement on the book website.

Problems

1. Show that the gravitational wave spacetime (16.2) has three Killing vectors: $(0, 1, 0, 0)$, $(0, 0, 1, 0)$, and $(1, 0, 0, 1)$.

2. Consider a Gaussian wave packet with $f(t-z) = a\exp[-(t-z)^2/\sigma^2]$.
   - **(a)** Draw a spacetime diagram showing a z-t slice of spacetime with $x = y = 0$. Shade the region where the wave packet has a size greater than $a/2$. Show the world line of the test mass at the origin.
   - **(b)** Draw a graph of the distance between the two test masses initially at rest in the given frame, one at the origin and the other at a distance L_* along the x-axis. What is the maximum value of the change in that distance?

3. Consider the gravitational wave in (16.2) and two test masses, one at the origin and the other at a location (X, Y, Z) in the Cartesian coordinates used in (16.2). Show that the change in distance between the masses produced by the wave is given by
$$\delta L(t) = \frac{1}{2}\int_0^{L_*} d\lambda h_{ij}\left(t - n^z\lambda\right) n^i n^j,$$
Here $n^i = (X/L_*, Y/L_*, Z/L_*)$ is the unit tangent vector to the straight-line path between the test masses and L_* is the unperturbed distance between them.

4. [C] Calculate the displacement $\delta x^i_{(\ell)}(\lambda)$ in the path of a light ray between two test masses from that of a flat-space straight line. Assume a gravitational wave of the form (16.2) having a definite frequency ω.

5. An observer is riding on one of the test particles discussed in Section 16.2 holding a cup of coffee filled to the brim. The size of the coffee cup is much less than the wavelength of the gravitational wave. Is there any danger that the coffee will spill because of the passage of the gravitational wave? If so, estimate how close to the top the observer can fill the cup and not have it spill.

6. The equation for an ellipse is $x^2/a^2 + y^2/b^2 = 1$, where a is the semimajor axis and b is the semiminor axis if $a > b$. Show that an initial circle of test particles distorts into an ellipse according to (16.13) to lowest order in a and compute the semimajor and semiminor axes as a function of time.

7. In Section 16.3 we produced a gravitational wave with $\times$ polarization by rotating the $+$ polarization (16.2) by 45°. Show that a rotation by an arbitrary angle θ doesn't give another independent solution but rather one that could be written as a superposition of $+$ and $\times$. This is one way of seeing that there are only two linearly independent polarizations of a gravitational wave.

8. [P] **(a)** In a linearly polarized electromagnetic wave, the electric field oscillates along one fixed direction in space. What pattern of motion is produced in a ring of test

charges like those in Figure 16.2 by an electromagnetic wave propagating in the z-direction that is polarized in the x-direction and normally incident on the plane of the ring? (Neglect the magnetic forces on the charges.)

(b) Is there a combination of the two gravitational wave polarizations that would produce the same motion?

9. [C] *Circularly Polarized Gravitational Waves* If a linearly polarized electromagnetic wave with a given frequency is added to a wave of the same amplitude and frequency propagating in the same direction but polarized in a perpendicular direction and 90° out of phase, the result is a circularly polarized wave in which the tip of the electric field vector moves in a circle at any one position in space. By analogy, the superposition of a + polarized gravitational plane wave with another of the same amplitude, frequency, and propagation direction but with × polarization and 90° out of phase is called a *circularly polarized* gravitational wave. Show that a circularly polarized plane gravitational wave with frequency ω that is normally incident on an ellipse of test particles causes each test particle to rotate in a small circle such that the elliptical pattern rotates with a constant angular frequency. What is that angular frequency?

10. *Interference Pattern* Suppose at the detector D in Figure 16.3 the electric fields of the two light waves that have traveled along the different arms of the interferometer have the forms $a \sin[\omega(t - L_{(x)})]$ and $a \sin[\omega(t - L_{(y)})]$, respectively. Show that if these are combined (added), the intensity of the resulting wave (proportional to the square of the amplitude) has the form of the interference pattern discussed in Figure 16.4.

11. [S] *Energy Density in a Newtonian Gravitational Field* Consider assembling a system of N particles of mass M_A at assigned positions $\vec{x}_A, A = 1, \ldots, N$. The Newtonian potential energy of the system W is the total potential energy of all the particles found by bringing them one by one from infinity in the potential of the particles already assembled. Show that this is

$$W = -\frac{1}{2}\sum_{A \neq B} \frac{G M_A M_B}{|\vec{x}_A - \vec{x}_B|}$$

and that the corresponding formula for a continuum distribution of mass with density $\mu(\vec{x})$ is

$$W = -\frac{1}{2}\int d^3x \int d^3x' \frac{G\mu(\vec{x})\mu(\vec{x}')}{|\vec{x} - \vec{x}'|} = \frac{1}{2}\int d^3x \mu(\vec{x})\Phi(\vec{x}).$$

Use the Newtonian field equation (3.18) to eliminate $\mu(\vec{x})$ from this expression and then the divergence theorem to write this as

$$W = -\frac{1}{8\pi G}\int d^3x [\vec{\nabla}\Phi(\vec{x})]^2 = -\frac{1}{8\pi G}\int d^3x [\vec{g}(\vec{x})]^2 \equiv \int d^3x \epsilon_{\text{Newt}}(\vec{x}),$$

where $\epsilon_{\text{Newt}}(\vec{x})$ is the energy density of a Newtonian gravitational field.

12. Show that for a wave traveling at the speed of light the flux of energy across a surface is the momentum density multiplied by c^2. Show that the magnitude of the momentum density is the energy density divided by c.

13. The LIGO gravitational wave detector expects to detect gravitational waves at frequencies of ~ 200 Hz that cause a dimensionless strain of $\delta L/L \sim 10^{-21}$. What is the flux of energy of such waves incident on Earth? If they come from 20 Mpc away, how fast was their source losing energy to gravitational waves when they were emitted? How far away would the Sun have to be to produce the same flux in electromagnetic radiation?

14. [E] The binary star system ι Boo is located about 11.7 parsecs from Earth in the direction of the constellation Boötes. (1 parsec = 3.09×10^{18} cm.) The two stars orbit each other with a period of approximately 6.5 h. A gravitational wave detector in the vicinity of Earth detects gravitational radiation from this source with a strain of $\delta L/L \sim 10^{-21}$. *Estimate* the energy flux in this radiation at the Earth and compare to that of the Sun in electromagnetic radiation if it were located the same distance away. *Comment*: Gravitational wave detectors contemplated on Earth can't make this detection because the frequency of the wave is too low, but detectors in space might be able to do it.

CHAPTER 17

The Universe Observed

Cosmology is the part of science concerned with the structure and evolution of the universe on the largest scales of space and time. As mentioned in Chapter 1, gravity governs the structure of the universe on these scales and determines its evolution. General relativity is thus central to cosmology, and cosmology is one of the most important applications of general relativity.

Our understanding of the universe on the largest scales of space and time has increased dramatically in recent years—both observationally and theoretically. This book does not have enough space to survey the wealth of observational detail that is available, and it does not assume the breadth of physics necessary to analyze all the processes that are important for the structure and evolution of the universe. We therefore concentrate on the role of relativistic gravity in cosmology, introducing only the most basic observational facts and working out the simplest theoretical models.

The next three sections sketch the three basic observational facts about our universe on the largest distance scales that will guide the construction of cosmological models in the next chapter:

- The universe consists of stars and gas in gravitationally bound collections of matter called galaxies, diffuse radiation, dark matter of unknown character,[1] and vacuum energy.
- The universe is expanding.
- Averaged over large distance scales, the universe is *isotropic*—the same in one direction as in any other—and *homogeneous*—the same in one place as in any other. The densities of galaxies, radiation, and vacuum energy are uniform.

17.1 The Composition of the Universe

The visible matter in the universe is mostly contained in galaxies—gravitationally bound collections of stars, gas, and dust (Figure 17.1). A typical galaxy has about

[1] Beginning here and for the rest of this chapter and the next two, we use the usual terminology of cosmology, in which zero rest-mass particles (photons, gravitons, etc.) are referred to as *radiation*, and nonzero rest-mass particles (protons, neutrons, electrons, etc.) are called *matter*. Neutrinos with very small masses behave approximately like radiation in some circumstances and matter in others, neither of which will be of concern in this text. Outside these cosmological chapters we employ the usual convention of general relativity, where both kinds of particles are called *matter*. The intent is not to confuse, but to conform to contemporary usage.

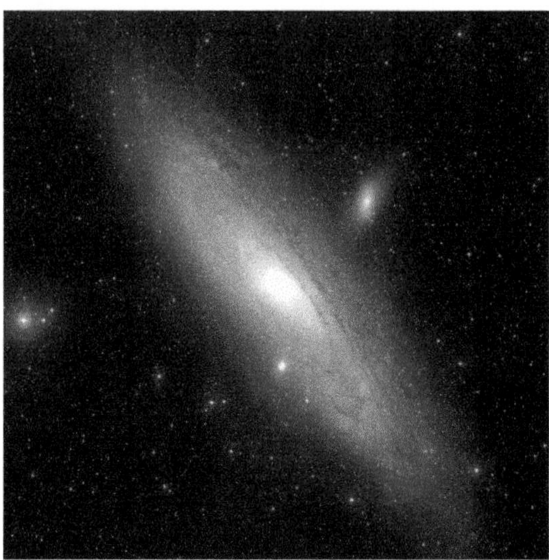

FIGURE 17.1 The Andromeda galaxy. This gravitationally bound collection of stars and dust is the nearest large galaxy to our own. The gravitational attraction holding the visible stars and dust in their orbits about the center is mostly due to unseen dark matter. (See Figure 17.4.)

10^{11} stars and a total mass of $10^{12}M_{\odot}$. There are very roughly 10^{11} galaxies in the part of the universe that is, in principle, accessible to our observations today (Figure 17.2). If the visible matter in galaxies were smoothed out uniformly over the largest scales, it would correspond at the present moment to a density of approximately

$$\rho_{\text{visible}}(t_0) \sim 10^{-31}\ \text{g/cm}^3, \tag{17.1}$$

roughly one proton per cubic meter.

Besides galaxies, the universe also contains radiation consisting of zero rest mass particles—photons, perhaps some neutrinos, and gravitational waves. This radiation, traveling at the speed of light, is not clustered in gravitationally bound clumps as is the matter that makes up the galaxies. The detected radiation with the greatest energy density is the *cosmic background radiation*—the electromagnetic radiation left over from the hot big bang. To impressive experimental accuracy, this has a blackbody spectrum with a temperature of $2.725 \pm .001$ K today. (See Figure 17.3.) This very precise fit to a blackbody spectrum is one of the strongest pieces of evidence for a big bang. The peak of this spectrum lies in the microwave band, and for that reason this radiation is often referred to as the cosmic *microwave* background (CMB). The density of the cosmic background radiation, like all blackbody radiation with a temperature of 2.725 K [cf. (18.24)], is

$$\rho_r(t_0) \sim 10^{-34}\ \text{g/cm}^3. \tag{17.2}$$

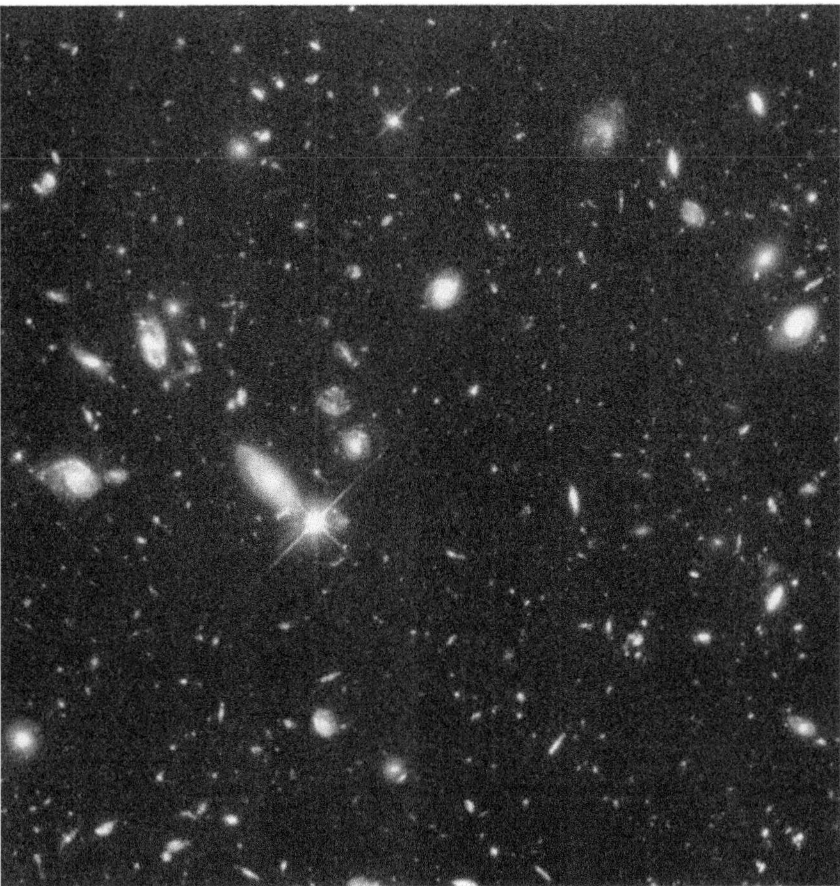

FIGURE 17.2 The Hubble deep field. Perhaps no single picture is more convincing that we live in a universe of galaxies than this image from the Hubble space telescope. This image covers just a narrow region of the sky with an angular size $\frac{1}{30}$ of that subtended by the full Moon. The region is so small that just a few foreground stars in our own galaxy are visible. However, the picture includes very faint objects whose light has been traveling to us a long time over great distances. The light from the most distant galaxies may have been emitted less than 1 billion years after the big bang—less than 6% of the present age.

Today the energy density in this background radiation is much less than the average density of the matter in the galaxies. But, as we will see in the next chapter, earlier in the universe it was the other way around.

There is considerable evidence that most of the mass in the universe is neither in the luminous matter in galaxies nor in the radiation detected so far. Mass can be detected by its gravitational influence even if it cannot be seen directly. The simplest evidence for unseen "dark" matter comes from "weighing" spiral galaxies. A spiral galaxy (Figure 17.1) is a disk of stars and dust rotating about a central

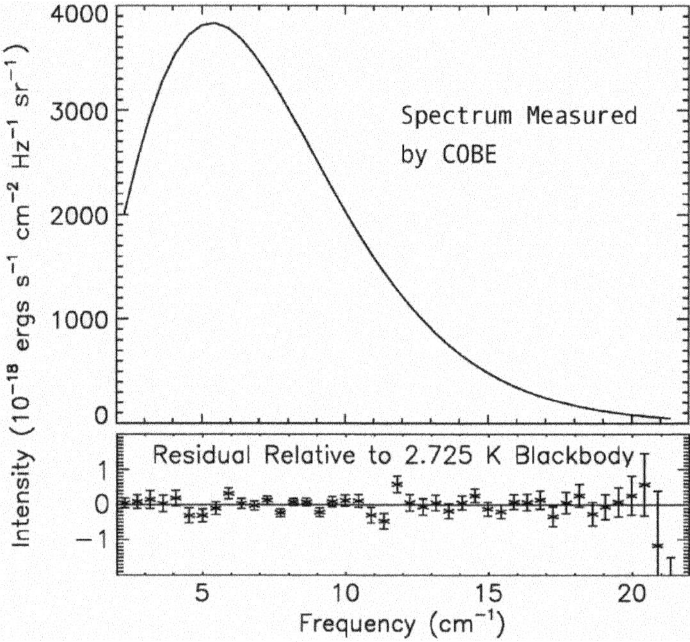

FIGURE 17.3 The spectrum of the cosmic background radiation observed by the FIRAS spectrophotometer on the COBE satellite (Fixsen et al., 1996). The horizontal axis is frequency measured in inverse centimeters. The vertical axis is the intensity measured in energy per unit time, per unit area, per unit solid angle, per unit frequency. The lower plot shows the data plotted as the difference between the observed intensity and that predicted by the Planck radiation law for a 2.725 K blackbody, as well as the experimental error bars. The upper plot is the Planck radiation law fitted to that data. Note the difference in intensity scale between the top and bottom plots. The error bars would be smaller than the width of the line in the top plot. This spectrum is a closer fit to the Planck blackbody spectrum than from any experiment done on earth.

nucleus. By measuring the Doppler shifts in the 21-cm line of neutral hydrogen, the velocities of clouds of this gas in the disk can be mapped as a function of the distance r from the center of the galaxy (Figure 17.4). One would expect the velocity $V(r)$ at radius r to be related to the mass $M(r)$ interior to that radius by a relation roughly like

$$\frac{GM(r)}{r^2} = \frac{V^2(r)}{r}. \tag{17.3}$$

Outside a radius that contains most of the mass, $V(r)$ should, therefore, fall off as $r^{-1/2}$. But this is not seen. Rather, in almost all cases, $V(r)$ remains approximately constant as far out as can be measured. (See, e.g., Figure 17.4.) This implies that, even in the outer reaches of the galaxy, $M(r)$ is growing $\propto r$. The inference is plain that almost every galaxy contains a halo of dark, unseen matter perhaps 10 times the mass seen in visible light.

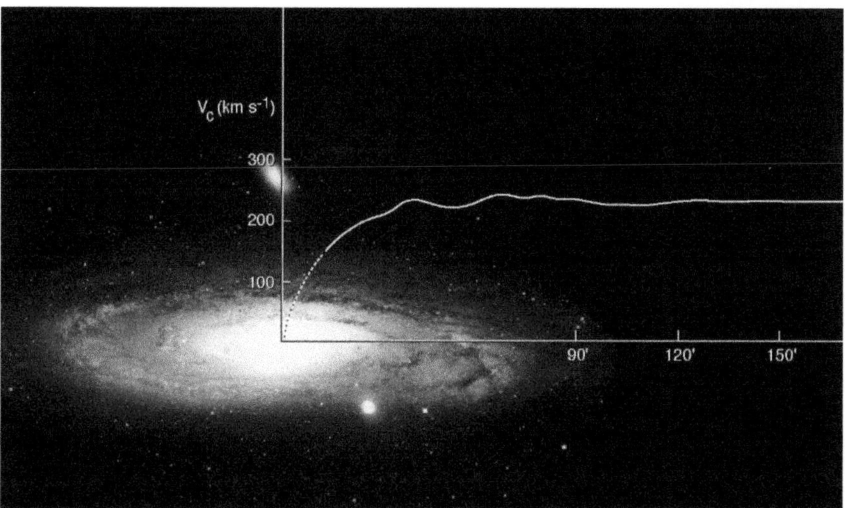

FIGURE 17.4 The rotation curve of the Andromeda galaxy (M31) (from Cram et al., 1980). By mapping the Doppler shift of the 21-cm line of hydrogen, the velocity of rotation of stars in the galaxy can be approximately determined as a function of distance r from its center, measured here in minutes of arc of angular separation as viewed from Earth. This rotation curve, like that of many other galaxies, does not fall off for large r as $\propto r^{-1/2}$, as would be expected if most of the mass were concentrated at the center. Rather, it is level as far out as the rotation of visible matter can be detected, indicating that the mass interior is still growing with r. This is evidence for a halo of dark matter perhaps 10 times as massive as the matter that is visible.

There are many other pieces of evidence that the visible matter and detectable radiation comprise only a small fraction of the mass in the universe, perhaps as little as a few percent. The nature of the "missing mass" is a central problem for cosmology. Speculations range from black holes, dim stars, and other gravitationally bound clumps to new species of particles. Even empty space could have an energy density. In Newtonian mechanics, a constant can be added to the potential energy without any observable effect because the equations of motion are unchanged. But in general relativity *all* energy curves spacetime, and the curvature produced by a constant vacuum energy could be detected. If empty space does have an energy density, it can be detected, if by no other means, by its effect on the expansion of the universe itself. We rely on gravity to detect dark energy.

17.2 The Expanding Universe

The spectra of starlight from galaxies outside our local group are redshifted as illustrated in Figure 17.5. Interpreted as a Doppler effect in flat spacetime, this redshift means that the galaxies outside the local group are all moving away from

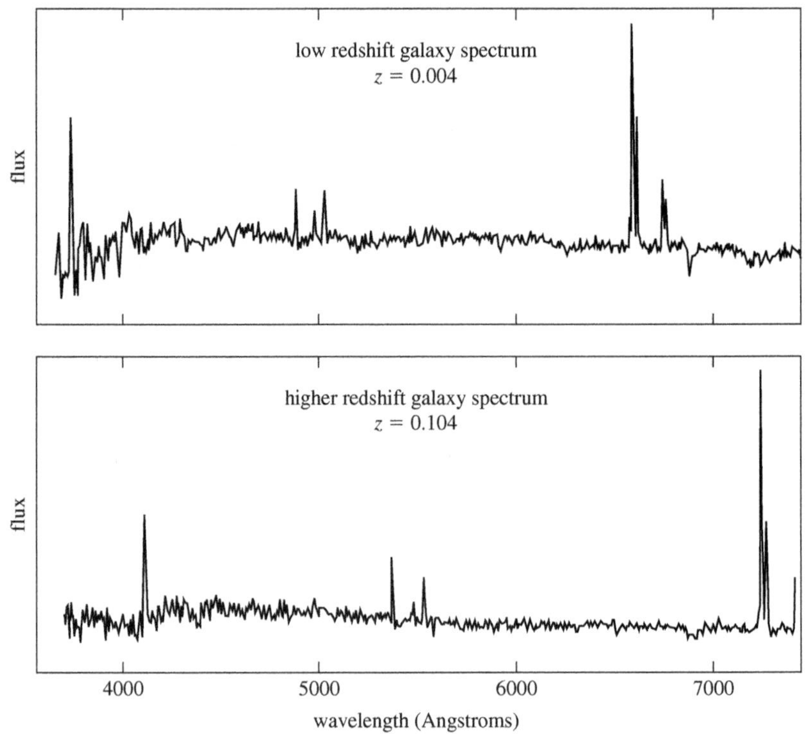

FIGURE 17.5 The cosmological redshift. The spectra of two galaxies are shown as a plot of received intensity vs. wavelength in Angstroms (1 Angstrom $= 10^{-8}$ cm). The bright lines of the two spectra correspond when shifted by $\Delta\lambda/\lambda \equiv z = .1$. That is the cosmological redshift.

us with a velocity V related to the shift in wavelength $\Delta\lambda/\lambda$ by the Doppler formula

$$V/c = \Delta\lambda/\lambda \equiv z, \tag{17.4}$$

where we have introduced the usual astronomical designation for redshift, z, and assumed $V \ll c$.

For galaxies sufficiently close that their distance can be measured, a simple linear relation between the velocity of recession V and the distance d is observed, called Hubble's law:

Hubble's Law

$$V = H_0 d. \tag{17.5}$$

The constant H_0 is called the Hubble constant. Observations that will be described shortly determine it to be

The Hubble Constant

$$H_0 = 72 \pm 7 (\text{km/s})/\text{Mpc}. \tag{17.6}$$

BOX 17.1 Distance Scales in Cosmology

This list of distances from the Earth is intended to give you a rough feel for the distance scales involved in cosmology. The distances are quoted in terms of the *parsec* —the standard distance unit in cosmology (1 pc = 3.09 $\times$ 10^{18} cm = 3.26 light-years). However, the range of scales is such that it is useful to deal with microparsecs (μpc), parsecs (pc), kiloparsecs (kpc), megaparsecs (Mpc), and gigaparsecs (Gpc). The numbers in this table are all rough order of magnitudes, and in some cases (indicated by ~) where typical scales are quoted the variation can be larger than an order of magnitude.

- Distance to the Sun: 5 μpc
- Distance to the nearest star: 1 pc
- Distance to the galactic center: 10 kpc
- Distance to a galaxy in the local group of about 30 galaxies (e.g., Andromeda at 725 kpc): 50–1000 kpc
- Distance to the nearest large cluster (the Virgo cluster of several thousand galaxies): 20 Mpc
- Distance scale of largest structures in the distribution of galaxies: ~ 100 Mpc
- Distance to the edge of the visible universe: 14 Gpc

The distance unit used here, the megaparsec (Mpc), is a conventional one for intergalactic distances. One parsec is 3.08×10^{18} cm $=$ 3.26 light-years, and a megaparsec is a million parsecs. To get a better feeling for the scale of cosmological distances, see Box 17.1.

The recession of galaxies away from us does not imply that we are at the center of the universe. Indeed, Hubble's law implies that there is no center that can be deduced from the expansion itself. Observers in another galaxy would see every other galaxy receding from *them* according to Hubble's law with the same constant, H_0. This can be seen qualitatively from the pictures in Figure 17.6.

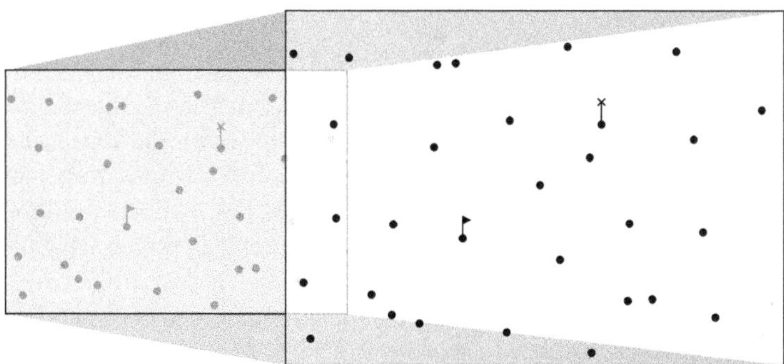

FIGURE 17.6 The pattern of dots on the right was obtained by expanding that on the left by a factor of approximately 20%, keeping the size of the dots the same. Pick any point (a "galaxy") in the box at left, for instance the one labeled by the flag, and measure the distance from it to a few other points. Those distances are increased by 20% in the box on the right. Pick any other galaxy, for example the one labeled by the cross, and do the same thing finding the same increase. Observers on each galaxy see the distances of all others increasing; no galaxy is the center of the expansion. You can make the expansion yourself with a copy machine (Problem 6).

More quantitatively, let's write Hubble's law in vector form

$$\vec{V} = H_0\vec{d}, \tag{17.7}$$

where $\vec{d}$ is the displacement vector from us to any galaxy. Consider a particular galaxy located at $\vec{d}_{\rm them}$ with velocity $\vec{V}_{\rm them}$. Observers in this galaxy will measure a velocity $\vec{V} - \vec{V}_{\rm them}$ for a galaxy we see moving with velocity $\vec{V}$, assuming $|\vec{V}| \ll c$. From (17.7)

$$\vec{V} - \vec{V}_{\rm them} = H_0(\vec{d} - \vec{d}_{\rm them}). \tag{17.8}$$

But $\vec{d} - \vec{d}_{\rm them}$ is the displacement vector from their galaxy. Observers in that galaxy therefore see an expansion governed by (17.5), just as we do. Neither galaxy is the center of the expansion.

Hubble's law is a phenomenological relationship between redshift and distance. It holds for galaxies far enough away that the expansion velocity dominates any velocities they might have acquired from the gravitational attraction of other galaxies nearby. But the galaxies for which it holds must not be so far away that the effects of spacetime curvature become important or that the universe expands significantly in the time it takes their light to travel to us. The Hubble constant is inferred by measuring the redshifts and distances of galaxies satisfying these criteria. (The redshift when these assumptions don't hold is discussed in the next chapter.) In the following we describe in a bit more detail how the relevant distances are determined.

It's not an easy task to measure the distance to the galaxies needed to establish Hubble's law. Distances to nearby stars can be determined precisely. However the cosmological redshift can be measured accurately only for galaxies sufficiently far away that the recession velocity dominates their local motions due to the attraction of other galaxies that may be nearby. The following discussion gives a greatly simplified description of how those nearby distances are connected to the ones far away.

Distance to nearby stars can be determined by triangulation using the Earth's orbit as a baseline (Figure 17.7). The Hipparcos astrometric satellite determined the distances to approximately 120,000 stars, mostly within the solar neighborhood, of which about 15,000 within approximately 100 pc of the Sun have distances determined to better than 10%. That 100 pc is less than a ten-millionth of the size of the visible universe, but these distances are the ones on which all the others are based.

The key to determining distances beyond those that can be found by triangulation is the idea of a *standard candle*. A standard candle is an object whose luminosity can be inferred from a physical property that can be independently determined. Consider stars as a possible example. The luminosity L of a nearby star can be determined from its apparent brightness f (energy flux at Earth) and distance d determined by triangulation using the inverse square law:

$$f = \frac{L}{4\pi d^2}. \tag{17.9}$$

Measure the flux, f, know the distance, d, infer the luminosity, L.

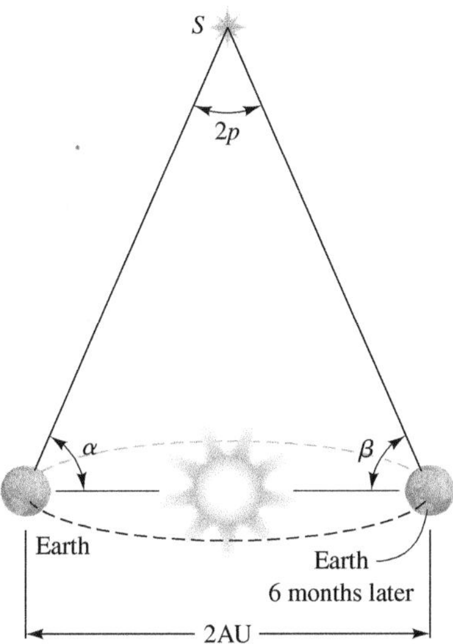

FIGURE 17.7 Triangulation using the Earth's orbit as a baseline can be used to determine the distances to nearby stars. The angular position of the star is observed from two points on the Earth's orbit giving the angles α and β and therefore the *parallax* p by the relation $2p \equiv \pi - (\alpha + \beta)$. For the small angles that actually occur, the distance is $d = (AU)/p$, where the astronomical unit AU is the semi-major axis of the Earth's orbit. (If p is measured in seconds of arc, the distance in parsecs is given by $d = 1/p$ pc—the original definition of this unit.) Parallaxes of greater than 10 milliarcsec were measured to better than 10% for about 15,000 stars by the Hipparcos astrometric satellite, thus surveying the solar neighborhood out to a distance of order 100 pc. This figure is greatly exaggerated. Even at the 70,000 AU distance to the nearest star, α-Centauri, the point S would be several kilometers above the top of the page were the figure drawn to scale.

Suppose it turned out that all stars of a certain blue color to which distances could be determined by triangulation had the *same* luminosity. These blue stars would be standard candles. Once identified by their color, their luminosity would be known—calibrated by the triangulation measurements of nearby blue stars. The distance of blue stars too far away to be determined directly could be found by using the inverse square law in reverse. Measure f, know L, infer d. Unfortunately, blue stars are not good standard candles; their luminosity varies with age and composition among other things. But the idea of the distance ladder is the same. Use distances at one step to calibrate a standard candle that can be used to determine distances in the next step.

The true story of cosmological distances is one of consistency among many different interlocking ladders based on many different kinds of standard candles (and indeed standard rulers as well). It is not possible to review all that here. Rather, we will just illustrate the idea by describing a three-step ladder that takes us out to the largest distances measured.

In preparation for this discussion, here is as good a place as any to introduce the astronomical measures of luminosity and apparent brightness. Astronomers use logarithmic measures of luminosity and flux called *absolute magnitude* and *apparent magnitude*, respectively. Without entering into the complexities of how these are precisely defined, you can take them to be given by the following relations for the purposes of this text. The *absolute magnitude* M is related to the total luminosity L by the formula

$$M = -2.5 \log_{10}(L/L_{\odot}) + 4.74, \tag{17.10}$$

where $L_{\odot}$ is the solar luminosity 3.85×10^{33} erg/s. There are various other kinds of magnitude to measure the luminosity in the different wavelength bands that can be observed. The apparent magnitude m is related to the flux f by

$$m = -2.5 \log_{10}(f/f_{\odot \text{ at 10 pc}}) + 4.74, \tag{17.11}$$

where $f_{\odot \text{ at 10 pc}}$ is the flux of the Sun at the Earth if it were 10 pc away—3.21×10^{-7} erg/(cm$^2 \cdot$ s). The apparent magnitude of an object equals its absolute magnitude if it is 10 pc away. The difference $m - M$ is a logarithmic measure of f/L. This called the *distance modulus* because, were space flat, $m - M = 5 \log_{10}(d/10 \text{ pc})$, where d is the distance away. That is the content of the flat space inverse square law in astronomical notation.

We now describe three different standard candles that together make one distance ladder to faraway galaxies and calibrate Hubble's law.

- *The Main Sequence*. As mentioned before, individual stars of a given color are not good standard candles. But the statistical properties of the relation between their luminosities and colors can be used as a standard candle. Figure 17.8 shows a plot of the luminosity-color relationship for about 14,000 stars whose distances were accurately determined by the Hipparcos satellite and used to infer their luminosities from their apparent brightnesses. The main sequence is the broad swath from upper left to lower right consisting of stars that are burning hydrogen to make helium. Suppose apparent brightnesses and colors are measured for stars in a cluster too far away to determine its distance by triangulation. The physical association in a cluster means all its stars are at approximately the same distance from us. The distance at which the inferred main sequence of the cluster matches the main sequence determined by triangulation measurements is the distance to the cluster. Measurements of stars whose distances were determined by triangulation have calibrated a standard candle—the main sequence—that can be used to determine the distance to faraway clusters.
- *Cepheid Variable Stars*. Cepheid variable stars are massive, bright, yellow stars with surface temperatures not unlike the Sun, but much brighter in luminosity. The luminosities of Cepheid variables in sufficiently nearby clusters can be found from the distances determined in the previous step in the distance ladder. (There are also a few parallax measurements for Cepheids.) These show a well defined empirical relation between the absolute luminosity of Cepheids and the periods of their variation that is shown in Figure 17.9. This correlation makes Cepheid variables a standard candle. Identified by their color and variability,

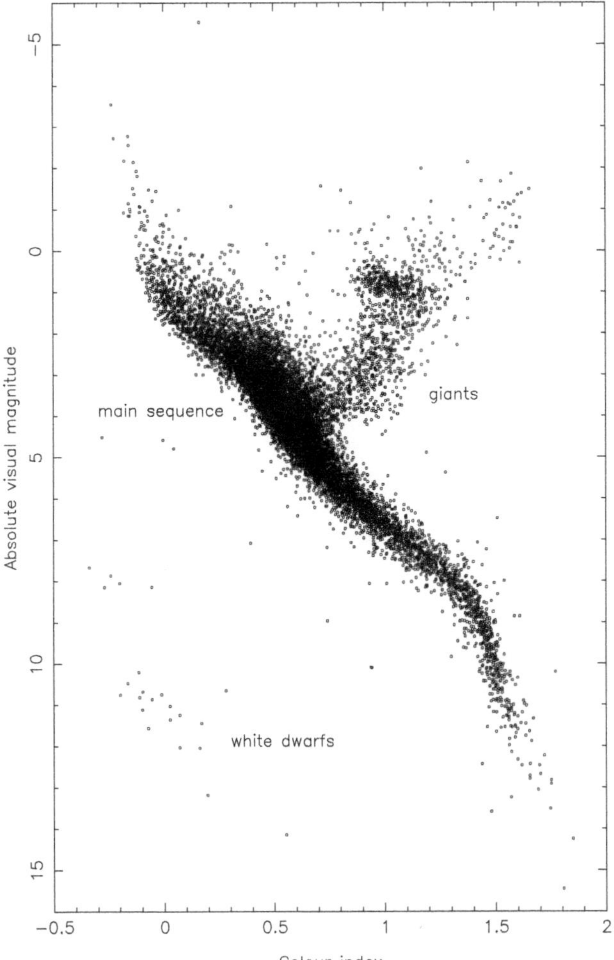

FIGURE 17.8 A Hertzsprung-Russell diagram showing the main sequence. The observed relationship between luminosity and color is shown for approximately 14,000 stars in the solar neighborhood whose distances are determined from parallaxes measured by the Hipparcos satellite (cf. Figure 17.7). The luminosity of a star is inferred from its distance and apparent brightness using the inverse square law (17.9). Luminosity in a range of wavelengths around visible light is plotted vertically in terms of absolute visual magnitude. This is a logarithmic measure of luminosity similar to the magnitude defined in (17.10) but referring to the limited range of wavelengths observed. Brighter stars have smaller magnitudes. The horizontal axis is a measure of the star's color and, therefore, roughly of its surface temperature. Bluer, hotter stars are to the left; redder cooler stars are to the right. The Sun has an absolute visual magnitude of 4.82 and a color index of .88 on the scales used. The main sequence is the broad swath from upper left to lower right consisting of stars burning hydrogen to make helium.

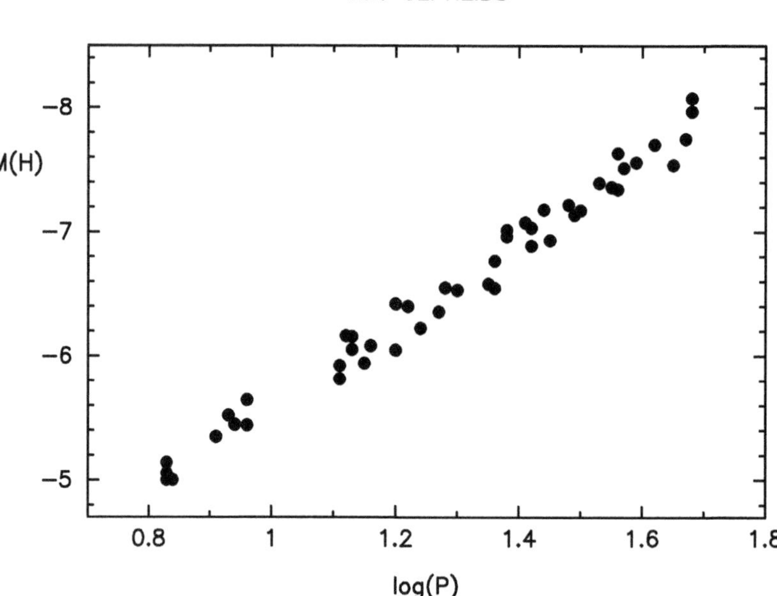

FIGURE 17.9 The period-luminosity relationship for Cepheid variable stars in the Large Magellanic Cloud—a satellite galaxy of our Milky Way (Persson et al., 2002). The figure shows a plot of the absolute magnitudes of stars in the infrared H-band centered about 1.6 μm versus the log (base 10) of their periods in days. The close correlation between period and luminosity makes Cepheid variables standard candles. The luminosity of a Cepheid variable can be determined from its period, and its distance can be determined from that and its apparent brightness using the inverse square law.

their distances can be inferred from a measurement of their apparent brightness and period. Calibrated by main sequence matching, Cepheid variables can be used to determine distances to galaxies too far away for the measurements of the main sequence of clusters to be feasible. Similarly, Cepheid variables form the basis for a next step in the distance ladder—they provide distances from which absolute luminosities for even brighter objects can be calibrated.

- *Type Ia Supernovae.* Supernovae are catastrophic explosions of stars whose peak brightness can rival that of the whole host galaxy (Figure 17.10). They are, therefore, detectable at great distances. They typically rise to peak brightness over a period of a few weeks and die away more slowly over a period of a few months. They come in several varieties with different mechanisms for the explosion.

  Type Ia supernovae (SNIa) are not the brightest supernovae that result from the collapse of a massive star that has exhausted its thermonuclear fuel (recall the discussion at the beginning of Chapter 12). Type Ia supernovae result when a normal star in mutual orbit around a white dwarf sheds mass on its much more compact companion. As the mass of the white dwarf nears the maximum mass that can be supported by the pressure of degenerate electrons

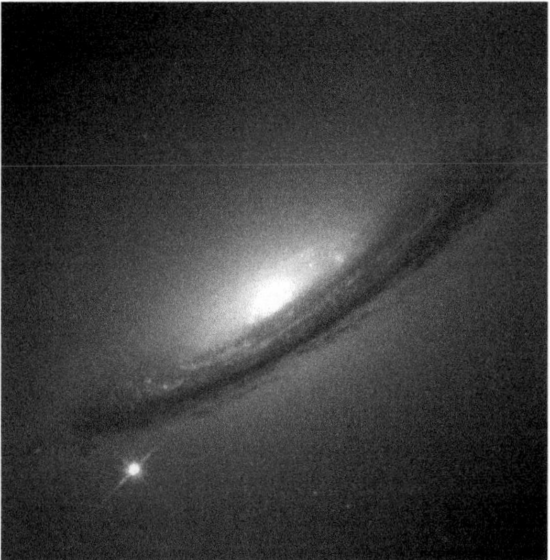

FIGURE 17.10 Supernova 1994D in the outskirts of the galaxy NGC 4526. This example of a type Ia supernova shows that at peak brightness they rival the cores of galaxies in luminosity. That is why they can be detected to great distances. At approximately 20 Mpc away, this galaxy is "nearby" on the distance scales of the universe. How long ago did the supernova go off?

(see Box 12.1 on p. 257), the star erupts in a powerful nuclear burning wave that releases enough energy to completely disrupt it and create the explosion. Since the maximum mass of a white dwarf is fixed ($\approx 1.4 M_{\odot}$), there is some similarity in basic mechanism between one SNIa and the next and, consequently, some similarity in their peak luminosities.

The peak luminosity of SNIa's can be determined from their apparent peak magnitudes if they occur in galaxies whose distances are known from Cepheid variable stars. The peak luminosity is similar from one SNIa to another, and there is an even tighter empirical correlation between peak brightness and time it takes that brightness to decay. SNIa can therefore be used as standard candles. The relation between peak brightness and decay time that was calibrated with Cepheid variables can be used to extend the distance scale beyond the range that Cepheid variables can be detected.

The result of this detailed astronomy and three-step distance scale is shown in Figure 17.11. This is a plot of the recession velocities vs. distance for a sample of galaxies whose distances are determined by SNIa standard candles. The evident linear relationship fixes the Hubble constant H_0 at 72 ± 7 (km/s)/Mpc. That's how fast our universe is expanding right now.

Once the constant H_0 is determined from galaxies that are not too far away, Hubble's law can be used to determine the distance to even further objects. Measure the redshift, then determine the recession velocity from (17.4) and the dis-

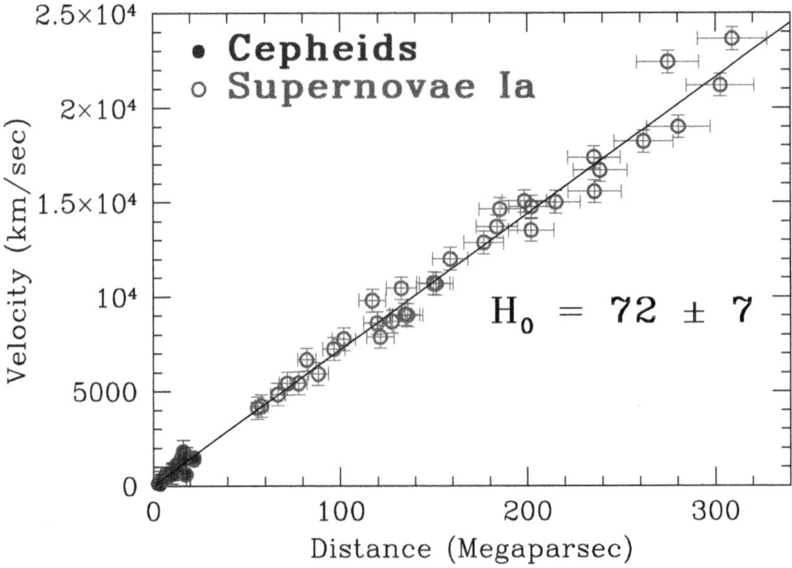

FIGURE 17.11 A Hubble diagram (Freedman et al., 2001). This figure shows the velocity of recession of galaxies as determined from their red shifts plotted against their distances, as determined by measurements of Type Ia supernovae. The nearer galaxies represented by closed circles in the lower left-hand corner contain both Type Ia supernovae and Cepheid variable stars. The distance to these galaxies can therefore be determined from the Cepheid variables and used to calibrate the SNIa's as standard candles. This relation then determines the distance to the more distant SNIa's represented by open circles. There is an approximate linear relation between velocity and distance, which is Hubble's law with a Hubble constant $H_0 = 72 \pm 7$ (km/s)/Mpc.

tance from (17.5). For larger distances the redshift can still be used as a distance indicator, but Hubble's law must be corrected for the curvature of the universe and its evolution over time, as we will see in Chapter 19.

17.3 Mapping the Universe

How is the detectable matter and radiation in the universe organized on the largest distance scales? How did this organization change over time? To answer such questions, the location and distribution of matter and radiation in the universe must be mapped. This is not easy. The distances are vast; the times are long. We live approximately 13 billion years after the big bang. In rough terms this means that, using light, we could in principle have observed at most a region of the universe that is approximately 26 billion light-years ($\sim 10^{23}$ km!) in diameter. In this section we present a few maps of this very large place. These maps provide compelling evidence that on the largest scales the universe is *isotropic*—the same in one direction as in any other—and *homogeneous*—the same in one place as in any other.

The Evidence for Isotropy

Mapping the Radiation

The most straightforward maps to make are of properties of the matter versus angular position on the sky. The most important of these is the map of the temperature of the cosmic background radiation.

If we naively extrapolate the Hubble expansion backward in time keeping the velocity constant, we arrive at a singular state at a time (*the Hubble time*),

$$t_H \equiv 1/H_0 \sim 14 \text{ billion years,} \tag{17.12}$$

when the distances between galaxies go to zero. Of course, there is no reason to believe this constant velocity extrapolation because the velocity of the galaxies should be changing with time according to the dynamic laws of gravity. Neither is there any reason to believe that there were galaxies at such an early time. But, in fact, we will see in the next chapter that general relativity *does* predict such a singularity at a comparable time. This is the big bang—a time of infinite density, infinite temperature, and infinite spacetime curvature. Near this early time matter and radiation had not yet condensed into galaxies but were in equilibrium with each other in a smooth, hot fluid described by a temperature.

As the universe expanded, both matter and radiation cooled. Some hundreds of thousands of years after the big bang the temperature dropped enough that previously free electrons combined with nuclei to make neutral, transparent matter—mostly hydrogen and helium. Light emitted at that time when the temperature was approximately 3000 K has been traveling to us ever since and forms the cosmic background radiation. The intervening expansion has cooled the radiation to a temperature of 2.73 K above absolute zero. A map of the temperature of this radiation on the sky is as close as we can come to a picture of the universe at the big bang. Figure 17.12 shows a series of three maps on different temperature scales of the cosmic background radiation from measurements made by the Cosmic Background Explorer (COBE) satellite. The top picture shows the temperature distribution at a resolution well above a millikelvin. It is completely uniform at 2.73 K! The next shows the same data at a resolution of a millikelvin. It shows an anisotropy entirely attributable to Doppler shifts due to our motion relative to a frame in which the radiation is nearly exactly isotropic. This effect of motion is subtracted out of the bottom picture, whose resolution is a microkelvin. Across the center of the picture one sees the contribution of the radiation from our own galaxy. The remaining patches are fluctuations of only tens of millionths of a kelvin. Evidently the early universe was remarkably *isotropic*—the same in one direction as in any other. The fluctuations away from exact isotropy are important, however, because they are signatures of density fluctuations that in the intervening time grew by gravitational attraction to become the galaxies we see today.

Mapping the Matter

A map of the angular positions of galaxies on the sky gives us a map of the universe today in much the same way that the map of the temperature of the cosmic

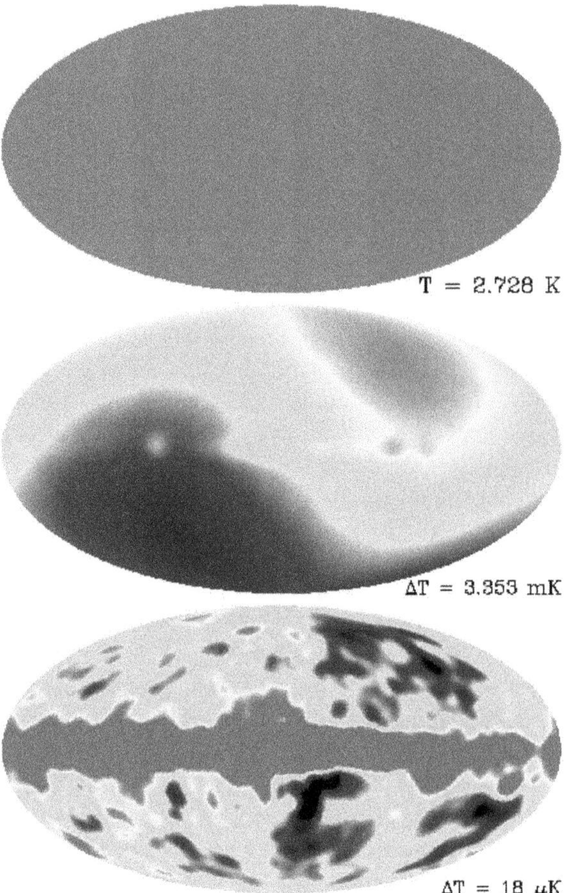

FIGURE 17.12 The temperature of the cosmic background radiation as measured by the COBE satellite (Bennett et al., 1996). Three maps of the temperature on different scales are shown. The maps are in galactic coordinates, so the plane of the galaxy is the equator. At the top is a picture that could be a map at a temperature resolution well above a millikelvin. The temperature is completely uniform at 2.73 K. The middle map shows the temperature at a resolution of millikelvin. There is a dipole anisotropy attributable to the motion of the solar system with respect to the rest frame of the background radiation. When this anisotropy is subtracted out, the bottom figure is obtained, which shows the remaining anisotropies at a resolution of a microkelvin. The dark strip through the center is due to residual radiation from the galaxy. The remaining variations correspond to fluctuations at the time the radiation was emitted, some hundreds of thousands of years after the big bang.

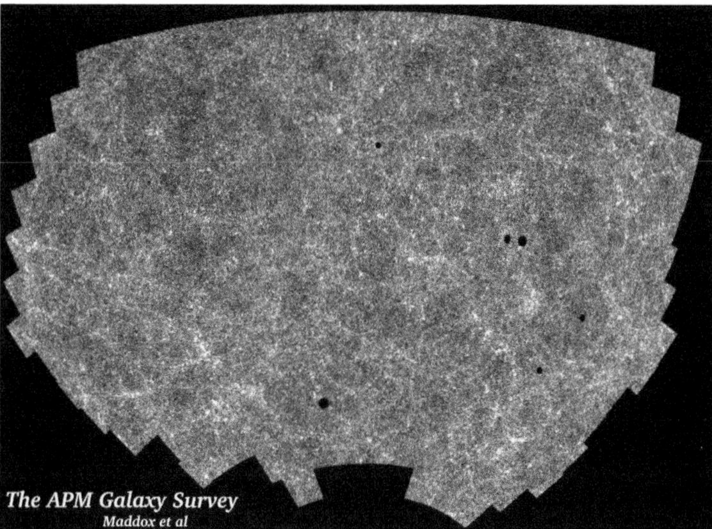

FIGURE 17.13 The angular distribution of galaxies from the APM survey (Maddox et al., 1990). The figure shows roughly one-quarter of the sky divided up into small squares shaded according to the number of galaxies counted in each square. (More is whiter.) On these large angular scales there are about as many galaxies in one direction as in any other.

background radiation gave us a picture of the very early universe. Figure 17.13 shows the map of the galaxies obtained by the A(utomatic) P(late) M(achine) survey. A large region of the sky was divided into small areas and the number of galaxies brighter than a certain limit was counted in each area. The grey scale on the figure indicates that number. Much more structure is apparent in this picture than in Figure 17.12. This is structure that emerged over the intervening time from the clumping action of gravity. Yet, averaged over large scales this picture is much the same in one direction as in any other. The universe today is isotropic.

The Evidence for Homogeneity

An isotropic universe (the same in one direction as in any other) is not necessarily homogeneous (the same in one place as in any other).[2] We might be in the center of a universe where the density of galaxies decreases rapidly away from us in all directions and still see the distribution as isotropic. However, a three-dimensional map of the distribution in space reveals that the distribution of galaxies is approximately homogeneous on distance scales above several hundred megaparsecs.

The construction of such a map is a large project. To probe the largest distances, only the redshift itself is available as an indicator of distance through

[2]Neither is a homogeneous universe necessarily isotropic. Suppose one kind of galaxy were observed to be moving in one direction relative to the others. That velocity could be the same everywhere in the universe (homogeneous), but the direction along the motion would be preferred.

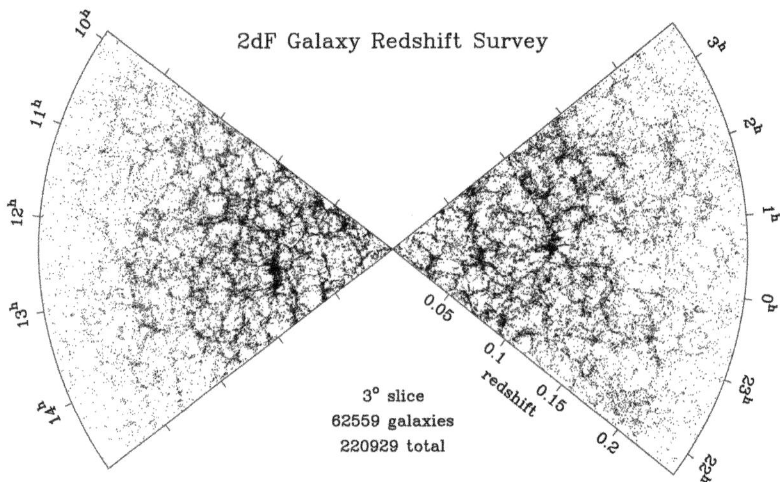

FIGURE 17.14 The 2dF redshift survey. This shows a map of the locations of 62,559 of the 220,929 galaxies whose angular positions and redshifts were measured in the 2dF Galaxy Redshift Survey (Colless et al. 2001). The radial location is given in units of redshift in accord with Hubble's law. It might seem that the galaxies are thinning out at larger redshifts, but that is merely a selection effect. Only the brightest galaxies can be seen far away, so there are fewer galaxies total that can be seen at large redshifts than smaller. In the inner regions, where the survey is more complete, voids, knots, and filaments are evident, but on the largest scales the galaxy distribution is much the same in one part of the universe as another. That's homogeneity.

Hubble's law. Spectra and location are needed for many thousands of galaxies. The further the survey goes, the more galaxies must be measured because their number grows roughly with the cube of the distance from us. Two large surveys in progress at the time of writing are the 2dF survey and the Sloan Digital Sky survey. Preliminary results of the 2dF survey are shown in Figure 17.14. Certainly, this distribution exhibits structure. On small scales there are knots and filaments where galaxies are concentrated and voids where there are very few galaxies. Yet on larger scales—above differences in z of about .02—the inner part of the distribution is much the same in one place as in any other. This is compelling evidence that the universe is *homogeneous* on large-distance scales. We do not seem to be at a special place in it.

Problems

1. A distant galaxy has a redshift $z = \Delta\lambda/\lambda$ of .2. According to Hubble's law, how far away was the galaxy when the light was emitted if the Hubble constant is 72 (km/s)/Mpc?

2. [E] Parallaxes greater than $.005''$ can be measured for about 120,000 stars in the immediate solar neighborhood. *Estimate* the number of stars per cubic pc in the solar neighborhood.

3. [P] Planck's radiation law specifies the energy dE in a blackbody gas at temperature T that is incident in a small time dt, on a small area dA, from a small solid angle $d\Omega$ about the normal direction, in a small frequency range $d\omega$, as

$$dE = \frac{\hbar\omega^3}{4\pi^3 c^2} \frac{1}{\exp(\hbar\omega/k_B T) - 1} dt\, dA\, d\Omega\, d\omega.$$

Assuming that the data in the top part of Figure 17.3 fit a blackbody spectrum, calculate the temperature of the radiation. (Note that the frequency plotted in Figure 17.3 is $\omega/2\pi c$.) (*Hint*: There are several ways to do this, some easier than others.)

4. [E] The distance to the Andromeda galaxy (M31) is 725 kpc. Use the data in Figure 17.4 to *estimate* the mass in $M_\odot$'s inside a sphere extending $150'$ from Andromeda's center.

5. [E,S] Use the data from the 2dF Survey in Figure 17.14 to *estimate* the distance scale above which the universe is approximately homogeneous.

6. *Expansion by Copy Machine* For a more compelling demonstration that the expansion of the universe doesn't have a center, try the following experiment related to Figure 17.6. Take a transparency, such as used with overhead projectors, and cover it approximately uniformly with dots representing galaxies, as in one of the boxes in Figure 17.6. Copy these dots onto another transparency at 20% expansion. (Real galaxies won't expand in size, so ignore the copier's expansion of the dots.) Line up the transparencies so the position of one galaxy after the expansion is on top of its position before. The rest will be seen to be moving outward in all directions. Try the same thing with a different galaxy. What do you see?

7. Radio signals are received from the vicinity of a star exactly like our Sun that has an apparent magnitude of 3.9. How long ago were these signals sent to us?

8. The main sequence of a distant cluster of stars is approximately fit by the relation (apparent magnitude) = 6 (color index) +18. Assuming stars are of the same kind as those in the solar neighborhood, approximately how far away is this cluster?

9. A Cepheid variable star is observed with an apparent magnitude of 22 and a period of 25 days. How far away is this star?

CHAPTER 18

Cosmological Models

The observations described in the last chapter show our universe to be approximately homogeneous and isotropic on spatial distance scales above several hundred megaparsecs. The simplest cosmological models enforce these symmetries exactly as a first approximation. For instance, the matter in galaxies and the radiation are approximated by smooth density distributions that are exactly uniform in space. Similarly, the geometry of spacetime incorporates the homogeneity and isotropy of space exactly. These simplifying assumptions define the *Friedman–Robertson–Walker* (FRW) family of cosmological models, which are the subject of this chapter.

18.1 Homogeneous, Isotropic Spacetimes

Let's begin with the spacetime geometry of a homogeneous, isotropic cosmological model. A homogeneous, isotropic spacetime is one for which the geometry is spherically symmetric about any one point in space (isotropic) and the same at one point in space as at any other (homogeneous). The homogeneity and isotropy are symmetries of *space* and not of *spacetime*. Homogeneous, isotropic spacetimes have a family of preferred three-dimensional spatial slices on which the three-dimensional geometry is homogeneous and isotropic. (Spacelike surfaces as three-dimensional slices of four-dimensional spacetime were discussed in Section 7.9.)

The Flat Robertson–Walker Metric

The simplest example of a homogeneous, isotropic cosmological geometry is described by the line element

Line Element for Flat FRW Model

$$\boxed{ds^2 = -dt^2 + a^2(t)(dx^2 + dy^2 + dz^2)} \tag{18.1}$$

where $a(t)$ is a function of the time coordinate t called the *scale factor*. The metric (18.1) is a homogeneous, isotropic cosmological model because its spacetime can be divided up as $ds^2 = -dt^2 + dS^2$ into time and homogeneous, isotropic spatial geometries with line element dS^2. The geometries of the t = const. spacelike surfaces are described by the line element

$$dS^2 = a^2(t)(dx^2 + dy^2 + dz^2). \tag{18.2}$$

At any one instant t, new coordinates $X = a(t)x$, $Y = a(t)y$, $Z = a(t)z$ can be introduced so that (18.2) takes the form $dS^2 = dX^2 + dY^2 + dZ^2$. The geometry of each $t =$ const. spatial slice is thus flat three-dimensional space—manifestly homogeneous and isotropic. Equation (18.1) is called the *flat Robertson–Walker metric*, not because the spacetime is flat but because the geometry of the spatial slices is flat. It's a *Friedman*–Robertson–Walker (FRW) model when the scale factor obeys the Einstein equation.

The distributions of galaxies and radiation in a FRW model are smoothed out into a *cosmological fluid*. An individual galaxy may be thought of as a particle in this fluid located by the three coordinates x^i at any time. The velocity dx^i/dt of a galaxy must vanish in the approximations of the FRW models; otherwise, it would establish a preferred direction contradicting the assumption of isotropy. The coordinates (x, y, z) are therefore *comoving*—an individual galaxy has the same coordinates x^i for all time. In a similar way the x^i define the rest frame of the radiation—the one in which the CMB temperature exhibits no dipole anisotropy of the kind illustrated in Figure 17.12.

If $a(t)$ increases in time, the line element (18.1) describes an expanding universe. To see that, consider the world lines of a pair of galaxies separated by coordinate intervals Δx, Δy and Δz. The *coordinate* distance between them, $d_{\text{coord}} \equiv (\Delta x^2 + \Delta y^2 + \Delta z^2)^{1/2}$, remains fixed in time. But the physical distance between them on a surface of constant time, $d(t)$, is defined by the metric (18.2) and given by

$$d(t) = a(t)d_{\text{coord}}. \tag{18.3}$$

This increases with time if $a(t)$ does. That is the sense in which (18.1) describes an expanding universe. It's perhaps natural to ask, Where is the universe expanding from? That kind of question doesn't make much sense in a universe of infinite spatial extent—without boundary—as Example 18.1 helps to show.

Example 18.1. What's expanding? Into What? From Where? Imagine a large lump of dough embedded approximately uniformly with raisins and being baked into raisin bread. As it bakes, the loaf expands. This homey example has some analogies with cosmology. Instead of dough mediating interactions between raisins, think of gravity mediating interactions between galaxies. Just as we view the universe from one galaxy, imagine viewing the universe of baking dough from one raisin. What's expanding is the distance between raisins, and the density of the dough is decreasing.

A realistic, finite, loaf of raisin bread expands into the surrounding space because it has a boundary, i.e., because it's not homogeneous on large distance scales. However, if the loaf was infinite in extent in all directions, without a boundary, exactly homogeneous, it wouldn't make sense to talk about it expanding "into" something. It's just expanding. In a similar way it wouldn't make sense to talk about it expanding "from" somewhere. The expansion would look the same viewed from any raisin—each raisin would see all the other raisins receding radially from it. (Recall the discussion of Hubble's law in Section 17.2.)

So far there is not a shred of observational evidence that the universe of galaxies is contained within a boundary, although it could be. We could assume that the universe is contained within an as yet unseen boundary with an as yet unknown center, but that assumption would not affect the predictions of our observations in the interior and complicate the calculations of them. It's simplest and most elegant to assume that the observed homogeneity and isotropy extend over the whole universe. In an exactly homogeneous model there can be no center and no boundary. In that context it doesn't make sense to talk about the universe expanding into something or from somewhere.

The flat Robertson–Walker metric is not the only homogeneous isotropic spacetime geometry. Any line element of the form

$$ds^2 = -dt^2 + a^2(t)\, d\mathcal{L}^2 \tag{18.4}$$

is a homogeneous, isotropic spacetime if $d\mathcal{L}^2$ is the line element of a time-independent, homogeneous, isotropic, three-dimensional *space*.[1] These are known collectively as *Robertson–Walker metrics.* Flat three-dimensional space is one example, but there are also two curved possibilities. One is the three-dimensional surface of a sphere in four-dimensional Euclidean space. However, we will defer discussion of these curved FRW models until Section 18.6 for two reasons: (1) The flat FRW models are the basis of a simple discussion, of which very little needs to be changed to include the curved cases. (2) The flat Robertson–Walker metric is not simply an academic exercise. The data to be discussed in the next chapter suggest that it is the flat homogeneous, isotropic model that most closely represents our universe. The simplest model is also the most realistic.

18.2 The Cosmological Redshift

The flat Robertson–Walker geometry (18.1) is time dependent. The energy of a particle will change as it moves in this geometry similarly to the way it would if it moved in a time-dependent potential. For a photon, whose energy is proportional to frequency, that change in energy is the cosmological redshift. Let's now derive the simple relation that gives its form.

To begin let's rewrite the line element (18.1) in polar coordinates:

$$ds^2 = -dt^2 + a^2(t)\left[dr^2 + r^2(d\theta^2 + \sin^2\theta\, d\phi^2)\right]. \tag{18.5}$$

Pick the origin of these coordinates to coincide with our location. Consider an observer in a galaxy a coordinate distance $r = R$ away, as illustrated in Fig-

[1] In fact, coordinates can always be chosen so the most general homogeneous, isotropic spacetime has the form (18.4). A supplement on the book website demonstrates this.

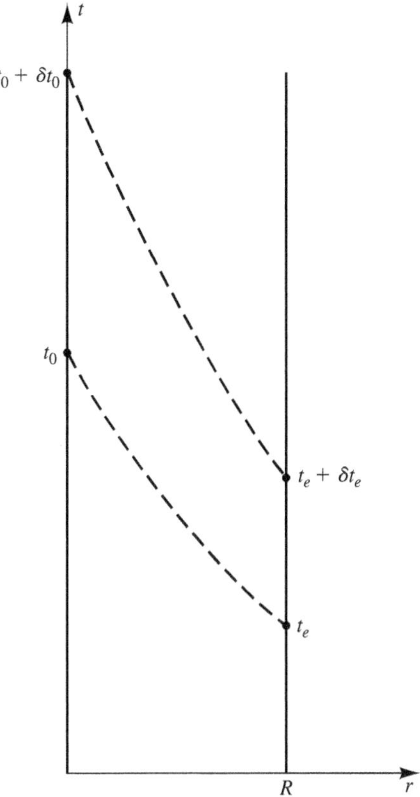

FIGURE 18.1 Two observers in an expanding universe exchanging light signals. The observers are at rest in the surfaces of homogeneity; their world lines are vertical in this t-r spacetime diagram. The observer at $r = R$ emits two light rays at times t_e and $t_e + \delta t_e$ which are received by an observer at $r = 0$ at times t_0 and $t_0 + \delta t_0$. Because the universe is expanding, the curves followed by light rays are not straight lines in (t, r) coordinates. In particular, the interval δt_0 between the times when two photons are received at $r = 0$ is greater than the interval δt_e between the times when they are emitted. That is the cosmological redshift.

ure 18.1. (Remember that the coordinate r is comoving, so $r = R$ labels the same galaxy for all time.) Suppose the observer in the distant galaxy emits a photon with frequency ω_e at time t_e which we receive at the present time t_0. What is the frequency, ω_0, of the received photon?

The pulse of light emitted by the distant observer travels on a radial null curve for which

$$ds^2 = 0 = -dt^2 + a^2(t)\, dr^2. \tag{18.6}$$

In the time between emission at t_e and reception at t_0, the pulse traveled a spatial coordinate distance R. Integrating (18.6) gives

$$R = \int_{t_e}^{t_0} \frac{dt}{a(t)}. \tag{18.7}$$

This relation connects the times of emission and reception with the coordinate distance traveled.

Suppose the observer in the distant galaxy emits a series of pulses spaced by equal short-time intervals δt_e, that is, with (circular) frequency $\omega_e = 2\pi/\delta t_e$. The time interval δt_0 between the pulses at reception can be calculated from (18.7) because all pulses travel the same spatial coordinate separation R:

$$\int_{t_e+\delta t_e}^{t_0+\delta t_0} \frac{dt}{a(t)} = R = \int_{t_e}^{t_0} \frac{dt}{a(t)}. \tag{18.8}$$

Assuming δt_e and δt_0 are small, the integral on the left differs from that on the right by just a small extension of the upper limit and a small contraction of the lower one. The net change must vanish, namely,

$$\frac{\delta t_0}{a(t_0)} - \frac{\delta t_e}{a(t_e)} = 0, \tag{18.9}$$

which in terms of frequencies $\omega = 2\pi/\delta t$ means

$$\boxed{\frac{\omega_0}{\omega_e} = \frac{a(t_e)}{a(t_0)}.} \tag{18.10}$$

Although derived for a spatially flat FRW model, this relation holds in any of the other homogeneous, isotropic models that will be discussed in Section 18.6 (Problem 21). In an expanding universe where $a(t)$ grows with t, the ratio $a(t_e)/a(t_0)$ will be less than 1 and the received frequency ω_0 less than the emitted one ω_e. That is the cosmological redshift. As the universe expands, the frequency of a photon decreases, and its wavelength increases linearly with the scale factor $a(t)$. Thus,

Cosmological Red Shift

$$\boxed{1 + z \equiv \frac{\lambda_0}{\lambda_e} = \frac{\omega_e}{\omega_0} = \frac{a(t_0)}{a(t_e)}.} \tag{18.11}$$

Astronomers call z *the redshift*, as in the phrase "the most distant quasar has a z of 6.6." No redshift at all corresponds to $z = 0$—the redshift of the present moment.

As a special case, consider a galaxy a small distance away at the time of reception so that its coordinate separation R is small. Its distance at reception is, from (18.3),

$$d = a(t_0)R. \tag{18.12}$$

Any light ray from the galaxy travels along a null path ($ds^2 = 0$) to us. From (18.6) the coordinate time Δt that it travels is approximately $\Delta t = a(t_0)R +$

(terms of order R^2). (It doesn't matter whether $a(t_0)$ or $a(t_e)$ is used in this expression—the difference would only be of order R^2.) The time of travel is therefore also d, and $t_e = t_0 - d$, both neglecting R^2 corrections. Evaluating (18.10) for small d gives the fractional change in wavelength:

$$z \equiv \frac{\Delta\lambda}{\lambda} = \left[\frac{\dot{a}(t_0)}{a(t_0)}\right] d \qquad (d \text{ small}), \tag{18.13}$$

where $\dot{a} \equiv da/dt$. This is Hubble's law [cf. (17.4) and (17.5) in $c = 1$ units], and (18.13) gives the connection of the Hubble constant to the geometry of spacetime

$$\boxed{H_0 \equiv H(t_0) \equiv \frac{\dot{a}(t_0)}{a(t_0)}.} \tag{18.14}$$

Hubble Constant

The Hubble constant H_0 we measure today is the value of (18.14) at the present age t_0. Figure 18.2 gives a simple geometric construction of the Hubble constant from the scale factor. Note that observers living at a different time would measure a different Hubble constant. The Hubble constant is not constant in time, but it is a constant in the sense of being a number describing the expansion of the universe from our perspective.

Although usually quoted in units of (km/s)/Mpc, the Hubble constant has the dimensions of an inverse time, as (18.14) makes clear. The inverse of H_0 is called the *Hubble time*, t_H. Its value in units of a billion years (1 Gyr $= 10^9$ yr $=$ 1 billion years) is

$$t_H \equiv \frac{1}{H_0} = 9.78\, h^{-1}\text{ Gyr}, \tag{18.15}$$

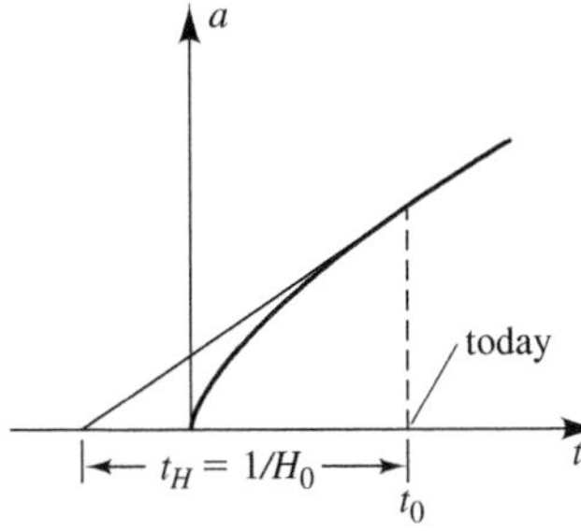

FIGURE 18.2 The Hubble constant related to the scale factor. Equation (18.14) implies that the Hubble constant at time t_0 is inversely related to the intercept of the tangent to $a(t)$ at time t_0. The Hubble time $t_H = 1/H_0$ is an upper bound and a rough estimate of the age of decelerating FRW models for which $\ddot{a}(t) < 0$. That is the case for models dominated by any combination of matter and radiation (Problem 23).

where h is the Hubble constant measured in units of 100 (km/s)/Mpc, namely,

$$h \equiv \frac{H_0}{100\ (\mathrm{km/s})/\mathrm{Mpc}}. \tag{18.16}$$

(Cosmologists like to write their equations in terms of h because, although the Hubble constant is one of the most well-determined parameters characterizing our universe, it is not all that well determined. The value we are using in this text is $h \approx .72$.) The Hubble time is a convenient unit of time in cosmology. As Figure 18.2 shows, it gives a rough estimate of the age of the universe.

18.3 Matter, Radiation, and Vacuum

The FRW models assume a simple model for the cosmological fluid consisting of three noninteracting components mentioned in the last chapter—pressureless matter, radiation, and vacuum. The fluid of galaxies is well approximated by a pressureless gas—*dust* in the parlance of relativists—because the typical random motions of galaxies $\sim 10^2$ km/s give a thermal energy much less than the rest energy. Radiation includes the cosmic background photons but also, for example, neutrino species with zero or small enough rest masses that they move relativistically today. Finally, we allow for an energy of the vacuum. These components did interact in the very early universe, but it is not a bad approximation to assume that they were independent over most of history.

The First Law of Thermodynamics for FRW Models

The entire evolution of a homogeneous isotropic universe is contained in the scale factor $a(t)$. For example, the evolution of the matter, radiation, and vacuum densities in the universe can be derived from it. All these quantities depend only on t because the universe is homogeneous. They are connected to $a(t)$ by the first law of thermodynamics—an expression of local energy conservation. This states that, for any change $d(\Delta\mathcal{V})$ in a volume $\Delta\mathcal{V}$ containing a fixed number of particles, the change in the total energy in the volume is the work done on it minus the heat that flows out of it. Heat flow in any direction would violate the assumption of isotropy. Put differently, the heat flow is zero because homogeneity implies that the temperature T depends only on time, so that no one place is hotter or colder than any other at a moment of time. The first law of thermodynamics for the cosmological fluid in a volume $\Delta\mathcal{V}$ then connects infinitesimal changes in that volume $d(\Delta\mathcal{V})$ to the corresponding infinitesimal change in its total energy $d(\Delta E)$, namely

$$d(\Delta E) = -p\,d(\Delta\mathcal{V}). \tag{18.17}$$

Here, p is the pressure exerted by matter in the volume, and the energy in the volume ΔE is $\rho\Delta\mathcal{V}$, where ρ is the total energy density.

Fixed coordinate intervals Δx, Δy, Δz define a volume in which the number of particles remains fixed because the coordinates (x, y, z) are comoving. The product of these coordinate intervals $\Delta\mathcal{V}_{\text{coord}}$ is not the physical size of the volume. That is given by (7.29) and (18.2) as

$$\Delta\mathcal{V} = a^3(t)\Delta\mathcal{V}_{\text{coord}}. \tag{18.18}$$

Substituting (18.18) in (18.17) and dividing by dt gives

$$\frac{d}{dt}(\rho a^3 \Delta\mathcal{V}_{\text{coord}}) = -p\frac{d}{dt}(a^3\Delta\mathcal{V}_{\text{coord}}). \tag{18.19}$$

Since $\Delta\mathcal{V}_{\text{coord}}$ is independent of time, it can be divided out, yielding

$$\boxed{\frac{d}{dt}[\rho(t)a^3(t)] = -p(t)\frac{d}{dt}[a^3(t)]} \tag{18.20}$$

First Law of Thermodynamics for Cosmology

as the form of the first law of thermodynamics appropriate in a homogeneous, isotropic cosmology. Let's see how it applies to the three kinds of energy assumed in the FRW models. In the epochs where different kinds of matter are not interacting, (18.20) applies to each kind separately.

Matter

As already mentioned, matter in the galaxies is well approximated by a pressureless gas. Since there is no internal energy, the energy density, ρ_m, is just the rest energy density of the galaxies. Then (18.20) becomes

$$\frac{d}{dt}(\rho_m a^3) = 0, \tag{18.21}$$

expressing the conservation of rest mass. Equivalently, ρ_m varies with the scale factor as

$$\rho_m(t) = \rho_m(t_0)\left[\frac{a(t_0)}{a(t)}\right]^3, \tag{18.22}$$

where t_0 is the present instant of time. Thus the overall time dependence of the matter density is determined by the scale factor $a(t)$.

Radiation

For a gas of blackbody radiation at temperature T, energy density ρ_r and pressure p_r are related by

$$p_r = \frac{1}{3}\rho_r, \tag{18.23}$$

and, in units where $c \neq 1$ and ρ_r is an energy density,

$$\rho_r = g\frac{\pi^2}{30}\frac{(k_B T)^4}{(\hbar c)^3}, \tag{18.24}$$

where k_B is Boltzmann's constant converting between kelvins and energy units (e.g., $k_B = 1.38 \times 10^{-16}$ erg/K) and g is the number of degrees of freedom of the zero rest mass particles making up the radiation. For photons $g = 2$ for the two possible polarizations. More realistically, including three species of neutrinos turns out to lead to a formula like (18.24) but with an effective g of about 3.4 that is valid for $k_B T <\sim 1$ MeV and will be adequate for our purposes (see, e.g., Kolb and Turner, 1990).

Inserting (18.23) into (18.20), we find that it can be easily integrated to give

$$\rho_r(t) = \rho_r(t_0)\left[\frac{a(t_0)}{a(t)}\right]^4, \tag{18.25}$$

or, equivalently in terms of the temperature $T(t)$ from (18.24),

$$T(t) = T(t_0)\left[\frac{a(t_0)}{a(t)}\right]. \tag{18.26}$$

Thus the time dependence of the radiation energy density and its temperature are fixed by the scale factor. Temperature varies inversely with the scale factor. If the universe began in a big bang where $a = 0$, the temperature began at infinity and cooled as the universe expanded. Many important large-scale properties of the matter in the universe, such as the primordial abundances of the elements, can be understood as arising from the cooling of an initial thermal equilibrium at very high temperature. Some of them are briefly described in Box 18.1.

Matter dominates radiation now, but the converse was true in the early universe. For any densities now there was some early time when $a(t)$ was smaller than now and ρ_r was bigger than ρ_m. The universe then was *radiation dominated.* From the numbers in (17.1) and (17.2) and the relations (18.22) and (18.25) this happened when $a(t_0)/a(t) \sim 10^3$, when the universe was about 10^{-3} of its present size. Thus, over most of its history to date, the universe's matter dominates radiation.

Vacuum

The energy density of the vacuum, ρ_v, is the final case. At the time of writing, there is no fundamental theory that fixes the value of the vacuum energy. Not even its sign is predicted. To avoid having to consider a multiplicity of cases, we will restrict attention in this text to a vacuum energy that is (i) constant in space and time and (ii) positive as indicated by present observations (Chapter 19). The first law of thermodynamics (18.20) then implies a pressure

$$p_v = -\rho_v, \tag{18.27}$$

BOX 18.1 The Thermal History of the Universe

The temperature is infinite at the big bang that begins the FRW models. At a sufficiently early moment the universe was hot enough that matter was dissociated into its most basic constituents. The abundances of the various kinds of elementary particles were then plausibly set by the conditions of thermal equilibrium. Consider by way of example the abundances of neutrino species. Neutrinos (ν's) and antineutrinos ($\bar{\nu}$'s) interact with electrons (e^-'s) and positrons (e^+'s) through reactions such as

$$\nu + \bar{\nu} \longleftrightarrow e^+ + e^-, \qquad \nu + e^- \longleftrightarrow \nu + e^-, \qquad \text{etc.} \tag{a}$$

If there were too many ν's and $\bar{\nu}$'s, they would annihilate by the first reaction. If there were too few, they would be produced by the annihilation of electrons and positrons. In thermal equilibrium the number density of these particles is such that these reactions balance. It is the same with all other kinds of particles at the high temperatures of sufficiently early moments.[a]

As the universe expands, the temperature drops according to (18.26). The number of reactions per unit time, Γ, drops with temperature for processes such as (a). The densities of particles are dropping because of expansion, energies per particle are declining, and cross sections typically decrease with energy. If the rate Γ drops below the expansion rate of the universe,

$$\Gamma(t) < \dot{a}(t)/a(t) \equiv H(t), \tag{b}$$

then the reaction is no longer fast enough to maintain thermal equilibrium in the face of the expanding universe. The abundances "freeze out" at their value when (b) happens. For neutrinos this happens at a temperature of about 1 MeV at a time of about 1 s after the big bang (see Example 18.2). A cosmic neutrino background with blackbody spectrum is then one of the predictions of the big bang, although currently it is not possible to detect it.

Understanding in detail the relic abundances of particles and nuclei left over from the big bang is a rich and fascinating subject that is important both for cosmology and for the theory of the elementary particles. The particle theory required to understand it in detail puts it beyond the scope of this book. However, we can mention three significant transitions in the thermal history of the universe.

Baryosynthesis ($T \sim 10^{14}$ GeV, $t \sim 10^{-34}$ s). Total baryon number is a conserved quantity for elementary particle interactions below the energy scale of 10^{14} GeV characteristic of the unification of the strong and electroweak forces. Protons and neutrons are familiar examples of baryons, each with a baryon number of $+1$. Antiprotons and antineutrons have baryon numbers of -1. Baryons and antibaryons are not equally represented in today's universe. If they were, there would be a lot more action from matter-antimatter annihilation than observations detect. Were baryon number conserved, this asymmetry could only be attributed to the initial condition of the universe. But above 10^{14} GeV, it may not be conserved, and the abundances of baryons and antibaryons could be set by the conditions of thermal equilibrium, just as the abundances of neutrino species were at much lower energy, as discussed before. Small differences between the interactions of matter and antimatter (for which there is evidence at accelerator energies) could have led to asymmetries in the abundances of baryons and antibaryons in thermal equilibrium. When the temperature dropped below 10^{14} GeV, these would have "frozen out," leading to the asymmetry observed today.

Nucleosynthesis ($T \sim .1$ MeV, $t \sim 3$ min). When the temperature drops through the range of a few tenths of a MeV, the free protons and neutrons combine to make light nuclei. The primordial abundances of the elements is thereby fixed—approximately 75% H, 24% ^{4}He by mass and much smaller but important fractions of other light elements. These abundances are sensitive to the number density of baryons and provide an important test of big bang cosmology. (See Box 19.1 on p. 400 for more details.)

Recombination ($T \sim .3$ eV, $t \sim 4 \times 10^5$ yr). When the temperature drops below a few tenths of an electron volt, electrons and nuclei combine to make atoms, in particular atomic hydrogen.[b] The universe then becomes transparent to radiation and the remaining photons constitute the cosmic background radiation.

Baryon number asymmetry, the primordial abundances of the elements, and the cosmic background radiation are relics of the thermal history of the universe. These relics yield a great deal of information about the early epochs in which they were made.

[a] Gravitons may not be in thermal equilibrium because the same coupling constant G that governs their interactions also governs the rate of expansion of the universe.

[b] To say that electrons and nuclei *re*combine is misleading because they were never combined before this moment, but it is the standard terminology, and this event is called recombination.

which is constant in space and time but is negative. A negative pressure is something like a tension in a rubber band. It takes work to expand the volume rather than work to compress it. The vacuum energy density is often written for historical reasons (in $c \neq 1, G \neq 1$ units) as

$$\rho_v = \frac{c^4 \Lambda}{8\pi G}, \tag{18.28}$$

where Λ is a constant with the dimensions of an inverse squared length called the *cosmological constant.* If there is a nonzero vacuum energy, it remains constant while the energy densities in matter and radiation decay away as the universe expands. The long-term future of a universe that expands indefinitely is dominated by vacuum energy.

Except for (18.24) and (18.28), the preceding discussion of matter, radiation, and vacuum did not commit to one or another system of units. In a familiar $\mathcal{MLT}$ system (e.g., g-cm-s), pressure has units of (force)/(area) $= (\mathcal{ML}/\mathcal{T}^2)/\mathcal{L}^2$, which are exactly the same as those of energy density, (energy)/(volume) $= \mathcal{M}(\mathcal{L}/\mathcal{T})^2/\mathcal{L}^3$. Temperature is traditionally measured in kelvins, for which Boltzmann's constant k_B provides the conversion to energy. In early universe cosmology, temperature is more conveniently measured in energy units, say, electron-volts. The next section introduces the Einstein equation for cosmology, for which ($c = G = 1$) geometrized ($\mathcal{L}$) units are convenient. (Recall Section 9.1.) There both pressure and energy density have units of $\mathcal{L}^{-2}$, whereas temperature has units of $\mathcal{L}$. In geometrized units, the relation (18.24) for the energy density of a radiation gas at temperature T becomes

$$\rho_r = g\frac{\pi^2}{30\ell_{Pl}^2}\left(\frac{T}{\ell_{Pl}}\right)^4, \tag{18.29}$$

where $\ell_{Pl} \equiv (G\hbar/c^3)^{1/2} = 1.62 \times 10^{-33}$ cm [cf. (1.6)] and T is measured in cm. Surprised to see the Planck length in a formula that does not involve gravity? It's really just $\hbar = \ell_{Pl}^2$ in geometrized units.

18.4 Evolution of the Flat FRW Models

Only one consequence of the Einstein equation is needed in addition to the first law of thermodynamics (18.20) to find the time behavior of the scale factor $a(t)$ for a flat FRW model universe. This is the relation (geometrized units)

Friedman Equation for a Flat FRW Model

$$\boxed{\dot{a}^2 - \frac{8\pi\rho}{3}a^2 = 0,} \tag{18.30}$$

where $\dot{a} \equiv da/dt$ and ρ is the energy density in matter, radiation, and vacuum. This dynamical equation is a special case of the *Friedman equation,* which we will meet in (18.63) and derive in Section 22.4 from the Einstein equation. Very roughly, (18.30) can be thought of as a statement that the kinetic energy of expansion just balances the potential energy of gravitational self-attraction for a flat FRW universe. This is discussed in greater detail in Section 18.7.

Before solving (18.30) for how the scale factor depends on time, just evaluating it at the present time t_0 and dividing by $a^2(t_0)$ yields an important connection between the total present density ρ_0 of a flat FRW model and the Hubble constant H_0 defined by (18.14), namely,

$$H_0^2 - \frac{8\pi\rho_0}{3} = 0. \tag{18.31}$$

The present density of a flat FRW model is called the *critical density* and has the value

$$\boxed{\rho_{\text{crit}} \equiv \frac{3H_0^2}{8\pi} = 1.88 \times 10^{-29} h^2 \text{ g/cm}^3.} \tag{18.32}$$

Critical Density

(We'll see later what is "critical" about it.) This total density is divided up among the densities of matter, radiation, and vacuum energy. The relative fractions at the present are conventionally denoted by[2]

$$\Omega_m \equiv \frac{\rho_m(t_0)}{\rho_{\text{crit}}}, \qquad \Omega_r \equiv \frac{\rho_r(t_0)}{\rho_{\text{crit}}}, \qquad \Omega_v \equiv \frac{\rho_v(t_0)}{\rho_{\text{crit}}}, \tag{18.33}$$

where $\Omega_m + \Omega_r + \Omega_v \equiv \Omega = 1$ for these flat models.

Another point to note about the dynamical equation (18.30) is that it determines $a(t)$ only up to a multiplicative constant. If $a(t)$ is a solution, then $Ka(t)$ is a solution for any positive constant K. This reflects the fact that the form of the line element (18.1) is unchanged by sending $a(t)$ into $Ka(t)$ and also sending x^i into x^i/K. This arbitrariness of normalization is a special property of the flat FRW models and won't hold for the spatially curved ones, whose finite radius of spatial curvature would be changed by such a transformation. There are various ways of fixing the arbitrary normalization of the scale factor for the flat FRW models. For our discussion let's normalize the scale factor to be unity at the present time—$a(t_0) = 1$. No physical results can depend on this choice.

An immediate advantage of choosing the normalization $a(t_0) = 1$ is a simple expression for the total density as a function of scale factor

$$\rho(a) = \rho_{\text{crit}}\left(\Omega_v + \frac{\Omega_m}{a^3} + \frac{\Omega_r}{a^4}\right) \qquad (a(t_0) = 1). \tag{18.34}$$

This is correct at the present moment when $a = 1$ from (18.33) and correct at other values of a, because of the way densities vary with a as required by the first law of thermodynamics [cf. (18.22), (18.25), and (18.28)]. The dynamical equation (18.30) can then be written

[2] The notation Ω_Λ for vacuum energy is also widely used at the time of writing.

$$\frac{1}{2H_0^2}\dot{a}^2 + U_{\text{eff}}(a) = 0, \tag{18.35}$$

where

$$U_{\text{eff}}(a) \equiv -\frac{1}{2}\left(\Omega_v a^2 + \frac{\Omega_m}{a} + \frac{\Omega_r}{a^2}\right) \qquad (a(t_0) = 1). \tag{18.36}$$

Equation (18.35) is like the expression for a conserved energy (in this case zero) in Newtonian mechanics. Solving for the evolution of a flat FRW model is the same as solving for the motion of a zero-energy particle in Newtonian mechanics moving in one dimension in the effective potential U_{eff}. The form of $U_{\text{eff}}(a)$ is illustrated in Figure 18.3.

Equation (18.35) can be straightforwardly solved for matter, radiation, and vacuum separately with the following results for the scale factor normalized to unity today. These can be verified by substituting them back in the equation.

- *Matter dominated,* $\Omega_m = 1, \Omega_r = 0, \Omega_v = 0$:

$$a(t) = (t/t_0)^{2/3}. \tag{18.37}$$

- *Radiation dominated,* $\Omega_m = 0, \Omega_r = 1, \Omega_v = 0$:

$$a(t) = (t/t_0)^{1/2}. \tag{18.38}$$

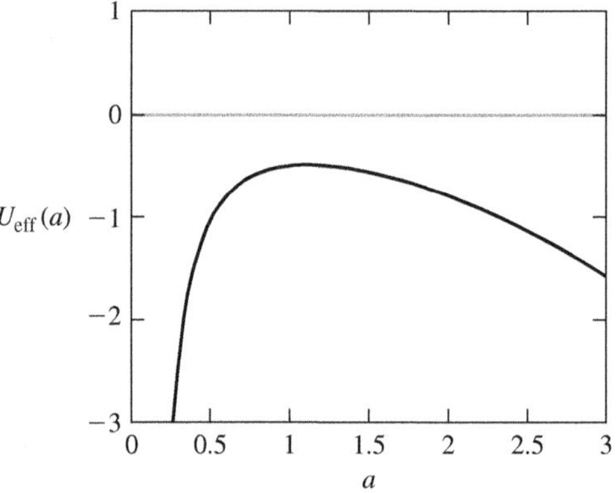

FIGURE 18.3 The effective potential for a flat FRW model. The figure shows the effective potential $U_{\text{eff}}(a)$ defined in (18.36) for an illustrative FRW model with equal amounts of vacuum, radiation, and matter today: $\Omega_v = \Omega_r = \Omega_m = 1/3$. (Our universe has much less radiation and more vacuum energy. For its effective potential, see the first of the plots in Figure 18.9.) The evolution of the scale factor $a(t)$ is the same as the Newtonian motion of a zero-energy particle in this effective potential. The universe starts from a big bang at $a = 0$, decelerates until $a \approx 1$ (today), and then accelerates forever. Initially it is radiation dominated, later matter dominated while still decelerating, but eventually vacuum dominated during its accelerating phase.

- *Vacuum dominated*, $\Omega_m = 0$, $\Omega_r = 0$, $\Omega_v = 1$:

$$a(t) = e^{H(t-t_0)}, \tag{18.39}$$

where

$$H^2 \equiv \frac{8\pi\rho_v}{3} = \frac{\Lambda}{3}. \tag{18.40}$$

[If you're worried that this definition of H might be confused with the Hubble constant, note that H *is* the Hubble constant for an expansion of the form (18.39) [cf. (18.14)] and in this case it is constant in time.]

In all three cases, the universe expands without limit as t increases. In the radiation- and matter-dominated cases, the universe begins with a singularity where $a = 0$ at $t = 0$. This is a physical singularity because a physical quantity—the density—becomes infinite then. The moment $t = 0$ is the big bang. In the vacuum-dominated case, a goes to zero at $t = -\infty$. Whether that is a singularity or not is less clear because the density ρ_v is constant, but in any case our universe has some matter and radiation in it and had a big bang singularity.[3]

Figure 18.4 illustrates the evolution that results when all three kinds of matter are present, as in our universe. The evolution proceeds through stages where the various types of energy are dominant. Initially the universe is radiation dominated, but the density in radiation dies away faster [cf. (18.25) or (18.34)] than that of matter and vacuum, so eventually matter dominates. Matter density decays too

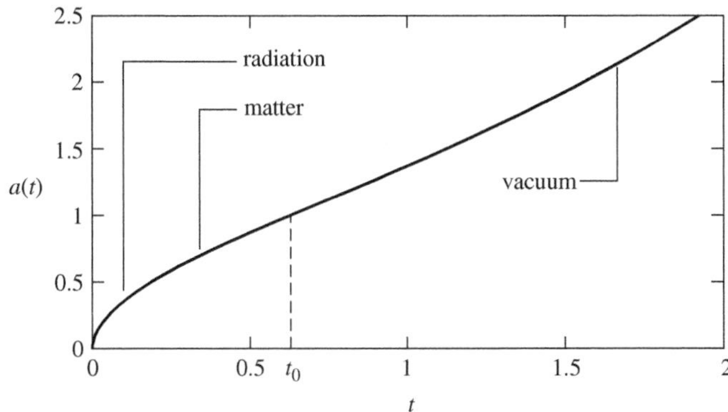

FIGURE 18.4 Stages of evolution in a flat FRW model. The various stages in the expansion of a flat FRW model are shown in this figure for the $\Omega_r = \Omega_v = \Omega_m = \frac{1}{3}$ illustrative model whose effective potential is shown in Figure 18.3. At early times the expansion is radiation dominated ($a(t) \propto t^{1/2}$), later matter dominated ($a(t) \propto t^{2/3}$), and finally vacuum dominated ($a(t) \propto \exp(Ht)$). The present age, t_0, when $a(t_0) = 1$ with our choice of normalization, is at the time of the dotted vertical line. This curve was calculated with the *Mathematica* program for general FRW models available on the book website.

[3] In fact, $a = 0$ is not a singularity in the pure vacuum case. Just the vanishing of a is not enough to ensure a singularity. Think of slicing a globe along lines of constant latitude. The radius of the slices goes to zero at the north pole, but the geometry is not singular there.

[cf. (18.22)], so that eventually only the constant vacuum energy ρ_v is important, with its associated exponential expansion.

Example 18.2. The First Few Minutes. The primoidial abundances of the elements are set at the time after the big bang when the temperature (in energy units) drops below $\sim .1$ MeV and thermonuclear reactions that can alter the abundances effectively cease. (See Boxes 18.1 and 19.1 on pages 375 and 400 for more discussion.) *When* does that happen? As mentioned before, the early universe is radiation dominated. The energy density is well approximated as a function of temperature by (18.29) with $g = 3.4$. The scale factor $a(t)$ is proportional to $t^{1/2}$. Equation (18.30) connects the scale factor to the energy density, and (18.29) connects the energy density to the temperature, giving

$$\frac{1}{4t^2} = \frac{8\pi}{3}\rho(t) = 2.75g\frac{1}{\ell_{Pl}^2}\left(\frac{T}{\ell_{Pl}}\right)^4. \tag{18.41}$$

This gives a connection between time in centimeters and temperature in centimeters as appropriate for geometrized units. When worked out in units of time in seconds and temperature in MeV, this is

$$t = 1.3\left(\frac{1\text{ MeV}}{T}\right)^2 \text{s} \qquad \text{(radiation dominated, } g \approx 3.4\text{)}. \tag{18.42}$$

Thus, $T \sim .1$ MeV corresponds to $t \sim 130$ s, or the first few minutes.

18.5 The Big Bang and Age and Size of the Universe

As we'll see in Section 18.7 and Chapter 19, like the flat cases, the curved FRW cosmologies that could model our universe began with a *big bang*—a moment in time at which the scale factor $a(t)$ vanishes and the geometry of the universe is singular. The singular nature of the big bang is apparent from (18.22) and (18.25). The densities of matter and radiation are infinite when $a = 0$. In Section 22.3 we will see quantitatively how the curvature of spacetime blows up at the big bang as well.

The big bang is not an explosion that happened at one point in space (see Example 18.1 on p. 367). Densities of matter and radiation diverge at all values of (x, y, z) at one value of t consistent with homogeneity. The big bang occurred at every place in space at one moment in time. The notion of a geometry of spacetime breaks down at a singularity, along with the predictive power of the laws of geometry, such as Einstein's equation. As far as making predictions in physics is concerned, the universe began at the big bang. (See Box 18.2 for more on this.) For this reason the big bang is conventionally assigned the time $t = 0$.

We are living later in the universe at some time denoted by t_0 (following the usual convention in cosmology that a subscript 0 refers to the present). This time

BOX 18.2 What Came Before the Big Bang?

The singularities formed in gravitational collapse are hidden inside black holes if the cosmic censorship conjecture is correct. (See p. 275 and Section 15.1.) But the big-bang singularity is visible in our past; we see the light from only a short time later in the cosmic background radiation today. The big bang forces the issue of singularities in general relativity and, in particular, the question in the title of this box.

Both the meaning of and the answer to this question depend on the theoretical context in which it is asked. In the context of the FRW models of this chapter, the big bang is a singular moment of infinite density and also infinite curvature, as we will see in Section 22.3. The singularity theorems of general relativity suggest that this singularity is not just an artifact of the high symmetry of the FRW models. Rather, in a broader context, there is a big-bang singularity in *any* general relativistic cosmological model that is compatible with present observations under reasonable assumptions on the matter, for instance, the positivity of energy (Problem 28). Perhaps matter did not obey these assumptions in the early universe, and the big bang was only a "bounce" of a very small size from an earlier recontracting phase (Problem 27). However, at the time of writing, there is little theoretical motivation for this, and singularities seem inevitable.

The classical idea of spacetime breaks down at a singularity. Consequently, the classical theory of spacetime—general relativity—has no meaningful way of determining what happened before the big bang from events after it, in particular, from observations today. In the context of general relativistic big-bang cosmology, there is no way of posing the question in the title, much less answering it. It's simplest to say that time began at the big bang.

But in a yet broader context, classical general relativity is only an approximation to a quantum theory of gravity, and its singularities signal regimes where the classical theory breaks down and the predictions of a quantum theory become important. As discussed on p. 11, when densities reach the Planck density $\rho_{Pl} \equiv (\hbar c/\ell_{Pl})/\ell_{Pl}^3 \sim 10^{94}\ \text{g/cm}^3$, significant quantum fluctuations in the geometry of spacetime can be expected. Such densities and higher are reached at singularities. In quantum gravity, spacetime geometry becomes a quantum variable, generally fluctuating and without definite value. There is no one geometry to supply a meaning to "before" and "after." Asking what happens before the big bang in quantum gravity is unlikely to make sense because the classical notion of time breaks down at a singularity.

since the big bang is the *age of the universe*. As discussed before, the Hubble time t_H is a rough estimate (and with certain assumptions an upper bound) on the age. To fix the age more precisely requires more observational input to determine *which* FRW model best fits our universe, as discussed in Section 18.7.

Example 18.3. Age and Hubble Constant in a Flat FRW Model. Suppose observations are reported that show us to be living in a flat FRW model with the line element (18.1) for which matter was the dominant component of energy. From (18.37), the scale factor is $a(t) \propto t^{2/3}$. Equation (18.14) connects the Hubble constant to the age

$$t_0 = \frac{2}{3H_0} = \frac{2}{3}t_H. \tag{18.43}$$

Assuming the Hubble constant of 72 (km/s)/Mpc favored by observations at the present time, this would mean an age of approximately 9 Gyr [cf. (18.15)]. However, the age of the oldest stars in our galaxy is approximately 12 Gyr, so our universe cannot correspond to a spatially flat, matter-dominated FRW model. Something must be wrong with those observations that show matter dominance.

How big is the universe? The trivial answer is "infinite" if the geometry is given by (18.1) because the volume of the $t = \text{const.}$ flat spatial slices is infinite. We'll see some FRW models in the next section with finite spatial volumes, but a more important question is, How big is the observable universe—the part we can see?

Even in principle, only that part of cosmological spacetime can be observed from which signals can travel to us at or less than the speed of light since the time of the big bang. Put differently, we can obtain information in principle only about events in the region bounded by the big bang and our past light cone. (Indeed, most of our information about distant parts of the universe is about events *on* our past light cone because it reaches us by electromagnetic radiation.) A very rough measure of the spatial radius of the region from which information can be obtained is the *Hubble distance* d_H defined to be ct_H, which is 2998 h^{-1} Mpc. To find a more accurate measure it is very convenient to introduce coordinates for the FRW models such that radial light rays move on 45° lines, just as we did for black holes in Chapter 12. Coordinates with this property can be introduced by defining a new time coordinate η such that

$$dt = a(t)\, d\eta. \tag{18.44}$$

In cosmology, this is sometimes called *conformal time*. Then, for example, the line element for a flat FRW model (18.5) becomes

$$ds^2 = a^2(\eta)[-d\eta^2 + dr^2 + r^2(d\theta^2 + \sin^2\theta\, d\phi^2)]. \tag{18.45}$$

Radial light rays move on curves where $ds^2 = a^2(\eta)(-d\eta^2 + dr^2) = 0$, that is, on the 45° lines in an η-r spacetime diagram.

Figure 18.5 shows an η-r spacetime diagram of a FRW model with the origin at $r = 0$ chosen to coincide with our world line. Past light cones of an observer at two different times are shown. (One of these could be our past light cone were the time t_0.) The largest radius r_{horiz} from which a light ray could have reached the observer in the time from the big bang is given by (18.7) with $t_e = 0$. Specifically,

$$r_{\text{horiz}}(t) = \int_0^t \frac{dt'}{a(t')}. \tag{18.46}$$

The comoving radius $r_{\text{horiz}}(t)$ divides those particles of the cosmological fluid from which the observer could have received information at time t, from those from which information could not have been received. The three-surface in spacetime with that radius is called the observer's *particle horizon*,[4] or *horizon* for short, when the context is unambiguously cosmological. Note that the radius of

[4] An *event* horizon separates regions of events by the spacetime's causal properties. A *particle* horizon separates spacetime regions by whether particles in them can be seen by a given observer. We can't see over the horizon on the surface of the earth, and we can't see beyond our particle horizon in cosmology.

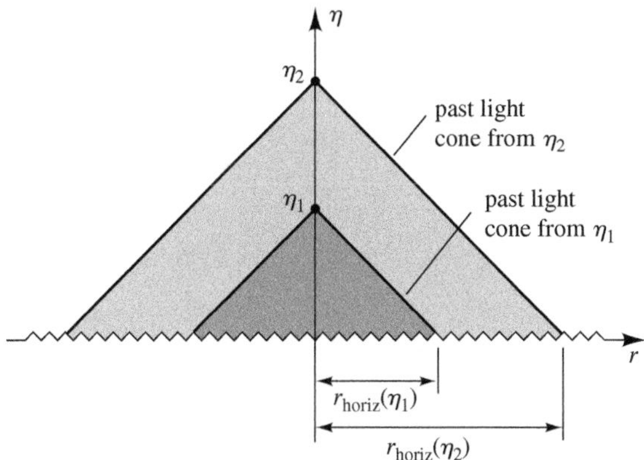

FIGURE 18.5 Light cones and horizons in a flat, matter-dominated FRW model. The figure shows an η-r spacetime diagram of a flat FRW spacetime using the conformal time coordinate defined by (18.44). The big bang is the line $\eta = 0$. Suppose our world line is the η-axis at $r = 0$. Two events are shown along that world line at an interval of η from the big bang and twice that interval. An observer at an event can receive signals from all points inside the past light cone but not from outside. The coordinate distance from the origin to the tick is the largest comoving radius $r_{\text{horiz}}(\eta)$ visible from conformal time η, whose size at the time of the observations is given by (18.46). As time goes on, the visible region increases in size.

that three-surface is constant in time for a given time of observation, as shown in Figure 18.5, but its value depends on the time t of the observation. As t increases, the size of the horizon grows because there is more time since the big bang for information to reach the observer. The physical distance to the horizon at the time of the observation is

$$d_{\text{horiz}}(t) = a(t)r_{\text{horiz}}(t) = a(t)\int_0^t \frac{dt'}{a(t')}. \tag{18.47}$$

Distance to the Horizon

Despite being denoted by the letter d, d_{horiz} is the physical *radius*, not the diameter, at the present moment of the region that is in principle visible. It's denoted by d_{horiz} because it is the *distance* to the horizon.

We are observers living at a time t_0 after the big bang. The radius today of the region from which we could have received information is $d_{\text{horiz}}(t_0)$. This turns out to be roughly 14 Gpc for the FRW model that best fits our universe at the present time (Problem 33). That is the radius of the region from which information could be received *in principle*. The radius of the universe visible in light is smaller because the early universe was opaque as we will see in Section 19.2.

Example 18.4. Horizon Size in Flat, Matter-Dominated, and Radiation-Dominated FRW Models. Evaluating (18.47) with (18.37) and (18.38) gives the horizon size for matter-dominated and radiation-dominated flat FRW models as

$$d_{\text{horiz}}(t) = 3t \qquad \text{(matter dominated)}, \tag{18.48a}$$

$$d_{\text{horiz}}(t) = 2t \qquad \text{(radiation dominated)}. \tag{18.48b}$$

For example, combining (18.48a) with (18.43) gives for the present horizon size of a matter dominated universe

$$d_{\text{horiz}}(t_0) = 2t_H \approx 8 \text{ Gpc} \qquad \text{(matter dominated)} \tag{18.49}$$

when t_H is expressed in units of length for a Hubble constant of 72 (km/s)/Mpc. The discrepancy between this number and the 14 Gpc mentioned above arises because the "best buy" FRW model has significant vacuum energy as well as matter, as we will see in the next chapter.

18.6 Spatially Curved Robertson–Walker Metrics

The flat Robertson–Walker line element (18.1) is one example of a homogeneous, isotropic cosmological spacetime geometry, but not the only one. The general Robertson–Walker line element for a homogeneous isotropic universe has the form (18.4)

$$ds^2 = -dt^2 + a^2(t)\, d\mathcal{L}^2, \tag{18.50}$$

where $d\mathcal{L}^2$ is the line element of a homogeneous, isotropic three-dimensional *space*. There are only three possibilities for this. Flat space with $d\mathcal{L}^2 = dx^2 + dy^2 + dz^2$ is one possibility already discussed. Let's now look at the other two.

Closed FRW Models

The surface of a unit sphere in a fictitious four-dimensional, flat Euclidean space is a homogeneous, isotropic spatial geometry called the *three-sphere*. (We choose a *unit* three-sphere because the overall scale is eventually to be described by the scale factor $a(t)$ in (18.50)). Using rectangular coordinates $X^\alpha = (W, X, Y, Z)$ for the fictitious space, this three-sphere is the surface

$$\delta_{\alpha\beta} X^\alpha X^\beta = W^2 + X^2 + Y^2 + Z^2 = 1. \tag{18.51}$$

A point in this surface is conveniently labeled by the generalization of polar angles to four dimensions:

$$X = \sin\chi \sin\theta \cos\phi, \qquad Z = \sin\chi \cos\theta,$$

$$Y = \sin\chi \sin\theta \sin\phi, \qquad W = \cos\chi, \tag{18.52}$$

where the range of the three polar angles (χ, θ, ϕ) is given by $0 \le \chi \le \pi$, $0 \le \theta \le \pi$, and $0 \le \phi \le 2\pi$. To see that (χ, θ, ϕ) specify a point on the three-surface (18.51), substitute (18.52) into that equation, note that it is satisfied for any choice of these angles, and further that any point on the surface corresponds to one set of angles in the given ranges. The line element on the three-surface can be worked out by inserting (18.52) into

$$ds^2 = \delta_{\alpha\beta} dX^\alpha dX^\beta = dX^2 + dY^2 + dZ^2 + dW^2 \qquad (18.53)$$

appropriate for a four-dimensional Euclidean space. This gives

$$d\mathcal{L}^2 = d\chi^2 + \sin^2\chi \, (d\theta^2 + \sin^2\theta \, d\phi^2) \qquad \text{(closed)}. \qquad (18.54)$$

This is the metric of a homogeneous isotropic three-dimensional space—the three-sphere.

This space is *closed* with a finite volume and no boundary, as is evident from its definition as a three-dimensional sphere in four-dimensional flat space. It is the three-dimensional analog of the surface of a sphere that has finite area but no boundaries. Indeed, we can compute the spatial volume of a spatial slice of the FRW model with (18.50) and (18.54) by integrating (7.29) over the whole range of coordinates:

$$V(t) = \int_0^{2\pi} d\phi \int_0^{\pi} d\theta \int_0^{\pi} d\chi \, a^3(t) \sin^2\chi \, \sin\theta = 2\pi^2 a^3(t). \qquad (18.55)$$

As the universe expands the volume gets bigger, and if it recontracts it gets smaller.

Open FRW Models

The remaining possible homogeneous, isotropic geometry for three-dimensional space has already been discussed as an example of a three-surface in Example 7.12 on p. 160. It is the geometry of a Lorentz hyperboloid—the three-surface in flat four-dimensional spacetime that is the analog of the three-surface of a sphere in flat four-dimensional Euclidean space. Using rectangular coordinates $X^\alpha = (T, X, Y, Z)$ to label the points in a fictitious flat spacetime with line element

$$ds^2 = \eta_{\alpha\beta} dX^\alpha dX^\beta = -dT^2 + dX^2 + dY^2 + dZ^2, \qquad (18.56)$$

the equation of a unit hyperboloid [cf. (7.74) with $a = 1$] can be reexpressed as

$$\eta_{\alpha\beta} X^\alpha X^\beta = -T^2 + X^2 + Y^2 + Z^2 = -1. \qquad (18.57)$$

(Capital letters are used for the coordinates, as in the analogous equation (18.53), to emphasize that this flat spacetime is only a convenient way of displaying a three-dimensional geometry as an embedded surface and has nothing to do with the cosmological spacetime geometry.) As the similarity between their equations shows, the three-surface (18.57) in a flat spacetime with metric $\eta_{\alpha\beta}$ is the analog of the sphere (18.51) in a Euclidean space with metric $\delta_{\alpha\beta}$. Any two points on the sphere can be mapped into one another by a combination of rotations leaving both the equation of the sphere (18.51) and the line element of the embedding geometry (18.53) unchanged. Any two points on the hyperboloid can be mapped into each other by a combination of Lorentz boosts and rotations that leave the equation of the hyperboloid (18.57) and the line element of flat spacetime unchanged. The geometries of both surfaces are, therefore, homogeneous and isotropic.

The analogs of polar coordinates (18.52) for the hyperboloid are

$$X = \sinh\chi \sin\theta \cos\phi, \qquad Z = \sinh\chi \cos\theta,$$
$$Y = \sinh\chi \sin\theta \sin\phi, \qquad W = \cosh\chi, \tag{18.58}$$

where the ranges are $0 \leq \chi < \infty$, $0 \leq \theta < \pi$, $0 \leq \phi < 2\pi$. Equation (18.57) defining the hyperboloid is satisfied by (18.58), and inserting these relations into (18.56) gives the line element on the hyperboloid that we found in (7.76):

$$d\mathcal{L}^2 = d\chi^2 + \sinh^2\chi\,(d\theta^2 + \sin^2\theta\, d\phi^2) \qquad \text{(open)}. \tag{18.59}$$

Spatial slices of this FRW model have infinite volume. These models are, therefore, called *open*.

The General FRW Metric

The conventional names *flat*, *closed*, and *open* have been used to distinguish the three possible homogeneous and isotropic geometries for spaces (18.5), (18.54), and (18.59), respectively. It might be better to distinguish them by their spatial curvature. Chapter 21 introduces quantitative measures of curvature. Homogeneity requires that the spatial curvature be the same at each point of these geometries. The flat case has zero spatial curvature everywhere, the closed case has constant positive spatial curvature, and the open case has negative constant spatial curvature. Figure 18.6 contains some embedding diagrams of two-surfaces in these

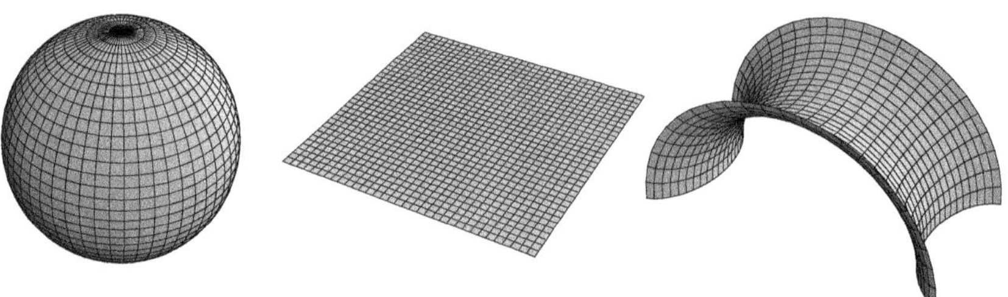

FIGURE 18.6 Three embedding diagrams of the possible homogeneous and isotropic geometries for space in FRW cosmological models. The first two figures are embeddings of a t = const., $\theta = \pi/2$ two-surface in the closed and flat FRW metrics (18.54) and (18.1) constructed as described in Section 7.7. They are the sphere (closed) and plane (flat). These are constant positive-curvature and zero-curvature surfaces, respectively, as we will see in Chapter 21. A t = const., $\theta = \pi/2$ slice of the open FRW metric can't be embedded as an axisymmetric surface in three-dimensional flat space. (Try it!) That surface has constant negative curvature and the embedding shown is of a limited piece of it. (It doesn't matter which piece because the geometry is homogeneous.) If you want to know how it was constructed, work through Problem 30.

possible homogeneous and isotropic geometries of space, which suggest the different kinds of spatial curvature.

The three possible line elements for homogeneous, isotropic cosmological models can be summarized as follows:

$$ds^2 = -dt^2 + a^2(t)\left[d\chi^2 + \begin{Bmatrix} \sin^2\chi \\ \chi^2 \\ \sinh^2\chi \end{Bmatrix} (d\theta^2 + \sin^2\theta\, d\phi^2)\right] \quad \begin{Bmatrix} \text{closed} \\ \text{flat} \\ \text{open} \end{Bmatrix}. \tag{18.60}$$

Here, the radial coordinate r in (18.5) has been written as χ to emphasize the similarity with the other cases. Another representation of all three is obtained by replacing the coordinate χ in all three cases by a new radial coordinate r as follows:

$$r = \sin\chi \quad \text{(closed)}, \qquad r = \chi \quad \text{(flat)}, \qquad r = \sinh\chi \quad \text{(open)}. \tag{18.61}$$

The three line elements in (18.60) can then be written in a unified form as

$$ds^2 = -dt^2 + a^2(t)\left[\frac{dr^2}{1-kr^2} + r^2(d\theta^2 + \sin^2\theta\, d\phi^2)\right], \tag{18.62}$$

Metric for Homogeneous Isotropic Cosmological Models

where $k = +1, 0, -1$ for closed, flat or open universes, respectively.

The Robertson–Walker metrics summarized in (18.62) or (18.60) each describe the time evolution of a homogeneous, isotropic space that gets larger in time as $a(t)$ increases and smaller as $a(t)$ decreases. All information about the evolution of the universe is contained in this one function determined by the Einstein equation, as discussed next.

18.7 Dynamics of the Universe

The Friedman Equation

Only a small modification of the dynamical equation (18.30) for the flat FRW models is needed to generalize it to include spatial curvature. The general relation is:

$$\dot{a}^2 - \frac{8\pi\rho}{3}a^2 = -k. \tag{18.63}$$

Friedman Equation

Here $\dot{a} \equiv da/dt$ and k is the constant appearing in (18.62), whose value ± 1 or 0 specifies the curvature of the spatial geometry of a FRW model. General relativity thus connects the time evolution of the universe to its spatial geometry. The relation (18.63) is called the *Friedman equation*. We will derive it from the Einstein equation in Section 22.4 but it has a simple intuitive motivation, given in Example 18.5.

Example 18.5. Equation of Motion of a Sphere of Pressureless Matter. Equation (18.63) has a simple interpretation for the case of pressureless matter. Imagine a spherical shell of test particles surrounding the origin in the general FRW spacetime (18.62). Label the shell by a comoving radial coordinate $r = r_s$. Take $r_s \ll 1$ so that the factor of $1 - kr^2$ can be replaced by 1 inside the shell to a good approximation. The proper radius $R(t)$ of the shell and the proper volume $V(t)$ inside are then given by [cf. (18.62) and (7.29)]

$$R(t) = a(t)r_s, \qquad V(t) = \frac{4}{3}\pi\,[a(t)r_s]^3 = \frac{4\pi}{3}R^3(t). \tag{18.64}$$

Since the energy density $\rho(t)$ in this homogeneous universe is a constant in space, the mass of matter inside the shell is

$$M(t) = \rho(t)V(t) = \frac{4\pi\rho(t)}{3}R^3(t). \tag{18.65}$$

Multiplying (18.63) by $\frac{1}{2}r_s^2$ and using (18.64) and (18.65) recasts the Friedman equation in the following form:

$$\frac{1}{2}\dot{R}^2(t) - \frac{M(t)}{R(t)} = -\frac{1}{2}kr_s^2. \tag{18.66}$$

Since r_s is constant in time (a comoving coordinate) the right-hand side of (18.66) is constant. When multiplied by the rest mass of the shell of test particles, (18.66) is the Newtonian expression for the conserved energy of the shell. The first term corresponds to the Newtonian kinetic energy, the second to the Newtonian gravitational potential energy. The Friedman equation predicts the same motion for the universe as for a shell in Newtonian theory for pressureless matter. It was necessary to assume that the matter was pressureless to get this result. Otherwise there would be pressure contributions to the mass $M(t)$ in Newtonian theory not present in the Friedman equation.

Dividing the Friedman equation by $a^2(t)$ and evaluating at the present moment, t_0, yields the generalization of (18.31) to curved FRW models:

$$H_0^2 - \frac{8\pi\rho_0}{3} = -\frac{k}{a_0^2}. \tag{18.67}$$

If the present energy density ρ_0 is larger than the critical density ρ_{crit} defined in (18.32), the universe is positively curved ($k = +1$) and closed. If ρ_0 is less than

ρ_{crit}, it is negatively curved ($k = -1$) and open. To be spatially flat ($k = 0$) the present density ρ_0 must equal the critical density. The two parameters H_0 and ρ_0 thus determine whether the universe is open, closed, or flat. We'll discuss their determination in the next chapter.

It is standard in cosmology to measure the present total density relative to the critical density by introducing the dimensionless parameter

$$\Omega \equiv \rho_0/\rho_{\text{crit}}, \tag{18.68}$$

as we did for the individual components of the cosmological fluid in (18.33). Positively curved FRW models have $\Omega > 1$, flat models have $\Omega = 1$, and negatively curved models have $\Omega < 1$. Flat models (zero curvature) are thus just on the borderline between open (negative curvature) and closed (positive curvature). That is why ρ_{crit} is called the critical density. If we could measure the present density and determine Ω, we would know the spatial curvature of the universe. But the amount of vacuum energy is inaccessible to local experiment and the unknown amount of dark matter in the universe described in the previous chapter makes it impossible to determine the matter density just by taking a census of the visible part. Instead, the next chapter discusses how to determine Ω by measuring the geometry of the universe.[5]

The solutions of the Friedman equation (18.63) can be exhibited in terms of elementary functions for spatially curved models for the three different components of the cosmological fluid separately, as Example 18.6 of a matter-dominated universe shows. (See also Problems 18 and 19.) However, our universe contains matter, radiation, and possibly vacuum energy together. We now turn to the general case, both qualitatively and quantitatively.

Example 18.6. Matter-Dominated FRW Cosmological Models. Over most of history, from several hundred thousand years after the big bang until approximately the present, matter has made the dominant contribution to the energy density driving the evolution of the universe. Matter-dominated FRW models with no radiation and no vacuum energy are, therefore, particularly appropriate as an example. For this special case, solutions to the Friedman equation can be found in closed form, as exhibited by the parametric equations following and as illustrated in Figure 18.7 (Problem 17). First, let's consider the positive curvature models:

$$\left.\begin{aligned} a(\eta) &= \frac{\Omega}{2H_0(\Omega-1)^{3/2}}(1-\cos\eta) \\ t(\eta) &= \frac{\Omega}{2H_0(\Omega-1)^{3/2}}(\eta-\sin\eta) \end{aligned}\right\} \qquad k=+1, \quad \Omega>1. \tag{18.69}$$

These closed models expand from a big bang singularity at $\eta = 0$ where $a = 0$, $t = 0$, and $\rho = \infty$. They reach a maximum volume at $\eta = \pi$ and then recollapse

[5] For a glimpse of what is coming, recall the discussion in Box 2.2 on p. 17.

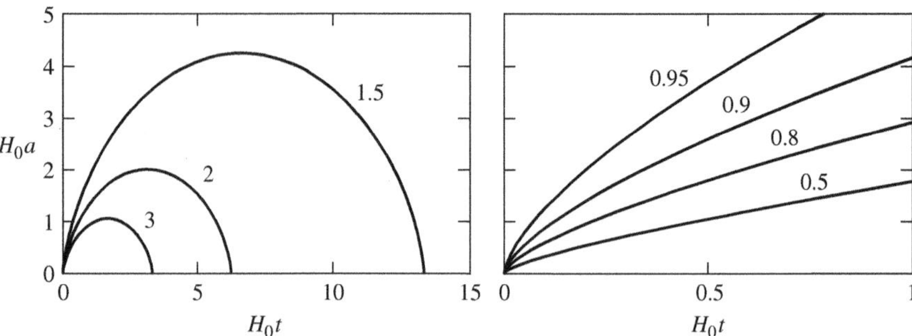

FIGURE 18.7 Matter-dominated FRW models. The time dependence of the scale factor $a(t)$ given in (18.69) and (18.71) is exhibited for closed models on the left and open ones on the right. Each curve is labeled by its value of $\Omega = \Omega_m$. Note that the vertical axes have the same range in both cases, but the horizontal axes have different ranges. All models expand from a big bang at $t = 0$. The positive-curvature ones reach a maximum radius and then recollapse to a singular big crunch. Negative-curvature models expand forever. As $\Omega \to 1$, the curves approach the vertical axis in each case.

to a "big crunch" singularity at $\eta = 2\pi$ where again $a = 0$. Their total duration is therefore $(\pi\Omega/H_0)(\Omega - 1)^{-3/2}$. The maximum spatial volume is, from (18.55)

$$V_{\max} = \frac{2\pi^2}{H_0^3}\left[\frac{\Omega}{(\Omega - 1)^{3/2}}\right]^3. \tag{18.70}$$

The negative curvature models are similar:

$$\left.\begin{aligned} a(\eta) &= \frac{\Omega}{2H_0(1-\Omega)^{3/2}}(\cosh\eta - 1) \\ t(\eta) &= \frac{\Omega}{2H_0(1-\Omega)^{3/2}}(\sinh\eta - \eta) \end{aligned}\right\} \qquad k = -1, \quad \Omega < 1. \tag{18.71}$$

These models also expand from a big bang singularity at $t = 0$, but continue expanding forever. Both models decelerate as they expand. That can be understood as gravitational attraction slowing down the expansion. The negatively curved models "escape" the attraction and expand forever; the positively curved models do not.

What about the spatially flat, zero-curvature, model with $\Omega = 1$, which lies on the border between the $\Omega > 0$ positively curved models and the $\Omega < 1$ negatively curved models? As Ω approaches 1, both the duration and the maximum spatial volume of the closed models approach infinity, and both for open and closed models the scale factor diverge at $\Omega = 1$. That is the correct answer as the limit of cases where the scale factor is proportional to a radius of curvature! Zero curvature *is* infinite radius of curvature. However, as explained on p. 377 the absolute magnitude of the scale factor for an exactly spatially flat model is arbitrary, and no physical quantity depends on it. Rescaling either (18.69) or (18.71) in the limit of $\Omega \to 1$ will give (18.37).

General Solution of the Friedman Equation

For both qualitative understanding and quantitative calculations of the FRW models, it is convenient to reexpress the Friedman equation (18.63) in terms of rescaled dimensionless variables. We use dimensionfull variables at the present moment, t_0, distinguished by a subscript 0 to define these rescalings. For instance, using the present value $a_0 \equiv a(t_0)$, a dimensionless measure of the scale factor can be defined by

$$\tilde{a}(t) \equiv a(t)/a_0. \tag{18.72}$$

Equation (18.11) shows that this rescaled variable is directly related to the redshift z of radiation coming from comoving galaxies at the time t by $\tilde{a} = 1/(1+z)$.

Similarly, the Hubble time, t_H, defined as the inverse of the present Hubble constant, H_0 [cf. (18.15)], can be used to define a dimensionless measure of time:

$$\tilde{t} \equiv t/t_H = H_0 t. \tag{18.73}$$

The critical density defined in (18.32) provides a convenient scale for densities. Present densities can be measured relative to this critical density by defining the various Ω's, as in (18.68) and (18.33). Thus, for example,

$$\rho_r(t) = \rho_{\text{crit}}\Omega_r/(\tilde{a}(t))^4, \qquad \text{etc.} \tag{18.74}$$

It is even convenient to introduce an Ω_c for curvature by defining

$$\Omega_c \equiv -k/(H_0 a_0)^2. \tag{18.75}$$

With this definition, the Friedman equation (18.63) evaluated at the present moment, t_0, reads

$$\Omega_r + \Omega_m + \Omega_v + \Omega_c = 1. \tag{18.76}$$

This elegant relation can be misleading since Ω_c can be negative in a closed universe, unlike all the other Ω's.

The payoff for all these redefinitions is a rescaled Friedman equation (18.63), which reads

$$\frac{1}{2}\left(\frac{d\tilde{a}}{d\tilde{t}}\right)^2 + U_{\text{eff}}(\tilde{a}) = \frac{\Omega_c}{2}. \tag{18.77}$$

where the effective potential U_{eff} [the same as in (18.36)] is defined by

$$U_{\text{eff}}(\tilde{a}) \equiv -\frac{1}{2}\left(\Omega_v \tilde{a}^2 + \frac{\Omega_m}{\tilde{a}} + \frac{\Omega_r}{\tilde{a}^2}\right), \tag{18.78}$$

and Ω_c is given in terms of the other Ω's by (18.76). Equations (18.77) and (18.78) reduce to (18.35) and (18.36), respectively, for a flat FRW model when the rescalings (18.72) and (18.73) are taken into account. As already discussed in that case, (18.77) is like the energy relation for a Newtonian particle moving in one dimension—$\tilde{a}$ is the coordinate, $\Omega_c/2$ is the energy.

To construct an FRW cosmological model, therefore, proceed as follows: (1) Specify the four parameters $H_0, \Omega_r, \Omega_m, \Omega_v$. (2) Use the last three to solve (18.77) for $\tilde{a}(\tilde{t})$ by writing $d\tilde{a}/[\Omega_c - 2\tilde{U}_{\text{eff}}(a)]^{1/2} = d\tilde{t}$ and doing the integral on both sides. (There is a *Mathematica* program on the book website to do that numerically.) (3) Undo the rescaling using H_0 to translate $\tilde{t}$ into t with (18.73) and find the value of a_0 from (18.75). The result for $a(t)$ is

$$a(t) = \frac{1}{H_0|\Omega_c|^{1/2}}\tilde{a}(H_0 t). \tag{18.79}$$

The solution to the rescaled Friedman equation (18.77) not only determines the scale factor $a(t)$ as a function of time; it also determines our location in time by fixing the present age t_0. The definition of $\tilde{a}$ in (18.72) implies that the present moment $\tilde{t}_0$ is when $\tilde{a}(\tilde{t}_0) = 1$. The present age is then $t_0 = \tilde{t}_0/H_0$ from (18.73).

An FRW model cosmology is, therefore, determined by four cosmological parameters

FRW Cosmological Parameters

$$\boxed{H_0, \qquad \Omega_r, \qquad \Omega_m, \qquad \Omega_v.} \tag{18.80}$$

These specify the present moment as well as the past history and future fate of the universe. Other properties of the universe are predicted as functions of these four parameters, as Example 18.7 illustrates. A goal of observational cosmology is to determine the values of these four parameters that specify our universe. A goal of theoretical cosmology is to explain why they have the values they do. Progress toward both these goals is discussed in the next chapter.

Example 18.7. The Age of the Universe as a Function of Cosmological Parameters. From the relation (18.73) between t and $\tilde{t}$, the age of the universe is

$$t_0 = \frac{1}{H_0}\tilde{t}_0(\Omega_r, \Omega_m, \Omega_v). \tag{18.81}$$

The dimensionless function $\tilde{t}_0(\Omega_r, \Omega_m, \Omega_v)$ is the value of $\tilde{t}$ at which $\tilde{a}(\tilde{t}_0) = 1$. This function can be determined by integrating the rescaled Friedman equation (18.77) to give

$$\tilde{t}_0(\Omega_r, \Omega_m, \Omega_v) = \int_0^1 d\tilde{a}\left[\Omega_c(\Omega_r, \Omega_m, \Omega_v) + \Omega_r\tilde{a}^{-2} + \Omega_m\tilde{a}^{-1} + \Omega_v\tilde{a}^2\right]^{-1/2}, \tag{18.82}$$

where Ω_c is given in terms of the other Ω's by (18.76). For our universe $\Omega_r \approx 8 \times 10^{-5}$ and so has little effect on the age. A contour plot of $\tilde{t}_0$ as function of the other Ω's is shown in Figure 18.8. The age of the oldest stars is about 12 billion

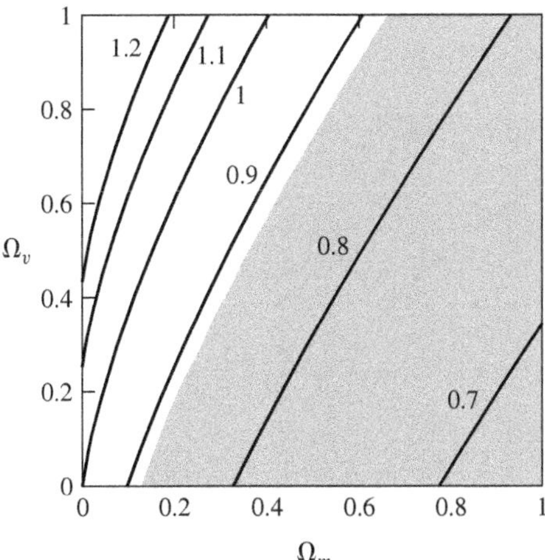

FIGURE 18.8 The age of the universe as a function of Ω_v and Ω_m. The figure shows contour lines of constant age t_0 expressed as a multiple of the Hubble time t_H. The age of the oldest stars is approximately 12 Gyr; the universe must be older. The unshaded region defines the ranges of Ω_v and Ω_m consistent with this limit when $h = .72$.

years. The universe must be older, and that fact puts one constraint on the values of the cosmological parameters.

The qualitative behavior of the scale factor with time for the various values of the FRW cosmological parameters can be read from plots like those in Figure 18.9 showing the relationship between effective potential $U_{\text{eff}}(\tilde{a})$ and Ω_c. In thinking about these, it's important to remember that *both* U_{eff} and Ω_c depend on the FRW parameters Ω_v, Ω_m, and Ω_r.

- *Open and Flat FRW Models* ($\Omega \leq 1$). Since $\Omega_c \equiv 1 - \Omega$ is nonnegative and the effective potential U_{eff} is negative, there are no turning points where $\dot{a} = 0$. All negative- or zero-curvature FRW models ($k = 0, -1$) therefore begin with a big bang singularity at $a = 0$ and expand forever.
- *Closed FRW Models* ($\Omega > 1$). Since $\Omega_c \equiv 1 - \Omega$ is negative, these may or may not have turning points, depending on whether or not the top of the potential is above the $\Omega_c/2$ line. If it isn't, the FRW model starts at a big bang singularity at $a = 0$ and expands forever. If it is, there are two turning points, and the FRW model can be one of two types: It can expand from a big bang singularity at $a = 0$, hit the smaller turning point at a maximum radius, and then recollapse to a singularity at $a = 0$ (the big crunch). The other possibility is a model that collapses from large values of a, "bounces" at the larger of the two turning points, and reexpands forever without ever becoming singular. The possibility corresponding to a given set of Ω's is determined by whether $\tilde{a} = 1$ (the present) is below the smallest turning point (recollapse) or above the largest one (bounce). Observations rule out a bounce for our universe (Problem 25) with the dust-radiation-vacuum model we have assumed.

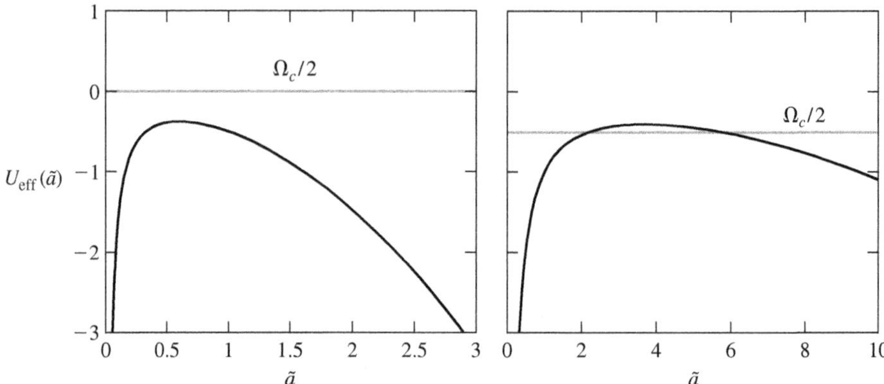

FIGURE 18.9 The effective potential for evolution of curved FRW models. This figure shows two examples of the effective potential (18.78) and its relation to the value of $\Omega_c/2$. The figure at left shows the potential for cosmological parameters approximating those of our universe, $\Omega_v \approx .7$, $\Omega_m \approx .3$, $\Omega_r \approx 5 \times 10^{-5}$, so that $\Omega_c \approx 0$. There are no turning points, so the universe starts with a big bang and expands forever. The figure at the right shows the relation when $\Omega_v = .02$, $\Omega_m = 2$, and $\Omega_r = 0$. There are two turning points. The universe starts from a big bang as $\tilde{a} = 0$ and expands to today at $\tilde{a} = 1$, turns around at the first turning point, and recollapses to a big crunch, following a trajectory similar to those at the left of Figure 18.7. Similarly, there are other positively curved models (not shown) for which the potential intersects the line $\Omega_c/2$, and $\tilde{a} = 1$ lies on the far side of the potential barrier. For such models, the universe collapses from a large radius, bounces at the larger turning point, and reexpands. These models have no big bang singularity. The regions of cosmological parameters corresponding to these behaviors are shown in Figure 18.10.

Figure 18.10 is a key diagram in contemporary cosmology. It shows the regimes of the behaviors discussed here in the plane of the least certain cosmological parameters Ω_v and Ω_m (Problem 26).

Of all the FRW models that could correspond to our universe, one feature stands out—the big bang. General relativity predicts that the universe began in a singular state of infinite density, and, as we will see in Section 22.3, of infinite curvature as well. This big bang singularity is not an artifact of the high symmetry of the FRW cosmological models but a feature of any general relativistic cosmological model compatible with present observations and reasonable assumptions on the nature of matter. (See Box 18.2 on p. 381.)

As it expands from the big bang, a curved FRW model passes through various stages identified by the dominant form of energy driving the evolution, much like the flat models illustrated in the left of Figure 18.4. Initially, the contribution of the spatial curvature term $\Omega_c/2$ in (18.77) is dwarfed by the radiation energy density. All FRW models with radiation begin alike. Later, when the radiation density has died away and the matter density has decreased, the spatial curvature term $\Omega_c/2$ becomes important and can lead to recollapse to a big crunch, as in the matter-

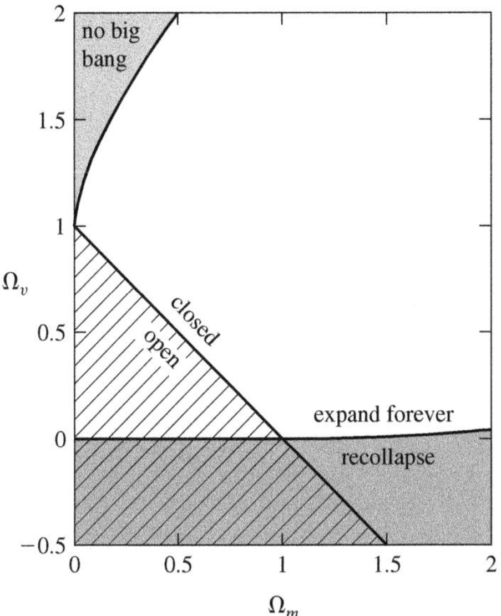

FIGURE 18.10 FRW models in the Ω_m-Ω_v plane. The figure shows the various regions of the Ω_m-Ω_v plane corresponding to various kinds of FRW models, assuming that Ω_r is very small, as it is in our universe. Flat models lie along the diagonal line $\Omega_v = 1 - \Omega_m$. Open models lie below this line; closed models lie above it. The lower shaded region contains parameters for which the universe expands from a big bang to a maximum size and then recollapses to a big crunch. Models above this region expand forever. The shaded region at the upper left contains values of the parameters for which the universe collapses from a large radius to a minimum one and then reexpands, never undergoing a big bang.

dominated models illustrated in the left of Figure 18.7. Any particular model can be calculated using the *Mathematica* program at the book website.

Problems

1. [S] Initially the raisins in Example 18.1 on p. 367 can be located by Cartesian coordinates (x, y, z) in the flat Euclidean space occupied by the dough. Continue to label the points occupied by the raisins with the same coordinates they started with so the coordinates (x, y, z) are comoving. Express the line element of flat Euclidean space in terms of these comoving coordinates and a scale factor $a(t)$ assuming the expansion is homogeneous and isotropic. Sketch the qualitative behavior of $a(t)$ between the start of baking and its completion. How would $a(t)$ look if the dough were a badly behaved spherical soufflé? If the dough were contained in a spherical boundary of radius R initially, what would be the equation of the boundary as a function of time? What would be the area of the boundary at the end of baking?

2. Suppose that the scale factor describing the expansion of the universe is

$$a(t) = (t/t_*)^{1/2},$$

where t_* is a constant and t is the proper time from the singularity. Suppose that the present age of the universe is 14 Gy.

(a) What would be the value (in yr^{-1}) of the Hubble constant observed today?

(b) At what age in years would the temperature of the microwave background be 3000 K?

3. Consider a flat FRW model whose metric is given by (18.1). Show that, if a particle is shot from the origin at time t_* with a speed V_* as measured by a comoving observer (constant x, y, z), then asymptotically it comes to rest with respect to a comoving frame. Express the comoving coordinate radius at which it comes to rest as an integral over $a(t)$ and discuss the conditions under which it is finite.

4. [S] Suppose the present value of the Hubble constant is 72 (km/s)/Mpc and that the universe is at critical density. A photon is emitted from our galaxy now. What is the redshift of this photon when it is received in another galaxy 10 billion years in the future, assuming it continues to be matter dominated?

5. [S] The cosmic background radiation has been propagating to us since the universe became transparent at a temperature of approximately 3000 K. Its temperature today is 2.73 K. What is the redshift z of the radiation?

6. [S] A type Ia supernova has a redshift of $z = 1.1$. The observed brightness rises and falls on a timescale of two months. (More precisely let's say the difference in times between when the supernova is at half peak brightness is two months.) What is the timescale for the rise and fall in the supernova's rest frame as would be seen by a hypothetical observer close to the supernova and at rest with respect to it?

7. Consider a galaxy whose light we see today at time t_0 that was emitted at time t_e. Show that the present proper distance to the galaxy (along a curve of constant t_0) is

$$d = a(t_0) \int_{t_e}^{t_0} dt/a(t).$$

8. In Section 9.2 the redshift of a photon in the Schwarzschild geometry was derived using the conservation law arising from time-translation symmetry. Show that the cosmological red shift (18.10) can be derived from the *space* translation symmetry of the metric (18.1) in a similar way.

9. [E] *Estimate* in centimeters the size of the universe visible today at the time the CMB radiation last interacted with matter at a temperature of approximately 3000 K.

10. [E, C] As the universe expands, the horizon grows. *Estimate* the time it has to grow for one new galaxy to come within the horizon, assuming the universe was matter dominated over the whole of its history.

11. **(a)** Equation (18.69) gives the scale factor as a function of time for closed, matter-dominated FRW models. Show that, if the parameter η that occurs there is used as a time coordinate, the FRW metric takes the form

$$ds^2 = a^2(\eta)\left[-d\eta^2 + d\chi^2 + \sin^2\chi\,(d\theta^2 + \sin^2\theta\, d\phi^2)\right].$$

(b) Draw an η-χ spacetime diagram indicating the big bang, the big crunch, and the past light cone of a comoving observer at the origin at the moment of maximum expansion.

(c) Is there time before the big crunch for the observer to receive information from all parts of this spatially finite universe, or are there parts of it he or she is doomed never to see?

(d) Could an observer traverse the entire circumference of the universe in the time between the big bang and the big crunch?

12. *Legislating the value of π* There is a story that a bill was introduced in a state legislature to declare the value of π to be some constant other than 3.14159 . . . . Could the bill's author have been correct? Is there some other geometry of three-dimensional space where the ratio of the circumference, C, to the radius R has a constant value for all circles that is different from $2 \times 3.14159\ldots$? (A circle in this context means the locus of points a given distance—the radius—from a given point called the center.) [*Hint:* Think why this problem is in a chapter on cosmology.]

13. Suppose the total spatial volume of a closed, matter dominated, FRW model is 10^{12} Mpc^3 at its moment of maximum expansion. What is the duration of this universe from the big bang to the big crunch in years?

14. [B] Box 2.2 on p. 17 described how objects of a given size would subtend different angles at a given distance in positively curved, flat, and negatively curved spatial geometries. Calculate these the angle subtended by an object of size s a distance d away in each of the FRW spacetimes, assuming for simplicity that the scale factor is independent of time. (The next chapter deals with the expansion.) Do your results confirm the statements in the box?

15. [S] Consider a homogeneous, isotropic, cosmological model described by the line element

$$ds^2 = -dt^2 + \left(\frac{t}{t_*}\right)\left[dx^2 + dy^2 + dz^2\right]$$

where t_* is a constant.

(a) Is this model open, closed, or flat?

(b) Is this a matter-dominated universe? Explain.

(c) Assuming the Friedman equation holds for this universe, find $\rho(t)$.

16. The scale factor $a(t)$ of any FRW model can be expanded about the present moment in the form

$$a(t) = a(t_0)\left[1 + H_0(t - t_0) - (1/2)q_0 H_0^2 (t - t_0)^2 + \cdots\right],$$

where q_0 is called the *deceleration parameter*. Explain why the coefficient of the first term is the Hubble constant and evaluate q_0 in terms of the cosmological parameters.

17. [S] Verify that (18.69) and (18.71) solve the Friedman equation (18.63) for a matter-dominated universe.

18. *Radiation Dominated FRW Models.* Solve the Friedman equation to exhibit the scale factor as a function of time for FRW models that are radiation dominated from start to finish. Express your answers in terms of H_0 and $\Omega = \Omega_r$.

19. *De Sitter Space.* Solve the Friedman equation (18.63) for the scale factor as a function of time for closed FRW models that have only vacuum energy ρ_v. Do these models have an initial big bang singularity?

20. [C] Find a closed-form solution to the dynamical equation for the flat FRW models (18.35) in the case when there is no radiation $\Omega_r = 0$, but both vacuum energy and matter are present. Express your answer in terms of H_0, Ω_m, and $\Omega_v = 1 - \Omega_m$. How large would Ω_v have to be for the universe to be accelerating ($\ddot{a} > 0$) at the present time? Find an explicit expression for the age of the universe t_0 as a function of H_0 and Ω_v.

21. [S] Show that the formula for the cosmological redshift (18.10) holds in nonspatially flat FRW models.

22. [S] Equation (18.47) for the present size of the horizon was derived for a flat FRW model. Show that the same formula holds for all FRW models.

23. **(a)** Show that for FRW models with any combination of matter and radiation but no vacuum energy, the curve of $a(t)$ curves downward, i.e., has negative second derivative. Show that this means that $1/H_0$ is always larger than the age t_0.

(b) Show that this is *not* always the case if there is a nonzero vacuum energy.

24. *The Einstein Static Universe* Consider a closed ($k = +1$) FRW model containing a matter density ρ_m, a vacuum energy density corresponding to a positive cosmological constant Λ, and no radiation.

(a) Show that for a given value of Λ, there is a critical value of ρ_m for which the scale factor does not change with time. Find this value.

(b) What is the spatial volume of this universe in terms of Λ?

(c) If ρ_m differs slightly from this value, the scale factor will vary in time. Does the evolution remain close to the static universe or diverge from it?

Comment: This is the Einstein static universe for which Einstein originally introduced the cosmological constant.

25. [E] *Estimate* the smallest value of the Ω_v that would allow the universe to bounce at a small radius but still reach a temperature $T \sim 10^{10}$ K such that nucleosynthesis could occur. Assume $\Omega_r = 8 \times 10^{-5}$ and $\Omega_m = .3$.

26. [A] Assuming that $\Omega_r = 0$ as is approximately true for our universe, find the algebraic relations between Ω_v and Ω_r that determine the boundaries in Figure 18.10 dividing the various behaviors of the FRW models.

27. [C] *Bouncing Universes*

(a) Show that for any form of the effective potential $U_{\text{eff}}(a)$ defined in (18.77), there is an equation of state $p = p(\rho)$ that will produce it. Find (parametric) expressions for p and ρ in terms of $U_{\text{eff}}(a)$.

(b) Sketch a potential $U_{\text{eff}}(a)$ that would give rise to a closed bouncing universe—one that eternally oscillates between a maximum and minimum volume. What properties does your potential have to have so that it has no detectable effect on the past evolution of the universe between today and say a radiation temperature of $k_B T \sim 10$ MeV just above that when nuclei were synthesized in the big bang? (See Box 19.1 on p. 400 for more on that, but that information is not necessary to work the problem.)

(c) Show that generally the combination $\rho + 3p$ for this hypothetical matter will be negative at very high densities.

Comment: The result in part (c) can be turned around to say that if $\rho + 3p$ is always positive, there is a big bang singularity—an example of a singularity theorem (Problem 28). No known form of matter has negative pressure *or* energy density below nuclear densities.

28. *FRW Singularity Theorem* Show in the context of the FRW models that if the combination $\rho + 3p$ is always positive, then there will be a big bang singularity sometime in the past.

29. [S] We don't know much about the vacuum energy. Suppose it were negative. Show that then every FRW model which contains some matter or radiation would recollapse to end in a big crunch.

30. [N] *Embedding A Slice of an Open FRW Universe*

(a) Show that a whole $t = \text{const.}$, $\theta = \pi/2$ slice of the open FRW metric in (18.60) can't be embedded as an axisymmetric surface in flat three-dimensional space.

(b) The following is a simple axisymmetric metric with constant negative curvature:

$$d\Sigma^2 = du^2 + \cosh^2 u \, d\phi^2.$$

Show that this can be embedded as an axisymmetric surface in flat three-dimensional space but only for a limited range of u starting at $u = 0$. Find the upper limit of this range, and exhibit the embedding diagram.

Comment: Minding's theorem in differential geometry says that all constant-curvature surfaces have the same local geometry. The surface in (b) is, therefore, an embedding of *a piece* of the surface discussed in (a). It doesn't matter which piece since the geometry is homogeneous. This is the surface shown in Figure 18.6.

31. Evaluate (18.82) to find the age of an FRW model that is matter dominated from start to finish as a function of H_0 and $\Omega = \Omega_m$. For given H_0, which are older, open models or closed models? Does your analytic answer agree with Figure 18.8?

32. Express the present distance to the particle horizon $d_{\text{horiz}}(t_0)$ in terms of the cosmological parameters by an integral formula analogous to (18.82) for the age.

33. [N] Evaluate the formula for the present distance to the horizon obtained in Problem 32 for the cosmological parameters $\Omega_v = .7$, $\Omega_r = 8 \times 10^{-5}$, $\Omega_m = .3$, $H_0 = 72$ (km/s)/Mpc, which best characterize our universe at this time. Express your answer in Gpc.

CHAPTER

19 Which Universe and Why?

Which of the four-parameter family of FRW cosmological models best fits our universe and why? Those are two central questions for observation and theory in cosmology that will be briefly introduced in this chapter. Of the four parameters $H_0, \Omega_r, \Omega_m, \Omega_v$ that define an FRW model [cf. (18.80)], only two are determined by the observations described so far. The first is the Hubble constant $H_0 = 72 \pm 7$ (km/s)/Mpc found from measurements of the redshifts and distances to galaxies as described in Section 17.2. The second is the ratio Ω_r of energy density in radiation to critical density. The energy density in cosmic background photons is known from the radiation's temperature through (18.24) and corresponds to

$$\Omega_{CMB} = 2.5 \times 10^{-5} h^{-2} \approx 5 \times 10^{-5} \tag{19.1}$$

for $h \equiv H_0/[100\,(\mathrm{km/s})/\mathrm{Mpc}] = .72$. Including massless neutrinos makes $\Omega_r \approx 8 \times 10^{-5}$. Gravitons are uncertain but probably make only a small contribution.

The number density of baryons (particles such as protons and neutrons) can be determined accurately from the observed primordial abundances of the elements and the theory of how they were synthesized in the big bang (see Box 19.1.) The corresponding Ω is

$$\Omega_{\text{baryon}} \approx .04. \tag{19.2}$$

BOX 19.1 Big Bang Nucleosynthesis

There were no atoms or nuclei in the very early universe. Earlier than a second after the big bang the radiation temperature was above the MeV binding energies of nuclei and way above the keV binding energies of atoms [cf. (18.42)]. Any atoms or nuclei present would have been quickly broken up into their constituent electrons, protons, and neutrons. How, when, and where were the elements in today's stars and planets made? Nearly all elements above Li in the periodic table were made by later thermonuclear burning in stars [cf. Section 12.1]. But significant amounts of the isotopes of H, He, and Li were made by nuclear reactions in the first few minutes after the big bang when the temperature had dropped to a value at which they could survive. The relative abundances synthesized in the minutes after the big bang depend on the cosmological parameters of the universe. These relative primordial abundances can be measured today in places where they are presumed to have changed little—places such as old stars, low mass local galaxies, and the edges of distant galaxies where some elements are detected by their absorption of light from more distant quasars. These measurements are some of the most important probes of physics at the big bang and our universe's cosmological parameters.

The relative abundances predicted from big bang nucleosynthesis depend on the composition of matter when the process started and the subsequent competition between the rates of nuclear reactions and the rate of expansion of the universe as briefly described in Box 18.1 on p. 375. At the start, less than a second after the big bang, matter was a hot soup of free protons, neutrons, electrons, positrons, photons, and neutrinos maintained in equilibrium by the strong and electroweak interactions. In that epoch these reactions were typically much faster than the expansion rate of the universe, and even a time as short as a second was sufficient for equilibrium to be established.

Conditions for equilibrium fix the initial relative abundances of protons and neutrons. For example, neutrons can decay *into* protons by reactions like $n \rightarrow p + e^- + \bar{\nu}$. But neutrons can also be made *from* protons by reactions like $\bar{\nu} + p \rightarrow n + e^+$. The relative number of neutrons and protons in equilibrium is just such that the number of neutrons made is balanced by the number destroyed.

Equilibrium does not fix the total number of neutrons and protons. Baryon number is conserved at and below MeV energies relevant here, when protons and neutrons are the only baryons around. The process of nucleosynthesis therefore depends on the total baryon number in the early universe. A convenient local measure of the baryon number is the ratio η of baryon number density to photon number density. That turns out to be roughly constant between the time of nucleosynthesis and now,[a] and is therefore directly related to the value of Ω_{baryon}.

Starting in the early universe with a value of η and the relative abundances from equilibrium, the evolution of nuclear abundances as the universe expands and cools can be predicted with high precision. When the temperature drops by about a factor of 10 below the binding energies of typical nuclei, protons and neutrons combine to make nuclei, and nuclei fuse with other nuclei to make more nuclei. The process is over after 3 min, by which time the temperature has become too low for any nuclear reactions and the primoidial abundances of the elements are fixed. The calculations are complex, but the precision results are simply displayed in the accompanying figure together with the error boxes set by observations at the present time. (Note the three very different scales!)

[a] Try showing this from the fact that the number density of photons in a blackbody gas is proportional to its temperature cubed.

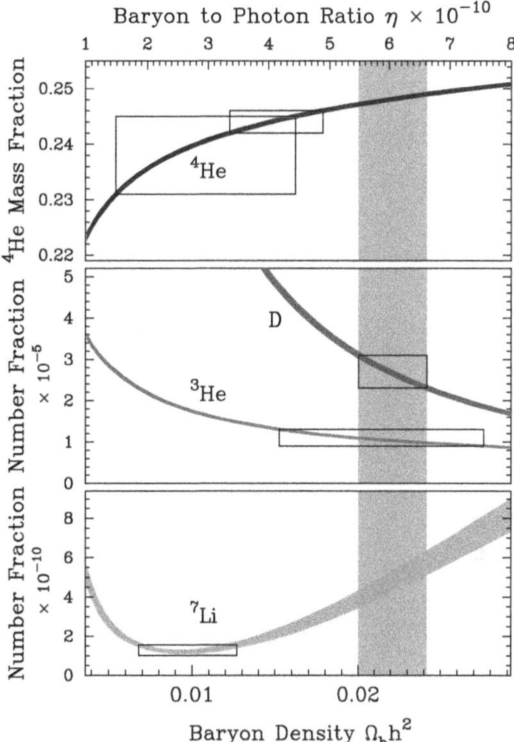

^{4}He—the lightest stable nucleus after hydrogen—is the winner in big bang nucleosynthesis. Over a wide range of baryon numbers, the universe emerges from the big bang with 76% matter in hydrogen by mass and 24% in helium. Much smaller amounts of deuterium, D, ^{3}He, and ^{7}Li are synthesized. However, in contrast to ^{4}He, the abundances of these elements *are* sensitive to η, or, equivalently, to Ω_{baryon}, as the figure shows. Since it is known that significant systematic errors are possible in these measurements, the rough agreement between the inferred values of Ω_{baryon} is usually counted as confirmation for the theory of big bang nucleosynthesis and is one of the strongest pieces of evidence for the big bang itself. The D abundance highlighted above that is measured from the absorption of light from quasars by D in the edges of the disks of intervening galaxies is perhaps the most reliable and gives

$$\Omega_{\text{baryon}} \approx .04, \qquad \text{(a)}$$

with $h \sim .7$.

This value supplies a lower bound for Ω_m, which could be considerably larger if there is nonbaryonic dark matter.

But beyond these known constituents, dark matter and vacuum energy are only detectable through their gravitational effects. To determine Ω_m and Ω_v, the spacetime geometry of the universe must be measured on large scales through a study of how matter moves through it. The following section describes two illustrative ways of doing that—one based on observations of distant supernovae and the other on observations of the cosmic background radiation already described briefly in Box 2.2 on p. 17. At the time of writing the combination of these two measurements as well as others are consistent with $\Omega_v \approx .7$ and $\Omega_m \approx .3$. This would give "best-buy" cosmological parameters at the time of writing of $H_0 \approx 72$ (km/s)/Mpc, $\Omega_r \approx 8 \times 10^{-5}$, $\Omega_m \approx .3$, and $\Omega_v \approx .7$. Remarkably, those numbers are consistent with the universe being spatially flat—right on the borderline between positive and negative spatial curvature.

19.1 Surveying the Universe

Redshift-Magnitude Relation

The redshift of a distant source directly measures the value of the scale factor when the light was emitted relative to its value today [cf. (18.11)]. The apparent brightness of a source is connected to its distance by the inverse square law (or its generalizations to curved geometry) and, therefore, to the time the light was emitted. Hence, measurements of redshift and magnitude for standard candles can probe the scale factor as a function of past time and determine cosmological parameters.

In Section 17.2 the flat-space inverse square law (17.9) was used to connect the observed flux from a standard candle to its distance. When coupled with Hubble's law, $z = H_0 d$ (in the $c = 1$ units used throughout this chapter), this gives a predicted connection between the flux f and red shift z of the form

$$\frac{f}{L} = \frac{H_0^2}{4\pi z^2}. \tag{19.3}$$

This is an example of a *redshift-magnitude relation* because in astronomy flux is measured in apparent magnitudes [cf. (17.11)]. The derivation of this relation assumed a flat geometry for space and neglected any evolution for that geometry. Those are appropriate approximations for nearby sources from which light takes only a short time to reach us, traveling a distance that is small compared to that over which space might be curved and small enough that Hubble's law holds.

However, for standard candles that are further away, deviations from (19.3) can be expected, arising from spatial curvature of the universe. Deviations can also be expected if the light from a standard candle travels to us over a time during which the expansion of the universe is significant. Observations of those deviations can be used to measure cosmological parameters.

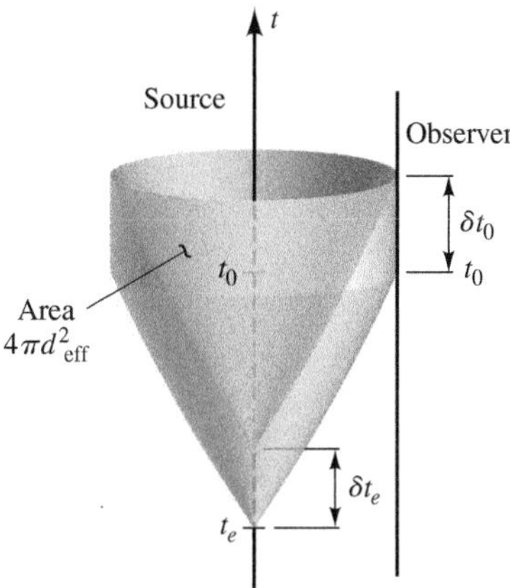

FIGURE 19.1 The relation between flux and distance in an FRW model. The world line of a source emitting at luminosity L runs up the center of this diagram. The world line of a comoving observer measuring flux from this source is at right. Both are at rest in the surfaces of homogeneity. The light emitted by the source over an interval δt_e at time t_e arrives at the observer over an interval δt_0 at time t_0 distributed over a sphere whose area, $4\pi d^2_{\rm eff}$, defines the effective distance, $d_{\rm eff}$. At reception the photons are redshifted from their source frequency and arriving less frequently than they were emitted. Each effect reduces the usual inverse square law for the flux by a factor of $1+z$, yielding (19.4).

Figure 19.1 shows the geometry necessary for deriving the revised redshift-magnitude relation. The world line of the source emitting with luminosity L (energy/second) is at the center of the diagram. Consider the photons of frequency ω_e emitted by the source over a time interval δt_e at time t_e. An observer located a certain comoving distance away measures the flux of radiation at time t_0. At reception, the emitted photons have been redshifted to a frequency ω_0, spread over a time interval δt_0, and spread over a sphere whose area is determined by the distance the observer is away. Define an effective distance $d_{\rm eff}$ such that the area of the sphere is $4\pi d^2_{\rm eff}$. This is not the distance from the source to the observer unless space is flat. The connections $\omega_0 = \omega_e/(1+z)$ and $\delta t_0 = (1+z)\delta t_e$ follow from (18.11) and (18.9). The energy flux f is, therefore, *reduced* from its flat space value because the photons have lower energy than at emission and arrive less frequently than they were emitted. Thus,

$$f = \frac{L}{4\pi d^2_{\rm eff}} \frac{1}{(1+z)^2}. \tag{19.4}$$

To find the redshift-magnitude relation for an FRW model we only have to express $d_{\rm eff}$ in terms of the redshift z and the cosmological parameters.[1] A flat, matter-dominated FRW model ($\Omega_r = 0$, $\Omega_m = 1$, $\Omega_v = 0$) provides a simple example. The geometry is described by the line element (18.5) and the scale factor is $a(t) =$

[1]It is conventional to define a *luminosity distance* $d_L = d_{\rm eff}(1+z)$ so that the inverse square law holds in its flat space form, but we prefer to keep the redshift and spatial geometry factors separate.

$(t/t_0)^{2/3}$ [cf. (18.37) with the normalization $a(t_0) = 1$]. Suppose the observer is a comoving coordinate distance $r = R$ away from the emitting galaxy. Since the spatial geometry is flat, the area of a sphere at time t_0 is just $4\pi a^2(t_0)R^2$, so $d_{\rm eff} = a(t_0)R = R$. The light travels over R in the time between t_e and t_0 on a null curve for which $-dt^2 + a(t)^2\,dr^2 = 0$. Thus, just as in (18.6) and (18.7), integrating this relation gives

$$d_{\rm eff} = a(t_0)\int_{t_e}^{t_0}\frac{dt}{a(t)} = 3t_0\left[1-\left(\frac{t_e}{t_0}\right)^{1/3}\right] = 3t_0\left[1-\left(\frac{a(t_e)}{a(t_0)}\right)^{1/2}\right]$$
$$= \frac{2}{H_0}\left(1-\frac{1}{\sqrt{1+z}}\right). \tag{19.5}$$

All the connections in the first line follow from $a(t) = (t/t_0)^{2/3}$. Those in the last follow from the definition of the redshift (18.11) and the Hubble constant (18.43). In the flat geometry, $d_{\rm eff}$ coincides with the present distance of the source d_0. Therefore, for small values of z this connection between redshift and distance reduces to Hubble's law—$z = H_0 d_0$—but for larger values the effects of spacetime curvature become important.

Inserting (19.5) for $d_{\rm eff}$ into the inverse square law (19.4) gives the following relation between the observable quantities flux, redshift, and Hubble constant:

Redshift-Magnitude Relation

$$\frac{f}{L} = \frac{H_0^2}{16\pi}\frac{1}{(1+z)[(1+z)^{1/2}-1]^2} \qquad \begin{pmatrix}\text{flat, matter}\\ \text{dominated}\end{pmatrix}. \tag{19.6}$$

This reduces correctly to (19.3) when z is small but differs from it significantly when $z \approx 1$ and larger. Were the apparent magnitudes of standard candles measured at larger and larger redshifts well fit by (19.6), we would conclude that the universe is spatially flat and matter dominated.

It is algebraically more complicated but no more difficult in principle to work out the relation between flux and luminosity for the general FRW model with nonzero values of all Ω's and nonflat spatial curvature. The inverse square law (19.4) still holds; only the effective distance, $d_{\rm eff}$, is different. To find $d_{\rm eff}$ in terms of the redshift and cosmological parameters, start with the line elements for the three types of spatial curvature written in the form (18.60). The effective distance $d_{\rm eff}$ is defined so that $4\pi d_{\rm eff}^2$ is the area of the sphere over which light from the emitting galaxy spreads in the time it travels to us. It is thus given in terms of the coordinate distance χ that the light rays travel by

$$d_{\rm eff} = a(t_0)\begin{Bmatrix}\sin\chi\\ \chi\\ \sinh\chi\end{Bmatrix} = \frac{1}{H_0|\Omega_c|^{1/2}}\begin{Bmatrix}\sin\chi\\ \chi\\ \sinh\chi\end{Bmatrix} \qquad \begin{Bmatrix}\text{closed}\\ \text{flat}\\ \text{open}\end{Bmatrix}. \tag{19.7}$$

(Recall that χ is just a different notation for r in the spatially flat case.) Here, (18.75) was used to express $a(t_0)$ in terms of cosmological parameters. It remains only to express χ in terms of them. In all three line elements (18.60), a radial light ray moves on a curve where $ds^2 = -dt^2 + a^2(t)\,d\chi^2 = 0$. The coordinate distance χ traveled is thus related to the time of emission and reception by an expression just like (18.7), with R replaced by χ. This can be written

$$\chi = \int_{t_e}^{t_0} \frac{dt}{a(t)} = \int_{a(t_e)}^{a(t_0)} \frac{da}{a\dot{a}(a)}. \tag{19.8}$$

Rewrite this expression in terms of the dimensionless variables $\tilde{a}$ and $\tilde{t}$ introduced in (18.72) and (18.73). Use the rescaled Friedman equation (18.77) to express $\dot{a}$ in terms of $\tilde{a}$ and the Ω's. Express $a(t_e)/a(t_0)$ in terms of the redshift z using (18.11). Assuming that the universe is expanding between time of emission and the present ($\dot{a} > 0$), the result is

$$\chi(\Omega_r, \Omega_m, \Omega_v) = |\Omega_c|^{1/2} \int_{(1+z)^{-1}}^{1} \frac{d\tilde{a}}{\tilde{a}[\Omega_c - 2U_{\text{eff}}(\tilde{a})]^{1/2}}, \tag{19.9}$$

where Ω_c and $U_{\text{eff}}(\tilde{a})$ are given in terms of the Ω's by (18.76) and (18.78), respectively. Inserting (19.9) into (19.7) and that into (19.4) gives the connection between flux and luminosity for a source at a redshift z as a function of the cosmological parameters. Converting flux and luminosity to apparent and absolute magnitude gives the redshift-magnitude relation.

The use of Type Ia supernovae as standard candles was described in Section 17.2. Figure 19.2 shows the redshift-magnitude from the combined data of Riess et al. (1998) and Perlmutter et al. (1999) together with a few representative curves of the kind (19.4) for various values of Ω_m and Ω_v (denoted by Ω_Λ in the figure). For values of z comparable to unity, the deviations from the flat-space inverse square law become significant and yield information about cosmological parameters. The data in this figure are evidence for a nonzero vacuum energy and its associated cosmological constant.

Cosmological Parameters from CMB Anisotropies

The tiny temperature anisotropies of the cosmic background information that are illustrated in Figure 17.12 contain a wealth of cosmological information. If measured accurately enough, they can determine all parameters characterizing the FRW model that best fits our universe. It is too detailed to go into all this here, but showing how the CMB anisotropies can answer the question of whether the universe is open or closed illustrates the idea.

The anisotropies of the CMB arise from temperature fluctuations in the radiation when it last scattered from matter at a temperature of approximately 3000 K, when electrons and nuclei combine to make neutral atoms (see Box 18.1 on p. 375). Afterward the universe was transparent to radiation. CMB photons arriving today started on their way to us from all over a last-scattering sphere (see Figure 19.3). The anisotropies in the CMB seen today (Figure 17.12) are images of the temperature fluctuations on this last-scattering surface. Their angular sizes depend on their physical size at this time of last-scattering. But they also depend

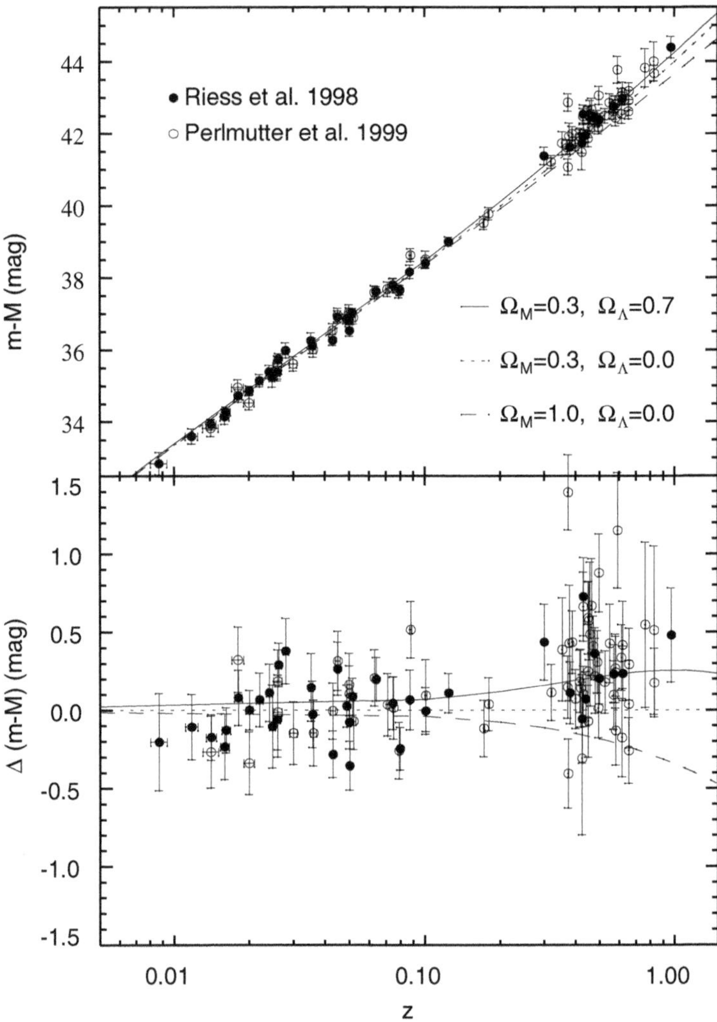

FIGURE 19.2 The redshift-magnitude relation for Type Ia supernovae measured by the High-Z SN Search Team (Riess et al. 1998) and the Supernova Cosmology Project (Perlmutter et al. 1999). The horizontal axis is the redshift z. The vertical axis is the difference between the peak apparent magnitude of the supernova and its observed peak absolute magnitude calibrated, as described in Section 17.2. This distance modulus is a logarithmic measure of f/L, as described on p. 356. Three representative theoretical curves are shown for various values of Ω_m and Ω_v, here denoted by Ω_Λ. [The small value of Ω_r [cf. (19.1)] does not affect these curves much.] The top and bottom boxes show the same data, but in the bottom box the data are plotted in terms of a difference between the observations and the predictions of an $\Omega_m = .3$, $\Omega_v = 0$ model. The bottom curve is the FRW prediction for $\Omega_m = 1$, $\Omega_v = 0$ that was calculated in (19.6). The data do not favor this matter-dominated flat FRW model but rather one with a nonzero value for the cosmological constant.

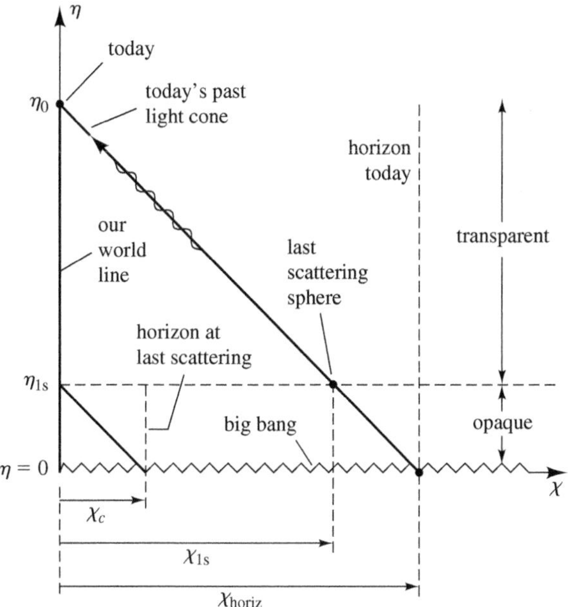

FIGURE 19.3 η-χ spacetime diagram of an FRW model. Radial light rays move on 45° lines when conformal time η, defined by (18.44), is used instead of t. Three important times are shown—the present moment, η_0, the time of last scattering of CMB photons, η_{ls}, and the big bang at $\eta = 0$. (The vertical axis is not to scale; realistically η_{ls} is only a few percent of η_0.) The intersection of the present moment's backward light cone with the big bang defines the coordinate radius of the present particle horizon, χ_{horiz}. The intersection with $\eta = \eta_{ls}$ defines the coordinate radius, χ_{ls}, of the last-scattering surface. This is the sphere from which CMB photons originate that travel along our past light cone to reach us today. Information could be received today in principle from any point with $\chi < \chi_{horiz}$ consistent with causality. But the universe is opaque to electromagnetic radiation before η_{ls}, so χ_{ls} defines the largest volume of the universe visible in light. (Neutrinos or gravitons would allow us to see earlier.) The coordinate radius χ_c defines the largest distance light could travel between the time of the big bang and the last-scattering time, which is just the radius of the horizon at η_{ls}.

on the geometry of the universe through which the light has been propagating in the 13 Gyr since then. Maps of the temperature fluctuations such as Figure 17.12 are a picture of this last-scattering surface processed through the geometry and evolution of an FRW model. Therein lies the possibility of using the anisotropies to determine cosmological parameters.

The connection between the angular sizes of CMB anisotropies and the geometry of the universe was described qualitatively in Box 2.2 on p. 17. We can now make this connection quantitative. For simplicity, suppose that the size Δs of some feature in the CMB at last-scattering is known from a theory of its origin. Let's calculate the angular size $\Delta\phi_s$ of the image of this feature today in a general FRW model assuming that $\Delta\phi_s$ is small. Write the FRW line elements in the form (18.60). Assume that our world line is at $\chi = 0$ and label the sphere of last-scattering by $\chi = \chi_{\rm ls}$. Assume that the coordinates are oriented so the feature lies in the equatorial plane $\theta = \pi/2$ of that surface. Since light from the extremities of the feature propagates to us along radial lines of constant ϕ, the angle $\Delta\phi_s$ is the angular coordinate length ϕ on the equator of an interval whose physical length is Δs. Specifically,

$$a(t_{\rm ls})\Delta\phi_s \begin{Bmatrix} \sin\chi_{\rm ls} \\ \chi_{\rm ls} \\ \sinh\chi_{\rm ls} \end{Bmatrix} = \Delta s \qquad \begin{Bmatrix} \text{closed} \\ \text{flat} \\ \text{open} \end{Bmatrix}, \tag{19.10}$$

where $t_{\rm ls}$ is the time of last scattering. Writing $a(t_{\rm ls}) = a(t_0)[a(t_{\rm ls})/a(t_0)] = a(t_0)/(1 + z_{\rm ls})$, one finds

$$\Delta\phi_s = (1 + z_{\rm ls})\frac{\Delta s}{d_{\rm eff}}, \tag{19.11}$$

where $d_{\rm eff}(z_{\rm ls}, H_0, \Omega_r, \Omega_m, \Omega_v)$ is given by (19.7) and $z_{\rm ls}$ is the redshift of the surface of last-scattering. This redshift could depend on the parameters of the cosmological model because rates of the processes involved in recombination are competing with the expansion of the universe (see Box 18.1 on p. 375). However, detailed calculations put the redshift of the last-scattering surface at $1 + z_{\rm ls} \approx 1100$, largely independent of cosmological parameters and corresponding to a temperature of about 3000 K. We will assume these values in what follows. Thus, the angle $\Delta\phi_s$ subtended today by a feature of length Δs on the last-scattering surface becomes a function of the cosmological parameters through $d_{\rm eff}$.

The universe is approximately matter-dominated over the period that the CMB radiation has been propagating to us. The case of a purely matter-dominated universe ($\Omega_v = \Omega_r = 0$) permits a straightforward analysis of the relation between angular size and spatial curvature. The effective distance $d_{\rm eff}$ in (19.11) is given by (19.7) and (19.9). For a matter-dominated universe at any value of $\tilde{a}$ contributing to (19.9), the denominator is larger for a closed universe ($\Omega_m > 1$) than for an open one ($\Omega_m < 1$). Correspondingly, the angle χ defined by (19.9) is smaller for a closed universe than for an open one. Since $\sin\chi < \sinh\chi$ for positive χ, it follows from (19.11), (19.10) and (19.7) that $d_{\rm eff}$ is smaller for a closed universe than for an open one, for a given z. In (19.11) this leads to

$$\Delta\phi_s^{\rm open} < \Delta\phi_s^{\rm flat} < \Delta\phi_s^{\rm closed}. \tag{19.12}$$

Thus, by measuring the angular size of features of the cosmic background of known physical size, it is in principle possible to tell whether the universe is open, flat, or closed.

Realistically the CMB's anisotropies do not have a definite size but rather a spectrum of them (see Figure 17.12 and the one in Box 2.2 on p. 17). Information about cosmological parameters is contained in the statistics of these observed angular sizes. The central quantity is the *correlation function* of the temperature anisotropies, $C(\theta)$, defined as follows: let $\Delta T(\vec{n})/T$ be the fractional deviation of CMB temperature from its mean value in the direction of a unit vector $\vec{n}$. Take two vectors $\vec{n}$ and $\vec{n}'$ that make a fixed angle θ with each other. The correlation function $C(\theta)$ is defined by averaging the product of the two $\Delta T/T$'s over the sky. Explicitly,

$$C(\theta) \equiv \left\langle \frac{\Delta T(\vec{n})}{T} \frac{\Delta T(\vec{n}')}{T} \right\rangle, \tag{19.13}$$

where the angle brackets denote the all-sky average over $\vec{n}$ and $\vec{n}'$ keeping $\vec{n} \cdot \vec{n}' = \cos\theta$.

Information in the correlation function is often most efficiently extracted through its multipole expansion in Legendre polynomials:

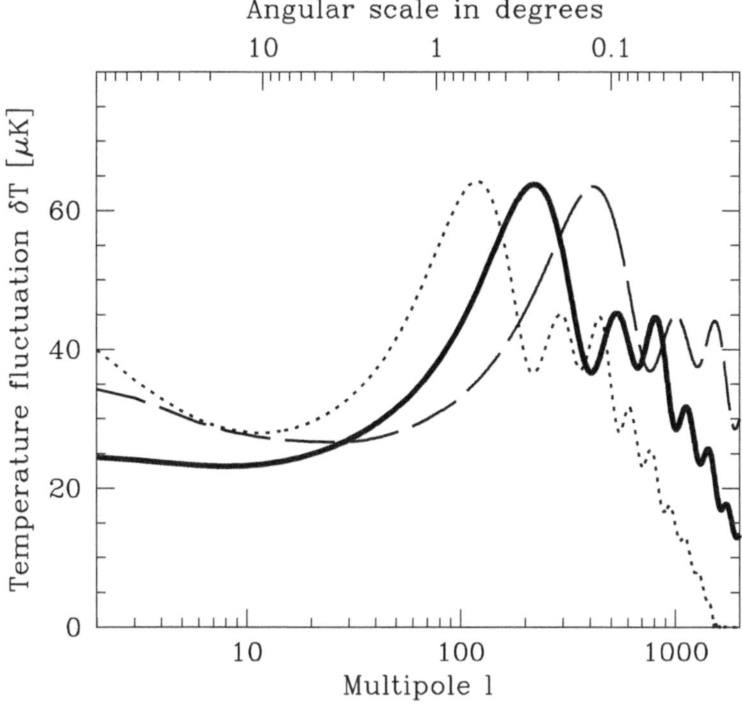

FIGURE 19.4 Theoretical predictions of the angular spectrum of temperature fluctuations in the CMB. The vertical axis is $\delta T \equiv T[\ell(\ell+1)C_\ell/2\pi]^{1/2}$. This is a measure of the temperature fluctuations in a given multipole ℓ (bottom scale) or on a corresponding angular scale (upper scale). All three curves assume $\Omega_{\text{baryon}} = .04$, $\Omega_m = .23$, and $h = .72$, as well as a common spectrum of fluctuations at the time of last scattering. They differ only in the value of Ω_v. The solid curve corresponds to $\Omega_v = .7$ (approximately flat), the dashed curve has $\Omega_v = 0$ (open), and the dotted curve, $\Omega_v = 1$ (closed). The largest peak in the open model occurs at a lower angular scale (higher ℓ) than in the approximately flat case whose angular scale is yet lower than the closed case, all as expected from (19.12). Measurements of the CMB anisotropies can thus determine whether the universe is open or closed and other cosmological parameters as well.

$$C(\theta) = \sum_{\ell=0}^{\infty} \frac{2\ell+1}{4\pi} C_\ell P_\ell(\cos\theta), \tag{19.14}$$

thus defining coefficients $C_\ell, \ell = 0, 1, 2, \ldots$.[2] If there are prominent features characterized by an angular size $\Delta\theta$ in radians, then the C_ℓ's will be enhanced for a value of ℓ inversely related to $\Delta\theta$. Equation (19.12) indicates that, for example, a given feature should show up at *lower* ℓ in a flat universe than it would in an open one. Figure 19.4 shows some theoretical predictions of the C_ℓ's, which illustrates just this effect. In this way, measurements of the anisotropy of the cosmic background radiation can determine whether we live in an open, closed, or flat universe. At the time of writing, the evidence from experiments such as that illustrated in Box 2.2 on p. 17 are consistent with the spatial geometry of the universe being flat.

19.2 Explaining the Universe

The evidence of the observations is that our universe is approximately homogeneous and isotropic on scales above several hundred megaparsecs and that it is close to being spatially flat—on the borderline between the open and closed FRW models. Further, the observations show our early universe to be even more homogeneous and isotropic than the universe today. A picture of remarkable simplicity thus emerges on the largest scales of space and time. These successes of observational cosmology have inevitably raised the question of *why* our universe has these simple special properties.

A homogeneous, isotropic, spatially flat universe is not the only cosmological model allowed by the Einstein equation. Zero spatial curvature FRW models, for instance, are but one point in a continuum of possibilities ranging from high negative spatial curvature to high positive spatial curvature. The Einstein equation also permits many *in*homogeneous, *an*isotropic cosmologies quite unlike the universe we live in. Which solution describes our universe depends on its initial condition, and ultimately cosmology requires a theory of this initial condition. At the big bang, where quantum gravity is important [cf. (1.6)], an initial condition means a quantum wave function for the universe. The subject of *quantum cosmology* concerned with that, however, is well outside the scope of this text.

Causality and Horizons

Physical processes that take place over the course of the history of the universe can help explain why it is the way we see it today. For example, clustering by the ever attractive force of gravity explains how tiny density fluctuations in the early

[2] If you haven't encountered them, Legendre polynomials are a series of orthogonal polynomials discussed in almost every text in electromagnetism or quantum mechanics: $P_0 = 1$, $P_1 = \cos\theta$, $P_2 = (3\cos^2\theta - 1)/2, \ldots$.

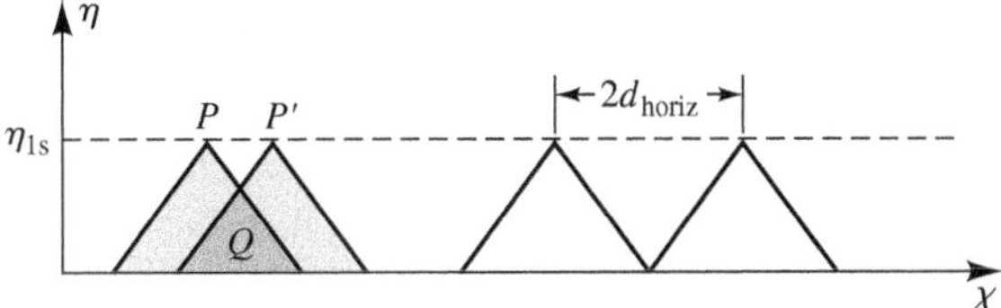

FIGURE 19.5 Causal contact at last scattering. Events in a spacetime region Q can influence things at points P and P' on the last scattering surface because Q is in the past light cone of both. However, if the points are separated by a physical distance larger than twice the horizon radius at last scattering $d_{horiz}(t_{ls})$, then the past light cones do not overlap and no event since the big bang could have influenced things at both.

universe, whose signatures are the minute temperature anisotropies of the CMB, can be amplified to eventually produce the condensations of matter that are the galaxies and stars we see today.

However, there is a fundamental obstacle to explaining the universe by any dynamical process that occurs over its history, which is illustrated in Figure 19.5. Light can travel only a finite distance since the big bang, and any *causal* physical process can act only over a volume of this radius. For the remarkable isotropy of the cosmic background radiation to be explained by any physical process, the whole of the last-scattering surface visible today would have to have been in causal contact at the time $t_{ls} \sim 400{,}000$ yr (Problem 9) when radiation last scattered from matter. Whether that's the case depends on how the universe expanded before that time.

Figure 19.5 shows the relationship between regions that could have been in causal contact at the time of last scattering. Figure 19.3 shows their relation to the visible universe today. The radius of the region that could have been in causal contact at last scattering is the radius of the particle horizon there, $d_{horiz}(t_{ls})$ [cf. (18.47)]. Since the universe is matter dominated at the time of last scattering (Problem 8) and since spatial curvature is unimportant in the effective potential (18.78) at a redshift of 1100, where $\tilde{a} = 1/(1+z)$ is very small, the horizon size can be estimated by assuming a spatially flat FRW model that was matter dominated for the whole of its history *before* t_{ls}. In this case, $d_{horiz}(t)$ is given by (18.48a) and

$$\begin{pmatrix} \text{radius of region} \\ \text{in causal contact} \\ \text{at last-scattering} \end{pmatrix} \approx 3t_{ls}. \tag{19.15}$$

The angular size such a region would subtend today on the sky can be calculated from (19.11) with $\Delta s = 6t_{ls}$ and is a few degrees. Thus, no physical mechanism acting before last scattering can explain the remarkable isotropy of the cosmic background radiation *if* the universe was matter dominated before that. Including radiation and vacuum energy does not change this conclusion, but matter in the very early universe is not necessarily well modeled by the simple assumptions of the FRW models.

Inflation

The early universe is the realm of high energy physics, and the extrapolation of the simple FRW model of noninteracting matter, radiation, and vacuum is unlikely to be valid there. Suppose the universe had a period when the scale factor increased exponentially like (18.39),

$$a(t) \propto e^{Ht}, \tag{19.16}$$

for some constant H. Such an exponentially rapid increase in the scale factor is called *inflation*. Even a short, early period of inflationary expansion can help explain why the universe is the way it is today. Here are three ways it does this:

- *Increase in Horizon Size.* Consider the increase in the horizon size, Δd_{horiz}, that accumulates in an inflationary epoch that lasts between a time t_s and t_f for an interval $\Delta t = t_f - t_s$. From (18.47) and (19.16), this is

$$\Delta d_{\text{horiz}} = e^{Ht_f} \int_{t_s}^{t_f} dt' e^{-Ht'} = \frac{e^{H\Delta t} - 1}{H}. \tag{19.17}$$

BOX 19.2 A Mechanism for Inflation

This box very crudely describes one of several ways in which a rapid inflationary expansion could have been produced in the very early universe ($t \approx 10^{-34}$ s). At that early time, matter is more accurately described in terms of quantum fields and their expectation values than in terms of particles and radiation. Consider just a single scalar field, denote its expectation value by ϕ, and assume it to be a function only of t, $\phi = \phi(t)$, consistent with the observed approximate homogeneity of the universe. It turns out that $\phi(t)$ evolves as though it were the position of a particle in an effective potential $V_\phi(\phi)$, describing how the field interacts with itself, such as that shown here.

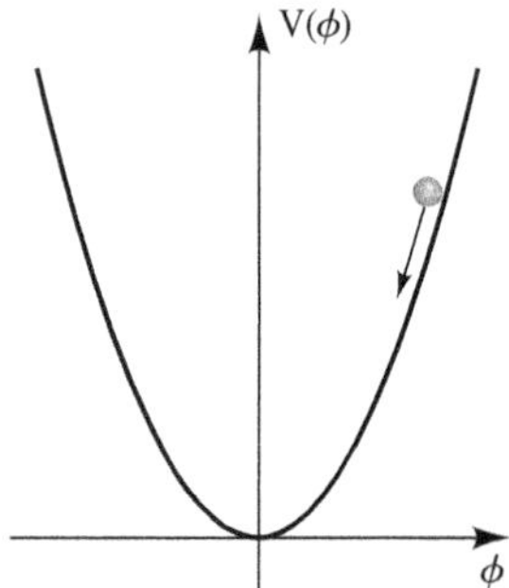

Suppose the field starts from rest at a value ϕ_*. Its potential energy, $V_\phi(\phi_*)$, is momentarily unchanging in time and acts like a vacuum energy. A flat universe inflates like (18.39) with $H^2 = 8\pi V_\phi(\phi_*)/3$, as in (18.40). As discussed in the text, if H is characterized by the energy scales of the unification of strong and electroweak forces, the inflation will be rapid. As the field rolls down, potential energy is converted into kinetic energy, which can be dissipated by various mechanisms, among them the expansion of the universe. The field winds up near zero energy and the inflationary expansion ends. A transition has been made from an inflating early universe to the universe we see today.

As mentioned in the text, inflation helps explain why the universe is homogeneous and isotropic. But the mechanism described here also helps explain the spectrum of *in*homogeneities that are observed in the present large-scale distribution of galaxies. Because of quantum fluctuations, the field may not be exactly homogeneous but can start rolling down the potential at slightly different times in different places. The resulting differences in the field at different places would lead to density fluctuations that are consistent with those observed in the CMB at last scattering and in the statistics of the distribution of galaxies today.

The horizon thus expands exponentially rapidly in an inflationary epoch. Even a value of $H\Delta t \sim 60$ would be enough to put the whole of the observable universe in causal contact at the time of last scattering.

- *Spatial Flatness.* Inflation also drives the universe toward $\Omega = 1$ and thus predicts that our universe is spatially flat, consistent with current observations. To see this, define $\Omega(t)$ to be the ratio of the total energy density $\rho(t)$ at time t to the critical density then. During an inflationary phase described by (19.16), the critical density is $\rho_{\text{crit}} = 3H^2/(8\pi)$, where $H(t) \equiv \dot{a}(t)/a(t)$ [cf. (18.14)]. Then, just rewriting the Friedman equation (18.63) using the inflationary $a(t)$ in (19.16) gives

$$\Omega(t) - 1 \propto ke^{-2Ht}. \tag{19.18}$$

  Thus, *very* quickly inflation drives the universe to $\Omega = 1$. This is one of inflation's most important predictions and is currently consistent with observations.

- *Homogeneity and Isotropy.* Like the inflation of an initially irregular balloon, inflation stretches the spatial size of initial inhomogeneities and helps explain why the universe is homogeneous and isotropic on the distance scales we can observe today. Of course, for any fixed duration of inflation, there are some large inhomogeneities that would not be stretched out in the time available, but an inflationary expansion goes a long way toward explaining the observed homogeneity and isotropy.

An exponential expansion such as (19.16) is characteristic of a vacuum energy [cf. (18.39)]. The vacuum energy today, as revealed by observations such as those summarized in Figure 19.2, may be important for the expansion of the late universe but is negligible on the scale of elementary particle physics energies (Problem 2) and unimportant for the evolution of the very early universe. But elementary particle interactions themselves could generate an inflationary expansion by mechanisms such as that described in Box 19.2 on p. 412. In typical models these mechanisms operate at energy scales of 10^{14} GeV and higher, approaching those characterizing the unification of the fundamental forces other than gravity. Energy scales of order 10^{14} GeV are achieved at *very* early times of order 10^{-34} s (18.42) and lead to values of H^{-1} of comparable magnitude. Only a very tiny period of inflationary expansion of this order of magnitude in duration, and at this very early time is needed to generate a horizon bigger than the visible universe, drive the universe to $\Omega = 1$, and stretch out significant initial irregularities to scales much bigger than we could observe them. That's one explanation of why observations show our universe to be approximately homogeneous and isotropic and on the borderline between positive and negative spatial curvature.

Problems

1. [B] The radio source 3C345 discussed in Box 4.3 on p. 61 has a redshift of $z = .595$. The angular velocity of the outward moving cloud $C2$ is approximately .47 mas/yr.

Assuming (contrary to fact) that the cloud is moving transverse to the line of sight, what velocity would be seen by an observer on station at 3C345 taking a flat ($k = 0$), matter-dominated FRW model of the universe with $h = .72$?

2. [S] Could the observed vacuum mass-energy density in the universe be a consequence of quantum gravity? One obstacle to such an explanation is the great difference in scale between observed vacuum mass density ρ_v and the Planck mass density $\rho_{Pl} \equiv c^5/\hbar G^2$ [cf. (1.6)] that might be expected on dimensional grounds to characterize quantum gravitational phenomena (Chapter 1).

(a) Show that ρ_{Pl} is the correct combination of $\hbar$, G, and c with the dimensions of mass density.

(b) Evaluate the ratio ρ_v/ρ_{Pl}.

3. [S] Show that the expression (19.5) for the effective distance d_{eff} in a flat, matter-dominated FRW model follows from (19.7) and (19.9).

4. [A] For small z the redshift-magnitude relation is given by the inverse square law (19.3). This is the first term in an expansion in z of the form

$$\frac{f}{L} = \frac{H_0^2}{4\pi z^2}(1 + \text{const.}\ z + \cdots).$$

Find the constant and express it in terms of the cosmological parameters. Sketch redshift-magnitude curves that have both $\Omega_m = .3$ and $\Omega_r = 0$ for the two values $\Omega_v = 0$ and $\Omega_v = .7$.

5. [S] Show that the effective distance, d_{eff}, defined in (19.7) of a galaxy in a spatially flat universe at redshift z can be written as

$$d_{\text{eff}} = \int_0^z dz'/H(z')$$

where $H(z')$ is the value of the Hubble constant when light from a galaxy with redshift z' was emitted.

6. *Standard Rulers* Suppose a certain kind of galaxy always had a fixed size. It then could be used as a standard ruler—from its angular size, its distance could be computed. Derive the redshift–angular size relation that is analogous to the redshift-magnitude relation for a flat, matter-dominated FRW universe. Show that there is a certain redshift beyond which the angular size of the object increases with redshift and find its value. Does this mean that objects will get brighter the further they are from us?

7. [C] *Number Counts of Galaxies* Suppose a census was taken of the number of galaxies $N_{\text{gal}}(Z)$ with a redshift *less* than a particular value Z. Assume the number density of galaxies $n_{\text{gal}}(t)$ is uniform in space, but changing in time. What is the prediction of a flat, matter-dominated, FRW model for how $N_{\text{gal}}(Z)$ depends on Z? Express your answer in terms of Z, the Hubble constant, and the present density of galaxies $n_{\text{gal}}(t_0)$. (*Comment:* Counting galaxies is another route to determining cosmological parameters, but the further away they are, the dimmer they are and the harder to count.)

8. Assume that our universe is characterized by the cosmological parameters $H_0 = 72$ (km/s)/Mpc, $\Omega_m = .3$, $\Omega_r = 8 \times 10^{-5}$, $\Omega_v = .7$. Also assume that last scattering occurs at a redshift of 1100.

(a) Compare the temperature at last scattering with the temperature T at which $k_B T$ is equal to the binding energy of hydrogen.

(b) Before what redshift z is the universe radiation dominated?

9. Calculate the age of our universe at the time of last scattering.

10. *Measure the Cosmological Constant in the Laboratory?* It's not possible to measure the matter density ρ_m in a laboratory of the typical size found on Earth. The FRW approximation that the matter is smoothly distributed breaks down on those scales. But a fundamental vacuum energy could be exactly uniform and therefore in principle detectable in a laboratory experiment. Calculate how two test particles in a freely falling laboratory would move relative to each other in the presence of a vacuum energy corresponding to $\Omega_v = 1$. Estimate the time scale for significant relative motion and the size of their relative acceleration assuming they start 1 cm apart. Is laboratory detection feasible?

PART III

The Einstein Equation

The Einstein equation governing the geometry of curved spacetime, which is the basic equation of general relativity, is introduced and solved to exhibit some of the geometries described previously, calculate the production of gravitational radiation, and analyze the structure of relativistic stars.

CHAPTER 20

A Little More Math

This last part of the book introduces the Einstein equation—the basic equation of general relativity in much the same way that Maxwell's equations are the basic equations of electromagnetism. Geometries such as the Schwarzschild geometry or those of the FRW cosmological models are particular solutions of the Einstein equation. Just three new mathematical ideas are needed to give an efficient and standard discussion of the Einstein equation: a more precise definition of vectors, the notion of dual vectors, and the covariant derivative. These mathematical concepts are introduced in this chapter.

20.1 Vectors

The introduction of vectors in curved spacetime in Section 7.8 was mathematically imprecise even if physically accurate. This section gives a more precise definition of vectors.[1] In particular, it gives a definition in terms of directional derivatives of what we meant in Section 7.8 by directions defined locally.

Vectors in flat spacetime were defined as directed line segments in Section 5.1. But there is another completely equivalent way of introducing vectors in flat spacetime by identifying them with directional derivatives. To recall the definition of the directional derivative of a function, consider a function $f(x^\alpha)$ and a curve $x^\alpha(\sigma)$. The directional derivative along the curve at the point labeled by σ is defined by

$$\frac{df}{d\sigma} = \lim_{\epsilon \to 0}\left[\frac{f(x^\alpha(\sigma+\epsilon)) - f(x^\alpha(\sigma))}{\epsilon}\right] = \frac{dx^\alpha}{d\sigma}\frac{\partial f}{\partial x^\alpha}. \tag{20.1}$$

The vector $\mathbf{t}$ with coordinate basis components

$$t^\alpha = \frac{dx^\alpha}{d\sigma} \tag{20.2}$$

is a tangent vector to the curve. (For the timelike curves followed by particles, $\mathbf{t}$ is the four-velocity $\mathbf{u}$ if σ is the proper time.) The directional derivative at the point labeled by σ is therefore specified by $\mathbf{t}$, and we can write

$$\frac{d}{d\sigma} = t^\alpha \frac{\partial}{\partial x^\alpha}. \tag{20.3}$$

[1]Specifically, four-vectors, but recall that the four was to be dropped in chapters after Chapter 7.

Thus, to every directional derivative there corresponds a vector. Conversely, to every vector **t**, there corresponds a directional derivative given by (20.3) along the curve $x^\alpha(\sigma) = x^\alpha(0) + t^\alpha\sigma$. *Vectors and directional derivatives are thus in one-to-one correspondence*, and we may write generally for any vector **a** the corresponding directional derivative

Vectors as Directional Derivative

$$\mathbf{a} \equiv a^\alpha \frac{\partial}{\partial x^\alpha}. \tag{20.4}$$

Thus, vectors in flat spacetime *could* have been defined as directional derivatives instead of directed line segments. Given a directional derivative (20.4), we could identify the components and construct the directed line segment to which they correspond. But it isn't necessary to do that. All the usual rules of vector algebra follow directly from (20.4). For instance, the rule that the components of the sum of two vectors is the sum of their components follows from the linearity of (20.4) in the components a^α. From (20.4) it also follows that the partial derivatives $\partial/\partial x^\alpha$ are coordinate basis vectors since the components (20.2) are the coordinate basis components of **t**. You can think of $\partial/\partial x^\alpha$ as just another notation for coordinate basis vectors.

Vectors cannot be defined as directed straight-line segments in curved spacetime for reasons given in Section 7.8. (How do you add straight-line segments in a curved spacetime?) However, the definition of vectors as directional derivatives *does* generalize to curved spacetime. Equations (20.1)–(20.4) hold in curved spacetime, and all the usual rules of vector algebra follow from them. From now on we'll think of vectors as directional derivatives. The linear space of directional derivatives is the *tangent space* referred to informally in Section 7.8.

Some find it unsettling to think of a vector as a differential operator and prefer to think of the notion of direction as defined by infinitesimal line segments, as described in Section 7.8. That is acceptable for physics, but it is the partial derivative that gives a precise mathematical meaning to the notion of infinitesimal line segments. Example 20.1 illustrates how useful the definition in terms of directional derivatives can be.

Example 20.1. Transforming from One Coordinate Basis to Another. How are the coordinate basis components of a vector **a** in one coordinate system related to those in another? Equation (20.4) and the algebra of partial derivatives provide a direct answer. Suppose x^α is one set of coordinates, x'^α another, and the connection between them $x'^\alpha(x^\beta)$ is known. Then

$$\mathbf{a} = a^\alpha \frac{\partial}{\partial x^\alpha} = a^\alpha \frac{\partial x'^\beta}{\partial x^\alpha}\frac{\partial}{\partial x'^\beta} \equiv a'^\beta \frac{\partial}{\partial x'^\beta}. \tag{20.5}$$

The transformation law between the coordinate basis components a^α in the coordinates x^α and the coordinate basis components a'^α in the coordinates x'^α follows from (20.5):

$$a'^\beta = \frac{\partial x'^\beta}{\partial x^\alpha} a^\alpha. \tag{20.6a}$$

The inverse transformation from the connection $x^\alpha(x'^\beta)$ is obtained just by interchanging primed and unprimed quantities in (20.6a):

$$a^\beta = \frac{\partial x^\beta}{\partial x'^\alpha} a'^\alpha. \tag{20.6b}$$

20.2 Dual Vectors

Linear Maps from Vectors to Real Numbers

Dual Vectors Defined

A *dual vector* $\boldsymbol{\omega}$ is a linear map from vectors to real numbers.[2] The real number to which a dual vector $\boldsymbol{\omega}$ maps a vector $\mathbf{a}$ is denoted by $\omega(\mathbf{a})$. (We don't use boldface ω here since the result of the map is a number.) A map is *linear* if, for any two vectors $\mathbf{a}$ and $\mathbf{b}$ and any two numbers α and β,

$$\omega(\alpha\mathbf{a} + \beta\mathbf{b}) = \alpha\omega(\mathbf{a}) + \beta\omega(\mathbf{b}). \tag{20.7}$$

A linear map from a vector $\mathbf{a}$ to a real number must also be a linear map from the vector's components a^α into the same real number. Assuming a zero vector is mapped to zero, the most general linear map of components to real numbers has the form

$$\omega(\mathbf{a}) = \omega_\alpha a^\alpha \tag{20.8}$$

for numbers ω_α, called the *components* of the dual vector $\boldsymbol{\omega}$.

Example 20.2. The Gradient. The gradient of a function $f(x^\alpha)$ provides the simplest example of a dual vector. We saw in (20.1) that the derivative of a function in the direction specified by a vector $\mathbf{t}$ is

$$\frac{\partial f}{\partial x^\alpha} t^\alpha. \tag{20.9}$$

The derivatives of a function $f(x^\alpha)$ thus specify a linear map from any vector $\mathbf{t}$ into the real number (20.9). That map is a dual vector called the *gradient of* f, whose components are $\partial f/\partial x^\alpha$. The gradient dual vector is conventionally denoted by $\boldsymbol{\nabla} f$.

Consider, for instance, the function $g(x) = -t^2 + x^2 + y^2 + z^2$, which gives the square of the distance of the point at x^α from the origin in flat spacetime. The gradient of g has the components $\partial g/\partial x^\alpha = (-2t, 2x, 2y, 2z)$.

A set of four linearly independent dual vectors $\{\mathbf{e}^\alpha\}$ (the curly brackets mean "set of") constitute a *basis* for all dual vectors. Any dual vector $\boldsymbol{\omega}$ is some linear combination of the basis dual vectors, namely,

$$\boldsymbol{\omega} = \omega_\alpha \mathbf{e}^\alpha. \tag{20.10}$$

[2] Alternative names for dual vectors are *one-forms* and *covectors*.

The numbers ω_α are the *components* of the dual vector in the basis $\{\mathbf{e}^\alpha\}$. The particular dual-vector basis that gives the components introduced in (20.8) is related to the basis for *vectors* $\{\mathbf{e}_\alpha\}$ that defined the components a^α in the same expression by

$$e^\alpha(\mathbf{e}_\beta) \equiv \delta^\alpha_\beta. \tag{20.11}$$

Here, δ^α_β is the Kronecker-δ, defined to be 1 when $\alpha = \beta$ and zero otherwise:

$$\delta^\alpha_\beta \equiv \begin{cases} 1 & \alpha = \beta, \\ 0 & \alpha \neq \beta. \end{cases} \tag{20.12}$$

The $\{\mathbf{e}^\alpha\}$ that satisfies (20.11) is called the basis of dual vectors that is *dual to* the basis of vectors $\{\mathbf{e}_\alpha\}$.

To see that the definition (20.11) reproduces the components of (20.8), just write out $\omega(\mathbf{a})$ using (20.11) and (20.10):

$$\omega(\mathbf{a}) = \omega_\alpha e^\alpha(a^\beta \mathbf{e}_\beta) = \omega_\alpha a^\beta e^\alpha(\mathbf{e}_\beta) = \omega_\alpha a^\beta \delta^\alpha_\beta = \omega_\alpha a^\alpha. \tag{20.13}$$

Do you find it difficult to keep track of all these definitions? The situation is about to become simpler.

The Correspondence Between Vectors and Dual Vectors

Any vector $\mathbf{a}$ specifies a linear map from other vectors $\mathbf{b}$ to real numbers through the scalar product

$$a(\mathbf{b}) \equiv \mathbf{a} \cdot \mathbf{b}. \tag{20.14}$$

Thus, to every vector there corresponds a dual vector. In a coordinate basis, utilizing (20.8) for the left-hand side of (20.14) gives

$$a(\mathbf{b}) \equiv a_\alpha b^\alpha = \mathbf{a} \cdot \mathbf{b} = g_{\beta\alpha} a^\beta b^\alpha = g_{\alpha\beta} a^\beta b^\alpha. \tag{20.15}$$

But, since (20.15) holds for *any* vector $\mathbf{b}$,

Lowering an Index

$$\boxed{a_\alpha = g_{\alpha\beta} a^\beta.} \tag{20.16}$$

Equation (20.16) specifies a correspondence between the *vector* with coordinate basis components a^α and the *dual vector* with components a_α in the basis of dual vectors dual to the coordinate basis.

This connection can be inverted by introducing the matrix inverse of $g_{\alpha\beta}$, called the *inverse metric*. The inverse metric is denoted by $g^{\alpha\beta}$ and is defined by the usual connection between a matrix and its inverse:

Inverse Metric

$$\boxed{g^{\alpha\gamma} g_{\gamma\beta} = \delta^\alpha_\beta.} \tag{20.17}$$

Multiplying both sides of (20.16) by the inverse metric $g^{\gamma\alpha}$ and using (20.12) gives (on relabeling the free indices)

$$a^{\alpha} = g^{\alpha\beta} a_{\beta}. \tag{20.18}$$

Raising an Index

Even simpler relations hold in an orthonormal basis, where $\eta_{\hat{\alpha}\hat{\beta}}$ replaces $g_{\alpha\beta}$ in the defining relation (20.15). For example, one has

$$a_{\hat{0}} = -a^{\hat{0}}, \qquad a_{\hat{1}} = a^{\hat{1}}, \qquad a_{\hat{2}} = a^{\hat{2}}, \qquad a_{\hat{3}} = a^{\hat{3}}. \tag{20.19}$$

The one-to-one connection between vectors and dual vectors supplied by the metric is the reason that we use the same boldface notation (e.g., **a**) for both. This same connection is the reason that physical quantities can be described either as vectors or dual vectors. The momentum **p** of a particle passing through an observer's laboratory (Section 5.6) can be described either by the vector components $p^{\hat{\alpha}}$ with respect to the orthonormal basis of the laboratory or by the dual-vector components $p_{\hat{\alpha}}$. One can be computed from the other using (20.16) or (20.18) with $\eta_{\hat{\alpha}\hat{\beta}}$ replacing $g_{\alpha\beta}$. The component $p^{\hat{0}}$ is the energy the observer would measure and $p_{\hat{0}}$ is minus the energy. This redundancy in description is the reason we did not need to introduce dual vectors before, but they will be *very* convenient in discussing curvature.

Since there is no physical distinction between representing a quantity such as momentum as a vector or dual vector, and since mathematically the representations are in one-to-one correspondence, it is convenient in physics to think of dual-vector components as just a different kind of component of the corresponding vector. In mathematical terms they can be *identified*. Thus, instead of referring to p_{α} as the components of the dual vector that correspond to the vector with components p^{α}, we refer to p_{α} and p^{α} as upper and lower components[3] of the vector **p**. *From now on we will refer just to vectors and their components.*

Example 20.3. Practice Raising and Lowering Indices. Try the following exercise to test whether you can raise and lower indices correctly. Consider two dimensions, where the indices $A, B, \ldots$ range over 1 and 2 and the metric is

$$g_{AB} = \begin{pmatrix} F & 1 \\ 1 & 0 \end{pmatrix} \tag{20.20}$$

for some constant F. Consider the following vectors:

$$a_A = (1, 0), \qquad b_A = (0, 1), \qquad c^A = (1, 0), \qquad d^A = (0, 1). \tag{20.21}$$

Find a^A, b^A, c_A, d_A, $\vec{a} \cdot \vec{b}$, $\vec{a} \cdot \vec{c}$, and $\vec{a} \cdot \vec{d}$. The answers are at the bottom of the page.

---

[3]Upper and lower are sometimes called upstairs and downstairs, or contravariant and covariant, respectively. One of the author's students uses the mnemonic "co is low, that's all you need to know" to remember the names of indices.

The components of the inverse metric are $g^{11} = 0$, $g^{12} = g^{21} = 1$, $g^{22} = -F$. Then $a^A = (0, 1)$, $b^A = (1, -F)$, $c_A = (F, 1)$, $d_A = (1, 0)$, $\vec{a} \cdot \vec{b} = 1$, $\vec{a} \cdot \vec{c} = 1$, and $\vec{a} \cdot \vec{d} = 0$.

Working with Bases and Dual Bases

The relations (20.16) and (20.18) mean that the scalar product between two vectors **a** and **b** can be written in a variety of ways:

$$\mathbf{a}\cdot\mathbf{b} = g_{\alpha\beta}a^\alpha b^\beta = a_\alpha b^\alpha = a^\alpha b_\alpha = g^{\alpha\beta}a_\alpha b_\beta. \tag{20.22}$$

These relations and others throughout this section can be summarized by a modification of the first of the rules for the summation convention of Section 7.3.

1′ Indices on vectors can be either upstairs or downstairs related by the operations of raising and lowering indices. Indices can be raised or lowered in an equation to give an equivalent equation provided (1) free indices on both sides are raised or lowered together so that free indices still balance, and (2) when one dummy summation index is raised, its repeated partner must be lowered, and vice versa, so that repeated indices always occur in upper-lower pairs.

The elements of the basis $\{\mathbf{e}^\alpha\}$ dual to a given basis $\{\mathbf{e}_\alpha\}$ are vectors that satisfy

$$\mathbf{e}^\alpha\cdot\mathbf{e}_\beta = \delta^\alpha_\beta \tag{20.23}$$

as a consequence of (20.11) and (20.14). The vectors $\{\mathbf{e}^\alpha\}$ and the vectors $\{\mathbf{e}_\alpha\}$ can be used to "project out" the various components of a vector **a** as follows:

Projecting Vector Components

$$a^\alpha = \mathbf{e}^\alpha\cdot\mathbf{a}, \qquad a_\alpha = \mathbf{e}_\alpha\cdot\mathbf{a}. \tag{20.24}$$

(Note how the usual rules for balancing indices on both sides of an equation help in remembering these formulas.) To check just the first of these relations, write

$$\mathbf{e}^\alpha\cdot\mathbf{a} = \mathbf{e}^\alpha\cdot(a^\beta\mathbf{e}_\beta) = a^\beta(\mathbf{e}^\alpha\cdot\mathbf{e}_\beta) = \delta^\alpha_\beta a^\beta = a^\alpha. \tag{20.25}$$

The other follows similarly.

In particular, the components of a vector **a** in an orthonormal basis can be found by projecting onto the basis vectors [cf. (5.82), (20.25)]

$$a^{\hat{\alpha}} = \mathbf{e}^{\hat{\alpha}}\cdot\mathbf{a}, \qquad a_{\hat{\alpha}} = \mathbf{e}_{\hat{\alpha}}\cdot\mathbf{a}. \tag{20.26}$$

If the components a^α of a vector **a** are known in a coordinate basis, then its components in a given orthonormal basis can be computed from (20.26) if the coordinate basis components, $(\mathbf{e}_{\hat{\alpha}})^\alpha$, of the orthonormal basis vectors and the basis dual to them, $(\mathbf{e}^{\hat{\alpha}})_\alpha$, are known. For then, using $\mathbf{a}\cdot\mathbf{b} = a_\alpha b^\alpha$,

$$a^{\hat{\alpha}} = (\mathbf{e}^{\hat{\alpha}})_\alpha a^\alpha, \qquad a_{\hat{\alpha}} = (\mathbf{e}_{\hat{\alpha}})^\alpha a_\alpha. \tag{20.27}$$

(In this clear but perhaps pedantic notation, $(\mathbf{e}_{\hat{2}})^1$ means the 1 coordinate basis component of the orthonormal basis vector $\mathbf{e}_{\hat{2}}$, etc., and we regard $\hat{\alpha}$ as a distinct index from α as far as the summation convention is concerned.)

TABLE 20.1 Bases and Dual Bases

General relations for any basis $\{\mathbf{e}_\alpha\}$, the basis $\{\mathbf{e}^\alpha\}$ dual to it, and a vector $\mathbf{a}$:

$$\mathbf{e}^\alpha \cdot \mathbf{e}_\beta = \delta^\alpha_\beta,$$
$$\mathbf{a} = a^\alpha \mathbf{e}_\alpha, \qquad \mathbf{a} = a_\alpha \mathbf{e}^\alpha,$$
$$a^\alpha = \mathbf{e}^\alpha \cdot \mathbf{a}, \qquad a_\alpha = \mathbf{e}_\alpha \cdot \mathbf{a}.$$

The inverse metric $g^{\alpha\beta}$ is the matrix inverse of $g_{\alpha\beta}$:

$$g^{\alpha\gamma} g_{\gamma\beta} = \delta^\alpha_\beta$$

Relations for a *coordinate basis*:

$$\mathbf{e}_\alpha \cdot \mathbf{e}_\beta = g_{\alpha\beta}, \qquad \mathbf{e}^\alpha \cdot \mathbf{e}^\beta = g^{\alpha\beta},$$
$$\mathbf{e}_\alpha = g_{\alpha\beta}\mathbf{e}^\beta, \qquad \mathbf{e}^\alpha = g^{\alpha\beta}\mathbf{e}_\beta,$$
$$a_\alpha = g_{\alpha\beta}a^\beta, \qquad a^\alpha = g^{\alpha\beta}a_\beta.$$

Relations for an *orthonormal basis*:

$$\mathbf{e}_{\hat\alpha} \cdot \mathbf{e}_{\hat\beta} = \eta_{\hat\alpha\hat\beta}, \qquad \mathbf{e}^{\hat\alpha} \cdot \mathbf{e}^{\hat\beta} = \eta^{\hat\alpha\hat\beta},$$
$$\mathbf{e}_{\hat\alpha} = \eta_{\hat\alpha\hat\beta}\mathbf{e}^{\hat\beta}, \qquad \mathbf{e}^{\hat\alpha} = \eta^{\hat\alpha\hat\beta}\mathbf{e}_{\hat\beta},$$
$$a_{\hat\alpha} = \eta_{\hat\alpha\hat\beta}a^{\hat\beta}, \qquad a^{\hat\alpha} = \eta^{\hat\alpha\hat\beta}a_{\hat\beta}.$$

Relations between basis vectors and dual-basis vectors and the components in each are summarized in Table 20.1.

Example 20.4. Bases and Dual Bases in Skew Rectangular Coordinates. Consider skew rectangular coordinates (x, y) for the flat plane, where the x- and y-axes make an angle ψ with each other, as illustrated in Figure 20.1. The flat-space line element in these coordinates is

$$dS^2 = dx^2 + 2\cos\psi\, dx\, dy + dy^2. \tag{20.28}$$

The coordinate basis vectors $\mathbf{e}_x$ and $\mathbf{e}_y$ point along the coordinate axes as shown and satisfy $\mathbf{e}_A \cdot \mathbf{e}_B = g_{AB}$ (the indices A and B ranging over 1 and 2.) [Recall the defining relation (7.56).] From (20.28) this means that $\mathbf{e}_x$ and $\mathbf{e}_y$ are unit vectors making an angle ψ with each other. The dual basis vectors $\mathbf{e}^x$ and $\mathbf{e}^y$ satisfy $\mathbf{e}^A \cdot \mathbf{e}_B = \delta^A_B$—conditions which determine their magnitudes and directions, as shown. The magnitude of $\mathbf{e}^x$ follows from one of these relations, $\mathbf{e}^x \cdot \mathbf{e}_x = 1$ and the familiar expression for the inner product in terms of the magnitude of the vectors and the angle between them $\mathbf{e}^x \cdot \mathbf{e}_x = |\mathbf{e}^x||\mathbf{e}_x|\cos(\pi/2 - \psi) = 1$. Noting that $|\mathbf{e}_x| = 1$, this gives $|\mathbf{e}^x| = 1/(\sin\psi)$, always greater than unity. The vector $\mathbf{e}^y$ has the same magnitude. The upper components of a vector $\mathbf{a}$ are the coefficients necessary to get a linear combination of $\mathbf{e}_x$ and $\mathbf{e}_y$ add up to $\mathbf{a}$ and the lower components are the coefficients necessary to get a linear combination of $\mathbf{e}^x$ and $\mathbf{e}^y$ to add up to $\mathbf{a}$. Their construction is illustrated in Figure 20.1.

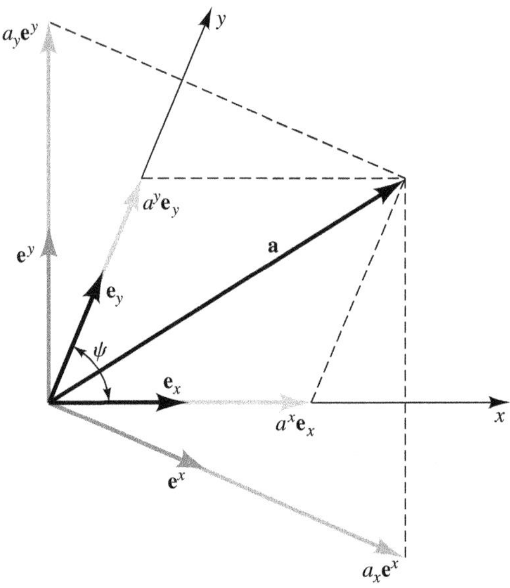

FIGURE 20.1 Skew rectangular coordinates for the plane. The figure shows x- and y-axes making an angle of ψ with each other. The coordinate basis vectors $\mathbf{e}_x$ and $\mathbf{e}_y$ pointing along the coordinate axes are shown. Also shown are the vectors $\mathbf{e}^x$ and $\mathbf{e}^y$ making up the basis dual to the basis $(\mathbf{e}_x, \mathbf{e}_y)$ and related to it by (20.23). An arbitrary vector $\mathbf{a}$ can be resolved into components using either of these bases defining two sets of components, (a^x, a^y) and (a_x, a_y).

Example 20.5. Normal Vectors. The orientation of a two-dimensional surface in three-dimensional space is expressed by the normal vector at each point. In a similar way, the orientation of a three-surface in four-dimensional space is expressed in terms of a normal four-vector at each point (Section 7.9). When the three-surface is defined by an equation of the form

$$f(x^\alpha) = \text{const.}, \tag{20.29}$$

the gradient of f provides one normal vector $\mathbf{n}$, whose lower components are

$$n_\alpha = \frac{\partial f}{\partial x^\alpha}. \tag{20.30}$$

A small displacement $\boldsymbol{\delta}\mathbf{x}$ in the surface with components δx^α doesn't change the value of f, meaning $\delta f = (\partial f/\partial x^\alpha)\delta x^\alpha = 0$ or [cf. (7.68)]

$$n_\alpha \delta x^\alpha = \mathbf{n} \cdot \boldsymbol{\delta}\mathbf{x} = 0. \tag{20.31}$$

That is, $\mathbf{n}$ is orthogonal (or normal) to displacements in the surface and vectors that are tangent to it. This construction doesn't necessarily yield a *unit* normal vector, but any other normal vector will be proportional to this one. As an example, a surface of constant value of a coordinate x^0 has a normal with lower components: $n_\alpha = (1, 0, 0, 0)$. Another example is the Lorentz hyperboloid defined by

$-t^2+r^2 = -a^2$ in (7.74). The normal from (20.30) is $n_\alpha = (-2t, 2r, 0, 0)$ which is proportional to the normal already found in (7.78).

20.3 Tensors

More Linear Maps of Vectors

If linear maps from vectors to real numbers are useful in physics, why not linear maps from *pairs* of vectors to real numbers? These ideas do prove useful. The general notion is called a *tensor*. We use boldface letters to denote tensors, e.g., **t**, as we have every other nonscalar quantity.

The metric is a tensor **g** that defines a linear map of two vectors into the number that is their inner product:

$$g(\mathbf{a}, \mathbf{b}) \equiv \mathbf{a} \cdot \mathbf{b} = g_{\alpha\beta} a^\alpha b^\beta. \tag{20.32}$$

Other important examples related to curvature and the energy density of matter will be encountered in the next chapters.

Tensor Defined

More generally, a *tensor* of *rank* r is a linear map from r vectors into real numbers.[4] A vector is, therefore, a tensor of rank 1. Conventionally, a function $f(x)$ is called a *scalar* when contrasting it with vectors and other tensors. The metric is a second-rank tensor. A third-rank tensor **t** is a linear map from any three vectors **a**, **b**, and **c** into real numbers that can be represented as

$$t(\mathbf{a}, \mathbf{b}, \mathbf{c}) = t_{\alpha\beta}{}^{\gamma} a^\alpha b^\beta c_\gamma. \tag{20.33}$$

The numbers $t_{\alpha\beta}{}^{\gamma}$ are the *components* of the tensor. Here we chose to represent the vectors **a** and **b** using their upper components and the vector **c** with lower components. The same number could equally well be expressed in terms of all upper components, namely,

$$t(\mathbf{a}, \mathbf{b}, \mathbf{c}) = t_{\alpha\beta\gamma} a^\alpha b^\beta c^\gamma, \tag{20.34}$$

and in many other ways.

The connection between the components $t_{\alpha\beta\gamma}$ and $t_{\alpha\beta}{}^{\gamma}$ is found by substituting (20.16) in the form $c_\gamma = g_{\gamma\delta} c^\delta$ into (20.33) and equating the result to (20.34). The resulting connection is

$$t_{\alpha\beta\gamma} = g_{\gamma\delta} t_{\alpha\beta}{}^{\delta}. \tag{20.35}$$

Indices of tensors can, therefore, be lowered just like vectors [cf. (20.16)]; by a similar argument they can also be raised [cf. (20.18)].

Example 20.6. Raising Indices on the Metric Tensor. To raise one index on the metric tensor $g_{\alpha\beta}$, the sum $g^\alpha{}_\beta \equiv g^{\alpha\gamma} g_{\gamma\beta}$ needs to be evaluated. However,

[4] A more general mathematical notion would be to consider linear maps from m vectors and n dual vectors to real numbers. However, once vectors and dual vectors have been identified, as here, there is no further generality in preserving this distinction.

from the definition of the inverse metric (20.17), this is the Kronecker delta:

$$g^{\alpha}{}_{\beta} = \delta^{\alpha}_{\beta}. \tag{20.36}$$

To raise the second index we construct $g^{\beta\gamma} g^{\alpha}{}_{\gamma}$. However, using (20.36), $g^{\beta\gamma} g^{\alpha}{}_{\gamma} = g^{\beta\gamma} \delta^{\alpha}_{\gamma} = g^{\alpha\beta}$, just as it must be if the notation is to be consistent!

Example 20.7. Practice Raising and Lowering Indices. Test your ability to raise and lower indices on tensors by working the following example, which continues from Example 20.3 with the same two dimensions and the same metric. Consider a tensor with components $t_{11} = G$, $t_{12} = 1$, $t_{21} = -1$, $t_{22} = 0$. Calculate $t^{A}{}_{B}$, $t_{A}{}^{B}$, t^{AB}, and $t^{A}{}_{A}$. The answers are at the bottom of the page.

Equation (20.35) obeys the usual rules for balancing free indices and dummy indices that were described in Section 7.3, as extended on p. 424. These also suggest how tensors can be thought of not just as maps between vectors and numbers, but also as maps between tensors and other tensors. By combining a vector **a** with a third-rank tensor **t**, a second-rank tensor can be formed with components

$$t_{\alpha\beta\gamma} a^{\gamma}, \tag{20.37}$$

and by combining two vectors **a** and **b** with the tensor **t**, we get a vector **v** with components

$$v_{\alpha} = t_{\alpha\beta\gamma} a^{\beta} b^{\gamma}. \tag{20.38}$$

The number of free indices in such expressions is the rank of the resulting tensor.

A very simple way of constructing tensors is to take products of vectors. For example, from three vectors **u**, **v**, **w**, we can form the third-rank tensor **s** whose components are

$$s^{\alpha\beta\gamma} = u^{\alpha} v^{\beta} w^{\gamma}, \tag{20.39}$$

and so forth.

Summing upper and lower indices in pairs is an operation called *contraction*, which reduces the rank of a tensor by two. For example, the last two indices of the third-rank tensor $t_{\alpha\beta}{}^{\gamma}$ can be contracted to get a vector **w** with the components

$$w_{\alpha} = t_{\alpha\beta}{}^{\beta}. \tag{20.40}$$

Contraction is thus defined in a basis, but the result is basis independent, as working Problem 9 shows.

In the following the rows in the matrices are labeled by the first index, the columns by the second.

$t^{A}{}_{B} = \begin{pmatrix} -1 & 0 \\ G+F & 1 \end{pmatrix}$, $t_{A}{}^{B} = \begin{pmatrix} 1 & G-F \\ 0 & -1 \end{pmatrix}$, $t^{AB} = \begin{pmatrix} 0 & -1 \\ 1 & G \end{pmatrix}$, and $t^{A}{}_{A} = 0$.

Converting from One Basis to Another

Not infrequently we have the components of a tensor in one basis but want them in another. This subsection discusses two examples of such transformations—changing from a coordinate basis to an orthonormal basis and changing from one coordinate basis to another. These transformations can easily be remembered by recalling that taking products of vectors, as in (20.39), is one way of forming a tensor. In that case transforming each vector and forming the transformed product gives the transformation of the tensor. The transformation rules for the general case have the same form. We will illustrate these rules with a generic second-rank tensor **t**; the generalizations to tensors of higher rank should be evident.

Converting from a Coordinate Basis to an Orthonormal Basis

Compute in a coordinate basis; interpret in an orthonormal basis. That has been a frequently used route to understanding in this text. To convert the coordinate basis components $t_{\alpha\beta}$ of a second-rank tensor **t** to the components $t_{\hat{\alpha}\hat{\beta}}$ in an orthonormal basis, we first need to know the coordinate basis components $(\mathbf{e}_{\hat{\alpha}})^{\alpha}$ of the orthonormal basis vectors. We then project the tensor onto the coordinate basis generalizing (20.27) as follows:

$$t_{\hat{\alpha}\hat{\beta}} = (\mathbf{e}_{\hat{\alpha}})^{\alpha}(\mathbf{e}_{\hat{\beta}})^{\beta} t_{\alpha\beta}. \tag{20.41}$$

Transforming between Coordinate Bases

Equations (20.6) give the transformation rules between the coordinate basis components of a vector in two different coordinate systems, x^{α} and x'^{α}. The transformation rule for the components of a tensor follows from this and the invariance of the action of tensors on vectors. Take, for example, the transformation rule for the lower components of a vector **a**. The scalar product $\mathbf{a} \cdot \mathbf{b}$ of **a** with any other vector **b** can be written $a_{\alpha}b^{\alpha}$ from (20.15). But it could equally well be written in terms of the coordinate basis in another coordinate system as $a'_{\alpha}b'^{\alpha}$. The two numbers must be equal. From this and (20.6b)

$$\mathbf{a} \cdot \mathbf{b} = a'_{\beta}b'^{\beta} = a_{\alpha}b^{\alpha} = a_{\alpha}\frac{\partial x^{\alpha}}{\partial x'^{\beta}}b'^{\beta}. \tag{20.42}$$

But since **b** is an arbitrary vector, this implies

$$a'_{\beta} = \frac{\partial x^{\alpha}}{\partial x'^{\beta}}a_{\alpha}, \tag{20.43}$$

which is the transformation rule for lower components.

The transformation rules for a general tensor follow from generalizations of this argument applied to expressions such as (20.32) and (20.34). But the rules are most easily remembered as the application of the transformation rules for vectors (20.6) and (20.43) to each index separately. For example, the rule for transforming the metric is

$$g'_{\alpha\beta}(x') = \frac{\partial x^{\gamma}}{\partial x'^{\alpha}}\frac{\partial x^{\delta}}{\partial x'^{\beta}}g_{\gamma\delta}(x). \tag{20.44}$$

As another example,

$$t'^{\alpha}{}_{\beta} = \frac{\partial x'^{\alpha}}{\partial x^{\gamma}} \frac{\partial x^{\delta}}{\partial x'^{\beta}} t^{\gamma}{}_{\delta}. \tag{20.45}$$

A fact that emerges clearly from these connections is that if the components of a tensor vanish in one basis, then they vanish in all bases. That is useful in proving many tensor relations. If a relation is between tensors then it must be satisfied as a relation between components in *any* basis. Thus, if a relation is known to be between tensors, and can be shown to be satisfied in one special basis, then it must be true in all bases.

Not every object with indices is a tensor. Coordinates x^{α} are not components of a tensor. The Christoffel symbols are not components of a tensor. They vanish in a freely falling frame but are nonzero in other bases. They may define a linear map from vector components into a real number in one basis, but that number is different in different bases. That means they don't define a map from vectors to a single real number, as the definition of a tensor requires.

20.4 The Covariant Derivative

The Derivative of Vectors

The partial derivative of a function f is a vector $\boldsymbol{\nabla} f$ with components $(\boldsymbol{\nabla} f)_{\alpha} = \partial f/\partial x^{\alpha}$, as we saw in Example 20.2. But the many vectorial equations of classical physics, such as the dynamical equations of fluid mechanics and electromagnetism, suggest that it would be useful to be able to differentiate vectors as well as functions. We would expect the derivative of a vector $\mathbf{v}$ to be a second-rank tensor $\boldsymbol{\nabla}\mathbf{v}$ with components $\nabla_{\alpha} v^{\beta}$—one index for the direction of the vector and one for the direction of the derivative. However, there is a basic problem to be overcome before such a derivative can be defined. The derivative of a vector will naturally involve the difference between vectors at nearby spacetime points. But, as stressed in Section 7.8, subtraction, addition, etc., of vectors are operations defined *only at one point.* Vectors at two different points live in two different tangent spaces. To define derivatives of vectors, we need to transport vectors from one spacetime point to another. A careful examination of the flat space case will show how to do that.

Figure 20.2 shows the construction of the derivative of a vector field in flat space. We consider the vectors $\mathbf{v}(x^{\alpha})$ and $\mathbf{v}(x^{\alpha} + dx^{\alpha})$ at two nearby spacetime points connected by an infinitesimal displacement $dx^{\alpha} = t^{\alpha}\epsilon$ along the vector $\mathbf{t}$ defining the direction of the derivative. To construct the derivative, the vector $\mathbf{v}(x^{\alpha} + t^{\alpha}\epsilon)$ is first transported parallel to itself back to the point x^{α} to give the vector $\mathbf{v}_{\parallel}(x^{\alpha})$. There it is in the tangent space of x^{α}, and $\mathbf{v}(x^{\alpha})$ can be subtracted from it by the familiar parallelogram rule. *Parallel transport* is thus the key notion in defining a derivative of vectors.

Parallel Transport

Parallel transport can also be defined in a local inertial frame in curved spacetime because, locally, a local inertial frame is equivalent to flat spacetime. We are

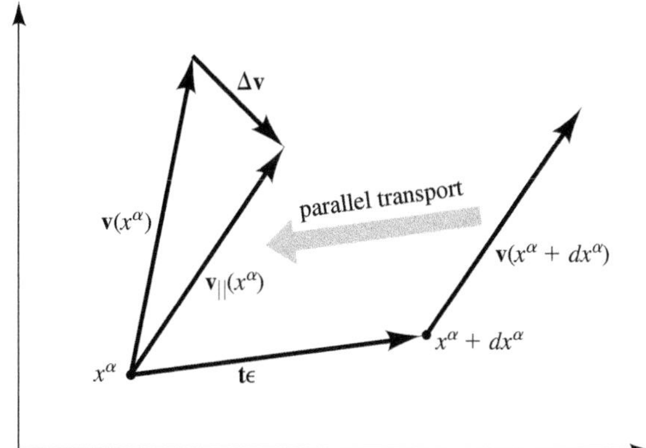

FIGURE 20.2 The derivative of a vector in flat space or in a freely falling frame in curved space. Two vectors of a vector field $\mathbf{v}(x^\alpha)$ are shown at two nearby points, x^α and $x^\alpha + dx^\alpha$, in spacetime. The two points are separated by a displacement $dx^\alpha = t^\alpha \epsilon$ along a vector t^α. To construct the difference between the vectors at x^α and $x^\alpha + dx^\alpha$, the vector $\mathbf{v}(x^\alpha + dx^\alpha)$ is first transported *parallel to itself* back to x^α to give the vector $\mathbf{v}_{\|}(x^\alpha) = [\mathbf{v}(x^\alpha + dx^\alpha)]_{\|\text{transported to } x^\alpha}$. The difference $\Delta\mathbf{v}(x^\alpha) = \mathbf{v}_{\|}(x^\alpha) - \mathbf{v}(x^\alpha)$ can then be constructed by the usual parallelogram rule. The limit $\Delta\mathbf{v}/\epsilon$ as $\epsilon \to 0$ defines the derivative of $\mathbf{v}$ in the direction of $\mathbf{t}$ at x^α.

thus led to the following definition of the *covariant derivative* of a vector field $\mathbf{v}(x^\alpha)$ in the direction $\mathbf{t}$ in curved spacetime:

$$\boxed{\nabla_{\mathbf{t}}\mathbf{v}(x^\alpha) = \lim_{\epsilon\to 0} \frac{[\mathbf{v}(x^\alpha + t^\alpha\epsilon)]_{\|\text{trans to } x^\alpha} - \mathbf{v}(x^\alpha)}{\epsilon}.} \tag{20.46}$$

Covariant Derivative

In the rectangular coordinates of flat space or of a local inertial frame (LIF) in curved spacetime (Section 7.4), the components v^α do not change as they are parallel transported (see Figure 20.2). Evaluating (20.46) in such coordinates is just like evaluating the derivative of a function (20.1):

$$(\nabla_{\mathbf{t}}\mathbf{v})^\alpha = t^\beta \frac{\partial v^\alpha}{\partial x^\beta} \qquad \text{(LIF)}. \tag{20.47}$$

For the tensor $\nabla\mathbf{v}$, we therefore have

$$\nabla_\beta v^\alpha = \frac{\partial v^\alpha}{\partial x^\beta} \qquad \text{(LIF)}. \tag{20.48}$$

(The LIFs have been added as a reminder that the formulas hold only in a local inertial frame at the point x^α where the formula is evaluated.)

However, even in flat space, (20.48) is not valid in curvilinear coordinates. As the case of polar coordinates illustrated in Figure 20.3 shows, the components of a vector change under parallel transport. The changes in the components result from

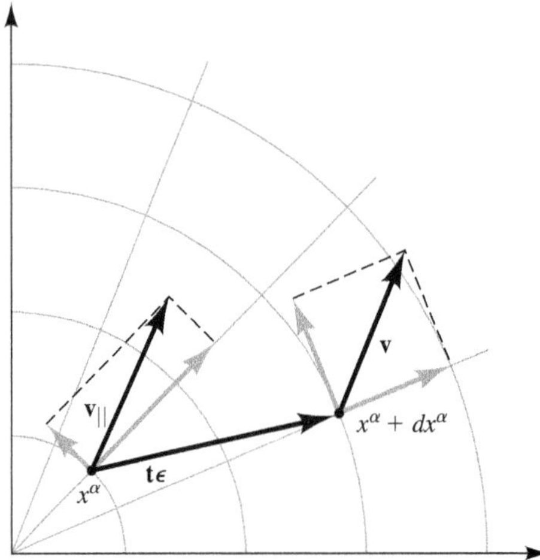

FIGURE 20.3 Components change under parallel transport. The figure shows the operation of parallel transport in a flat plane that was considered in Figure 20.2, but this time using polar coordinates. The vector $\mathbf{v}(x^\alpha + dx^\alpha)$ and the result $\mathbf{v}_{\|}(x^\alpha)$ of parallel transporting it to x^α along a displacement $dx^\alpha = \epsilon t^\alpha$ are illustrated. The resolution of the vectors into polar coordinate components at each point is shown. The vector doesn't change under parallel transport, but its components in polar coordinates do.

the changes in the angles the vector makes with the basis vectors. The changes in components will, therefore, be linear in the components themselves. In general, therefore, to first order in the displacement $dx^\alpha = \epsilon t^\alpha$, the components $v_{\|}^\alpha(x^\beta)$ are the sum of two terms—the components v^α at the displaced position and the changes in those components resulting from the change in the basis vectors during parallel transport, namely,

$$v_{\|}^\alpha(x^\delta) = v^\alpha(x^\delta + \epsilon t^\delta) + \widetilde{\Gamma}^\alpha_{\beta\gamma}(x^\delta)v^\gamma(x^\delta)(\epsilon t^\beta) \tag{20.49}$$

for yet to be determined coefficients $\widetilde{\Gamma}^\alpha_{\beta\gamma}$. By taking components of (20.46), we get the following general formula for the components of the covariant derivative:

$$\nabla_\beta v^\alpha = \frac{\partial v^\alpha}{\partial x^\beta} + \widetilde{\Gamma}^\alpha_{\beta\gamma} v^\gamma . \tag{20.50}$$

Roughly speaking, the first term comes from the change in the vector field from x^α to $x^\alpha + dx^\alpha$, and the second from the change in the basis vectors. Both terms are basis dependent, but the sum is basis independent as its construction (20.46) demonstrates.

Efficient calculation requires a formula for $\widetilde{\Gamma}^\alpha_{\beta\gamma}$. We could obtain one such formula by transforming (20.48) from coordinates of a local inertial frame to general coordinates. However, the resulting formula is not much use except to show that the $\widetilde{\Gamma}^\alpha_{\beta\gamma}$ are symmetric in β and γ (Problem 10). That is because we are not usu-

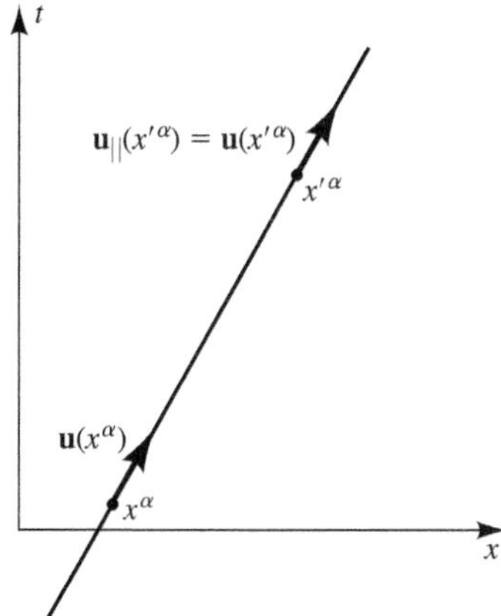

FIGURE 20.4 Geodesics and parallel propagation. In flat spacetime, as shown here, or in a local inertial frame, a geodesic is a straight line with the property that when a tangent vector $\mathbf{u}$ at x^{α} is parallel-propagated to x'^{α}, the parallel-propagated vector $\mathbf{u}_{\|}(x'^{\alpha})$ coincides with the tangent vector $\mathbf{u}(x'^{\alpha})$ there. More briefly, a geodesic is a curve whose tangent vector is parallelly propagated along itself. That same property characterizes geodesics in curved spacetime.

ally given the local inertial frames; rather we are given the metric and have to find them (see Section 8.4).

However, the coefficients $\widetilde{\Gamma}^{\alpha}_{\beta\gamma}$ can be found from something we already know—the equation for a geodesic. In a local inertial frame a geodesic is a straight line. A straight line can be defined either as a curve of extremal distance or as a curve whose unit tangent vector is propagated parallel to itself (see Figure 20.4). If $\mathbf{u}$ is that unit tangent vector, its covariant derivative in its own direction must vanish [cf. (20.46)]:

$$(\boldsymbol{\nabla}_{\mathbf{u}}\mathbf{u})^{\alpha} = u^{\beta}\left(\frac{\partial u^{\alpha}}{\partial x^{\beta}} + \widetilde{\Gamma}^{\alpha}_{\beta\gamma}u^{\gamma}\right) = 0, \tag{20.51}$$

where $u^{\alpha} = dx^{\alpha}/d\tau$. But we already know the geodesic equation (8.15), which can be written in a coordinate basis as

$$u^{\beta}\left(\frac{\partial u^{\alpha}}{\partial x^{\beta}} + \Gamma^{\alpha}_{\beta\gamma}u^{\gamma}\right) = 0, \tag{20.52}$$

where $\Gamma^{\alpha}_{\beta\gamma}$ are the Christoffel symbols expressed in terms of the metric by (8.19). Multiplying that equation by the inverse metric $g^{\epsilon\alpha}$, using its definition (20.17), and relabeling the free indices, the following explicit expression for the Γ's emerges:

$$\Gamma^{\alpha}_{\beta\gamma} = \frac{1}{2}g^{\alpha\delta}\left(\frac{\partial g_{\delta\beta}}{\partial x^{\gamma}} + \frac{\partial g_{\delta\gamma}}{\partial x^{\beta}} - \frac{\partial g_{\beta\gamma}}{\partial x^{\delta}}\right). \tag{20.53}$$

Thus, since there are geodesics with u^{α} pointing in any direction at one point, the coefficients $\widetilde{\Gamma}^{\alpha}_{\beta\gamma}$ defining the covariant derivative in a coordinate basis are identical with the Christoffel symbols. (Whence the labored notation $\widetilde{\Gamma}^{\alpha}_{\beta\gamma}$; but now erase the tildes in the previous expressions to get true formulas in terms of the Christoffel symbols.) The critical reader may worry that (20.52) involves only the symmetric part of the Christoffel symbols, but, as mentioned before, it is possible to prove that the $\widetilde{\Gamma}^{\alpha}_{\beta\gamma}$ have this symmetry (Problem 10). The following formula for the coordinate basis components of the second-rank tensor $\boldsymbol{\nabla}\mathbf{v}$ sums it all up:[5]

Covariant Derivative in a Coordinate Basis

$$\nabla_\alpha v^\beta = \frac{\partial v^\beta}{\partial x^\alpha} + \Gamma^\beta_{\alpha\gamma} v^\gamma \qquad \begin{pmatrix}\text{coordinate}\\ \text{basis}\end{pmatrix}. \tag{20.54}$$

This argument can now be turned around to give an elegant version of the geodesic equation in terms of the covariant derivative. A geodesic is a curve whose tangent vector $\mathbf{u}$ obeys

Geodesic Equation

$$\boldsymbol{\nabla}_{\mathbf{u}}\mathbf{u} = 0. \tag{20.55}$$

Example 20.8. The Acceleration of a Stationary Observer in the Schwarzschild Geometry. In an inertial frame of special relativity or a local inertial frame in general relativity, the acceleration four-vector of a particle can be defined by its coordinate basis components as

$$a^\alpha = \frac{du^\alpha}{d\tau} \qquad \text{(LIF only)}, \tag{20.56}$$

where $\mathbf{u}$ is the particle's four-velocity and τ is the proper time along its world line. But, even in flat space, (20.56) is not correct in a general coordinate system.[6] The correct and general definition of acceleration employs the correct and general way to differentiate a vector—the covariant derivative. Specifically, acceleration is defined generally by

$$\mathbf{a} \equiv \boldsymbol{\nabla}_{\mathbf{u}}\mathbf{u}. \tag{20.57}$$

This reduces to (20.56) in a local inertial frame because there $\boldsymbol{\nabla}_{\mathbf{u}} = u^\alpha \nabla_\alpha = u^\alpha(\partial/\partial x^\alpha) = (dx^\alpha/d\tau)(\partial/\partial x^\alpha) = d/d\tau$.

A stationary observer who remains at a fixed value of (r, θ, ϕ) in the spacetime of a Schwarzschild black hole is accelerating. Rocket thrust is needed to maintain

[5] There is a formula for the covariant derivative in an orthonormal basis—or, indeed, in any basis (see Problem 14). For simplicity we'll stick to computing covariant derivatives in a coordinate basis. Their components in an orthonormal basis can be found by projecting on the orthonormal basis vectors, as in (20.41).

[6] If you don't believe this, try to use (20.56) in spherical coordinates to compute the acceleration of a particle moving in a straight line at constant speed. You should get zero, but you don't.

a fixed position in space; the alternative is falling into the black hole.[7] To illustrate how to use the covariant derivative, we will calculate the components of the acceleration of a stationary observer, as defined by (20.57).

The Schwarzschild coordinate components of the normalized four-velocity of a stationary observer at radius r are [cf. (9.16)]:

$$u^\alpha = (u^t, \vec{0}) = [(1 - 2M/r)^{-1/2}, 0, 0, 0]. \tag{20.58}$$

Using (20.54) to evaluate (20.57), we have

$$a^\alpha \equiv u^\beta \nabla_\beta u^\alpha = u^t \nabla_t u^\alpha = u^t \left(\frac{\partial u^\alpha}{\partial t} + \Gamma^\alpha_{t\gamma} u^\gamma \right) = u^t \left(\frac{\partial u^\alpha}{\partial t} + \Gamma^\alpha_{tt} u^t \right) \tag{20.59}$$

since $\mathbf{u}$ has only a t component. Since the components u^α are independent of time, this reduces to

$$a^\alpha = \Gamma^\alpha_{tt} (u^t)^2. \tag{20.60}$$

Appendix B shows that the only nonvanishing Christoffel symbol that enters into (20.60) is $\Gamma^r_{tt} = (1 - 2M/r)(M/r^2)$. Thus,

$$a^\alpha = (0, M/r^2, 0, 0). \tag{20.61}$$

The acceleration points in the radial direction—the direction of the force necessary to keep the observer from falling into the black hole. Its components are finite at the horizon $r = 2M$, but the true measure of finiteness is its length,

$$(\mathbf{a} \cdot \mathbf{a})^{1/2} = \left(1 - \frac{2M}{r}\right)^{-1/2} \frac{M}{r^2}, \tag{20.62}$$

which diverges at $r = 2M$. Infinite acceleration is required to remain stationary at the horizon of a black hole. (See also the discussion in Box 12.2 on p. 261.)

A simple application of the covariant derivative is to derive the formulas familiar from basic electromagnetism or fluid mechanics for gradient, divergence, and curl in various curvilinear orthogonal coordinate systems in flat three-dimensional space. The general formulas for all such coordinate systems are worked out in Box 20.1 on p. 437.

Working with the Covariant Derivative

The idea of the covariant derivative can be extended to functions by extending the notation for the gradient. We write, for instance,

$$\nabla_\alpha f \equiv \frac{\partial f}{\partial x^\alpha}, \qquad \boldsymbol{\nabla}_{\mathbf{u}} f \equiv u^\alpha \frac{\partial f}{\partial x^\alpha}. \tag{20.63}$$

[7] In Newtonian mechanics the acceleration of a stationary particle outside of a mass is zero; in general relativity it is nonzero. In both cases "acceleration" is a measure of the deviation of a particle's trajectory from a geodesic. But the geometry assumed in Newtonian mechanics is different from the curved spacetime of general relativity.

The idea of covariant derivative can be extended from vectors to other tensors by enforcing Leibniz' rule. Consider two vectors, v^α and w^β, and the second-rank tensor, $v^\alpha w^\beta$, that is their product. Leibniz' rule would say

$$\nabla_\gamma (v^\alpha w^\beta) = v^\alpha (\nabla_\gamma w^\beta) + (\nabla_\gamma v^\alpha) w^\beta . \tag{20.64}$$

From (20.64) and (20.54) we have, immediately, the rule for differentiating a second-rank tensor

$$\nabla_\gamma t^{\alpha\beta} = \frac{\partial t^{\alpha\beta}}{\partial x^\gamma} + \Gamma^\alpha_{\gamma\delta} t^{\delta\beta} + \Gamma^\beta_{\gamma\delta} t^{\alpha\delta} . \tag{20.65}$$

Roughly, this could be summarized as the following instruction: differentiate the components and add terms with Γ's for each index one by one of the same form as for differentiating vectors in (20.54).

Since the covariant derivative of a vector defined in (20.54) is a tensor, we just have to lower one index [cf. (20.16)] to get $\nabla_\alpha v_\beta$. But we can derive a handy formula for these components by using Leibniz' rule (20.64) with the indices α and β contracted:

$$\nabla_\gamma \left(v_\alpha w^\alpha \right) = \left(\nabla_\gamma v_\alpha \right) w^\alpha + v_\alpha \left(\nabla_\gamma w^\alpha \right) . \tag{20.66}$$

The inner product, $v_\alpha w^\alpha$, is a scalar, so the usual Leibniz' rule using *partial* derivatives may be applied to the left-hand side of (20.66). The last term on the right is given by (20.54). The result is

$$\boxed{\nabla_\alpha v_\beta = \frac{\partial v_\beta}{\partial x^\alpha} - \Gamma^\gamma_{\alpha\beta} v_\gamma , \qquad \begin{pmatrix}\text{coordinate}\\ \text{basis}\end{pmatrix}} \tag{20.67}$$

This also generalizes to tensors, for instance,

$$\nabla_\gamma t^\alpha{}_\beta = \frac{\partial t^\alpha{}_\beta}{\partial x^\gamma} + \Gamma^\alpha_{\gamma\delta} t^\delta{}_\beta - \Gamma^\delta_{\gamma\beta} t^\alpha{}_\delta , \tag{20.68}$$

and so forth.

Example 20.9. Practice with Covariant Derivatives. Test your ability to calculate covariant derivatives by working the following exercise involving the geometry on the surface of a two-dimensional sphere [cf. (2.15)]:

$$dS^2 = a^2 (d\theta^2 + \sin^2\theta \, d\phi^2) \tag{20.69}$$

and a vector $\vec{v}$ with components $v^A = (0, 1)$. Calculate the four components of $\nabla_A v^B$ and then calculate the two quantities $\nabla_\theta \nabla_\phi v^\theta$ and $\nabla_\phi \nabla_\theta v^\theta$ to test whether covariant derivatives commute. The answers are at the bottom of the page.

You first need the Christoffel symbols. The only nonvanishing ones are $\Gamma^\theta_{\phi\phi} = -\sin\theta\cos\theta$ and $\Gamma^\phi_{\theta\phi} = \Gamma^\phi_{\phi\theta} = \cot\theta$. The only nonvanishing first derivatives of $\vec{v}$ are $\nabla_\phi v^\theta = -\sin\theta\cos\theta$ and $\nabla_\theta v^\phi = \cot\theta$. The requested second derivatives are $\nabla_\phi \nabla_\theta v^\theta = 0$ and $\nabla_\theta \nabla_\phi v^\theta = \sin^2\theta$. Covariant derivatives do not commute.

BOX 20.1 Gradient, Divergence, and Curl

The equations of electromagnetism, fluid mechanics, and many other areas of classical physics make use of a three-dimensional vector calculus employing the gradient $\vec{\nabla} f$ and Laplacian $\vec{\nabla}^2 f$ of functions together with the divergence $\vec{\nabla} \cdot \vec{V}$ and curl $\vec{\nabla} \times \vec{V}$ of vector fields. Explicit forms for these derivatives are given in many texts for useful coordinate systems, such as Cartesian, cylindrical, polar, parabolic, etc.—typically the 11 coordinate systems in which Laplace's equation separates. The covariant derivative provides a unified picture of all these derivatives and a direct route to the explicit forms in special coordinate systems.

We are concerned with three-dimensional *flat* space labeled by orthogonal coordinates (x^1, x^2, x^3). The metric is thus diagonal:

$$dS^2 = g_{11}(dx^1)^2 + g_{22}(dx^2)^2 + g_{33}(dx^3)^2, \qquad \text{(a)}$$

where g_{11}, g_{22}, and g_{33} are known functions of x^1, x^2, x^3. The coordinate basis vectors are denoted as usual by $\vec{e}_i$ and the dual-basis vectors, by $\vec{e}^{\,i}$. However, the basis most used in classical physics is neither of these, but rather it is an orthonormal basis $\vec{e}_{\hat{i}}$, with the three vectors pointing along the three coordinate lines. This is possible because the coordinates are orthogonal, and for the same reason there is a simple relationship among the different basis vectors. For instance, [cf. Example 7.9 on p. 156]

$$\vec{e}_{\hat{1}} = \vec{e}_1/(g_{11})^{1/2} = \vec{e}^{\,1}(g_{11})^{1/2}, \qquad \text{(b)}$$

with similar relations for directions 2 and 3. Then the inner products of the various sets of basis vectors satisfy the rules summarized in Table 20.1. The components of a vector $\vec{V}$ are connected by similar rescalings, which can be found from its expansion in terms of basis vectors

$$\vec{V} = V^k \vec{e}_k = V_k \vec{e}^{\,k} = V^{\hat{k}} \vec{e}_{\hat{k}} = V_{\hat{k}} \vec{e}^{\,\hat{k}}. \qquad \text{(c)}$$

For example, $V^{\hat{1}} = (g_{11})^{1/2} V^1$.

As we have defined it [cf. (20.9)], the gradient of a function is a vector with components $(\vec{\nabla} f)_i = \partial f/\partial x^i$. That is,

$$\vec{\nabla} f = \frac{\partial f}{\partial x^k} \vec{e}^{\,k}. \qquad \text{(d)}$$

When rewritten in terms of the unit vectors using (b),

$$\vec{\nabla} f = \frac{1}{(g_{11})^{1/2}} \frac{\partial f}{\partial x^1} \vec{e}_{\hat{1}} + \frac{1}{(g_{22})^{1/2}} \frac{\partial f}{\partial x^2} \vec{e}_{\hat{2}} + \frac{1}{(g_{33})^{1/2}} \frac{\partial f}{\partial x^3} \vec{e}_{\hat{3}}. \qquad \text{(e)}$$

This gives an expression for the gradient in an arbitrary orthogonal coordinate system.

The divergence of a vector $\vec{V}$ is the scalar $\nabla_i V^i \equiv \vec{\nabla} \cdot \vec{V}$. From (20.54) we have

$$\vec{\nabla} \cdot \vec{V} = \frac{\partial V^i}{\partial x^i} + \Gamma^i_{ik} V^k. \qquad \text{(f)}$$

A little calculation from (a) and the formula for the Γs (20.53) shows (Problem 21)

$$\Gamma^i_{ik} = \frac{1}{\sqrt{g}} \frac{\partial \sqrt{g}}{\partial x^k}, \qquad \text{(g)}$$

where g is defined by

$$g \equiv g_{11} g_{22} g_{33} = \det(g_{ij}). \qquad \text{(h)}$$

(With $g = \det(g_{ij})$, (g) turns out to hold for nondiagonal metrics as well.) Using this and writing (f) in terms of the components in the orthonormal basis, we find

$$\vec{\nabla} \cdot \vec{V} = \frac{1}{g^{1/2}} \left[\frac{\partial}{\partial x^1} \left(g_{22}^{1/2} g_{33}^{1/2} V^{\hat{1}} \right) + \frac{\partial}{\partial x^2} \left(g_{33}^{1/2} g_{11}^{1/2} V^{\hat{2}} \right) + \frac{\partial}{\partial x^3} \left(g_{11}^{1/2} g_{22}^{1/2} V^{\hat{3}} \right) \right]. \qquad \text{(i)}$$

This is a general formula for the divergence. A general formula for the Laplacian follows immediately from (i) and (e):

$$\vec{\nabla}^2 f \equiv \nabla_i \nabla^i f = \frac{1}{\sqrt{g}} \frac{\partial}{\partial x^i} \left(\sqrt{g}\, g^{ij} \frac{\partial f}{\partial x^j} \right). \qquad \text{(j)}$$

This form turns out to be valid whether or not the coordinate system is orthogonal.

The general derivative of a vector $\nabla_i V^j$ is a second-rank tensor. However, a special feature of three dimen-

BOX 20.1 (*continued*)

sions is that an antisymmetric second-rank tensor can be associated with a vector through the alternating tensor

$$g^{-1/2}\epsilon^{ijk}. \tag{k}$$

Here, ϵ^{ijk} is antisymmetric in the interchange of any two indices and equals +1 if (i, j, k) is an even permutation of $(1, 2, 3)$. In particular, ϵ^{ijk} vanishes if any two indices are equal. One can check (Problem 22) that this definition is basis-independent.

The curl of a vector $\vec{V}$ is

$$\left(\vec{\nabla} \times \vec{V}\right)^i = g^{-1/2}\epsilon^{ijk}\nabla_j V_k. \tag{l}$$

The covariant derivative is given by (20.67). But because the Γ's are symmetric and ϵ^{ijk} is antisymmetric, they do not enter (l). Thus,

$$\left(\vec{\nabla} \times \vec{V}\right)^i = g^{-1/2}\epsilon^{ijk}\left(\frac{\partial V_k}{\partial x^j} - \frac{\partial V_j}{\partial x^k}\right). \tag{m}$$

This leads to a very simple formula for the curl in terms of the components of $\vec{\nabla}$ in the orthonormal basis:

$$\left(\vec{\nabla} \times \vec{V}\right)^{\hat{1}} = \left(\frac{g_{11}}{g}\right)^{1/2}\left[\frac{\partial}{\partial x^3}\left(g_{22}^{1/2}V^{\hat{2}}\right) - \frac{\partial}{\partial x^2}\left(g_{33}^{1/2}V^{\hat{3}}\right)\right]. \tag{n}$$

To get the formulas for the other components, simply cyclically permute (1, 2, 3) in (n).

You might like to check that (e), (i), (j), and (n) give standard formulas in familiar cases, e.g., polar coordinates,

$$dS^2 = dr^2 + r^2\, d\theta^2 + r^2 \sin^2\theta\, d\phi^2, \tag{o}$$

or cylindrical coordinates

$$dS^2 = d\rho^2 + \rho^2\, d\phi^2 + dz^2. \tag{p}$$

The covariant derivative of the metric vanishes:

$$\boxed{\nabla_\gamma g_{\alpha\beta} = 0.} \tag{20.70}$$

This important property follows immediately because it clearly holds in a local inertial frame, where all first derivatives of the metric vanish. However, it could also be worked out explicitly in a general coordinate system from the expressions for the covariant derivative in terms of the Christoffel symbols (Problem 17).

The covariant derivative was constructed by comparing a vector at one point with one parallel propagated from a neighboring point along a curve [cf. (20.46)]. Thus, a vector is parallel propagated along a curve if its covariant derivative along the curve vanishes:

Parallel Propagation

$$\boxed{\begin{pmatrix}\mathbf{v}\text{ is parallel} \\ \text{propagated} \\ \text{along } x^\alpha(\sigma)\end{pmatrix} \Leftrightarrow (\nabla_{\mathbf{t}}\mathbf{v} = 0),} \tag{20.71}$$

where **t** is a tangent vector, $t^\alpha = dx^\alpha/d\sigma$.

Example 20.10. The Equations of a Gyroscope. We have already seen in (20.55) how the geodesic equation can be elegantly stated using the covariant derivative:

$$\nabla_{\mathbf{u}}\mathbf{u} = 0. \tag{20.72}$$

From (20.71) this is the statement that the four-velocity is parallel propagated along a geodesic. The equation of motion for the spin, **s**, of a gyroscope free from external forces (14.6) can be similarly compactly stated:

$$\nabla_{\mathbf{u}}\mathbf{s} = 0. \tag{20.73}$$

The equation means the spin of the gyro is parallel-transported along the geodesic it follows.

Constant vectors are not defined by constant components except in rectangular coordinates. Rather, they are vector fields that don't change if parallel-transported in any direction:

$$\begin{pmatrix}\text{a vector}\\ \text{field is constant}\end{pmatrix} \Leftrightarrow (\nabla_\alpha v^\beta = 0). \tag{20.74}$$

We can illustrate this idea as well as give an example of a calculation utilizing covariant derivatives by calculating the constant vector fields in the plane the hard way—in polar coordinates.

Example 20.11. Constant Vector Fields in the Two-Dimensional Plane. To find the constant vector fields in a flat plane, solve the four equations

$$\nabla_A v^B = 0, \tag{20.75}$$

where $A, B = 1, 2$. It is easy to do this in rectangular coordinates but more instructive to do it in polar coordinates. Equation (8.2) displays the metric in polar coordinates (r, ϕ), and the Christoffel symbols are given in (8.17). Writing out (20.75), the four equations are

$$\nabla_r v^r = \frac{\partial v^r}{\partial r} = 0, \tag{20.76a}$$

$$\nabla_\phi v^r = \frac{\partial v^r}{\partial \phi} - r v^\phi = 0, \tag{20.76b}$$

$$\nabla_r v^\phi = \frac{\partial v^\phi}{\partial r} + \frac{1}{r} v^\phi = 0, \tag{20.76c}$$

$$\nabla_\phi v^\phi = \frac{\partial v^\phi}{\partial \phi} + \frac{1}{r} v^r = 0. \tag{20.76d}$$

The first shows that v^r is a function only of ϕ, $v^r = g(\phi)$. The third leads to $\partial(r v^\phi)/\partial r = 0$, which implies $v^\phi = f(\phi)/r$ for some function $f(\phi)$. The remaining two equations imply $f' = -g$ and $g' = f$. The solution to these is

$g = A\cos(\phi - \phi_*)$ and $f = -A\sin(\phi - \phi_*)$ for constants A and ϕ_*. Thus, the possible constant vector fields have components

$$v^r = A\cos(\phi - \phi_*), \qquad v^\phi = -\frac{A}{r}\sin(\phi - \phi_*). \tag{20.77}$$

The meaning of this becomes clearer if we consider components not in the coordinate basis associated with polar coordinates, but rather in the orthonormal basis whose unit vectors $\vec{e}_{\hat{r}}, \vec{e}_{\hat{\phi}}$ point along the coordinate lines. Then

$$v^{\hat{r}} = A\cos(\phi - \phi_*) \qquad v^{\hat{\phi}} = A\sin(\phi_* - \phi). \tag{20.78}$$

These are vector fields of length A that everywhere make an angle ϕ_* with the x-axis. These are all the possible constant vector fields in the plane.

20.5 Freely Falling Frames Again

In Section 3.1 the inertial frames of Newtonian mechanics were constructed by parallel-transporting an initial choice of directions for the three coordinate axes along the straight-line path of a free particle in flat space. In Section 4.3 the inertial frames of special relativity were constructed in the same way. In Section 8.4 the same construction was applied to give the freely falling frames of general relativity—a specification of a *local* inertial frame all along a geodesic and the closest analogy possible to the global inertial frames of flat spacetime. However, in describing the construction of a freely falling frame, we did not give a quantitative explanation of how its axes changed along the geodesic defining its origin. Such an explanation will be useful to define curvature in the next chapter, and, with our understanding of the covariant derivative, we are now in a position to give it.

Consider the geodesic of a freely falling observer $x^\alpha(\tau)$ defining the origin of a freely falling frame. A set of orthonormal basis vectors $\{\mathbf{e}_{\hat{\alpha}}(\tau)\}$ define the axes of the frame all along the geodesic. These vectors will be the coordinate basis vectors for the frame. The four-velocity $\mathbf{u}(\tau)$ is the basis vector $\mathbf{e}_{\hat{0}}$ defining the time direction. Three mutually orthogonal vectors also orthogonal to $\mathbf{u}$ can be picked at one point along the geodesic to define the spatial directions. The axes at other points are found by parallel-propagating these vectors along the observer's geodesic. Thus, the orthonormal basis vectors along the axes of a freely falling frame satisfy [cf. (20.71)]

$$\boldsymbol{\nabla}_{\mathbf{u}}\mathbf{e}_{\hat{\alpha}} = 0. \tag{20.79}$$

The equation for $\mathbf{e}_{\hat{0}}$ is satisfied automatically because the world line of the observer is a geodesic [cf. (20.55)]. The remaining three equations determine how the spatial vectors $\{\mathbf{e}_{\hat{i}}\}$ change along that geodesic. Their directions could be said to be defined by the spins of gyroscopes because they also are parallel-propagated along the geodesic [cf. (20.73)].

Example 20.12. Freely Falling Frames in the Schwarzschild Geometry. Consider an observer who falls freely from infinity in the Schwarzschild spacetime described in Chapter 9 with the metric (9.9).

Suppose the observer starts from rest at infinity and falls radially inward. The observer follows an $e = 1$, $\ell = 0$ geodesic whose four-velocity is given by (9.36):

$$u^\alpha = ((1 - 2M/r)^{-1}, -(2M/r)^{1/2}, 0, 0), \tag{20.80}$$

where, as usual, $x^\alpha = (t, r, \theta, \phi)$. One component of the observer's frame is, therefore, $\mathbf{e}_{\hat{0}} \equiv \mathbf{e}_{\hat{\tau}} = \mathbf{u}(\tau)$. This and the other three vectors $\mathbf{e}_{\hat{i}}$ must be normalized and mutually orthogonal and satisfy (20.79) as well. Initially, when the observer is at rest, the three vectors can be chosen to lie along the r-, θ-, and ϕ-directions; accordingly, we denote them by $\mathbf{e}_{\hat{1}} \equiv \mathbf{e}_{\hat{r}}$, $\mathbf{e}_{\hat{2}} \equiv \mathbf{e}_{\hat{\theta}}$, and $\mathbf{e}_{\hat{3}} \equiv \mathbf{e}_{\hat{\phi}}$ to remind ourselves of this. Spherical symmetry dictates that $\mathbf{e}_{\hat{\theta}}$ and $\mathbf{e}_{\hat{\phi}}$ remain oriented along the θ- and ϕ-directions as the laboratory falls. The components of $\mathbf{e}_{\hat{r}}$ are then determined by orthogonality to the other vectors and normalization. The result is

$$u^\alpha = (\mathbf{e}_{\hat{\tau}})^\alpha \equiv (\mathbf{e}_{\hat{0}})^\alpha = ((1 - 2M/r)^{-1}, -(2M/r)^{1/2}, 0, 0), \tag{20.81a}$$

$$(\mathbf{e}_{\hat{r}})^\alpha \equiv (\mathbf{e}_{\hat{1}})^\alpha = (-(2M/r)^{1/2}(1 - 2M/r)^{-1}, 1, 0, 0), \tag{20.81b}$$

$$(\mathbf{e}_{\hat{\theta}})^\alpha \equiv (\mathbf{e}_{\hat{2}})^\alpha = (0, 0, 1/r, 0), \tag{20.81c}$$

$$(\mathbf{e}_{\hat{\phi}})^\alpha \equiv (\mathbf{e}_{\hat{3}})^\alpha = (0, 0, 0, 1/(r \sin\theta)). \tag{20.81d}$$

Symmetry enabled us to find these vectors of the freely falling frame without using (20.79), but it is an instructive exercise to check that it is satisfied (Problem 25). For example, to check that $\boldsymbol{\nabla}_{\mathbf{u}}\mathbf{e}_{\hat{1}} = 0$ one would have to write out $u^\beta \nabla_\beta (\mathbf{e}_{\hat{1}})^\alpha$ using (20.81a) for u^β, (20.81b) for the components of $\mathbf{e}_{\hat{1}}$, and (20.54) for the covariant derivative.

Problems

1. [S] Show explicitly that the transformation rule (20.6a) leads to the transformation of vector components under a Lorentz boost (4.33) given in (5.9).

2. [S] **(a)** Evaluate

$$\frac{\partial x^\beta}{\partial x'^\alpha}\frac{\partial x'^\alpha}{\partial x^\gamma}.$$

   (b) Use this result to show explicitly that the transformation law (20.6b) is the inverse of (20.6a).

3. [S] Use the transformation (7.2) connecting rectangular coordinates (t, x, y, z) for flat space to polar coordinates (t, r, θ, ϕ) to find the explicit transformation laws giving the components (a^t, a^x, a^y, a^z) of a vector $\mathbf{a}$ in terms of the components $(a^t, a^r, a^\theta, a^\phi)$ and the components (a_t, a_x, a_y, a_z) in terms of $(a_t, a_r, a_\theta, a_\phi)$.

4. In the Schwarzschild geometry consider the following function:

$$f(x) = (5t^2 - 2r^2)/(2M)^2,$$

where t and r are the usual Schwarzschild coordinates in which the metric has the form (9.9). Find the coordinate basis components $(\boldsymbol{\nabla} f)^\alpha$ of the gradient of f.

5. Equation (20.81) gives the upstairs coordinate basis components of a set of four vectors $\{\mathbf{e}_{\hat{\alpha}}\}$ constituting an orthonormal frame in the Schwarzschild geometry.
   (a) Verify explicitly that this is an orthonormal set of vectors.
   (b) Find the downstairs coordinate basis components of each of these vectors.
   (c) Find the upstairs coordinate basis components of the basis $\mathbf{e}^\alpha$ that is dual to the given set of basis vectors.
   (d) Consider a vector $\mathbf{a}$ with upstairs coordinate basis components

$$a^\alpha = (4, 3, 0, 0)$$

   at the point $(0, 3M, 0, 0)$. Find the components $a^{\hat{\alpha}}$ and $a_{\hat{\alpha}}$ of this vector in the given orthonormal frame.

6. For the basis of dual vectors $\{\mathbf{e}^\alpha\}$ that is dual to a basis of vectors $\{\mathbf{e}_\alpha\}$, work out $e^\alpha(\mathbf{a})$ and $a(\mathbf{e}^\alpha)$ in terms of the components of the vector $\mathbf{a}$ in the basis $\{\mathbf{e}_\alpha\}$.

7. Consider a set of *coordinate* basis vectors $\{\mathbf{e}_\alpha\}$ and the associated dual basis $\{\mathbf{e}^\alpha\}$ defined by the relations (20.11) or (20.23).
   (a) Show that basis vectors $\{\mathbf{e}_\alpha\}$ and dual-basis vectors $\{\mathbf{e}^\alpha\}$ are related to each other by $\mathbf{e}_\alpha = g_{\alpha\beta}\mathbf{e}^\beta$ and $\mathbf{e}^\alpha = g^{\alpha\beta}\mathbf{e}_\beta$.
   (b) Show that $\mathbf{e}^\alpha \cdot \mathbf{e}^\beta = g^{\alpha\beta}$.

8. [S] At a point, the coordinate-basis vectors $\{\mathbf{e}_\alpha\}$ in one system of coordinates x^α must be linear combinations of the coordinate-basis vectors $\{\mathbf{e}'_\alpha\}$ in another system of coordinates x'^α. Find the explicit transformation rule.

9. Show that the operation of contraction, as exemplified by (20.40), is basis independent by showing that if carried out in a another system of coordinates $x'^\alpha = x'^\alpha(x^\beta)$, the components of w^α transform correctly as a consequence of the transformation law for tensors.

10. [A] Equation (20.48) gives the expression for the components of the second-rank tensor that results from covariant differentiation in a local inertial frame where all the Γs vanish. Use the transformation law for tensors, (20.45), to obtain an expression for the Γs in a general coordinate system. Use this result to show that $\Gamma^\alpha_{\beta\gamma}$ is symmetric in β and γ.

11. Work out the expression for the covariant derivative $\nabla_\gamma t_{\alpha\beta}$ analogous to (20.65) and (20.68).

12. [A] Following Example 20.9, work out all the components of $\nabla_A w^B$ and $\nabla_A \nabla_B w^C$ for the vector $w^A = (1, 0)$.

13. In Example 20.8 the acceleration four-vector of a stationary observer in the Schwarzschild geometry *could* have been computed using

$$a_\alpha = u^\beta \nabla_\beta u_\alpha$$

and formula (20.67). Show that the same result, (20.61), could have been obtained this way.

14. *Covariant Derivative in an Arbitrary Basis* Let $x^\alpha(\sigma)$ be a curve and $\mathbf{t}(\sigma)$ be the unit tangent vector to the curve at σ. Show that the components of the covariant derivative $\boldsymbol{\nabla}_{\mathbf{t}}\mathbf{v}$ of a vector $\mathbf{v}$ in the direction $\mathbf{t}$ can be written in an arbitrary basis $\{\mathbf{e}_\alpha\}$ as

$$(\boldsymbol{\nabla}_{\mathbf{t}}\mathbf{v})^\alpha = \frac{dv^\alpha}{d\sigma} + \widetilde{\Gamma}^\alpha_{\beta\gamma} v^\beta t^\gamma.$$

where

$$\widetilde{\Gamma}^\alpha_{\beta\gamma} = \mathbf{e}^\alpha \cdot \boldsymbol{\nabla}_{\mathbf{e}_\gamma}\mathbf{e}_\beta.$$

These are called *Ricci rotation coefficients*. Show that they reduce to the Christoffel symbols when $\{\mathbf{e}_\alpha\}$ is a coordinate basis.

15. [S] *Null Geodesics with Nonaffine Parametrization* As we showed in Section 8.3, when the tangent vector to a null geodesic $\mathbf{u}$ is parametrized with an affine parameter λ, it obeys the geodesic equation

$$\boldsymbol{\nabla}_{\mathbf{u}}\mathbf{u} = 0.$$

Show that even if a nonaffine parameter is used,

$$\boldsymbol{\nabla}_{\mathbf{u}}\mathbf{u} = -\kappa\mathbf{u}$$

for some function κ of the parameter λ.

16. *Surface Gravity of a Black Hole* In the geometry of a spherical black hole, the Killing vector $\boldsymbol{\xi} = \partial/\partial t$ corresponding to time translation invariance is tangent to the null geodesics that generate the horizon. If you worked Problem 15, you will know that this means

$$\boldsymbol{\nabla}_{\boldsymbol{\xi}}\boldsymbol{\xi} = -\kappa\boldsymbol{\xi}$$

for a constant of proportionally κ, which is called the *surface gravity* of the black hole. Evaluate this relation to find the value of κ for a Schwarzschild black hole in terms of its mass, M. Be sure to use a coordinate system that is nonsingular on the horizon such as the Eddington-Finkelstein coordinates discussed in Section 12.1.

17. Show explicitly that the covariant derivative of the metric vanishes by working it out using expression (20.65) or analogous expressions for other components of the covariant derivative (for example that worked out in Problem 11) and the explicit expression for the Γ's in (20.53).

18. *Killing's equation* In Section 8.2 a Killing vector corresponding to a symmetry of a metric was defined in a coordinate system in which the metric was independent of one coordinate, x^1. The components of the corresponding Killing vector $\boldsymbol{\xi}$ are then

$$\xi^\alpha = (0,\ 1,\ 0,\ 0).$$

By explicit calculation show that

$$\nabla_\alpha \xi_\beta + \nabla_\beta \xi_\alpha = 0.$$

This is Killing's equation. It is a general characterization of Killing vectors in the sense that any solution corresponds to a symmetry of the metric.

19. [C] A three-surface $f(x^\alpha) = 0$ is null if its normal $\ell_\alpha = \partial f/\partial x^\alpha$ is a null vector (Section 7.9). Show that these normal vectors satisfy $\boldsymbol{\nabla}_{\ell}\boldsymbol{\ell} = \kappa\boldsymbol{\ell}$ where κ is some function of the x^α. Show that the equation for null geodesics (8.42) can be put in this form by using a parameter that is not an affine parameter, thus ℓ is a tangent to a null geodesic.

20. **(a)** Show that the three Killing vectors $\partial/\partial x$, $\partial/\partial y$, and $\boldsymbol{\eta} \equiv -y(\partial/\partial x) + x(\partial/\partial y)$ in Example 8.6 satisfy Killing's equation from Problem 18.
(b) Show that in polar coordinates on the plane, $\boldsymbol{\eta} = \partial/\partial\phi$.
(c) Show that the rotational symmetry about a point that is not the origin corresponds to a Killing vector that is a linear combination with constant coeeficients of $\partial/\partial x$, $\partial/\partial y$, and $\boldsymbol{\eta}$.

21. [B] Derive formula (g) in Box 20.1. For simplicity you can just consider the case of a diagonal metric although the result is general.

22. [B] Demonstrate that the alternating tensor defined in (k) in Box 20.1 transforms correctly as a third-rank tensor under coordinate transformations. *Hint:* The definition of the determinant of the matrix is given by

$$\det(A) = \epsilon^{ijk} A_{1i} A_{2j} A_{3k}.$$

23. [B, A] In three-dimensional flat space, parabolic coordinates (μ, ν, ϕ) are defined by

$$x = \mu\nu\cos\phi,$$
$$y = \mu\nu\sin\phi,$$
$$z = (\mu^2 - \nu^2)/2.$$

(a) Sketch the lines of constant μ and constant ν in the $\phi = 0$ plane.
(b) Find the flat space line element in the coordinates (μ, ν, ϕ).
(c) Work out the expressions for grad, div, curl and the Laplacian in these coordinates.

24. [S] Use formulas (20.72) and (20.73) to show that if the spin $\mathbf{s}$ of a free gyro starts out orthogonal to its four-velocity $\mathbf{u}$, it remains orthogonal.

25. [A] Show that the basis vectors of the freely falling frame (20.81) in the Schwarzschild geometry are indeed parallel-propagated along the geodesic of the freely falling observer by showing explicitly that they each satisfy (20.79).

26. Show that the orthonormal basis of the freely falling frame (20.81) at Schwarzschild coordinate radius r is connected to the orthonormal basis of a stationary observer at that point by a Lorentz boost. Find the velocity of that boost. *Comment:* This is a special case of the general result that any two orthonormal bases are connected by a Lorentz trans-formation, cf. Problem 7.23.

CHAPTER 21

Curvature and the Einstein Equation

Previous chapters have described particular spacetime geometries that occur in general relativity and the motion of test particles and light rays in them. Although it was mentioned that the presence of matter produces spacetime curvature, an equation that spells out in quantitative detail how this happens has not been written down. The central content of general relativity is just such an equation. It has the schematic form

$$\begin{pmatrix}\text{a measure of local}\\ \text{spacetime curvature}\end{pmatrix} = \begin{pmatrix}\text{a measure of}\\ \text{matter energy density}\end{pmatrix}. \tag{21.1}$$

This relation, called the *Einstein equation* (or *Einstein's equation*), is the field equation of general relativity in the way that Maxwell's equations are the field equations of electromagnetism. Maxwell's equations relate the electromagnetic field to its sources—charges and currents. Einstein's equation relates spacetime curvature to *its* source—the mass-energy of matter. The analogy goes further. Maxwell's equations are eight second-order partial differential equations for the electromagnetic potentials. Einstein's equation is a set of ten second-order partial differential equations for the metric coefficients $g_{\alpha\beta}(x)$. An important difference is that Maxwell's equations are linear but the Einstein equation is nonlinear.

In this chapter we give a very brief introduction to the Einstein equation. We consider the equation in the absence of matter sources (the vacuum Einstein equation) in this chapter and include matter sources in the next one. Even the vacuum Einstein equation has important implications. Just as the field of a static point charge and electromagnetic waves are solutions of the source free Maxwell's equations, the Schwarzschild geometry and gravitational waves are solutions of the vacuum Einstein equation.

21.1 Tidal Gravitational Forces

Our first task is to find the "measure of spacetime curvature" in (21.1). To do this let's consider thought experiments by which an observer could use the motion of test particles to measure curvature in principle.

The motion of a *single* test particle reveals nothing about spacetime curvature. Imagine studying that motion in a frame falling freely with the particle. In a freely falling frame, the test particle remains at rest. Its motion is indistinguishable from that of a test particle in flat spacetime. One test particle is not enough to detect curvature.

The motion of at least two test particles is needed to detect spacetime curvature. The example of astronauts in the space shuttle following the motion of two ping-pong balls (Example 6.3) suggests how to do this. By studying the *relative* motion of two nearby ping-pong balls over time the astronauts could detect the spacetime curvature produced by the Earth. Figures 21.1 and 21.2 show a more general idealized example. Two observers are in freely falling laboratories. One is in empty space and the other is falling toward the surface of a planet. Both observers start out with a circular pattern of test particles at rest with respect to them. After a while, the observer moving toward the planet will notice that the pattern changes. In a Newtonian description of the motion, the particle closest to the center of the planet will accelerate more than the observer and be seen to pull away. The particle furthest from the center will accelerate less and also pull away. The bodies on the sides moving toward the center of the planet will move closer to the observer. The net result is a distortion of the circular pattern into an ellipse, as shown. This distortion is a measure of local spacetime curvature. By contrast, the circular pattern remains unchanged for the observer in empty space.

To find the equation that governs the relative motion of two nearby particles, let's first consider this question in Newtonian theory. In an inertial frame the equation of motion for the position $\vec{x}(t)$ of the first particle moving in a gravitational potential $\Phi(x)$ is

$$\frac{d^2x^i}{dt^2} = -\delta^{ij}\frac{\partial\Phi(x^k)}{\partial x^j}. \tag{21.2}$$

(The δ^{ij} is included so that the indices balance.) Let $\vec{\chi}$ be the *separation vector*, which measures the relative separation of the second particle from the first so that the position of the second is $x^i(t) + \chi^i(t)$. If the particles are nearby, the length of $\vec{\chi}$ will be small. The equation of motion for the second particle is

$$\frac{d^2(x^i + \chi^i)}{dt^2} = -\delta^{ij}\frac{\partial}{\partial x^j}\Phi(x^k + \chi^k). \tag{21.3}$$

Expand the right-hand side of (21.3) to linear order in χ^j—a valid expansion because $|\vec{\chi}|$ is small. To make the expansion, just note that $\partial\Phi/\partial x^j$ is a function of the x^k and use Taylor series for that function, namely

$$\frac{\partial\Phi(x^i + \chi^i)}{\partial x^j} = \frac{\partial\Phi(x^i)}{\partial x^j} + \frac{\partial}{\partial x^k}\left(\frac{\partial\Phi(x^i)}{\partial x^j}\right)\chi^k + \cdots. \tag{21.4}$$

Subtract (21.2) from (21.3) using this expansion to find the following equation for the separation vector χ^i:

Newtonian Deviation Equation

$$\frac{d^2\chi^i}{dt^2} = -\delta^{ij}\left(\frac{\partial^2\Phi}{\partial x^j\partial x^k}\right)_{\vec{x}}\chi^k. \tag{21.5}$$

This is called the *Newtonian deviation equation*. Given the separation of two nearby particles at one time, and its rate of change, (21.5) can be used to calculate

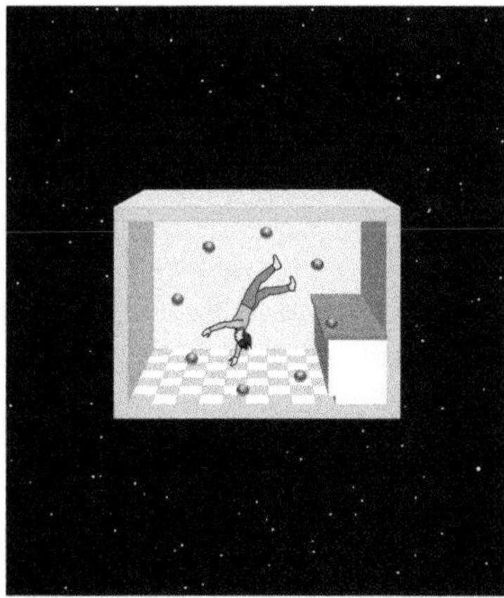
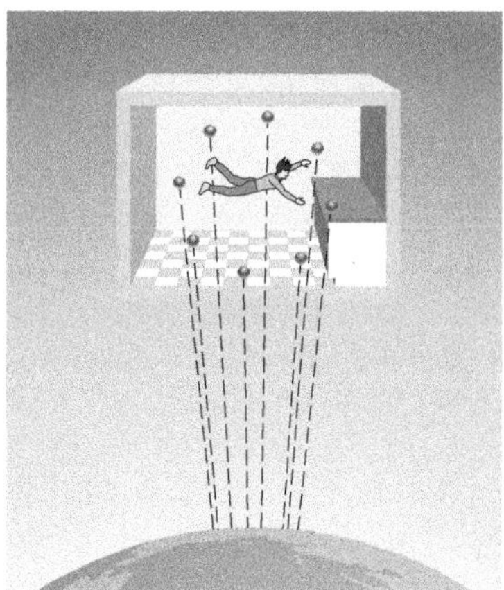

FIGURE 21.1 Two observers in freely falling laboratories. The laboratory at left is in empty space. The laboratory at right is falling toward the surface of a curvature-producing planet. Each is surrounded by a cloud of test particles initially at rest in the laboratory and arranged in a circle. In the laboratory in empty space, the relative acceleration of observer and test particles remains zero. Using a Newtonian description of motion in a frame in which the center of the planet is at rest, the observer and each test particle in the falling laboratory are accelerating toward the planet's center. The test particles further away from the center of the planet are accelerating slightly less than the ones closer. The effect on the pattern of test particles is shown in the next figure.

FIGURE 21.2 The two observers of Figure 21.1 after a little time. The unaccelerated particles in empty space have remained in a circle. The different accelerations of the test particles falling toward the planet have distorted the initial circle into an ellipse. In this way the freely falling observer can detect tidal gravitational accelerations if the laboratory is big enough to allow measurable differential accelerations between the test particles.

the separation for later times as long as $|\vec{\chi}|$ remains small. The tensor whose components are $\partial^2\Phi/\partial x^i \partial x^j$ measures the differential accelerations and determines the forces that tend to pull nearby particles apart or bring them closer together. These are called *tidal gravitational forces*, and the tensor is called the *tidal acceleration tensor*. (See Box 21.1 for the reason.)

Example 21.1. Tidal Acceleration Outside a Spherical Mass. The Newtonian gravitational potential outside a spherically symmetric distribution of mass is ($G = 1$ units) [cf. (3.13)]

$$\Phi = -M/r, \tag{21.6}$$

BOX 21.1 Tides from Tidal Forces

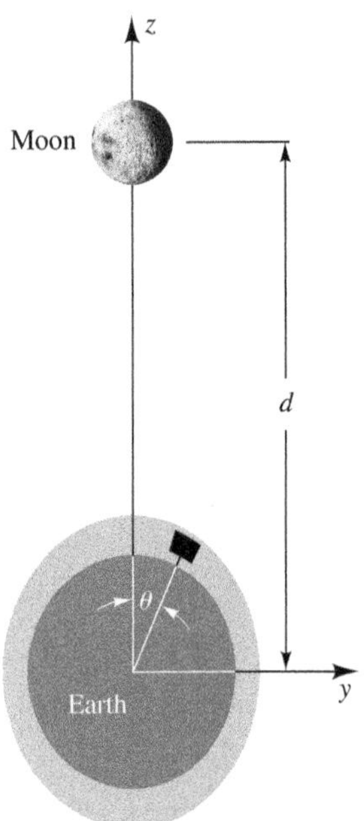

The gravitational pull of the Moon produces the daily tides as the Earth rotates under the resulting distortion of the surface of the oceans. The tides can be seen to be consequences of the tidal gravitational forces exerted by the Moon at the Earth. A simplification of the real situation is illustrated in the figure, in which the Moon rotates around a much more massive Earth at rest at the origin of an inertial frame (x, y, z). At the instant shown, the Moon is located a distance d away along the z-axis. The Newtonian gravitational potential of the Moon is

$$\Phi_{\text{Moon}}(x, y, z) = -\frac{GM_{\text{Moon}}}{[x^2 + y^2 + (z-d)^2]^{1/2}}. \tag{a}$$

The second derivatives of the potential at the origin, which determine the tidal gravitational accelerations at the Earth through (21.5), are

$$\left(\frac{\partial^2\Phi}{\partial x^i \partial x^j}\right)_0 = \frac{GM_{\text{Moon}}}{d^3}\,\text{diag}(1, 1, -2). \tag{b}$$

Consider an element of ocean of mass m located in the y-z plane at a displacement from the origin $\vec{r} = r(0, \sin\theta, \cos\theta)$, where θ is the usual polar angle measured from the z-axis. Using (b) to evaluate the tidal gravitational acceleration on the right-hand side of (21.5), we find ($r \ll d$)

$$F^i_{\text{tidal}} = \frac{GmM_{\text{Moon}}}{d^2}\left(\frac{r}{d}\right)(0, -\sin\theta, +2\cos\theta). \tag{c}$$

Along the positive z-axis ($\theta = 0$) the force is in the $+z$-direction, while along the negative z-axis ($\theta = \pi$) it has the same magnitude but points in the opposite direction. The tidal gravitational forces thus pull the oceans *away* from the center of the Earth on both sides of the z-axis. By contrast, the tidal forces push the ocean *toward* the center along the x- or y-axes. The result is the two-humped tidal bulge shown in the accompanying figure, which produces *two* high tides a day as the Earth rotates underneath. (More realistically, the tidal bulge does not point exactly at the Moon but in a direction so the high tide lags the time the Moon is in the zenith. Can you think why?)

where $r = (x^2 + y^2 + z^2)^{1/2}$ is the distance from the center of symmetry. Evaluating the tidal gravitational acceleration tensor that enters into (21.5) using the rectangular coordinates assumed in that equation gives

$$a_{ij} \equiv -\frac{\partial^2 \Phi}{\partial x^i \partial x^j} = -(\delta_{ij} - 3n_i n_j)\frac{M}{r^3}, \tag{21.7}$$

where $n^i \equiv x^i/r$ are the components of a unit vector in the radial direction. In an orthonormal basis $(\vec{e}_{\hat{r}}, \vec{e}_{\hat{\theta}}, \vec{e}_{\hat{\phi}})$ oriented along the coordinate directions of polar coordinates (r, θ, ϕ), the nonvanishing components of the tidal acceleration tensor are

$$a_{\hat{r}\hat{r}} = \frac{2M}{r^3}, \qquad a_{\hat{\theta}\hat{\theta}} = a_{\hat{\phi}\hat{\phi}} = -\frac{M}{r^3}. \tag{21.8}$$

An observer falling freely and radially might employ just such a basis where the components of the deviation equation (21.5) become

$$\frac{d^2\chi^{\hat{r}}}{dt^2} = +\frac{2M}{r^3}\chi^{\hat{r}}, \quad \frac{d^2\chi^{\hat{\theta}}}{dt^2} = -\frac{M}{r^3}\chi^{\hat{\theta}}, \quad \frac{d^2\chi^{\hat{\phi}}}{dt^2} = -\frac{M}{r^3}\chi^{\hat{\phi}} \tag{21.9}$$

An object falling towards the central mass is stretched in the radial direction and compressed in the transverse directions by tidal gravitational forces.

Example 21.2. Detecting the Earth's Gravitational Field from Inside the Space Shuttle. A space shuttle fallls freely and radially towards the Earth. Astronauts inside use Newtonian physics to analyze the relative motion of two ping pong balls to detect the Earth's gravitational field as in Example 6.3. The balls start at relative rest at $t = 0$, when radius R is crossed, separated radially by a small distance s. The separation vector is thus $\vec{\chi} = s\vec{e}_{\hat{r}}$ initially and its time derivative is zero then. The subsequent evolution of $\vec{\chi}$ is determined by the deviation equations (21.9). Over a short period of time t after $t = 0$ the solution to (21.9) is

$$\vec{\chi}(t) \approx s[1 + (M/R^3)t^2]\vec{e}_{\hat{r}}. \tag{21.10}$$

This is only valid over times t small enough that $|\vec{\chi}(t)|$ remains small and r changes little. For such times, the change $\delta s(t)$ in separation between the balls is

$$\delta s(t)/s \approx (2\pi t/P)^2 \tag{21.11}$$

where P is the period of a circular orbit at radius R. This coincides with the estimate (6.15) in Examaple 6.3. The Earth's gravitational field can be detected through the deviation of the balls.

Interestingly, the field equation for Newtonian gravity (3.18)

$$\nabla^2\Phi = 4\pi G\mu, \tag{21.12}$$

can be expressed in terms of the second derivatives of Φ, which define the tidal gravitational accelerations (21.5). In particular,

$$\nabla^2\Phi = \delta^{ij}\left(\frac{\partial^2\Phi}{\partial x^i \partial x^j}\right) = 4\pi G\mu. \tag{21.13}$$

There is an analogous connection in general relativity.

21.2 Equation of Geodesic Deviation

This section derives the generalization of the Newtonian deviation equation (21.5) to a four-dimensional curved spacetime equation called the *equation of geodesic deviation*. The analogs of $\partial^2\Phi/\partial x^i \partial x^j$ will give us a local measure of spacetime curvature.

The object to study is the separation four-vector $\boldsymbol{\chi}$ giving the infinitesimal displacement between two nearby geodesics, as shown in Figure 21.3. It connects events at the same proper time on both geodesics. The origin of proper time on the neighboring geodesic is arbitrary, and correspondingly there are a variety of ways of defining the separation vector. We could require, for example, that at one initial moment it satisfies $\boldsymbol{\chi} \cdot \mathbf{u} = 0$. However it is fixed initially, the important question is how it changes with τ.

To find the equation for the evolution of the separation vector $\boldsymbol{\chi}$ in general relativity that is analogous to (21.5) in Newtonian theory, it is necessary to compute the second derivative of $\boldsymbol{\chi}$ with respect to the observer's proper time τ. The derivative of a *function* $f(x)$ evaluated on the observer's world line $x^\alpha(\tau)$ with respect to proper time τ is given by

$$\frac{df}{d\tau} = \frac{\partial f}{\partial x^\alpha}\frac{dx^\alpha}{d\tau} = u^\alpha \frac{\partial f}{\partial x^\alpha} \equiv \boldsymbol{\nabla}_{\mathbf{u}} f. \tag{21.14}$$

Here, $u^\alpha = dx^\alpha/d\tau$ is the observer's four-velocity, and $\boldsymbol{\nabla}_{\mathbf{u}} f$ is the covariant derivative along $\mathbf{u}$ [cf. (20.63)]. The derivative with respect to τ of a *vector* like $\boldsymbol{\chi}$ is given by a similar *covariant* derivative along $\mathbf{u}$,

$$\mathbf{v} \equiv \boldsymbol{\nabla}_{\mathbf{u}}\boldsymbol{\chi}, \tag{21.15}$$

and the second derivative with respect to τ—the acceleration of the separation vector—is given by

$$\mathbf{w} \equiv \boldsymbol{\nabla}_{\mathbf{u}}\boldsymbol{\nabla}_{\mathbf{u}}\boldsymbol{\chi}. \tag{21.16}$$

We only need to evaluate (21.16) explicitly to find the equation for $\boldsymbol{\chi}$ analogous to the Newtonian (21.5).

The formula for the covariant derivative of a vector (20.54) gives an explicit expression for the components of the vectors $\nabla_{\mathbf{u}}\boldsymbol{\chi}$ and $\nabla_{\mathbf{u}}\mathbf{v}$ in any coordinate basis:

$$v^\alpha \equiv (\nabla_{\mathbf{u}}\boldsymbol{\chi})^\alpha = u^\beta \nabla_\beta \chi^\alpha = \frac{d\chi^\alpha}{d\tau} + \Gamma^\alpha_{\beta\gamma} u^\beta \chi^\gamma, \tag{21.17}$$

$$w^\alpha \equiv (\nabla_{\mathbf{u}}\mathbf{v})^\alpha = u^\delta \nabla_\delta v^\alpha = \frac{dv^\alpha}{d\tau} + \Gamma^\alpha_{\delta\gamma} u^\delta v^\gamma. \tag{21.18}$$

Here, expressions such as $u^\beta(\partial\chi^\alpha/\partial x^\beta)$ have been written as $d\chi^\alpha/d\tau$ using (21.14).

To derive an expression for the acceleration $\nabla_{\mathbf{u}}\nabla_{\mathbf{u}}\boldsymbol{\chi}$ of the separation vector, (1) substitute (21.17) for v^α into (21.18) and carry out the necessary differentiations. (2) Expand the geodesic equation (8.14) for $x^\alpha(\tau) + \chi^\alpha(\tau)$ to first order in $\chi^\alpha(\tau)$ similarly to the expansion of the Newtonian (21.3). (3) Subtract from this the geodesic equation for $x^\alpha(\tau)$ to find an expression that can be used to eliminate $d^2\chi^\alpha/d\tau^2$ from the results of (1). This is a good exercise in keeping careful track of indices, but it is sufficiently complicated that we defer the details to the book website. It is clear from the forms of (21.17) and (21.18) that the result will be linear in χ^α, be proportional to two factors of u^α, and involve first derivatives of the Christoffel symbols and products of them. The result is the following equation:

Equation of Geodesic Deviation

$$\boxed{(\nabla_{\mathbf{u}}\nabla_{\mathbf{u}}\boldsymbol{\chi})^\alpha = -R^\alpha{}_{\beta\gamma\delta} u^\beta \chi^\gamma u^\delta,} \tag{21.19}$$

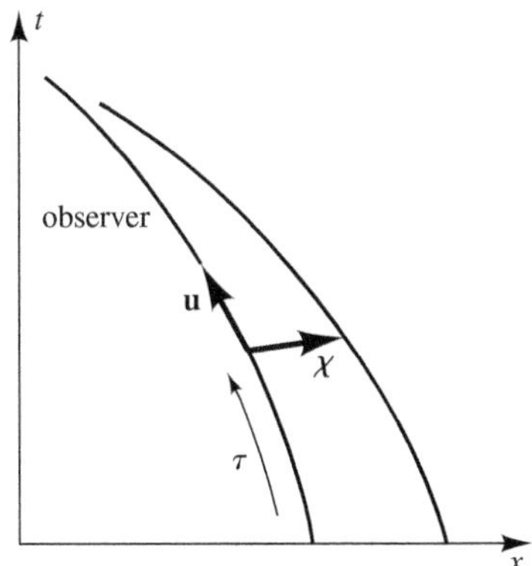

FIGURE 21.3 Geodesic deviation. This is a spacetime diagram illustrating the measurements defining tidal gravitational accelerations or spacetime curvature described in Figures 21.1 and 21.2. The figure shows the world line of the observer and of just one of the nearby test particles becoming closer as time moves on. Both particles are moving on geodesics since they are freely falling. The four-vector $\boldsymbol{\chi}$ giving the infinitesimal displacement from observer to test particle is called the separation vector. The acceleration of the separation vector along the world line of the observer is a quantitative measure of spacetime curvature.

where

Riemann Curvature

$$R^{\alpha}{}_{\beta\gamma\delta} = \frac{\partial\Gamma^{\alpha}_{\beta\delta}}{\partial x^{\gamma}} - \frac{\partial\Gamma^{\alpha}_{\beta\gamma}}{\partial x^{\delta}} + \Gamma^{\alpha}_{\gamma\epsilon}\Gamma^{\epsilon}_{\beta\delta} - \Gamma^{\alpha}_{\delta\epsilon}\Gamma^{\epsilon}_{\beta\gamma}. \tag{21.20}$$

Equation (21.19) is called the *equation of geodesic deviation* and is the general relativistic generalization of the Newtonian (21.5), as Table 21.1 makes clear. The quantity $R^{\alpha}{}_{\beta\gamma\delta}$ is, therefore, the sought-for measure of spacetime curvature—a rank-four tensor called the *Riemann curvature tensor* or *Riemann curvature* for short. It is a tensor because, from (21.19), it supplies a linear map from the vectors $\mathbf{u}$ and $\boldsymbol{\chi}$ to the acceleration vector on the left-hand side. Equation (21.20) gives its *coordinate basis* components. Its detailed properties are discussed in the next section.

The equation of geodesic deviation takes an even simpler form when written in a freely falling frame following a geodesic with four-velocity $\mathbf{u}$. Let $\mathbf{e}_{\hat{\alpha}}$ and $\mathbf{e}^{\hat{\alpha}}$ be the basis vectors and dual-basis vectors, respectively, of the freely falling frame satisfying [cf. (20.79)]

$$\nabla_{\mathbf{u}}\mathbf{e}_{\hat{\alpha}} = 0, \qquad \nabla_{\mathbf{u}}\mathbf{e}^{\hat{\alpha}} = 0, \qquad \text{(freely falling frame)}. \tag{21.21}$$

TABLE 21.1 Newtonian Gravity and General Relativity

	Newtonian Gravity	General Relativity
Basic field quantity	Gravitational potential $\Phi(x^i, t)$	Metric $g_{\alpha\beta}(x)$
Equation of motion (involving the first derivative of field quantity)	Newton's law $\dfrac{d^2x^i}{dt^2} = -\delta^{ij}\dfrac{\partial\Phi}{\partial x^j}$	Geodesic equation $\dfrac{d^2x^\alpha}{d\tau^2} = -\Gamma^{\alpha}_{\beta\gamma}\dfrac{dx^\beta}{d\tau}\dfrac{dx^\gamma}{d\tau}$
Equation of motion for the deviation between two particles (involving second derivatives)	Newtonian deviation $\dfrac{d^2\chi^i}{dt^2} = -\delta^{ij}\left(\dfrac{\partial^2\Phi}{\partial x^j\partial x^k}\right)\chi^k$	Geodesic deviation $\dfrac{d^2\chi^{\hat{\alpha}}}{d\tau^2} = -R^{\hat{\alpha}}{}_{\hat{\tau}\hat{\beta}\hat{\tau}}\chi^{\hat{\beta}}$
Quantity determining the acceleration of the deviation	Tidal forces $\dfrac{\partial^2\Phi}{\partial x^i\partial x^j}$	Riemann curvature $R^{\alpha}{}_{\beta\gamma\delta} = \partial\Gamma^{\alpha}_{\beta\delta}/\partial x^\gamma - \partial\Gamma^{\alpha}_{\beta\gamma}/\partial x^\delta + \Gamma^{\alpha}_{\gamma\epsilon}\Gamma^{\epsilon}_{\beta\delta} - \Gamma^{\alpha}_{\delta\epsilon}\Gamma^{\epsilon}_{\beta\gamma}$
Field equation in a vacuum	Laplace's equation $\nabla^2\Phi = \delta^{ij}\dfrac{\partial^2\Phi}{\partial x^i\partial x^j} = 0$	Vacuum Einstein equation $R_{\beta\delta} = R^{\alpha}{}_{\beta\alpha\delta} = 0$

The basis vectors for a freely falling frame in the Schwarzschild geometry are worked out in Example 20.12, for instance. To convert (21.19) to a freely falling frame, follow (20.41) and multiply both sides by $(\mathbf{e}^{\hat{\alpha}})_\alpha$, thereby implying a sum over α. Since the $\mathbf{e}^{\hat{\alpha}}$ satisfy (21.21) they can be moved inside the covariant derivatives on the left-hand side of (21.19), as in the following example:

$$(\mathbf{e}^{\hat{\alpha}})_\alpha(\boldsymbol{\nabla}_{\mathbf{u}}\boldsymbol{\chi})^\alpha = \mathbf{e}^{\hat{\alpha}} \cdot \boldsymbol{\nabla}_{\mathbf{u}}\boldsymbol{\chi} = \boldsymbol{\nabla}_{\mathbf{u}}(\mathbf{e}^{\hat{\alpha}} \cdot \boldsymbol{\chi}) = d\chi^{\hat{\alpha}}/d\tau. \tag{21.22}$$

The second equality follows from $\boldsymbol{\nabla}_{\mathbf{u}}\mathbf{e}^{\hat{\alpha}} = 0$; the third follows because $\mathbf{e}^{\hat{\alpha}} \cdot \boldsymbol{\chi}$ is a scalar function. Then, noting that $u^{\hat{\alpha}} \equiv (\mathbf{e}_{\hat{\tau}})^{\hat{\alpha}} = (1, \vec{0})$, we have[1]

$$\boxed{\frac{d^2\chi^{\hat{\alpha}}}{d\tau^2} = -R^{\hat{\alpha}}{}_{\hat{\tau}\hat{\beta}\hat{\tau}}\chi^{\hat{\beta}} \qquad \begin{pmatrix}\text{freely falling}\\ \text{frame}\end{pmatrix},} \tag{21.23}$$

where [cf. (20.41)]

$$R^{\hat{\alpha}}{}_{\hat{\beta}\hat{\gamma}\hat{\delta}} = R^{\alpha}{}_{\beta\gamma\delta}(\mathbf{e}^{\hat{\alpha}})_\alpha(\mathbf{e}_{\hat{\beta}})^\beta(\mathbf{e}_{\hat{\gamma}})^\gamma(\mathbf{e}_{\hat{\delta}})^\delta \tag{21.24}$$

are the components of the Riemann tensor in the freely falling frame.

With an equation as seemingly complex as the equation of geodesic deviation, it may be reassuring to see how easily it reduces to (21.5) in the Newtonian limit. To take that limit it is necessary only to use the static weak field metric (6.20) ($c = 1$ units)

$$ds^2 = -(1 + 2\Phi)\,dt^2 + (1 - 2\Phi)(dx^2 + dy^2 + dz^2) \tag{21.25}$$

to evaluate (21.23) to first order in the Newtonian potential Φ, which is small in the Newtonian limit. Since the curvature vanishes to zeroth order in Φ, only the leading order basis vectors are needed. These coincide with the coordinate basis vectors for (21.25) when the center of the freely falling frame is a particle moving at much less than the speed of light [cf. (20.81)]. Thus, to leading order we can write (21.23) as

$$\frac{d^2\chi^i}{dt^2} = -R^i{}_{tjt}\chi^j. \tag{21.26}$$

To agree with (21.5), $R^i{}_{tjt}$ must equal $\partial^2\Phi/\partial x^i\partial x^j$ when calculated using (21.25). Does it? It is not too difficult to get the answer directly from (21.20), or you can find it in Appendix B. Recognize that the answer is needed only to leading order in the small perturbation of the metric produced by Φ. In leading order,

[1] To avoid potential confusion, it is useful to review a few notational conventions relevant for (21.23): the proper time τ along the central geodesic is also the time coordinate in the freely falling frame. It appears as a parameter on the left-hand side of the equation and as a label of an orthonormal basis vector $\mathbf{e}_{\hat{\tau}}$ pointing in that direction [cf. (20.81a)]. Just as $p^{\hat{\tau}}$ indicates a component of the momentum along $\mathbf{e}_{\hat{\tau}}$, the $\hat{\tau}$'s on the right-hand side of (21.23) indicate components of the Riemann tensor in this direction. Their repetition does not indicate summation.

the Γ's will be proportional to Φ. The last two terms in (21.20) will therefore be negligible compared to the first two. The second of these vanishes because it is a time derivative and Φ is time-independent. There remains ($c = 1$)

$$R^i{}_{tjt} = \frac{\partial \Gamma^i_{tt}}{\partial x^j} = \delta^{ik} \frac{\partial^2 \Phi}{\partial x^k \partial x^j}. \tag{21.27}$$

Inserting (21.27) into (21.26), the Newtonian deviation equation (21.5) is recovered.

The computation of the 20 independent components of the Riemann curvature for even a modestly complex metric like the Kerr metric of a rotating black hole can be very laborious if done by hand. There are a number of computer packages that do the algebra for you. A simple one, *Curvature and the Einstein Equation*, is available on the book website as a *Mathematica* notebook. A hard copy showing the calculation of various curvature quantities for the Schwarzschild metric is in Appendix C. Appendix B contains tables of curvature quantities, including the Riemann curvature, for most of the metrics discussed in this book.

21.3 Riemann Curvature

Properties

The form of the Riemann curvature (21.20) simplifies considerably in a local inertial frame (LIF) of a point (see Section 7.4). There $g_{\alpha\beta} = \eta_{\alpha\beta}$ and the first derivatives of the metric vanish. For example, the last two terms in (21.20) vanish identically, and the first two terms simplify (Problem 6). Lowering the first index of $R^\alpha{}_{\beta\gamma\delta}$ as in (20.35) gives:

$$R_{\alpha\beta\gamma\delta} = \frac{1}{2}\left(\frac{\partial^2 g_{\alpha\delta}}{\partial x^\beta \partial x^\gamma} - \frac{\partial^2 g_{\alpha\gamma}}{\partial x^\beta \partial x^\delta} - \frac{\partial^2 g_{\beta\delta}}{\partial x^\alpha \partial x^\gamma} + \frac{\partial^2 g_{\beta\gamma}}{\partial x^\alpha \partial x^\delta}\right) \quad \text{(LIF)}. \tag{21.28}$$

The Riemann curvature has a number of symmetries that are true in general but most straightforwardly demonstrated from (21.28) (Problem 7):

$$R_{\alpha\beta\gamma\delta} = -R_{\beta\alpha\gamma\delta}, \tag{21.29a}$$

$$R_{\alpha\beta\gamma\delta} = -R_{\alpha\beta\delta\gamma}, \tag{21.29b}$$

$$R_{\alpha\beta\gamma\delta} = +R_{\gamma\delta\alpha\beta}, \tag{21.29c}$$

$$R_{\alpha\beta\gamma\delta} + R_{\alpha\delta\beta\gamma} + R_{\alpha\gamma\delta\beta} = 0. \tag{21.29d}$$

These symmetries show that all $4\times4\times4\times4$ components of the Riemann curvature are not independent of each other. In fact, there are only 20 independent components (Problem 7). If you worked through Problem 7.9, you know that, unlike the first derivatives, there are 20 combinations of the *second* derivatives of the metric that cannot be made to vanish by a coordinate transformation. The components of the Riemann curvature are these 20 combinations.

The dimensions of the Riemann curvature are (length)$^{-2}$, and the length scale L for which typical components are $\sim 1/L^2$ is called the *curvature scale*. For example, for a two-dimensional sphere of radius a we expect the components of the curvature to be $\sim 1/a^2$. In general relativity mass produces curvature, and so we expect curvatures to be $\sim$ (mass scale)/(distance scale)^3. The Schwarzschild geometry provides a simple example, as we will see in the next section.

Curvature of the Schwarzschild Geometry

The idea of curvature can be simply illustrated by the Schwarzschild geometry described in Chapter 9. Consider an observer who falls freely and radially from infinity with a laboratory of test particles, as in Figure 21.1. Suppose, for simplicity, that the laboratory starts from rest at infinity and falls radially. The basis vectors for this freely falling frame were worked out in (20.81). Nonvanishing components of the Riemann curvature in this freely falling frame turn out to be (Problem 8)

$$R_{\hat{\tau}\hat{r}\hat{\tau}\hat{r}} = -2M/r^3, \tag{21.30a}$$

$$R_{\hat{\theta}\hat{\phi}\hat{\theta}\hat{\phi}} = +2M/r^3, \tag{21.30b}$$

$$R_{\hat{\tau}\hat{\theta}\hat{\tau}\hat{\theta}} = R_{\hat{\tau}\hat{\phi}\hat{\tau}\hat{\phi}} = +M/r^3, \tag{21.30c}$$

$$R_{\hat{r}\hat{\theta}\hat{r}\hat{\theta}} = R_{\hat{r}\hat{\phi}\hat{r}\hat{\phi}} = -M/r^3. \tag{21.30d}$$

Any other nonvanishing components can be worked out from these using the symmetries (21.29). (The repeated Greek indices in these expressions do not indicate summation but are just component labels for directions in the freely falling frame.)

With (21.30) components of the equation of geodesic deviation (21.23) take the simple form

$$\frac{d^2\chi^{\hat{r}}}{d\tau^2} = +\frac{2M}{r^3}\chi^{\hat{r}}, \qquad \frac{d^2\chi^{\hat{\theta}}}{d\tau^2} = -\frac{M}{r^3}\chi^{\hat{\theta}}, \qquad \frac{d^2\chi^{\hat{\phi}}}{d\tau^2} = -\frac{M}{r^3}\chi^{\hat{\phi}}. \tag{21.31}$$

Coincidently these have the same form as the Newtonian deviation equation (21.9) in a spherically symmetric gravitational potential. Equations (21.31) show that an observer who falls into a Schwarzschild black hole is stretched in the radial direction and compressed in the transverse directions. At the Schwarzschild radius $r = 2M$, the tidal accelerations are finite. This is another way of seeing that the singularity in the metric in Schwarzschild coordinates is a coordinate singularity and not a real singularity. Indeed, the tidal gravitational accelerations at the horizon are of order $\sim 1/M^2$, which can be very small if the black hole is large.

The radius $r = 0$ is a different story. There the tidal forces become infinite and the observer (or anything else) will be destroyed as $r = 0$ is approached. The radius $r = 0$ is a real singularity!

Example 21.3. Space Pirates. You are the unfortunate victim of space pirates who dispose of their captives by dropping them in their spacesuits radially into the

$3 \times 10^6\ M_\odot$ black hole at the center of our galaxy (Figure 13.4). Understanding general relativity, the pirates know that their victims will never return from beyond the horizon. On your way in, you console yourself with the thought that at least you will have about a minute after crossing the horizon to see the interior of a black hole. But will you survive the tidal stresses even to make it across the horizon? If so, how close to the singularity will you get before being stretched apart in the radial direction and crushed in the transverse ones? You estimate that you could stand about 10^4 N (roughly 2000 lb, or 1 ton) of force pulling you apart. The tidal forces at a radius r can be estimated from the geodesic deviation equation (21.31), where the separation vector extends from your feet to your head. The force is m times the acceleration of the separation vector on the left-hand side. Putting back the factors of G and c, the tidal force at radius r from the right-hand side is roughly $(GmM/r^2)(h/r)$, where M is the mass of the black hole and h is your height. Assuming $m \sim 80$ kg and $h \sim 2$ m, this gives .1 N at the black-hole horizon $r = 2GM/c^2$. You will, therefore, easily cross the horizon. The tidal force reaches 10^4 N at $r \sim 2 \times 10^5$ km. You will pass through most of the radius from the horizon to the singularity before being destroyed, but it will go by very quickly.

21.4 The Einstein Equation in Vacuum

Having found the local measure of gravitational curvature, we are now in a position to introduce the Einstein equation. As with the Newtonian case (21.13), the field equations for gravity in a vacuum involve an object formed by taking a sum over the quantities that describe geodesic deviation. In the relativistic case this sum defines the ten components of the *Ricci curvature*,

$$R_{\alpha\beta} \equiv R^{\gamma}{}_{\alpha\gamma\beta} \tag{21.32}$$

(sum over repeated indices). The Ricci curvature can be expressed directly in terms of the Christoffel symbols by

Ricci Curvature

$$\boxed{R_{\alpha\beta} = \frac{\partial \Gamma^{\gamma}_{\alpha\beta}}{\partial x^{\gamma}} - \frac{\partial \Gamma^{\gamma}_{\alpha\gamma}}{\partial x^{\beta}} + \Gamma^{\gamma}_{\alpha\beta}\Gamma^{\delta}_{\gamma\delta} - \Gamma^{\gamma}_{\alpha\delta}\Gamma^{\delta}_{\beta\gamma}.} \tag{21.33}$$

The Einstein equation in vacuum, the relativistic generalization of $\nabla^2\Phi = 0$, is

Vacuum Einstein Equation

$$\boxed{R_{\alpha\beta} = 0.} \tag{21.34}$$

It is not difficult to show from (21.33) that $R_{\alpha\beta}$ is symmetric in α and β (Problem 14). The equations $R_{\alpha\beta} = 0$ are, therefore, ten second-order partial differential equations for the ten metric coefficients $g_{\alpha\beta}$. This would seem to be a nice balance between equations and unknowns until you realize that no set of field equations should determine the $g_{\alpha\beta}$ uniquely. There should always be the freedom to make four independent coordinate transformations. In fact, there turn out to be four differential identities relating the $R_{\alpha\beta}$, so that there are really only six independent equations corresponding to the six independent metric degrees of freedom. These identities, comprising the Bianchi identity, are exhibited in the next chapter.

Unlike those of Newtonian gravity or electromagnetism, the field equations of general relativity are nonlinear. Flat spacetime is one solution of the vacuum Einstein equation (21.34)—all the Γs vanish in usual rectangular coordinates; therefore, so does $R_{\alpha\beta}$. However, finding other solutions, either analytically or numerically, is in general a difficult task. Indeed it is a lengthy calculation just to verify that the Schwarzschild geometry quoted in (9.9) is a solution of the Einstein equation. It is instructive to work through this calculation once by hand to see what is involved following the leads in Problems 16 and 17. However, for efficiency, algebraic computing programs such as *Mathematica* are hard to beat. There is one available at the text website for calculating the Christoffel symbols, Riemann curvature, and Ricci curvature for any metric. We illustrate their power with the following example.

Example 21.4. Solving the Vacuum Einstein Equation to find the Schwarzschild Metric. Suppose you were trapped on a desert island and had forgotten the form of the Schwarzschild metric (9.9)—the most general solution of the vacuum Einstein equation outside of a static (no moving parts), spherically symmetric distribution of mass. How could you solve the vacuum Einstein equation to find it again?[2]

Your first step would be to use the spherical symmetry and time independence to simplify the metric. You might reason as follows: The metric of a static source should be independent of a time coordinate t; further, there should be no $dx^i dt$ terms in the line element because these change sign under $t \to -t$. Spherical symmetry implies that space—the $t = \text{const.}$ hypersurfaces—can be thought of a nested family of two-spheres, each labeled by a radial coordinate r, which can be defined in terms of the sphere's area by $r = [(\text{area})/4\pi]^{1/2}$. You introduce the familiar polar angles θ and ϕ on each sphere, so its intrinsic geometry is given by the line element $d\Sigma^2 = r^2(d\theta^2 + \sin^2\theta\, d\phi^2)$ as it would be if embedded in flat space [cf. (2.15)]. Further, you choose the polar coordinates on all spheres so that a rotation of the whole spacetime transforms the coordinates θ and ϕ in the same way on all spheres. This means there can be no $d\theta\, dr$ or $d\phi\, dr$ terms in the line element because they would not be spherically symmetric. The result

[2]The circumstances surrounding the discovery of the Schwarzschild solution had some similarities with this imagined one, although more tragic. In the spring and summer of 1915, Schwarzschild was serving with the German army on the eastern front during World War I. He contracted a fatal illness and died in May 1916. It was during this illness in December 1915 that he discovered the Schwarzschild solution, which was published just a few months after Einstein had published the basic equations of general relativity.

of this rough argument (which can be promoted to a more rigorous one with a little work) is that you decide that the line element outside a static, spherically symmetric source of curvature can be put in the general form

$$ds^2 = -e^{\nu(r)}dt^2 + e^{\lambda(r)}dr^2 + r^2(d\theta^2 + \sin^2\theta\, d\phi^2) \tag{21.35}$$

containing just two unknown functions of r. Writing $g_{tt}(r)$ and $g_{rr}(r)$ in terms of $\nu(r)$ and $\lambda(r)$ in this way is just tradition. The important point is that ten unknown metric functions of four variables have been reduced to two unknown functions of one variable, and ten nonlinear partial differential equations will be reduced to two nonlinear ordinary differential equations—enormous simplifications.

The next step is to write out the Einstein equation $R_{\alpha\beta} = 0$ using the metric (21.35). Luckily you have been marooned with your laptop containing the *Mathematica* notebook *Curvature and the Einstein Equation* supplied on the website for this book, and you can calculate any curvature quantity you want.[3] The program outputs the Ricci curvature $R_{\alpha\beta}$ but also the combination of it, $G_{\alpha\beta} \equiv R_{\alpha\beta} - (1/2)g_{\alpha\beta}R^{\gamma}_{\gamma}$, called the Einstein tensor. (This is the left-hand side of the Einstein equation with sources discussed in the next chapter.) The Einstein tensor must vanish if the Ricci tensor vanishes and vice versa. Although it isn't necessary, you use (20.41) to project the components of $G_{\alpha\beta}$ on an orthonormal basis along the coordinate axes constructed as in Example 7.9 because you remember that was how results are displayed in Appendix B of your text. You only need two of the equations $G_{\hat{\alpha}\hat{\beta}} = 0$ to solve for $\nu(r)$ and $\lambda(r)$. You pick the simplest, secure in the knowledge that the others will be automatically satisfied since the components of the Einstein equation are not all independent. The simplest two are:

$$G_{\hat{t}\hat{t}} = e^{-\lambda}\left(\frac{\lambda'}{r} - \frac{1}{r^2}\right) + \frac{1}{r^2} = 0, \tag{21.36a}$$

$$G_{\hat{r}\hat{r}} = e^{-\lambda}\left(\frac{\nu'}{r} + \frac{1}{r^2}\right) - \frac{1}{r^2} = 0, \tag{21.36b}$$

where a prime denotes an r-derivative.

The first of the equations (21.36) is an ordinary differential equation for λ alone, which can be rewritten

$$\frac{d}{dr}(re^{-\lambda(r)}) = 1, \tag{21.37}$$

whose solution is

$$e^{-\lambda(r)} = 1 + A/r \tag{21.38}$$

for some constant A. The combination $G_{\hat{t}\hat{t}} + G_{\hat{r}\hat{r}} = 0$ shows that $\nu' = -\lambda'$, so that $\nu(r) = -\lambda(r) + B$, where B is another constant. The result for $g_{tt}(r)$ is

$$e^{\nu(r)} = e^{B}(1 + A/r). \tag{21.39}$$

[3] You can see a hard copy of this notebook applied to just this problem in Appendix C. You can also find the answer in Appendix B.

Matching the metric to the standard flat metric in polar coordinates (7.4) at large r shows $B = 0$. Rewriting the constant A as $-2M$ gives the Schwarzschild metric, where M is identified with the total mass, as discussed in Section 9.1.

The Schwarzschild geometry is more general than this derivation. Even if the source is changing in time, it is the only spherically symmetric solution of the vacuum Einstein equation. This result—called Birkhoff's theorem—was important for the understanding of spherical collapse in Chapter 12. You are led through a derivation of it in Problem 18.

21.5 Linearized Gravity

Until now we have presented various interesting metrics solving the Einstein equation. The Einstein equation comprises ten nonlinear, partial differential equations for ten metric coefficients, $g_{\alpha\beta}(x)$. There is currently no general technique for solving such systems and no such thing (yet) as a "general solution." Rather, there are a large variety of sometimes powerful techniques for solving the equations in particular circumstances—typically those involving symmetries such as the spherical symmetry of the Schwarzschild geometry discussed earlier.[4] However, unlike the general nonlinear case, it *is* possible to give a complete analysis of the solutions of the Einstein equation for spacetimes whose geometries differ only slightly from flat spacetime. Examples are the static, weak-field metric (21.25) discussed in Chapter 6, and the linear gravitational wave spacetimes discussed in Chapter 16. We next linearize the vacuum Einstein equation and show how to solve it in general circumstances.

The Linearized Vacuum Einstein Equation

In Minkowski coordinates (t, x, y, z), the metric for flat spacetime is $g_{\alpha\beta} = \eta_{\alpha\beta}$, where $\eta_{\alpha\beta} = \text{diag}(-1, 1, 1, 1)$. Metrics of geometries that are *close* to flat can therefore be written [cf. (16.1)]

$$g_{\alpha\beta}(x) = \eta_{\alpha\beta} + h_{\alpha\beta}(x)\,, \tag{21.40}$$

Metric Perturbations

where the $h_{\alpha\beta}(x)$ are small quantities called *metric perturbations*.

The linearized Einstein equation is obtained by inserting (21.40) into $R_{\alpha\beta} = 0$ and expanding it to first order in $h_{\alpha\beta}(x)$. The first term on the left-hand side of this expansion is the Ricci curvature of flat spacetime, which vanishes. The second term is the first-order perturbation in the Ricci curvature $\delta R_{\alpha\beta}$, which is linear in $h_{\alpha\beta}(x)$. The linearized vacuum Einstein equation is, thus,

$$\delta R_{\alpha\beta} = 0. \tag{21.41}$$

This is a set of ten *linear*, partial differential equations for $h_{\alpha\beta}(x)$.

To find an explicit formula for how $\delta R_{\alpha\beta}$ depends on $h_{\alpha\beta}(x)$, use the expression for the Ricci curvature in terms of the Christoffel symbols (21.33) and the ex-

[4]For a whole book on such techniques, see Kramer et al. (1980).

pression for the Christoffel symbols in terms of the metric (20.53). In the lowest, or zeroth, order, the perturbations $h_{\alpha\beta}(x)$ are neglected, the Christoffel symbols vanish, and so does the curvature because all the components of $\eta_{\alpha\beta}$ are constant.

The first-order perturbations in Christoffel symbols are

$$\delta\Gamma^{\gamma}_{\alpha\beta} = \frac{1}{2}\eta^{\gamma\delta}\left(\frac{\partial h_{\delta\alpha}}{\partial x^{\beta}} + \frac{\partial h_{\delta\beta}}{\partial x^{\alpha}} - \frac{\partial h_{\alpha\beta}}{\partial x^{\delta}}\right). \tag{21.42}$$

The perturbation in the Ricci curvature is easy to compute because the last two terms in (21.33) are quadratic in $h_{\alpha\beta}$ and negligible in the linear approximation. Thus, we have

$$\delta R_{\alpha\beta} = \frac{\partial \delta\Gamma^{\gamma}_{\alpha\beta}}{\partial x^{\gamma}} - \frac{\partial \delta\Gamma^{\gamma}_{\alpha\gamma}}{\partial x^{\beta}}. \tag{21.43}$$

Substituting (21.42) into (21.43), the linearized Einstein equation in vacuum (21.41) becomes:

$$\delta R_{\alpha\beta} = \frac{1}{2}\left[-\Box h_{\alpha\beta} + \partial_{\alpha}V_{\beta} + \partial_{\beta}V_{\alpha}\right] = 0. \tag{21.44}$$

The quantities that appear in this expression are as follows: the operator $\Box$ is the flat-space wave operator (sometimes called the d'Alembertian)

Wave Operator

$$\Box \equiv \eta^{\alpha\beta}\partial_{\alpha}\partial_{\beta} = -\frac{\partial^2}{\partial t^2} + \vec{\nabla}^2 \tag{21.45}$$

and ∂_{α} is a shorthand for $\partial/\partial x^{\alpha}$. The vector V_{α} is the particular combination of perturbations

$$V_{\alpha} \equiv \partial_{\gamma}h^{\gamma}_{\alpha} - \frac{1}{2}\partial_{\alpha}h^{\gamma}_{\gamma}, \tag{21.46}$$

where[5]

$$h^{\gamma}_{\alpha} = \eta^{\gamma\delta}h_{\delta\alpha}. \tag{21.47}$$

The latter equation is an example of a general relation useful in linearized gravity. *To linearized accuracy, indices on perturbations can be raised and lowered with the flat-space metric.* Including the first-order corrections to $g^{\alpha\beta}$ following from (21.40) when raising an index would result in negligible *second*-order corrections to h^{α}_{β} and $\delta R_{\alpha\beta}$.

Example 21.5. The Newtonian Limit. A first application of the equations of linearized gravity is to verify that the static, weak-field metric (21.25) satisfies the linearized vacuum Einstein equation when Φ is independent of t and satisfies the

[5]Strictly speaking, we should write $h^{\beta}{}_{\alpha}$ for the mixed indices because it is the first index that is raised in (21.47) and the order of mixed indices generally *does* make a difference. However, there is no difference when, as here, $h_{\alpha\beta}$ is symmetric. Then $h^{\beta}{}_{\alpha} = h_{\alpha}{}^{\beta}$ and since there is no danger of confusion, we write h^{β}_{α} for either, following the usual conventions of the subject.

vacuum equation of Newtonian gravity [cf. (3.18)]:

$$\nabla^2 \Phi(x^i) = 0. \tag{21.48}$$

Comparing the metric (21.25) with (21.40) gives the following metric perturbations:

$$h_{tt} = -2\Phi, \qquad h_{ti} = h_{it} = 0, \qquad h_{ij} = -2\delta_{ij}\Phi, \tag{21.49a}$$

and

$$h^i_{\ j} = -2\delta^i_{\ j}\Phi, \qquad h^\beta_{\ \beta} = -4\Phi. \tag{21.49b}$$

It then follows directly that $V_\alpha = 0$ identically, and the linearized Einstein equation (21.44) reduces to the Newtonian equation (21.48).

Choosing Coordinates—Gauge

As has been stressed many times, coordinates are arbitrary. By careful choice of coordinates the solution of the linearized Einstein equation can be simplified. Indeed, it is essential to impose *some* conditions on the coordinates. The ten equations $\delta R_{\alpha\beta} = 0$ can't determine the $h_{\alpha\beta}(x)$ uniquely because their values could be changed without changing the geometry by changing the coordinates.

In (21.40) we assumed a coordinate system in which the flat metric takes the form $\eta_{\alpha\beta} = \text{diag}(-1, 1, 1, 1)$ and $h_{\alpha\beta}$ is an as yet unknown perturbation of it. However, that assumption does not uniquely fix the coordinates. We can still make small changes in the coordinates that leave $\eta_{\alpha\beta}$ unchanged but make small changes in the $h_{\alpha\beta}(x)$. Such changes preserve the form of (21.40) but change the functional form of the $h_{\alpha\beta}$.

To see this, consider a change in coordinates of the form

$$x'^\alpha = x^\alpha + \xi^\alpha(x), \tag{21.50}$$

where $\xi^\alpha(x)$ are four arbitrary functions whose derivatives are of the same small size as the metric perturbations $h_{\alpha\beta}(x)$. Under a change of coordinates, the metric generally transforms as we found in (20.44):

$$g'_{\alpha\beta}(x') = \frac{\partial x^\gamma}{\partial x'^\alpha}\frac{\partial x^\delta}{\partial x'^\beta} g_{\gamma\delta}(x). \tag{21.51}$$

From (21.50), we have $x^\alpha = x'^\alpha - \xi^\alpha(x^\beta) = x'^\alpha - \xi^\alpha(x'^\beta)$, the last equality being accurate to first order in ξ^α. Indeed, generally we can substitute x'^α for x^α or vice versa in any first-order expression. For example,

$$\frac{\partial x^\alpha}{\partial x'^\beta} = \delta^\alpha_\beta - \frac{\partial \xi^\alpha}{\partial x'^\beta} = \delta^\alpha_\beta - \frac{\partial \xi^\alpha}{\partial x^\beta}. \tag{21.52}$$

TABLE 21.2 Gauge Transformations in Linearized Gravity and Electromagnetism

	Linearized Gravitation	Electromagnetism
Basic "potentials"	Linearized metric perturbation $h_{\alpha\beta}(x)$	Vector and scalar potentials $\left(\Phi(t,\vec{x}),\ \vec{A}(t,\vec{x})\right)$
Field quantities	Linearized Riemann curvature $\delta R_{\alpha\beta\gamma\delta}(x)$	Electric and magnetic fields $\vec{E}(t,\vec{x}),\ \vec{B}(t,\vec{x})$
Gauge transformation leading to new potentials but the same fields	$h_{\alpha\beta} \to h_{\alpha\beta} - \partial_\alpha \xi_\beta - \partial_\beta \xi_\alpha$	$\vec{A} \to \vec{A} + \nabla\Lambda$ $\Phi \to \Phi - \partial\Lambda/\partial t$
Example of a gauge condition	Lorentz gauge $\partial_\beta h^\beta_\alpha - \frac{1}{2}\partial_\alpha h^\beta_\beta = 0$	Lorentz condition $\vec{\nabla}\cdot\vec{A} + \partial\Phi/\partial t = 0$
Field equations simplified by the gauge condition	$\Box h_{\alpha\beta} = 0$	Maxwell's equations $\Box\vec{A} = 0$ $\Box\Phi = 0$

After a little work, we find from (21.51) that a metric of the form (21.40) transforms into a metric of the same form but with new perturbations given by

Gauge Transformation

$$h'_{\alpha\beta} = h_{\alpha\beta} - \partial_\alpha \xi_\beta - \partial_\beta \xi_\alpha . \tag{21.53}$$

The transformations (21.53) are often called *gauge transformations* because of their analogy with gauge transformations in electromagnetism (see Table 21.2).

Since the $\xi^\alpha(x)$ are four arbitrary functions (though small), we can choose them to simplify the form of the transformed $h_{\alpha\beta}(x)$. In particular, we can choose them so that the four conditions

$$V'_\alpha(x) = 0 \tag{21.54}$$

are satisfied and (21.44) reduces to $\delta R'_{\alpha\beta} = -(\frac{1}{2})\Box h'_{\alpha\beta} = 0$. But we have not yet solved for the perturbations $h_{\alpha\beta}$. So we might as well assume that we already have a system of coordinates where (21.54) is true! Thus, dropping the prime, the

Einstein equation $\delta R_{\alpha\beta} = 0$ becomes the wave equation

$$\boxed{\Box h_{\alpha\beta}(x) = 0,} \tag{21.55}$$

Linearized Einstein Equation

together with the gauge conditions

$$\boxed{V_\alpha(x) \equiv \partial_\beta h^\beta_\alpha(x) - \tfrac{1}{2}\partial_\alpha h^\beta_\beta(x) = 0.} \tag{21.56}$$

Lorentz Gauge Condition

If you have studied electromagnetism, you will recognize an analogy between (21.56) and the Lorentz gauge condition used there. This analogy is spelled out in Table 21.2. For this reason the four conditions in (21.56) are often called the *Lorentz gauge conditions.* Equations (21.55) and (21.56) are the two basic equations of linearized gravity.

Example 21.6. The Gravitational Wave Metric Satisfies the Linearized Einstein Equation. We can now verify that the gravitational wave metric (16.2a) that was the subject of Chapter 16 indeed solves the linearized Einstein equation (21.55) in the Lorentz gauge (21.56). A quick check shows $V_\alpha = 0$, either because it vanishes identically or because $\partial_t f(t-z) = -\partial_z f(t-z)$. Thus, the Lorentz gauge condition (21.56) is satisfied. Equation (21.55) is satisfied because $f(z-t)$ is a solution of the wave equation. The perturbation (16.2a) thus solves the linearized Einstein equation.

Example 21.6 shows that the gravitational wave metric exhibited in (16.2a) satisfies the linearized Einstein equation. We now show how that metric could be found by solving the equations of linearized gravity directly.

Solving the Wave Equation

Metric perturbations are determined by two equations: the wave equation (21.55) and the gauge condition (21.56). Solving the wave equation is a problem that occurs in many areas of physics, and we briefly review how to go about it.

To keep things simple, consider first the flat-space wave equation for a scalar $f(x)$:

$$\Box f(x) \equiv \eta^{\alpha\beta}\frac{\partial^2 f}{\partial x^\alpha \partial x^\beta} = -\frac{\partial^2 f}{\partial t^2} + \vec{\nabla}^2 f = 0. \tag{21.57}$$

We can solve this by Fourier transforms. First, try a solution of the form

$$f(x) = ae^{i\mathbf{k}\cdot\mathbf{x}}, \tag{21.58}$$

where

$$\mathbf{k}\cdot\mathbf{x} = -k^t t + \vec{k}\cdot\vec{x}. \tag{21.59}$$

Here, $\mathbf{x}$ is the flat-space position four-vector with components $x^\alpha = (t, \vec{x})$ and $\mathbf{k}$ is a constant wave four-vector. Inserting (21.58) into (21.57), we find

$$\Box f = -\mathbf{k} \cdot \mathbf{k} f = 0, \tag{21.60}$$

where $\mathbf{k} \cdot \mathbf{k}$ is computed in the flat-space scalar product. Equation (21.60) vanishes if $\mathbf{k} \cdot \mathbf{k} = 0$, that is, when $\mathbf{k}$ is a null vector. Null vectors can be written in the form

$$\mathbf{k} = \left(|\vec{k}|, \vec{k}\right), \tag{21.61}$$

and we frequently use the standard notation $\omega_{\vec{k}}$ for the frequency $k^t = |\vec{k}|$.

Physical solutions are obtained by taking the real part of (21.58), which is of the form

$$|a| \cos(\mathbf{k} \cdot \mathbf{x} + \delta) = |a| \cos\left(-\omega_{\vec{k}} t + \vec{k} \cdot \vec{x} + \delta\right). \tag{21.62}$$

This represents a wave with frequency $\omega_{\vec{k}} = |\vec{k}|$ and wavelength $2\pi/|\vec{k}|$, traveling in the direction of the vector $\vec{k}$ with speed $\omega_{\vec{k}}/|\vec{k}| = 1$. Thus, we learn that solutions of the wave equation are waves propagating with the speed of light.

The general solution of the wave equation (21.57) may be built up by superposing waves like (21.58) with definite wave vectors:

$$f(x) = \int d^3k \, a(\vec{k}) e^{i\mathbf{k}\cdot\mathbf{x}}. \tag{21.63}$$

The integral is over all values of $\vec{k}$, e.g., k^x from $-\infty$ to $+\infty$, and the $a(\vec{k})$ are arbitrary complex amplitudes. To get real solutions that represent quantities in physics, take the real part of (21.63).

More Gauge

Each component of the gravitational wave perturbation satisfies the flat-space wave equation according to (21.55). Thus, for a wave of definite wave vector $\vec{k}$,

$$h_{\alpha\beta}(x) = a_{\alpha\beta} e^{i\mathbf{k}\cdot\mathbf{x}}, \tag{21.64}$$

where $a_{\alpha\beta}$ is a symmetric 4×4 matrix of constants giving the amplitudes of the various components of the wave. These amplitudes are not arbitrary. In addition to the wave equation, the metric perturbations must satisfy the gauge condition (21.56). However, these conditions still allow for further coordinate transformations, which can be used to simplify the matrix $a_{\alpha\beta}$. Transformations of the form (21.53) for $\xi_\alpha(x)$ that don't disturb the conditions $V_\alpha(x) = 0$ are allowed. Inserting (21.53) into $V_\alpha(x) = 0$, noting that $h_{\alpha\beta}(x)$ has already been assumed to satisfy this condition, we find

$$\Box \xi_\alpha(x) = 0 \tag{21.65}$$

as the condition on $\xi_\alpha(x)$ that leaves the Lorentz gauge conditions satisfied. But, since the $h_{\alpha\beta}(x)$ also satisfy the wave equation [cf. (21.55)], these transformations can be used to make any four of the $h_{\alpha\beta}$ vanish identically. We choose to set

$$h_{ti} = 0, \tag{21.66a}$$

$$h^{\beta}_{\beta} = 0, \tag{21.66b}$$

or $a_{ti} = a^{\beta}_{\beta} = 0$.

Using (21.66), the four conditions $V_\alpha = 0$, where V_α is given by (21.56), now imply

$$V_t = \frac{\partial h^t_t}{\partial t} = +i\omega_k a_{tt} e^{i\mathbf{k}\cdot\mathbf{x}} = 0,$$

$$V_i = \frac{\partial h^j_i}{\partial x^j} = ik^j a_{ji} e^{i\mathbf{k}\cdot\mathbf{x}} = 0. \tag{21.67}$$

(There is no implied sum in the first of these.) From this we learn that

$$a_{tt} = 0, \tag{21.68a}$$

$$k^j a_{ij} = 0. \tag{21.68b}$$

The last condition means that gravitational waves are transverse, just as are electromagnetic waves.

Of the original ten $a_{\alpha\beta}$, only two are left. The four time components vanish because of (21.66a) and (21.68a). Equations (21.66b) and (21.68b) are a total of four additional conditions. Thus, there are only two independent $a_{\alpha\beta}$. The easiest way to write them explicitly is to orient the spatial coordinates so that one axis—say the z-axis—is along the direction of propagation of the wave. Then $\vec{k} = (0, 0, \omega)$. The transversality condition then implies that all the components a_{zi} vanish. All that's left is the 2×2 symmetric matrix in the x-y subspace whose trace must vanish because of (21.66b). Thus, with this careful choice of coordinates, the *most general* solution of the linearized Einstein equation with definite wave number is

$$h_{\alpha\beta}(x) = \begin{array}{c} \\ t \\ x \\ y \\ z \end{array}\!\!\begin{array}{c} \begin{array}{cccc} t & x & y & z \end{array} \\ \begin{pmatrix} 0 & 0 & 0 & 0 \\ 0 & a & b & 0 \\ 0 & b & -a & 0 \\ 0 & 0 & 0 & 0 \end{pmatrix} \end{array} e^{i\omega(z-t)}. \tag{21.69}$$

Transverse Traceless Gauge

This choice of coordinates in which the transverse and traceless conditions are represented explicitly is called *transverse-traceless gauge*, or TT-gauge for short.

Equation (21.69) is exactly the general form of the gravitational wave spacetimes discussed in Chapter 16. It is (16.17) when that wave has a definite frequency ω. The parts of (21.69) proportional to a and b represent the two different polarizations of the gravitational wave. The part proportional to a is usually called the $+$ (plus) polarization and the part proportional to b is called the $\times$ (cross) polarization, although these are basis-dependent distinctions. The general solution of the linearized Einstein equation is a superposition of waves of the form (21.69) with different values of ω, different directions of propagation, and different amplitudes for the two kinds of polarization.

Transforming to Transverse-Traceless Gauge

There is a simple and useful algorithm for taking a plane wave such as (21.64), with all components nonvanishing but satisfying the Lorentz gauge condition and transforming it into TT-gauge. To exhibit it, consider for simplicity a wave propagating in the z-direction so that $\vec{k} = (0, 0, \omega)$. The gauge transformations that effect the conditions (21.66) are of the form $\xi^\alpha(x) \propto \exp[i\omega(z - t)]$ and depend only on z and t. From the expression for the transformation, (21.53), it follows that the components in the x-y submatrix of $h_{\alpha\beta}$ are unchanged. Therefore, to obtain the result of the gauge transformation, first simply set all the components equal to zero that are not in the x-y submatrix. The trace of the remainder will vanish automatically because of the Lorentz gauge condition and can, therefore, be freely subtracted out. The simple algorithm for transforming a general perturbation $h_{\alpha\beta}$ satisfying the wave equation and Lorentz condition to TT-gauge is, therefore, to *set all the nontransverse parts of the metric equal to zero and subtract out the trace from the remaining diagonal elements to make it traceless.* In the example of a wave propagating in the z- direction this would give

$$h^{TT}_{\alpha\beta} = \begin{matrix} \\ t \\ x \\ y \\ z \end{matrix} \begin{matrix} t & x & y & z \\ \left(\begin{matrix} 0 \\ 0 \\ 0 \\ 0 \end{matrix}\right. & \begin{matrix} 0 \\ \frac{1}{2}\left(h_{xx} - h_{yy}\right) \\ h_{xy} \\ 0 \end{matrix} & \begin{matrix} 0 \\ h_{xy} \\ \frac{1}{2}\left(h_{yy} - h_{xx}\right) \\ 0 \end{matrix} & \left.\begin{matrix} 0 \\ 0 \\ 0 \\ 0 \end{matrix}\right) \end{matrix} . \tag{21.70}$$

This is exactly of the form (21.69).

Problems

1. Why do Newtonian tidal gravitational forces outside a mass distribution always squeeze in some directions and expand in others? (*Hint*: The Newtonian gravitational potential satisfies Laplace's equation.)

2. [B, C] *The Shape of the Tides* This problem concerns the shape of the tides raised by the Moon in Newtonian gravity. (See Box 21.1 on p. 448.) Consider the freely falling frame following the center of mass of the Earth in its mutual orbit with the Moon. (Neglect the slower motion of the Earth around the Sun and the rotation of the Earth.) Assume the surface of the solid Earth is a sphere, which is covered with a worldwide ocean.

(a) Explain why the surface of the ocean should be at an equal total gravitational potential.

(b) Find a gravitational potential Φ_{tidal} that will reproduce the tidal gravitational force of the Moon given in (c) in Box 21.1 and the gravitational force of the Earth on an ocean fluid element of mass m according to

$$\vec{F}_{\text{tidal}} = -m\vec{\nabla}\Phi_{\text{tidal}}.$$

(c) Find the difference $\delta h(\theta, \phi)$ between the depth of the ocean in the presence of the Moon and in its absence caused by the tidal gravitational force of the Moon. Use the usual polar angles with the z-axis pointing toward the Moon. Express your answer in terms of the mass of the Earth, the mass of the Moon, the distance between them, and the distance from the center of the Earth to the surface of the ocean were the Moon not present.

(d) Estimate the expected height of the ocean tides from your result in part (c).

(e) Answer the question at the end Box 21.1.

3. [E] A meter stick falls radially into the center of a Newtonian gravitational attraction produced by one solar mass located at a point. *Estimate* the distance from the point at which the meter stick would break or be crushed.

4. Show that if χ is a separation vector obeying the equation of geodesic deviation, then $\chi + C\mathbf{u}$ is another separation vector also obeying the equation of geodesic deviation (21.19), where C is any constant.

5. Fill in the details in the derivation of (21.27) for the Riemann curvature component $R^i{}_{tjt}$ in the Newtonian limit.

6. Derive expression (21.28) for the Riemann curvature in a local inertial frame from its definition (21.20).

7. **(a)** Derive the symmetries (21.29) from the form of the Riemann curvature in a local inertial frame (21.28).

(b) Use these symmetries to show that the Riemann curvature has 20 independent components.

8. [A] Calculate $R_{\hat{t}\hat{r}\hat{t}\hat{r}}$ for the Schwarzschild metric in the frame of the freely falling observer described in (20.81). To do this, first calculate $R_{\alpha\beta\gamma\delta}$ in the Schwarzschild coordinate basis, and then use (21.24) to get the components in the freely falling frame. Does your answer agree with (21.30a)?

9. [C] *Are We Already in a Black Hole?* Measurements of the velocities of galaxies indicate that the Milky Way (our own galaxy) is falling toward the Andromeda galaxy and that these two, together with other members of the local group of galaxies, are falling toward a "great attractor" in the direction of the Hydra-Centaurus supercluster of galaxies. What observations would be necessary to determine whether or not we are already in a black hole falling toward its center? To discuss this question you may assume that the great attractor is spherically symmetric.

10. *A Uniform Gravitational Field* Calculate the Riemann curvature for the metric

$$ds^2 = -(1 + gx)^2 dt^2 + dx^2 + dy^2 + dz^2,$$

thereby showing this spacetime is flat. Find a coordinate transformation that puts this metric into the usual Minkowski form.

11. Problem 8.12 introduced the two-dimensional hyperbolic plane and claimed it has constant negative curvature. Does it? Calculate $R \equiv R^{\alpha}_{\alpha}$ for this two-dimensional geometry to find out.

12. [C, A] **(a)** For the wormhole metric (7.39), calculate the components of the Riemann curvature in an orthonormal basis whose vectors point along the (t, r, θ, ϕ) coordinate axes.

(b) Show that a stationary observer at the wormhole throat feels no tidal gravitational forces.

(c) Show that an observer moving radially through the throat with speed V, as measured by a stationary observer at any point along its trajectory, experiences tidal gravitational forces proportional to V^2.

(d) How do these tidal forces depend on the radius of the throat? What combination of b and V would make for a survivable trip through the wormhole?

13. [S] The Ricci curvature was defined by a particular sum of the components of the Riemann curvature in (21.32). Show that if any pair of indices of $R_{\alpha\beta\gamma\delta}$ are summed using the inverse metric (e.g., $g^{\alpha\delta} R_{\alpha\beta\gamma\delta}$), the result is either zero or a multiple of the Ricci curvature.

14. [S] Starting from its definition (21.33) or from equations (21.32) and (21.28), show that the Ricci curvature $R_{\alpha\beta}$ is symmetric in α and β.

15. Calculate R_{AB} for the metric on the sphere

$$ds^2 = a^2\left(d\theta^2 + \sin^2\theta \, d\varphi^2\right)$$

where A, B, etc., range over 1 and 2, $x^1 = \theta$, $x^2 = \varphi$, and $a =$ constant. This problem can easily be done using the *Mathematica* program for computing curvature on the book website. However, work through this problem by hand to make sure you understand what the *Mathematica* program is doing.

16. Check by hand three of the nonvanishing Christoffel symbols for the Schwarzschild metric given in Appendix B.

17. [A] *The Schwarzschild geometry satisfies the Einstein equation* Insert the Christoffel symbols for the Schwarzschild geometry given in Appendix B into (21.33) and evaluate. You should find $R_{\alpha\beta} = 0$ identically for the each of the ten possible combinations of α and β, thus proving that the Schwarzschild geometry is a solution of the empty-space Einstein equation. For a shorter problem do just the diagonal cases.

18. [C] *Birkhoff's Theorem*

(a) In Example 21.4, invariance under $t \to -t$ was used to exclude a $g_{rt}\, dr\, dt$ from the line element representing the geometry outside a *static* spherically symmetric distribution of stress energy. However, in a dynamic situation such as spherically symmetric collapse, that argument no longer holds. Then the most general spherically symmetric line element is of the form

$$ds^2 = -A(r,t)\, dt^2 + 2B(r,t)\, dr\, dt + C(r,t)\, dr^2 + r^2\left(d\theta^2 + \sin^2\theta \, d\phi^2\right).$$

Show that a transformation of the form

$$t \to t + f(r,t),$$

for some $f(r,t)$, can be used to eliminate the $dr\,dt$ term leaving the most general spherically symmetric metric in the form

$$ds^2 = -e^{\nu(r,t)}\,dt^2 + e^{\lambda(r,t)}\,dr^2 + r^2\left(d\theta^2 + \sin^2\theta\,d\phi^2\right).$$

(b) Using the expressions for the Einstein tensor in Appendix B, show that the equation

$$G_{\hat{r}\hat{t}} = 0$$

implies that $\lambda(r,t)$ is independent of time, and use the remaining components of the Einstein equation to show

$$\nu(r,t) = -\lambda(r) + h(t)$$

for some $h(t)$.

(c) Use these results to conclude that the Schwarzschild geometry is the most general asymptotically flat, spherically symmetric solution of the Einstein equation, dynamic or not. This is called Birkhoff's theorem.

19. [C] *Static Weak Field Metric Derived* This problem shows that the static, weak-field metric (21.25) is the most general such solution of the linearized, vacuum Einstein equation.

(a) For a time-independent *static* source (no velocities) argue that it is possible to choose coordinates so the metric perturbations $h_{\alpha\beta}$ are unchanged by $t \to -t$ and that this means $h_{it} = h_{ti} = 0$.

(b) Show that the residual gauge freedom analogous to that discussed in the subsection "More Gauge" can be used to make h_{ij} diagonal without affecting either $h_{it} = 0$ or the Lorentz gauge condition.

(c) Show that then (21.25) is the unique asymptotically flat solution of the equations of linearized gravity.

20. [S] Calculate the Ricci tensor for the five-dimensional metric (a) in Box 7.3 on p. 157. (*Hint*: No computation is needed.)

21. Carry out the steps leading to expression (21.44) for the perturbation of the Ricci curvature in linearized gravity.

22. [S] Equation (21.40) gives the metric in linearized gravity. Work out the inverse metric to first order in $h_{\alpha\beta}$.

23. Evaluate all components of the linearized Riemann curvature for the gravitational wave metric (16.2a).

24. A linearized gravitational wave in the + polarization is normally incident on a plane containing a circle of test particles such as shown in Figure 16.2. Work out what happens to the test particles when the gravitational wave passes by using the equation of geodesic deviation.

25. Show explicitly using (21.28) and (21.53) that the linearized Riemann curvature found in Problem 23 is invariant under gauge transformations.

26. Consider the metric perturbation:

$$h_{\alpha\beta}(x) = \begin{pmatrix} c & 0 & 0 & 0 \\ 0 & -c & 0 & 0 \\ 0 & 0 & a & 0 \\ 0 & 0 & 0 & -a \end{pmatrix} \sin k(x-t)$$

for arbitrary constants c and a.

(a) Show how this solves the linearized Einstein equation $\delta R_{\alpha\beta} = 0$.

(b) Find the functions $\xi_\alpha(x)$ that transform it into a form where $c = 0$. What is the new value of a?

CHAPTER 22

The Source of Curvature

The previous chapter went part of the way toward introducing the Einstein equation, which describes how the mass-energy of matter curves spacetime:

$$\begin{pmatrix} \text{a measure of local} \\ \text{spacetime curvature} \end{pmatrix} = \begin{pmatrix} \text{a measure of} \\ \text{matter energy density} \end{pmatrix}. \tag{22.1}$$

That discussion focused on the *vacuum* Einstein equation, for which the right-hand side of (22.1) vanishes. The Ricci curvature $R_{\alpha\beta}$ was the measure of curvature on the left-hand side when there are no matter sources on the right. This chapter completes the description of the Einstein equation by finding the correct measure of energy density to go on the right-hand side of (22.1) and the more general measure of spacetime curvature appropriate for the left. A density is a quantity per unit spatial volume, such as rest-mass density, charge density, number density, energy density, etc. The chapter begins by discussing how *densities* are represented in special and general relativity.

22.1 Densities

The flat spacetime of special relativity is the context of this section. The usual rectangular coordinates (t, x, y, z) in which the metric is $g_{\alpha\beta} = \eta_{\alpha\beta} = \text{diag}(-1, 1, 1, 1)$ are used throughout. The discussion aims at the correct relativistic description of the density of energy, which is the density of a component of the energy-momentum four-vector. But we begin with a simpler case, the correct relativistic description of the number density—the density of a scalar.

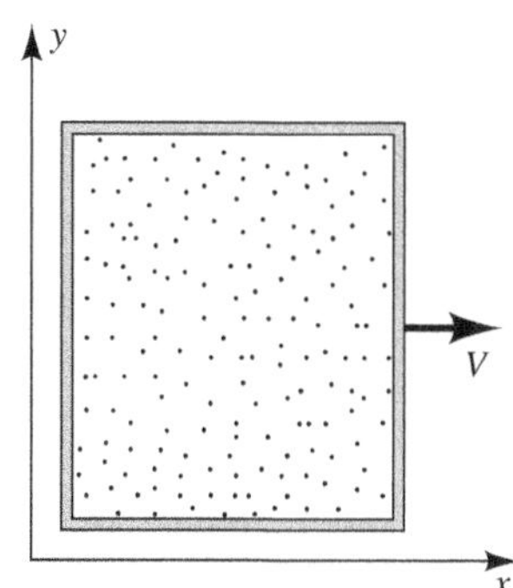

FIGURE 22.1 A box containing $\mathcal{N}$ particles moving with speed V along the x-axis. The number density is higher than in the same box at rest because the box is Lorentz-contracted and its volume is smaller.

Number Density

Consider the situation illustrated in Figure 22.1. A box containing $\mathcal{N}$ particles is moving along one of its dimensions with speed[1] V. In its rest frame the volume of the box is $\mathcal{V}_*$, so the rest number density n of particles inside is[2]

$$n = \mathcal{N}/\mathcal{V}_*. \tag{22.2}$$

[1]This chapter deals with both speed V and three-volume $\mathcal{V}$. Don't mix them up.

[2]It would be consistent with previous usage to put subscript $_*$'s on quantities like n that are defined in the rest frame. But that would be inconsistent with the usual usage in general relativity, which is followed here.

What will be the number density N in the frame in which the box is moving with speed V? In that frame, the length of the box is Lorentz-contracted by a factor $(1-V^2)^{1/2}$ in the direction it is moving. The volume is, therefore, smaller by this same factor:

$$\mathcal{V} = \mathcal{V}_*(1-V^2)^{1/2}. \tag{22.3}$$

The total number of particles inside the box $\mathcal{N}$ is the same in all frames, so the number density N in the frame where the box is moving is

$$N = \frac{\mathcal{N}}{\mathcal{V}} = \frac{n}{\sqrt{1-V^2}}. \tag{22.4}$$

The number density N is thus nu^t, where n is the *rest* number density and u^t is the time component of the four-velocity of the box [cf. (5.28)]. This shows that the number density N is the time component of the four-vector $n\mathbf{u}$. This *number-current four-vector*[3] is[4]

Number Current Four-vector

$$\mathbf{N} = n\mathbf{u}. \tag{22.5}$$

The number-current four-vector $\mathbf{N}$ has components $N^\alpha = (N, \vec{N})$, where the spatial parts are

$$\vec{N} = n\vec{u} = \frac{n\vec{V}}{\sqrt{1-V^2}}. \tag{22.6}$$

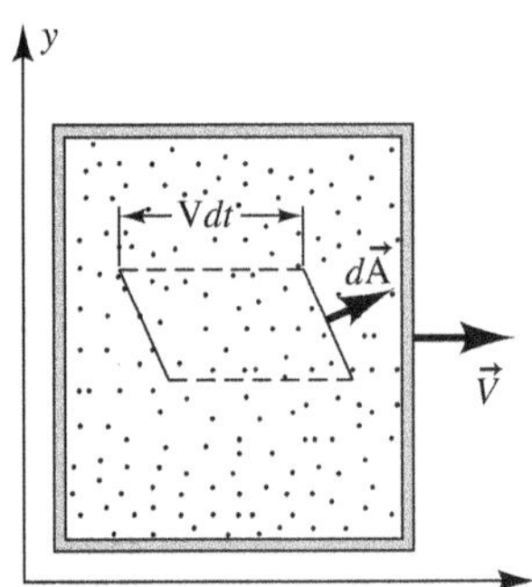

FIGURE 22.2 The number of particles crossing the surface $d\vec{A}$ in time dt is the number of particles in the illustrated tube of length $V\,dt$ and cross-sectional area $d\vec{A}\cdot\vec{V}/V$.

These form the *number current density*. Given an element of area $d\vec{A}$, $\vec{N}\cdot d\vec{A}$ is the number of particles flowing across the area per unit time. To see that, look at Figure 22.2. The number of particles flowing across the area $d\vec{A}$ is the same as the number in a volume with a cross sectional area $d\vec{A}\cdot\vec{V}/V$ and a length $V\,dt$. That is, $\vec{N}\cdot d\vec{A}$.

By taking the volume $\mathcal{V}_*$ to be very small, the number density and number current density can be defined at a point in spacetime—$N(x)$ and $\vec{N}(x)$. In general, these will vary from point to point in spacetime, but they must vary so that the number of particles is conserved:

$$\frac{\partial N}{\partial t} + \nabla\cdot\vec{N} = 0. \tag{22.7}$$

To see that (22.7) represents the conservation of the number of particles, integrate it over any volume of space $\mathcal{V}$ and use the divergence theorem to find

$$\frac{d}{dt}\int_{\mathcal{V}} N\,d^3x + \int_{\partial\mathcal{V}} \vec{N}\cdot d\vec{A} = 0. \tag{22.8}$$

[3] More accurately but more clumsily it is the number-(number-current) four-vector.

[4] Various forms of the letter "N" are employed for several different quantities in this chapter: n for number density in the rest frame, $\vec{n}$ for the normal to a two-surface, N for number density in a general frame, $\vec{N}$ for number current, $\mathbf{n}$ for the normal to a three-surface, $\mathbf{N}$ for number-current four-vector, and $\mathcal{N}$ for total number. It's important not to get these mixed up.

The first integral is the total number of particles in the volume $\mathcal{V}$. The second integral is over the surface $\partial\mathcal{V}$ bounding the volume and is the rate at which particles cross the surface—the number flux. Equation (22.8) says that the time rate of change of the number of particles inside is minus the net rate at which particles flow out through the surface. That's conservation.

The conservation of number (22.7) can be expressed elegantly in terms of the number-current four-vector $N^\alpha = (N, \vec{N})$ as

$$\boxed{\frac{\partial N^\alpha}{\partial x^\alpha} = 0.} \tag{22.9}$$

Number Conservation

Here the usual sum over repeated indices and rectangular coordinates are assumed.

The lesson of this discussion is that densities of scalar quantities such as number density are the time components of a four-vector whose spatial components are the corresponding current density. There is a more geometrical way of seeing why this is the case. A density associates a scalar with an element of three-dimensional volume. A three-dimensional volume is a *three-surface* in four-dimensional space, as discussed in Section 7.9 (cf. Figure 7.8). The orientation of that surface in spacetime is specified by a normal four-vector $\mathbf{n}$, so that an element of three-volume is $\mathbf{n}\Delta\mathcal{V}$. To get a scalar quantity associated with the volume, a four-vector current is needed to form a scalar product with the normal vector. For example, the number of particles $\Delta\mathcal{N}$ in the three-volume $\mathbf{n}\Delta\mathcal{V}$ is

$$\Delta\mathcal{N} = \mathbf{N}\cdot(\mathbf{n}\Delta\mathcal{V}) = N^\alpha n_\alpha \Delta\mathcal{V}. \tag{22.10}$$

This relation shows that one can think of a spatial density in four-dimensional terms as a *flux* of the number four-current through an element of spacelike three-surface. Densities are fluxes in timelike directions through spacelike three-surfaces; currents are fluxes in spacelike directions through timelike three-surfaces.

Example 22.1. The Charge-Current Four-Vector. Number density illustrates the idea of the density of a scalar, but the discussion applies equally well to other scalar quantities. Electric charge is a useful example. Charge density ρ_{elec} and electric current density $\vec{J}_{\text{elec}}$ together make up a four-vector current density

$$J^\alpha = \left(\rho, \vec{J}_{\text{elec}}\right). \tag{22.11}$$

Conservation of charge is expressed by

$$\frac{\partial J^\alpha}{\partial x^\alpha} = \frac{\partial \rho_{\text{elec}}}{\partial t} + \vec{\nabla}\cdot\vec{J}_{\text{elec}} = 0 \tag{22.12}$$

analogous to (22.9). The charge inside a volume $\mathbf{n}\Delta\mathcal{V}$ is given by the analog of (22.10).

Densities of Energy and Momentum

The densities of energy and momentum are the sources of spacetime curvature that occur on the right-hand side of the Einstein equation (22.1). Equation (22.10) shows how a density-current four-vector is needed to associate a scalar quantity with a volume $n_\alpha \Delta\mathcal{V}$. But energy and momentum are not scalars. Rather, they are different components of the energy-momentum four-vector p^α. To associate a four-vector Δp^α with a three-volume $n_\alpha \Delta\mathcal{V}$, an object with two indices $T^{\alpha\beta}$ is needed so that we can write

$$\Delta p^\alpha = T^{\alpha\beta} n_\beta \Delta\mathcal{V}. \tag{22.13}$$

The quantity $T^{\alpha\beta}$ is a second-rank tensor called the *energy-momentum-stress tensor* (accurate but too long), or sometimes the *energy-momentum tensor* (could be confused with energy-momentum four-vector), or sometimes (as here) just the *stress-energy tensor* (gets most of the components).

To understand what the components of the stress-energy tensor are, consider a particular inertial frame in flat spacetime and a three-dimensional volume $\Delta\mathcal{V}$ at rest in that frame. That volume is part of a $t =$ const. three-surface in spacetime whose normal can be chosen to be $n_\alpha = (1, 0, 0, 0)$ [cf. (20.30)]. With that choice of normal, (22.13) becomes

$$\Delta p^\alpha = T^{\alpha t} \Delta\mathcal{V}. \tag{22.14}$$

The energy density is $\epsilon \equiv \Delta p^t/\Delta\mathcal{V} = T^{tt}$, and the momentum density is $\vec{\pi} \equiv \Delta p^i/\Delta\mathcal{V} = T^{it}$. Thus we understand the significance of four of the components of the stress-energy tensor:

$$T^{tt} = (\text{energy density}) \equiv \epsilon, \tag{22.15a}$$

$$T^{it} = \begin{pmatrix}\text{momentum density}\\ \text{in direction } i\end{pmatrix} \equiv \pi^i, \tag{22.15b}$$

each as would be measured by an observer at rest in the inertial frame under discussion.

A simple illustration of the stress-energy tensor is provided by the moving box of particles in Figure 22.1. Suppose that the particles inside are all at rest with respect to the box and have rest mass m. In the inertial frame where the box is moving with speed V, the energy of each particle in the box is $m\gamma$, where $\gamma = (1 - V^2)^{-1/2}$. The *energy density* is the number density (22.4) times this energy. Thus,

$$\epsilon \equiv T^{tt} = mn\,\gamma^2 = mnu^t u^t, \tag{22.16a}$$

and similarly the momentum density is

$$\pi^i \equiv T^{it} = mn\,\gamma^2 V^i = mnu^i u^t. \tag{22.16b}$$

From (22.16) it is easy to guess that the expression for the stress-energy tensor of the particles inside the box is

$$T^{\alpha\beta} = mnu^\alpha u^\beta \equiv \mu u^\alpha u^\beta, \tag{22.17}$$

where $u^\alpha = (\gamma, \gamma \vec{V})$ is the four-velocity of the box [cf. (5.28)] and $\mu \equiv mn$ is the rest-mass density. Equation (22.17) illustrates an important property of stress-energy tensors in general. They are symmetric: $T^{\alpha\beta} = T^{\beta\alpha}$.

What is the meaning of the $T^{\alpha j}$ components of the energy-momentum tensor? The answer can't be found by looking at a spacelike three-surface of constant t because these components don't enter (22.14). The answer can be found by considering a timelike three-surface. Consider a three-volume spanned by coordinate intervals Δy, Δz, and Δt. The unit normal to this three-surface pointing in the x-direction is $n_\alpha = (0, 1, 0, 0)$. The analog of (22.14) is then

$$\Delta p^\alpha = T^{\alpha x} \Delta y \Delta z \Delta t. \tag{22.18}$$

The time component of this equation gives

$$T^{tx} = \frac{\Delta p^t}{\Delta A \Delta t}, \tag{22.19}$$

where ΔA is the area $\Delta y \Delta z$. The component T^{tx} is thus the flux of energy in the x-direction. A flux of energy is the same thing as a momentum density so that generally $T^{tx} = T^{xt}$. A simple example is a box of particles already discussed (Figure 22.2). The amount of energy that crosses a surface with area dA extending in the y-and z-directions in a time dt is (energy flux) $dA\ dt$ = (energy density)$V\ dA\ dt$ = (momentum density) $dA\ dt$ since for each of the particles in the box, (energy)$V = m\gamma\, V$ = (momentum). The symmetry $T^{tx} = T^{xt}$ is explicit in (22.17).

The spatial parts of (22.18) can be written in the revealing way

$$T^{ix} = \frac{\Delta p^i / \Delta t}{\Delta A}. \tag{22.20}$$

The numerator $\Delta p^i / \Delta t$—a rate of change of momentum—is a force. Equation (22.20) says that T^{ix} is the ith component of the force per unit area exerted across a surface whose normal lies in the x-direction. More generally we can write for the components of the force $\vec{F}$ exerted across an area ΔA with normal $\vec{n}$:

$$\Delta F^i = T^{ij} n_j \Delta A. \tag{22.21}$$

Thus,

$$T^{ij} = \begin{pmatrix} i\text{th component of the force per unit} \\ \text{area exerted across a surface with} \\ \text{normal in direction } j \end{pmatrix}. \tag{22.22}$$

Stress Tensor

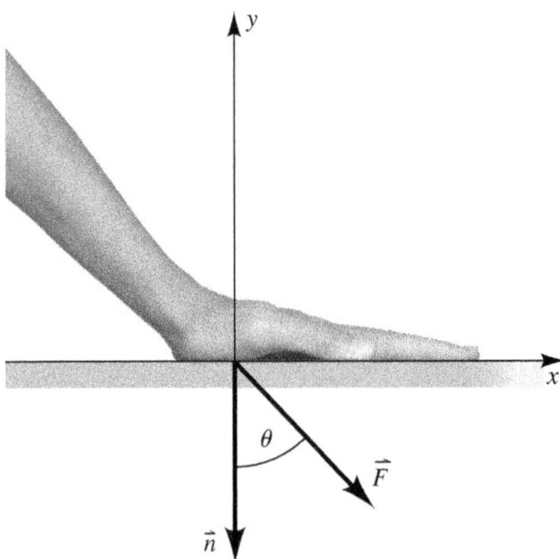

FIGURE 22.3 Push on a table with your hand, exerting a force $\vec{F}$ that makes an angle θ with the direction normal to the surface. You are exerting a *stress* across the surface of the table described by a stress tensor T^{ij}. As explained in Example 22.2, since the normal points in the $-y$-direction, $-T^{xy}A$ is the force you exert in the x-direction and $-T^{yy}A$ is the component of force you exert in the y-direction, where A is the area of your palm.

In classical mechanics a force per unit area is called a *stress*, and T^{ij} is the *stress tensor.*

Example 22.2. Pushing on a Table. You push on a table with a force $\vec{F}$, making an angle θ with the vertical (Figure 22.3). What are the components of the stress you are exerting, assuming it is evenly distributed over your palm of area A?

Equation (22.21) implies

$$F\sin(\theta) = F^x = T^{xj}n_j A = -T^{xy}A, \tag{22.23a}$$

$$-F\cos(\theta) = F^y = T^{yj}n_j A = -T^{yy}A. \tag{22.23b}$$

Thus,

$$T^{xy} = -(F/A)\sin(\theta), \qquad T^{yy} = +(F/A)\cos(\theta) \tag{22.24}$$

are the only relevant components of the stress.

Example 22.3. Pressure. Pressure in a fluid is the simplest example of a stress—a force per unit area. In a fluid at rest, the force exerted across a surface is always along its normal and the same for all orientations. The stress tensor is, therefore, diagonal, with all the diagonal values equal to the pressure p:

$$T^{ij} = p\delta^{ij}, \tag{22.25}$$

so that (22.21) becomes

$$\Delta F^i = pn^i \Delta A. \tag{22.26}$$

Equations (22.15), (22.19), and (22.22) complete our understanding of the components of the energy-momentum tensor. In summary, in the inertial frame under discussion, the components of $T^{\alpha\beta}$ are

$$T^{\alpha\beta} = \left(\begin{array}{c|c} \text{energy density} & \text{energy flux} \\ \hline \text{mom. density} & \text{stress tensor} \end{array}\right). \tag{22.27}$$

Components of the Stress-Energy Tensor

The stress-energy tensor is symmetric, $T^{\alpha\beta} = T^{\beta\alpha}$. Energy flux is the same thing as momentum density, as discussed earlier. The symmetry of the stress-tensor was exhibited in the specific case of a box of gas [cf. (22.17)], and can be demonstrated more generally in a variety of ways (see Problems 4 and 5). The stress-energy tensor $T^{\alpha\beta}$ is the correct relativistic description of energy density to go on the right-hand side of the Einstein equation (22.1).

Example 22.4. Energy Density Measured by an Observer. What matter energy density is measured by an observer moving through spacetime with a four-velocity $\mathbf{u}_{\text{obs}}$? To answer this question consider a little volume $\Delta\mathcal{V}$ in the observer's rest frame. That volume is an element of spacelike three-surface (Section 7.9) with normal $-\mathbf{u}_{\text{obs}}$ if we use the same convention for its direction as in (22.14). The energy-momentum four-vector of the matter contained in that volume is, from (22.13) (with some indices raised and lowered),

$$\Delta p_\alpha = T_{\alpha\beta}(-u^\beta_{\text{obs}})\Delta\mathcal{V}. \tag{22.28}$$

The energy ΔE in the volume measured by the observer is [cf. (7.53)]

$$\Delta E = -\Delta \mathbf{p} \cdot \mathbf{u}_{\text{obs}} = -\Delta p_\alpha u^\alpha_{\text{obs}} = T_{\alpha\beta} u^\alpha_{\text{obs}} u^\beta_{\text{obs}} \Delta\mathcal{V}. \tag{22.29}$$

The energy density measured by the observer is thus

$$\left(\begin{array}{c}\text{energy density measured by} \\ \text{an observer with four-velocity } \mathbf{u}_{\text{obs}}\end{array}\right) = T_{\alpha\beta} u^\alpha_{\text{obs}} u^\beta_{\text{obs}}. \tag{22.30}$$

This is just a special case of the general statement that observations made by an observer are components in the orthonormal basis associated with the observer's laboratory (Sections 5.6 and 7.8). In the present example the measured energy density is $T_{\hat{0}\hat{0}}$, where $\mathbf{e}_{\hat{0}} = \mathbf{u}_{\text{obs}}$, and (22.30) is a particular case of (20.41).

22.2 Conservation of Energy-Momentum

Conservation of Energy-Momentum in Flat Spacetime

In flat space, energy of matter is conserved; momentum of matter is conserved. These conservation laws can be expressed, like the conservation of number (22.7) and charge (22.12), in terms of the components of the stress-energy. Since each *component* of the energy-momentum four-vector is conserved, the conservation laws for energy and momentum take the form

Conservation of Energy and Momentum in Flat Space

$$\frac{\partial T^{\alpha\beta}}{\partial x^\beta} = 0 \qquad \left(\begin{array}{c}\text{flat}\\ \text{spacetime}\end{array}\right). \tag{22.31}$$

The free index α means that (22.31) represents four separate conservation laws—one for energy and three for the components of momentum. Let's write out these four relations separately.

Take the t component of (22.31). Using the identifications in (22.27), this is

$$\frac{\partial \epsilon}{\partial t} + \vec{\nabla} \cdot \vec{\pi} = 0, \tag{22.32}$$

where ϵ is the energy density and $\vec{\pi}$ is the energy flux or current. As discussed earlier, this is the same thing as momentum density. Integrated over a small spatial volume as in (22.8), the first term in (22.32) is the rate of change of energy inside. The second term is the flux of energy out of the volume. Equation (22.32) shows that these are opposite, meaning energy is conserved.

The three spatial components of (22.31) read

$$\frac{\partial \pi^i}{\partial t} = -\frac{\partial T^{ij}}{\partial x^j} \equiv \phi^i. \tag{22.33}$$

A time rate of change of momentum is a force. Equation (22.33) defines a component of a *force density* that we have denoted ϕ^i. Equation (22.33) is, therefore, the equation of motion for the fluid—an expression of $\vec{F} = m\vec{a}$ for a continuum. Integrating ϕ^i over a small fixed spatial volume gives the force acting on that volume, and (22.33) shows the connection to the stresses acting on it. That connection can be made clearer by using the divergence theorem to write the total force F^i acting *on* a volume $\mathcal{V}$ as

$$F^i = \int_{\mathcal{V}} d^3x\, \phi^i = -\int_{\mathcal{V}} d^3x \frac{\partial T^{ij}}{\partial x^j} = -\int_{\partial\mathcal{V}} dA\, n_j^{(\text{out})} T^{ij} = \int_{\partial\mathcal{V}} dA\, n_j^{(\text{in})} T^{ij}, \tag{22.34}$$

where $\partial\mathcal{V}$ is the boundary of $\mathcal{V}$ and $\vec{n}^{(\text{out})}$ and $\vec{n}^{(\text{in})}$ are the outward- and inward-pointing normals to $\partial\mathcal{V}$, respectively. The term on the right is the sum of all the forces exerted across the bounding surface from the outside to the inside. That is equal to the total force on the volume, as the equation shows.

Example 22.5. Pressure Force on a Volume. The simple example of fluid pressure illustrates (22.33). In a fluid $T^{ij} = \delta^{ij} p$, where p is the pressure [cf. (22.25)]. Consider a cube of fluid with sides of length L oriented along the x-, y-, z-axes and one corner at position (x, y, z). The pressure force exerted on the face in the y-z plane at x will be $p(x, y, z)L^2$ in the x-direction. The force across the other y-z face at $x + L$ will be $p(x + L, y, z)L^2$ in the negative x-direction. The net force is

$$\begin{aligned} F^x &= L^2[p(x, y, z) - p(x + L, y, z)] \\ &\approx -L^3 \frac{\partial p}{\partial x} \end{aligned} \tag{22.35}$$

for small L. Since the force density is $\phi^x = F^x/L^3$, this is exactly (22.33) when $T^{ij} = \delta^{ij} p$.

Perfect Fluids

A simple example of a stress-energy tensor is that for a *perfect fluid*. As seen in (22.25) the stress tensor T^{ij} for a fluid is diagonal, with equal pressures for the diagonal elements. A fluid is said to be perfect when heat conduction, viscosity, or other transport or dissipative processes are negligible. In an inertial frame where it is at rest, a perfect fluid is characterized by its energy density ρ, pressure p, and its stress-energy is

$$T^{\alpha\beta} = \mathrm{diag}(\rho, p, p, p). \tag{22.36}$$

(The symbol ϵ was used for the energy density of anything in any frame [cf. (22.15a)], but the symbol ρ is reserved for the energy-density of a fluid in its rest frame.) A pressure is a force per unit area with usual units $(\mathcal{ML}/\mathcal{T}^2)/\mathcal{L}^2$, which are the same as those of energy density $(\mathcal{ML}^2/\mathcal{T}^2)/\mathcal{L}^3$. In geometrized units that are often convenient, where mass is measured in units of length, both energy density and pressure have units of $1/\mathcal{L}^2$. Typically, pressure and density are related by an equation of state. For example, in Section 18.3 the gas of galaxies was modeled by a fluid with $p = 0$. The cosmic microwave background radiation was modeled by a perfect fluid with $p = \rho/3$ [cf. (18.23)].

Typically a fluid is not at rest but flowing with a four-velocity $\mathbf{u}(x)$ differing from one point to the next. The stress-energy tensor will, therefore, depend on $\mathbf{u}(x)$ as well as $\rho(x)$ and $p(x)$. The most general possible form that can be constructed from $\mathbf{u}$ and the metric $\eta^{\alpha\beta}$ without involving derivatives is $T^{\alpha\beta} = Au^\alpha u^\beta + B\eta^{\alpha\beta}$. The coefficients A and B are determined by the requirement that the stress-energy must reduce to (22.36) in the frame of an observer at rest with respect to the fluid where $u^\alpha = (1, \vec{0})$. This implies

$$T^{\alpha\beta} = (\rho + p)u^\alpha u^\beta + \eta^{\alpha\beta} p. \tag{22.37}$$

Perfect Fluid Stress-Energy

We use this perfect fluid stress-energy to model matter in diverse situations—inside neutron stars, sources of gravitational radiation, the gas of galaxies, the cosmic background radiation, and vacuum energy. The stress-energy (22.17) is an example of a perfect fluid where the pressure vanishes—called *dust* in general relativity.

Example 22.6. The Stress-Energy of the Vacuum. In Section 18.3 the idea of a vacuum energy, ρ_{vac}, constant in both space and time was discussed. Consistency with the first law of thermodynamics required a negative vacuum pressure, $p_{\text{vac}} = -\rho_{\text{vac}}$, at least for the homogeneous, isotropic cosmological models under consideration. From (22.37) this implies that the form of the vacuum stress-energy tensor is

Vacuum Stress-Energy

$$T^{\alpha\beta}_{\text{vac}} = -\rho_{\text{vac}}\eta^{\alpha\beta} = -\frac{\Lambda}{8\pi G}\eta^{\alpha\beta}, \tag{22.38}$$

where Λ is the *cosmological constant.* Indeed, that is the only possible form of stress-energy tensor that depends only on a constant. In curved spacetime this becomes $T^{\alpha\beta}_{\text{vac}} = -(\Lambda/8\pi G)g^{\alpha\beta}$.

Local Conservation of Energy-Momentum in Curved Spacetime

The discussion of stress-energy in the previous section has been in the context of the flat spacetime of special relativity. But the stress-energy is to serve as a source of spacetime curvature in the Einstein equation. A straightforward but not unique way of generalizing a flat spacetime stress-energy tensor to curved spacetime is to replace the flat metric by $g_{\alpha\beta}$. For example, for a perfect fluid the obvious generalization is

Perfect Fluid Stress-Energy

$$\boxed{T^{\alpha\beta} = (\rho + p)u^\alpha u^\beta + g^{\alpha\beta}p.} \tag{22.39}$$

This reduces to (22.37) in a local inertial frame.

However, the conservation equation (22.31) is no longer satisfied, nor should it be. What is satisfied is the natural generalization of (22.31) to curved spacetime:

Local Conservation of Energy Momentum

$$\boxed{\nabla_\beta T^{\alpha\beta} = 0,} \tag{22.40}$$

where ∇_β is the covariant derivative. This relation is called the *local conservation of energy-momentum* because it reduces to the conservation law (22.31) in a local inertial frame. However, it is not a conservation law like (22.31), nor should it be. The energy of matter is not conserved in the presence of dynamic spacetime curvature but changes in response to it. The most familiar example is the cosmic microwave background radiation. As the universe expands, the energy density and temperature of the radiation *decrease*. Equation (22.40) describes how they decrease, as the following example illustrates.

Example 22.7. $\nabla_\alpha T^{\alpha\beta} = 0$ in a Homogeneous, Isotropic Cosmology. The geometry of a homogeneous, isotropic, spatially flat cosmological model is summarized by the line element [cf. (18.1)]

$$ds^2 = -dt^2 + a(t)^2(dx^2 + dy^2 + dz^2), \tag{22.41}$$

where $a(t)$ is the scale factor whose time dependence describes the expansion of the universe. The source of curvature is a homogeneous, isotropic perfect fluid of matter, radiation, and vacuum energy whose stress-energy $T^{\alpha\beta}$ is given by (22.39). (This kind of model was discussed in detail in Chapter 18, but the point being made in this example can be appreciated without that discussion.)

Consistent with homogeneity, the energy density $\rho(t)$ and pressure $p(t)$ are functions only of time. Consistent with homogeneity and isotropy, the four-velocity of the fluid has only a time component, $u^\alpha = (1, \vec{0})$. Thus, the nonvanishing components of $T^{\alpha\beta}$ are, from (22.39),

$$T^{tt} = \rho(t), \quad T^{ij} = g^{ij} p(t) = \delta^{ij}[\, p(t)/a(t)^2]. \tag{22.42}$$

The t component of the local conservation equation (22.40) can be written out using (20.65) as

$$\nabla_\beta T^{t\beta} = \frac{\partial T^{t\beta}}{\partial x^\beta} + \Gamma^t_{\beta\gamma} T^{\gamma\beta} + \Gamma^\beta_{\beta\gamma} T^{t\gamma} = 0. \tag{22.43}$$

The relevant Christoffel symbols are easily calculated directly form (22.41) and are listed in Appendix B. The only nonvanishing Γ's are

$$\Gamma^t_{ij} = a\dot{a}\,\delta_{ij}, \qquad \Gamma^i_{jt} = \frac{\dot{a}}{a}\delta^i_j, \tag{22.44}$$

where a dot denotes differentiation with respect to t. The three terms in the local conservation equation (22.43) become

$$\dot{\rho} + 3\frac{\dot{a}}{a}p + 3\frac{\dot{a}}{a}\rho = 0, \tag{22.45}$$

or, equivalently,

$$\frac{d}{dt}\left(\rho a^3\right) = -p\frac{da^3}{dt}. \tag{22.46}$$

Consider the fluid in a small coordinate volume $\Delta\mathcal{V}_{\text{coord}} = \Delta x \Delta y \Delta z$ that occupies a physical volume $\Delta\mathcal{V} = a^3(t)\Delta\mathcal{V}_{\text{coord}}$, which increases over time because of the expansion of the universe that increases $a(t)$. The energy in that volume is

$\Delta E(t) = \rho(t)\Delta\mathcal{V}(t)$. Multiplying (22.46) by $\Delta\mathcal{V}_{\text{coord}}\,dt$, it becomes

$$d(\Delta E(t)) = -p(t)d(\Delta\mathcal{V}(t)). \tag{22.47}$$

This is the first law of thermodynamics expressing how the energy of an element of fluid decreases by the work done against pressure while expanding. (For a more thorough discussion see p. 372.)

22.3 The Einstein Equation

The Einstein equation relating curvature to density of mass-energy is a fundamental equation of classical physics. It cannot be derived, for there is no more fundamental classical theory to derive it from. However, its form can be motivated by a few arguments, which are given in this section.

The necessary ingredients to form the Einstein equation are already in hand. On the right-hand side of (22.1) is the measure of matter energy density—the stress-energy $T_{\alpha\beta}$. On the left-hand side is a measure of curvature. The Ricci curvature $R_{\alpha\beta}$ is one such measure. But another rank-two symmetric tensor that can be formed from it and the metric is $g_{\alpha\beta}R$, where

$$R = R^{\gamma}{}_{\gamma} = g^{\gamma\delta}R_{\gamma\delta} \tag{22.48}$$

is called the *Ricci curvature scalar*. Thus, a candidate for the relation between curvature and stress-energy is

$$R_{\alpha\beta} + \lambda\, g_{\alpha\beta}R = \kappa\, T_{\alpha\beta} \tag{22.49}$$

for as yet undetermined constants κ and λ.

Consistency with the local conservation law (22.40) determines the combination of $R_{\alpha\beta}$ and $g_{\alpha\beta}R$ that stands on the left-hand side of the Einstein equation. Applying ∇^{β} to the right-hand side of (22.49) gives zero. Applying it to the left-hand side of (22.49) therefore must also give zero. This is true for one and only one combination of $R_{\alpha\beta}$ and $g_{\alpha\beta}R$. The particular combination follows from the *Bianchi identity*:

Bianchi Identities

$$\nabla_{\beta}(R^{\alpha\beta} - \tfrac{1}{2}g^{\alpha\beta}R) = 0. \tag{22.50}$$

These four relations are satisfied for any metric $g_{\alpha\beta}$ one cares to choose. Their validity can be established by working through Problem 13 or accepted as a result of differential geometry. The only left-hand side of the Einstein equation consistent with local conservation of the right-hand side is, therefore, $R_{\alpha\beta} - (1/2)g_{\alpha\beta}R$ giving $\lambda = -\frac{1}{2}$ in (22.49).

Still undetermined is the value of κ. This must be proportional to the gravitational coupling constant G. The Newtonian limit fixes its precise value at $8\pi G$ (in the units where $c = 1$ used throughout this chapter), as will be seen in the next section.

The Einstein equation is thus

$$R_{\alpha\beta} - \tfrac{1}{2} g_{\alpha\beta} R = 8\pi G T_{\alpha\beta}. \tag{22.51}$$

Einstein Equation

In $c \neq 1$ units the factor $8\pi G$ is replaced by $(8\pi G)/c^4$. In geometrized units, where mass is measured in units of length and $G = 1$, it is just 8π.

It is conventional to define the *Einstein curvature tensor* by

$$G_{\alpha\beta} \equiv R_{\alpha\beta} - \tfrac{1}{2} g_{\alpha\beta} R. \tag{22.52}$$

Einstein Curvature

Then the Einstein equation can be written in the shorter form:

$$G_{\alpha\beta} = 8\pi G T_{\alpha\beta} \tag{22.53}$$

or in the even shorter form:

$$\mathbf{G} = 8\pi G \mathbf{T}. \tag{22.54}$$

The astute reader may have noticed that there is one other term that could be added to the left-hand side of the Einstein equation consistent with local conservation of $T_{\alpha\beta}$. This is a term of the form $\Lambda g_{\alpha\beta}$ for some constant Λ. Adding it to the left-hand side doesn't affect local conservation because the covariant derivative of the metric is zero [cf. (20.70)]. Indeed, Einstein did just this when he introduced such a term, calling Λ the cosmological constant. However, the modern practice is to identify this term with the stress-energy of the vacuum (if any) and include it on the right-hand side as a contribution to the stress-energy tensor of the form $T^{\text{vac}}_{\alpha\beta} = -(\Lambda/8\pi G) g_{\alpha\beta}$ as in Example 22.6. That is how it was treated in the chapters on cosmology [cf. (18.28)], and that is how it will be treated here.

The Einstein equation reduces to $R_{\alpha\beta} = 0$ when $T_{\alpha\beta} = 0$. To see this, put $T_{\alpha\beta} = 0$ in (22.51) and multiply it by $g^{\alpha\beta}$. Use (22.48) and the definition of the inverse metric (20.12) to find the result $R = 0$. The Einstein equation is then just

$$R_{\alpha\beta} = 0 \quad \text{when } T_{\alpha\beta} = 0, \tag{22.55}$$

which is the vacuum Einstein equation (21.34) used in the previous chapter.

The Einstein equation relates the Ricci curvature of spacetime to the stress-energy of matter. Its components are ten partial differential equations for the metric coefficients $g_{\alpha\beta}(x)$ given the matter sources $T_{\alpha\beta}(x)$. They are analogous to Maxwell's equations, which determine the electromagnetic potentials given the charge and current densities. Unlike Maxwell's equations, the differential equations of Einstein's theory are nonlinear. Nonlinearity makes them much more difficult to solve than Maxwell's equations.

The ten equations are not all independent. They are related by the four Bianchi identities (22.50). Thus, there are only six independent equations. This is the correct number because the metric can be changed by transformations of the four coordinates. Only six metric functions or combinations should be determined by the basic field equation for gravity, and six is exactly the number of independent differential relations in the Einstein equation.

Example 22.8. The Einstein Equation for Homogeneous, Isotropic Cosmological Models. The Robertson–Walker homogeneous isotropic cosmological models discussed in Chapter 18 provide the simplest realistic example of writing out the components of the Einstein equation.

The metric of the Robertson–Walker models is [cf. (18.62)]

$$ds^2 = -dt^2 + a^2(t)\left[\frac{dr^2}{1-kr^2} + r^2\left(d\theta^2 + \sin^2\theta\, d\phi^2\right)\right], \tag{22.56}$$

where $k = \pm 1, 0$ is a constant prescribing the spatial curvature. The stress-energy of the cosmological perfect fluid of matter, radiation, and vacuum was described in Example 22.7.

The components of the Einstein tensor for the metric (22.56) can be worked out using the *Mathematica* notebook *Curvature and the Einstein Equation* on the book website and are listed in Appendix B.

The results are most elegantly expressed in an orthonormal basis $\{\mathbf{e}_{\hat{\alpha}}\}$ of an observer moving with the fluid, where $\mathbf{e}_{\hat{0}} = \mathbf{u}$ and the other three basis vectors are oriented along the directions of the (r, θ, ϕ) coordinate lines. In this basis the components of the stress-energy tensor are, by definition, just those in (22.36)—$T_{\hat{\alpha}\hat{\beta}} = \mathrm{diag}(\rho, p, p, p)$. The coordinate basis components of the Einstein tensor can be projected into this orthonormal basis by working out the components of the basis vectors as in Example 7.9 and carrying out the projection as in (20.41). The result is

$$G_{\hat{t}\hat{t}} = \frac{3}{a^2}\left(k + \dot{a}^2\right) = 8\pi\rho, \tag{22.57a}$$

$$G_{\hat{r}\hat{r}} = G_{\hat{\theta}\hat{\theta}} = G_{\hat{\phi}\hat{\phi}} = -\left[2\frac{\ddot{a}}{a} + \frac{1}{a^2}\left(k + \dot{a}^2\right)\right] = 8\pi p. \tag{22.57b}$$

Here a dot means derivative with respect to t. All other components of the Einstein equation vanish identically.

The first of these, (22.57a), is the Friedman equation (18.63), from which we derived the properties of the FRW cosmological models in Chapter 18. What of the rest? They are consequences of (22.57a) and the first law of thermodynamics (22.46). To see this, multiply (22.57a) by a^3 and differentiate with respect to time. Evaluate $d(\rho a^3)/dt$ with (22.46) to get (22.57b). Alternatively, we may say that the Einstein equation *implies* the first law of thermodynamics (22.46). Equations (22.57a) and (22.46) are just the two equations used throughout Chapter 18.

It is clear from any of the equations (22.57) that the big bang of the FRW models at $a = 0$ is a singularity not only in pressure and density, but in the curvature of spacetime as well.

22.4 The Newtonian Limit

General relativity must reproduce the inverse square law of Newtonian gravity in the limit of small spacetime curvature produced by matter sources having velocities small compared to the velocity of light. Put differently, the Einstein equation must reduce to the Newtonian field equation (3.18) in this limit.

The conclusion of Example 21.5 was that the *vacuum* Einstein equation reduces to the vacuum Newtonian equation $\nabla^2\Phi = 0$ [cf. (21.48)] for the static, weak-field metric [cf. (21.25)]:

$$ds^2 = -(1+2\Phi)dt^2 + (1-2\Phi)(dx^2 + dy^2 + dz^2). \tag{22.58}$$

(Here, as throughout this chapter, $c = 1$ units are used.) Let's see what happens when the Einstein equation with sources (22.53) is evaluated in the same linear approximation, with the same metric, with the stress-energy $T_{\alpha\beta}$ of nonrelativistic matter.

Rest energy dominates the stress-energy of nonrelativistic matter in a frame where the matter is moving with typical velocities, V, that are small compared to the velocity of light. We are assuming that the rest energy density μ is small, so that it produces only a slight spacetime curvature consistent with (22.58) with a small Φ. But we also assume that the kinetic energy proportional to μV^2 and potential energy proportional to $\mu\Phi$ are smaller still and negligible in comparison.[5] The $T^{\alpha\beta}$ of (22.17) with $u^\alpha \approx (1, \vec{0})$ is, therefore, a good first approximation to the stress-energy of nonrelativistic matter. The only significant component of $T^{\alpha\beta}$ is

$$T^{tt} = \mu + (\text{terms of order } \mu\Phi \text{ and } \mu V^2). \tag{22.59}$$

All other components of $T^{\alpha\beta}$ are of order μV^2 at largest and negligible. Since $g_{tt} = -1$ in leading order, the leading approximation to $T_{\alpha\beta}$ will be $T_{tt} \approx \mu$, with other components being negligible. This stress-energy is the right-hand side of the Einstein equation (22.53).

Using the *Mathematica* program on the book website, the Einstein tensor, $G_{\alpha\beta}$, can be evaluated for the Newtonian metric (22.58) to first order in the small values of Φ. The result, which is listed in Appendix B, is

$$G_{tt} = 2\nabla^2\Phi + (\text{terms of order } \Phi^2), \tag{22.60}$$

with all other components of $G_{\alpha\beta}$ of order Φ^2.

[5] Another way to see this is to put back in the factors of c so that V is replaced by V/c and Φ by Φ/c^2.

TABLE 22.1 Newtonian Gravity and General Relativity Compared

	Newtonian Gravity	General Relativity
What mass does	Produces a field Φ causing a force on other masses $\vec{F} = -m\vec{\nabla}\Phi$	Curves spacetime $ds^2 = g_{\alpha\beta}(x)dx^\alpha dx^\beta$
Motion of a particle	Newton's law of motion $\frac{d^2x^i}{dt^2} = -\delta^{ij}\frac{\partial\Phi}{\partial x^j}$	Geodesic equation $\frac{d^2x^\alpha}{d\tau^2} = -\Gamma^\alpha_{\beta\gamma}\frac{dx^\beta}{d\tau}\frac{dx^\gamma}{d\tau}$
Field equation	Newtonian field equation $\nabla^2\Phi = 4\pi G\mu$	The Einstein equation $R_{\alpha\beta} - \frac{1}{2}g_{\alpha\beta}R = 8\pi G T_{\alpha\beta}$

Substituting (22.59) (with indices lowered) and (22.60) into the Einstein equation (22.53) gives

$$\nabla^2\Phi = 4\pi G\mu, \tag{22.61}$$

which is the Newtonian field equation relating rest mass density μ to gravitational potential Φ. This is the Newtonian gravitational field equation (3.18). If the constant of proportionality between the left- and right-hand side of the Einstein equation [κ in (22.49)] were not known, this recovery of the Newtonian limit would determine it.

The recovery of the Newtonian gravitational field equation (22.61) from the Einstein equation completes the demonstration that Newtonian gravity is an approximation to general relativity appropriate for small curvatures and nonrelativistic matter sources. The other part of the demonstration—that the geodesic equation implies the Newtonian equation of motion—was given in Section 6.6. In particular, in this approximation, general relativity implies the familiar inverse square law for gravitational forces. The more than 300 years of successful applications of Newtonian gravity to the mechanics of the solar system are thus incorporated as approximate predictions of general relativity but with small corrections, such as the precession of the perihelion of Mercury. Newtonian gravity is not wrong, it is a nonrelativistic approximation to a relativistic theory of gravity—general relativity.

With the formulation of the Einstein equation we have fulfilled our pledge made in Chapter 6 to exhibit a theory of gravity consistent with special relativity. There are analogies between the two theories, as Table 22.1, which completes Table 6.1, shows. But general relativity is qualitatively different from Newtonian theory in its view of space and time.

Problems

1. *Four Dimensional Divergence Theorem and Conservation Laws* Consider a cube in flat spacetime with sides oriented along the (t, x, y, z)-axes of an inertial frame. Suppose that the cube's dimensions are Δt, Δx, Δy, and Δz in the respective directions. Show that for any vector $\mathbf{v}(x)$,

$$\int_{\mathcal{V}_4} d^4x \frac{\partial v^\alpha}{\partial x^\alpha} = \int_{\partial\mathcal{V}_4} d^3x (n_\alpha v^\alpha),$$

where $\mathcal{V}_4$ is the four-volume of the cube, $\partial\mathcal{V}_4$ is the three-surface boundary, and $\mathbf{n}$ is an appropriately chosen normal. Discuss how the normal should be chosen on the spacelike and timelike parts of the cube so that this relation is true. Show that when $\partial v^\alpha/\partial x^\alpha = 0$ and the cube extends over all *space*, the right-hand side of this relation implies a conservation law, and find the conserved quantity that is the same on the two bounding spacelike surfaces.

2. A cube of mass M with sides of length L is at rest on an inclined plane whose surface makes an angle θ with the horizontal. What are the components of the stress T^{ij} exerted by the cube

   (a) In rectangular coordinates oriented along the plane and perpendicular to it?

   (b) In rectangular coordinates oriented horizontally and vertically?

   (c) Does the stress you calculated in (b) give the correct force in the plane?

3. *The Law of Atmospheres* Assume the atmosphere is a perfect fluid gas of molecules with mass m, where the pressure, p, number density, n, and temperature, T, are related by

$$p = nk_BT,$$

where k_B is Boltzmann's constant. Using Newtonian gravity, find how the pressure p varies with height z when pressure forces are in equilibrium with gravitational forces. Assume the atmosphere has a constant temperature T, that the pressure at sea level is p_{sea}, that the heights of interest are small compared to the Earth's radius, and that the Earth's gravitation supplies all the force on the particles of the gas.

4. *The Stress Tensor Is Symmetric* Calculate the torque about its center exerted on a small cube of side L assuming that T^{xy} and T^{yx} are the only nonzero components of the stress tensor but that the stress tensor is not symmetric, $T^{xy} \neq T^{yx}$. Consider smaller and smaller cubes made of the same-density material. How does the net torque vary with smaller and smaller pieces? Can you see any reason from this variation why the stress tensor has to be symmetric?

5. A box of gas is at rest. The molecules of the gas are uniformly distributed throughout the box and are moving with a distribution of momenta $f(\vec{p})$ so that $f(\vec{p})d^3p$ is the number of molecules per unit volume with momentum in the range d^3p centered on $\vec{p}$. Suppose $f(\vec{p})$ is *isotropic*, meaning it depends only on $|\vec{p}|$.

   (a) Argue that the stress tensor for the gas is

$$T^{\alpha\beta} = \int d^3p \frac{f(\vec{p})}{m} p^\alpha p^\beta,$$

where p^α is the four-momentum considered as a function of $\vec{p}$ and m is the rest mass of the molecule.

(b) Calculate T^{ij} and show it is diagonal with all diagonal entries equal.

(c) Find the pressure and energy density in the gas. Assuming the distribution is peaked about ultrarelativistic momenta, find the equation of state of the gas.

6. [P] *The Stress Energy for Electromagnetism* This problem concerns the stress-energy tensor for the electromagnetic field in flat spacetime. Standard results are quoted in $c = 1$, SI units, which involve the defined factors $\mu_0 \equiv 4\pi \times 10^{-7}$ and $\epsilon_0 \equiv 1/\mu_0$. You can also use Gaussian units simply by making the replacements $\mu_0 \to 4\pi$ and $\epsilon_0 \to 1/(4\pi)$.

In electrodynamics the energy density ϵ is given in a vacuum by

$$\epsilon = \frac{1}{2}\left(\epsilon_0 \vec{E}^2 + \frac{1}{\mu_0}\vec{B}^2\right),$$

the energy flux is given by the Poynting vector

$$\vec{S} = \frac{1}{\mu_0}(\vec{E} \times \vec{B}),$$

and the stress is given by the Maxwell stress-tensor:

$$T^{ij} = \epsilon_0\left(-E^i E^j + \frac{1}{2}\delta^{ij}\vec{E}^2\right) + \frac{1}{\mu_0}\left(-B^i B^j + \frac{1}{2}\delta^{ij}\vec{B}^2\right).$$

(In comparing with other possible formulas you may have seen, remember that $c^2 = 1 = 1/(\epsilon_0\mu_0)$. Also watch for sign changes in the definition of the stress tensor.)

(a) Put these together to form the stress energy tensor $T^{\alpha\beta}$ for the electromagnetic field.

(b) Show explicitly from Maxwell's equations that this $T^{\alpha\beta}$ is conserved, i.e., satisfies the four equations (22.31).

7. [S] Consider the cube described in Example 22.5. Show that the surface integral in (22.34) gives the net pressure force acting on the cube.

8. [S] Show that the stress-energy of the vacuum defined by (22.38) satisfies the local conservation law (22.40).

9. [S] Show that all observers measure the same energy density of the vacuum no matter how they are moving through spacetime.

10. *The Weak Energy Condition* Consider stress-energy tensors that in a local inertial frame have the form $T_{\alpha\beta} = \text{diag}(A, B, C, D)$.

(a) What condition on A, B, C, and D must be satisfied so that any observer will see positive energy density no matter how fast that observer is moving with respect to the frame in which the stress-energy tensor is given?

(b) The vacuum stress-energy tensor (22.39) is of this form with negative values of B, C, and D. Is there some frame where an observer would see negative energy density?

11. Reinforce the argument given in Section 16.5 that there is no local gravitational energy by showing that there is no stress energy tensor that can be constructed from the metric and its first derivatives that reduces to zero when space is flat.

12. [C] *Symmetry Implies Conservation* The results of the following problem are general, but to keep the algebra manageable, restrict attention to metrics of the form

$$ds^2 = -A^2dt^2 + B^2\,dx^2 + C^2\,dy^2 + D^2\,dz^2,$$

where A, B, C, and D are functions of (t, x, y, z).

(a) Show that a relation such as

$$\nabla_\alpha J^\alpha = 0$$

can be written in the form

$$\frac{\partial(fJ^\alpha)}{\partial x^\alpha} = 0$$

for some function f specified by the metric and therefore implies a conservation law for the current J^α.

(b) When spacetime has a symmetry there is an associated Killing vector satisfying

$$\nabla_\alpha \xi_\beta + \nabla_\beta \xi_\alpha = 0.$$

(See Problem 20.18, to be led through a demonstration.) Show that

$$J^\alpha \equiv \xi_\beta T^{\alpha\beta}$$

is a conserved current.

13. [A, C] *Proving the Bianchi Identity* In a local inertial frame the Bianchi identities (22.50) read

$$\partial_\alpha R^\alpha_\beta - \tfrac{1}{2}\partial_\beta R = 0,$$

where $\partial_\alpha = \partial/\partial x^\alpha$. Use (21.20) and (21.28) to demonstrate these identities as follows:

(a) Use (21.20) to demonstrate that only terms containing *third* derivatives of the metric survive in the local inertial frame.

(b) Use (21.28) to evaluate the combinations of third derivatives of the metric that occur in the above expression for the Bianchi identities and show that they cancel.

14. [C, N] *Warp Drive Requires Negative Energy Density* This problem concerns the Alcubierre warp drive spacetime, discussed in Section 7.4, whose line element is given in (7.24).

(a) Calculate the components of the normal n_α to a surface of constant t.

(b) Modify the *Mathematica* program *Curvature and the Einstein Equation* available on the book website to show that

$$T_{\alpha\beta}n^\alpha n^\beta = -\frac{1}{8\pi}\frac{V_s^2(y^2+z^2)}{(2r_s)^2}\left(\frac{df}{dr_s}\right)^2.$$

This is the energy density measured by observers at rest with respect to the surfaces of constant t. The fact that it is negative means that the warp drive spacetime can't be supported by classical matter with positive energy density.

15. *Wormholes Require Negative Energy Density* Recall the wormhole geometry (7.39). Calculate the components of the stress-energy $T^{\hat{\alpha}\hat{\beta}}$ that would be needed for this geometry to be a solution of the Einstein equation in the orthonormal basis used in that example. Show that the energy density (as measured by a stationary observer) required is *negative*. Since all realistic matter described classically has *positive* energy density, it is impossible to construct a wormhole like (7.39) by classical means.

16. [A, N] *Embedding Constant Negative Curvature Surfaces in Euclidean Space*

(a) Compute the scalar curvature of the *two*-surface that is a $\theta = \pi/2$ slice of the spatial geometry of an open universe, as represented in (18.60). Show that the scalar curvature is constant over the surface and negative.

(b) Show that the two-dimensional geometry

$$d\Sigma^2 = du^2 + \cosh^2 u \, d\phi^2$$

has constant negative curvature as well. By Minding's theorem in differential geometry, this surface must have the same local geometry as the slice of an open universe in (a).

(c) Find an embedding of the surface in (b) in three-dimensional flat space. Does it look like a potato chip? (This part is the same as part (b) of Problem 18.30 if you worked that.)

CHAPTER 23

Gravitational Wave Emission

As described in Chapter 16, gravitational waves provide a window on the universe of astronomical phenomena that is different from any in the electromagnetic spectrum. Mass in many different varieties of motion is a source of propagating ripples in spacetime curvature. Sources of gravitational radiation are, therefore, widespread in the universe. Regions of rapidly varying, strong spacetime curvature, such as those that occur at the big bang or in gravitational collapse to black holes, will produce gravitational waves copiously. But even the motion of a pair of stars in orbit about one another will produce some radiation. Indeed, as described in Section 23.7, the first experimental detection of the effects of gravitational radiation was through the decay in the orbital period of a pair of neutron stars.

Gravitational wave detectors on Earth and in space are the instruments necessary to explore the universe with gravitational waves. The workings of some of them were sketched in Chapter 16. But to interpret their observations, and predict what they might see, it's necessary to solve the Einstein equation for the gravitational radiation produced by given sources. Predicting the gravitational radiation from strong-curvature, rapidly varying sources is a problem generally tractable only by numerical simulation of the fully nonlinear Einstein equation—a subject well beyond the scope of this book. However, some insight into the production of gravitational waves can be obtained from examining the more tractable problem of the small ripples in spacetime emitted by weak, nonrelativistic sources. That problem is treated in this chapter.

23.1 The Linearized Einstein Equation with Sources

The Einstein equation (22.51),

$$R_{\alpha\beta} - \tfrac{1}{2} g_{\alpha\beta} R = 8\pi T_{\alpha\beta}, \tag{23.1}$$

relates the stress-energy of mass in motion to propagating ripples in spacetime curvature. We will solve the Einstein equation assuming that the waves produced by the source $T_{\alpha\beta}$ are so weak that the metric can be written as a small perturbation

$h_{\alpha\beta}(x)$ of the metric of flat spacetime $\eta_{\alpha\beta}$. This means

Small Perturbations of Flat Spacetime

$$\boxed{g_{\alpha\beta}(x) = \eta_{\alpha\beta} + h_{\alpha\beta}(x),} \tag{23.2}$$

assuming rectangular (t, x, y, z) coordinates for flat spacetime, where $\eta_{\alpha\beta} = \mathrm{diag}(-1, 1, 1, 1)$. Weak means $|h_{\alpha\beta}| \ll 1$ for all α and β. We will evaluate the left-hand side of Einstein equation (23.1) to first order in the metric perturbation, just as we did in Section 21.5 to obtain the linearized *vacuum* Einstein equation. We will equate that to the right-hand side to obtain the equations of linearized gravity with weak sources. Eventually, we will also assume that all velocities in the source are small compared to the velocity of light. That will mean that the stress energy will be dominated by the rest-mass density μ:

$$T^{\alpha\beta} = \mu u^\alpha u^\beta. \tag{23.3}$$

However, we don't assume that the source is static—if we did there would be no gravitational radiation! We will postpone introducing the specific form (23.3) as long as possible so that many of our intermediate results are more general.

As we described in Section 21.5, small changes in coordinates (gauge transformations) can be used to impose four gauge conditions on the metric perturbation $h_{\alpha\beta}(x)$. We continue to use Lorentz gauge (21.56). The four Lorentz gauge conditions can be written in a simple form by introducing the "trace-reversed" amplitude

$$\bar{h}_{\alpha\beta} \equiv h_{\alpha\beta} - \tfrac{1}{2}\eta_{\alpha\beta} h, \tag{23.4}$$

where[1] $h \equiv h^\gamma{}_\gamma$. Then the Lorentz gauge condition (21.56) becomes

Lorentz Gauge

$$\boxed{\frac{\partial \bar{h}^{\alpha\beta}}{\partial x^\beta} = 0.} \tag{23.5}$$

Equation (21.44) shows that, to first order in the metric perturbation $h_{\alpha\beta}(x)$, the Ricci curvature on the left-hand side of the Einstein equation (23.1) has the simple form $\delta R_{\alpha\beta} = (-\frac{1}{2})\Box h_{\alpha\beta}$ in Lorentz gauge. Here, $\Box$ is the *flat-space* wave operator defined in (21.45), namely, $\Box = -\partial^2/\partial t^2 + \vec{\nabla}^2$. The linearized curvature scalar δR [cf. (22.48)] is, therefore, $\delta R = (-\frac{1}{2})\Box h$. The linearization of the Einstein equation (23.1) is then

Linearized Einstein Equation for Weak Sources

$$\boxed{\Box \bar{h}_{\alpha\beta} = -16\pi T_{\alpha\beta}.} \tag{23.6}$$

[1]In linearized gravity, indices are raised with the flat metric $\eta^{\alpha\beta}$ as discussed in Chapter 21; [cf. (21.47)].

It should be stressed that (23.6), like (23.5) and (23.2), holds only for the rectangular coordinate components of the metric perturbations and stress-energy tensor where the flat-space metric is $\eta_{\alpha\beta}$.

Equation (23.6) shows that each component of $\bar{h}_{\alpha\beta}(x)$ obeys a separate flat-space wave equation with source of the form

$$-\frac{\partial^2 f(x)}{\partial t^2} + \vec{\nabla}^2 f(x) = j(x). \tag{23.7}$$

The solution of the wave equation for $f(x)$ with a given source $j(x)$ is a standard problem in physics—familiar, for example, from the theory of electromagnetic waves. The solution is reviewed in the next section.

23.2 Solving the Wave Equation with a Source

First consider the case when the source $j(x)$ in (23.7) is a δ-function located at an event at a definite time and a definite location in space. Because the wave equation is linear, more general sources can be built up by adding waves from such δ-function sources (see Figure 23.1). For convenience let's begin by putting this spacetime event at the origin so that

$$j(x) \equiv \delta(t)\,\delta(x)\,\delta(y)\,\delta(z) = \delta(t)\,\delta^{(3)}(\vec{x}). \tag{23.8}$$

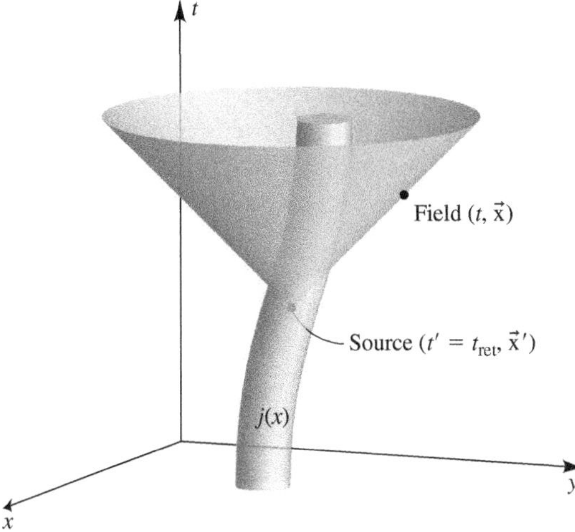

FIGURE 23.1 The solution of the wave equation (23.7) for a general source $j(x)$ may be built up by superposing δ-function point sources at spacetime events $(t', \vec{x}')$ weighted by the strength $j(t', \vec{x}')$. The cone in this spacetime diagram shows the wave from $(t', \vec{x}')$ moving outward at the speed of light. The value of the wave at $(t, \vec{x})$ depends on the source at the retarded time $t_{\text{ret}} \equiv t - |\vec{x} - \vec{x}'|$.

The spherical symmetry of this source implies that the particular solution of (23.7) it produces, which we call $g(t, \vec{x})$, can depend only on t and $r = |\vec{x}|$. Away from the origin at $r \neq 0$, therefore, $g(t, r)$ must satisfy

$$-\frac{\partial^2 g}{\partial t^2} + \frac{1}{r^2}\frac{\partial}{\partial r}\left(r^2 \frac{\partial g}{\partial r}\right) = 0. \tag{23.9}$$

It is not difficult to check that a solution of this equation is

$$g(t, r) = \frac{1}{r}\Big[O(t - r) + I(t + r)\Big] \tag{23.10}$$

for any two functions $I(\cdot)$ and $O(\cdot)$. The solution $O(t - r)/r$ represents a wave moving outward from the source with the speed of light; the part $I(t + r)/r$ represents a wave imploding inward on the source with the same speed. That is the most general possible physical situation, and (23.10) is the most general possible mathematical solution. However, until we learn how to construct devices that focus gravitational waves, it is only the outgoing wave part that is relevant physically. That is called the *retarded*, or *causal*, solution. The retarded wave is received *after* the event that is its source. We consider only the outgoing, retarded solutions O in (23.10).

It remains to find the function $O(t - r)$ that is the solution of (23.9) with the δ-function source (23.8). The singular δ-function is difficult to work with directly, so we integrate both sides of (23.7) over a small spatial volume of radius ϵ that contains the source (23.8), finding

$$\int_\epsilon d^3x \left[-\frac{\partial^2 g(x)}{\partial t^2} + \vec{\nabla}^2 g(x)\right] = \int_\epsilon d^3x\, \delta(t)\, \delta^{(3)}(\vec{x}) = \delta(t). \tag{23.11}$$

The solution g diverges as $1/r$ for small r from (23.10), but the volume element in (23.11) is decreasing as $4\pi r^2$. The integral over the $\partial^2 g/\partial t^2$ term goes to zero as $\epsilon \to 0$. Thus, the entire contribution in the limit comes from the volume integral of $\vec{\nabla}^2 g$, which can be transformed into a surface integral using the divergence theorem, giving the following:

$$\lim_{\epsilon \to 0} \int_\epsilon d^3x \vec{\nabla}^2 g = \lim_{\epsilon \to 0} \int_\epsilon d\vec{A} \cdot \vec{\nabla} g = \delta(t). \tag{23.12}$$

The first integral is over the volume of a sphere of radius ϵ and the second is over its surface. Inserting $g(t, r) = O(t - r)/r$, the integral and limit are easy to evaluate, giving

$$-4\pi\, O(t) = \delta(t). \tag{23.13}$$

The solution $g(t, r)$ to the wave equation with δ-function source at the origin and outgoing wave (retarded) boundary conditions is, therefore,

$$g(t, r) = -\frac{\delta(t - r)}{4\pi r}. \tag{23.14}$$

Note that the solution is confined to the light cone $t = r$ as expected and as illustrated in Figure 23.1.

If the source $j(x')$ is distributed over many spacetime points $(t', \vec{x}')$, simply sum the contributions from each of them weighted by $j(t', \vec{x}')$ to obtain the solution to (23.7):

$$f(t, \vec{x}) = \int dt'\, d^3x'\, g(t - t', \vec{x} - \vec{x}')\, j(t', \vec{x}'). \tag{23.15}$$

The effect of the δ-function in (23.14) is to evaluate the time integral in (23.15) at the retarded time $t' = t_{\text{ret}} \equiv t - |\vec{x} - \vec{x}'|$. Thus,

$$\boxed{f(t, \vec{x}) = -\frac{1}{4\pi} \int d^3x' \frac{[j(t', \vec{x}')]_{\text{ret}}}{|\vec{x} - \vec{x}'|},} \tag{23.16}$$

General Solution to the Wave Equation

where $[\ \cdot\ ]_{\text{ret}}$ means that the argument should be evaluated at the retarded time.

Equation (23.16) is the general solution of the wave equation with source and outgoing wave boundary conditions. A case of special interest is when the source varies harmonically in time with frequency ω, e.g.,

$$j(t, \vec{x}) = j_\omega(\vec{x}) \cos(\omega t), \tag{23.17}$$

and the corresponding wavelength $\lambda = 2\pi/\omega$ is large compared to the characteristic dimensions of the source, R_{source}:

$$\lambda \gg R_{\text{source}} \qquad \begin{pmatrix} \text{long wavelength} \\ \text{approximation} \end{pmatrix}. \tag{23.18}$$

Long wavelengths mean low frequencies and characteristic velocities $V_{\text{source}} \sim \omega R_{\text{source}} \ll 1$. In simple sources, long wavelengths mean low velocities.

In the long-wavelength approximation there is an especially simple formula for the solution a long distance away, $r \gg R_{\text{source}}$, as we now show. Inserting (23.17) into the general solution (23.16) gives

$$f(t, \vec{x}) = -\frac{1}{4\pi} \int d^3x' \frac{j_\omega(\vec{x}') \cos[\omega(t - |\vec{x} - \vec{x}'|)]}{|\vec{x} - \vec{x}'|}. \tag{23.19}$$

For a large distance r from the source, $|\vec{x} - \vec{x}'|$ can be replaced by r in the denominator. The long-wavelength approximation means that the same replacement can be made in the cosine because $(2\pi/\lambda)|\vec{x} - \vec{x}'|$ will not change much as $\vec{x}'$ varies over the source if (23.18) holds. Thus, far away from a source whose size is much smaller than a wavelength, the solution is asymptotically

$$f(t, \vec{x}) \xrightarrow[r \to \infty]{} -\frac{1}{4\pi r} \int d^3x'\, j(t - r, \vec{x}') \qquad \begin{pmatrix} \text{long wavelengths} \\ \text{large } r \end{pmatrix}. \tag{23.20}$$

The subscript ω has been dropped from j in (23.20) because it does not matter whether the source varies exactly harmonically as long as it contains only frequencies low enough for the long wavelength approximation (23.18) to be valid. Equation (23.20) is an outgoing spherical wave whose amplitude is determined by the integral of the source over space at the retarded time.

23.3 The General Solution of Linearized Gravity

The results of the preceding discussion for the solutions of the wave equation can be immediately applied to the case of linearized gravity. That is because the linearized Einstein equation (23.6) is a set of ten flat-space wave equations for the components of $\bar{h}_{\alpha\beta}$ with a separate source for each. Equivalently, by raising all indices, it is a set of ten wave equations for $\bar{h}^{\alpha\beta}$ with $-16\pi T^{\alpha\beta}$ as the source. From (23.16) the general solution of the linearized Einstein equation is, therefore,

Solution of Linearized Einstein Equation

$$\bar{h}^{\alpha\beta}(t,\vec{x}) = 4\int d^3x' \frac{\left[T^{\alpha\beta}(t',\vec{x}')\right]_{\text{ret}}}{|\vec{x}-\vec{x}'|}, \tag{23.21}$$

where, again, $[\ \cdot\ ]_{\text{ret}}$ means the argument is evaluated at the retarded time $t' = t_{\text{ret}} \equiv t - |\vec{x}-\vec{x}'|$. Like (23.6), from which it came, this relation holds only for the components of the metric perturbations and stress-energy in the usual rectangular coordinates of an inertial frame.

The metric perturbation (23.21) satisfies the wave equation with source (23.6). But that is not the only requirement for a metric perturbation to solve the linearized Einstein equation. It must also satisfy the Lorentz gauge condition (23.5). However, (23.5) is automatically satisfied by the solution of the wave equation (23.21) as a consequence of the flat-space conservation of stress-energy (22.31), as the following calculation shows. Insert the solution (23.21) into the Lorentz gauge condition (23.5) to find

$$\frac{\partial \bar{h}^{\alpha\beta}(t,\vec{x})}{\partial x^\beta} = 4\int d^3x' \left[\frac{\partial T^{\alpha t}(t_{\text{ret}},\vec{x}')}{\partial t}\frac{1}{|\vec{x}-\vec{x}'|} + \frac{\partial}{\partial x^i}\left(T^{\alpha i}(t_{\text{ret}},\vec{x}')\frac{1}{|\vec{x}-\vec{x}'|}\right)\right]. \tag{23.22}$$

To evaluate this remember that $t_{\text{ret}} \equiv t - |\vec{x}-\vec{x}'|$ is itself a function of x^i and x'^i. Use the identity $\partial|\vec{x}-\vec{x}'|/\partial x^i = -\partial|\vec{x}-\vec{x}'|/\partial x'^i$ to replace appropriate derivatives with respect to x^i by derivatives with respect to x'^i. Integrate by parts (noting that surface terms outside the matter vanish) to obtain the following expression (Problem 3):

$$\frac{\partial \bar{h}^{\alpha\beta}(t,\vec{x})}{\partial x^\beta} = 4\int d^3x' \frac{1}{|\vec{x}-\vec{x}'|}\left(\frac{\partial T^{\alpha\beta}(t',\vec{x}')}{\partial x'^\beta}\right)_{t'=t_{\text{ret}}}. \tag{23.23}$$

In linearized gravity the metric perturbations $h^{\alpha\beta}(x)$ and the stress-energy $T^{\alpha\beta}(x)$ are of comparably small size. The linearized local conservation law (22.40) is just the flat-space conservation law (22.31) in this approximation. The right-hand side of (23.23) therefore vanishes, and the Lorentz gauge condition (23.5) is automatically satisfied by (23.21).

Example 23.1. A Little Rotation. The general solution of the linearized Einstein equation (23.21) provides the most direct route to understanding how the gravitomagnetic effects of rotation described in Chapter 14 arise in general relativity. Consider for simplicity a time-independent, spherically symmetric distribution of nonrelativistic matter. Suppose it is rotating uniformly with an angular velocity $\vec{\Omega}$ that is constant throughout the interior and slow enough that the body is not significantly rotationally distorted. As in other nonrelativistic situations we have worked on, the stress-energy is dominated by the rest energy of the matter and well approximated by $T^{\alpha\beta} = \mu u^\alpha u^\beta$, where $\mu(x)$ is the rest-energy density and $u^\alpha(x)$ is the four-velocity. When the rotational velocities are small compared to the velocity of light, the most important components of the stress-tensor accurate to linear order in Ω are [cf. (5.28)]:

$$T^{tt}(\vec{x}) = \mu(r), \qquad T^{ti}(\vec{x}) = T^{it}(\vec{x}) = \mu(r)V^i(\vec{x}), \tag{23.24}$$

where $\vec{V}(\vec{x}) = \vec{\Omega} \times \vec{x}$ is the three-velocity. All other components are of order Ω^2. The component T^{tt} is the source of the Newtonian perturbations of flat space exhibited for instance in (21.49) (Problem 1). The components T^{ti} are the source of gravitomagnetic effects that depend on the *velocity* of the matter as well as the distribution of mass. Specifically, from (23.21),

$$h^{it} = h^{ti} = \bar{h}^{ti} = 4\int d^3x' \frac{\mu(r')V^i(\vec{x}')}{|\vec{x} - \vec{x}'|}. \tag{23.25}$$

To evaluate this integral choose rectangular coordinates (x, y, z) with the z-axis oriented along $\vec{\Omega}$. Then $\vec{\Omega} = \Omega\vec{e}_z$ and

$$V^x = -\Omega y, \qquad V^y = \Omega x, \qquad V^z = 0. \tag{23.26}$$

It is simplest to first evaluate (23.25) at a large spatial distance r from the rotating body using the familiar expansion

$$\frac{1}{|\vec{x} - \vec{x}'|} = \frac{1}{r} + \frac{\vec{x} \cdot \vec{x}'}{r^3} + \cdots, \tag{23.27}$$

valid for $r \equiv |\vec{x}| \gg |\vec{x}'|$. The contribution from the first term vanishes because the integrand is odd under $\vec{x}' \to -\vec{x}'$. The second term (Problem 4) yields the result

$$h^{xt} = h^{tx} = -\frac{2yJ}{r^3}, \qquad h^{yt} = h^{ty} = \frac{2xJ}{r^3}, \qquad h^{zt} = h^{tz} = 0, \tag{23.28}$$

where J is the magnitude of the Newtonian angular momentum

$$\vec{J} = \int d^3x \big[\vec{x} \times (\mu \vec{V})\big] = \vec{e}_z \int d^3x \, \mu \Omega (x^2 + y^2). \tag{23.29}$$

This is exactly the form of the rotational perturbation quoted in (14.25). The result is more general than its derivation. It holds for any r outside the body provided it is rotating uniformly (Problem 2).

The perturbations (23.28) do not describe a gravitational wave. They fall off like $1/r^2$ at large r rather than the $1/r$ that characterizes gravitational radiation. Axisymmetric rotation in general is an example of a highly symmetrical motion that does not produce gravitational radiation.

23.4 Production of Weak Gravitational Waves

Gravitational Waves Far from Their Source

Equation (23.21) is the general solution of linearized gravity for a prescribed source of stress-energy assuming outgoing waves. Equation (23.20) gives the asymptotic form of solutions of the wave equation in the long-wavelength approximation. In this section we apply these results to the calculation of the gravitational waves a large distance from a weak source assuming, in particular, that its velocities are slow. More concretely we assume that $r \gg R_{\text{source}}$ and $\lambda \gg R_{\text{source}}$, where R_{source} is the characteristic size of the source and $\lambda = 2\pi/\omega$ is the wavelength associated with the characteristic frequency of variation of the source ω. Applying (23.20) to (23.21) then gives for the asymptotic gravitational wave amplitudes:

$$\bar{h}^{\alpha\beta}(t, \vec{x}) \xrightarrow[r\to\infty]{} \frac{4}{r} \int d^3x' \, T^{\alpha\beta}(t - r, \vec{x}') \qquad \begin{pmatrix} \text{weak source} \\ \text{long wavelengths} \\ \text{large } r \end{pmatrix}. \tag{23.30}$$

Over a limited range of angle about any one direction, the wave from (23.30) is approximately a plane wave at large r. This means that Chapter 16's analysis of polarization, energy flux, and the response of detectors for plane waves can be applied here. That analysis requires only the spatial components of the metric perturbation $\bar{h}^{ij}(x)$. The sources of these spatial components in (23.30) are the quantities $\int d^3x \, T^{ij}(t - r, x)$. They can be put in a more useful form by using the flat-space conservation law (22.31) obeyed by $T^{\alpha\beta}$ to this linear order. One component of this is

$$\frac{\partial T^{tt}}{\partial t} + \frac{\partial T^{kt}}{\partial x^k} = 0. \tag{23.31}$$

Differentiate this equation with respect to time, and use the symmetry $T^{tk} = T^{kt}$ and the conservation law (22.31) once again to find

$$\frac{\partial^2 T^{tt}}{\partial t^2} = -\frac{\partial}{\partial t}\left(\frac{\partial T^{tk}}{\partial x^k}\right) = -\frac{\partial}{\partial x^k}\left(\frac{\partial T^{tk}}{\partial t}\right) = +\frac{\partial^2 T^{k\ell}}{\partial x^k \partial x^\ell}. \tag{23.32}$$

Multiply both sides of this equation by $x^i x^j$ and integrate over space. The integral over the right-hand side can be carried out by parts; the surface terms vanish because the source is bounded. The result is the identity

$$\int d^3x\, T^{ij}(x) = \frac{1}{2}\frac{d^2}{dt^2}\int d^3x\, x^i x^j T^{tt}(x). \tag{23.33}$$

Long wavelengths mean low velocities as we mentioned in Section 23.2. In that limit we assume that the stress-energy tensor has the form (23.3) with non-relativistic velocities. The energy density $T^{tt}(x)$ will then be dominated by the rest-mass density $\mu(x)$, and the integral in (23.33) defines the *second mass moment*[2] $I^{ij}(t)$:

$$\boxed{I^{ij}(t) \equiv \int d^3x\, \mu(t, \vec{x}) x^i x^j.} \tag{23.34}$$

Second Mass Moment

The gravitational wave metric perturbation far from a weak, nonrelativistic source in the long-wavelength approximation becomes

$$\boxed{\bar{h}^{ij}(t, \vec{x}) \xrightarrow[r\to\infty]{} \frac{2}{r}\ddot{I}^{ij}(t-r) \qquad \begin{pmatrix}\text{weak source}\\ \text{long wavelengths}\\ \text{large } r\end{pmatrix},} \tag{23.35}$$

Gravitational Wave Amplitude at Large Distance

where a dot means a derivative with respect to t.

Equation (23.35) is more general than its derivation. The assumptions of the derivation would not cover self-gravitating systems like a pair of stars in mutual orbit, no matter how accurately that orbit was approximated by Newtonian theory. That is because *all* perturbations of flat space are neglected on the right-hand side of the wave equation (23.6). But Newtonian perturbations of flat space are the source of motion in a self-gravitating system and cannot be neglected. However, result (23.35) depends only on the motion of the mass sources, not on how that motion was produced. It turns out that the formula holds to a good approximation for weak sources that are slowly moving because of Newtonian gravity even though its derivation does not.[3] Relying on that fact, in the next section and Example 23.2 we apply (23.35) to binary stars.

[2]Despite the use of the letter I, the second mass moment is not exactly the same as the moment of inertia tensor,

$$\mathcal{J}^{ij} = \int d^3x\, \mu(\vec{x})[\delta^{ij}r^2 - x^i x^j],$$

which is important for the motion of rigid bodies in mechanics, although one tensor can be constructed from the other.

[3]The derivation, which involves keeping track of the first-order *non*linearities in general relativity, can be found in advanced texts, for example, Misner, Thorne, and Wheeler (1970).

Example 23.2. Estimating Gravitational Wave Amplitudes: The Binary Star System ι Boo. The binary star system ι Boo is located about 11.7 pc from Earth in the direction of the constellation Boötes. It consists of a $1M_\odot$ star and a $.5M_\odot$ star so close together they are in contact and revolving around each other with an orbital period $P = 6.5$ hr. What order of magnitude fractional strain sensitivity $\delta L/L$ [cf. (16.19)] would be needed in a gravitational wave detector to receive waves from this source?

The result (23.35) can be used to make simple order-of-magnitude estimates of the gravitational wave amplitude far from weak, nonrelativistic sources that are needed to answer this kind of question. To keep the estimate simple, assume that both stars have the same mass M and are each moving about their center of mass in a circular orbit of radius R and period P. (The error made by not taking into account that one mass is roughly half of the other isn't important for a rough order-of-magnitude estimate.) A rough estimate of the typical component of the second mass moment is then $I^{ij} \sim MR^2$. Two time derivatives add two factors of the period to yield

$$\ddot{I}^{ij} \sim MR^2/P^2. \tag{23.36}$$

The radius of the orbit R is related to the mass and period by

$$\frac{V^2}{R} = \frac{(2\pi R/P)^2}{R} = \frac{M}{(2R)^2}.$$

Thus, from (23.35) we estimate for the gravitational wave amplitude a distance r from the binary system

$$\bar{h}^{ij} \sim \left(\frac{M}{r}\right)\left(\frac{M}{P}\right)^{2/3}, \tag{23.37}$$

or, putting back the units,

$$\bar{h}^{ij} \sim 10^{-21}\left(\frac{M}{M_\odot}\right)^{5/3}\left(\frac{1\text{ h}}{P}\right)^{2/3}\left(\frac{100\text{ pc}}{r}\right). \tag{23.38}$$

For the parameters of ι Boo, this gives $\bar{h}^{ij} \sim 10^{-21}$, and this is the fractional strain sensitivity needed for a detector to see the gravitational waves from it [cf. (16.19)]. In the next section we will carry out a detailed evaluation of (23.35) for binary star systems and see just how good this very rough estimate is. However, the factor of 10^{-21} captures in one number the difficulty in detecting gravitational waves because ι Boo is one of the *brightest* binary star sources of gravitational waves at Earth. The LIGO detector described in Section 16.4 is not sensitive in the relevant frequency range of $\sim 10^{-5}$ Hz. The ι Boo system would be one of a handful of brightest binary star sources that might be seen by a detector in space.

TABLE 23.1 Production of Linearized Gravitational and Electromagnetic Waves

	Linearized gravitation ($c = G = 1$)	Electromagnetism ($c = 1$)
Field equation	Einstein equation with $g_{\alpha\beta} = \eta_{\alpha\beta} + h_{\alpha\beta}$	Maxwell's equations
Basic potentials	Linearized metric perturbations $h_{\alpha\beta}(x)$	Vector and scalar potentials $\left(\Phi(x), \vec{A}(x)\right)$
Sources	Stress-energy $T_{\alpha\beta}$	Charge and current $(\rho_{\text{elec}}, \vec{J})$
Lorentz gauge	$\dfrac{\partial \bar{h}^{\alpha\beta}}{\partial x^\alpha} = 0$	$\dfrac{\partial \Phi}{\partial t} + \vec{\nabla} \cdot \vec{A} = 0$
Wave equation with source	$\Box \bar{h}_{ij} = -16\pi T_{ij}$	$\Box \vec{A} = -\mu_0 \vec{J}$
General solution	$\bar{h}^{ij} = 4 \displaystyle\int d^3x' \frac{[T^{ij}]_{\text{ret}}}{\lvert\vec{x} - \vec{x}'\rvert}$	$\vec{A} = \dfrac{\mu_0}{4\pi} \displaystyle\int d^3x' \frac{[\vec{J}]_{\text{ret}}}{\lvert\vec{x} - \vec{x}'\rvert}$
Large r, long-wavelength approximation	$\bar{h}^{ij} = \dfrac{2[\ddot{I}^{ij}]_{\text{ret}}}{r}$ $I^{ij} = \displaystyle\int d^3x\, \mu x^i x^j$	$\vec{A} = \dfrac{\mu_0}{4\pi}\dfrac{[\dot{\vec{p}}]_{\text{ret}}}{r}$ $\vec{p} = \displaystyle\int d^3x\, \rho_{\text{elec}} \vec{x}$
Time-averaged radiated power	$\dfrac{dE}{dt} = \dfrac{1}{5}\langle \dddot{I}_{ij} \dddot{I}^{ij} \rangle$	$\dfrac{dE}{dt} = \dfrac{\mu_0}{6\pi}\langle \ddot{\vec{p}}^{\,2} \rangle$

The angle brackets, $\langle \cdot \rangle$, denote an average over a time longer than the characteristic period of the source. The equations in the linearized gravitation column are in $c = G = 1$ geometrized units. The electromagnetic equations are in SI units with $c = 1$, where $\mu_0 \equiv 4\pi \times 10^{-7}$. To convert this column to Gaussian units with $c = 1$, replace μ_0 by 4π.

Analogies with Electromagnetism

There is a close analogy between the theory of the production of weak gravitational waves and the theory of the production of electromagnetic waves. If you have had a course in electromagnetism, that analogy should be helpful; it is laid out in Table 23.1. (The last row of that table refers to the total radiated power, to be discussed in Section 23.6.) If you have not had a course in electromagnetism, skip to the next section.

The similarities between linearized gravity and electromagnetism can be traced to the fact that the basic field equations of both theories reduce to the wave equation with source in an appropriate gauge. The chief differences arise because gravity is a tensor field, but electromagnetism is described by vector fields. In both cases the long-wavelength approximation can be systematically pursued to give an expansion of the $1/r$ part of the field far away from the source (the radiation field) in powers of $(R_{\text{source}}/\lambda)$. Equation (23.20) and its consequent (23.35) are just the first terms in such expansions; the others arise from a large r expansion of the cosine in (23.19). For both gravity and electromagnetism the coefficients in these expansions are proportional to time derivatives of the multipole moments—integrals of the source multiplied by powers of x^i. In each case there are two families of multipole moments: electric (charge) moments and magnetic (charge current) moments in electromagnetism, and mass moments and mass current moments in gravity. There are no monopole (zero factors of x^i) contributions to the radiation field in either electromagnetism or gravity. Put differently, there are no spherically symmetric electromagnetic or gravitational waves. That is because charge is conserved in electromagnetism and the mass of matter is conserved to linear order in gravity, where the flat-space conservation law (22.31) holds (even though mass is radiated away in the next (quadratic) order, as we will see). The leading term in electromagnetism is, therefore, electric dipole radiation, as shown in Table 23.1. However, the dipole moment of a mass distribution $\int d^3x\, \mu x^i$ is simply the total mass times the center of mass position. The center of mass position can be made to vanish by an appropriate choice of coordinates; therefore, there is no mass moment dipole gravitational radiation (Problem 6). The analog of the magnetic moment in gravity is the angular momentum—an integral of one power of x^i times the mass-current (cf. Example 23.1). But angular momentum of matter is also conserved in the linear approximation, so there is no gravitational radiation from this multipole either. Therefore, the leading gravitational effect is quadrupolar, as we will see in more detail in Section 23.6. This difference means that although the leading approximation to the amplitude of an electromagnetic wave is proportional to $(\omega R_{\text{source}})$, that of a gravitational wave is proportional to $(\omega R_{\text{source}})^2$, where ω and R_{source} are the characteristic frequency and size of the source, respectively.

23.5 Gravitational Radiation from Binary Stars

There are as many sources of gravitational radiation in the universe as there are mass distributions with nonuniformly time-varying second-mass moments. However, were we to single out one typical source for detailed analysis, it would be two stars moving in orbit about one another under their mutual gravitational attraction. Approximately two-thirds of all stars are in such binary systems. Gravitational radiation from some nearby binaries should be detectable by receivers in space as the case of ι Boo considered in Example 23.2 showed. Observations of the decay of the orbit of a binary pulsar system due to its gravitational radiation were the first detection of the effects of gravitational radiation (Section 23.7).

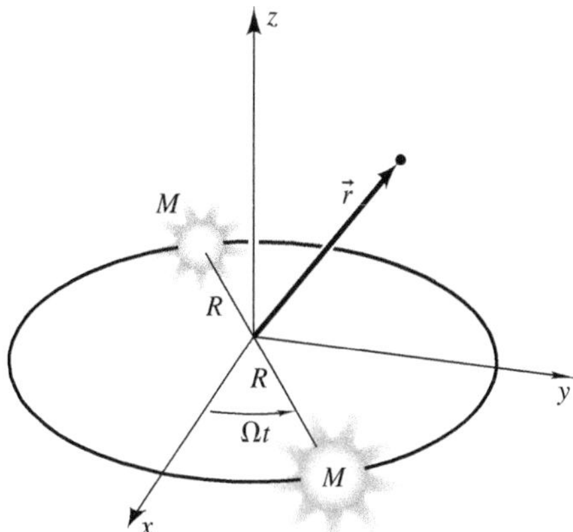

FIGURE 23.2 A binary star system. Two stars of equal mass M are in orbit about each other in the x-y plane under their mutual gravitational attraction. The orbit is circular with radius R and the orbital frequency is Ω. We are interested in the gravitational radiation they produce a long way away in any direction $\vec{r}$.

As discussed before, the weak-source, low-velocity approximation turns out to give a good approximation to the gravitational radiation from binary star systems, even though our derivation of it does not strictly apply to self-gravitating systems. To see how this goes, let's consider in detail the simplest possible case of a binary pair illustrated in Figure 23.2. Two stars of equal mass M are in a circular orbit of radius R about their center of mass. Assuming a Newtonian analysis is sufficiently accurate, the radius of the orbit is related to its period P by Newton's law ($G = 1$ units):

$$\frac{V^2}{R} = \frac{1}{R}\left(\frac{2\pi R}{P}\right)^2 = \frac{M}{(2R)^2}. \tag{23.39}$$

Defining the orbital frequency Ω by $2\pi/P$, we have

$$R = \left(\frac{M}{4\Omega^2}\right)^{1/3} = \left(\frac{MP^2}{16\pi^2}\right)^{1/3}, \tag{23.40}$$

which is Kepler's law for this binary system.[4]

A few simple dimensional estimates are in order before proceeding with a detailed calculation. First, the ratio of source size to wavelength from (23.40) is

$$R/\lambda \sim (M/P)^{1/3}. \tag{23.41}$$

But a limit on the period is provided by the obvious condition that the radius of the orbit, R, is larger than the radii of the stars, R_*. Again from (23.40), this implies

$$P > 4\pi R_*(R_*/M)^{1/2}. \tag{23.42}$$

[4]This is not (3.24), which is valid when one mass is much heavier than the other. You may be familiar with this in the more general form $\Omega^2 a^3 = M_{\text{tot}}$, where a is the semimajor axis of the elliptical orbit and M_{tot} is the sum of the masses of the binary pair.

Then from (23.41)

$$R/\lambda \lesssim (M/R_*)^{1/2}. \tag{23.43}$$

But R_* is always greater than $2M$ (the Schwarzschild radius) and typically is much greater. The long-wavelength approximation is thus easily valid for any realistic binary systems except those that are coalescing.

With these estimates behind us, let's return to a detailed evaluation of the wave amplitude (23.35). The assumed geometry is shown in Figure 23.2. We take

$$x(t) = R\cos(\Omega t), \qquad y(t) = R\sin(\Omega t), \qquad z(t) = 0 \tag{23.44}$$

for the trajectory of one of the masses. The components of the second mass moment, including both masses, are then easily evaluated from (23.34):

$$\begin{aligned} I^{xx} &= 2MR^2\cos^2(\Omega t) = MR^2\left[1+\cos(2\Omega t)\right], \\ I^{xy} &= 2MR^2\sin(\Omega t)\cos(\Omega t) = MR^2\sin(2\Omega t), \\ I^{yy} &= 2MR^2\sin^2(\Omega t) = MR^2\left[1-\cos(2\Omega t)\right]. \end{aligned} \tag{23.45}$$

The other components, I^{zz}, I^{zx}, and I^{zy}, are all zero. Inserting this in (23.35) we have

$$\bar{h}^{ij} \xrightarrow[r\to\infty]{} -\frac{8\Omega^2 MR^2}{r}\begin{pmatrix} \cos\left[2\Omega(t-r)\right] & \sin\left[2\Omega(t-r)\right] & 0 \\ \sin\left[2\Omega(t-r)\right] & -\cos\left[2\Omega(t-r)\right] & 0 \\ 0 & 0 & 0 \end{pmatrix}. \tag{23.46}$$

The frequency of the emitted radiation is thus *twice* the orbital frequency.

Using Kepler's law (23.40) to eliminate R from (23.46) gives a peak gravitational wave amplitude of the same form as in the rough estimate (23.37) we made for the ι Boo system. However, the factor of approximately $1/5$ by which the estimate (23.37) differs from this more detailed analysis shows that one shouldn't base the design of several hundred million dollar experiments on such rough calculations.

The asymptotic gravitational wave amplitude (23.46) looks spherically symmetric because there is no angular dependence of any of the functions involved. In fact, it contains all the information on how the polarization of the wave and the energy flux vary in different directions. These are easy to calculate because, for large r, the wave (23.46) is well approximated by a plane wave in a small solid angle about any particular direction. The analysis of polarization and energy flux for plane waves from Chapter 16 can, therefore, be applied.

In Chapter 16 we saw that plane gravitational waves had two types of polarization exhibited explicitly in transverse traceless (TT) gauge [cf. (21.69)]. Using the transverse traceless gauge, we also found the flux of energy in any one polarization time averaged over a period [cf. (16.22)]. The approximate plane wave in any one direction from (23.46) is not necessarily in TT gauge but can easily be put in that gauge using the algorithm discussed in Chapter 21 [cf. (21.70)]:

Briefly, just make the nontransverse components zero and subtract out the trace. The transformation to TT gauge is different in different directions, and that is how the angular properties of the radiation emerges. We illustrate this by considering just two directions.

Example 23.3. Gravitational Radiation from a Binary Star System in Two Directions. *Normal to the Orbital Plane.* Fix attention on the radiation propagating in the z-direction perpendicular to the plane of the orbit, as illustrated in Figure 23.2. Equation (23.46) is already in transverse-traceless form for a wave propagating in the z-direction. The wave is an equal superposition of the two linear polarizations exhibited in (21.69) 90° out of phase.[5]

The time-averaged energy flux, f_{GW}, in one linear polarization of a plane gravitational wave was given in Section 16.5 as $f_{GW} = \omega^2 a^2/(32\pi)$ [cf. (16.22)], where ω is the frequency and a is the amplitude. From (23.46) we see that the frequency of the wave, ω, is 2Ω and its amplitude, a, is $-8(\Omega^2 MR^2)/r$ in either of the two linear polarizations. The gravitational wave luminosity, L_{GW}, (energy/time) radiated into a solid angle,[6] $d\Omega_{sa}$, about the z-direction is $r^2 d\Omega_{sa} f_{GW}$. The time-averaged angular differential luminosity, $dL_{GW}/d\Omega_{sa}$, in the z-direction from both of the two polarizations is then

$$\left(\frac{dL_{GW}}{d\Omega_{sa}}\right)_{z\text{-direction}} = 2(r^2)\frac{(2\Omega)^2}{32\pi}\left(\frac{8(\Omega^2 MR^2)}{r}\right)^2$$
$$= \frac{16}{\pi}(\Omega^3 MR^2)^2 = \frac{16}{\pi}4^{1/3}\left(\frac{\pi M}{P}\right)^{10/3}. \tag{23.47}$$

Here, the leading factor of 2 is due to the equal contributions of the two linear polarizations in (23.46). The Kepler's law connection (23.40) between the radius of the orbit and its period has been used to arrive at the last expression for the differential luminosity.

In the Orbital Plane. The radiation propagating in the x-direction is represented by (23.46) but not in transverse-traceless gauge. Making the longitudinal xx and xy components of $\bar{h}^{ij}$ zero and subtracting out the trace as in (21.70) gives

$$\bar{h}^{ij}_{TT} \xrightarrow[r\to\infty]{} \frac{4\Omega^2 MR^2}{r}\begin{pmatrix} 0 & 0 & 0 \\ 0 & \cos[2\Omega(t-r)] & 0 \\ 0 & 0 & -\cos[2\Omega(t-r)] \end{pmatrix} \tag{23.48}$$

for the transverse-traceless form appropriate to the x-direction. This is linear polarization, one of the two represented in (21.69).

[5]In fact, this is *circular* polarization, in which a transverse ellipse of particles, such as shown in Figure 16.2, rotates with the angular frequency 2Ω in response to the wave. The individual particles in the ellipse do not rotate around its center. Rather, they rotate in small circles displaced from the center in such a way that the whole pattern rotates. (See Problem 16.9.)

[6]The clumsy notation $d\Omega_{sa}$ is used here so you don't get the solid angle mixed up with the orbital frequency, Ω.

Since there is only one contributing linear polarization and its amplitude is smaller by a factor of 2 than the wave in the z-direction discussed earlier, the energy flux in the x-direction is smaller by a factor of 8 than in the z-direction (23.47):

$$\left(\frac{dL_{GW}}{d\Omega_{\text{sa}}}\right)_{x\text{-direction}} = \frac{2}{\pi}(\Omega^3 M R^2)^2 = \frac{2}{\pi}4^{1/3}\left(\frac{\pi M}{P}\right)^{10/3}. \tag{23.49}$$

Thus, both polarization and energy flux vary with angle. The complete angular distribution of radiated power can be found by working through Problem 9.

23.6 The Quadrupole Formula for the Energy Loss in Gravitational Waves

Gravitational waves carry energy away from a radiating system. By calculating the energy flux in different directions, much as we did in the two special cases in Example 23.3, and integrating over a solid angle, a useful expression for the total rate of energy loss in gravitational radiation in the weak-field, long-wavelength approximation can be derived. This is called the *quadrupole formula* for total gravitational-wave radiated power for reasons that will become clear shortly. We skip over the derivation, not because it is difficult, but only because it is long. A supplement giving it is available on the book website. In fact, we can anticipate the form of the quadrupole formula just from a few simple facts we already know.

Equation (23.35) gives the gravitational wave amplitude far from the source in terms of the second time derivative of the second mass moment, I^{ij}. We expect the expression for the energy flux to be quadratic in the wave amplitude, and the expression for the plane wave energy flux (16.22) confirms this. The luminosity L_{GW} (total radiated power) in gravitational radiation should, therefore, be quadratic in I^{ij} and its time derivatives. The number of time derivatives can be determined by dimensional analysis. In geometrized units $L_{GW} = d(\text{energy})/d(\text{time})$ is dimensionless. The *third* time derivative of I^{ij} is dimensionless; therefore, L_{GW} must be a quadratic combination of $\dddot{I}^{ij}$. Alternatively, we can note that the wave amplitude in (23.35) is proportional to $\ddot{I}^{ij}$ and there is an additional factor of ω^2 in the energy flux (16.22), making one more time derivative for each of the two factors of $\ddot{I}^{ij}$. L_{GW} also behaves as a scalar under rotations in space and so must be a quadratic scalar combination of $\dddot{I}^{ij}$. The only two possibilities are $\dddot{I}_{ij}\dddot{I}^{ij}$ or $(\dddot{I}^k{}_k)^2$. The precise combination is fixed by the fact that there is no radiation from a spherically symmetric system and, therefore, no energy loss. For a spherically symmetric system x, y, and z are all equivalent, and $I^{ij} \propto \delta^{ij}$. The combination

Mass Quadrupole Moment

$$\boxed{I\!\!\!-^{ij} \equiv I^{ij} - \frac{1}{3}\delta^{ij} I^k{}_k,} \tag{23.50}$$

called the *quadrupole moment tensor*,[7] vanishes for spherical symmetry. L_{GW} must therefore be proportional to $\dddot{I}_{ij}\dddot{I}^{ij}$. The factor turns out to be $\frac{1}{5}$. The quadrupole formula is, thus,

$$L_{GW} = \frac{1}{5}\left\langle \dddot{I}_{ij}\dddot{I}^{ij} \right\rangle. \tag{23.51}$$

Total Power Radiated in Gravitational Waves

Here, $\langle \; \cdot \; \rangle$ denotes the time average over a period. This expression is in geometrized units. In $\mathcal{MLT}$ ($c \neq 1$, $G \neq 1$) units the gravitational wave luminosity, L_{GW} (energy/time), is given by

$$L_{GW} = \frac{1}{5}\frac{G}{c^5}\left\langle \dddot{I}_{ij}\dddot{I}^{ij} \right\rangle. \tag{23.52}$$

If you have studied electromagnetism, it may be helpful to note that (23.51) is the gravitational analog of the expression for the radiated power (see the last row in Table 23.1) although in the electromagnetic case, it is the dipole rather than the quadrupole moment that supplies the leading term.

The quadrupole formula can be immediately applied to find the power radiated in gravitational waves by a binary system. The components of I^{ij} are given in (23.45). The trace, $I^k_k = 2MR^2$, is independent of time in this case; the time derivatives of I_{ij} and I_{ij} therefore coincide. Take the third time derivative of each component of I^{ij} in (23.45) and sum their squares. Average over a period to obtain a factor of $1/2$. The result for L_{GW} is

$$L_{GW} = \frac{128}{5} M^2 R^4 \Omega^6. \tag{23.53}$$

This can be expressed in terms of the period P using Kepler's law (23.40) for R with the result

$$L_{GW} = \frac{128}{5} 4^{1/3} \left(\frac{\pi M}{P}\right)^{10/3} = 1.85 \times 10^3 \left(\frac{M}{P}\right)^{10/3}. \tag{23.54}$$

(The factor of 1.85×10^3 shows that dimensional estimates ignoring factors of 2, π, etc., can be quite far off sometimes.) We can write this in $\mathcal{MLT}$ units by inserting the relevant factors of G and c. From Appendix A, $M \to GM/c^2$, $t \to ct$, and $P \to cP$. Thus,

$$L_{GW} = \frac{128}{5} 4^{1/3} \frac{c^5}{G} \left(\frac{\pi G M}{c^3 P}\right)^{10/3}, \tag{23.55}$$

which numerically is

$$L_{GW} = 1.9 \times 10^{33} \left(\frac{M}{M_\odot}\frac{1\,\text{h}}{P}\right)^{10/3} \frac{\text{erg}}{\text{s}}. \tag{23.56}$$

[7] There are a number of different conventions for the quadrupole moment tensor, all proportional to (23.50).

The luminosity of the Sun in electromagnetic radiation is 3.9×10^{33} erg/s. Binary stars with typical stellar masses and short periods are, therefore, not so extraordinarily faint in gravitational radiation. It is only that the weak coupling of gravity to matter makes the radiation hard to detect.

23.7 Effects of Gravitational Radiation Detected in a Binary Pulsar

Gravitational radiation will reduce the energy and angular momentum of an orbiting binary system and, in particular, change the orbital period P (called P_b in Section 11.3). We can evaluate the rate of change for the equal-mass circular orbit example considered in the previous section. In the Newtonian approximation adequate for this nonrelativistic system, its energy is [cf. Figure 23.2]

$$E_{\text{Newt}} = 2\left(\frac{1}{2}MV^2\right) - \frac{M^2}{2R}, \tag{23.57}$$

where V is the orbital speed. Using Newton's law (23.39) to relate V to R and Kepler's law (23.40) to relate R to P gives

$$E_{\text{Newt}} = -\frac{M^2}{4R} = -\frac{1}{4}M\left(\frac{4\pi M}{P}\right)^{2/3}. \tag{23.58}$$

The Newtonian energy is negative because the orbiting stars are bound. Reducing the energy E_{Newt} will therefore *decrease* the period P. Smaller P means more negative (lower) energy. Differentiating (23.58) with respect to t and equating dE_{Newt}/dt to $-L_{GW}$ in (23.54) gives the following formula for the rate of decrease of P:

$$\frac{dP}{dt} = -\frac{96}{5}\pi 4^{1/3}\left(\frac{2\pi M}{P}\right)^{5/3}. \tag{23.59}$$

Working out the numbers, this is

$$\frac{dP}{dt} = -3.4 \times 10^{-12}\left(\frac{M}{M_\odot}\frac{1\text{ h}}{P}\right)^{5/3}. \tag{23.60}$$

This is a dimensionless quantity.

For the binary pulsar PSR B1913+16 discussed in Section 11.3, the mass of both the pulsar and its unseen companion is about $1.4M_\odot$, and the orbital period is 7.75 h. The predicted decrease in the orbital period because of gravitational radiation can be estimated from (23.60) as of order of 10 μs per year, although the actual orbit is not circular as assumed there. Yet, so precise are the measurements of the arrival times of the signals from the pulsar that the effect of this slow decrease in the orbital period can be detected. Timing measurements over an epoch of many years gave dP/dt for PSR B1913+16 as $(-2.422 \pm .006) \times 10^{-12}$ on

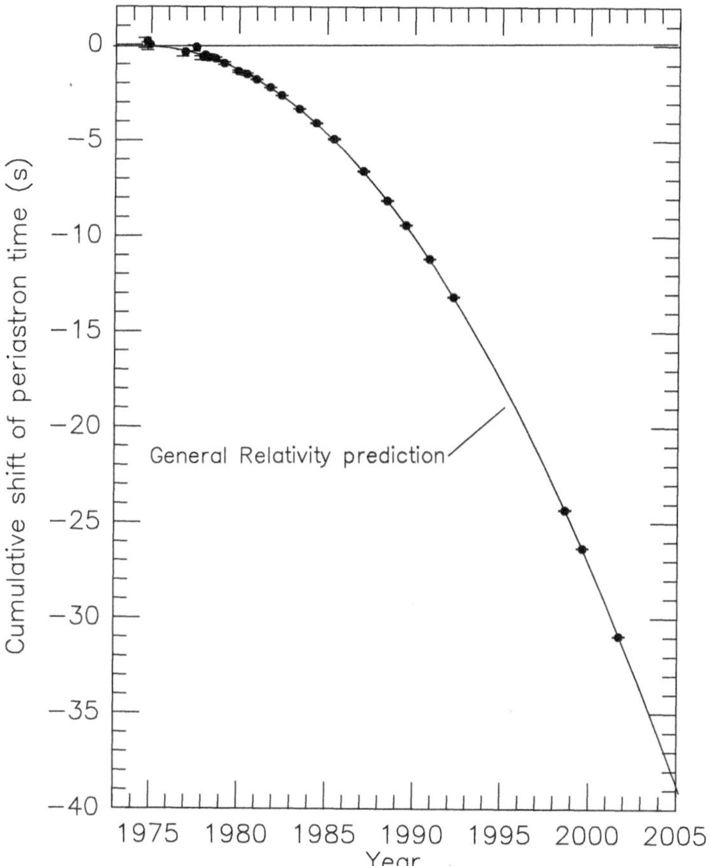

FIGURE 23.3 The detection of the effects of gravitational radiation in a binary pulsar. The binary pulsar PSR B1913+16 is a pair of neutron stars in mutual orbit about one another. The emission of gravitational radiation reduces the orbital period [cf. (23.59)]. One measure of the decrease in orbital period is the steady change with time of the time in the orbit of the periastron—the position of the pulsar's closest approach to its companion star. The figure shows the cumulative value of this shift as measured by J. Taylor and J. Weisberg at the Arecibo radio telescope in Puerto Rico (Figure 11.9) over several decades. The points are their data points. The solid line is the shift predicted by general relativity as a consequence of the emission of gravitational waves. The agreement is better than a third of a percent. The effects of gravitational waves have thus been detected in the universe, although gravitational waves have not yet been directly received at Earth.

July 7, 1984, about 6 h after midnight GMT. The results of several decades of careful timing measurements by Taylor and Weisberg for the change in orbital period are shown in Figure 23.3. The agreement of the observed decrease in orbital period with the predictions of Einstein's general relativity for the decrease due to gravitational radiation is to better than a $\frac{1}{3}$% accuracy. The effects of gravitational radiation have thus been detected.

23.8 Strong Source Expectations

Just as important as understanding how to use a relation like (23.35) to estimate or calculate in detail the amplitudes of gravitational waves is understanding when not to use it. The expression was derived under two important assumptions. (1) Small (weak) curvature was assumed everywhere in spacetime, which permitted the use of linearized gravity. (2) Nonrelativistic matter whose energy is dominated by rest energy and whose velocities are much less than light was assumed. This allowed the long-wavelength approximation to the general solution of the linearized Einstein equation (23.21). However, the most significant gravitational wave sources in the universe do not meet these criteria. Neither curvatures nor velocities are typically small in the spacetimes of two colliding black holes or a supernova collapse.

Accurate calculation of the gravitational radiation from events generating strong, rapidly varying spacetime curvature generally requires detailed numerical simulation. However, the quadrupole formula can be used as guide to a rough order-of-magnitude, dimensional analysis of the amount of radiation one might expect, as in Example 23.4.

Example 23.4. Estimating the Gravitational Radiation from Merging Black Holes. There is evidence that every sufficiently massive galaxy contains a black hole at its center. There is also evidence that every galaxy merges with another at least once in its lifetime, which could lead to the coalescence of their central black holes. What peak gravitational wave luminosity could be expected from such an event? The collision of two black holes of comparable masses M is not characterized by weak curvatures, but simple dimensional estimates based on the quadrupole formula give some idea of what to expect. All scales in this problem are determined by the value of M—the mass scale, the size of the black holes, and the time scale for collapse [cf. (9.40)]. Since the radiated power L_{GW} is dimensionless in geometrized units, we expect it to be of order of magnitude unity. Perhaps more cautiously, we might put $L_{GW} \sim \varepsilon$, where ε is an efficiency factor depending on the geometry of the merger that is not too many orders of magnitude less than unity. Estimating the various contributors to the quadrupole formula (23.51) yields the same result. Converting to $\mathcal{MLT}$ units (see Appendix A) gives

$$L_{GW} \sim \varepsilon(c^5/G) \sim \varepsilon 10^{59} \text{ erg/s}. \qquad (23.61)$$

At the time of writing, preliminary numerical simulations suggest that ε is of order 10^{-2}. To appreciate how large the luminosity (23.61) is, compare it with typical optical luminosities. The luminosity of the Sun is $\sim 10^{33}$ erg/s, the total luminosity of a large galaxy is $\sim 10^{44}$ erg/s, the luminosity of a large radio source can range up to 10^{48} erg/s, and the luminosity of the brightest gamma-ray bursts are about $\sim 10^{52}$ erg/s. Indeed, in certain simple situations c^5/G is roughly the maximum possible physical luminosity (Problem 18). The merger of two $10^9 M_\odot$ black holes at the centers of two colliding galaxies might produce the peak luminosities of (23.61) over time scales of days, thereby becoming transiently the brightest event in the universe (Problem 19). Numerical simulation can put such rough expectations on firmer ground, and gravitational wave detectors in space can check them observationally.

Care must be taken in making estimates such as these to use the mass that is contributing the time-dependent quadrupole moment. For example, in the spherically symmetric gravitational collapse to a billion-solar-mass black hole, the factor ε is exactly zero. Estimates are useful but are no substitute for detailed calculation.

Problems

1. In Example 21.5 we showed that the equations of vacuum linearized gravity were satisfied for the time-independent static, weak field, metric (21.25) when Φ satisfies the vacuum Newtonian equation $\nabla^2\Phi = 0$. Show the same thing for the equations of linearized gravity with nonrelativistic sources when Φ satisfies the Newtonian equation with sources $\nabla^2\Phi = 4\pi\mu$ (geometrized units).

2. [C] Equation (23.28) for the metric perturbation produced by the slow and uniform rotation of a spherical body was derived only for large values of r compared to the size of the source. Show that it holds for all values of r outside the rotating body.

3. Work through the details of deriving the Lorentz gauge condition (23.23) from (23.22).

4. Spell out all the steps in the derivation of the metric outside a slowly rotating body (23.28) from (23.25).

5. [E] Would a nuclear explosion halfway around the Earth produce a gravitational wave of sufficient amplitude to be detected by the LIGO gravitational wave receiver? To answer this question, *estimate* the amplitude h that might be expected from such an explosion and compare with the rough sensitivity $h \sim 10^{-22}$ expected of the advanced LIGO detectors. (A large nuclear explosion is 20 megatons of TNT. One megaton of TNT $= 4.2 \times 10^{22}$ erg.)

6. [C] *No Dipole Gravitational Radiation*

(a) In Section 23.4 we did not discuss the large r behavior of the $\bar{h}^{t\alpha}$ parts of the metric perturbations. Show that in the long-wavelength approximation, these are given by

$$\bar{h}^{t\alpha} = 4P^\alpha/r,$$

where P^α is the total energy momentum four-vector of the matter. To simplify your discussion, you may assume that the stress-energy tensor of the matter has the nonrelativistic form $T^{\alpha\beta} = \mu u^\alpha u^\beta$ with all velocities much less than unity.

(b) Show that

$$\bar{h}^{ti} = 4\dot{p}^i/r,$$

where $\vec{p}$ is the mass dipole moment

$$\vec{p} = \int d^3x\, \mu(x)\vec{x}.$$

What other important quantity in Newtonian mechanics is the mass dipole moment connected to?

(c) Argue that by a Lorentz transformation to a new inertial frame, the mass dipole term can be made to vanish and, therefore, that there is no contribution to gravitational radiation. Find the velocity and direction of the Lorentz transformation.

7. Spell out the detailed steps in the derivation of the the large-distance gravitational wave amplitude (23.35) from (23.30).

8. What combination of the two polarizations $+$ and $\times$ is the gravitational radiation emerging at an angle of $45°$ with respect to the axis perpendicular to the plane of the circular orbit of two equal mass stars?

9. [C] *Angular Distribution of Radiated Gravitational Wave Power from a Binary Star System* In Example 23.3, the time-averaged power radiated in gravitational waves was calculated for a binary star system for two directions—one normal to and one in the plane of the orbit. This problem aims at calculating the complete angular distribu-tion (the "antenna pattern"). The time-averaged distribution will be symmetric about the axis of rotation because the time-averaged source is axisymmetric. It is, therefore, necessary to calculate only the power radiated in a direction making an angle θ with the z-axis, which can be conveniently taken to lie in the y-z plane. Proceed as follows:

(a) Rotate the spatial coordinates about the x-axis by an angle θ so that the new z-axis makes an angle θ with the old one. Transform the gravitational wave amplitude (23.46) to this coordinate system.

(b) Put the approximate plane wave propagating in the new z-direction in TT gauge.

(c) Calculate the power radiated in the new z-direction, thereby getting the radiated power as a function of θ. Check that your answer agrees with the two special cases in Example 23.3. Draw a rough plot of the antenna pattern.

(d) If you integrate the angular distribution of radiated power to get the total radiated power, do you get the answer from the quadrupole formula quoted in (23.53)?

10. Two equal masses M are at the ends of a massless spring of unstretched length L and spring constant k. The masses are started oscillating in line with the spring with an amplitude A so that their center of mass remains fixed. Calculate the amplitude of gravitational radiation a long distance away from the center of mass of the spring as a function of the angle θ from the axis of the spring to lowest nonvanishing order in A. Analyze the polarization of the radiation. Calculate the angular distribution of power radiated in gravitational waves.

11. A particle of mass m moves along the z-axis according to $z(t) = (1/2)gt^2$ (g is a constant) between times $t = -T$ and $t = +T$ and is otherwise moving with constant speed. Calculate the gravitational wave metric perturbations at a large distance L along the positive z-axis in the Lorentz gauge used in (23.21). Find the same perturbation in the TT-gauge appropriate for the z-axis and evaluate the power in gravitational waves emitted along the positive x-axis.

12. What is the longest period a binary consisting of two neutron stars in circular orbit, each with $1.4M_\odot$, could have *now* and coalesce before the end of the universe (assuming that it has about 15 billion more years to go)?

13. *Angular Momentum Loss Through Gravitational Radiation* In Newtonian physics an axisymmetric body rotating rigidly about a principal axis with an angular velocity

Ω has a kinetic energy E and angular momentum along the axis J given by

$$E = \frac{1}{2} I \Omega^2, \qquad J = I\Omega,$$

where I is the moment of inertia about that axis. Assuming that this is true for linearized gravity (it is), calculate the average rate over a period at which the binary star system discussed in Section 23.5 is losing angular momentum in gravitational radiation.

14. [C] *Gravitational Radiation Reaction* A particle of mass m moves because of an applied force and radiates gravitational radiation. Suppose the velocity of the particle is much less than the velocity of light so that nonrelativistic kinematics applies. Show that the rate at which the particle loses energy in gravitational waves is the same, in a time-averaged sense, as if it were acted on by a gravitational radiation reaction force

$$F_i^{\text{rad. react.}}(t) = -\frac{2}{5} m \frac{d^5 \bar{I}_{ik}(t)}{dt^5} x^k(t),$$

where x^i are usual rectangular coordinates giving the particle's position and $\bar{I}_{ij}$ is the quadrupole moment defined in (23.50). If you are familiar with electromagnetism, compare this force with the radiation reaction force in electrodynamics.

15. [E] Lunar laser ranging measurements of the position of the Moon relative to the Earth lead to the inference that the length of the day is increasing by 2 millisec per century. *Estimate* whether gravitational radiation from the Earth is an important or negligible contribution to this slowdown in the Earth's rotation rate.

16. A steel beam of mass M and length L, much longer than it is wide, rotates about an axis through its center of mass perpendicular to its length with an angular frequency Ω. Under what conditions is the quadrupole formula for the total power applicable? Assuming it is, use it to calculate the power radiated in gravitational radiation. If the beam were contained in a drag-free satellite, what would be the predicted decrease in angular frequency in one year of rotation?

17. [E] When a small body of mass m falls from rest into a large black hole of mass M, there is a burst of gravitational radiation. *Estimate* the duration of the burst, the peak gravitational wave luminosity, and also the total power radiated as a fraction of the small body's rest mass. What is the peak gravitational wave luminosity produced by a $10M_\odot$ black hole falling into the $\sim 10^6 M_\odot$ black hole at the center of our galaxy (Section 13.2)? How does this compare with the optical luminosity of the whole galaxy?

18. [E] *Maximum Luminosity* Imagine a point source of radiation at the center of a spherical star of radius R. Suppose that the luminosity of the source is L (energy/time) and is steady in time. Use Newtonian physics to estimate the maximum luminosity as follows:

(a) Calculate the energy density in radiation in the interior of the star.

(b) Estimate the maximum L above which the star would be inside its Schwarzschild radius.

(c) Compare the maximum luminosity to the estimate of the luminosity from two merging black holes (23.61).

19. [E, C] *Gravitational Waves from Merging Supermassive Black Holes* Suppose for simplicity that (1) every galaxy contains a $10^9 M_\odot$ black hole, (2) that every galaxy merges once in its lifetime, and (3) that when they do, the black holes in their cores coalesce. Consider a detector built to detect the gravitational waves from such events. Even though they do not really apply, use the results of linearized gravity to:

(a) Estimate the frequency range in which the detector would have to operate.

(b) Estimate the strain sensitivity that would be necessary to see mergers out to the edge of the visible universe, ~ 1 Gpc in radius.

(c) Estimate the duration of such events in usual time units.

(d) Estimate the rate at which such events would be detected.

CHAPTER

24

Relativistic Stars

Most stars support themselves against the collapsing force of gravity by the pressure of hot gas. In steady state the energy lost to radiation is supplied by the thermonuclear reactions that combine nuclei and release energy. The energy from our Sun, for example, results mainly from energy released in reactions where four hydrogen nuclei (four protons) combine to make one helium nucleus.

As described in Chapter 12, eventually the core of a star may run out of thermonuclear fuel. It can then evolve to one of two endstates: (1) ongoing gravitational collapse leading to a black hole or (2) a star supported against gravity by a nonthermal source of pressure. The first possibility—collapse to a black hole—was described in Chapter 12. This chapter returns to the second possibility realized in nature by white dwarf stars and neutron stars.

Unlike black holes, which can be understood entirely in the context of general relativity, an understanding of the stars at the endstate of stellar evolution requires almost all of the rest of physics in some way. For instance, the simplest examples of nonthermal pressure by which the Earth (and indeed ordinary objects such as tables and chairs) are supported against gravity involve understanding the properties of solids. That is why this chapter is at the end of the book.

To make this point more emphatically, imagine boring from the surface of a neutron star to its center. Beneath an ocean of hydrogen and helium lies a solid crust whose properties are determined by the quantum electronic forces between atoms. These are the same forces that determine the properties of ordinary solids, but here operating at densities much greater than any available on Earth.

Dig further and we enter a region where the properties of the matter are determined by the Pauli exclusion principle applied to relativistic electrons kept in equilibrium with protons and neutrons by the weak interactions. These nucleons are bound in neutron-rich nuclei unlike any found naturally on Earth.

Dig yet deeper and we enter a region where almost all the electrons have vanished. The nuclei have dissolved to yield superfluid nuclear matter, whose properties are determined by the strong interactions, but here operating at densities beyond those in ordinary nuclei or accessible in terrestrial laboratories.

This passage from surface to center goes through regimes where condensed matter physics, relativistic statistical mechanics, weak interaction physics, nuclear physics, and strong interaction physics are central to understanding. Quantum mechanics is important throughout, and gravity ties it all together.

We cannot hope to review the range of physics necessary for a complete understanding of the equilibrium endstates of stellar evolution in this book. However, we can isolate the essential role of gravitational physics and discuss the overall structure of these stars because of one central fact: Gravitation is the only long-range force operating. The forces governing the properties of the matter are all short range. Nuclear forces, for example, operate over distances of order 10^{-13} cm; the radius of a typical neutron star is of order 10 km; white dwarfs are even larger. Electromagnetic forces can be long range for matter with a net electric charge. But the matter from which neutron stars and white dwarfs are made is electrically neutral, so long-range electromagnetic forces are effectively screened.

This difference in ranges means that, to an excellent approximation, the properties of the matter relevant for the gross structure of neutron stars and white dwarfs can be summarized by an *equation of state* relating the pressure p of an ideal matter fluid to its energy density ρ. The job of understanding the equilibrium endstates of stellar evolution can, therefore, be divided into two parts: (1) calculating the equation of state of matter at the end of thermonuclear evolution, and (2) calculating how stars made from this matter are held together by gravity. In the following we merely report on the results of (1) but derive (2). We begin with the very simplest example of a nonthermal source of pressure.

24.1 The Power of the Pauli Principle

A simple but very important example of a nonthermal source of pressure is the Fermi pressure arising from the Pauli exclusion principle. The Pauli principle restricts the quantum states allowed to half-integral spin identical particles (such as electrons, protons and neutrons) referred to collectively as *fermions*. The Pauli principle prohibits any two identical fermions from occupying the same quantum state. The Pauli principle is crucial for the structure of atoms and their chemical properties. It has the particular consequence that in the lowest-energy state of an atom, the electrons are not in the lowest-energy level near the nucleus; instead, they are arranged in higher-energy-level *shells*. In effect, the Pauli principle supports the outer electrons in an atom against the attractive electric force of the nucleus. We will see how the Pauli principle can support a star against the attractive force of gravity.

To understand the operation of the Pauli principle, think about a gas of spin-$\frac{1}{2}$ identical fermions in a box. Assume that the fermions are at *zero temperature* so that the gas is in its lowest-possible energy state. Such a gas of fermions is said to be *degenerate*. Even at zero temperature, even if there is no interaction poten-tial between the fermions, there is a pressure arising from the Pauli principle. The essential idea can be understood in one dimension.[1]

First, consider the states of a *single* fermion of mass m moving in a one-dimensional box that extends from $x = 0$ to $x = \mathcal{L}$, as illustrated in Figure 24.1. For simplicity, assume for the moment that the fermion is *nonrelativistic*. The

[1]The following discussion assumes some very elementary quantum mechanics. If it is unfamiliar, you can take a look at the supplement on the book website, where more details are filled in. Alternatively, you might want to review a basic quantum mechanics text or just assume the result and skip to the next section.

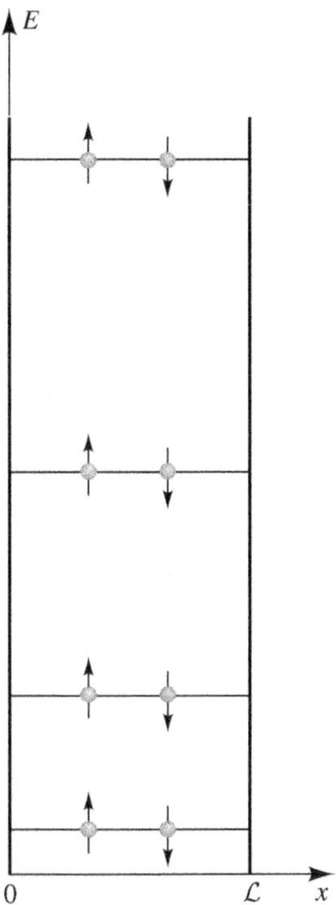

FIGURE 24.1 The energy levels of a one-dimensional box of length $\mathcal{L}$ containing eight noninteracting fermions. The energy levels available to a single particle are discrete. The lowest energy state of eight fermions has two fermions with opposite spins in the lowest level, two in the next lowest, and so on. That is the lowest-energy configuration of the eight particles in which no two fermions occupy the same state, as required by the Pauli exclusion principle. If the particles in the box did not obey the exclusion principle, the lowest-energy state would have all eight fermions in the lowest single-particle level.

possible energy levels in the box are quantized

$$E_k = \frac{1}{2m}\left(\frac{k\pi\hbar}{\mathcal{L}}\right)^2 \equiv \frac{p_k^2}{2m}, \qquad k = 1, 2, \ldots \tag{24.1}$$

where p_k is the magnitude of the momentum equal to

$$p_k \equiv \frac{k\pi\hbar}{\mathcal{L}}, \qquad k = 1, 2, \ldots . \tag{24.2}$$

These turn out to be the allowed values for momentum, whether the particle is nonrelativistic or not.

The lowest-energy state of the box containing one particle is achieved by putting that particle in the lowest single-particle energy eigenstate $k = 1$. But the lowest-energy state of a box with $\mathcal{N}$ particles has to take account of the Pauli principle. Suppose, for simplicity, that $\mathcal{N}$ is even and imagine filling the box one particle at a time, putting the added particle into the lowest available single-particle energy eigenstate. A spin-$\frac{1}{2}$ particle has two possible spin states. The first two particles can be put into the lowest-energy ($k = 1$) single-particle state, but the next two must go into the next higher ($k = 2$) single-particle state, and so on. The magnitude of the momentum of the highest state filled that contains the last two particles is called the *Fermi momentum* p_F. Thus, the lowest-energy state of a box with $\mathcal{N}$ particles has the total energy

$$\mathcal{E} = \sum_{k=1}^{\mathcal{N}/2} (2E_k). \tag{24.3}$$

Sums such as those in (24.3) can be replaced by integrals for the very large values of $\mathcal{N}$ that might characterize a realistic gas or star. It is convenient to write them as integrals over the magnitude of the momentum p from 0 to p_F. Equation (24.2) shows that there is one state for every interval $\pi\hbar/\mathcal{L}$ in p. Thus, for the sums of interest,

$$2\sum_{k=1}^{\mathcal{N}/2} F(p_k) \approx 2\frac{\mathcal{L}}{\pi\hbar}\int_0^{p_F} dp\, F(p) \tag{24.4}$$

for any function $F(p)$. The factor of $(\mathcal{L}/\pi\hbar)$ is thus the density of states in magnitude of momentum.

We live in three spatial dimensions, not one. The energy eigenstates of a free particle moving in a cubical box of size $\mathcal{L}$ can be characterized by the magnitudes of the three components of the momentum $(p^x,\ p^y,\ p^z)$, each quantized according to a rule such as (24.2). The density of states in the three-dimensional momentum space spanned by positive values of p^x , p^y, and p^z is $(\mathcal{L}/\pi\hbar)^3$. In the lowest-energy state of a system of $\mathcal{N}$ fermions, all these states are filled out to some momentum space radius p_F.

We can immediately illustrate these ideas by finding the Fermi momentum p_F for a gas of $\mathcal{N}$ identical fermions. The total number $\mathcal{N}$ must be two times the sum of all the states with positive p^x , p^y, p^z inside the sphere of radius p_F:

$$\mathcal{N} = 2\left(\frac{\mathcal{L}}{\pi\hbar}\right)^3 \frac{1}{8}\int_0^{p_F} 4\pi p^2\, dp. \tag{24.5}$$

Here, the integral over the octant of momentum space with positive p's has been written as one-eighth of the integral over the whole sphere. This gives the following connection between the number density $n = \mathcal{N}/\mathcal{L}^3$ and p_F:

$$n = \frac{p_F^3}{3\pi^2\hbar^3}. \tag{24.6}$$

If we introduce the Fermi wavelength $\lambda_F \equiv 2\pi\hbar/p_F$, the number density can be written $n = 8\pi/3\lambda_F^3$, showing that one fermion is confined to a volume of characteristic size λ_F. This is the relation that we used to estimate the maximum mass of stars supported by Fermi pressure in Box 12.1 on p. 257. We will shortly supply a more quantitative derivation of this maximum mass.

In a similar manner the energy density ρ of the gas can be calculated by weighting the integral in (24.5) by the energy as a function of momentum—$E(p) = (m^2c^4 + p^2c^2)^{1/2}$. (We retain the factors of c for comparison with other formulas you may have seen.) The result will depend on the upper limit of integration p_F, but that can be reexpressed in terms of n using (24.6). The integral is not difficult and is particularly simple in the nonrelativistic limit, where $E(p) \approx mc^2 + p^2/2m$, and in the extreme relativistic limit, where $E(p) \approx pc$. In these limits

$$\rho = mc^2 n + \frac{3}{10}(3\pi^2)^{2/3}\left(\frac{\hbar^2}{m}\right) n^{5/3} \qquad \text{(nonrelativistic)}, \qquad (24.7a)$$

$$\rho = \frac{3}{4}(3\pi^2)^{1/3}(\hbar c)n^{4/3} \qquad \text{(relativistic)}. \qquad (24.7b)$$

To calculate the pressure we can use the first law of thermodynamics at zero temperature. This relates the change $\Delta\mathcal{E}$ of the total energy in the box to a small change $\Delta\mathcal{V}$ in its volume, keeping the number of fermions, $\mathcal{N}$, fixed. Specifically,

$$\Delta\mathcal{E} = -p\Delta\mathcal{V}, \qquad (24.8)$$

where p is the pressure. (Be careful not to get p for pressure mixed up with p for momentum—both notations are standard.) Noting that $\mathcal{E} = \rho\mathcal{V}$ and $\mathcal{V} = \mathcal{N}/n$,

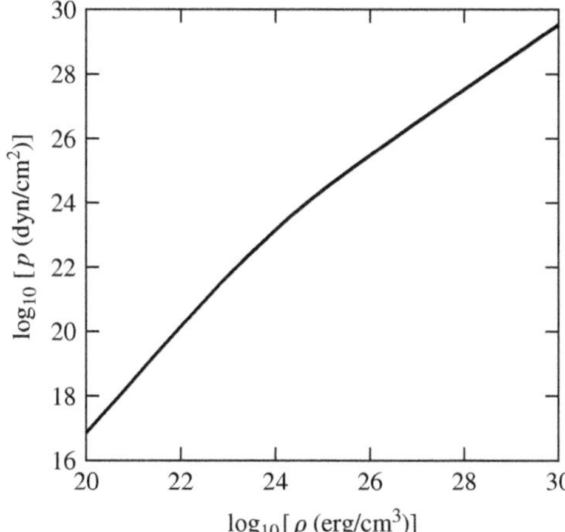

FIGURE 24.2 The equation of state of a gas of noninteracting electrons at zero temperature. Pressure p in dyn/cm^2 is plotted vertically and energy density ρ in ergs/cm^3 is plotted horizontally on this log-log plot. (These are the same units, just expressed differently). This illustrates how pressure can arise from the Pauli exclusion principle. The pressure changes gradually from the equation of state for a nonrelativistic gas of electrons to one where the electrons are relativistic.

this gives $p = n(d\rho/dn) - \rho$, which together with (24.7), leads to

Equation of State of Degenerate Fermions

$$p = \frac{1}{5}(3\pi^2)^{2/3}\left(\frac{\hbar^2}{m}\right)n^{5/3} \qquad \text{(nonrelativistic)}, \tag{24.9a}$$

$$p = \frac{1}{4}(3\pi^2)^{1/3}(\hbar c)n^{4/3} \qquad \text{(relativistic)}. \tag{24.9b}$$

The number density n can be eliminated between the expressions (24.7) for the energy density ρ and the expressions (24.9) for the pressure to give the equation of state $p = p(\rho)$ of noninteracting fermions at zero temperature. The results are shown graphically in Figure 24.2. In the following we will use this simple equation of state to illustrate how this nonthermal source of pressure can be balanced by the attractive forces of relativistic gravity to give a family of equilibrium stars.

24.2 Relativistic Hydrostatic Equilibrium

This section exhibits the equations of structure that determine the properties of a relativistic star in equilibrium between the compressive force of gravity and a pressure that resists compression. We assume an equation of state $p = p(\rho)$ is given, such as the one described in the previous section.

For simplicity, we will study only nonrotating, spherically symmetric stars. The geometry outside the star is then described by the familiar Schwarzschild metric (9.9),

$$ds^2 = -\left(1 - \frac{2M}{r}\right)dt^2 + \left(1 - \frac{2M}{r}\right)^{-1}dr^2 + r^2\left(d\theta^2 + \sin^2\theta\, d\phi^2\right), \tag{24.10}$$

where M is the total mass of the star. However, to determine the structure of the star, the metric inside is needed as well.

Example 21.4 showed that coordinates (t, r, θ, ϕ) can be chosen so that any spherically symmetric, time-independent metric can be put in the form

$$ds^2 = -e^{\nu(r)}dt^2 + e^{\lambda(r)}dr^2 + r^2\left(d\theta^2 + \sin^2\theta\, d\phi^2\right) \tag{24.11}$$

for some functions $\nu(r)$ and $\lambda(r)$ that depend on r alone. The Schwarzschild metric has this form. A metric of this form will hold inside the star as well as outside.

We assume that the matter making up the star is a perfect fluid with stress-energy tensor of the form described in Section 22.2. The stress-energy given in (22.39) depends on the energy density ρ, the fluid pressure p, and the fluid four-velocity $\mathbf{u}$. Since the star's matter is static, the spatial part of $\mathbf{u}$ vanishes, and the normalized $\mathbf{u}$ has the form $u^\alpha = (e^{-\nu/2}, \vec{0})$. The pressure is connected to the density by the equation of state $p = p(\rho)$. In the geometrized units used in almost all this chapter, both pressure and energy density have units of inverse length squared (cf. the discussion on p. 479).

Between metric and matter there are thus a total of three unknown functions for which the Einstein equation must be solved—$\nu(r)$, $\lambda(r)$, and $\rho(r)$. We will not write out the components of the Einstein equation $G_{\alpha\beta} = 8\pi T_{\alpha\beta}$ explicitly. Rather, we will exhibit three independent equations that follow from combining these equations in various ways (see Problem 4). We begin with a useful redefinition:

$$e^{-\lambda(r)} \equiv 1 - \frac{2m(r)}{r}. \tag{24.12}$$

This just replaces one as yet unknown function, $\lambda(r)$, by another, $m(r)$. The new function $m(r)$ has a constant value outside the star—its total mass, M. The three equations for $m(r)$, $\rho(r)$, and $\nu(r)$ are

Equations of Structure for Spherical Relativistic Stars

$$\frac{dm(r)}{dr} = 4\pi r^2 \rho(r), \tag{24.13a}$$

$$-\frac{dp(r)}{dr} = [\rho(r) + p(r)]\left(\frac{m(r) + 4\pi r^3 p(r)}{r^2(1 - 2m(r)/r)}\right), \tag{24.13b}$$

$$\frac{1}{2}\frac{d\nu(r)}{dr} = -\frac{1}{\rho(r) + p(r)}\frac{dp(r)}{dr} = \frac{m(r) + 4\pi r^3 p(r)}{r^2(1 - 2m(r)/r)}. \tag{24.13c}$$

Here, p is always understood to be related to ρ by the equation of state. These three equations are collectively referred to as the *equations of structure* for spherical relativistic stars. We will see how they determine the distributions of pressure, density, and geometry inside a star in the next section.

To better understand these equations, it is instructive to look at their nonrelativistic limit. Putting back the factors of G and c as in Appendix A sends

$$m \to \frac{Gm}{c^2}, \qquad \rho \to \frac{G\rho}{c^4}, \qquad p \to \frac{Gp}{c^4}. \tag{24.14}$$

The rest-mass density, μc^2, dominates the energy density ρ in the nonrelativistic limit. Comparing (24.11) with the static, weak field metric (6.20) shows that in that limit $\nu(r)$ becomes $2\Phi(r)/c^2$, where $\Phi(r)$ is the Newtonian gravitational potential inside the star. Inserting these results in the equations of structure (24.13) and taking the leading order in $1/c$ gives, for their nonrelativistic limit,

$$\frac{dm(r)}{dr} = 4\pi r^2 \mu(r), \tag{24.15a}$$

$$-\frac{dp(r)}{dr} = \mu(r)\frac{Gm(r)}{r^2}, \tag{24.15b}$$

$$\frac{d\Phi(r)}{dr} = -\frac{1}{\mu(r)}\frac{dp(r)}{dr} = \frac{Gm(r)}{r^2}. \tag{24.15c}$$

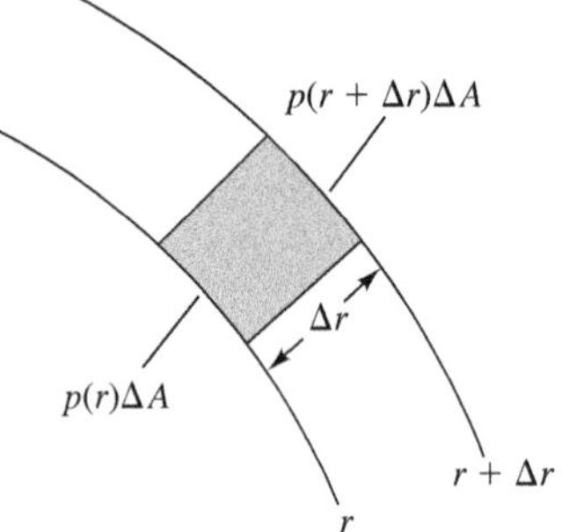

FIGURE 24.3 The pressure forces on a small fluid element inside a spherical star. A fluid element has dimension Δr in the radial direction and an area ΔA in the directions transverse to r. The radial pressure force on the outer face is $-p(r+\Delta r)\Delta A$. The pressure force on the inner face is $+p(r)\Delta A$. The net pressure force pointing in the radial direction is $-p(r+\Delta r)\Delta A + p(r)\Delta A = -(dp/dr)\Delta\mathcal{V}$, where $\Delta\mathcal{V} = \Delta A \Delta r$ is the volume of the fluid element. In hydrostatic equilibrium this net pressure force will be balanced by gravitational attraction toward the center.

The first equation determines the rest mass, $m(r)$, interior to the radius, r, in terms of the rest mass density, $\mu(r)$. The next equation expresses the balance between pressure forces and gravitational forces in the equilibrium star. To see that, consider a small cubical volume $\Delta\mathcal{V}$ located at a radius r (see Figure 24.3). The volume has one side of size Δr oriented along the radius; the two sides transverse to the radius make an area ΔA. The net pressure force pointing outward on the volume is the difference between the pressure forces on the two transverse faces:

$$-p(r+\Delta r)\Delta A + p(r)\Delta A \approx -\frac{dp}{dr}\Delta\mathcal{V}. \tag{24.16}$$

This must balance the inward gravitational force on the mass in the volume:

$$(\mu\Delta\mathcal{V})\frac{Gm(r)}{r^2}. \tag{24.17}$$

That balance is the content of (24.15b). When multiplied by the mass in the volume, (24.15c) equates the gravitational force derived from the gravitational potential with that derived from Newton's theorem for spherical bodies (recall Example 3.1). Accepting Newton's theorem, it could be regarded as the definition of the gravitational potential.

The general relativistic equations (24.13) have roughly the same interpretation. The first (24.13a) defines a quantity $m(r)$ that behaves like a mass interior to the radius r in determining the equilibrium. It cannot be exactly a mass because, although there is an energy density of matter, there is no notion of a local energy density in general relativity that includes an energy for gravity, as was discussed in Section 16.5. Nevertheless, the value of $m(r)$ at the surface radius R is the mass M in the exterior Schwarzschild geometry and the total mass of the star as measured from infinity. Equation (24.13c) determines $\nu(r)$ and, therefore, the metric component $g_{tt}(r)$. Equation (24.13b) is the relativistic equation of hydrostatic equilibrium expressing the balance between pressure and gravitational forces on each small volume inside the star.

A comparison of the right-hand side of (24.13b) with its Newtonian limit (24.15b) shows that, roughly speaking, in general relativity the forces of gravitational attraction are stronger than the Newtonian $m(r)/r^2$. The contributions of the pressure, for example, *add* to $m(r)$ and $\rho(r)$. The denominator is *smaller* than r^2 by a factor of $1-2m(r)/r$. These differences have important consequences for relativistic stars.

24.3 Stellar Models

The equations of relativistic hydrostatic equilibrium (24.13) are a system of first-order, ordinary differential equations that can be integrated by standard numerical algorithms for a given equation of state. We now describe how to carry out that integration in detail. We first consider the two equations (24.13a) and (24.13b) for $m(r)$ and $\rho(r)$, which form a closed system by themselves.

1. Begin at the center of the star $r = 0$ with a value for the central density ρ_c. The central pressure p_c is determined by the equation of state, $p_c = p(\rho_c)$. The central value of $m(r)$ must be zero; otherwise spacetime would not be locally flat with a spatial metric of the form $dS^2 = dr^2 + r^2(d\theta^2 + \sin^2\theta\, d\phi^2)$ in the neighborhood of the center (Problem 5).
2. Integrate the coupled equations of structure, (24.13a) and (24.13b), outward with these boundary conditions. Equation (24.13b) shows that $dp/dr < 0$, so the pressure drops steadily, as illustrated in Figure 24.4. So does the density when $dp/d\rho > 0$ (as in Figure 24.2). (In fact, this is a property of *any* equation of state, as is explained in Section 24.6.) The mass, $m(r)$, rises steadily. Eventually, a radius R is reached where the pressure vanishes, $p(R) = 0$. That is the surface of the star, where no weight of external matter is needed to hold the last small volume of fluid in place. The value of $m(R)$ at the surface is the total mass of the star M.
3. Repeat this process for all values of ρ_c from zero to infinity to find the family of spherical stars made of matter with the given equation of state. These form a one-parameter family, with ρ_c as the parameter. The masses, $M(\rho_c)$, and radii, $R(\rho_c)$, of these stars are functions of ρ_c.
4. To complete the calculation of the spacetime geometry inside the star, (24.13c) can be integrated inward[2] from the surface value $\log(1 - 2M/R)$ to find $\nu(r)$ in the interior. The metric inside the star is given by (24.11), with $\nu(r)$ determined by this calculation and $\lambda(r)$ by (24.12).

[2] In practice we integrate (24.13c) outward with the other equations, starting with $\nu(0) = 0$, for example, and then add a constant so that it matches the Schwarzschild geometry at the star's surface.

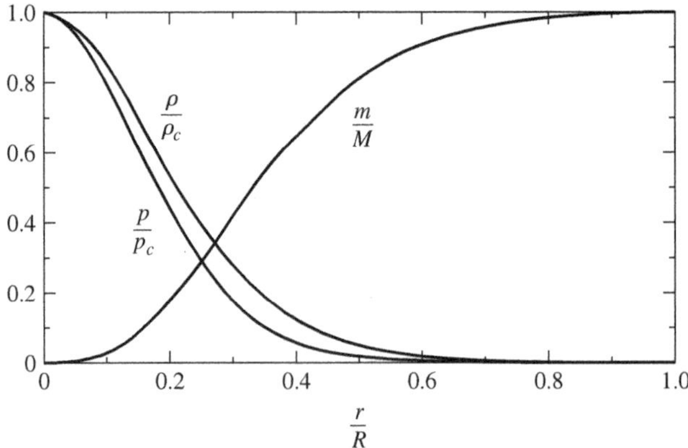

FIGURE 24.4 The structure of a stellar model. The figure shows $\rho(r)$, $p(r)$, and $m(r)$ obtained by integrating the relativistic equations of structure (24.13) using the equation of state of degenerate electrons supplemented by the rest energy of an equal number of protons. The particular model starts with a central density of $\rho_c = 1 \times 10^{10}$ g/cm^3 and a corresponding central pressure of $p_c = 1.06 \times 10^{28}$ dyn/cm^2. As the integration moves outward, pressure, $p(r)$, and density, $\rho(r)$, fall, and the mass function, $m(r)$, rises. The pressure vanishes at a radius $R = 1314$ km, which is the radius of the surface of the star. The mass $M = m(R)$ is $1.42\ M_\odot$. This is a model near the peak of the sequence of models illustrated in Figure 24.5.

The results of such a calculation are illustrated in Figure 24.5 for an equation of state of matter consisting of equal numbers of protons and electrons. The pressure is assumed to be supplied by degenerate electrons obeying (24.9). Most of the energy density comes from the rest energy of the protons. This approximates the physics inside realistic white dwarfs. There is a one-parameter family of stars labeled by their central density, ρ_c. The plot shows the total mass, M, and Schwarzschild coordinate radius, R, for each star in the sequence. An important result of this calculation is that there is an upper limit of the total mass that can be supported by the Fermi pressure of degenerate electrons of about $1.4 M_\odot$. This is the *Chandrasekhar mass*, whose value was estimated roughly in Box 12.1.

Chandrasekhar Mass

The family of stars exhibited in Figure 24.5 is a reasonable approximation to realistic white dwarfs. These stars are not very relativistic. The characteristic ratio GM/c^2R is a few parts in a thousand at the most. However, the pressure of degenerate electrons is a good approximation to the source of pressure in matter in stars at the endstate of stellar evolution only for densities below about 10^{11} g/cm^3. It is to higher densities and more relativistic objects that we now turn.

24.4 Matter in Its Ground State

The equation of state for free degenerate fermions used to construct the family of stars illustrated in Figure 24.5 is the answer to the question: "What is the lowest energy state (the ground state) of $\mathcal{N}$ *noninteracting* fermions in a box?" More

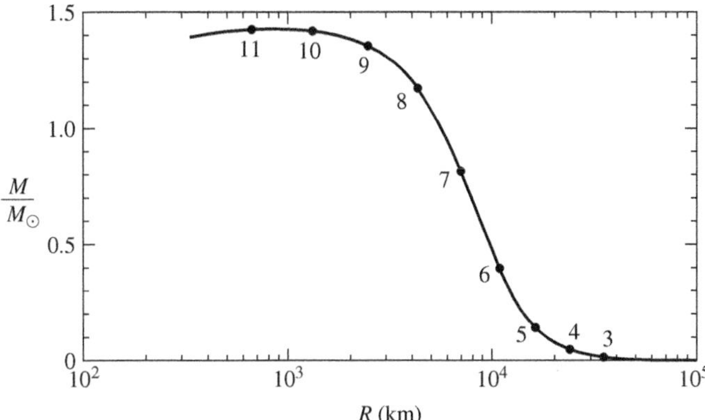

FIGURE 24.5 Mass vs. radius for the nonrotating stars supported by the Fermi pressure of zero temperature electrons arising from the Pauli principle. The equation of state discussed in Section 24.1 was used for the electrons supplemented by the rest energy of an equal number of accompanying protons. This approximates the matter in white dwarfs. The curve is the one-parameter family of possible stars parametrized by the central density ρ_c. Values of $\log_{10}[\rho_c(\text{g/cm}^3)]$ are indicated along the curve. There is an upper limit of approximately $1.4M_\odot$ to the amount of this matter that can be supported by Fermi pressure against gravitational collapse, which is called the Chandrasekhar mass. This family of stellar models approximates realistic white dwarfs.

generally, we ask for the equation of state of the ground state of matter, *including* realistic forces such as the strong and weak interactions, which become important at sufficiently high densities. That is a much harder question to answer theoretically, but some of the physics involved is described qualitatively in Box 24.1. The family of spherical stars constructed from this realistic ground state matter is a good approximation to the endstates of stellar evolution at central densities above 10^{14} g/cm^3. The overall properties of these stars, such as their mass and radius, depend almost entirely on the properties of the matter above this density, where an ideal fluid is a good approximation, with an equation of state relating pressure and density.

It is often convenient (and conventional) to summarize an equation of state by the quantity

$$\gamma \equiv \frac{\rho + p}{p}\frac{dp}{d\rho}. \tag{24.18}$$

That's because some useful simple models yield equations of state for which γ is a constant. For instance, the equation of state of degenerate fermions discussed in Section 24.1 has $\gamma = \frac{5}{3}$ when the fermions are nonrelativistic, making a smooth transition to $\gamma = \frac{4}{3}$ when they become relativistic (Problem 13). The dimensionless quantity γ is a measure of the stiffness of the equation of state. A larger γ means a larger increase in pressure for given increase in energy density and a stiffer equation of state.

BOX 24.1 The Ground State of Matter

What is the lowest-energy state of 10^{27} protons and an equal number of electrons in a box, say, 1 m on a side? That lowest-energy state is called the *ground state* of matter under those conditions. The ground state will have zero temperature, so there is no energy in thermal motion. Electrons will be bound to nuclei to make atoms, and atoms will be bound together to make a solid. But the ground state does not consist of 10^{27} hydrogen atoms. Protons can be combined with electrons to make neutrons, and protons and neutrons can be combined to make nuclei, which are more bound (lower energy) than their constituents separately. Never mind how the nuclear reactions necessary for these transitions could be made to happen; we are investigating a question of principle. The atom with the lowest mass per nucleon has a ^{56}Fe nucleus. That is mostly because ^{56}Fe is close to being the most bound nucleus. (See the curve of binding energy vs. nucleon number in Figure 12.1.) The ground state of an initial configuration of 10^{27} protons and an equal number of electrons is, therefore, a lump of solid ^{56}Fe at zero temperature. Since the density of iron at low pressure is ~ 7.9 g/cm^3, the lump is approximately 20 cm on a side and fits nicely into the 1 m box at zero pressure.

Now imagine making the box smaller and smaller (eventually compressing the iron) to find the ground state at higher density. Shrink the box more slowly than the cooling time so that it continues to be at zero temperature and any reactions between nuclei that can take place will take place. The energy density ρ will rise, and so will the pressure p. In this way we could imagine determining the equation of state $p = p(\rho)$ for matter in its ground state at all densities. The principal features of that equation of state are described qualitatively in the following discussion. The results of quantitative calculations are summarized in Figure 24.6 in terms of the *stiffness parameter*, γ, defined in (24.18).

The lowest densities, where the ground state consists of atoms bound into a solid, are unimportant for the overall structure of realistic stars for two reasons: First, the description of the matter in terms of individual atoms would be valid only for a tiny region near the surface of such stars. Second, ground state matter is not a realistic approximation to the actual surface conditions of white dwarf and neutron stars. We may, therefore, safely begin the discussion at higher densities, where the physics is simpler.

As the material in the box is compressed to smaller volumes, the energies of the electrons inside rise—their momenta p varying very roughly as $p \sim \hbar/\lambda$, where λ is the size of the volume to which a typical electron is confined [cf. (24.2)]. By a density $\rho \sim 10^4$ g/cm^3 they are no longer bound in atoms. The dominant source of pressure is the Fermi pressure of this gas of approximately free electrons arising from the Pauli principle, as described in Section 24.1. The energy density is the rest energy of the nuclei to an excellent approximation.

The stiffness parameter γ begins at 5/3 for a nonrelativistic electron gas and drops to 4/3 as the electrons become relativistic (energies $\sim m_e c^2 \sim .5$ MeV) at about $\rho \sim 10^6$ g/cm^3. Shortly above $\rho \sim 10^6$ g/cm^3, the typical electron energy reaches the neutron-proton mass difference ($m_n c^2 - m_p c^2 = 1.3$ MeV). Roughly at that energy it becomes energetically favorable for protons bound in nuclei to absorb an electron and become a neutron through the weak interaction $e^- + p \to n + \nu$. The equation of state softens as the electrons disappear and the nuclei become increasingly neutron-rich and proton-poor. At a density of about 4×10^{11} g/cm^3, the matter becomes so neutron-rich that the most energetic neutrons become unbound from the nuclei. This phenomenon is called *neutron drip*, and γ drops precipitously as compressional energy goes into releasing neutrons rather than supplying pressure.

As the compression proceeds further, the density of neutrons *in between* nuclei increases, and eventually equals the density of neutrons *in* nuclei. The nuclei then merge to form a uniform fluid consisting mainly of neutrons with a few percent of protons and enough electrons to ensure electrical neutrality. This is the *neutron matter* from which neutron stars are mostly made.

In this regime the neutrons are packed together at separations that become comparable to and eventually smaller than the range of 10^{-13} cm, over which nuclear (strong interaction) forces operate. These now supply the dominant source of pressure rather than electrons, which are essentially absent.

Calculating the equation of state in these regimes is no easy matter. The system of nucleons (protons and neutrons) is strongly coupled; eventually at higher densities it is fully relativistic and coupled to other elementary particles, such as π's Λ's, Δ's, K's, etc. Figure 24.6 shows the results of one calculation. The essential qualitative feature is the steady rise of γ with energy density, as a consequence of the repulsive part of nuclear forces so that the equation of state becomes increasingly stiff until it is well above nuclear density.

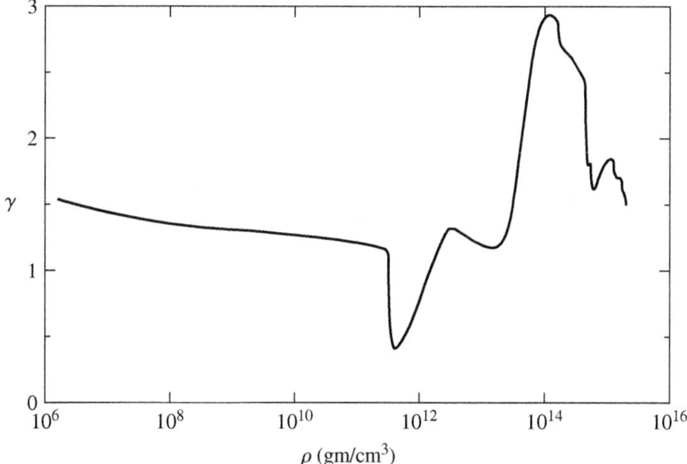

FIGURE 24.6 The equation of state of matter in its ground state. This figure shows γ defined by (24.18) resulting from combining two calculations: the calculations of Harrison and Wheeler (1965) below nuclear densities and the calculation of Glendenning (1985) above nuclear densities. The graph starts at 10^6 g/cm^3, where the zero-temperature electrons supplying most of the pressure are on their way to being relativistic. γ drops slowly through $4/3$ as more and more electrons are absorbed by protons to make neutrons. At approximately 4×10^{11} g/cm^3, the equation of state suddenly softens as neutrons drip off neutron-rich nuclei. Nuclear forces and neutrons become increasingly important as sources of pressure above this density. The equation of state becomes stiffer and stiffer until well above nuclear densities.

One calculation of γ as a function of density for ground state matter is shown in Figure 24.6. The family of spherical stars that results from using this equation of state to integrate the relativistic equations of structure, as described in the previous section, is shown in Figure 24.7.[3] Above densities of order 10^{14} g/cm^3, almost all electrons have combined with protons to make neutrons; for this reason the stars above this central density are called *neutron stars*. The strong interactions are important for their properties, and as a consequence the calculations at the very highest densities shown are somewhat uncertain. Nevertheless, several important features emerge clearly. First, these stars are very compact, with masses of order of a solar mass and radii of order 10 km. The neutrons supplying the pressure are approximately 1000 times more massive than the electrons that supply the pressure in white dwarfs, and neutron star radii are roughly 1000 times smaller (Problem 14). Second, since $GM/Rc^2 \sim .1$, relativistic gravity is important for the structure of neutron stars. Third, even repulsive strong interaction forces are not enough to support an arbitrary amount of matter in its ground

[3]Below densities of 10^{11} g/cm^3, Figure 24.7 differs in detail from Figure 24.5. That is because even though degenerate electrons supply the pressure in both cases, the nuclear reactions assumed to have taken place to lower the energy to the ground state have not taken place in realistic white dwarfs. A different assumption for the rest energy of nuclei is made in Figures 24.4 and 24.7.

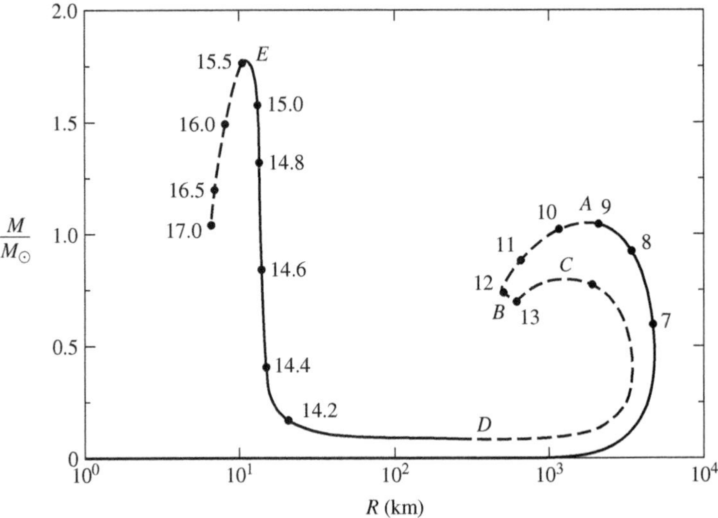

FIGURE 24.7 Mass vs. radius for the nonrotating equilibrium endstates of stellar evolution calculated from the equation of state of matter in its ground state described in Box 24.1 and summarized in Figure 24.6. The curve is the one-parameter family of equilibrium endstates of stellar evolution parametrized by the central density, ρ_c. Values of $\log_{10}[\rho_c(\text{g/cm}^3)]$ are indicated along the curve. The extrema are labeled by $A, B, C, \ldots$. The curve begins at the origin with $M \propto R^3$, but this low-density part of the curve is indistinguishable from the horizontal axis on the scales of the figure. There are two regions of stable equilibria indicated by solid lines. Stars with densities below the first maximum of the mass at A are supported by the Fermi pressure of electrons arising from the Pauli exclusion principle. The other family of stable equilibria lies between the second minimum of the mass at D and the third maximum at E. They consist mostly of neutron nuclear matter and are, therefore, called *neutron stars*. The repulsive forces between nucleons are the dominant source of the pressure. The dotted parts of the curve are unstable configurations, which will not exist in nature.

state against gravity; there is a maximum neutron star mass of approximately a few solar masses. That maximum mass is used to distinguish neutron stars from black holes, as described in Section 13.1. However, not all the stars shown in Figure 24.7 can exist in nature, because not all of them are stable. It is, therefore, to the question of stability that we now turn.

24.5 Stability

It is not enough for a star to be in equilibrium to exist in nature. It must be a *stable* equilibrium.

Figure 24.8 recalls the idea of stable and unstable equilibria in classical mechanics for a particle of mass m moving in one dimension in a potential $V(x)$. The maxima and minima of the potential are equilibrium positions where the force on the particle vanishes. *Minima* of the potential, such as A in Figure 24.8, are *stable*

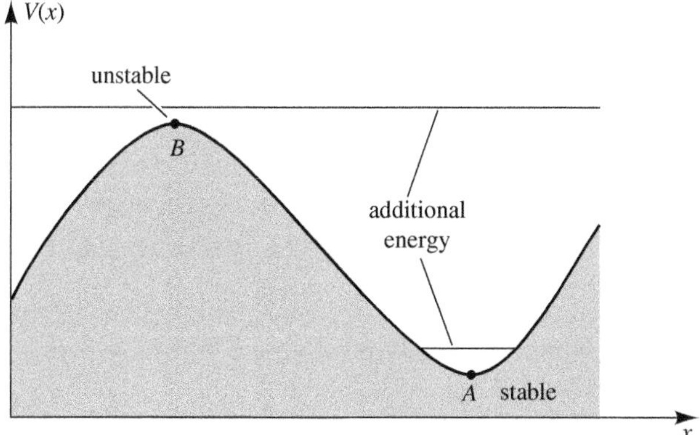

FIGURE 24.8 Newtonian stable and unstable equilibria. The figure shows a potential governing the motion of a particle in one dimension. There are two equilibria at points A and B, where the force $-dV/dx$ on the particle vanishes. If the particle is placed at rest at one of these points, it will stay there. However, if the particle is given a little additional kinetic energy (horizontal lines), its behavior is very different in the two cases. The particle at A will execute small oscillations about equilibrium. The particle at B will move exponentially quickly in one direction or the other away from the equilibrium position. *Minima* like A are *stable. Maxima* like B are *unstable.*

equilibria. Give a particle resting there a bit more energy, and it will oscillate about that equilibrium, never getting very far away if the additional energy is small. Small displacements $\xi(t)$ away from equilibrium will obey the equation of motion obtained by expanding the equation of motion, $m\ddot{x} = -dV/dx$, in ξ and keeping only the lowest term. That is

$$m\frac{d^2\xi}{dt^2} = -\left(\frac{\partial^2 V}{\partial x^2}\right)_{x_A} \xi. \tag{24.19}$$

Solutions to this equation vary harmonically like $\xi(t) \propto \exp(\pm i\omega_A t)$ (of course only the real part gives the displacement), where

$$\omega_A^2 \equiv \frac{1}{m}\left(\frac{\partial^2 V}{\partial x^2}\right)_{x_A} . \tag{24.20}$$

Since A is a minimum, ω_A^2 is positive, ω_A is real, and the solutions will oscillate with bounded amplitude. That is stability.

In contrast, a *maximum* of the potential, such as B in Figure 24.8, is an *unstable* equilibrium. Give a small additional energy to a particle resting there, and it will move further and further away from the equilibrium. More explicitly, the ω_B^2 defined in analogy to (24.20) will be negative and ω_B will be imaginary. Solutions to (24.19) will, therefore, behave like $\xi \propto \exp(\pm|\omega_B|t)$—one growing in time and the other decaying. A pencil standing on its tip is a simple example of an unstable equilibrium and these kinds of motions. There is a decaying motion in which the pencil coasts to the vertical and stops, remaining in equilibrium. But realistic pencils will be subject to uncontrollable perturbations, for example, from air currents, producing *both* growing and decaying motions. Any growing perturbation, however small initially, will eventually dominate the decaying ones, and the pencil will fall over. *A single growing perturbation is enough to demonstrate the instability of a mechanical system.*

The analysis of the stability of stars is qualitatively similar; the only real difference is that the star has many more possible modes of oscillation. These can be decomposed into normal modes with definite frequencies. If even one of these squared frequencies is negative, the star will be unstable and will not survive in nature.[4]

The Einstein equation would have to be solved for small oscillations about equilibrium to actually calculate the normal modes of a star and exhibit their frequencies. That calculation is not so very difficult, but it yields much more information than is needed just to answer the question of whether the star is stable or unstable. For that, just a few general facts about the modes are needed—indeed, facts that are common with many similar vibrating systems, such as strings and drums. We now describe these facts, beginning with the simple case of a vibrating string in Example 24.1.

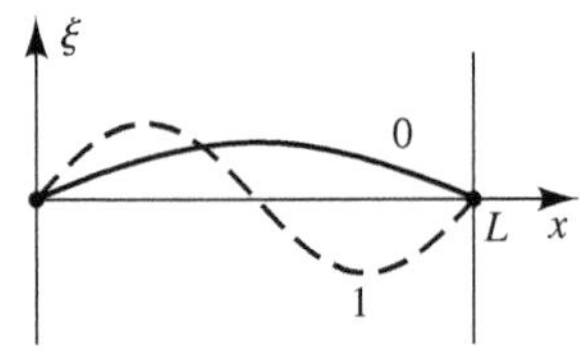

FIGURE 24.9 Two normal modes of a vibrating string fixed at its endpoints. Plotted are the amplitudes $\xi_n(x)$ of the lowest two modes of a vibrating string of length L given explicitly in (24.24).

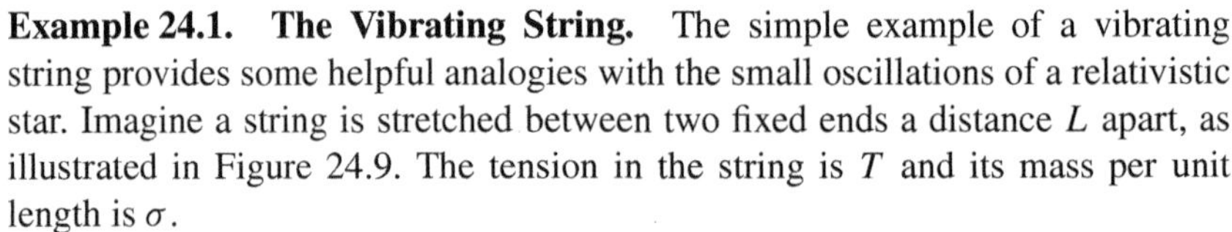

Example 24.1. The Vibrating String. The simple example of a vibrating string provides some helpful analogies with the small oscillations of a relativistic star. Imagine a string is stretched between two fixed ends a distance L apart, as illustrated in Figure 24.9. The tension in the string is T and its mass per unit length is σ.

The straight stretched string is an equilibrium configuration—unchanging in time—and analogous to the equilibrium of a relativistic star. If the string is plucked, it will vibrate with an amplitude $\xi(t, x)$ that obeys the wave equation[5]

$$\sigma \frac{\partial^2 \xi}{\partial t^2} = T \frac{\partial^2 \xi}{\partial x^2}. \tag{24.21}$$

The equation for the radial oscillations of a relativistic star is similar in character, although more complex. The normal modes of vibration of the string are the solutions to (24.21) that have harmonic time dependence $\xi(t) \propto \exp(\pm i\omega t)$. The fixed end boundary conditions can be satisfied only for the discrete spectrum of frequencies

$$\omega_n^2 = (n+1)^2 \left(\frac{T}{\sigma}\right)\left(\frac{\pi}{L}\right)^2, \qquad n = 0, 1, 2 \ldots, \tag{24.22}$$

[4]In some cases the time scale of an instability can be so long that the star *does* survive for an astrophysically interesting time. But such *secular instabilities* are not among the spherical modes we are about to analyze.

[5]Review your basic mechanics book or work through Problem 15 if this or anything else in the following standard discussion seems unfamiliar.

at which an integral number of half-wavelengths can fit between the ends. The corresponding normal modes are

$$\xi(t,x) = e^{\pm i\omega_n t}\xi_n(x), \tag{24.23}$$

where

$$\xi_n(x) = A_n \sin[(n+1)\pi x/L] \tag{24.24}$$

for some constant complex amplitude A_n. Frequencies have been labeled so that mode n has n nodes (not counting the ends).

As long as T is positive, there is a restoring force for any displacement from the straight equilibrium, the squared frequencies of all modes are positive, and the equilibrium is stable. But imagine what would happen if T could become negative, so that compressive tension became an expansive force, as with a spring compressed from both ends. The straight configuration is still an equilibrium, but now the smallest disturbance will grow. The squared frequencies from (24.22) are now all negative, and all modes are unstable. As T varies from positive to negative, the straight configuration passes from being a stable equilibrium to an unstable one.

It turns out that modes in which the displacement of the fluid is purely radial control the stability of spherical stars. Therefore, we focus on these radial modes where the motion of the fluid is in and out in the radial direction. Not unlike the vibrating string, a spherical oscillation is described by giving the radial displacement $\xi(r,t)$ of a fluid element located at radius r in the unperturbed star as a function of time. The possible displacements can be analyzed into a discrete spectrum of normal modes with definite frequencies ω_n, $n = 0, 1, 2, 3, \dots$. In each mode the displacement is

$$\xi_n(r)e^{\pm i\omega_n t}. \tag{24.25}$$

Stable modes have $\omega_n^2 > 0$, have real frequencies, and oscillate. Unstable modes have $\omega_n^2 < 0$, imaginary frequencies, and can grow exponentially quickly in time. However they behave, these modes conserve energy. It takes a little energy to start them, but afterward that energy is conserved. A few of the lowest mode functions $\xi_n(r)$ are illustrated schematically in Figure 24.10. The modes must vanish at the center because it remains fixed during the oscillation. At the surface they must preserve the condition that the pressure vanishes there. The lowest-frequency mode has no nodes other than the one at the center. Mode n has n nodes.

Imagine having carried out the calculation of the frequencies of the normal modes of each star in the one-parameter family of the stars illustrated in Figure 24.7, with the equation of state of matter in its ground state. The result would be the squared frequencies of each normal mode $\omega_n^2(\rho_c)$, $n = 0, 1, 2, \dots$, as a function of the central density ρ_c that parametrizes the family. Figure 24.11 shows schematically how the curves of $\omega_n^2(\rho_c)$ vs. ρ_c might look.

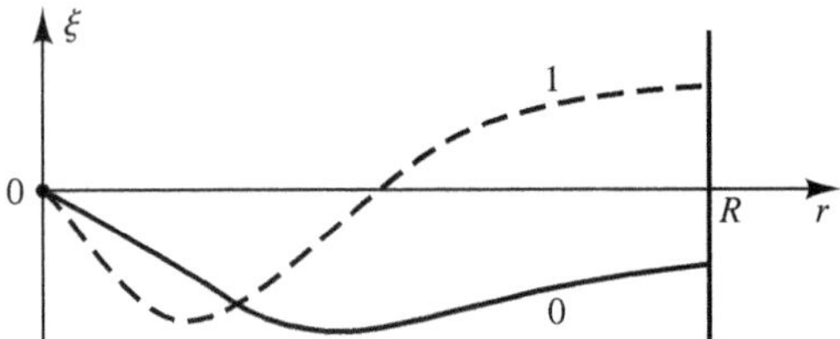

FIGURE 24.10 A schematic plot of the two lowest radial modes of a spherical star. The horizontal axis is the radius from the center of the star ranging from 0 at the center to the star's radius, R, at the surface. The vertical axis shows the maximum displacement of a fluid element at radius r. The normalizations are arbitrary because a given mode may have any small amplitude. They are shown large on this plot for clarity. All modes vanish at the center, which remains fixed during the oscillation. The solid line represents the lowest (0) mode. This has no zeros (nodes) other than the center. When the center is being compressed (negative ξ near $r = 0$), the surface is displaced inward. The dotted line shows the next (1) mode, which has one node. When the center is being compressed, the surface is displaced outward.

As ρ_c varies the squared frequencies of the modes change, and modes can change from being stable to unstable and vice versa. A mode n changes from being stable to unstable at a central density where ω_n^2 changes from being positive to negative and from being unstable to stable when the change is the other way around. In either case, a mode has zero frequency when it changes stability. But a zero-frequency mode is something special. It is a time-*independent* displacement.

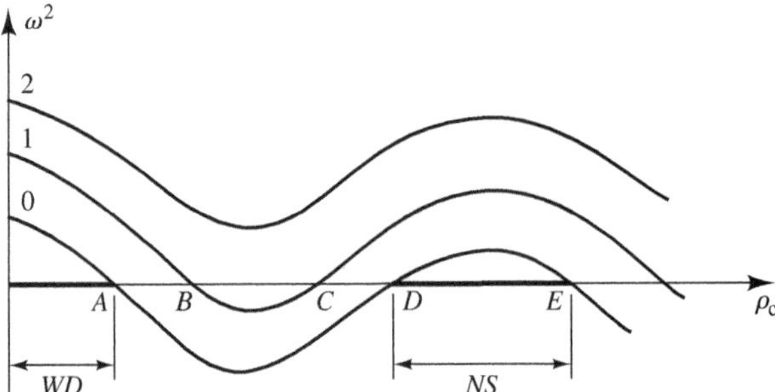

FIGURE 24.11 A schematic representation of the squared frequencies of the lowest three radial modes of a family of nonrotating stars shown in Figure 24.7 as a function of central density ρ_c with 0, 1, and 2 nodes, as labeled. At low densities all modes have positive squared frequency and are stable. At the density A the lowest mode's squared frequency turns negative, and the mode becomes unstable. At B the second mode also becomes unstable. These two modes return to stability in succession at C and D. The lowest mode again becomes unstable at E. Zero-frequency modes that characterize a change in stability occur at these central densities. These are displacements between equilibrium configurations that can occur only at extrema of the M vs. R curve in Figure 24.7, which are labeled by A, B, etc. Stars are stable only when all their modes are stable. This happens in the range of densities from 0 to A corresponding to white dwarfs and D to E corresponding to neutron stars.

It is a displacement from one equilibrium configuration to another—a displacement *along the* $M(R)$ *curve* shown in Figure 24.7.

A displacement corresponding to a zero-frequency mode cannot occur just anywhere along the sequence of equilibrium configurations. A zero-frequency mode has zero energy, and so the displacement must conserve total mass-energy. Zero-frequency modes—and, therefore, changes in stability—can occur only at the *extrema* of the M vs. R curve, where mass is unchanged to first order in a small displacement that changes R. Conversely, at any extremum there is a time-independent displacement between equilibria that conserves energy. Zero-frequency modes and extrema are, therefore, in correspondence. The stability of the family of equilibrium endstates of stellar evolution can therefore be analyzed entirely from the M vs. R curve in Figure 24.7, as we now show.

Begin at low central densities ρ_c, where the equilibrium configurations may be presumed stable (positive squared frequencies for all modes), and follow how the modes change stability with increasing ρ_c, as illustrated in Figure 24.11. The first extremum where a change in stability occurs is at the peak in $M(R)$, labeled A in Figure 24.7. There the lowest mode becomes unstable. Stars with central densities below A are a stable family supported over most of this range by the Fermi pressure of degenerate electrons and are qualitatively similar to the stars in Figure 24.5.

Stars with central densities above A but below that of the next extremum (the minimum at B) are unstable and will not exist in nature. There are two possibilities for the zero modes at B: (1) The lowest mode could return to stability, or (2) as illustrated in Figure 24.11, the second mode could become unstable along with the first. We can tell which happens by how the two different modes change the radius of the star.

Consider a zero-frequency mode for which the displacement along the family of equilibrium configurations *increases* the central density. The mode function, $\xi(r)$, must therefore be negative near $r = 0$ (compression). Thus, the lowest mode, which has no nodes except at $r = 0$, must be negative at the surface corresponding to a decrease in the radius R of the star (see Figure 24.10). In contrast, the second mode with one node would increase R. But we observe from Figure 24.7 that, at B, the radius R is *increasing* with increasing central density. Therefore, it is not the lowest mode that is changing stability but the next-lowest, with one node, which is becoming unstable as shown in Figure 24.11. Following this kind of analysis (Problem 16), we find that the only other stable regime is from the minimum at D to the maximum at E. These are stars with central densities from about 1.2×10^{14} g/cm^3 to 3×10^{15} g/cm^3. These stars, made mostly from the neutron matter described in Box 24.1, are called *neutron stars*.

Neutron stars are general relativistic objects with $GM/c^2R \sim .1$ (see Figure 24.7). Theoretical extrapolation of the equation of state beyond 3×10^{15} g/cm^3 yields no evidence of further families of stable stars. Neutron stars and white dwarfs are, therefore, the two possible stable equilibrium endstates of stellar evolution. White dwarfs can be detected from their optical properties; neutron stars can be detected from the pulsar phenomenon among other means (see Box 24.2).

BOX 24.2 Pulsars

The observational signature of a pulsar is a continuing series of short radio-frequency pulses of electromagnetic radiation spaced on average by precise periods typically of order of a second. These radio signals originate from rotating, magnetized neutron stars.

Neutron stars could easily have inherited enough angular momentum in their formation in a supernova collapse to be spinning around once per second. Even at this rate they are not spinning rapidly in the sense that centrifugal forces are much less than gravitational ones (Problem 18). The inertia of a compact spinning solar mass makes it a very accurate clock. A narrowly beamed beacon rotating with the neutron star could produce the observed pulses as its beam sweeps over the Earth. But what is the mechanism for producing such a beam?

A neutron star inherits not only angular momentum in its formation but also a magnetic field. Neutron star material is highly conducting, so much so that any magnetic field threading the material will be frozen in for a time much longer than the age of the pulsar. In the process of collapse, magnetic flux is approximately conserved inside the conducting matter, and ordinary stellar magnetic fields of a few gauss can be amplified to fields of 10^{12} gauss by the compression. The rotating neutron star can, therefore, be highly magnetized. An observer a long distance from the star would see it as a rotating magnetic moment. If the moment is not aligned along the rotation axis as shown, then it will change in time due to the rotation, and electromagnetic waves will be emitted. The energy departing in these waves slows the rotation of the star. The pulsar is such an accurate clock that the consequent minute decrease in its rotational period can be measured and is consistent with such large values of the magnetic field.

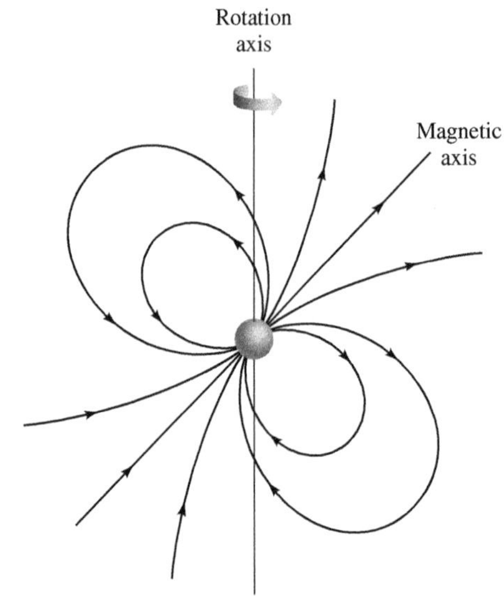

This same magnetic field is the origin of a plasma surrounding the neutron star. Inside the highly conducting interior the rotating magnetic field gives rise to an electric field. The force on an electron must approximately vanish in steady state so that the two terms in the Lorentz force law must cancel—$\vec{E} = -\vec{V}/c \times \vec{B}$, where $\vec{V} = \vec{\Omega} \times \vec{r}$, with $\vec{\Omega}$ being the star's angular velocity. The tangential component of the electric field must be continuous across the star's surface. This means that there will be an electric field outside the star, which in general will also have a component normal to the surface. This field is strong enough to pull electrons out of atoms and create a plasma rotating with the star. Charged particles of the plasma flow outward along the open field lines shown and are trapped along the closed ones. The trapped particles rotate with the neutron star out to the distance from the rotation axis where their speed would equal the velocity of light. The mechanism for the emission of radio pulses is still imperfectly understood, but one idea is that instabilities in this plasma could create concentrations of charge that could radiate coherently because of their rotation. The relativistic beaming effect discussed in Section 5.5 would naturally produce a narrow beam, which could be the beacon observed. In this way, as in many others, neutron stars are natural laboratories for physics at extreme conditions.

24.6 Bounds on the Maximum Mass of Neutron Stars

Figure 24.7 shows that there is a maximum nonrotating mass of about $2M_\odot$ that can be supported against gravitational collapse by the pressure forces of matter in its ground state. This maximum mass of neutron stars is an important number for astrophysics because it is used observationally to distinguish black holes from neutron stars, as discussed in Section 13.1. However, as already mentioned, results such as those shown in Figure 24.7 rely on challenging theoretical calculations of the equation of state in regimes of density beyond those of ordinary nuclei and, therefore, beyond many checks by experiments on Earth. Thus, it is important that general relativity *by itself* provides a bound on the maximum mass of neutron stars, assuming a detailed knowledge of the equation of state up to the density of ordinary nuclei but making only general assumptions on its properties beyond. That bound is the subject of this section. We will not try to derive the best bound possible but rather indicate the basic reasons for it.

The bound on the maximum mass follows from the equations of relativistic hydrostatic equilibrium (24.13) and a few general assumptions on the equation of state, which we now spell out:

- We assume the matter is described by an equation of state $p = p(\rho)$ with (1) $\rho > 0$, (2) $p > 0$, and (3) $dp/d\rho > 0$. Positive energy (1) is a very general property of matter. Properties (2) and (3) are necessary for matter to be microscopically stable. If the pressure were not positive or if the pressure did not increase with compression, it would be energetically favorable to compress any small volume to an even smaller one, leading to collapse on small scales.
- We assume that the equation of state is known up to some fiducial density ρ_0. Nuclei ranging from helium to uranium have nearly the same density at their cores. This density is called *nuclear density* and has the value 2.9×10^{14} g/cm^3. Up to this density, theoretical calculations can be compared with actual nuclei so it is a reasonable choice for the density ρ_0, below which we can say we know the equation of state of matter in its ground state with some confidence.

Under these assumptions, we can derive a bound on the maximum mass, as follows. The right-hand side of the equation of hydrostatic equilibrium (24.13b) is always positive. Each term is positive, and $1 - 2m(r)/r$ can never be negative or the region inside r would be inside a black hole. It follows that the pressure decreases with radius, and since $dp/d\rho > 0$ by assumption, the density also decreases with radius. A star can, therefore, be divided into a *core* with $\rho \geq \rho_0$, where the equation of state is unknown in detail, and an *envelope* with $\rho < \rho_0$, where the equation of state is known. Denote by r_0 the radius of the core and by M_0 the value of $m(r)$ at that radius. We informally refer to M_0 as the *mass of the core*. From (24.13a) we have

$$M_0 = \int_0^{r_0} dr\, 4\pi r^2 \rho(r) \geq \int_0^{r_0} dr\, 4\pi r^2 \rho_0, \tag{24.26}$$

because the density at any point inside the core is larger than at its surface. Thus,

$$M_0 \geq \tfrac{4}{3}\pi r_0^3 \rho_0. \tag{24.27}$$

On the other hand, the core can't be inside its own Schwarzschild radius, or it would be a black hole. Thus,

$$\frac{2M_0}{r_0} < 1. \tag{24.28}$$

The two constraints (24.27) and (24.28) are illustrated in Figure 24.12. There is a maximum core mass, which is

$$M_0 \leq \frac{1}{2}\left(\frac{3}{8\pi\rho_0}\right)^{1/2} = 8.0M_\odot \left(\frac{2.9 \times 10^{14}\ \mathrm{g/cm^3}}{\rho_0}\right)^{1/2}. \tag{24.29}$$

However, the bound in (24.29) is not the best that can be made. If you work through Problem 19, you will find that the largest value of $2M_0/r_0$ for a constant density core is $\frac{8}{9}$, not 1. Just this one improvement is enough to reduce the $8M_\odot$ bound on the core mass in (24.29) to $6.7M_\odot$, still not optimum. The total bounding mass is the sum of the mass of the core and the mass of the envelope. The mass of the envelope can be found by integrating the equations of structure as described in Section 24.3 but starting from r_0 and M_0 instead of the center. It turns out that the optimum bound on the core is about $6.5M_\odot$, to which the envelope makes a small but nonnegligible addition for values of ρ_0 around nuclear densities. Detailed calculations (Hartle 1978) show that, under the preceding assumptions, the optimum general relativistic bound on maximum mass of neutron stars—core plus envelope—is $6.7M_\odot$, assuming the equation of state summarized in Figure 24.6 up to $\rho_0 = 2.9 \times 10^{14}\ \mathrm{g/cm^3}$. By assuming further restrictions on

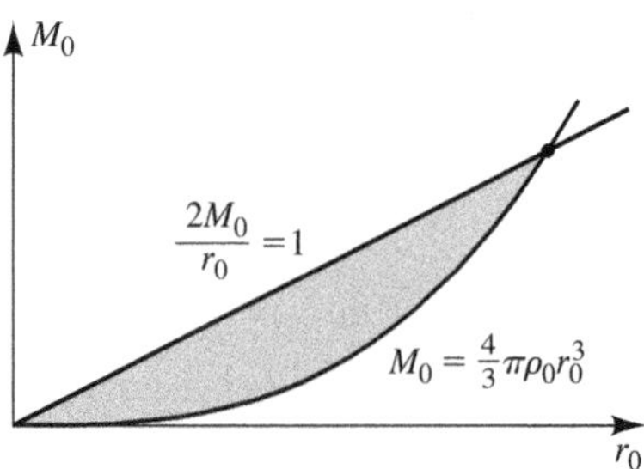

FIGURE 24.12 A bound on the mass of nonrotating, equilibrium endstates of stellar evolution. The shaded region shows the allowed values of mass M_0 and radius r_0 for the cores of neutron stars where the density is above the fiducial density, ρ_0, below which the equation of state is accurately known. The mass must be greater than the lower curve, which is the mass-radius relation for uniform-density cores with the minimum possible density, ρ_0. The mass must be less than the upper curve that would make the configuration a black hole. The maximum core mass occurs at the dot where the curves intersect. For ρ_0 near nuclear densities, this is the maximum possible neutron star mass to a good approximation.

the equation of state, such as a velocity of sound less than the velocity of light, the bound can be driven down to about $4M_\odot$.

This upper bound on the mass of spherical neutron stars independent of the properties of matter above nuclear densities is what gives confidence to the identifications of solar-mass-scale black holes in binary X-ray sources discussed in Section 13.1. If the analysis of the orbit of the binary from the radial velocity and spectrum of one star reveals a compact source of X-rays with a mass greater than the upper bound described previously, it cannot be a neutron star. It must be a black hole.

Problems

1. [S] White dwarfs can have surface temperatures of 10^5 K, which is hot by everyday measures. Is this temperature large enough that approximation of a degenerate gas of electrons at *zero* temperature will break down?

2. *The One-Dimensional Gas of Degenerate Fermions* Suppose that we lived in one dimension, not three. A gas of $\mathcal{N}$ degenerate fermions in a "box" of length $\mathcal{L}$ would be characterized by an energy per unit length, ρ, a force on the walls, p, and a number per unit length, n. (The notation stresses the analogy with three dimensions.)

(a) Evaluate the sum in (24.3) to find the total energy in the box for large values of $\mathcal{N}$. Find ρ as a function of n.

(b) Find the force on the walls p as a function of n.

3. *No Equation of State for Bosons* Continuing the one-dimensional example from the previous problem, what would be the ground-state energy of $\mathcal{N}$ bosons (particles not restricted by the Pauli principle) in a box of length $\mathcal{L}$? Would there be a relation between p and n that is independent of $\mathcal{N}$ when it is large?

4. [A] *Deriving the Equations of Structure from the Einstein Equation* Use the metric (24.11) and the stress-energy of a perfect fluid (22.37) to derive the three equations of hydrostatic equilibrium (24.13) from Einstein equation $G_{\alpha\beta} = 8\pi T_{\alpha\beta}$ and the local conservation of stress-energy $\nabla_\beta T^{\alpha\beta} = 0$ which follows from it. (*Hint*: You can use any combination of the equations you choose but the equations $\nabla_\beta T_r^\beta = 0$, $G_{\hat{t}\hat{t}} = 8\pi T_{\hat{t}\hat{t}}$, and $G_{\hat{r}\hat{r}} = 8\pi T_{\hat{r}\hat{r}}$ in the orthonormal basis pointing along the coordinate axes involve the least algebra. Those components of the Einstein tensor are given in Appendix B for the metric (24.11).)

5. Argue that the metric

$$d\Sigma^2 = \alpha^2 dr^2 + r^2 d\phi^2,$$

where α is a constant greater than 1, represents the geometry on the two-dimensional surface of a cone. What is the opening angle of the cone? (*Comment*: This is an example of a geometry that is not locally flat at $r = 0$.)

6. [C] Incompressible matter with a constant fixed density ρ is inconsistent with special relativity because it could be used to send signals faster than the speed of light. (How?) But it does provide a simple example of relativistic hydrostatic equilibrium. Integrate the equations of structure (24.13a) and (24.13b) to find the pressure as a function of radius for a star of total mass M made out of material for which ρ is a given constant. Plot the mass vs. radius relation for these stars. Is there a maximum mass?

7. [A] **(a)** Find the metric in the interior of the family of spherical, constant-density stars whose structure was solved for in Problem 6. Carefully discuss the junction conditions between the interior and the exterior of the star.

(b) Show that the geometry of a $t =$ const. surface inside a constant-density star is the same locally as that of a homogeneous spatial surface of a closed FRW universe discussed in Chapter 18.

(c) Is the volume inside the star bigger or smaller for a given surface area than it would be in flat space?

8. Using the results of Problem 7, find and plot an embedding diagram for a $t =$ const., $\theta =$ const., two-dimensional slice of the spacetime geometry of a constant-density star.

9. [C] **(a)** In the text we took the central density ρ_c to parametrize families of non-rotating stars, as, for example, in Figure 24.7. But they could equally well have been parametrized by the central pressure p_c. Stars made from the hypothetical constant-density material discussed in Problem 6 all have the same given density.

(b) For these constant-density stars show that for each value of the central pressure p_c, there is a unique mass, M, and radius, R.

(c) What is the largest red shift [cf. (9.20)] from the surface that is exhibited in this family of spherical stars with constant density, ρ? To what central pressure, p_c, does it correspond?

10. A spherical distribution of matter with a $p = \rho$ equation of state is contained within a spherical shell of area $4\pi R^2$ and negligible mass. (The shell is not made of realistic matter.)

(a) Find a simple solution of the equations of hydrostatic equilibrium (24.13) in which the distributions of pressure and density are inverse powers of the radius.

(b) What is the total mass of this distribution?

(c) What pressure does the shell have to exert?

11. [E, B] *Estimate* the densities in g/cm^3 in which, for matter in its ground state, the energy of a typical electron (1) exceeds typical atomic binding energies $\sim$ 10 keV, (2) exceeds the electron rest mass $\sim$ .5 MeV, and (3) exceeds the neutron-proton mass difference $\sim$ 1.3 MeV.

12. [B] An electrically neutral gas of highly relativistic *free* electrons, protons, and neutrons is maintained in equilibrium by the reactions

$$e^- + p \leftrightarrow n + \nu,$$

so that no electrons are absorbed on the average and no neutrons decay. How are the number densities of electrons, protons, and neutrons related to each other?

13. [S] Evaluate γ defined by (24.18) for the equation of state of degenerate fermions given in (24.7) and (24.9) for both the nonrelativistic and relativistic limits.

14. [E] *Estimating the Radius of Neutron Stars* Estimate the radius of a neutron star by assuming that the degeneracy pressure of free neutrons supplies the pressure that holds it up, and then work Problem 12.2.

15. [S] Starting from the wave equation (24.21), derive the frequencies and shapes of the normal modes of a vibrating string that are given in (24.22) and (24.24), respectively.

16. [S] Work through the changes in stability of the modes that occur at the extrema of the mass vs. radius relation for the family of stars in Figure 24.7. Assume that the curves of squared frequency vs. central density never cross. Show that the changes in stability are as illustrated in Figure 24.11 and that there are only two ranges of stable equilibrium stars, as shown.

17. *Stable Equilibria Beyond Neutron Stars?* A theorist proposes a new equation of state for matter above nuclear densities and wonders whether it might lead to a new kind of ultra-high-density endstates to stellar evolution beyond neutron and white dwarf stars. You use the equation of state and the equations of this chapter to calculate the mass-radius relationship of stars with central density greater than nuclear density that is shown here. The curve represents stable neutron stars (NS) at the lowest densities but then spirals around at higher densities. Assuming the lowest-density, largest-radius stars shown are stable, will there be a new family of stable equilibria?

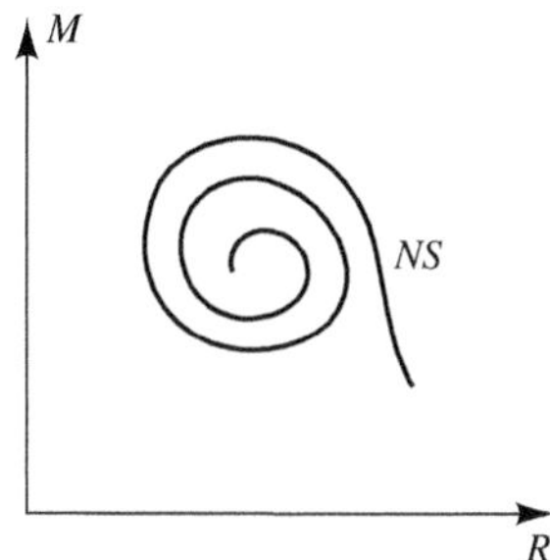

18. [E] Using Newtonian physics, *estimate* the ratio of centrifugal forces to gravitational forces at the surface of a neutron star rotating with a period of 1 s.

19. *Refining the Upper Bound on the Maximum Mass of Neutron Stars* Use the results of Problem 9 to show that, in the notation of Section 24.6, mass and radius of the core must satisfy $M_0/r_0 < \frac{4}{9}$. Assuming that ρ_0 is nuclear density, find a bound on the mass of the core for stars with central densities that are above this value.

APPENDIX A

Units

A.1 Units in General

To understand something of units, imagine the problem of communicating the predictions of our physical theories to intelligent aliens living near a distant star. You can send a message saying that the mass of the proton is approximately 1835 times the mass of the electron; that ratio of masses is a dimensionless number that can be transmitted as bits. But a message saying that the mass of a proton is 1.672×10^{-27} kg will make no sense. The aliens don't know what a kilogram is. Nor can we explain it to them exactly because it is defined as the mass of the block of metal kept in the Bureau International des Poids et Mesures, in Sèvres outside Paris. You could send a message saying that the standard kilogram was *approximately* 5.980×10^{26} proton masses because that is a dimensionless ratio between the mass of the international kilogram and the proton mass. That will be a less interesting message because it is not about the predictions of physical laws, but rather about how humans organize those predictions.

The predictions of fundamental physical theories are reducible to dimensionless numbers. Units are introduced for *convenience*, and the number and system of units varies considerably with the notion of convenience. For example, today the second is *defined* as the time required for exactly 9,192,631,770 cycles in the transition between the two lowest energy states of a cesium atom, and the meter is *defined* to be 1/(299,792,458) of one of those seconds—both definitions involving defined dimensionless numbers. We use hours, minutes, and seconds partly because of tradition, but also because it would be inconvenient to talk about lectures that were 28 trillion cycles of a cesium transition long. Were it convenient, we *could* introduce a unit to measure the areas of circles that differs from that used to measure the area of rectangles. Suppose a 1-cm radius circle is defined to have an area of 1 Archimedes. Then there would be the conversion 1 Archimedes equals 3.14159265... cm^2. That would add one more unit but not much convenience, so there is little motivation for the Archimedes. But from the perspective of special relativity the use of separate units to measure spacelike and timelike distances is not so very different (Section 4.6). When a dimensionless ratio is between a measured quantity and a *standard*, then it is convenient to introduce a unit for the standard as in the case of the international kilogram.

Accepted physical theory plays an important role in the choice of units. The second could not be defined as above if we did not have confidence from the many successes of atomic theory that all cesium atoms were identical. Confidence in special relativity is behind the definition of the meter in terms of time units because that theory asserts that the velocity of light is the same in all inertial frames.

Progress in experiment also plays a role in determining what units are used. One reason we have separate units for mass, length, and time is that there once were separate standards for these quantities. The second was once defined as a certain fraction of the mean solar day, and the meter by the distance between two marks on a particular bar. When it became possible to measure the frequency of atomic transitions more accurately than the solar day it made sense to change the definition of the unit of time to the one we have today. Future progress could change the present situation. For example, with confidence in the equality of gravitational and inertial mass, general relativity, and access to precise enough mea-surements, the kilogram could be defined as the mass of a sphere such that a test mass completes a circular orbit of radius 1 m in some defined number of days. (At current accuracies that would be approximately 8.90 days from Kepler's law (3.24)). Newton's gravitational constant would then be a defined quantity, rather than a measured one, just like the velocity of light is today. Indeed by defining mass as an appropriate multiple of the inverse period squared of such an orbit G could be made equal to 1.

A.2 Units Employed in This Book

This text employs three systems of units for mechanics and the special and general theories of relativity that are convenient in different circumstances. For the traditional mass-length-time ($\mathcal{MLT}$) system we use the (cgs) units of gram, centimeter, and second. These are standard in astrophysics for most of the applications we consider. The units convenient for special relativity are a mass-length ($\mathcal{ML}$) system, in which the velocity of light is unity ($c = 1$). The gram and the centimeter are used for these units. The units convenient for general relativity are a length ($\mathcal{L}$) system called *geometrized units*, in which $G = 1$ and $c = 1$ where mass, length, and time all have units of length.

Tables A.1 and A.2 show how to convert various quantities between $\mathcal{MLT}$ units, $\mathcal{ML}$ ($c = 1$) units, and $\mathcal{L}$ ($G = c = 1$) units. The tables can be used in two ways: To convert from $\mathcal{MLT}$ units to either of the other systems, multiply by the indicated factor in the last column. For instance, to convert mass in grams to mass in centimeters use the first line of Table A.2 to find

$$M(\text{in cm}) = (G/c^2)M(\text{in g}) = .742 \times 10^{-28} M(\text{in g}). \tag{A.1}$$

To convert equations back to $\mathcal{MLT}$ units from either of the other two systems, replace quantities by the expressions in the last column with c and G restored. For instance, the equation giving the escape velocity of a particle from a Schwarzschild coordinate radius R outside a spherical black hole of mass M is [cf. (9.42)] $V_{\text{escape}} = (2M/R)^{1/2}$ in geometrized units. To find the same relation in $\mathcal{MLT}$ units we find from Table A.1 that V_{escape} should be replaced by V_{escape}/c and from

Table A.2 that M should be replaced by GM/c^2. This gives

$$\frac{V_{\text{escape}}}{c} = \left(\frac{2GM}{c^2R}\right)^{1/2}, \quad \text{or} \quad V_{\text{escape}} = \left(\frac{2GM}{R}\right)^{1/2}. \tag{A.2}$$

TABLE A.1 Mass-Length and Mass-Length-Time Units

Quantity	Typical symbol	$\mathcal{ML}$ unit	$\mathcal{MLT}$ unit	Conversion $\mathcal{MLT} \to \mathcal{ML}$
Mass	m	$\mathcal{M}$	$\mathcal{M}$	m
Length	L	$\mathcal{L}$	$\mathcal{L}$	L
Time	t	$\mathcal{L}$	$\mathcal{T}$	ct
Spacetime distance	s	$\mathcal{L}$	$\mathcal{L}$	s
Proper time	τ	$\mathcal{L}$	$\mathcal{T}$	$c\tau$
Energy	E	$\mathcal{M}$	$\mathcal{M}(\mathcal{L}/\mathcal{T})^2$	E/c^2
Momentum	p	$\mathcal{M}$	$\mathcal{M}(\mathcal{L}/\mathcal{T})$	p/c
Velocity	V	dimensionless	$\mathcal{L}/\mathcal{T}$	V/c

TABLE A.2 Geometrized and Mass-Length-Time Units

Quantity	Typical symbol	Geometrized unit	$\mathcal{MLT}$ unit	Conversion $\mathcal{MLT} \to$ geom.
Mass	M	$\mathcal{L}$	$\mathcal{M}$	GM/c^2
Length	L	$\mathcal{L}$	$\mathcal{L}$	L
Time	t	$\mathcal{L}$	$\mathcal{T}$	ct
Spacetime distance	s	$\mathcal{L}$	$\mathcal{L}$	s
Proper time	τ	$\mathcal{L}$	$\mathcal{T}$	$c\tau$
Energy	E	$\mathcal{L}$	$\mathcal{M}(\mathcal{L}/\mathcal{T})^2$	GE/c^4
Momentum	p	$\mathcal{L}$	$\mathcal{M}(\mathcal{L}/\mathcal{T})$	Gp/c^3
Angular momentum	J	$\mathcal{L}^2$	$\mathcal{M}(\mathcal{L}^2/\mathcal{T})$	GJ/c^3
Power (luminosity)	L	dimensionless	$\mathcal{ML}^2/\mathcal{T}^3$	GL/c^5
Energy density	ϵ	$\mathcal{L}^{-2}$	$\mathcal{M}/(\mathcal{LT}^2)$	$G\epsilon/c^4$
Momentum density (energy flux)	$\vec{\pi}$	$\mathcal{L}^{-2}$	$\mathcal{M}/(\mathcal{L}^2\mathcal{T})$	$G\vec{\pi}/c^3$
Pressure (stress)	p	$\mathcal{L}^{-2}$	$\mathcal{M}/(\mathcal{LT}^2)$	Gp/c^4
Energy of an orbit per unit mass	e	dimensionless	$(\mathcal{L}/\mathcal{T})^2$	e/c^2
Angular momentum of an orbit per unit mass	ℓ	$\mathcal{L}$	$\mathcal{L}^2/\mathcal{T}$	ℓ/c
Planck's constant	$\hbar$	$\mathcal{L}^2$	$\mathcal{M}(\mathcal{L}^2/\mathcal{T})$	$G\hbar/c^3$

APPENDIX B

Curvature Quantities

The following tables give useful quantities for the simplest of the geometries considered in the text. Specifically they give the metric, Christoffel symbols, Riemann curvature, and Einstein curvature. These are enough to form the geodesic equations and the Einstein equation.

Only nonzero components are shown, and only those nonzero components sufficient to construct the rest by summetries. For instance, we don't give both Γ^t_{tr} and Γ^t_{rt}, since $\Gamma^\alpha_{\beta\gamma}$ is symmetric in β and γ. Similarly other nonzero components of the Riemann curvature can be found from the ones displayed by making use of the symmetries in (21.29).

In each case one coordinate system is used and the Christoffel symbols are given in that coordinate basis. Both the Christoffel symbols and coordinate basis components of the curvature quantities can be computed using the *Mathematica* program *Curvature and the Einstein Equation* on the book website. However, curvature quantities are quoted in an orthonormal basis. This gives simpler expressions for highly symmetric metrics, and ones that are not singular at coordinate singularities. Since all the metrics considered are diagonal, we use an orthonormal basis whose vectors point along the coordinate directions. The coordinate components of these basis vectors are easily calculated from the metric according to the prescription in Example 7.9. Specifically,

$$(\mathbf{e}_{\hat{0}})^\alpha = [(-g_{00})^{-1/2}, 0, 0, 0],$$
$$(\mathbf{e}_{\hat{1}})^\alpha = [0, (g_{11})^{-1/2}, 0, 0], \quad \text{etc.} \qquad \begin{pmatrix}\text{diagonal}\\ \text{metrics}\end{pmatrix}$$

Components in this coordinate basis and the orthonormal basis are connected by (20.41), which in the case of the Einstein curvature reads

$$G_{\hat{\alpha}\hat{\beta}} = (e_{\hat{\alpha}})^\alpha \, (e_{\hat{\beta}})^\beta \, G_{\alpha\beta}$$

which, for these simple diagonal metrics, reduces to a simple prescription, e.g.,

$$G_{\hat{0}\hat{1}} = (-g_{00})^{-1/2} \, G_{01} (g_{11})^{-1/2} \text{, etc.} \qquad \text{(diagonal metrics).}$$

The analogous relation for the Riemann curvature is given in (21.25). Inverting these relations allows the coordinate basis components to be computed from the orthonormal basis components given.

The Ricci curvature components and the Ricci curvature scalar can be found from the Riemann curvature components in an orthonormal basis by

$$R_{\hat{\alpha}\hat{\beta}} = \eta^{\hat{\gamma}\hat{\delta}} R_{\hat{\alpha}\hat{\gamma}\hat{\beta}\hat{\delta}}, \qquad R = \eta^{\hat{\alpha}\hat{\beta}} R_{\hat{\alpha}\hat{\beta}}.$$

Schwarzschild Geometry

- **Metric (Schwarzschild coordinates):**

$$ds^2 = -\left(1 - \frac{2M}{r}\right) dt^2 + \left(1 - \frac{2M}{r}\right)^{-1} dr^2 + r^2(d\theta^2 + \sin^2\theta\, d\phi^2)$$

- **Christoffel Symbols:**

$\Gamma^t_{tr} = (M/r^2)(1 - 2M/r)^{-1}$

$\Gamma^r_{tt} = (M/r^2)(1 - 2M/r)$

$\Gamma^r_{rr} = -(M/r^2)(1 - 2M/r)^{-1}$

$\Gamma^r_{\theta\theta} = -(r - 2M)$

$\Gamma^r_{\phi\phi} = -(r - 2M)\sin^2\theta$

$\Gamma^\theta_{r\theta} = 1/r$

$\Gamma^\theta_{\phi\phi} = -\cos\theta\,\sin\theta$

$\Gamma^\phi_{r\phi} = 1/r$

$\Gamma^\phi_{\theta\phi} = \cot\theta$

- **An Orthonormal Basis:**

$(e_{\hat{t}})^\alpha = \left[(1 - 2M/r)^{-1/2}, 0, 0, 0\right]$

$(e_{\hat{r}})^\alpha = \left[0, (1 - 2M/r)^{1/2}, 0, 0\right]$

$(e_{\hat{\theta}})^\alpha = [0, 0, 1/r, 0]$

$(e_{\hat{\phi}})^\alpha = [0, 0, 0, 1/(r\sin\theta)]$

- **Riemann Curvature:**

$R_{\hat{t}\hat{r}\hat{t}\hat{r}} = -2M/r^3$

$R_{\hat{\theta}\hat{\phi}\hat{\theta}\hat{\phi}} = +2M/r^3$

$R_{\hat{t}\hat{\theta}\hat{t}\hat{\theta}} = R_{\hat{t}\hat{\phi}\hat{t}\hat{\phi}} = +M/r^3$

$R_{\hat{r}\hat{\theta}\hat{r}\hat{\theta}} = R_{\hat{r}\hat{\phi}\hat{r}\hat{\phi}} = -M/r^3$

- **Einstein Curvature**

$G_{\hat{\alpha}\hat{\beta}} = 0$

Spherically Symmetric Geometries

- **Metric:**

$$ds^2 = -e^{\nu(r,t)}dt^2 + e^{\lambda(r,t)}dr^2 + r^2\,(d\theta^2 + \sin^2\theta d\phi^2)$$

- **Christoffel Symbols:** (a prime denotes a partial derivative with respect to r; a dot denotes a partial derivative with respect to t)

$\Gamma^t_{tt} = \dot{\nu}/2$

$\Gamma^t_{tr} = \nu'/2$

$\Gamma^t_{rr} = e^{\lambda-\nu}\dot{\lambda}/2$

$\Gamma^r_{tt} = e^{\nu-\lambda}\nu'/2$

$\Gamma^r_{tr} = \dot{\lambda}/2$

$\Gamma^r_{rr} = \lambda'/2$

$\Gamma^r_{\theta\theta} = -e^{-\lambda}r$

$\Gamma^r_{\phi\phi} = -e^{-\lambda}r\sin^2\theta$

$\Gamma^\theta_{r\theta} = 1/r$

$\Gamma^\theta_{\phi\phi} = -\cos\theta\,\sin\theta$

$\Gamma^\phi_{r\phi} = 1/r$

$\Gamma^\phi_{\theta\phi} = \cot\theta$

- **An Orthonormal Basis:**

$(e_{\hat{t}})^\alpha = \left[e^{-\nu/2}\,,0\,,0\,,0\right]$
$(e_{\hat{r}})^\alpha = \left[0\,,e^{-\lambda/2}\,,0\,,0\right]$
$(e_{\hat{\theta}})^\alpha = [0\,,0\,,1/r\,,0]$
$(e_{\hat{\phi}})^\alpha = [0\,,0\,,0\,,1/(r\sin\theta)]$

- **Riemann Curvature:**

$R_{\hat{t}\hat{r}\hat{t}\hat{r}} = e^{-\lambda}\left[2\nu'' + (\nu')^2 - \lambda'\nu'\right]/4 - e^{-\nu}(2\ddot{\lambda} + \dot{\lambda}^2 - \dot{\nu}\dot{\lambda})/4$
$R_{\hat{t}\hat{\theta}\hat{t}\hat{\theta}} = R_{\hat{t}\hat{\phi}\hat{t}\hat{\phi}} = e^{-\lambda}\nu'/(2r)$
$R_{\hat{t}\hat{\theta}\hat{r}\hat{\theta}} = R_{\hat{t}\hat{\phi}\hat{r}\hat{\phi}} = e^{-(\nu+\lambda)/2}\dot{\lambda}/(2r)$
$R_{\hat{r}\hat{\theta}\hat{r}\hat{\theta}} = R_{\hat{r}\hat{\phi}\hat{r}\hat{\phi}} = e^{-\lambda}\lambda'/(2r)$
$R_{\hat{\theta}\hat{\phi}\hat{\theta}\hat{\phi}} = (1 - e^{-\lambda})/r^2$

- **Einstein Curvature:**

$G_{\hat{t}\hat{t}} = e^{-\lambda}(-1 + e^{\lambda} + r\lambda')/r^2$
$G_{\hat{t}\hat{r}} = e^{-(\nu+\lambda)/2}\dot{\lambda}/r$
$G_{\hat{r}\hat{r}} = e^{-\lambda}(1 - e^{\lambda} + r\nu')/r^2$
$G_{\hat{\theta}\hat{\theta}} = G_{\hat{\phi}\hat{\phi}} = e^{-\lambda}\left[2\nu'' + (\nu')^2 + 2(\nu' - \lambda')/r - \nu'\lambda'\right]/4 - e^{-\nu}\left[2\ddot{\lambda} + \dot{\lambda}^2 - \dot{\lambda}\dot{\nu}\right]/4$

Friedman-Robertson-Walker (FRW) Geometries

- **Metric:**

$$ds^2 = -dt^2 + a^2(t)\left[\frac{dr^2}{1 - kr^2} + r^2\left(d\theta^2 + \sin^2\theta\, d\phi^2\right)\right]$$

- **Christoffel Symbols:**

$\Gamma^t_{rr} = a\dot{a}/(1 - kr^2)$
$\Gamma^t_{\theta\theta} = r^2 a\dot{a}$
$\Gamma^t_{\phi\phi} = r^2\sin^2\theta\, a\dot{a}$
$\Gamma^r_{rr} = kr/(1 - kr^2)$
$\Gamma^r_{\theta\theta} = -r(1 - kr^2)$
$\Gamma^r_{\phi\phi} = -r(1 - kr^2)\,\sin^2\theta$
$\Gamma^r_{tr} = \dot{a}/a$
$\Gamma^\theta_{r\theta} = 1/r$
$\Gamma^\theta_{\phi\phi} = -\cos\theta\,\sin\theta$
$\Gamma^\theta_{t\theta} = \dot{a}/a$
$\Gamma^\phi_{r\phi} = 1/r$
$\Gamma^\phi_{\theta\phi} = \cot\theta$
$\Gamma^\phi_{t\phi} = \dot{a}/a$

- **An Orthonormal Basis:**

$(e_{\hat{t}})^\alpha = [1, 0, 0, 0]$
$(e_{\hat{r}})^\alpha = [0, \sqrt{1 - kr^2}, 0, 0]/a$
$(e_{\hat{\theta}})^\alpha = [0, 0, 1/r, 0]/a$
$(e_{\hat{\phi}})^\alpha = [0, 0, 0, 1/(r\sin\theta)]/a$

- **Riemann Curvature:**
$R_{\hat{t}\hat{r}\hat{t}\hat{r}} = R_{\hat{t}\hat{\theta}\hat{t}\hat{\theta}} = R_{\hat{t}\hat{\phi}\hat{t}\hat{\phi}} = -\ddot{a}/a$
$R_{\hat{r}\hat{\theta}\hat{r}\hat{\theta}} = R_{\hat{r}\hat{\phi}\hat{r}\hat{\phi}} = R_{\hat{\theta}\hat{\phi}\hat{\theta}\hat{\phi}} = (k+\dot{a}^2)/a^2$

- **Einstein Curvature:**
$G_{\hat{t}\hat{t}} = 3\left(k+\dot{a}^2\right)/a^2$
$G_{\hat{r}\hat{r}} = G_{\hat{\theta}\hat{\theta}} = G_{\hat{\phi}\hat{\phi}} = -\left(k+\dot{a}^2+2a\ddot{a}\right)/a^2$

Static, Weak Field Geometry

- **Metric:** $\Phi = \Phi(\vec{x}) = \Phi(x,y,z)$

$$ds^2 = -\left(1+\frac{2\Phi}{c^2}\right)(cdt)^2 + \left(1-\frac{2\Phi}{c^2}\right)\left(dx^2+dy^2+dz^2\right) .$$

- **Christoffel Symbols:** (to linear order in Φ/c^2):

$$\Gamma^t_{tx} = \frac{1}{c^2}\frac{\partial\Phi}{\partial x}, \qquad \Gamma^x_{tt} = \frac{\partial\Phi}{\partial x},$$
$$\Gamma^x_{xx} = -\frac{1}{c^2}\frac{\partial\Phi}{\partial x}, \quad \Gamma^x_{xy} = -\frac{1}{c^2}\frac{\partial\Phi}{\partial y}, \quad \Gamma^x_{yy} = \frac{1}{c^2}\frac{\partial\Phi}{\partial x}$$

plus cyclic permutations of (x,y,z).

- **An Orthonormal Basis:** (to leading order in $1/c$):

$$\left(e_{\hat{0}}\right)^\alpha = (1/c,0,0,0), \quad \left(e_{\hat{1}}\right)^\alpha = (0,1,0,0),$$
$$\left(e_{\hat{2}}\right)^\alpha = (0,0,1,0), \quad \left(e_{\hat{3}}\right)^\alpha = (0,0,0,1)$$

- **Riemann Curvature:** (to linear order in Φ/c^2):

$$R_{\hat{t}\hat{x}\hat{t}\hat{x}} = \frac{1}{c^2}\frac{\partial^2\Phi}{\partial x^2}, \qquad R_{\hat{t}\hat{x}\hat{t}\hat{y}} = \frac{1}{c^2}\frac{\partial^2\Phi}{\partial x\,\partial y},$$
$$R_{\hat{x}\hat{y}\hat{x}\hat{y}} = \frac{1}{c^2}\left(\frac{\partial^2\Phi}{\partial x^2}+\frac{\partial^2\Phi}{\partial y^2}\right), \quad R_{\hat{x}\hat{y}\hat{x}\hat{z}} = \frac{1}{c^2}\frac{\partial^2\Phi}{\partial y\,\partial z}$$

plus cyclic permutations of (x,y,z).

- **Einstein Tensor:** (to linear order in Φ/c^2):

$$G_{\hat{t}\hat{t}} = \frac{2}{c^2}\nabla^2\Phi .$$

Linearized Gravity

- **Metric:**

$$g_{\alpha\beta}(x) = \eta_{\alpha\beta} + h_{\alpha\beta}(x)$$

- **Christoffel Symbols:**

$$\Gamma^{\alpha}_{\beta\gamma} = \frac{1}{2}\,\eta^{\alpha\delta}\left(\frac{\partial h_{\delta\beta}}{\partial x^{\gamma}} + \frac{\partial h_{\delta\gamma}}{\partial x^{\beta}} - \frac{\partial h_{\beta\gamma}}{\partial x^{\delta}}\right)$$

- **An Orthonormal Basis:** (same as the coordinate basis to zeroth order)
- **Riemann Curvature:**

$$R_{\alpha\beta\gamma\delta} = \frac{1}{2}\left(\frac{\partial^2 h_{\alpha\delta}}{\partial x^{\beta}\partial x^{\gamma}} + \frac{\partial^2 h_{\beta\gamma}}{\partial x^{\alpha}\partial x^{\delta}} - \frac{\partial^2 h_{\alpha\gamma}}{\partial x^{\beta}\partial x^{\delta}} - \frac{\partial^2 h_{\beta\delta}}{\partial x^{\alpha}\partial x^{\gamma}}\right)$$

- **Einstein Curvature:**

$$G_{\alpha\beta} = \frac{1}{2}\left(-\Box\,\bar{h}_{\alpha\beta} + \frac{\partial V_{\alpha}}{\partial x^{\beta}} + \frac{\partial V_{\beta}}{\partial x^{\alpha}} - \eta_{\alpha\beta}\,\frac{\partial V^{\gamma}}{\partial x^{\gamma}}\right)$$

where $V_{\alpha} \equiv \partial\bar{h}^{\beta}_{\alpha}/\partial x^{\beta}$ with $\bar{h}^{\beta}_{\alpha} \equiv h^{\beta}_{\alpha} - (1/2)\delta^{\beta}_{\alpha}h^{\gamma}_{\gamma}$ and

$$\Box \equiv \eta^{\alpha\beta}\frac{\partial^2}{\partial x^{\alpha}\partial x^{\beta}} = -\frac{\partial^2}{\partial t^2} + \vec{\nabla}^2.$$

APPENDIX C

Curvature and the Einstein Equation

This is the *Mathematica* notebook *Curvature and the Einstein Equation* available from the book website. From a given metric $g_{\alpha\beta}$, it computes the components of the following: the inverse metric, $g^{\lambda\sigma}$, the Christoffel symbols or affine connection,

$$\Gamma^{\lambda}{}_{\mu\nu} = \frac{1}{2}\, g^{\lambda\sigma}(\partial_\mu\, g_{\sigma\nu} + \partial_\nu\, g_{\sigma\mu} - \partial_\sigma\, g_{\mu\nu}),$$

(∂_α stands for the partial derivative $\partial/\partial x^\alpha$), the Riemann tensor,

$$R^{\lambda}{}_{\mu\nu\sigma} = \partial_\nu\, \Gamma^{\lambda}{}_{\mu\sigma} - \partial_\sigma\, \Gamma^{\lambda}{}_{\mu\nu} + \Gamma^{\eta}{}_{\mu\sigma}\, \Gamma^{\lambda}{}_{\eta\nu} - \Gamma^{\eta}{}_{\mu\nu}\, \Gamma^{\lambda}{}_{\eta\sigma},$$

the Ricci tensor

$$R_{\mu\nu} = R^{\lambda}{}_{\mu\lambda\nu},$$

the scalar curvature,

$$R = g^{\mu\nu}\, R_{\mu\nu},$$

and the Einstein tensor,

$$G_{\mu\nu} = R_{\mu\nu} - \frac{1}{2}\, g_{\mu\nu}\, R.$$

You must input the covariant components of the metric tensor $g_{\mu\nu}$ by editing the relevant input line in this *Mathematica* notebook. You may also wish to change the names of the coordinates. Only the nonzero components of the above quantities are displayed as the output. All the components computed are in the *coordinate basis* in which the metric was specified.

■ Clearing the values of symbols:

First clear any values that may already have been assigned to the names of the various objects to be calculated. The names of the coordinates that you will use are also cleared.

```
In[1]:= Clear[coord, metric, inversemetric,
          affine, riemann, ricci, scalar, einstein, r, θ, ϕ, t]
```

■ Setting the dimension:

The dimension **n** of the spacetime (or space) must be set:

```
In[2]:= n = 4

Out[2]= 4
```

▪ Defining a list of coordinates:

The example given here is the Schwarzschild metric. The coordinate choice of Schwarzschild is appropriate for this spherically symmetric spacetime.

```
In[3]:= coord = {r, θ, ϕ, t}

Out[3]= {r, θ, ϕ, t}
```

You can change the names of the coordinates by simply editing the definition of **coord**, for example, to **coord = {x, y, z, t}**, when another set of coordinate names is more appropriate. In this program indices range over **1** to **n.** Thus for spacetime they range from 1 to 4 and x^4 is the same as x^0 used in the text.

▪ Defining the metric:

Input the metric as a list of lists, i.e., as a matrix. You can input the components of any metric here, but you must specify them as explicit functions of the coordinates.

```
In[4]:= metric = {{(1 - 2 m / r) ^ (-1), 0, 0, 0},
          {0, r^2, 0, 0}, {0, 0, r^2 Sin[θ]^2, 0}, {0, 0, 0, - (1 - 2 m / r)}}
```

Out[4]= $\left\{\left\{\frac{1}{1-\frac{2m}{r}}, 0, 0, 0\right\}, \{0, r^2, 0, 0\}, \{0, 0, r^2 \,\text{Sin}[\theta]^2, 0\}, \left\{0, 0, 0, -1+\frac{2m}{r}\right\}\right\}$

You can also display this in matrix form.

```
In[5]:= metric // MatrixForm
```

Out[5]//MatrixForm=

$$\begin{pmatrix} \frac{1}{1-\frac{2m}{r}} & 0 & 0 & 0 \\ 0 & r^2 & 0 & 0 \\ 0 & 0 & r^2\,\text{Sin}[\theta]^2 & 0 \\ 0 & 0 & 0 & -1+\frac{2m}{r} \end{pmatrix}$$

▪ Note:

It is important not to use the symbols, **i, j, k, l, s,** or **n** as constants or coordinates in the metric that you specify above. The reason is that the first five of those symbols are used as summation or table indices in the calculations done below, and **n** is the dimension of the space. For example, if **m** were used as a summation or table index below, then you would get the wrong answer for the present metric because the **m** in the metric would be treated as an index, rather than as the mass.

▪ Calculating the inverse metric:

The inverse metric is obtained through matrix inversion.

```
In[6]:= inversemetric = Simplify[Inverse[metric]]
```

Out[6]= $\left\{\left\{1-\frac{2m}{r}, 0, 0, 0\right\}, \left\{0, \frac{1}{r^2}, 0, 0\right\}, \left\{0, 0, \frac{\text{Csc}[\theta]^2}{r^2}, 0\right\}, \left\{0, 0, 0, \frac{r}{2m-r}\right\}\right\}$

This can also be displayed in matrix form:

```
In[7]:= inversemetric // MatrixForm
```

Out[7]//MatrixForm=

$$\begin{pmatrix} 1-\frac{2m}{r} & 0 & 0 & 0 \\ 0 & \frac{1}{r^2} & 0 & 0 \\ 0 & 0 & \frac{\text{Csc}[\theta]^2}{r^2} & 0 \\ 0 & 0 & 0 & \frac{r}{2m-r} \end{pmatrix}$$

▪ Calculating the Christoffel symbols:

The calculation of the components of the Christoffel symbols is done by transcribing the definition given earlier into the notation of *Mathematica* and using the *Mathematica* functions **D** for taking partial derivatives, **Sum** for summing over repeated indices, **Table** for forming a list of components, and **Simplify** for simplifying the result.

```
In[8]:= affine := affine = Simplify[Table[(1 / 2) * Sum[(inversemetric[[i, s]]) *
              (D[metric[[s, j]], coord[[k]] ] +
                D[metric[[s, k]], coord[[j]] ] - D[metric[[j, k]], coord[[s]] ]), {s, 1, n}],
           {i, 1, n}, {j, 1, n}, {k, 1, n}] ]
```

▪ Displaying the Christoffel symbols:

The nonzero Christoffel symbols are displayed below. You need not follow the details of constructing the functions that we use for that purpose. In the output the symbol Γ[1,2,3] stands for $\Gamma^1{}_{23}$. Because the Christoffel symbols are symmetric under interchange of the last two indices, only the independent components are displayed.

```
In[9]:= listaffine := Table[If[UnsameQ[affine[[i, j, k]], 0],
           {ToString[Γ[i, j, k]], affine[[i, j, k]]}] , {i, 1, n}, {j, 1, n}, {k, 1, j}]

In[10]:= TableForm[Partition[DeleteCases[Flatten[listaffine], Null], 2], TableSpacing → {2, 2}]
```

Out[10]//TableForm=

Γ[1, 1, 1]	$\frac{m}{2mr-r^2}$
Γ[1, 2, 2]	$2m-r$
Γ[1, 3, 3]	$(2m-r)\,\text{Sin}[\theta]^2$
Γ[1, 4, 4]	$\frac{m(-2m+r)}{r^3}$
Γ[2, 2, 1]	$\frac{1}{r}$
Γ[2, 3, 3]	$-\text{Cos}[\theta]\,\text{Sin}[\theta]$
Γ[3, 3, 1]	$\frac{1}{r}$
Γ[3, 3, 2]	$\text{Cot}[\theta]$
Γ[4, 4, 1]	$\frac{m}{-2mr+r^2}$

▪ Calculating and displaying the Riemann tensor:

The components of the Riemann tensor, $R^{\lambda}{}_{\mu\nu\sigma}$, are calculated using the definition given above.

```
In[11]:= riemann := riemann = Simplify[Table[
        D[affine[[i, j, l]], coord[[k]] ] - D[affine[[i, j, k]], coord[[l]] ] +
         Sum[affine[[s, j, l]] affine[[i, k, s]] - affine[[s, j, k]] affine[[i, l, s]],
           {s, 1, n}],
        {i, 1, n}, {j, 1, n}, {k, 1, n}, {l, 1, n}] ]
```

The nonzero components are displayed by the following functions. In the output, the symbol R[1, 2, 1, 3] stands for $R^{1}{}_{213}$, and similarly for the other components. You can obtain R[1, 2, 3, 1] from R[1, 2, 1, 3] using the antisymmetry of the Riemann tensor under exchange of the last two indices. The antisymmetry under exchange of the first two indices of $R_{\lambda\mu\nu\sigma}$ is not evident in the output because the components of $R^{\lambda}{}_{\mu\nu\sigma}$ are displayed.

```
In[12]:= listriemann := Table[If[UnsameQ[riemann[[i, j, k, l]], 0],
          {ToString[R[i, j, k, l]], riemann[[i, j, k, l]]}] ,
         {i, 1, n}, {j, 1, n}, {k, 1, n}, {l, 1, k - 1}]
```

```
In[13]:= TableForm[Partition[DeleteCases[Flatten[listriemann], Null], 2],
          TableSpacing → {2, 2}]
```

Out[13]//TableForm=

R[1, 2, 2, 1]	$\frac{m}{r}$
R[1, 3, 3, 1]	$\frac{m\,\text{Sin}[\theta]^2}{r}$
R[1, 4, 4, 1]	$\frac{2\,m\,(-2\,m+r)}{r^4}$
R[2, 1, 2, 1]	$\frac{m}{(2\,m-r)\,r^2}$
R[2, 3, 3, 2]	$-\frac{2\,m\,\text{Sin}[\theta]^2}{r}$
R[2, 4, 4, 2]	$\frac{m\,(2\,m-r)}{r^4}$
R[3, 1, 3, 1]	$\frac{m}{(2\,m-r)\,r^2}$
R[3, 2, 3, 2]	$\frac{2\,m}{r}$
R[3, 4, 4, 3]	$\frac{m\,(2\,m-r)}{r^4}$
R[4, 1, 4, 1]	$\frac{2\,m}{r^2\,(-2\,m+r)}$
R[4, 2, 4, 2]	$-\frac{m}{r}$
R[4, 3, 4, 3]	$-\frac{m\,\text{Sin}[\theta]^2}{r}$

▪ Calculating and displaying the Ricci tensor:

The Ricci tensor $R_{\mu\nu}$ was defined by summing the first and third indices of the Riemann tensor (which has the first index already raised).

```
In[14]:= ricci :=
          ricci = Simplify[Table[Sum[riemann[[i, j, i, l]], {i, 1, n}], {j, 1, n}, {l, 1, n}] ]
```

Next we display the nonzero components. In the output, R[1, 2] denotes R_{12}, and similarly for the other components.

```
In[15]:= listricci := Table[If[UnsameQ[ricci[[j, l]], 0],
            {ToString[R[j, l]], ricci[[j, l]]}] , {j, 1, n}, {l, 1, j}]

In[16]:= TableForm[Partition[DeleteCases[Flatten[listricci], Null], 2], TableSpacing → {2, 2}]

Out[16]//TableForm=
```

A vanishing table (as with the Schwarzschild metric example) means that the vacuum Einstein equation is satisfied.

▪ Calculating the scalar curvature:

The scalar curvature R is calculated using the inverse metric and the Ricci tensor. The result is displayed in the output line.

```
In[17]:= scalar = Simplify[Sum[inversemetric[[i, j]] ricci[[i, j]], {i, 1, n}, {j, 1, n}] ]

Out[17]= 0
```

▪ Calculating the Einstein tensor:

The Einstein tensor, $G_{\mu\nu} = R_{\mu\nu} - \frac{1}{2}\, g_{\mu\nu}\, R$, is found from the tensors already calculated.

```
In[18]:= einstein := einstein = Simplify[ricci - (1 / 2) scalar * metric]
```

The results are displayed in the same way as for the Ricci tensor earlier.

```
In[19]:= listeinstein := Table[If[UnsameQ[einstein[[j, l]], 0],
            {ToString[G[j, l]], einstein[[j, l]]}] , {j, 1, n}, {l, 1, j}]

In[20]:= TableForm[Partition[DeleteCases[Flatten[listeinstein], Null], 2],
          TableSpacing → {2, 2}]

Out[20]//TableForm=
```

A vanishing table means that the vacuum Einstein equation is satisfied!

▪ Acknowledgment

This program was kindly written by *Leonard Parker, University of Wisconsin, Milwaukee* especially for this text.

APPENDIX

D

Pedagogical Strategy

. . . as simple as possible, but not simpler.
attributed to A. Einstein

Physics first!
Anon.

The straightforward approach to teaching general relativity is to

1. develop the necessary mathematical concepts and tools,
2. motivate the Einstein equation and the requisite physical concepts,
3. solve the equation for the models of realistic physical situations, and
4. compare the predictions of the theory with experiment and observation.

The logic of this order is unassailable, and, by and large, it is the way the theory is presented in the classic expositions mentioned in the bibliography, as well as many excellent introductory texts. However, following this order in the limited time that is typically available and appropriate for a basic introductory course is difficult. There is a considerable body of beautiful, powerful, and straightforward mathematics that is necessary. But developing it takes time. Similarly, solving the nonlinear Einstein equation in any realistic situation can be a lengthy exercise. The length available for an introductory course is often not sufficient to present the subject in this logical way and also discuss its important applications. This book introduces general relativity in a different order. In this Appendix we present some pedagogical principles on which the present text is constructed.

D.1 Pedagogical Principles

Explore First, Derive Later

The simplest physically relevant solutions of the Einstein equation are presented *first*, without derivation, as spacetimes whose observational consequences are to be explored by the study of the motion of test particles and light rays in them. This brings the student to the physical phenomena as quickly as possible. It is the part of the subject most directly connected to classical mechanics, and requires the minimum of new mathematical ideas. Later the Einstein equation is introduced and solved to show where these geometries originate. Readers who have time to work through the entire text should understand both the important solutions and their origin. But those who stop earlier will at least have understood some of the basic phenomena for which curved spacetime is important.

Only the Simplest Examples

The simplest solutions of the Einstein equation are the most physically relevant. The Sun is approximately spherical, the universe is approximately homogeneous and isotropic, and detectable gravitational waves are weak and approximately planar. Only the simplest physically relevant spacetimes of general relativity are presented. Thus we discuss black holes with mass and angular momentum but not charge, spherical gravitational collapse quantitatively but nonspherical collapse only qualitatively, homogeneous, isotropic cosmologies but not anisotropic ones, weak gravitational waves in flat spacetime but not nonlinear waves or waves in curved spacetime, and spherical stars but not rotating ones.

Introduce New Math Only As Necessary

Mathematical ideas beyond those in the usual advanced calculus toolkit are introduced only as needed. Only a few additional tools are needed to understand a spacetime geometry and explore it through the motion of test particles and light rays. The basic concepts of metric, four-vector, and geodesics largely suffice. These are introduced in various chapters at the start of Parts 1 and 2 and are sufficient for all the development there. It is not necessary, for example, to develop a general theory of tensors in Parts 1 and 2, because only one tensor—the metric—is used. Tensors and the covariant derivative are introduced in Chapter 20. Quantitative measures of curvature are introduced in Chapter 21 as a prerequisite to understanding the Einstein equation.

Stress Physical Phenomena and Their Connection to Experiment and Observation

The Global Positioning System, the orbits of planets and light in the solar system, X-ray binaries, active galactic nuclei, neutron stars, gravitational lensing, gravitational waves, the large-scale structure of the universe, and the big bang are just some of the phenomena in the universe for which relativistic gravity is important. This book stresses the growing connection between general relativity and experiment and observation. Astrophysics and cosmology are home to many of these applications. However, *this is not a text on astrophysics or cosmology.* The connection between theory and observation is typically made by way of only the simplest type of model, and then often only in a qualitative way.

Classic Experiments but Not an Overview of Experiment

No contemporary exposition of general relativity would be complete without describing its experimental confirmation and application to astrophysics. But the inevitable downside to any discussion of experiment and observation is that it will become quickly dated. That is especially the case in gravitational physics, where the domain of application is growing rapidly at the time of writing and will grow even faster when the gravitational wave detectors now under construction come on line. For this reason the author has not tried to write an overview of the experimental situation, nor necessarily included the latest data, but rather has used classic examples that illustrate the basic methods.

D.2 Organization

Prerequisites

The main prerequisite is the introductory mechanics course that is typically a standard part of any undergraduate major in physics. Especially important are a grounding in the general principles of mechanics, conservation laws, orbits in the central force problem, and Lagrangian mechanics. An introduction to the variational principle for mechanics will be helpful although an abbreviated discussion is given in Chapter 3. Similarly an introduction to special relativity would be helpful but the discussion in Chapters 4–5 is self-contained. There are passing references to Maxwell's equations, and there are elementary applications of electromagnetism in the boxes, but a detailed course in the subject is not a prerequisite to tackling the main text.

The Three Parts

The book is divided into three parts. Part 1 introduces the idea that gravity is geometry and reviews the basic parts of Newtonian and special relativistic mechanics that are relevant for general relativity. Part 2 introduces the basic ideas of general relativity and then focuses on understanding the simplest black hole, cosmological, and gravitational wave spacetimes through a study of the motion of test particles and light rays in them. These geometries are presented and analyzed, not derived. They are derived in Part 3 after the mathematics of curvature and the Einstein equation are introduced. Part 3 goes on to give an elementary discussion of the production of gravitational waves and relativistic stars for which the Einstein equation is essential. Within each part the order of the topics is roughly by increasing sophistication—either of mathematical detail or physical concept or both.

Boxes

The discussion in the boxes is intended to extend and illustrate the basic ideas in the main text. Sometimes a box concerns a related idea (such as Penrose diagrams), sometimes a relevant experiment (such as a modern Michelson–Morley experiment), and sometimes an introduction to a complex phenomenon in which general relativity plays an important role (such as the electromagnetic extraction of energy from rotating black holes). Some of these extensions require modest parts of physics beyond the basic mechanics assumed for most of the main text. The discussion in such cases is typically more qualitative and abbreviated than the standard typical of the main text. The aim of the boxes is not to achieve an in-depth understanding of the subjects treated, but to illustrate some of the ramifications of the main development briefly and qualitatively. Depending on their preparation, students will find some boxes more difficult to understand than others, but it is not necessary to understand any box to understand the main text.

Mathematica Notebooks

Analyzing even the simplest of physical situations in general relativity can sometimes require messy algebra, or lead to differential equations lacking elementary closed-form solutions. To help with this, the following *Mathematica* notebooks are provided on the book website which do some standard algebra and solve some of the most important differential equations.

- Christoffel Symbols and Geodesic Equation
- Shape of Orbits in the Schwarzschild Geometry
- Friedman-Robertson-Walker Cosmological Models
- Curvature and the Einstein Equation

Web Supplements

Some conceptually simple results require lengthy derivations that tend to interrupt the main development. Conventionally these would be relegated to appendices. But to keep the book to a managable length these are housed in the book website.

D.3 Constructing Courses

The text contains more material than can be reasonably covered in a one-quarter ($\sim$ 30 hour lectures) or a one-semester ($\sim$ 45 hour lectures) course. A variety of course plans can therefore be constructed by selecting chapters, or parts of them, in various ways. The following chart shows how the various chapters depend on each other. Student preparation will determine where to start in the first part (Chapters 1 through 5) or how quickly to cover them. Chapters 6–9 introduce some basic ideas and techniques of general relativity. Any selection of later chapters that includes the earlier ones on which they depend could in principle form the basis of a course. For example, a focus on black holes would include Chapters 12–15. A focus on gravitational waves might include Chapters 16, 20–23. The author's own quarter course typically covers Chapters 1–10, 12–13, and then Chapters 17–19 on cosmology or Chapters 20–21 introducing the Einstein equation, depending on class interest.

The author has several times employed the text as the basis for an introductory graduate course. It works well for students who are seeing the general relativity for the first time, are more interested in applications than the general framework, or if there is limited time.

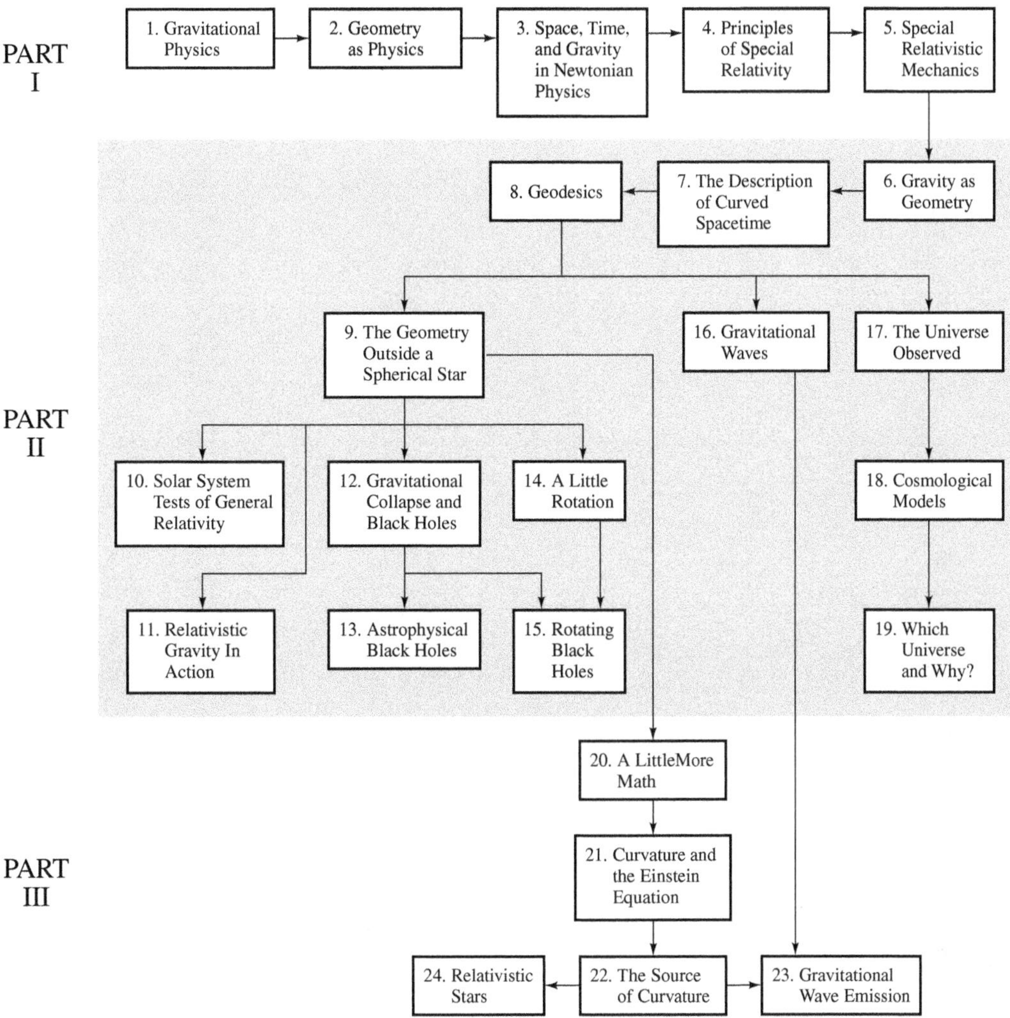
PART I
1. Gravitational Physics
2. Geometry as Physics
3. Space, Time, and Gravity in Newtonian Physics
4. Principles of Special Relativity
5. Special Relativistic Mechanics
PART II
6. Gravity as Geometry
7. The Description of Curved Spacetime
8. Geodesics
9. The Geometry Outside a Spherical Star
16. Gravitational Waves
17. The Universe Observed
10. Solar System Tests of General Relativity
12. Gravitational Collapse and Black Holes
14. A Little Rotation
18. Cosmological Models
11. Relativistic Gravity In Action
13. Astrophysical Black Holes
15. Rotating Black Holes
19. Which Universe and Why?
PART III
20. A LittleMore Math
21. Curvature and the Einstein Equation
24. Relativistic Stars
22. The Source of Curvature
23. Gravitational Wave Emission

Bibliography

This bibliography contains references to classic expositions of general relativity, works of a more specialized character, and to the papers cited in the text—chiefly the sources of the experimental or observational data described. No attempt has been made to be comprehensive or to cite the original sources of the ideas discussed. Readers interested in pursuing any subject mentioned in this book, including its history, are probably best advised to first consult one of the expositions below.

Classic General Expositions

Landau, L. and Lifshitz, E.M. (1962). *The Classical Theory of Fields*, Pergamon Press, London.

The 150 pages of this text devoted to general relativity give a concise introduction to the basics of the subject in the clear and straightforward Landau and Lifshitz style, although few applications are covered in any depth.

Misner, C.W., Thorne, K.S., and Wheeler, J.A. (1970). *Gravitation*, W.H. Freeman, San Francisco.

Over thirty years after its publication, *Gravitation* is still the most comprehensive treatise on general relativity. An authoritative and complete discussion of almost any topic in the subject can be found within its 1300 pages. It also contains an extensive bibliography with references to original sources. Written by three twentieth-century masters of the subject, it set the style for many later texts on the subject, including this one.

Taylor, E.F. and Wheeler, J.A. (1963). *Spacetime Physics*, W.H. Freeman, San Francisco.

A lively and insightful presentation of special relativity at an entry level stressing the spacetime point of view.

Wald, R. (1984). *General Relativity*, University of Chicago Press, Chicago.

General Relativity is a popular graduate textbook. The differential geometry necessary for general relativity is developed thoroughly, rigorously, and in a contemporary mathematical style. The book is notable for the care and clarity of its basic arguments as well as its treatment of a number of advanced topics including causal structure, asymptotic flatness, spinors, and quantum field theory in curved spacetime.

Weinberg, S. (1972). *Gravitation and Cosmology*, John Wiley & Sons, New York.

This book treats general relativity from the perspective of field theory and elementary particle physics stressing its applications to cosmology. The viewpoint is mainly nongeometrical, which contrasts with that of this text. However, this different viewpoint often contributes to understanding. *Gravitation and Cosmology* remains a valuable resource for gravitation theory in cosmology despite the great changes in observation that have taken place in thirty years since its publication.

More Specialized Texts

Hawking, S.W. and Ellis, G.F.R. (1973). *The Large Scale Structure of Space-Time*, Cambridge University Press, Cambridge.

Kolb, E.W. and Turner, M.S. (1990). *The Early Universe*, Addison-Wesley, Redwood City, CA.

Krolik, J.H. (1999). *Active Galactic Nuclei*, Princeton University Press, Princeton.

Shapiro, S.L. and Teukolsky, S.A. (1983). *Black Holes, White Dwarfs, and Neutron Stars*, Wiley-Interscience, New York.

Thorne, K.S., Price, R.H., and MacDonald, D.A. (1986). *Black Holes: The Membrane Paradigm*, Yale University Press, New Haven.

Will, C.A. (1993). *Theory and Experiment in Gravitational Physics*, Cambridge University Press, Cambridge, UK (revised edition).

History

Miller, A. I. (1981). *Albert Einstein's Special Theory of Relativity*, Addison-Wesley, Reading, MA.

Pais, A. (1982). *Subtle is the Lord . . .*, Oxford University Press, New York.

Further References Cited in the Text

Alcock, C. et al. (1997). "The MACHO Project Large Magellanic Cloud Microlensing Results from the First Two Years and the Nature of the Galactic Dark Halo," *Ap. J.*, **486**, 697.

Alcubierre, M. (1994), "The Warp Drive: Hyper-fast Travel within General Relativity," *Class. Quant. Grav.*, **11**, L73.

Anderson, J.D. and Williams, J.G. (2001). "Long Range Tests of the Equivalence Principle," *Class. Quant. Grav.*, **18**, 2447.

Bailey, J. et al. (1977). "Measurements of Relativistic Time Dilation for Positive and Negative Muons in Circular Orbit," *Nature*, **268**, 301.

Bennett, C.L., Banday, A.J, Gorski, K.M., Hinshaw, G., Jackson, P., Keegstra, P., Kogut, A., Smoot, G.F., Wilkinson, D.T., and Wright, E.L. (1996). "4-Yr COBE Cosmic Microwave Background Observations: Maps and Basic Results," *Ap.J.*, **464**, L1.

Biretta, J.A., Moore, R.L., and Cohen, M.H. (1986). "The Evolution of the Compact Radio Source in 3C 345. I. VLBI Observations," *Ap. J.*, **308**, 93.

Braginsky, V.B., and Panov, V.I. (1971). "Verification of the Equivalence of Inertial and Gravitational Mass," *Zh. Eksp. Theor. Fiz.*, **61**, 873 [*Sov. Phys. JETP*, **34**, 463 (1972).]

Brillet, A. and Hall, J.L. (1979). "Improved Laser Test of the Isotropy of Space," *Phys. Rev. Letters*, **42**, 549.

Brown, T.M. et al. (1989). "Inferring the Sun's Internal Angular Velocity from Observed p-Mode Frequency Splittings," *Ap. J.*, **343**, 526.

Campbell, W.W. and Trumpler, R. (1923). "Observation on the Deflection of Light in Passing Through the Sun's Gravitational Field Made During the Total Solar Eclipse of Sept. 21 1922," *Lick Observatory Bulletin*, No. 346, **11**, 41.

Chandrasekhar, S. (1983). *The Mathematical Theory of Black Holes*, Oxford University Press, Oxford.

Colless, M., Dalton, G., Maddox, S., Sutherland, W., et al. (2001). "The 2dF Galaxy Redshift Survey: Spectra and Redshifts," *MNRAS*, **328**, 1039–1063.

Colley, W.N., Tyson, J.A., and Turner, E.L. (1996). "Unlensing Multiple Arcs in 0024+1654: Reconstruction of the Source Image," *Ap. J.*, **461**, L83.

Cram, T.R., Roberts, M.S., and Whitehurst, R.N. (1980). "A Complete, High-Sensitivity 21-cm Hydrogen Line Survey of M31," *Astron. and Astrophys. Suppl.*, **40**, 215.

de Bernardis, P. et al. (2000). "A Flat Universe from High-Resolution Maps of the Cosmic Microwave Background Radiation," *Nature*, **404**, 955.

Dickey, J.O. et al. (1994). "Lunar Laser Ranging: A Continuing Legacy of the Apollo Program," *Science*, **265**, 482.

Feynman, R. (1965). *The Character of Physical Law*, MIT Press, Cambridge, MA.

Fixsen, D.J., Cheng, E.S., Gales, J.M., Mather, J.C., Shafer, R., and Wright. E. (1996). "The Cosmic Microwave Background Spectrum from the Full COBE FIRAS Data Set," *Ap.J.* **473**, 576.

Fomalont, E.B. and Sramek, R.A. (1975). "A Confirmation of Einstein's General Theory of Relativity by Measuring the Bending of Microwave Radiation in the Gravitational Field of the Sun," *Ap. J.*, **199**, 749.

Fomalont, E.B. and Sramek, R.A. (1977). "The Deflection of Radio Waves by the Sun," *Comm. Astrophys.*, **7**, 19.

Freedman, W. L., Madore, B.F., and Kennicutt, R.C. (2001). "The Hubble Space Telescope Key Project to Measure the Hubble Constant," *Ap. J.*, **553**, 47.

Ghez, A.M., Hornstein, S., Tanner, A., Morris, M., and Becklin, E.E. (2002). "Full 3-D Orbital Solutions for Stars Making a Close Approach to the Supermassive Black Hole at the Center of the Galaxy," in M.J. Rees Symposium "Making Light of Gravity," July 8, 2002 (unpublished).

Glendenning, P. (1985). "Neutron Stars Are Giant Hypernuclei?," *Ap. J.*, **293**, 470.

Gustavson, T.L., Bouyer, P., and Kasevich, M.A. (1997). "Precision Rotation Measurements with an Atom Interferometer Gyroscope," *Phys. Rev. Lett.*, **78**, 2046.

Hafele, J.C. and Keating R.E. (1972). "Around-the-World Atomic Clocks: Observed Relativistic Time Gains," *Science*, **177**, 168.

Harrison, B.K., Thorne, K.S., Wakano, M., and Wheeler, J.A. (1965). *Gravitation Theory and Gravitational Collapse*, University of Chicago Press, Chicago.

Hartle, J.B. (1978). "Bounds on the Mass and Moment of Inertia of Non-Rotating Neutron Stars," *Phys. Reports*, **46**, 201.

Hartle, J.B. et al. (1999). *Gravitational Physics: Exploring the Structure of Space and Time*, National Academies Press, Washington, DC.

Herrnstein, J.R., Moran, J.M., Greenhill, L.J., Diamond, P.J., Inoue, M., Nakai, N., Miyoshi, M., Henkel, C., and Riess, A. (1999). "A Geometric Distance to the Galaxy NGC 4258 from Orbital Motions in a Nuclear Gas Disk," *Nature*, **400**, 539.

Kramer, D., Stephani, H., MacCallum M., and Herlt, E. (1980). *Exact Solutions of Einstein's Field Equations*, Schmutzer, E., ed. Cambridge University Press, Cambridge.

Lebach, D., Corey, B., Shapiro, I., Ratner, M., Webber, J., Rogers, A., Davis, J., and Herring, T. (1995). "Measurement of the Solar Gravitational Deflection of Radio Waves Using Very-Long-Baseline Interferometry," *Phys. Rev. Letters*, **75**, 1439.

Maddox, S., Efstathiou, G., Sutherland, W., and Loveday, J. (1990). "The APM Galaxy Survey I," *MNRAS*, **243**, 692.

Marey, Étienne-Jules (1885). *La méthode graphique dans les sciences experimentales et principalement en physiologie et en medecine*, G. Masson, Paris.

Mather, J., Fixsen, D., Shafer, R., Mosier, C., and Wilkinson, D. (1999). "Calibrator Design for the COBE Far Infrared Absolute Spectrophotometer (FIRAS)," *Ap. J.*, **512**, 511.

Miyoshi, M., Moran, J.M., Herrnstein, J.R., Greenhill, L.J., Nakai, N., Diamond, P.J., and Inoue, M. (1995). "Evidence for a Massive Black Hole from High Rotation Velocities in a Sub-Parsec Region of NGC 4258," *Nature*, **373**, 127.

Orosz, J.A., Bailyn, C.D., McClintock, J.E., and Remillard, R.A. (1996). "Improved Parameters for the Black Hole Binary System X-ray Nova Muscae 1991," *Ap. J.*, **468**, 380.

Parkinson, B.W., and Spilker, J.J. eds. (1996). *Global Positioning System: Theory and Applications*, vols I and II, American Institute of Aeronautics and Astronautics, Washington, D.C.

Perlmutter, S. et al. (1999). "Measurements of Omega and Lambda from 42 High-Redshift Supernovae," *Ap.J.* **517**, 565–586.

Persson, S.E, Madore, B.F., Freedman, W.L., Krzeminski, W., Roth M., and Murphy, D.C. (2002). (to be published in Ap. J.).

Pound, R.V. and Rebka, G.A. (1960). "Apparent Weight of Photons," *Phys. Rev. Lett.* **4**, 337.

Pound, R.V. and Snider, J.L. (1964). "Effect of Gravity on Nuclear Resonance," *Phys. Rev. Lett.* **13**, 539.

Riess, A.G. et al. (1998). "Observational Evidence from Supernovae for an Accelerating Universe and a Cosmological Constant," *Astron.J.* **116**, 1009–1038.

Roberts, M. (1988). "How Much of the Universe Do We See?" In *Proceedings of the Bicentennial Commemoration of R.G. Boscovich*, Bossi, M. and Tucci, P., eds. Edizioni Unicopli, Milan.

Roll, P.G., Krotkov, R., and Dicke, R.H. (1964). "The Equivalence of Inertial and Passive Gravitational Mass," *Ann. Phys. (N.Y.)*, **26**, 442.

Saulson, P. (1994). *Fundamentals of Interferometric Gravitational Wave Detectors*, World Scientific, Singapore.

Shapiro, I.I., Reasenberg, R.D., MacNeil, P.E., Goldstein, R.B., Brenkle, J.P., Cain, D.L., Komarek, T., Zygielbaum, A.I., Cuddihy, W.F., and Michael, W.H., Jr. (1977). "The Viking Relativity Experiment," *J. Geophy. Res.* **82**, 4329.

Shapiro, I. (1990). "Solar System Tests of General Relativity," in *General Relativity and Gravitation 1989*, Ashby, N., Bartlett, D.F., and Wyss, W., eds. Cambridge University Press, Cambridge.

Su, Y., Heckel, B.R., Adelberger, E.G., Gundlach, J.H., Harris, M., Smith, G.L., and Swanson, H.E. (1994). "New Tests of the Universality of Free Fall," *Phys. Rev. D*, **50**, 3614.

Tanaka, Y., et al. (1995). "Gravitationally Redshifted Emission Implying an Accretion Disk and Massive Black-Hole in the Active Galaxy MCG:-6-30-15," *Nature*, **375**, 659.

Taylor, J.H. (1994). "Binary Pulsars and Relativistic Gravity," *Rev. Mod. Phys.*, **66**, 711.

Taylor, J.H. and Weisberg, J.M. (1989). "Further Experimental Tests of Relativistic Gravity Using the Binary Pulsar PSR 1913+16," *Ap. J.* **345**, 434.

Thorne, K.S. (1994). *Black Holes and Time Warps: Einstein's Outrageous Legacy*, W.W. Norton, New York.

Vessot, R.F.C. and Levine, M.W. (1979). "A Test of the Equivalence Principle Using a Space-Borne Clock," *Gen. Rel. and Grav.*, **10**, 181.

Wang, X., Tegmark, M., Zaldarriaga, M. (2002). "Is Cosmology Consistent?," *Phys. Rev. D*, **65**, 123001.

Williams, J.G., Newhall, X.X., and Dickey, J.O. (1996). "Relativity Parameters Determined from Lunar Laser Ranging," *Phys. Rev. D*, **53**, 6730.

Index

COORDINATE AND ORTHONORMAL BASES

- A set $\{\mathbf{e}_{\hat{\alpha}}\}$ of four *orthonormal* basis vectors satisfies

$$\mathbf{e}_{\hat{\alpha}}(x) \cdot \mathbf{e}_{\hat{\beta}}(x) = \eta_{\hat{\alpha}\hat{\beta}}.$$

- A set $\{\mathbf{e}_{\alpha}\}$ of four *coordinate* basis vectors associated with a set of coordinates x^{α} satisfies

$$\mathbf{e}_{\alpha}(x) \cdot \mathbf{e}_{\beta}(x) = g_{\alpha\beta}(x)$$

  where the line element has the form $ds^2 = g_{\alpha\beta}(x)dx^{\alpha}dx^{\beta}$.
- If the coordinate system is *orthogonal* ($g_{\alpha\beta}(x) = 0$ for $\alpha \neq \beta$), the coordinate basis components of an orthonormal basis pointing along the coordinate directions have the form

$$(\mathbf{e}_{\hat{0}})^{\alpha} = [(-g_{00})^{-1/2}, 0, 0, 0], \quad (\mathbf{e}_{\hat{1}})^{\alpha} = [0, (g_{11})^{-1/2}, 0, 0], \quad \text{etc.}$$

USEFUL NUMBERS

Conversion Factors

Velocity of light	$c \equiv 299792458$ m/s $\approx 3 \times 10^{10}$ cm/s
Boltzmann's constant	$k_B = 1.38 \times 10^{-16}$ erg/K $= 8.59 \times 10^{-5}$ eV/K
Second of arc	1 arcsec = $1''$ =4.85×10^{-6} rad
Light year	1 ly $= 9.46 \times 10^{17}$ cm
Parsec	1 pc $= 3.09 \times 10^{18}$ cm $= 3.26$ ly
Electron volt	1 eV = 1.60×10^{-12} erg $= 1.16 \times 10^{4}$ K
Erg (cgs unit of energy)	1 erg = 10^{-7} J
Dyne (cgs unit of force)	1 dyne = 10^{-5} N

Physical Constants

Gravitational constant	$G = 6.67 \times 10^{-8}$ dyn $\cdot$ cm^2/g^2
Stefan–Boltzmann constant	$\sigma = 5.67 \times 10^{-5}$ erg/(cm$^2 \cdot$ s $\cdot$ K^4)
Radiation constant	$a = 7.56 \times 10^{-15}$ erg/(cm$^3 \cdot$ K^4)
Mass of an electron	$m_e = 9.11 \times 10^{-28}$ g
Mass of a proton	$m_p = 1.67 \times 10^{-24}$ g
Planck's constant	$\hbar = 1.05 \times 10^{-27}$ erg $\cdot$ s